T0211649

Lecture Notes in Computer Science 12142

Valeria V. Krzhizhanovskaya ·
Gábor Závodszky · Michael H. Lees ·
Jack J. Dongarra · Peter M. A. Sloot ·
Sérgio Brissos · João Teixeira (Eds.)

Computational Science – ICCS 2020

20th International Conference
Amsterdam, The Netherlands, June 3–5, 2020
Proceedings, Part VI

 Springer

Editors
Valeria V. Krzhizhanovskaya ⓘ
University of Amsterdam
Amsterdam, The Netherlands

Michael H. Lees
University of Amsterdam
Amsterdam, The Netherlands

Peter M. A. Sloot ⓘ
University of Amsterdam
Amsterdam, The Netherlands

ITMO University
Saint Petersburg, Russia

Nanyang Technological University
Singapore, Singapore

João Teixeira
Intellegibilis
Setúbal, Portugal

Gábor Závodszky ⓘ
University of Amsterdam
Amsterdam, The Netherlands

Jack J. Dongarra ⓘ
University of Tennessee
Knoxville, TN, USA

Sérgio Brissos
Intellegibilis
Setúbal, Portugal

ISSN 0302-9743   ISSN 1611-3349 (electronic)
Lecture Notes in Computer Science
ISBN 978-3-030-50432-8   ISBN 978-3-030-50433-5 (eBook)
https://doi.org/10.1007/978-3-030-50433-5

LNCS Sublibrary: SL1 – Theoretical Computer Science and General Issues

This Springer imprint is published by the registered company Springer Nature Switzerland AG
The registered company address is: Gewerbestrasse 11, 6330 Cham, Switzerland

Preface

Twenty Years of Computational Science

Welcome to the 20th Annual International Conference on Computational Science (ICCS – https://www.iccs-meeting.org/iccs2020/).

During the preparation for this 20th edition of ICCS we were considering all kinds of nice ways to celebrate two decennia of computational science. Afterall when we started this international conference series, we never expected it to be so successful and running for so long at so many different locations across the globe! So we worked on a mind-blowing line up of renowned keynotes, music by scientists, awards, a play written by and performed by computational scientists, press attendance, a lovely venue... you name it, we had it all in place. Then corona hit us.

After many long debates and considerations, we decided to cancel the physical event but still support our scientists and allow for publication of their accepted peer-reviewed work. We are proud to present the proceedings you are reading as a result of that.

ICCS 2020 is jointly organized by the University of Amsterdam, NTU Singapore, and the University of Tennessee.

The International Conference on Computational Science is an annual conference that brings together researchers and scientists from mathematics and computer science as basic computing disciplines, as well as researchers from various application areas who are pioneering computational methods in sciences such as physics, chemistry, life sciences, engineering, arts and humanitarian fields, to discuss problems and solutions in the area, to identify new issues, and to shape future directions for research.

Since its inception in 2001, ICCS has attracted increasingly higher quality and numbers of attendees and papers, and 2020 was no exception, with over 350 papers accepted for publication. The proceedings series have become a major intellectual resource for computational science researchers, defining and advancing the state of the art in this field.

The theme for ICCS 2020, "Twenty Years of Computational Science", highlights the role of Computational Science over the last 20 years, its numerous achievements, and its future challenges. This conference was a unique event focusing on recent developments in: scalable scientific algorithms, advanced software tools, computational grids, advanced numerical methods, and novel application areas. These innovative novel models, algorithms, and tools drive new science through efficient application in areas such as physical systems, computational and systems biology, environmental systems, finance, and others.

This year we had 719 submissions (230 submissions to the main track and 489 to the thematic tracks). In the main track, 101 full papers were accepted (44%). In the thematic tracks, 249 full papers were accepted (51%). A high acceptance rate in the thematic tracks is explained by the nature of these, where many experts in a particular field are personally invited by track organizers to participate in their sessions.

ICCS relies strongly on the vital contributions of our thematic track organizers to attract high-quality papers in many subject areas. We would like to thank all committee members from the main and thematic tracks for their contribution to ensure a high standard for the accepted papers. We would also like to thank Springer, Elsevier, the Informatics Institute of the University of Amsterdam, the Institute for Advanced Study of the University of Amsterdam, the SURFsara Supercomputing Centre, the Netherlands eScience Center, the VECMA Project, and Intellegibilis for their support. Finally, we very much appreciate all the Local Organizing Committee members for their hard work to prepare this conference.

We are proud to note that ICCS is an A-rank conference in the CORE classification.

We wish you good health in these troubled times and hope to see you next year for ICCS 2021.

June 2020

Valeria V. Krzhizhanovskaya
Gábor Závodszky
Michael Lees
Jack Dongarra
Peter M. A. Sloot
Sérgio Brissos
João Teixeira

Organization

Thematic Tracks and Organizers

Advances in High-Performance Computational Earth Sciences: Applications and Frameworks – IHPCES

Takashi Shimokawabe
Kohei Fujita
Dominik Bartuschat

Agent-Based Simulations, Adaptive Algorithms and Solvers – ABS-AAS

Maciej Paszynski
David Pardo
Victor Calo
Robert Schaefer
Quanling Deng

Applications of Computational Methods in Artificial Intelligence and Machine Learning – ACMAIML

Kourosh Modarresi
Raja Velu
Paul Hofmann

Biomedical and Bioinformatics Challenges for Computer Science – BBC

Mario Cannataro
Giuseppe Agapito
Mauro Castelli
Riccardo Dondi
Rodrigo Weber dos Santos
Italo Zoppis

Classifier Learning from Difficult Data – CLD²

Michał Woźniak
Bartosz Krawczyk
Paweł Ksieniewicz

Complex Social Systems through the Lens of Computational Science – CSOC

Debraj Roy
Michael Lees
Tatiana Filatova

Computational Health – CompHealth

Sergey Kovalchuk
Stefan Thurner
Georgiy Bobashev

Computational Methods for Emerging Problems in (dis-)Information Analysis – DisA

Michal Choras
Konstantinos Demestichas

Computational Optimization, Modelling and Simulation – COMS

Xin-She Yang
Slawomir Koziel
Leifur Leifsson

Computational Science in IoT and Smart Systems – IoTSS

Vaidy Sunderam
Dariusz Mrozek

Computer Graphics, Image Processing and Artificial Intelligence – CGIPAI

Andres Iglesias
Lihua You
Alexander Malyshev
Hassan Ugail

Data-Driven Computational Sciences – DDCS

Craig C. Douglas
Ana Cortes
Hiroshi Fujiwara
Robert Lodder
Abani Patra
Han Yu

Machine Learning and Data Assimilation for Dynamical Systems – MLDADS

Rossella Arcucci
Yi-Ke Guo

Meshfree Methods in Computational Sciences – MESHFREE

Vaclav Skala
Samsul Ariffin Abdul Karim
Marco Evangelos Biancolini
Robert Schaback

Rongjiang Pan
Edward J. Kansa

Multiscale Modelling and Simulation – MMS

Derek Groen
Stefano Casarin
Alfons Hoekstra
Bartosz Bosak
Diana Suleimenova

Quantum Computing Workshop – QCW

Katarzyna Rycerz
Marian Bubak

Simulations of Flow and Transport: Modeling, Algorithms and Computation – SOFTMAC

Shuyu Sun
Jingfa Li
James Liu

Smart Systems: Bringing Together Computer Vision, Sensor Networks and Machine Learning – SmartSys

Pedro J. S. Cardoso
João M. F. Rodrigues
Roberto Lam
Janio Monteiro

Software Engineering for Computational Science – SE4Science

Jeffrey Carver
Neil Chue Hong
Carlos Martinez-Ortiz

Solving Problems with Uncertainties – SPU

Vassil Alexandrov
Aneta Karaivanova

Teaching Computational Science – WTCS

Angela Shiflet
Alfredo Tirado-Ramos
Evguenia Alexandrova

Uncertainty Quantification for Computational Models – UNEQUIvOCAL

Wouter Edeling
Anna Nikishova
Peter Coveney

Program Committee and Reviewers

Ahmad Abdelfattah
Samsul Ariffin
  Abdul Karim
Evgenia Adamopoulou
Jaime Afonso Martins
Giuseppe Agapito
Ram Akella
Elisabete Alberdi Celaya
Luis Alexandre
Vassil Alexandrov
Evguenia Alexandrova
Hesham H. Ali
Julen Alvarez-Aramberri
Domingos Alves
Julio Amador Diaz Lopez
Stanislaw
  Ambroszkiewicz
Tomasz Andrysiak
Michael Antolovich
Hartwig Anzt
Hideo Aochi
Hamid Arabnejad
Rossella Arcucci
Khurshid Asghar
Marina Balakhontceva
Bartosz Balis
Krzysztof Banas
João Barroso
Dominik Bartuschat
Nuno Basurto
Pouria Behnoudfar
Joern Behrens
Adrian Bekasiewicz
Gebrai Bekdas
Stefano Beretta
Benjamin Berkels
Martino Bernard

Daniel Berrar
Sanjukta Bhowmick
Marco Evangelos
  Biancolini
Georgiy Bobashev
Bartosz Bosak
Marian Bubak
Jérémy Buisson
Robert Burduk
Michael Burkhart
Allah Bux
Aleksander Byrski
Cristiano Cabrita
Xing Cai
Barbara Calabrese
Jose Camata
Mario Cannataro
Alberto Cano
Pedro Jorge Sequeira
  Cardoso
Jeffrey Carver
Stefano Casarin
Manuel Castañón-Puga
Mauro Castelli
Eduardo Cesar
Nicholas Chancellor
Patrikakis Charalampos
Ehtzaz Chaudhry
Chuanfa Chen
Siew Ann Cheong
Andrey Chernykh
Lock-Yue Chew
Su Fong Chien
Marta Chinnici
Sung-Bae Cho
Michal Choras
Loo Chu Kiong

Neil Chue Hong
Svetlana Chuprina
Paola Cinnella
Noélia Correia
Adriano Cortes
Ana Cortes
Enrique
  Costa-Montenegro
David Coster
Helene Coullon
Peter Coveney
Attila Csikasz-Nagy
Loïc Cudennec
Javier Cuenca
Yifeng Cui
António Cunha
Ben Czaja
Pawel Czarnul
Flávio Martins
Bhaskar Dasgupta
Konstantinos Demestichas
Quanling Deng
Nilanjan Dey
Khaldoon Dhou
Jamie Diner
Jacek Dlugopolski
Simona Domesová
Riccardo Dondi
Craig C. Douglas
Linda Douw
Rafal Drezewski
Hans du Buf
Vitor Duarte
Richard Dwight
Wouter Edeling
Waleed Ejaz
Dina El-Reedy

Amgad Elsayed
Nahid Emad
Chriatian Engelmann
Gökhan Ertaylan
Alex Fedoseyev
Luis Manuel Fernández
Antonino Fiannaca
Christos
   Filelis-Papadopoulos
Rupert Ford
Piotr Frackiewicz
Martin Frank
Ruy Freitas Reis
Karl Frinkle
Haibin Fu
Kohei Fujita
Hiroshi Fujiwara
Takeshi Fukaya
Wlodzimierz Funika
Takashi Furumura
Ernst Fusch
Mohamed Gaber
David Gal
Marco Gallieri
Teresa Galvao
Akemi Galvez
Salvador García
Bartlomiej Gardas
Delia Garijo
Frédéric Gava
Piotr Gawron
Bernhard Geiger
Alex Gerbessiotis
Ivo Goncalves
Antonio Gonzalez Pardo
Jorge
   González-Domínguez
Yuriy Gorbachev
Pawel Gorecki
Michael Gowanlock
Manuel Grana
George Gravvanis
Derek Groen
Lutz Gross
Sophia
   Grundner-Culemann

Pedro Guerreiro
Tobias Guggemos
Xiaohu Guo
Piotr Gurgul
Filip Guzy
Pietro Hiram Guzzi
Zulfiqar Habib
Panagiotis Hadjidoukas
Masatoshi Hanai
John Hanley
Erik Hanson
Habibollah Haron
Carina Haupt
Claire Heaney
Alexander Heinecke
Jurjen Rienk Helmus
Álvaro Herrero
Bogumila Hnatkowska
Maximilian Höb
Erlend Hodneland
Olivier Hoenen
Paul Hofmann
Che-Lun Hung
Andres Iglesias
Takeshi Iwashita
Alireza Jahani
Momin Jamil
Vytautas Jancauskas
João Janeiro
Peter Janku
Fredrik Jansson
Jirí Jaroš
Caroline Jay
Shalu Jhanwar
Zhigang Jia
Chao Jin
Zhong Jin
David Johnson
Guido Juckeland
Maria Juliano
Edward J. Kansa
Aneta Karaivanova
Takahiro Katagiri
Timo Kehrer
Wayne Kelly
Christoph Kessler

Jakub Klikowski
Harald Koestler
Ivana Kolingerova
Georgy Kopanitsa
Gregor Kosec
Sotiris Kotsiantis
Ilias Kotsireas
Sergey Kovalchuk
Michal Koziarski
Slawomir Koziel
Rafal Kozik
Bartosz Krawczyk
Elisabeth Krueger
Valeria Krzhizhanovskaya
Pawel Ksieniewicz
Marek Kubalcík
Sebastian Kuckuk
Eileen Kuehn
Michael Kuhn
Michal Kulczewski
Krzysztof Kurowski
Massimo La Rosa
Yu-Kun Lai
Jalal Lakhlili
Roberto Lam
Anna-Lena Lamprecht
Rubin Landau
Johannes Langguth
Elisabeth Larsson
Michael Lees
Leifur Leifsson
Kenneth Leiter
Roy Lettieri
Andrew Lewis
Jingfa Li
Khang-Jie Liew
Hong Liu
Hui Liu
Yen-Chen Liu
Zhao Liu
Pengcheng Liu
James Liu
Marcelo Lobosco
Robert Lodder
Marcin Los
Stephane Louise

Frederic Loulergue
Paul Lu
Stefan Luding
Onnie Luk
Scott MacLachlan
Luca Magri
Imran Mahmood
Zuzana Majdisova
Alexander Malyshev
Muazzam Maqsood
Livia Marcellino
Tomas Margalef
Tiziana Margaria
Svetozar Margenov
Urszula
    Markowska-Kaczmar
Osni Marques
Carmen Marquez
Carlos Martinez-Ortiz
Paula Martins
Flávio Martins
Luke Mason
Pawel Matuszyk
Valerie Maxville
Wagner Meira Jr.
Roderick Melnik
Valentin Melnikov
Ivan Merelli
Choras Michal
Leandro Minku
Jaroslaw Miszczak
Janio Monteiro
Kourosh Modarresi
Fernando Monteiro
James Montgomery
Andrew Moore
Dariusz Mrozek
Peter Mueller
Khan Muhammad
Judit Muñoz
Philip Nadler
Hiromichi Nagao
Jethro Nagawkar
Kengo Nakajima
Ionel Michael Navon
Philipp Neumann

Mai Nguyen
Hoang Nguyen
Nancy Nichols
Anna Nikishova
Hitoshi Nishizawa
Brayton Noll
Algirdas Noreika
Enrique Onieva
Kenji Ono
Eneko Osaba
Aziz Ouaarab
Serban Ovidiu
Raymond Padmos
Wojciech Palacz
Ivan Palomares
Rongjiang Pan
Joao Papa
Nikela Papadopoulou
Marcin Paprzycki
David Pardo
Anna Paszynska
Maciej Paszynski
Abani Patra
Dana Petcu
Serge Petiton
Bernhard Pfahringer
Frank Phillipson
Juan C. Pichel
Anna
    Pietrenko-Dabrowska
Laércio L. Pilla
Armando Pinho
Tomasz Piontek
Yuri Pirola
Igor Podolak
Cristina Portales
Simon Portegies Zwart
Roland Potthast
Ela Pustulka-Hunt
Vladimir Puzyrev
Alexander Pyayt
Rick Quax
Cesar Quilodran Casas
Barbara Quintela
Ajaykumar Rajasekharan
Celia Ramos

Lukasz Rauch
Vishal Raul
Robin Richardson
Heike Riel
Sophie Robert
Luis M. Rocha
Joao Rodrigues
Daniel Rodriguez
Albert Romkes
Debraj Roy
Katarzyna Rycerz
Alberto Sanchez
Gabriele Santin
Alex Savio
Robert Schaback
Robert Schaefer
Rafal Scherer
Ulf D. Schiller
Bertil Schmidt
Martin Schreiber
Alexander Schug
Gabriela Schütz
Marinella Sciortino
Diego Sevilla
Angela Shiflet
Takashi Shimokawabe
Marcin Sieniek
Nazareen Sikkandar
    Basha
Anna Sikora
Janaína De Andrade Silva
Diana Sima
Robert Sinkovits
Haozhen Situ
Leszek Siwik
Vaclav Skala
Peter Sloot
Renata Slota
Grazyna Slusarczyk
Sucha Smanchat
Marek Smieja
Maciej Smolka
Bartlomiej Sniezynski
Isabel Sofia Brito
Katarzyna Stapor
Bogdan Staszewski

Contents – Part VI

Meshfree Methods in Computational Sciences

Quantum Computing Workshop

Data Driven Computational Sciences

A High-Performance Implementation of Bayesian Matrix Factorization with Limited Communication

Tom Vander Aa[1]([✉]), Xiangju Qin[2,3], Paul Blomstedt[2,4], Roel Wuyts[1],
Wilfried Verachtert[1], and Samuel Kaski[2]

[1] ExaScience Life Lab at imec, Leuven, Belgium
tom.vanderaa@imec.be
[2] Department of Computer Science, Helsinki Institute
for Information Technology (HIIT), Aalto University, Espoo, Finland
[3] University of Helsinki, Helsinki, Finland
[4] F-Secure, Helsinki, Finland

Abstract. Matrix factorization is a very common machine learning technique in recommender systems. Bayesian Matrix Factorization (BMF) algorithms would be attractive because of their ability to quantify uncertainty in their predictions and avoid over-fitting, combined with high prediction accuracy. However, they have not been widely used on large-scale data because of their prohibitive computational cost. In recent work, efforts have been made to reduce the cost, both by improving the scalability of the BMF algorithm as well as its implementation, but so far mainly separately. In this paper we show that the state-of-the-art of both approaches to scalability can be combined. We combine the recent highly-scalable Posterior Propagation algorithm for BMF, which parallelizes computation of blocks of the matrix, with a distributed BMF implementation that users asynchronous communication within each block. We show that the combination of the two methods gives substantial improvements in the scalability of BMF on web-scale datasets, when the goal is to reduce the wall-clock time.

1 Introduction

Matrix Factorization (MF) is a core machine learning technique for applications of collaborative filtering, such as recommender systems or drug discovery, where a data matrix $\mathbf{R}$ is factorized into a product of two matrices, such that $\mathbf{R} \approx \mathbf{U}\mathbf{V}^\top$. The main task in such applications is to predict unobserved elements of a partially observed data matrix. In recommender systems, the elements of $\mathbf{R}$ are often ratings given by users to items, while in drug discovery they typically represent bioactivities between chemical compounds and protein targets or cell lines.

Bayesian Matrix Factorization (BMF) [2,13], formulates the matrix factorization task as a probabilistic model, with Bayesian inference conducted on the

© Springer Nature Switzerland AG 2020
V. V. Krzhizhanovskaya et al. (Eds.): ICCS 2020, LNCS 12142, pp. 3–16, 2020.
https://doi.org/10.1007/978-3-030-50433-5_1

unknown matrices U and V. Advantages often associated with BMF include robustness to over-fitting and improved predictive accuracy, as well as flexible utilization of prior knowledge and side-data. Finally, for application domains such as drug discovery, the ability of the Bayesian approach to quantify uncertainty in predictions is of crucial importance [9].

Despite the appeal and many advantages of BMF, scaling up the posterior inference for industry-scale problems has proven difficult. Scaling up to this level requires both data and computations to be distributed over many workers, and so far only very few distributed implementations of BMF have been presented in the literature. In [16], a high-performance computing implementation of BMF using Gibbs sampling for distributed systems was proposed. The authors considered three different distributed programming models: Message Passing Interface (MPI), Global Address Space Programming Interface (GASPI) and ExaS-HARK. In a different line of work, [1] proposed to use a distributed version of the minibatch-based Stochastic Gradient Langevin Dynamics algorithm for posterior inference in BMF models. While a key factor in devising efficient distributed solutions is to be able to minimize communication between worker nodes, both of the above solutions require some degree of communication in between iterations.

A recent promising proposal, aiming at minimizing communication and thus reaching a solution in a faster way, is to use a hierarchical embarrassingly parallel MCMC strategy [12]. This technique, called BMF with Posterior Propagation (BMF-PP), enhances regular embarrassingly parallel MCMC (e.g. [10,17]), which does not work well for matrix factorization [12] because of identifiability issues. BMF-PP introduces communication at predetermined limited phases in the algorithm to make the problem identifiable, effectively building one model for all the parallelized data subsets, while in previous works multiple independent solutions were found per subset.

The current paper is based on a realization that the approaches of [16] and [12] are compatible, and in fact synergistic. BMF-PP will be able to parallelize a massive matrix but will be the more accurate the larger the parallelized blocks are. Now replacing the earlier serial processing of the blocks by the distributed BMF in [16] will allow making the blocks larger up to the scalability limit of the distributed BMF, and hence decrease the wall-clock time by engaging more processors.

The main contributions of this work are:

- We combine both approaches, allowing for parallelization both at the algorithmic and at the implementation level.
- We analyze what is the best way to subdivide the original matrix into subsets, taking into account both compute performance *and* model quality.
- We examine several web-scale datasets and show that datasets with different properties (like the number of non-zero items per row) require different parallelization strategies.

The rest of this paper is organized as follows. In Sect. 2 we present the existing Posterior Propagation algorithm and distributed BMF implementation and we explain how to combine both. Section 3 is the main section of this paper, where we document the experimental setup and used dataset, we compare with related MF methods, and present the results both from a machine learning point of view, as from a high-performance compute point of view. In Sect. 4 we draw conclusions and propose future work.

2 Distributed Bayesian Matrix Factorization with Posterior Propagation

In this section, we first briefly review the BMF model and then describe the individual aspects of distributed computation and Posterior Propagation and how to combine them.

2.1 Bayesian Matrix Factorization

In matrix factorization, a (typically very sparsely observed) data matrix $\mathbf{R} \in \mathbb{R}^{N \times D}$ is factorized into a product of two matrices $\mathbf{U} \in \mathbb{R}^{N \times K} = (\mathbf{u}_1, \ldots, \mathbf{u}_N)^\top$ and $\mathbf{V} = (\mathbf{v}_1, \ldots, \mathbf{v}_D)^\top \in \mathbb{R}^{D \times K}$. In the context of recommender systems, $\mathbf{R}$ is a rating matrix, with N the number of users and D the number of rated items.

In Bayesian matrix factorization [2,13], the data are modelled as

$$p(\mathbf{R}|\mathbf{U}, \mathbf{V}) = \prod_{n=1}^{N} \prod_{d=1}^{D} \left[\mathcal{N} \left(r_{nd} | \mathbf{u}_n^\top \mathbf{v}_d, \tau^{-1} \right) \right]^{I_{nd}} \tag{1}$$

where I_{nd} denotes an indicator which equals 1 if the element r_{nd} is observed and 0 otherwise, and τ denotes the residual noise precision. The two parameter matrices $\mathbf{U}$ and $\mathbf{V}$ are assigned Gaussian priors. Our goal is then to compute the joint posterior density $p(\mathbf{U}, \mathbf{V}|\mathbf{R}) \propto p(\mathbf{U})p(\mathbf{V})p(\mathbf{R}|\mathbf{U}, \mathbf{V})$, conditional on the observed data. Posterior inference is typically done using Gibbs sampling, see [13] for details.

2.2 Bayesian Matrix Factorization with Posterior Propagation

In the *Posterior Propagation* (PP) framework [12], we start by partitioning $\mathbf{R}$ with respect to both rows and columns into $I \times J$ subsets $\mathbf{R}^{(i,j)}$, $i = 1, \ldots, I$, $j = 1, \ldots, J$. The parameter matrices $\mathbf{U}$ and $\mathbf{V}$ are correspondingly partitioned into I and J submatrices, respectively. The basic idea of PP is to process each subset using a hierarchical embarrassingly parallel MCMC scheme in three phases, where the posteriors from each phase are propagated forwards and used as priors in the following phase, thus introducing dependencies between the subsets. The approach proceeds as follows (for an illustration, see Fig. 1):

Phase (a): Joint inference for submatrices $(\mathbf{U}^{(1)}, \mathbf{V}^{(1)})$, conditional on data subset $\mathbf{R}^{(1,1)}$:

$$p\left(\mathbf{U}^{(1)}, \mathbf{V}^{(1)}|\mathbf{R}^{(1,1)}\right) \propto p\left(\mathbf{U}^{(1)}\right) p\left(\mathbf{V}^{(1)}\right) p\left(\mathbf{R}^{(1,1)}|\mathbf{U}^{(1)}, \mathbf{V}^{(1)}\right).$$

Phase (b): Joint inference in parallel for submatrices $(\mathbf{U}^{(i)}, \mathbf{V}^{(1)})$, $i = 2, \ldots, I$, and $(\mathbf{U}^{(1)}, \mathbf{V}^{(j)})$, $j = 2, \ldots, J$, conditional on data subsets which share columns or rows with $\mathbf{R}^{(1,1)}$, and using posterior marginals from phase (a) as priors:

$$p\left(\mathbf{U}^{(i)}, \mathbf{V}^{(1)} | \mathbf{R}^{(1,1)}, \mathbf{R}^{(i,1)}\right) \propto p\left(\mathbf{V}^{(1)} | \mathbf{R}^{(1,1)}\right) p\left(\mathbf{U}^{(i)}\right) p\left(\mathbf{R}^{(i,1)} | \mathbf{U}^{(i)}, \mathbf{V}^{(1)}\right),$$

$$p\left(\mathbf{U}^{(1)}, \mathbf{V}^{(j)} | \mathbf{R}^{(1,1)}, \mathbf{R}^{(1,j)}\right) \propto p\left(\mathbf{U}^{(1)} | \mathbf{R}^{(1,1)}\right) p\left(\mathbf{V}^{(j)}\right) p\left(\mathbf{R}^{(1,j)} | \mathbf{U}^{(1)}, \mathbf{V}^{(j)}\right).$$

Phase (c): Joint inference in parallel for submatrices $(\mathbf{U}^{(i)}, \mathbf{V}^{(j)})$, $i = 2, \ldots, I$, $j = 2, \ldots, J$, conditional on the remaining data subsets, and using posterior marginals propagated from phase (b) as priors:

$$p\left(\mathbf{U}^{(i)}, \mathbf{V}^{(j)} | \mathbf{R}^{(1,1)}, \mathbf{R}^{(i,1)}, \mathbf{R}^{(1,j)}, \mathbf{R}^{(i,j)}\right)$$

$$\propto p\left(\mathbf{U}^{(i)} | \mathbf{R}^{(1,1)}, \mathbf{R}^{(i,1)}\right) p\left(\mathbf{V}^{(j)} | \mathbf{R}^{(1,1)}, \mathbf{R}^{(1,j)}\right) p\left(\mathbf{R}^{(i,j)} | \mathbf{U}^{(i)}, \mathbf{V}^{(j)}\right).$$

Finally, the aggregated posterior is obtained by combining the posteriors obtained in phases (a)–(c) and dividing away the multiply-counted propagated posterior marginals; see [12] for details.

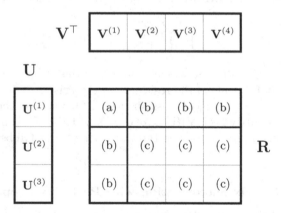

Fig. 1. Illustration of Posterior Propagation (PP) for a data matrix $\mathbf{R}$ partitioned into 3×4 subsets. Subset inferences for the matrices $\mathbf{U}$ and $\mathbf{V}$ proceed in three successive phases, with posteriors obtained in one phase being propagated as priors to the next one. The letters (a), (b), (c) in the matrix $\mathbf{R}$ refer to the phase in which the particular data subset is processed. Within each phase, the subsets are processed in parallel with no communication. Figure adapted from [12].

2.3   Distributed Bayesian Matrix Factorization

In [16] a distributed parallel implementation of BMF is proposed. In that paper an implementation the BMF algorithm [13] is proposed that distributes the rows and columns of $\mathbf{R}$ across different nodes of a supercomputing system.

Since Gibbs sampling is used, rows of $\mathbf{U}$ and rows of $\mathbf{V}$ are independent and can be sampled in parallel, on different nodes. However, there is a dependency between samples of $\mathbf{U}$ and $\mathbf{V}$. The communication pattern between $\mathbf{U}$ and $\mathbf{V}$ is shown in Fig. 2.

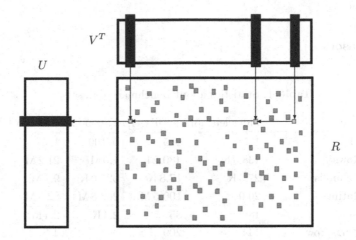

Fig. 2. Communication pattern in the distributed implementation of BMF

The main contribution of this implementation is how to distribute $\mathbf{U}$ and $\mathbf{V}$ to make sure the computational load is distributed as equally as possible and the amount of data communication is minimized. The authors of [16] optimize the distributions by analysing the sparsity structure of $\mathbf{R}$.

As presented in the paper, distributed BMF provides a reasonable speed-up for compute systems up to 128 nodes. After this, speed-up in the strong scaling case is limited by the increase in communication, while at the same computations for each node decrease.

2.4   Distributed Bayesian Matrix Factorization with Posterior Propagation

By combining the distributed BMF implementation with PP, we can exploit the parallelism at both levels. The different subsets of $\mathbf{R}$ in the parallel phases of PP are independent and can thus be computed on in parallel on different resources. Inside each subset we exploit parallelism by using the distributed BMF implementation with MPI communication.

3   Experiments

In this section we evaluate the distributed BMF implementation with Posterior Propagation (D-BMF+PP) and competitors for benchmark datasets. We first have a look at the datasets we will use for the experiments. Next, we compare our proposed D-BMF+PP implementation with other related matrix factorization methods in terms of accuracy and runtime, with a special emphasis on the benefits of Bayesian methods. Finally we look at the strong scaling of D-BMF+PP and what is a good way to split the matrix in to blocks for Posterior Propagation.

3.1   Datasets

Table 1. Statistics about benchmark datasets.

Name	Movielens [5]	Netflix [4]	Yahoo [3]	Amazon [8]
Scale	1–5	1–5	0–100	1–5
#Rows	138.5 K	480.2 K	1.0 M	21.2 M
#Columns	27.3 K	17.8 K	625.0 K	9.7 M
#Ratings	20.0 M	100.5 M	262.8 M	82.5 M
Sparsity	189	85	2.4 K	2.4 M
Ratings/row	144	209	263	4
#Rows/#cols	5.1	27.0	1.6	2.2
K	10	100	100	10
Rows/sec ($\times 1000$)	416	15	27	911
Ratings/sec ($\times 10^6$)	70	5.5	5.2	3.8

Table 1 shows the different webscale datasets we used for our experiments. The table shows we have a wide diversity of properties: from the relatively small Movielens dataset with 20M ratings, to the 262M ratings Yahoo dataset. Sparsity is expressed as the ratio of the total number of elements in the matrix (#rows × #columns) to the number of filled-in ratings (#ratings). This metric also varies: especially the Amazon dataset is very sparsely filled.

The scale of the ratings (either 1 to 5 or 0 to 100) is only important when looking at prediction error values, which we will do in the next section.

The final three lines of the table are not data properties, but rather properties of the matrix factorization methods. K is the number of latent dimensions used, which is different per dataset but chosen to be common to all matrix factorization methods. The last two lines are two compute performance metrics, which will be discussed in Sect. 3.4.

3.2   Related Methods

In this section we compare the proposed BMF+PP method to other matrix factorization (MF) methods, in terms of accuracy (Root-Mean-Square Error or RMSE on the test set), and in terms of compute performance (wall-clock time).

Stochastic gradient decent (SGD [15]), Coordinate Gradient Descent (CGD [18]) and Alternating Least Squares (ALS [14]) are the three most-used algorithms for non-Bayesian MF.

CGD- and ALS-based algorithms update along one dimension at a time while the other dimension of the matrix remains fixed. Many variants of ALS and CGD exist that improve the convergence [6], or the parallelization degree [11], or are optimized for non-sparse rating matrices [7]. In this paper we limit ourselves to comparing to methods, which divide the rating matrix into blocks, which is necessary to support very large matrices. Of the previous methods, this includes the SGD-based ones.

FPSGD [15] is a very efficient SGD-based library for matrix factorization on multi-cores. While FPSGD is a single-machine implementation, which limits its capability to solve large-scale problems, it has outperformed all other methods on a machine with up to 16 cores. NOMAD [19] extends the idea of block partitioning, adding the capability to release a portion of a block to another thread before its full completion. It performs similarly to FPSGD on a single machine, and can scale out to a 64-node HPC cluster.

Table 2. RMSE of different matrix factorization methods on benchmark datasets.

Dataset	BMF+PP	NOMAD	FPSGD
Movielens	0.76	0.77	0.77
Netflix	0.90	0.91	0.92
Yahoo	21.79	21.91	21.78
Amazon	1.13	1.20	1.15

Table 2 compares the RMSE values on the test sets of the four selected datasets. For all methods, we used the K provided in Table 1. For competitor methods, we used the default hyperparameters suggested by the authors for the different data sets. As already concluded in [12], BMF with Posterior Propagation results in equally good RMSE compared to the original BMF. We also see from the table that on average the Bayesian method produces only slightly better results, in terms of RMSE. However we do know that Bayesian methods have significant statistical advantages [9] that stimulate their use.

For many applications, this advantage outweighs the much higher computational cost. Indeed, as can be seem from Table 3, BMF is significantly more expensive, than NOMAD and FPSGD, even when taking in to account the speed-ups thanks to using PP. NOMAD is the fastest method, thanks to the aforementioned improvements compared to FPSGD.

Table 3. Wall-clock time (hh:mm) of different matrix factorization methods on benchmark datasets, running on single-node system with 16 cores.

Dataset	BMF+PP	BMF	NOMAD	FPSGD
Movielens	0:07	0:14	0:08	0:09
Netflix	2:02	4:39	0:08	1:04
Yahoo	2:13	12:22	0:10	2:41
Amazon	4:15	13:02	0:40	2:28

3.3 Block Size

The combined method will achieve some of the parallelization with the PP algorithm, and some with the distributed BMF within each block. We next compare the performance as the share between the two is varied by varying the block size. We find that blocks should be approximately square, meaning the number of rows and columns inside the each block should be more or less equal. This implies the number of blocks across the rows of the **R** matrix will be less if the **R** matrix has fewer rows and vice versa.

Figure 3 shows optimizing the block size is crucial to achieve a good speed up with BMF+PP and avoid compromising the quality of the model (as measured with the RMSE). The block size in the figure, listed as $I \times J$, means the R matrix is split in I blocks (with equal amount of rows) in the vertical direction and J blocks (with equal amount of columns) in the horizontal direction. The figure explores different block sizes for the Netflix dataset, which has significantly more rows than columns ($27\times$) as can be seen from Table 1. This is reflected by the fact that the data point with the smallest bubble area (20×3) provides the best trade-off between wall-clock time and RMSE.

We stipulate that the reason is the underlying trade-off between the amount of information in the block and the amount of compute per block. Both the amount of information and the amount of compute can optimized (maximized and minimized respectively) by making the blocks approximately squared, since both are proportionate to the ratio of the area versus the circumference of the block.

We will come back to this trade-off when we look at performance of BMF+PP when scaling to multiple nodes for the different datasets in Sect. 3.4.

3.4 Scaling

In this section we look at how the added parallelization of Posterior Propagation increases the strong scaling behavior of BMF+PP. Strong scaling means we look at the speed-up obtained by increasing the amount of compute nodes, while keeping the dataset constant.

The graphs in this section are displayed with a logarithmic scale on both axes. This has the effect that linear scaling (i.e. doubling of the amount of resources

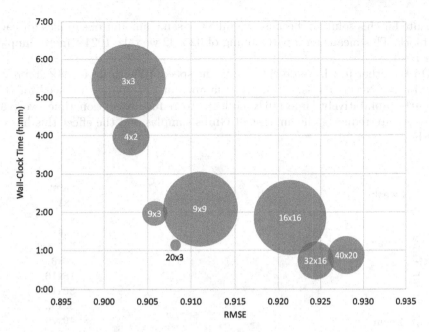

Fig. 3. Block size exploration: RMSE on test set and wall-clock time (hh:mm) for different block size on Netflix. Each bubble is the result for a different block size indicated as the number of blocks across the rows and the columns inside the bubble. The size of bubbles is an indication for the aspect ratio of the blocks inside PP. Smaller bubbles indicate the blocks are more square.

results in a halving of the runtime) is shown as a straight line. Additionally, we indicate Pareto optimal solutions with a blue dot. For these solutions one cannot reduce the execution time without increasing the resources.

We performed experiments with different block sizes, indicated with different colors in the graphs. For example, the yellow line labeled 16×8 in Fig. 4 means we performed PP with 16 blocks for the rows of $\mathbf{R}$ and 8 blocks for the columns of $\mathbf{R}$.

We have explored scaling on a system with up to 128 K nodes, and use this multi-node parallelism in two ways:

1. Parallelisation inside a block using the distributed version of BMF [16].
2. Parallelism across blocks. In phase (b) we can exploit parallelism across the blocks in the first row and first column, using up to $I + J$ nodes where I and J is the number of blocks in the vertical, respectively horizontal direction of the matrix. In phase (c) we can use up to $I \times J$ nodes.

General Trends. When looking at the four graphs (Fig. 4 and 5), we observe that for the same amount of nodes using more blocks for posterior propagation, increases the wall-clock time. The main reason is that we do significantly more compute for this solution, because we take the same amount of samples for each sub-block. This means for a partitioning of 32×32 we take $1024\times$ more samples than 1×1.

On the other hand, we do get significant speedup, with up to $68\times$ improvement for the Netflix dataset. However the amount of resources we need for this is clearly prohibitively large ($16\,K$ nodes). To reduce execution time, we plan to investigate reducing the amount of Gibbs samples and the effect this has on RMSE.

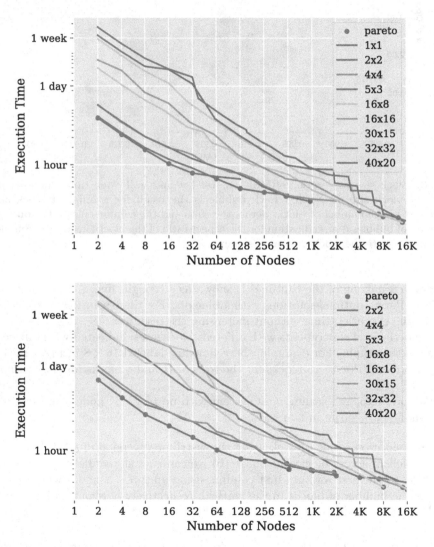

Fig. 4. Strong scaling results for the Netflix (top) and Yahoo (bottom) datasets. X-axis indicates the amount of compute resources (#nodes). Y-axis is wall-clock time. The different series correspond to different block sizes. (Color figure online)

Netflix and Yahoo. Runs on the Netflix and Yahoo datasets (Fig. 4) have been performed with 100 latent dimensions (K = 100). Since computational intensity (the amount of compute per row/column of **R**), is $O(K^3)$, the amount of compute versus compute for these experiments is high, leading to good scalability for a single block (1 × 1) for Netflix, and almost linear scalability up to 16 or even 64 nodes. The Yahoo dataset is too large to run with a 1 × 1 block size, but we see a similar trend for 2 × 2.

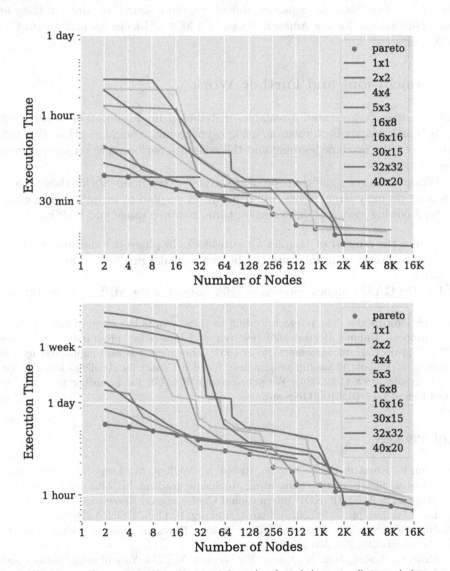

Fig. 5. Strong scaling results for the Movielens (top) and Amazon (bottom) datasets. X-axis indicates the amount of compute resources (#nodes). Y-axis is wall-clock time. The different series correspond to different block sizes.

Movielens and Amazon. For Movielens and Amazon (Fig. 5), we used $K = 10$ for our experiments. This implies a much smaller amount of compute for the same communication cost inside a block. Hence scaling for 1×1 is mostly flat. We do get significant performance improvements with more but smaller blocks where the Amazon dataset running on 2048 nodes, with 32×32 blocks is $20\times$ faster than the best block size (1×1) on a single node.

When the nodes in the experiment align with the parallelism across of blocks (either $I + J$ or $I \times J$ as explained above), we see a significant drop in the run time. For example for the Amazon dataset on 32×32 blocks going from 1024 to 2048 nodes.

4   Conclusions and Further Work

In this paper we presented a scalable, distributed implementation of Bayesian Probabilistic Matrix Factorization, using asynchronous communication. We evaluated both the machine learning and the compute performance on several web-scale datasets.

While we do get significant speed-ups, resource requirements for these speed-ups are extremely large. Hence, in future work, we plan to investigate the effect of the following measures on execution time, resource usage and RMSE:

- Reduce the number of samples for sub-blocks in phase (b) and phase (c).
- Use the posterior from phase (b) blocks to make predictions for phase (c) blocks.
- Use the GASPI implementation of [16], instead of the MPI implementation.

Acknowledgments. The research leading to these results has received funding from the European Union's Horizon2020 research and innovation programme under the EPEEC project, grant agreement No 801051. The work was also supported by the Academy of Finland (Flagship programme: Finnish Center for Artificial Intelligence, FCAI; grants 319264, 292334). We acknowledge PRACE for awarding us access to Hazel Hen at GCS@HLRS, Germany.

References

1. Ahn, S., Korattikara, A., Liu, N., Rajan, S., Welling, M.: Large-scale distributed Bayesian matrix factorization using stochastic gradient MCMC. In: Proceedings of the 21th ACM SIGKDD International Conference on Knowledge Discovery and Data Mining, pp. 9–18 (2015). https://doi.org/10.1145/2783258.2783373
2. Bhattacharya, A., Dunson, D.B.: Sparse Bayesian infinite factor models. Biometrika **98**, 291–306 (2011)
3. Dror, G., Koenigstein, N., Koren, Y., Weimer, M.: The Yahoo! music dataset and KDD-Cup'11. In: Dror, G., Koren, Y., Weimer, M. (eds.) Proceedings of KDD Cup 2011. Proceedings of Machine Learning Research, vol. 18, pp. 3–18. PMLR (2012). http://proceedings.mlr.press/v18/dror12a.html

4. Gomez-Uribe, C.A., Hunt, N.: The netflix recommender system: algorithms, business value, and innovation. ACM Trans. Manag. Inf. Syst. **6**(4), 13:1–13:19 (2015). https://doi.org/10.1145/2843948
5. Harper, F.M., Konstan, J.A.: The movielens datasets: history and context. ACM Trans. Interact. Intell. Syst. **5**(4), 19:1–19:19 (2015). https://doi.org/10.1145/2827872
6. Hsieh, C.J., Dhillon, I.S.: Fast coordinate descent methods with variable selection for non-negative matrix factorization. In: ACM SIGKDD International Conference on Knowledge Discovery and Data Mining (KDD) (2011)
7. Koren, Y., Bell, R., Volinsky, C.: Matrix factorization techniques for recommender systems. Computer **42**(8), 30–37 (2009). https://doi.org/10.1109/MC.2009.263
8. Lab, S.S.: Web data: Amazon reviews. https://snap.stanford.edu/data/web-Amazon.html. Accessed 18 Aug 2018
9. Labelle, C., Marinier, A., Lemieux, S.: Enhancing the drug discovery process: Bayesian inference for the analysis and comparison of dose-response experiments. Bioinformatics **35**(14), i464–i473 (2019). https://doi.org/10.1093/bioinformatics/btz335
10. Neiswanger, W., Wang, C., Xing, E.P.: Asymptotically exact, embarrassingly parallel MCMC. In: Proceedings of the Thirtieth Conference on Uncertainty in Artificial Intelligence, UAI 2014, pp. 623–632. AUAI Press, Arlington (2014). http://dl.acm.org/citation.cfm?id=3020751.3020816
11. Pilászy, I., Zibriczky, D., Tikk, D.: Fast ALS-based matrix factorization for explicit and implicit feedback datasets. In: Proceedings of the Fourth ACM Conference on Recommender Systems, pp. 71–78. ACM (2010)
12. Qin, X., Blomstedt, P., Leppäaho, E., Parviainen, P., Kaski, S.: Distributed Bayesian matrix factorization with limited communication. Mach. Learn. **108**(10), 1805–1830 (2019). https://doi.org/10.1007/s10994-019-05778-2
13. Salakhutdinov, R., Mnih, A.: Bayesian probabilistic matrix factorization using Markov chain Monte Carlo. In: Proceedings of the 25th International Conference on Machine Learning, pp. 880–887. ACM (2008). https://doi.org/10.1145/1390156.1390267
14. Tan, W., Cao, L., Fong, L.: Faster and cheaper: parallelizing large-scale matrix factorization on GPUs. In: Proceedings of the 25th ACM International Symposium on High-Performance Parallel and Distributed Computing (HPDC 2016), pp. 219–230. ACM (2016). https://doi.org/10.1145/2907294.2907297
15. Teflioudi, C., Makari, F., Gemulla, R.: Distributed matrix completion. In: Proceedings of the 2012 IEEE 12th International Conference on Data Mining (ICDM 2012), pp. 655–664 (2012). https://doi.org/10.1109/ICDM.2012.120
16. Vander Aa, T., Chakroun, I., Haber, T.: Distributed Bayesian probabilistic matrix factorization. Procedia Comput. Sci. **108**, 1030–1039 (2017). International Conference on Computational Science, ICCS 2017
17. Wang, X., Guo, F., Heller, K.A., Dunson, D.B.: Parallelizing MCMC with random partition trees. In: Advances in Neural Information Processing Systems, vol. 28, pp. 451–459. Curran Associates, Inc. (2015)

18. Yu, H.F., Hsieh, C.J., Si, S., Dhillon, I.: Scalable coordinate descent approaches to parallel matrix factorization for recommender systems. In: Proceedings of the 2012 IEEE 12th International Conference on Data Mining (ICDM 2012), pp. 765–774. IEEE Computer Society (2012). https://doi.org/10.1109/ICDM.2012.168
19. Yun, H., Yu, H.F., Hsieh, C.J., Vishwanathan, S.V.N., Dhillon, I.: NOMAD: non-locking, stochastic multi-machine algorithm for asynchronous and decentralized matrix completion. Proc. VLDB Endow. 7(11), 975–986 (2014). https://doi.org/10.14778/2732967.2732973

Early Adaptive Evaluation Scheme for Data-Driven Calibration in Forest Fire Spread Prediction

Edigley Fraga[✉], Ana Cortés, Andrés Cencerrado, Porfidio Hernández, and Tomàs Margalef[✉]

Universitat Autónoma de Barcelona, 08193 Cerdanyola del V. (Barcelona), Spain
edigley@gmail.com, {ana.cortes,andres.cencerrado,porfidio.hernandez,
tomas.margalef}@uab.cat
http://www.uab.cat

Abstract. Forest fires severely affect many ecosystems every year, leading to large environmental damages, casualties and economic losses. Established and emerging technologies are used to help wildfire analysts determine fire behavior and spread aiming at a more accurate prediction results and efficient use of resources in fire fighting. Natural hazards simulations need to deal with data input uncertainty and their impact on prediction results, usually resorting to compute-intensive calibration techniques. In this paper, we propose a new evaluation technique capable of reducing the overall calibration time by 60% when compared to the current data-driven approaches. This is achieved by means of the proposed adaptive evaluation technique based on a periodic monitoring of the fire spread prediction error ϵ estimated by the normalized symmetric difference for each simulation run. Our new strategy avoid wasting too much computing time running unfit individuals thanks to an early adaptive evaluation.

Keywords: Forest fires · Urgent computing · Data uncertainty · Data driven prediction

1 Introduction

Fire is a natural element of many ecosystems and even large wildfires are part of a defined disturbance regime [20]. For that reason, the challenge from both a prevention and a suppression point of view is to anticipate and reduce the spread potential of large wildfires, and the succeeding risk for lives, property and land use systems [7]. Wildfires have a relatively unpredictable nature as their spread can vary based on the flammable material and can differ by their extent and wind speeds. Forest fire prevention strategies for detection and suppression have

This research has been supported by MINECO-Spain under contract TIN2017-84553-C2-1-R and by the Catalan government under grant 2017-SGR-313.

© Springer Nature Switzerland AG 2020
V. V. Krzhizhanovskaya et al. (Eds.): ICCS 2020, LNCS 12142, pp. 17–30, 2020.
https://doi.org/10.1007/978-3-030-50433-5_2

improved significantly through the years, both due to technological innovations and the adoption of various skills and methods. Nowadays, wildfire researchers use technologies that integrate data on weather prediction, topography and other factors to predict how fires spread.

Nevertheless, wildfires still occur widely and represent a permanent threat whose consequences may be catastrophic both in terms of human fatalities, ecosystem degradation and economic losses. For example, years marked by intense drought and dry conditions contribute to the high levels of wildfire activity which could be turned into natural disasters. In the last years, unusually large wildfires severely damaged forests in Indonesia, Canada, United States, Spain, Chile, Portugal, and Australia, just to name a few. Actually, some studies suggests that over the past few decades, the number of wildfires has indeed increased [9,15,17,18].

In light of this situation, forest fire prediction, prevention and management measures have become increasingly important over the decades. Systems for wildfire prediction represent an essential asset to back up the forest fire monitoring and extinction phase, to predict forest fire risks and to help in the fire control planning and resource allocation.

When dealing with the extinction phase, an accurate prediction of the fire propagation is a critical issue to minimize its effects. Actually, in order to be used in a fire extinction activity, a wildfire spread prediction is a *hard-deadline-driven task*. For instance, a complex wildfire simulation that could accurately predict the perimeter of a wildfire, can drive firefighters to put firebreaks where they would be most effective to stop the fire propagation. In this particular case, an accurate prediction that comes up late compared to the actual event is useless to the task of fire suppression. These characteristics represent an urgent computing system, from which the simulation results are needed by relevant authorities in making timely informed decisions to mitigate financial losses, manage affected areas and reduce casualties [14]. The following three urgent computing requirements can be found in the mentioned forest fire propagation system:

a The computation operates under a strict deadline after which the computation results may give little practical value (*"late results are useless"*);
b The beginning of the event that demands the computation is unpredictable;
c The computation requires significant resource usage.

To fulfill those requirements, a solution must be *deadline-driven, on-demand provisioned* and *scalable*. To deal with those kind of applications, High Performance Computing (HPC) community usually rely on high-end clusters, on supercomputers or on distributed computing platforms. As deadline-based resource management like resource reservation schedulers and minimum latency dispatch are key-components to urgent computing procedures, much work have been conducted to address this issue [5,11–13,23]. In particular, HPC resources involved into simulation should be provisioned and managed in a way that the computation process will be finished within a defined time limit. Those are technologies that aim to support urgent computation dynamically and yet preserve overall machine utilization and the productivity of the other users working daily on the platforms being used.

The use of simulators for forest fire propagation models requires sufficient time for the processing, a precise fuel model data and an accurate knowledge of small and large-scale interaction of weather and topography [7]. To start a simulation, it is necessary a plethora of input parameters, which include spatial data describing the elevation, slope, aspect, and fuel type. Figure 1 illustrates the input data commonly used for fire simulations.

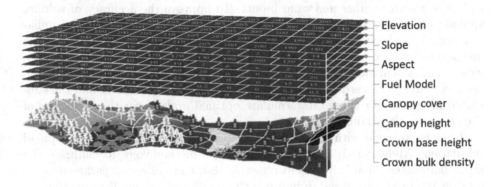

- Elevation
- Slope
- Aspect
- Fuel Model
- Canopy cover
- Canopy height
- Crown base height
- Crown bulk density

Fig. 1. Fuel and topographic grid data used for wildfire simulations.

In reality, the input data describing the actual scenario where the fire is taking place is usually subject to high levels of uncertainty that represents a serious drawback for the correctness of the prediction [22]. In this paper, we build up our work on a calibration phase that uses a *Genetic Algorithm* (GA) as an optimization technique, proposing a new evaluation strategy capable of reducing the overall calibration time by 60% when compared to the current state-of-the-art data-driven approach for the problem at hand [2]. The new adopted strategy avoid wasting too much computing time running unfit individuals thanks to an early adaptive evaluation for the fitness function.

The remainder of this document is organized as follows. Related works are discussed in Sect. 2 whereas Sect. 3 describes the fire spread simulation model and the two-stage prediction method employed to deal with input-data uncertainty. Section 4 presents the proposed evaluation strategy. A case study and computational experiments are presented in Sect. 5. Finally, Sect. 6 contains some concluding remarks.

2 Related Works

In a scenario with uncertainties regarding input data, to accurately predict a fire spread under a strict deadline constraint represents a challenge for any designed solution. Current state-of-the-art presents different approaches to tackle data-uncertainty problem, ranging from applying ensemble strategies to soften the uncertainty of input parameters effects to apply Kalman filter to certain input variables in order to tune their values [21]. Another approach is to resort on computing-intensive methods to relieve such data uncertainty effects, namely a Two-Stage prediction method, composed of a *Calibration* and a *Prediction* stage, wherein the former the input parameters values are adjusted to better reproduce the observed past behaviour of the fire, and the latter where those calibrated parameters are used to forecast the forest fire spread evolution [8].

With regard to fire simulators, FARSITE [10] is a well-known fire growth simulation modeling system which uses spatial information on topography and fuels along with weather and wind inputs. To improve the accuracy of wildfire spread predictions, Srivas [21] extended FARSITE to incorporate data assimilation techniques based on noisy and limited spatial resolution observations of the fire perimeter. The adjustment is calculated from the Kalman filter gain in an Ensemble Kalman filter, based on a Monte-Carlo implementation of the Bayesian update problem. Uncertainty on both the measured fire perimeter and the simulated fire perimeter is used to formulate optimal updates for the prediction of the spread of the wildfire.

In order to cope with the input data uncertainty related with fire spread simulation, Abdalhaq [1] proposed a two-stage methodology to calibrate the input parameters in an adjustment phase so that the calibrated parameters are used in the prediction stage to improve the quality of the predictions. Cencerrado [6] applied Genetic Algorithm as the calibration technique in the adjustment phase, which requires the execution of many simulations to generate the best calibrated set of input parameters. Similar work was also carried over by Méndez-Garabetti et al [16]. Cencerrado also devised one strategy based on *Decision Trees* to identify long running execution individuals of a fire spread simulation. Such strategy was the base for a classification method that allows to estimate in advance the execution time of a simulation given a certain set of input parameters.

More recently, Artés [2] proposed and evaluated a set of resource allocation policies to assign more computing resources to estimated long running executions and less resources to the fast ones, allowing to reduce the adjustment stage time to a more acceptable deadline. That was possible due to the use of a parallel version of FARSITE model that could reduce long running execution times by 35% [3]. To work in a time-constrained fashion, a hybrid MPI-OpenMP application based on the Master-Worker paradigm was developed to take advantage of the execution in a parallel HPC cluster environment. The Two-Stage framework has been proved to be a good methodology to deal with the input data uncertainties and it is leveraged in this current work.

3   Two-Stage Prediction Method

Usually, to predict forest fire behavior a simulator takes the initial state of the fire front perimeter (P_0) along with other parameters as input. As output, the simulator then returns the fire front spread prediction for a later instant in time ($\hat{P}_1$). After comparing the simulation result with the actual advanced fire front

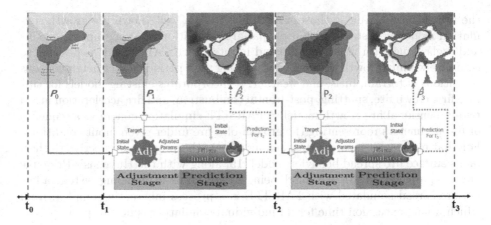

Fig. 2. Two-stage prediction method.

(P_1), the predicted fire line tends to differ from the actual one. Besides the natural phenomena modeling complexity uncertainty, the reason for this mismatch is that the classic scheme calculation is based solely on a single set of input parameters, affected by the aforementioned data uncertainty. To overcome this drawback, a simulator independent data-driven prediction scheme was proposed to calibrate model input parameters [8].

Introducing a previous adjustment stage (see Fig. 2), the set of input parameters is calibrated before every prediction step. Thus, the solution comes from reversing the problem, coming up with a parameter configuration such that the fire simulator would produce predictions that match the actual fire behavior. After detecting the simulator input that better reproduces the observed fire propagation, the same set of parameters is used to describe the conditions for the next prediction ($\hat{P}_2$), assuming that meteorological circumstances remain constant during the next prediction interval. Then, the final prediction becomes the result of a series of automatically adjusted input configurations. The process can be applied again for subsequent fire perimeters ($\hat{P}_3$), ($\hat{P}_4$), ($\hat{P}_5$) and so on.

In order to enhance the quality of the predictions, as a data-driven scheme, the two-stage method is applied continuously, providing calibrated parameters at different time intervals and taking advantage of observed fire behavior and helping to reduce the negative effects related to the input-data uncertainty. This approach has been proven to be appropriate in order to enhance the quality of the predictions. In particular, a Genetic Algorithm based adjustment technique gives accurate results [8] although not being able to give fast response times even when using multi-core allocation strategies, as showed in Sect. 5.

3.1 Forest Fire Spread Prediction Model Simulator

In the field of forest fire behavior modeling, there is a few fire propagation simulators, based on different physical models, whose main objective is to predict the fire evolution. Among those, FARSITE [10] is a well-known fire growth simulation modeling system which uses spatial information on topography (terrain) and fuels along with weather and wind inputs. With FARSITE it is possible to compute wildfire growth and behavior for long time periods under heterogeneous conditions of terrain, fuels, and weather. It incorporates existing models for surface fire, crown fire, spotting, post-frontal combustion, and fire acceleration into a two-dimensional fire growth model. A FARSITE simulation generates a sequence of fire perimeters representing the growth of a fire under given input conditions. For that purpose, it incorporates, among others, the simple but effective Rothermel's surface fire spread behavior model [19] along with the Huygens's Principle of wave propagation [10]. Although being a deterministic modeling system, a forest fire spread simulated with FARSITE is a process inherently complex, from which a long execution time for an individual simulation is not atypical.

3.2   Genetic Algorithm Implementation

The adjustment stage is based on a genetic algorithm implemented in a Master-Worker paradigm. The calibration starts from an initial random population of individuals, each one representing a scenario to be simulated. An individual is composed of a set of different genes that represent input variables such as dead fuel moisture, live fuel moisture, wind speed and direction, among others.

Each individual is simulated and it is evaluated comparing the predicted and the real fire propagation by estimating the fitness function (or prediction error function) based on the normalized symmetric difference between predicted and real burned areas. Eq. 1 defines how such difference is calculated, where *Real* is the area burned by the real fire at a certain time and *Pred* is the area burned by the predicted fire at the same time instant.

$$SymDiff = \frac{\bigcup(Real, Pred) - \bigcap(Real, Pred)}{Real}$$
$$= \frac{Misses + FalseAlarms}{Misses + Hits} \tag{1}$$

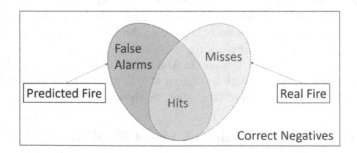

Fig. 3. Different categories present in forecast verification.

As illustrated in Fig. 3, the areas around the simulation map that have not been burned by neither the real fire nor the simulated fire are considered *Correct Negatives*. Areas that have been burned by both fires are called *Hits*. The areas that have been burned only in the real fire are called *Misses* whereas areas that have been burned only in the simulated fire are called *False Alarms*.

Due to the high parallelizable characteristics of GAs, its implementation in a Master-Worker paradigm is straightforward. One can take advantage of stable tools, platforms and programming language support to exploit the execution in a parallel HPC cluster environment. Figure 4 illustrates how the adjustment phase may be implemented in a Master-Worker paradigm, with one worker node per individual and the master node acting as a coordinator.

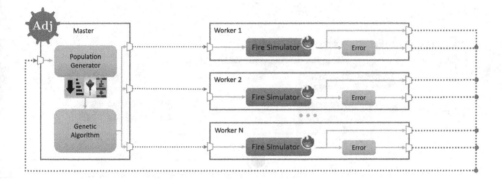

Fig. 4. Genetic Algorithm implemented in a Master-Worker paradigm.

In the current implementation, those individuals whose execution time is estimated to be longer than a preset value are discarded. In this type of situation, if the GA drives the system to a search space associated with parameter settings whose correspondent simulation takes longer than the preset value, this search space will never be considered. In the following section, a technique to overcome this drawback is proposed.

4   Early Adaptive-Evaluation Strategy

In order to overcome the slow time-to-result characteristic of the genetic algorithm, we propose an ***adaptive evaluation*** technique based on a periodic monitoring of the fire spread prediction error ε estimated by the normalized symmetric difference, as defined in Eq. 1, for each execution of the individuals. Figure 5a shows the steps done by the monitor process to determine whether the monitored individual should be early terminated or not. Figure 5b depicts two individuals being executed and monitored according to the monitor flow defined in Fig. 5a. The one described in Fig. 5b(I) terminates normally whereas the other described in Fig. 5b(II) is early terminated by the monitoring agent due to its unfitness based on its ongoing prediction error.

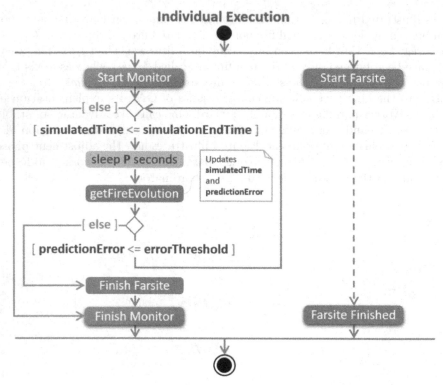

(a) Monitor activity diagram.

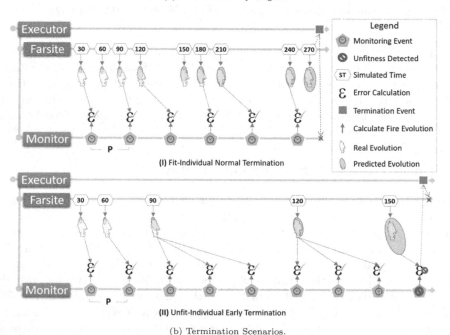

(b) Termination Scenarios.

Fig. 5. Individual termination strategies.

The reasoning behind the early termination is to avoid wasting computing time running individuals that are doomed to unfitness. If along its execution the monitor agent detects that the prediction deviates too much from the actual fire spread, it is considered safe to early terminate the individual. Each monitoring agent is launched with a prediction error threshold T, i.e. a maximum tolerable error above which the monitoring agent must early terminate the individual, and a monitoring period P, usually defined in the order of dozen of seconds.

For example, if we consider the canonical scenario of a fire prediction evolution described in Fig. 6, together with their corresponding normalized symmetric difference (prediction error ε) calculated as described in Eq. 1, we notice that as the prediction overspread beyond the real fire, the prediction error increases at a rate higher than the ones circumscribed to the real fire perimeter. Therefore, if at an earlier stage of the prediction a monitor configured with a threshold $T = 2.0$ detects a fire evolution similar to the one showed in Fig. 6f, then, the execution can be terminated as the individual is considered unfit to represent the real fire.

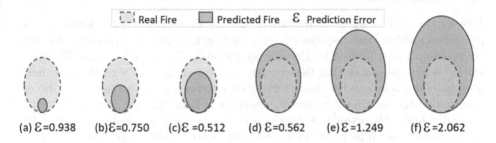

(a) ε=0.938   (b)ε=0.750   (c)ε =0.512   (d) ε=0.562   (e)ε =1.249   (f)ε =2.062

Fig. 6. normalized symmetric difference calculated along fire spread prediction.

In the current implementation, as described in Sect. 2, the strategy used to early terminate and then discard individuals was based on a deadline previously defined to avoid delaying an entire generation due to longer individuals. Or, even worse, previously filtering out those individuals whose execution time is estimated to be longer than a preset value. The problem regarding early termination is its impact on the population diversity, a crucial characteristic to the genetic algorithm's ability to continue fruitful exploration of the search space.

Another difference from previous works, where the normalized symmetric difference was directly used as the fitness function for the evaluation of the individual, is that, in this proposal, we use a weighted version of it, in order to take into account early terminations. The formula showed in Eq. 2 is a weighted version for the normalized symmetric difference in which $PredictionTime$ is the total time to be simulated whereas $SimulatedTime$ is the total time simulated until a normal or an early termination takes place.

$$fitness = \frac{PredictionTime}{SimulatedTime} \times SymDiff \qquad (2)$$

In Fig. 5b(I), in the scenario of a normal termination, for a total simulated time equals to 270 min, we can see that the simulated and prediction time are equals, then the fitness function is equals to the normalized symmetric difference. On the other hand, in the early termination scenario showed in Fig. 5b(II), the penalty is directly proportional to the simulated time left to a normal termination, i.e., $fitness = \frac{270}{150} \times SymDiff = 1.8 \times SymDiff$. The idea is to be able to compare unfinished individuals results against the actual burned area, considering how much has been simulated until the moment of the early termination: the sooner the termination, the greater should be the penalty. Therefore the main advantage of such strategy is to avoid wasting too much computing time running indubitably unfit individuals. Nevertheless, thanks to the weighted normalized symmetric difference those individuals can still be evaluated and be unlikely selected to the following generations, ensuring more diversity to the evolution process.

4.1 Adaptive-Evaluation and Deadline-Driven Strategies Comparison

Due to the *fork-and-join* characteristic of the evolution stream defined by a genetic algorithm, the response time of each generation and therefore for the entire evolution is limited by the slowest individual's execution time. Previous works [6] have identified that the execution time distribution of randomly generated individuals follows the characteristic of a power law, with a few small values dominating the distribution, complemented by a long tail representing the slowest individuals. Unfortunately, like in all unusual extreme events modelled by power laws, those infrequent occurrences are responsible for the worst damage to the calibration phase capacity to provide shorter response time.

Then, in order to decrease the overall calibration time, it is crucial to be able to *consistently* decrease the random individuals execution time. The violin plots in Fig. 9(a) compare the FARSITE execution time for a 1 h deadline-driven strategy and the adaptive-evaluation for a real forest fire scenario (see Sect. 5). Considering those two execution time distributions, it would be expected that the new strategy helps to calibrate the input data approximately three times faster than any strategy based solely on a deadline-driven approach set to 1 h per generation. In Sect. 5 we evaluate the adaptive strategy applied to the aforementioned real fire scenario.

5   Case Study and Computational Experiments

The Mediterranean area is one of the European regions most affected by forest fires during high risk seasons. The selected case study corresponds to a region within the Mediterranean coast that is frequently affected by forest fires. In particular, we used a wildfire that occurred in La Jonquera (North-East of Catalonia, Spain - see Fig. 7) in July 2012 that devastated over 13,000 ha and caused the death of two people. This wildfire scenario has been thoroughly studied and compared in previous works [2,4].

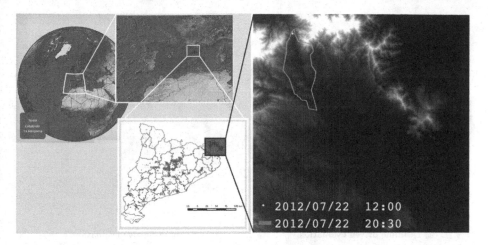

Fig. 7. La Jonquera' Fire (Catalonia - Spain), and its calibration perimeter

The simulated calibration time period was set to match exactly the real observed evolution that goes from 22 July 2012 at 12:00 to 22 July 2012 at 20:30, as depicted in Fig. 7. The genetic algorithm has been configured to evolve for 10 generations, each one with population size set-up to 100 individuals. Probabilities of crossover and mutation used in the executions are 0.7 and 0.3, respectively. Each individual is simulated and the resulting simulated fire perimeter is compared to the actual fire spread using the weighted normalized symmetric difference between the real burned area and the predicted one, as described in Sect. 4. Moreover, each monitoring agent is launched with a prediction error threshold set to 1.5. This implies that the maximum ongoing tolerable error above which the monitoring agent must early terminate the individual is 1.5. Or saying in other words: when the predicted area is one and a half times the size of the one defined by the real fire perimeter, the corresponding simulation will be terminated.

Figure 8 shows an overview of a single calibration execution using the adaptive-evaluation strategy and the deadline-driven approach set to 1 h per generation for a population of 64 individuals where each individual has been executed sequentially in a single core.

Confronting the two figures, we can notice that the overall calibration time is reduced from above 8 h to 1 h on average, representing an improvement of impressive 85%. When compared to the current best data-driven Hybrid MPI-OpenMP Parallel approach [2,4] dubbed TAC (*Time Aware Core allocation*), the calibration time is reduced from above 3 h to the same 1 h on average, resulting in a 60% time-to-result improvement. The TAC approach, which applies classification techniques to detect in advance those individuals that will last longer and allocate more cores to them, was implemented with strict time restrictions: a time lapse of 4 h was considered to carry out the calibration stage and the final prediction. This time window was split into two intervals: 3 h for the calibration, and 1 h for the simulation that will result in the final prediction.

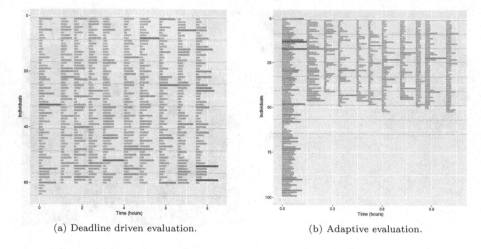

(a) Deadline driven evaluation.   (b) Adaptive evaluation.

Fig. 8. Deadline-driven and adaptive evaluation calibration executions overview.

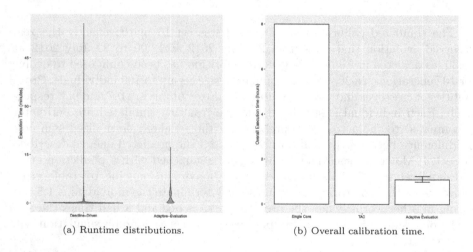

(a) Runtime distributions.   (b) Overall calibration time.

Fig. 9. Execution time comparisons between the different evaluation strategies.

Figure 9b shows the overall comparison between the three calibration strategies. In relation to the adaptive evaluation metric, the error bars represent the minimum and the maximum values obtained for 20 different calibrations whereas the intermediate point is the average value. For the other scenarios, the bar height represent the average value as reported by Artés et al. [2]. As expected, the adaptive evaluation indubitably outperforms the other two strategies, representing a major improvement even when compared to the TAC strategy.

6 Concluding Remarks

Forecasting forest fire spread is a complex problem by itself, considering the plethora of factors capable of influence in this natural phenomena. Apart from that, input data uncertainty represents an additional challenge to devise an accurate model. When applied to a real case scenario in production, there is also an extra key factor that challenge the forecast process, the response time. In such situation, late results are useless.

For this reason, any approach that focuses on accelerating forest fire spread prediction without losing accuracy is more than welcome. In this work, we propose a new strategy able to, at the same time, simplify implementation and improve response time of the critical adjustment stage of a Two-Stage Prediction Method, implemented as a High Performance Computing framework to deal with the input data uncertainties. In that methodology, the prediction phase is preceded by an adjustment stage, the latter being the most time consuming one. In this stage, a genetic algorithm is carried out to calibrate unknown parameters such as fuel humidity and meteorological data.

Previous works exploited heavily cluster environments, classification algorithms and core allocation techniques to be able to deliver acceptable response time under strict time constraints. In contrast, we resort to a monitoring strategy that, apart from being much simpler to implement and easier to productionize, managed to achieve better calibration response time results. In order to validate our proposed strategy, we analyzed a real forest fire scenario that took place in La Jonquera (Catalonia, Spain) in 2012. Results show that the new technique is capable of reducing the overall calibration time by 60% when compared to the current best data-driven time aware core allocation approach.

As future work we intend to propose and evaluate other fitness functions together with the early termination technique and its impact in the overall prediction accuracy.

References

1. Abdalhaq, B., Cortés, A., Margalef, T., Luque, E.: Enhancing wildland fire prediction on cluster systems applying evolutionary optimization techniques. Future Gener. Comput. Syst. **21**(1), 61–67 (2005)
2. Artés, T., Cencerrado, A., Cortés, A., Margalef, T.: Relieving the effects of uncertainty in forest fire spread prediction by hybrid MPI-OpenMP parallel strategies. Procedia Comput. Sci. **18**, 2278–2287 (2013)
3. Artés, T., Cortés, A., Margalef, T.: Large forest fire spread prediction. Procedia Comput. Sci. **80**(C), 909–918 (2016)
4. Brun, C., Artés, T., Cencerrado, A., Margalef, T., Cortés, A.: A high performance computing framework for continental-scale forest fire spread prediction. Procedia Comput. Sci. **108**, 1712–1721 (2017)
5. Brzoza-Woch, R., Konieczny, M., Kwolek, B., Nawrocki, P., et al.: Holistic approach to urgent computing for flood decision support. Procedia Comput. Sci. **51**, 2387–2396 (2015)

6. Cencerrado, A., Cortés, A., Margalef, T.: Genetic algorithm characterization for the quality assessment of forest fire spread prediction. Procedia Comput. Sci. **9**, 312–320 (2012)

7. Costa, P., Castellnou, M., Larrañaga, A., Miralles, M., Kraus, D.: Prevention of large wildfires using the fire types concept. European Forest Institute (January 2011)

8. Denham, M., Cortés, A., Margalef, T.: Computational steering strategy to calibrate input variables in a dynamic data driven genetic algorithm for forest fire spread prediction. In: Allen, G., Nabrzyski, J., Seidel, E., van Albada, G.D., Dongarra, J., Sloot, P.M.A. (eds.) ICCS 2009. LNCS, vol. 5545, pp. 479–488. Springer, Heidelberg (2009). https://doi.org/10.1007/978-3-642-01973-9_54

9. Dennison, P.E., Brewer, S., Arnold, J.D., Moritz, M.: Large wildfire trends in the western United States, 1984–2011. Geophys. Res. Lett. **41**, 2928–2933 (2014)

10. Finney, M.: FARSITE: fire area simulator-model. development and evaluation. USDA Forest Service Research Paper, RMRS-RP-4 (1998)

11. Knyazkov, K.V., Nasonov, D.A., Tchurov, T.N., Boukhanovsky, A.V.: Interactive workflow-based infrastructure for urgent computing. Procedia Comput. Sci. **18**, 2223–2232 (2013)

12. Kovalchuk, S.V., Smirnov, P.A., Maryin, S.V.: Deadline-driven resource management within urgent computing cyberinfrastructure. Procedia Comput. Sci. **18**, 2203–2212 (2013)

13. Leong, S.H., Frank, A., Kranzlmüller, D.: Leveraging e-infrastructures for urgent computing. Procedia Comput. Sci. **18**, 2177–2186 (2013)

14. Leong, S.H., Kranzlmüller, D.: Towards a general definition of urgent computing. Procedia Comput. Sci. **51**, 2337–2346 (2015)

15. Moreno, M.V., Conedera, M., Chuvieco, E., Pezzatti, G.B.: Fire regime changes and major driving forces in Spain from 1968 to 2010. Environ. Sci. Policy **37**, 11–22 (2014)

16. Méndez-Garabetti, M., Bianchini, G., Tardivo, L., Scutary, P.: Comparative analysis of performance and quality of prediction between ESS and ESS-IM. Electron. Notes Theor. Comput. Sci. **314**, 45–60 (2015)

17. Pausas, J.G., Fernández-Muñoz, S.: Fire regime changes in the western Mediterranean basin: from fuel-limited to drought-driven fire regime. Clim. Change **110**(1), 215–226 (2012)

18. Pausas, J.G., Llovet, J., Rodrigo, A.A., Vallejo, R.A.V.: Are wildfires a disaster in the Mediterranean basin? Int. J. Wildland Fire **17**, 713–723 (2008)

19. Rothermel, R.: A mathematical model for predicting fire spread in wildland fuels. USDA Forest Service research paper INT, Intermountain Forest & Range Experiment Station, Forest Service, U.S. Deptartment of Agriculture (1972)

20. San-Miguel-Ayanz, J., Moreno, J.M., Camia, A.: Analysis of large fires in European Mediterranean landscapes: lessons learned and perspectives. For. Ecol. Manag. **294**, 11–22 (2013)

21. Srivas, T., de Callafon, R.A., et al.: Data assimilation of wildfires with fuel adjustment factors in FARSITE using ensemble Kalman filtering. Procedia Comput. Sci. **108**, 1572–1581 (2017)

22. Thompson, M.P., Calkin, D.E.: Uncertainty and risk in wildland fire management: a review. J. Environ. Manag. **92**(8), 1895–1909 (2011)

23. Yoshimoto, K., Choi, D., Moore, R., Majumdar, A., Hocks, E.: Implementations of urgent computing on production HPC systems. Procedia Comput. Sci. **9**, 1687–1693 (2012)

Strategic Use of Data Assimilation for Dynamic Data-Driven Simulation

Yubin Cho[1,2], Yilin Huang[1(✉)], and Alexander Verbraeck[1]

[1] Faculty of Technology, Policy and Management, Delft University of Technology,
Delft, The Netherlands
{y.huang,a.verbraeck}@tudelft.nl
[2] To70 Aviation Consultants, The Hague, The Netherlands
yubin.cho@to70.nl

Abstract. Dynamic data-driven simulation (DDDS) incorporates real-time measurement data to improve simulation models during model run-time. Data assimilation (DA) methods aim to best approximate model states with imperfect measurements, where particle Filters (PFs) are commonly used with discrete-event simulations. In this paper, we study three critical conditions of DA using PFs: (1) the time interval of iterations, (2) the number of particles and (3) the level of actual and perceived measurement errors (or noises), and provide recommendations on how to strategically use data assimilation for DDDS considering these conditions. The results show that the estimation accuracy in DA is more constrained by the choice of time intervals than the number of particles. Good accuracy can be achieved without many particles if the time interval is sufficiently short. An over estimation of the level of measurement errors has advantages over an under estimation. Moreover, a slight over estimation has better estimation accuracy and is more responsive to system changes than an accurate perceived level of measurement errors.

Keywords: Dynamic Data-Driven Simulation · Data Assimilation · Particle Filters · Discrete-event simulation · Sensitivity analysis

1 Introduction

Simulation modeling has been widely used for studying complex systems [10–12]. In a highly evolving environment, classical simulation shows limitations in situational awareness and adaptation [8,9]. Dynamic Data-Driven Application Systems (DDDAS) is a relative new paradigm [4] proposed to integrate the computational and instrumental aspects of complex application systems offering more accurate measurements and predictions in real-time. A related concept is Dynamic Data-Driven Simulation (DDDS) [6,9], where Data Assimilation (DA) [3,14] is used to combine a numerical model with real-time measurements at simulation run-time. DA aims to obtain model states that best approximate the current and future states of a system with imperfect measurements [18].

© Springer Nature Switzerland AG 2020
V. V. Krzhizhanovskaya et al. (Eds.): ICCS 2020, LNCS 12142, pp. 31–44, 2020.
https://doi.org/10.1007/978-3-030-50433-5_3

Owing to disciplinary traditions, DA is predominantly used with simulation of continuous systems but less with discrete systems [7]. A few examples of the latter can be found e.g. in wildfire and transport simulations [5–7,26], and in agent-based simulations that predict the behavior of residents in buildings [21,22]. For DA in discrete systems simulations, the Sequential Monte Carlo (SMC) methods, a.k.a. Particle Filters (PFs), are commonly used [6,7,23,25]. Two major reasons are mentioned in literature. First, PFs methods are more suitable to DDDS than variational methods [15] since the models can easily incorporate the real-time data that arrives sequentially [23]. Second, the classical sequential methods such as Kalman Filter and its extensions rely on requirements that are difficult to fulfil by systems that exhibit non-linear and non-Gaussian behaviors which typically do not have analytical forms [7]. SMC or PFs are sample-based methods that use Bayesian inference, stochastic sampling and importance resampling to iteratively estimate system states from measurement data [7,23,25]. The probability distributions of interest are approximated using a large set of random samples, named particles, from which the outcomes are propagated over time [7,23,25].

In this paper, we study three common and critical conditions of DA using PFs for discrete-event simulation – the time interval of iterations, the number of particles and the level of measurement errors (or noises) – to understand the effect of these conditions on the estimation accuracy of system states. A number of works studied the conditions of DA for continuous systems such as meteorology, geophysics and oceanography [13,16,17,20]. But little is known for discrete-event simulation in this regard.

The time interval of assimilating measurement data and the number of particles in PFs are two critical conditions because they directly affect computational cost and estimation accuracy in DA. One recent research studied the effects of both conditions independently [24]. Our experiments also study their mutual influences, since they are two conditions that restrict one another given that the computational time is often limited between two successive iterations in DA. The level of measurement errors is another critical condition in DA. The actual level of measurement errors is rarely known in real world situations. What is included in DA algorithms is always the perceived level (or assumptions) of measurement errors. Our experimental setup imitates the actual level of measurement errors, and allows the study of the differences between the actual and perceived measurement errors, and their effects on estimation accuracy. In the following, we present the methodology used, discuss the experimental results and provide recommendations on future research.

2   Methodology

This research uses an $M/M/1$ single server queuing system with balking for the DA experiments. The real system is imitated with a sensing process that generates measurement data where errors (or noises) are introduced. The discrete-event simulation model is a perfect representation of the real system. The DA process uses PFs to iteratively construct probability distributions for particle weight calculation incorporating measurement data. The DA results are evaluated with regard to different time intervals Δt, the numbers of particles N and the levels of actual and perceived measurement errors ϵ and ϵ'.

2.1   Experimental Setup

The experimental setup consists of four components (cf. [7,24]): (1) Real System, (2) Measurement Model, (3) Simulation Model, and (4) Data Assimilation. The real system and the simulation model are implemented with Salabim[1]. The whole experimental setup is implemented in python[2].

Real System. The real system is represented by an ESP32 microcontroller, which (1) imitates the real $M/M/1$ queuing system with balking, and (2) generates the "sensor data" in real-time.

The queuing process has exponential inter-arrival times of jobs (or customers) with mean arrival rate λ, and exponential service times with mean service (or processing) rate μ. The queue has a limit of length L for balking [1]: when the queue reaches length L, no new job is appended to the queue. The state of the queuing system S_{real} at time t is denoted as

$$S_{t,real} := \{arrRate_{t,real}, procRate_{t,real}, queLen_{t,real}\}$$

where $arrRate_{t,real}$ is the mean arrival rate λ, i.e. the inter-arrival time $T_{arr,real} \sim Exp(arrRate_{t,real})$; $procRate_{t,real}$ is the mean processing rate μ, i.e. the processing time $T_{proc,real} \sim Exp(procRate_{t,real})$; and $queLen_{t,real} \in [0, L]$ is the queue length.

To imitate second order dynamics [8] in the queuing system, every 15 s the values of $arrRate_{t,real}$ and $procRate_{t,real}$ are updated stochastically from a uniform distribution as

$$arrRate_{t,real} \sim U(0, 20)$$
$$procRate_{t,real} \sim U(0, 20)$$

These are the two internal values (i.e. non observables) the data assimilation component needs to estimate for the simulation model. Two "observables" are measured from the real system:

$$\{numArr_{real}, numDep_{real}\}$$

the number of arrival $numArr_{real}$ at the queue, and the number of departure $numDep_{real}$ from the queue during a measurement period. These two variables are added with noises and then used for DA.

---

[1] https://www.salabim.org.

[2] https://github.com/yuvenious/ddds_queuing.

Measurement Model. The "real system" sends sensor data (a set of two values each time) $\{numArr_{real}, numDep_{real}\}$ through serial communications, and generates measurement data:

$$\{numArr_{measure}, numDep_{measure}\}$$

The measurement data at time t is denoted as

$$numArr_{t,measure} = numArr_{t,real} + error_{t,arr}$$
$$numDep_{t,measure} = numDep_{t,real} + error_{t,dep}$$

where $error_{t,arr}$ and $error_{t,dep}$ are the imitated measurement errors (or noises), sampled from Gaussian distributions $N \sim (0, \sigma^2)$ at time t. The variance σ can take one of the four values denoted by $\epsilon \cdot \Delta t^2$, where ϵ is the level of measurement errors during the sensing process: $\epsilon \in [0,3]$ represents the error levels from zero (0) to low (1), medium (2) till high (3). Δt is the time interval of DA. For example, if $\Delta t = 5$ then σ is set to be $[0, 5, 10, 15]$ in the experiments depending on the corresponding error levels. In addition, σ_{arr} and σ_{dep} are independent to each other in the experiments. As such, the joint probability can be obtained by the product of the two probabilities.

Note that in our experiments, the data assimilation process uses the perceived level of measurement errors ϵ' to represent the difference between the assumption of the level of measurement errors and their actual level. To our knowledge, these two are deemed as the same, i.e. $\epsilon = \epsilon'$, in previous works.

Simulation Model. The simulation model of the single server queuing system with balking has state $S_{t,sim}$ at time t denoted as

$$S_{t,sim} := \{arrRate_{t,sim}, procRate_{t,sim}, queLen_{t,sim}\}$$

where $arrRate_{t,sim}$ is the mean arrival rate; $procRate_{t,sim}$ is the mean processing rate; and $queLen_{t,real}$ is the queue length. In the simulation, the inter-arrival times and processing times also have exponential distributions, and the queue has maximum length L as in the "real system".

Each simulation replication is a particle in the DA. The state transition of a replication i (i.e. particle i) from time t to $t + \Delta t$ is denoted as

$$S_{t,sim}^i \longmapsto S_{t+\Delta t,sim}^i \ (i = 1, 2, \cdots, N)$$
$$S_{t,sim}^i := \{arrRate_{t,sim}^i, procRate_{t,sim}^i, queLen_{t,sim}^i\}$$

where N is the total number of particles. The simulation time is repeatedly advanced by time interval Δt, each time after the measurement data becomes available and when the calculations in the DA are completed. The measurement data is "compared with" the corresponding predicted values by the simulation model:

$$\{numArr_{sim}, numDep_{sim}\}$$

which are the number of arrival and the number of departure in the simulation.

Data Assimilation. At initialization $(t = 0)$, N sets of mean arrival rates and mean processing rates are sampled from uniform distribution $U(0, 20)$ for the N particles in the simulation, and each particle has equal weight:

$$\{arrRate^i_{0,sim}, procRate^i_{0,sim}\}$$

The simulation time t of each particle i then advances by Δt denoted as

$$S^i_{0,sim} \longmapsto S^i_{0+\Delta t,sim}$$

Iteratively, the simulation time t advances by Δt, and each simulation (replication, i.e. particle i) $S^i_{t,sim} \longmapsto S^i_{t+\Delta t,sim}$ is interpreted as the predictive distribution $p(x^i_{t+\Delta t} | x^i_t)$ of state variable $x \in S_{sim}$.

The importance weight w^i of each particle i is calculated by comparing the measurement data with the simulation (prediction). Each particle i is equally weighted at initialization: $w^i_0 = 1/N$. For the subsequent iteration steps, weights are calculated as:

$$w^i_{t+\Delta t} = p(measure_{t+\Delta t} \,|\, predict^i_{t+\Delta t}) \cdot w^i_t \quad \text{where}$$
$$measure_{t+\Delta t} = \{numArr_{t+\Delta t,measure}, numDep_{t+\Delta t,measure}\}$$
$$predict^i_{t+\Delta t} = \{numArr^i_{t+\Delta t,sim}, numDep^i_{t+\Delta t,sim}\}$$

As mentioned earlier, the level of measurement errors ϵ is used to imitate the measurement noises, $error_{arr}$ and $error_{dep}$, that are added into the measurement data. A different value (i.e. the level of perceived measurement errors ϵ') is used for the weight calculation of each particle, comparing the measurement data, $measure_{t+\Delta t}$ (or $measure_t$), with the prediction by the simulation, $predict^i_{t+\Delta t}$ (or $predict^i_t$). The conditional probability of $measure_t$ given $predict^i_t$, is interpreted as the conditional probability of the difference between the two, $measure_t - predict^i_t$, given the level of perceived measurement errors ϵ' (cf. [23] p.47):

$$p(measure_t \,|\, predict^i_t) = p(measure_t - predict^i_t \,|\, \epsilon')$$
$$= \frac{1}{\sigma'\sqrt{2\pi}} \cdot e^{-\frac{\left(measure_t - predict^i_t\right)^2}{2\sigma'^2}}$$
$$\text{where } \sigma' = \epsilon' \cdot \Delta t^2$$

In each iteration, $arrRate^i_{sim}$ and $depRate^i_{sim}$ of every particle i are resampled according to its weight w^i. This means a higher probability of resampling is given to a particle with a higher weight. As a result, the resampled particles are located nearby the highly weighted particles in the previous iteration.

For example, if the evaluated weight of particle i is $w^i_{t+\Delta t} = 0.6$ and $N = 1000$, then 600 new particles $(j = 1, 2, \cdots, 600)$ are subjected to resampling derived from particle i. In principle, $S^i_{t+\Delta t,sim}$ is assigned to $S^j_{t+\Delta t,sim}$ as

$$S^j_{t+\Delta t,sim} \longleftarrow \{arrRate^i_{t+\Delta t,sim}, procRate^i_{t+\Delta t,sim}, queLen^i_{t+\Delta t,sim}\}$$

But since all these resampled particles contain the identical state, different random seeds shall be used to prevent identical simulation runs. We also use Gaussian distributions to scatter the values of $arrRate^i_{t,sim}$ and $depRate^i_{t,sim}$. This additional treatment guarantees that the resampled particle j is close but different to the previous particle i to represent the dynamic change of the system.

$$arrRate^j_{t+\Delta t,sim} \sim N(arrRate^i_{t,sim}, arrRate^i_{t,sim}/10)$$

$$depRate^j_{t+\Delta t,sim} \sim N(depRate^i_{t,sim}, depRate^i_{t,sim}/10)$$

Thereafter, all resampled particles are evenly weighted: $w^j_{t+\Delta t} = 1/N$. These resampled particles are used for the next iteration $(t \leftarrow t + \Delta t)$.

The (aggregated) system state at time t can be estimated by the state of each particle and their corresponding weights as

$$S_{t,sim} = \frac{1}{N} \sum_i^N (S^i_{t,sim} \cdot w^i_t)$$

2.2   Sensitivity Analysis

In the experiments, three critical conditions in DA are investigated to study their effects on the estimation accuracy: (1) the time interval Δt, (2) the number of particles N, and (3) the level of measurement errors ϵ and the level of perceived measurement errors ϵ'. The time interval Δt determines the frequency of the DA steps, i.e. how often the measurement data is assimilated to the simulation which triggers the calculation of the subsequent predictive distributions. The number of particles N is the number of simulation replications used for the DA algorithm. It determines the "number of samples" used for the predictive distribution. The level of measurement errors ϵ is used to introduce noises in the measurement data, and the level of perceived measurement errors ϵ' is used in importance weight calculation. The experiments make combinations of the levels of actual and perceived measurement errors to study the effect.

Each DA experiment run lasts 50 s, during which $arrRate_{real}$ and $procRate_{real}$ change every 15 s in the "real system". The values of $numArr$ and $numDep$ are assimilated to the simulation model in the experiment using different time interval Δt which ranges from 1 to 5 s. The number of particles N for the DA varies from 10 to 2000. The measurement errors and perceived measurement errors are set to be different as will be further explained in the next section.

To compare the estimation accuracy of different DA experiment settings, distance correlation [2,19] is used to measure the association between the state variables of the "real system" and the simulated values:

$$0 \le dCor(S_{real}, S_{sim}) = \frac{dCov(S_{real}, S_{sim})}{\sqrt{dVar(S_{real})dVar(S_{sim})}} \le 1$$

$dCor$ is measured for each state variable. The overall distance correlation of the estimation is the mean of the individual distance correlations.

3   Experimental Results and Discussions

This section first presents the results regarding time interval and number of particles, as they produce related effects on computational cost and estimation accuracy. Since computational cost is often limited in practice, experiments are also made to show the trade-offs of the two. The second part of this section compares the effect of measurement errors with perceived measurement errors.

3.1   Time Interval and Number of Particles

The time interval Δt of iternation in DA is experimented ranging from 1 to 5 s. The number of particles N is set to be 1000 in those experiments ($\epsilon = 1$ and $\epsilon' = 1$). As shown in Fig. 1, when Δt decreases, the estimation accuracy $dCor$ increases significantly with narrower variances.

The number of particles N is experimented ranging from 10 to 2000 with different steps, as shown in Fig. 2, where $\Delta t = 1$, $\epsilon = 1$ and $\epsilon' = 1$. The estimation accuracy $dCor$ increases with narrower variances as more particles are used in the DA. However, when N exceeds 100, the increment in accuracy becomes slower. The Tuckey test (CI = 95%) is performed to compare the difference of $dCor$ between $N = 100$ and higher numbers of particles. The result shows that the increase in the number of particles above 400 in these experiments is no more effective in improving estimation accuracy.

Trade-Off Between Time Interval and Number of Particles. To understand the relation between the time interval Δt and number of particles N with regard to the estimation accuracy $dCor$, an extensive number of DA experiments are performed. The results are displayed in Fig. 3, where the X-axis shows the total number of simulation runs over one DA experiment. For example, if $\Delta t = 2$ s and $N = 1000$ in a DA experiment, then the number of total simulation runs within that experiment is $50/2 \cdot 1000 = 25000$. The Y-axis is the resulting $dCor$ of that experiment. Each dot in Fig. 3 hence represents one DA experiment, where the size of the dot (small to large) denotes the number of particles $N \in \{500, 1000, 1500, 2000\}$, and the color of the dot (blue to red) indicates the time interval $\Delta t \in \{1, 2, 3, 4, 5\}$ used in that DA experiment.

The result shows that when N increases (large dots) and Δt decreases (blue dots), thereby more simulation replications and iterations executed, the estimation accuracy improves and $dCor$ approaches to 1. Notably, there is hardly any red dots close to $dCor = 1$, and many large red dots (i.e. experiments with high numbers of particles and long time intervals) are located at where $dCor \leq 0.8$. This means, if Δt is too long, using a large number of particles increases computational cost *without* improvement in estimation accuracy. On the other hand, there are small blue dots (i.e. experiments with low numbers of particles and short time intervals) that are located close to $dCor = 1$. This indicates, if Δt is sufficiently short, good estimation accuracy can be achieved even though not many particles are used.

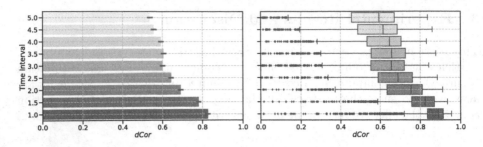

Fig. 1. Time interval Δt and estimation accuracy $dCor$ (N=1000)

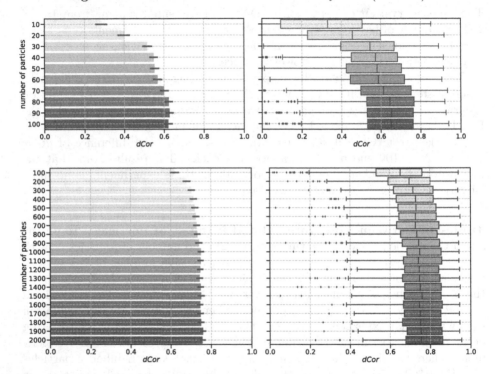

Fig. 2. Number of particles N and estimation accuracy $dCor$ (Δt=1)

To summarize the findings: while the number of particles is positively correlated and the time interval is negatively correlated to estimation accuracy in DA, the estimation accuracy is more constrained by the choice of time interval than the number of particles in the experiments. This implies that, given limited computational resources in DA applications, once the number of particles is sufficiently large, more computational resources can be allocated to shorten the time interval of iteration in DA to improve the estimation accuracy.

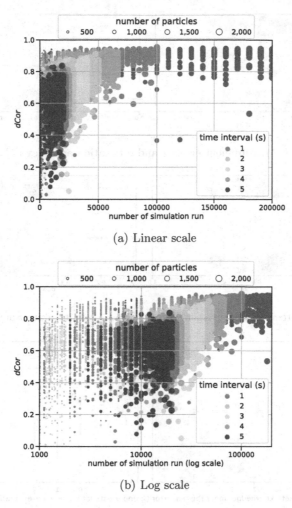

(a) Linear scale

(b) Log scale

Fig. 3. Trade-off between the time interval Δt and number of particles N (Color figure online)

3.2 Measurement Errors and Perceived Measurement Errors

In the experiments, the levels of measurement errors $\epsilon \in [0,3]$ are from zero (0) to low (1), medium (2) till high (3). The levels of perceived measurement errors ϵ' are represented in a similar manner. Different levels of measurement errors $\epsilon \in [0,3]$ are experimented first with perceived measurement errors $\epsilon' = 1$, $\Delta t = 1$ and $N = 400$. As shown in Fig. 4, when ϵ increases from zero to high, the estimation accuracy $dCor$ decreases with increasing variances.

The levels of perceived measurement errors $\epsilon' \in [1,4]$ are experimented with $\epsilon = 1$, $\Delta t = 1$ and $N = 400$. Figure 5 shows that a higher level of perceived measurement errors in DA does not seem to generate a clear pattern in relation with $dCor$. The variances of $dCor$ have slight reduction, however.

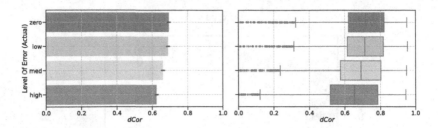

Fig. 4. Measurement errors ϵ and estimation accuracy $dCor$

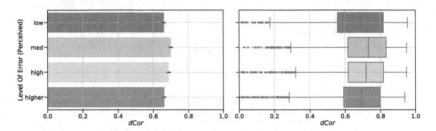

Fig. 5. Perceived measurement errors ϵ' and estimation accuracy $dCor$

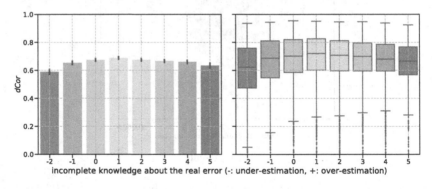

Fig. 6. Difference between perceived measurement errors and actual measurement errors $\epsilon' - \epsilon$ and estimation accuracy $dCor$

How does the difference between ϵ and ϵ' affect the estimation accuracy in DA? We further experiment this by sweeping $\epsilon \in \{0, 1, 2, 3\}$ and $\epsilon' \in \{1, 2, 3, 4, 5\}$ where $\Delta t = 1$ and $N = 400$. The results are shown in Fig. 6, where the X-axis shows the difference of perceived measurement errors and actual measurement errors by subtracting the value of the latter from the former, i.e. $x = \epsilon' - \epsilon$. For example, when the levels of measurement errors $\epsilon = 0$ and the levels of perceived measurement errors $\epsilon' \in \{1, 2, 3, 4, 5\}$, the results are plotted along $x \in \{1, 2, 3, 4, 5\}$; when $\epsilon = 3$ then the results are along $x \in \{-2, -1, 0, 1, 2\}$. This means, a negative x value indicates under estimation and a positive x indicates over estimation of the measurement errors.

The experimental results show that under estimation of the measurement errors ($x < 0$) leads to lower estimation accuracy $dCor$ in average, and over estimation ($x > 0$) often has higher $dCor$ than under estimation ($x < 0$). Perfect knowledge about measurement errors ($x = 0$) does not necessarily result in better $dCor$, while slight over estimation ($x = 1$) has better $dCor$ than perfect knowledge. In the cases when $x > 1$, $dCor$ gradually decreases again (see the slight right skew of the bars in Fig. 6) but it is no worse than the same levels of under estimation. In addition, $dCor$ has lower variances when over estimating the errors than under estimation, which is often a desired feature in DA.

To further illustrate the difference, we present and discuss another experiment that compares two cases: (a) perfect knowledge about measurement errors ($x = 0$); (b) slight over estimation of measurement errors ($x = 1$). The result is shown in Fig. 7. In both cases, the level of the actual measurement errors is *Low* ($\epsilon = 1, \Delta t = 2$ and $N = 1300$). The first case (a) has perceived measurement errors at level *Low* ($\epsilon' = 1$) while the second case (b) over estimates the measurement errors at level *Medium* ($\epsilon' = 2$). These two cases perform distinctly in estimating the queue length $queLen_{sim}$ in the simulation responding to the sudden change of the arrival rate $arrRate_{real}$ and processing rate $procRate_{real}$ at time $t = 15$ in the "real system". In case (a), the simulation can not well follow the trajectory of $queLen$ already in the first 15 s ($t : 0 \rightarrow 15$). Once the sudden change occurs at $t = 15$, $queLen$ diverges more and can catch up the system state again after 10 iterations in DA. In case (b), the simulation can follow the sudden change more responsively.

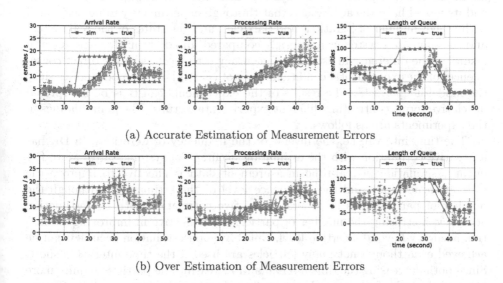

(a) Accurate Estimation of Measurement Errors

(b) Over Estimation of Measurement Errors

Fig. 7. Accurate estimation (a) vs over estimation (b) of measurement errors

The difference in response time in the two cases can be explained by the spread of particles, which are depicted as gray dots in Fig. 7. Note that the vertical spread of particles in case (a) is narrower than that in case (b). In case (a), only a few particles having a small deviation from the measurement can "survive" throughout the experiment. Particles are discarded when they are located far apart. Consequently, sudden and large changes in the system are not detected rapidly because of the restricted spread of particles. In case (b), as the particles spread wider, the aggregated result can quickly converge to the true value under sudden changes. Thus widespread particles are more tolerating and show more responsive estimation in detecting capricious system changes.

Given these observations in the experiments, we conclude that a pessimistic view on measurement errors has advantages over an optimistic view on measurement errors with respect to the resulting estimation accuracy in DA. In addition, a slight pessimistic view on measurement errors results in better estimation accuracy than an accurate view on measurement errors in the experiments. (This is rarely an intuitive choice in DA experimental setup.)

4   Conclusions and Future Work

The experiments presented in this paper study the effect of experimental conditions – namely the time interval of iterations, the number of particles and the level of measurement errors (or noises) – of data assimilation (DA) on estimation accuracy using an $M/M/1$ queuing system (which is implemented in discrete event simulation). The simulation model is constructed with perfect knowledge about the internal process of the system. The choice of a simple target system and its model have the advantages that thorough experiments can be performed with a high number of iterations and particles, and the states of the real system and the simulated system can be easily compared. In addition, the experimental results of the difference in estimation accuracy (or inaccuracy) are direct consequences of the experimental conditions but not (partly) due to model noises since the model is "perfect". The results of the experiments can thus be interpreted in relative terms contrasting different experimental setups. The main findings in the experiments are as follows.

The time interval, i.e. the inverse of the frequency of iterations, in DA has a negative correlation with the estimation accuracy of system states. More frequent assimilation of real-time measurement data is effective to improve the estimation accuracy and the confidence level of the estimation. Although the number of particles has in general a positive correlation with the estimation accuracy, increasing the number of particles is ineffective in improving estimation accuracy beyond a certain level. Notably, good estimation accuracy can be achieved even though not many particles are used if the time interval is short. Since both decreasing the time interval and increasing the particles require more computation, the former can be more cost effective when the number of particles is sufficiently large. With regard to measurement errors, an over estimation of the level of measurement errors leads to higher estimation accuracy than an under estimation in our experiments. A slight over estimation has better estimation accuracy and more responsive model adaptation to system states than an accurate estimation of measurement errors. An overly pessimistic view on measurement errors, however, deteriorates the estimation accuracy.

In this paper, the assimilation of real-time data to the simulation model is performed with fixed time intervals during an experiment run. An event based data assimilation approach and its effects can be an interesting future research direction. The experimental setups could also be dynamically configured during DA in real-time to achieve good estimation results.

References

1. Ancker, C., Gafarian, A.: Some queuing problems with balking and reneging. i. Oper. Res. **11**(1), 88–100 (1963)
2. Bickel, P.J., Xu, Y.: Discussion of brownian distance covariance. Ann. Appl. Stat. **3**(4), 1266–1269 (2009)
3. Bouttier, F., Courtier, P.: Data assimilation concepts and methods. ECMWF (European Centre for Medium-Range Weather Forecasts) (2002)
4. Darema, F.: Dynamic data driven applications systems: a new paradigm for application simulations and measurements. In: Bubak, M., van Albada, G.D., Sloot, P.M.A., Dongarra, J. (eds.) ICCS 2004. LNCS, vol. 3038, pp. 662–669. Springer, Heidelberg (2004). https://doi.org/10.1007/978-3-540-24688-6_86
5. Gu, F.: On-demand data assimilation of large-scale spatial temporal systems using sequential monte carlo methods. Simulation Modell. Pract. Theory **85**, 1–14 (2018)
6. Hu, X.: Dynamic data driven simulation. SCS M&S Magazine **5**, 16–22 (2011)
7. Hu, X., Wu, P.: A data assimilation framework for discrete event simulations. ACM Trans. Model. Comput. Simul. **29**(3), 171–1726 (2019). https://doi.org/10.1145/3301502
8. Huang, Y., Seck, M.D., Verbraeck, A.: Towards automated model calibration and validation in rail transit simulation. In: Sloota, P.M.A., van Albada, G.D., Dongarrab, J. (eds.) Proceedings of The 2010 International Conference on Computational Science. Procedia Computer Science, vol. 1, pp. 1253–1259. Elsevier, Amsterdam (2010)
9. Huang, Y., Verbraeck, A.: A dynamic data-driven approach for rail transport system simulation. In: Rossetti, M.D., Hill, R.R., Johansson, B., Dunkin, A., Ingalls, R.G. (eds.) Proceedings of The 2009 Winter Simulation Conference, pp. 2553–2562. IEEE, Austin (2009)
10. Huang, Y., Seck, M.D., Verbraeck, A.: Component based light-rail modeling in discrete event systems specification (DEVS). Simulation **91**(12), 1027–1051 (2015)
11. Huang, Y., Verbraeck, A., Seck, M.D.: Graph transformation based simulation model generation. J. Simul. **10**(4), 283–309 (2016)
12. Huang, Y., Warnier, M., Brazier, F., Miorandi, D.: Social networking for smart grid users - a preliminary modeling and simulation study. In: Proceedings of 2015 IEEE 12th International Conference on Networking, Sensing and Control, pp. 438–443 (2015). DOI: https://doi.org/10.1109/ICNSC.2015.7116077
13. Ma, C., et al.: Multiconstituent data assimilation with WRF-Chem/DART: Potential for adjusting anthropogenic emissions and improving air quality forecasts over eastern China. J. Geophys. Res.: Atmospheres **124**, 7393–7412 (2019). https://doi.org/10.1029/2019JD030421

14. Nichols, N.: Data assimilation: aims and basic concepts. In: Swinbank, R., Shutyaev, V., Lahoz, W.A. (eds.) Data Assimilation for the Earth System, pp. 9–20. Springer, Dordrecht (2003). https://doi.org/10.1007/978-94-010-0029-1_2

15. Petropoulos, G.P.: Remote Sensing of Surface Turbulent Energy Fluxes, chap. 3, pp. 49–84. CRC Press, Boca Raton (2008)

16. Ren, L., Nash, S., Hartnett, M.: Data assimilation with high-frequency (HF) radar surface currents at a marine renewable energy test site. C. Guedes Soares (Leiden: CRC Press/Balkema) pp. 189–193 (2015)

17. Shuwen, Z., Haorui, L., Weidong, Z., Chongjian, Q., Xin, L.: Estimating the soil moisture profile by assimilating near-surface observations with the ensemble kaiman filter (ENKF). Adv. Atmosph. Sci. **22**(6), 936–945 (2005)

18. Smith, P., Baines, M., Dance, S., Nichols, N., Scott, T.: Data assimilation for parameter estimation with application to a simple morphodynamic model. Math. Rep. **2**, 2008 (2008)

19. Székely, G.J., Rizzo, M.L., Bakirov, N.K., et al.: Measuring and testing dependence by correlation of distances. Ann. Stat. **35**(6), 2769–2794 (2007)

20. Tran, A.P., Vanclooster, M., Lambot, S.: Improving soil moisture profile reconstruction from ground-penetrating radar data: a maximum likelihood ensemble filter approach. Hydrol. Earth Syst. Sci. **17**(7), 2543–2556 (2013)

21. Wang, M., Hu, X.: Data assimilation in agent based simulation of smart environment. In: Proceedings of the 1st ACM SIGSIM Conference on Principles of Advanced Discrete Simulation, pp. 379–384. ACM (2013)

22. Wang, M., Hu, X.: Data assimilation in agent based simulation of smart environments using particle filters. Simulation Modell. Pract. Theory **56**, 36–54 (2015)

23. Xie, X.: Data assimilation in discrete event simulations. Ph.D. thesis, Delft University of Technology (2018)

24. Xie, X., van Lint, H., Verbraeck, A.: A generic data assimilation framework for vehicle trajectory reconstruction on signalized urban arterials using particle filters. Transport. Res. Part C: Emerg. Technol. **92**, 364–391 (2018)

25. Xie, X., Verbraeck, A., Gu, F.: Data assimilation in discrete event simulations: a rollback based sequential monte carlo approach. In: Proceedings of the Symposium on Theory of Modeling & Simulation, p. 11. Society for Computer Simulation International (2016)

26. Xue, H., Gu, F., Hu, X.: Data assimilation using sequential monte carlo methods in wildfire spread simulation. ACM Trans. Model. Comput. Simulation (TOMACS) **22**(4), 23 (2012)

PDPNN: Modeling User Personal Dynamic Preference for Next Point-of-Interest Recommendation

Jinwen Zhong[1,2], Can Ma[1(✉)], Jiang Zhou[1], and Weiping Wang[1]

[1] Institute of Information Engineering, Chinese Academy of Sciences, Beijing, China
{zhongjinwen,macan,zhoujiang,wangweiping}@iie.ac.cn
[2] School of Cyber Security, University of Chinese Academy of Sciences, Beijing, China

Abstract. Next Point of Interest (POI) recommendation is an important aspect of information feeds for Location Based Social Networks (LSBNs). The boom in LSBN platforms such as Foursquare, Twitter, and Yelp has motivated a considerable amount of research focused on POI recommendations within the last decade. Inspired by the success of deep neural networks in many fields, researchers are increasingly interested in using neural networks such as Recurrent Neural Network (RNN) to make POI recommendation. Compared to traditional methods like Factorizing Personalized Markov Chain (FPMC) and Tensor Factorization (TF), neural network methods show great improvement in general sequences prediction. However, the user's personal preference, which is crucial for personalized POI recommendation, is not addressed well in existing works. Moreover, the user's personal preference is dynamic rather than static, which can guide predictions in different temporal and spatial contexts. To this end, we propose a new deep neural network model called Personal Dynamic Preference Neural Network(PDPNN). The core of the PDPNN model includes two parts: one part learns the user's personal long-term preferences from the historical trajectories, and the other part learns the user's short-term preferences from the current trajectory. By introducing a similarity function that evaluates the similarity between spatiotemporal contexts of user's current trajectory and historical trajectories, PDPNN learns the user's personal dynamic preference from user's long-term and short-term preferences. We conducted experiments on three real-world datasets, and the results show that our model outperforms current well-known methods.

Keywords: Point-of-interest recommendation · User personal dynamic preference · Recurrent Neural Network

1 Introduction

Due to the prevalence of smart mobile devices, people frequently use Location-Based Social Networks (LBSNs) such as Foursquare, Twitter, and Yelp to post

© Springer Nature Switzerland AG 2020
V. V. Krzhizhanovskaya et al. (Eds.): ICCS 2020, LNCS 12142, pp. 45–57, 2020.
https://doi.org/10.1007/978-3-030-50433-5_4

check-ins and share their life experiences. Point-of-interest (POI) recommendation has become an important way to help people discover attractive and interesting venues based on their historical preferences. Many merchants also use POI recommendation as an important channel to promote their products.

POI recommendation has been extensively studied in recent years. In the literature, Markov Chains (MC) [1,2], collaborative filtering [3–6], and Recurrent Neural Networks [7–11] are three main approaches. These methods commonly focus on capturing either the user's short-term preferences or long-term preferences. User's short-term and long-term preferences are both important for achieving higher accuracy of recommendation [12]. The short-term preference is greatly influenced by time and POIs the user has visited recently, while the long-term preference reflects the user's personal habits thus all historically visited POIs need to be considered. However, the user's personal long-term preference, which is crucial for personalized POI recommendation, is not addressed well in existing work. Moreover, the user's personal preference is dynamic rather than static, which can guide predictions in different temporal and spatial contexts. To this end, we propose a new deep neural network model called Personal Dynamic Preference Neural Network (PDPNN). The core of the PDPNN model includes two parts: one part learns the user's personal long-term preferences from the historical trajectories, and the other part learns the user's short-term preferences from the current trajectory. By introducing a similarity function that evaluates the similarity between spatiotemporal contexts of user's current trajectory and historical trajectories, PDPNN learns the user's personal dynamic preference from user's long-term and short-term preferences.

We summarize our contributions in this paper as follows:

- We propose a new approach to better model dynamic user preference for their next POI from their current trajectory in conjunction with their trajectory context. Based on the attention-enhanced Long Short-Term Memory (LSTM) neural network model, we build a new neural network model named PDPNN for next POI recommendation.
- To accelerate training of trajectory contexts, we propose inner epoch cache mechanism to store the trajectory context output states during the process of model training. This reduces the complexity of the user's personal dynamic preference module from $O(n^2)$ to $O(n)$ in each training epoch.
- We conducted experiments on three real world datasets and the results show that our model outperforms current well known methods.

2   Related Work

Plenty of approaches have been proposed that focus on sequential data analysis and recommendation. Collaborative filtering based models, such as Matrix Factorization (MF) [3] and Tensor Factorization(TF) [4] are widely used for recommendation. These methods aim to cope with data sparsity and the cold start problem by capturing common short-term preferences among users, but cannot capture the user's personal long-term preferences. Other existing studies employ the properties of a Markov chain to capture sequential patterns, such as Markov Chain (MC) [1] and Factorizing Personalized Markov Chains (FPMC) [2]. FPMC models user's preference and sequential information jointly by combining factorization method and Markov Chains for next-basket recommendation. However, both MC and FPMC methods fall short in learning the long-term preference and the periodicity of the user's movement.

Recently, as a result of the success of deep learning in speech recognition, vision and natural language processing, Recurrent Neural Networks (RNNs) have been widely used in sequential item recommendation [7–11]. Spatial Temporal Recurrent Neural Networks (STRNN) [8] utilizes a RNN architecture and linear interpolation to learn the regularity of sequential POI transition. However, traditional RNNs suffer from the issues of vanishing gradients and error propagation when they learn long-term dependencies [13]. Therefore, special gating mechanisms such as Long Short-Term Memory network (LSTM) [14] have been developed and widely used in recent work [9–11]. By controlling access to memory cells, LSTMs can alleviate the problem of long-term dependencies.

Attention mechanism is a key advancement in deep learning in recent years, and it shows a promising performance improvement for RNNs [15–17]. By introducing the attention mechanism, Attention-based Spatio-Temporal LSTM network (ATST-LSTM) [9] can focus on the relevant historical check-in records in a check-in sequence selectively using the spatiotemporal contextual information. However, these models are designed to learn short-term preferences and are not well suited to learning personal long-term preferences. They commonly add user embedding to RNN outputs, and reduce the problem of user context learning to the problem of learning a static optimal user embedding representation. It is more difficult to learn long-term preferences with simple attention mechanisms, especially for users with only a small set of historical trajectories.

3 Model Description

3.1 Problem Formulation

Let $U = \{u_1, \ldots, u_m\}$ denote the user set and $P = \{p_1, \ldots, p_j\}$ denote the POI set, where $\|U\|$ and $\|P\|$ are the total numbers of users and POIs, respectively. Each POI p_k is associated with a geographic location $l_k = (l_a, l_o)$, where l_a and l_o denote the latitude and longitude of the POI location. For a user $u \in U$, a check-in behavior means u_i visits a POI p_k at time t_k, which is denoted as tuple (u, p_k, t_k). Any check-in sequence with all the time intervals of successive check-ins less than a threshold value T_{delta}, is called a trajectory. Obviously, a user's historical check-ins will be segmented into many trajectories. We denote $T_i^u = \{(u, p_1, t_1), \ldots, (u, p_n, t_n)\}$ as the i-th trajectory of the user u, with $t_k - t_{k-1} \leq T_{delta}$ for neighbor check-ins of the trajectory, and $\|T_i^u\|$ is the total number of check-ins. All the historical trajectories of the user u are denoted as $T^u = \{T_1^u, T_2^u, \ldots, T_n^u\}$.

When we are processing the i-th trajectory T_i^u, we need to take into account the previous historical trajectories of user u, which is called the trajectory context. We denote $C_i^u = \{T_1^u, T_2^u, \ldots, T_{i-1}^u\}$ as the context of the current trajectory T_i^u. When there is no ambiguity, we will omit the subscript u.

Formally, given a user u and all of their historical trajectories $T^u = \{T_1^u, T_2^u, \ldots, T_n^u\}$, the problem of making the next POI recommendation is to predict the next POI p_{n+1} that the user is most likely to visit.

3.2   PDPNN Model

Basic Framework. The PDPNN model receives current trajectory T_i and its trajectory context C_i as input. C_i is the previous trajectories of each trajectory T_i of the same user. The model learns short-term preference mainly from current trajectory T_i, and builds a personal dynamic preference from both T_i and its trajectory context C_i. The architecture of our proposed model PDPNN is shown in Fig. 1 :

- (I) Attention-enhanced recurrent neural network (ARNN) module: We use a LSTM network to capture short-term and long-term spatial-temporal sequence patterns. Additionally, an attention mechanism is introduced to capture the weight of all hidden states of the trajectory sequence. This component is used as an important part of the personal dynamic preference and the short-term preference learning module.
- (II) Personal dynamic preference learning module: The user's personal preferences is important for personalized POI recommendation, which is implicit in their historical trajectories. This module obtains a representation of personal dynamic preference from the user's current trajectory and trajectory context C_i. We first utilize the ARNN module to calculate the output hidden states of all trajectories in the trajectory context. Obviously, not every state can help to predict a POI, and the historical trajectories that are highly correlated with the current trajectory are supposed to have greater weight. To this end, we introduce attention mechanism into the module to measure the spatial temporal context similarity of the trajectory context and the current trajectory, and aggregate all the hidden states of the trajectory context as the personal dynamic preference representation.
- (III) Short-term preference learning module: The short-term preference, which is usually referred to as sequential preference, is commonly learned from the current trajectory. We simply apply the ARNN module described above to get the hidden state of the current trajectory T_i .
- (IV) Classifier: This is the final output component which unifies the user embedding output, the last hidden state from current trajectory and the aggregated attention from historical trajectories into a feature representation; after this it predicts the next POI. The output of this module is the probability vector of every POI that the user is likely to visit next. Cross-entropy loss and L2 regularization is used to measure the total loss in this module.

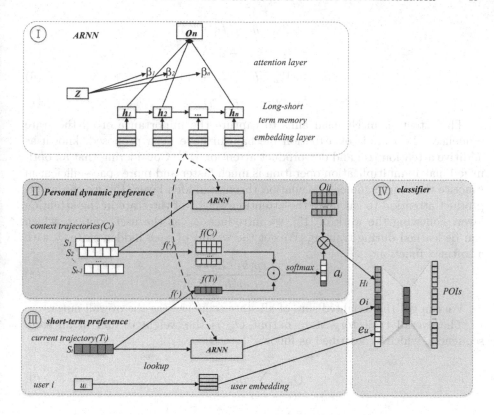

Fig. 1. The architecture of PDPNN model.

ARNN: ARNN is an attention-enhanced recurrent neural network, which receives a trajectory T^u as input and outputs the weighted sum of the hidden states of the user's trajectory via a attention mechanism. The ARNN model consists of an embedding layer, a recurrent layer and an attention layer.

Each input trajectory contains a sequence of POI identifier (id) p, time interval Δs, geographic distance Δt, longitude lo and latitude la. The input POI id is then transformed into a latent space vector e_i^p by the embedding layer. Input data I_t can be described as follows:

$$I_t = [e_t^p; \Delta s_t; \Delta t_t; la_t; lo_t] \tag{1}$$

where $\Delta s = \sqrt{(la_t - la_{t-1})^2 - (lo_t - lo_{t-1})^2}$, and $\Delta t = t_t - t_{t-1}$, the subscript $t \in [2, \|T^u\|]$.

We utilize a Long Short-Term Memory (LSTM) network in the recurrent layer. A LSTM neuron unit consists of an input gate i_t, an input gate f_t, and an output gate o_t. These parameters are explained in detail below:

$$f_t - \delta(W_f \cdot [h_{t-1}; I_t] + b_f) \tag{2}$$

$$i_t = \delta(W_i \cdot [h_{t-1}; I_t] + b_i) \tag{3}$$

$$\tilde{c}_t = \delta(W_c \cdot [h_{t-1}; I_t] + b_c) \tag{4}$$

$$c_t = f_t \odot c_{t-1} + i_c \odot \tilde{c}_t \tag{5}$$

$$o_t = \delta(W_o \cdot [h_{t-1}; I_t] + b_o) \tag{6}$$

$$h_t = o_t \odot tanh(c_t) \tag{7}$$

The attention mechanism aims to capture the importance of all the state sequences. Two kinds of attention mechanism have been proposed, known as additive attention [16] and dot-product attention [15]. Considering that an optimized matrix multiplication operation is much faster and more space-efficient in practice than an alignment calculation through hidden layer [15], we chose dot-product attention to calculate the attention weight of the state in the attention layer. Following the work of [15], we introduce Z_u as the user context, which can be learned during training. We get the weight of each hidden state β_t with a softmax function:

$$\beta_t = \frac{exp(h_t \cdot Z_u)}{\sum_{j=1}^{n} exp(h_j \cdot Z_u)} \tag{8}$$

We can get the weighted sum of all state sequences as the overall output.

The overall trajectory state output O_n is the weighted sum of all state sequences, which is described as follows:

$$O_n = \sum_{t=1}^{n} \beta_t \odot h_t \tag{9}$$

For the convenience of subsequent references, we summarize Eqs. (1)–(9) as function $ARNN(\cdot)$:

$$O_n = ARNN(T^u) \tag{10}$$

Personal Dynamic Preference Learning Module: Unlike routine RNN models that treat each separate trajectory as input, PDPNN attaches to each trajectory T_i the trajectory context $C_i = \{T_1, T_2, \ldots, T_{i-1}\}$ of the same user to produce the input.

Obviously, the length of the current trajectory processed by the padding operation is fixed, while the trajectory context can not be filled or truncated due to the large length changes. This represents a difficult problem in model training. To solve this problem, we separate the processing of trajectory context from the processing of the current trajectory. For each trajectory context C_i , we use the ARNN model to calculate the hidden state output of each trajectory T_j in C_i:

$$O_j^{(i)} = ARNN(T_j), j \in [1, i - 1] \tag{11}$$

Trajectories that are highly relevant to the current trajectory are considered to play a more important role in guiding POI predictions for all historical trajectories in the trajectory context. To this end, we introduce a correlation function

$f(\cdot)$ to calculate the weight of each historical trajectory, and use softmax to calculate the weight of all historical trajectory states:

$$w_{ij} = \frac{exp(f(T_i^u, T_j^u))}{\sum_{j=1}^{i-1} exp(f(T_i^u, T_j^u))}, j \in [1, i-1] \tag{12}$$

Function $f(\cdot)$ here can be any function used to measure the correlation of trajectories. We use the Jaccard Similarity of the POI set of the trajectory to measure the similarity between the trajectories in the paper:

$$f(T_i^u, T_j^u) = \frac{|P_i \cap P_j|}{|P_i \cup P_j|}), j \in [1, i-1] \tag{13}$$

where P_i and P_j are the POI sets of T_i^u and T_j^u, respectively.

Then we multiply the weight of the context trajectories and their corresponding output latent states. The user's personal dynamic preference H_i is represented as the weighted sum of the products above:

$$H_i = \sum_{j=1}^{i-1} w_{ij} \cdot O_j^{(i)} \tag{14}$$

Short-Term Preference Learning Module: The short-term preference learning module is mainly used to capture the sequential POI transition preferences of users. The module consists of a user embedding layer and an ARNN model, and receives a user and trajectories as input. Each trajectory is represented by a sequence of POI identifier, time interval, geographic distance, longitude and latitude as desribed in Eq. (1). Users are represented by a user identifier u in real world data, which can not precisely reflect the similarities and differences between users. The appropriate representation of a user can accurately measure the similarities between users, so that the model can learn the similarity of user behavior preferences according to degree of similarity between users. To this end, we use a fully connected network to embed user identifiers into latent space.

$$e_u = tanh(W_u \cdot u + b_u) \tag{15}$$

where parameter $W_u \in \mathbf{R}^{\|U\| \times d_u}$ and $b_u \in \mathbf{R}^{1 \times d_u}$.

The output of the short-term preference learning module can be described as follows:

$$O_i = ARNN(T_i) \tag{16}$$

Classifier: We consider the next POI recommendation as a multiclass classification problem. The classifier module concatenates the user embedding, the user's personal dynamic preference and the current trajectory state as the input:

$$Q = [H_i; O_i; e_u] \tag{17}$$

Then, we feed Q into a fully connected layer with softmax function to calculate the probability of each POI:

$$\hat{y}_t = softmax(sigmoid(W_s \cdot Q + b_s)) \tag{18}$$

where parameter $W_s \in \mathbf{R}^{\|P\| \times d_p}$ and b_s is the bias parameter.

We adopt the cross-entropy loss between the ground truth POI y_t and the predicted POI $\hat{y}_t$. The loss function can be calculated as follows:

$$L(I_1^{(1)}, \cdots I_1^{(T^{(1)}-1)} \cdots, I_N^{(1)}, I_N^{(T^{(N)}-1)}, \theta) =$$
$$-\frac{1}{N} \sum_{n=1}^{N} y_t log(\hat{y}_t) + (1 - y_t)log(1 - \hat{y}_t) + \frac{\lambda}{2} \|\Theta^2\| \tag{19}$$

where Θ denotes the parameter set of PDPNN, and λ is a pre-defined hyper parameter for the L2 regularization to avoid overfitting. To optimize the above loss function, we use Stochastic Gradient Descent to learn the parameter set Θ.

3.3   Model Training

Training the model with short-term preference and personal dynamic preference jointly is time-consuming, so we use pre-trained model and inner epoch cache method to accelerate the training process.

Pre-trained Model: The ARNN is the core component of the PDPNN model, and it takes most of the time required by the training process. In other words, if the ARNN component could be trained as quickly as possible, the training speed of the whole model would be accelerated. To this end, we try to pre-train the ARNN component in a more simple model, which connects the output of the ARNN and the classifier directly. Then, we load the pre-trained model and modify its network structure to adapt to our current model in training.

Inner Epoch Cache: Each input of the PDPNN model contains the current trajectory and the trajectory context, which makes the training process time consuming. For a certain user u with n trajectories $\{T_1, T_2, \ldots, T_n\}$, the PDPNN needs to process all of the n trajectories and the corresponding trajectory contexts. For each trajectory T_i , the PDPNN learns a short-term preference from T_i, and learns the user's personal dynamic preference from its trajectory context $C_i = \{T_1, T_2, \ldots, T_{i-1}\}$, which means the PDPNN needs to calculate the output state of C_i with ARNN component for i times. So, the whole training complexity of the n trajectories user u is $O(n^2)$.

As a part of trajectory contexts of $T_{i+1}, T_{i+2}, \ldots, T_n$, the i-th trajectory T_i is repeatedly calculated for $n - i$ times during each training epoch. The optimization we can intuitively think of is to eliminate the repeated calculations. However, trajectory context state values are updated during training synchronously, which means we can not simply cache the state value in the whole training process. Considering the increment of state value updates between batches is very small in each training epoch, if the trajectory context output states can be cached within one training epoch and updated between epochs, a large number of approximately repeated calculations can be eliminated. To this end, we apply the inner epoch cache to store the trajectory context output state value during model training, and the calculation of each user's personal dynamic preference can be reduced to linear complexity.

4   Experiment Analysis and Evaluation

4.1   Experiment Settings

Datasets: We conducted experiments on three publicly available LBSN datasets, NYC, TKY and CA. NYC and TKY [18] are two datasets collected from users sharing their check-ins on the Foursquare website in New York and Tokyo, respectively. CA is a subset of a Foursquare dataset [19], which includes long-term, global-scale check-in data. We chose the check-ins of users in California for the dataset in this paper. The check-in times of above datasets range from 2012 Apr to 2013 Sep. Each record contains an anonymous user identifier, time, POI-id, POI category, latitude and longitude of the check-in behavior. In order to alleviate the problem of data sparsity, following previous work [12], we expand the set of trajectories by adding the sub-trajectories of the original trajectories. For all datasets, we choose 90% of each user's trajectories as the training set, and the remaining 10% as testing data.

Table 1. The statistics of datasets.

Dataset	#Users	#check-ins	#location	#trajectories
LA	1,083	22,7428	38,333	31,941
TKY	2,293	57,3703	61,858	66,663
CA	4,163	48,3805	2,9529	47,276

Comparing Methods: We compare PDPNN with several representative methods for location prediction:

- RNN [7]: This is a basic method for POI prediction, which has been successfully applied in word embedding and ad click prediction.
- AT-RNN: This method empowers the RNN model with an attention mechanism, which has been successfully applied in machine translation and vision processing.
- LSTM [14]: This is a variant of the RNN model, which contains a memory cell and three multiplicative gates to allow long-term dependency learning.

- ATST-LSTM [9]: This is a state-of-the-art method for POI prediction, which applies an attention mechanism to a LSTM network.
- PDPNN: This is our approach, which learns a user's personal dynamic preference base on the current spatial and temporal context.

Parameter Settings: The key hyper parameters in PDPNN include: (1) the embedding dimension for POI and user, namely d_u and d_p; (2) the dimension d_h for the hidden state; (3) the regularization parameter λ. In general, the performance of the PDPNN increases with the above dimensions and gradually stabilizes when dimensions are large enough. In our experiments, we finally set $d_u = d_p = d_h = 200$. For the regularization parameter λ, we tried values in $\{1,0.1,0.01,0.001\}$ and $\lambda = 0.01$ turns out to have the best performance.

Table 2. Prediction performance comparison on three dataset.

Dataset	Method	Recall@1	Recall@5	Recall@10	Recall@20
LA	RNN	0.7901%	3.0536%	5.2317%	9.4597%
	AT-RNN	1.9218%	7.4098%	14.4352%	24.685%
	LSTM	6.8332%	25.71%	39.0562%	50.5659%
	ATST-LSTM	7.9650%	26.009%	39.9317%	51.5909%
	PDPNN	**8.5629%**	**28.2084%**	**41.0634%**	**52.0606%**
TKY	RNN	2.5639%	7.2056%	9.4411%	11.8282%
	AT-RNN	4.0859%	16.8172%	26.8203%	36.9498%
	LSTM	2.6271%	7.2182%	9.8958%	13.5965%
	ATST-LSTM	4.4143%	18.5475	**28.9233%**	**38.8317%**
	PDPNN	**4.8058%**	**18.5665%**	28.2918%	37.0445%
CA	RNN	0.5711%	2.4112%	4.3993%	8.0795%
	AT-RNN	6.5144%	17.8299%	24.6616%	32.0643%
	LSTM	2.0093%	7.0008%	10.7445%	15.4399%
	ATST-LSTM	6.6836%	17.7453%	24.4078%	32.1489%
	PDPNN	**7.1912%**	**18.6548%**	**25.5499%**	**32.9315%**

Metrics: To evaluate the performance of all methods for the POI recommendation problem, we employ a commonly used metric known as recall@N. The recall@N metric is popular in ranking tasks, which evaluates where the ground truth next POI appears in a ranked prediction list. A larger metric value indicates better performance.

4.2   Comparison of Recommendation Performance

We conducted experiments on a machine with Intel a Xeon CPU and a NVIDIA Tesla P4 GPU. The performance comparison of methods on three datasets is illustrated in Table 2. The RNN has a lower baseline performance than other methods with no extra optimization. The AT-RNN improves the performance greatly by 2 or 3 times over the RNN on all three datasets, which shows that the attention mechanism can alleviate the problem of long-term dependencies with gradient descent of RNN. The LSTM outperforms the RNN largely because of its memory and forget gate design, but fails to outperform the AT-RNN for the TKY and CA datasets. ATST-LSTM turns out to be the strongest baseline and shows significant improvement on the TKY and CA datasets. The reason is that ATST-LSTM works better at learning long-term dependencies through the gate and attention mechanisms. This shows that the attention mechanism is an effective supplement to LSTM models.

PDPNN outperforms the above baseline methods in almost all of the recall metrics. Compared with ATST-LSTM, the PDPNN obtains recall@1, recall@5 and recall@10 improvements of 7–10% on LA and CA, and recall@1 and recall@5 improvements of 5–7% on TKY. By creating an elaborate modeling of users' personal dynamic preference, the PDPNN can capture the intentions of user activities more accurately. This enables the PDPNN to achieve better performance in next POI recommendation. It is worth noting that the PDPNN has lower performance in recall@10 and recall@20 than the ATST-LSTM on the TKY dataset. The reason is that the PDPNN treats POI recommendation as a classification problem, and the objective function is to optimize the accuracy of recall@1. Besides, there may be another interesting reason that people's preferences vary widely from country to country.

4.3   Inner Epoch Cache Evaluation

To evaluate the efficiency of the inner epoch cache, we compare the average time consumption of models with and without the inner epoch cache during one epoch training. Figure 2 shows that the PDPNN with an inner epoch cache

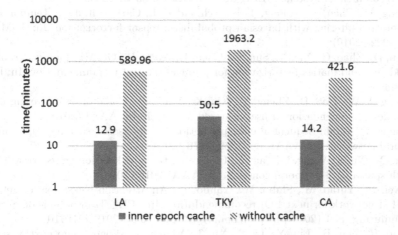

Fig. 2. Time-consuming comparison in one training epoch.

trains 45×, 38× and 30× faster than the model without a cache on LA, TKY and CA dataset, respectively. By introducing the inner epoch cache optimization in training, the per-epoch training time is reduced from more than one day to one hour on the TKY data set. It shows that the inner epoch cache can greatly accelerate the model training process.

5 Conclusion

In this paper, we have proposed a new approach to better model the user's personal dynamic preference of next POI from current trajectory supplemented with a trajectory context. Based on an attention empowered LSTM neural network, we build a new neural network model named PDPNN for next POI recommendation. Moreover, to accelerate the training of trajectory contexts, we proposed an inner epoch cache to store the trajectory context output state value during model training, and reduce the complexity of the user's personal dynamic preference module from $O(n^2)$ to $O(n)$ in each training epoch. We conduct experiments on three real world data set and show that our model outperforms current well-known methods.

Acknowledgment. This work was supported by Beijing Municipal Science & Technology Commission (Z191100007119003).

References

1. Gambs, S., Killijian, M.-O., del Prado Cortez, M.N.: Next place prediction using mobility markov chains. In: MPM (2012)
2. Steffen, R., Christoph, F., Lars, S.T.: Factorizing personalized Markov chains for nextbasket recommendation. In: WWW 811–820 (2010)
3. Yehuda Koren; Robert Bell; and Chris Volinsky: Matrix factorization techniques for recommender systems. IEEE Comput. **42**(8), 30–37 (2009)
4. Xiong, L., Chen, X., Huang, T.-K., Schneider, J., Carbonell, J.G.: Temporal collaborative filtering with bayesian probabilistic tensor factorization. In: SDM, pp. 211–222 (2010)
5. Lian, D., Zhao, C., Xie, X., Sun, G., Chen, E., Yong, R.: GeoMF, joint geographical modeling and matrix factorization for point-of-interest recommendation. In: KDD (2014)
6. Zheng, V.W., Cao, B., Zheng, Y., Xie, X., Yang, Q.: Collaborative filtering meets mobile recommendation, a user-centered approach. In: AAAI (2010)
7. Zhang, Y., et al.: Sequential click prediction for sponsored search with recurrent neural networks. In: AAAI, pp. 1369–1376 (2014)
8. Liu. Q., Wu, S., Wang, L., Tan, T.: Predicting the next location : a recurrent model with spatial and temporal contexts. In: AAAI 2016 (2016)
9. Liwei, H., Yutao, M., Shibo, W., Yanbo, L.: An attention-based spatio-temporal LSTM network for next POI recommendation. In: IEEE Transactions on Services Computing, p. 1 (2019). https://doi.org/10.1109/TSC.2019.2918310
10. Zhao, P., Zhu, H., Liu, Y., Li, Z., Xu, J., Victor, S.: Where to go next: a spatio-temporal LSTM model for next POI recommendation. In: AAAI 2019 (2019)

11. Yao, D., Zhang, C., Huang, J., Bi, J.: SERM: a recurrent model for next location prediction in semantic trajectories. In: CIKM (2017)
12. Jannach, D., Lerche, L., Jugovac, M.: Adaptation and evaluation of recommendations for short-term shopping goals. In: RecSys, Adaptation and evaluation of recommendations for short-term shopping goals. pp. 211–218 (2015)
13. Bengio, Y., Frasconi, P., Simard, P.: Learning long-term dependencies with gradient descent is difficult. IEEE Trans. Neural Netw. **5**(2), 157–166 (1994)
14. Schuster, M., Paliwal, K.P.: Bidirectional recurrent neural networks. IEEE Trans. Signal Process. **45**(11), 2673–2681 (1997)
15. Yang, Z., Yang, D., Dyer, C., He, X., Smola, A., Hovy, E.: Hierarchical attention networks for document classification. In: Proceedings of NAACL (2016)
16. Bahdanau, D., Cho, K., Bengio, Y.: Neural machine translation by jointly learning to align and translate. In: Proceedings of ICLR (2015)
17. Vaswani, A., et al.: Attention is all you need. In: Proceedings of NIPS (2017)
18. Yang, D., Zhang, D., Zheng, V.W., Yu, Z.: Modeling user activity preference by leveraging user spatial temporal characteristics in LBSNs. IEEE Trans. Syst. Man Cybern. Syst. (TSMC) **45**(1), 129–142 (2015)
19. Yang, D., Zhang, D., Qu, B.: Participatory cultural mapping based on collective behavior data in location based social networks. ACM Trans. Intell. Syst. Technol. (TIST) **7**(3), 30 (2016)

Cyber Attribution from Topological Patterns

Yang Cai[✉], Jose Andre Morales, and Guoming Sun

Carnegie Mellon University, 5000 Forbes Avenue, Pittsburgh, PA 15213, USA
ycai@cmu.edu, jose@josemorales.org

Abstract. We developed a crawler to collect live malware distribution network data from publicly available sources including Google Safe Browser and VirusTotal. We then generated a dynamic graph with our visualization tool and performed malware attribution analysis. We found: 1) malware distribution networks form clusters rather than a single network; 2) those cluster sizes follow the Power Law; 3) there is a correlation between cluster size and the number of malware species in the cluster; 4) there is a correlation between the number of malware species and cyber events; and finally, 5) infrastructure components such as bridges, hubs, and persistent links play significant roles in malware distribution dynamics.

Keywords: Cyber attribution · Malware · Malware distribution network · MDN · Dynamics · Graph · Security · Computer virus · Malicious software · Topology

1 Introduction

Similar to an epidemic virus spread, malicious files infect computer systems over a set of globally connected domains or IP addresses, which we call a malware distribution network (MDN) [4–7, 9–15]. In this paper, we study temporal topological structures of an MDN with subsets of connected domains as a malicious cluster (M-Cluster). We created a novel dataset over an eight-month period by crawling the transparency report repository of Google Safe Browsing as well as collected URL and malware file hash scanning results from VirusTotal [8, 17]. We analyzed the topological structural evolution and malware hosted on various domain servers of the three largest M-Clusters in an eight-month period. Our analysis revealed the layout of an M-Cluster as a *hub* and *bridge* structure. We further observed that the increase in size of an M-Cluster occured in parallel to an increase in discovered malware on the domain servers. One scenario in which the manifestation of an M-Cluster may occur is in conjunction with global events, for example, the 2017 Presidential Inauguration of the United States of America. Our M-Cluster analysis also revealed a consistent presence of multiple layers of URL redirection services, which, we believe, serves to obfuscate servers hosting malware. The contributions of this paper are: 1) observation and analysis of malware distribution networks as clusters with a bridge and hub construction; 2) correlation between size increases of M-Clusters and the presence of hosted malware; 3) the significant roles of persistent bridges and hubs in malware distribution dynamics; and 4) development of algorithms to identify hubs and bridges.

© Springer Nature Switzerland AG 2020
V. V. Krzhizhanovskaya et al. (Eds.): ICCS 2020, LNCS 12142, pp. 58–71, 2020.
https://doi.org/10.1007/978-3-030-50433-5_5

2   Literature Review

Dynamic graphs have been used in software engineering and operation research. Schiller and Strufe developed the framework for the analysis of dynamic graphs with DNA (Dynamic Network Analyzer) [2]. The topological properties of a dynamic graph include topological metrics of degree distribution (DD), connected components (C), assortativity (ASS), clustering coefficient (CC), rich-club connectivity (RCC), all-pairs-shortest paths (SP), and betweenness centrality (BC) [1]. Yu, et al. [26] studied the malware propagation dynamics of a single malware ConFlicker botnet. The authors tracked three top-domain layers and the growth of total compromised hosts by Android malware. The authors used the epidemic dynamics model to interpolate the malware distribution process. They discovered the Power Law distribution of ConFlicker botnet in the top three levers, i.e. ranking in botnet size of the malware versus probability of the distribution. This is perhaps the most comprehensive study of malware distribution at single botnet with a computational distribution model.

Here, we define a malware distribution network (MDN) as a *dynamic graph* whose vertex (nodes) and edge (links) sets change over time. We consider a dynamic graph at an initial state $M_0 = (V_0, E_0)$ and its development over time: M_0, M_1, M_2, ... The transition between two states M_i and M_{i+1} of the graph can be described by a set of updates T_{i+1}. The evolution of a dynamic graph over time is the result of a sequence of transitions.

$$M_0 \rightarrow M_1 \rightarrow M_2 \rightarrow M_3 \rightarrow ...$$

Given a malware distribution network (MDN), we have specific infrastructural measurements: *Inbound Hub Node* – a node that has more than m inbound links; *Outbound Hub Node* - a node that has more than n outbound links; *Bridge Node (Center Node)* – a node that connects to multiple hubs; *Sink Node* – a node that has only inbound links. *Root Node* – a node that has only outbound links; *Transition Node* – a node that has both inbound and outbound links; *Sink Node* – a node that has only inbound links. *Root Node* – a node that has only outbound links; *Transition Node* – a node that has both inbound and outbound links; *Persistent Link* - a link that stays active for a period of time p. Figure 1 shows an example of infrastructural components of an MDN.

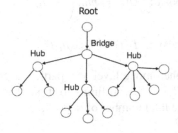

Fig. 1. Infrastructural components of an MDN

3 Semantic Graph Model

In this study, we embed semantic information into the dynamic graph of malware distribution networks. Graphs are represented by an augmented adjacency list data structure that is designed to capture both the dependencies of graph links and the mode of linkage types. We describe this data structure as a list of key–value pairs, whose keys are the top level domain of a website, denoted as a source, and key values are a pair <mode, destination> whereby destination is the top-level domain which is reported as being affected by the source. To place all of the top-level domains on the visualization, we used a Dynamic Behavioral Graph [22–24] to incorporate event frequencies, protocol types, packet contents and data flow information into one graph. In contrast to a typical Force-Directed Graph such as D3 [18], our model goes beyond the aesthetic layout of a graph to reveal the dynamic sequential patterns in a three-dimensional virtual space. In the model, the attraction force between a pair of nodes is calculated using the formula:

$$f_a = \frac{||x_j - x_i||^2}{\alpha T} \tag{1}$$

$$f_r = \frac{\beta}{||x_j - x_i||^2} \tag{2}$$

where: i and j are distinct nodes, α is the value of elasticity where a greater value increases the length of the edge. β is the coefficient for repulsion force. T is equal to the average time between each nodes' timestamps and $||x_i - x_j||$ is the distance between two nodes.

We use a gradient arc for displaying the direction of edges. The decrease of alpha value indicates the direction, with 1 at the source and 0 at the end. This novel visual representation also enables us to add the attributes to the edges [19–21].

Here, we enable digital pheromone deposit and decay on the edges of a network. The digital pheromones are stored on the connected edges over time. The digital pheromones also decay at a certain rate. The amount of pheromones at an edge at time t is:

$$\text{Deposit}: \quad D(t) = min\left(\sum_{i=0}^{N} u_i(t), M\right) \tag{3}$$

$$\text{Decay}: \quad D(t) = max(u_i(t) - rt, L) \tag{4}$$

where, $D(t)$ is the current pheromone level at a particular edge i between two nodes. M and L are the upper and lower bound limits to it. $u_i(t)$ is an individual pheromone deposit at time t, and N is the total number of deposits on that particular edge. 'r' is the linear decay rate. See Fig. 2.

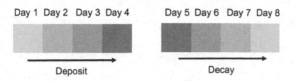

Day 1 Day 2 Day 3 Day 4   Day 5 Day 6 Day 7 Day 8

Deposit   Decay

Fig. 2. Pheromone deposit and decay representation of persistency of the malware distribution channels (connected edges in the graph).

4 Data Collection and Malware Attribution

The MDN and M-Clusters were built from our dataset collected from Google Safe Browsing (GSB) and VirusTotal.com (VT). The data set spans a period of eight months from 19 January to 25 September 2017. The collection start date was specifically chosen to capture data related to the 2017 U.S. Presidential Inauguration. The end date, unfortunately, resulted from the unavailability of GSB API services. The GSB service has been used to warn users not to visit potentially unsafe URLs. The GSB Transparency Report is an online resource providing statistics from the collected data repository. An API set was made available to automate the retrieval of data from the repository for any submitted URL. The API requires a URL as input and returns a report including the timestamp of the last visit, the source, and the destination of the transmission. However, the report does not contain specific malware information.

VirusTotal (VT), on the other hand, provides a scanning service to detect the presence of malicious code in files and URLs. VT provides specific malware information. However, it does not contain the source-destination data. Scanning is a combination of multiple commercial anti-malware products providing both static and heuristic-based data analysis. In this study, we used the academic API service to automate submission and result retrieval for large data sets.

The site *vk.net* was selected as the seed website based on a four-month observation of the site reliably appearing on GSB. The report, in JSON format, consisted of various statistics. The statistics of interest to us were labeled: *name, sendsToAttackSites, receivesTrafficFrom, sendsToIntermediary-Sites, lastVisitDate,* and *lastMaliciousDate.* An MN with no incoming edges for the current collection was relabeled to a Root Malicious Node (RMN). This node is unique to our MDN graphs as it cannot be determined from the GSB reports alone. It is revealed only if the MDN graph is completed.

5 Topological Dynamic Clusters

The malware distribution network is not a giant web. Instead, there are many clusters of subnetworks. Some are large; others are small. All of the clusters are dynamic. They formed for a period of time and then dissolved gradually. Figures 3, 4 and 5 are the top three clusters in size. Figure 6 shows an overview of the 8-month dataset of cluster sizes (nodes) evolved over time, where each curve represents a cluster whose nodes are more than 5 nodes. The first blue line between 19 January, 2017 and 1 April, 2017 was the biggest cluster.

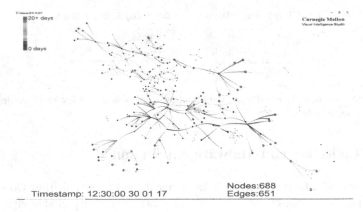

Timestamp: 12:30:00 30 01 17   Nodes:688
                                      Edges:651

Fig. 3. The biggest cluster on 01/30/2017 from the visualization

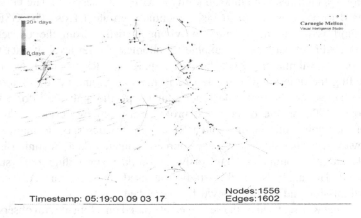

Timestamp: 05:19:00 09 03 17   Nodes:1556
                                      Edges:1602

Fig. 4. The second biggest cluster on 03/09/2017

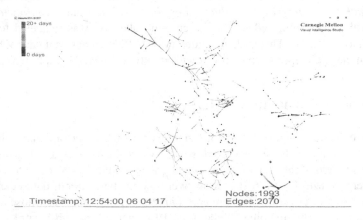

Timestamp: 12:54:00 06 04 17   Nodes:1993
                                      Edges:2070

Fig. 5. The third biggest cluster on 04/06/2017

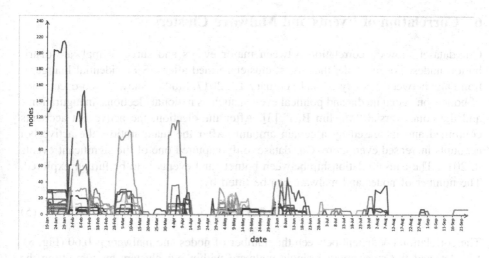

Fig. 6. The overview of the 9-month (1/19/2017–9/25/2017) dataset of cluster sizes (nodes) evolved over time, where each curve represents a cluster whose nodes are more than 5 nodes. The first blue line between 19 January, 2017 and 1 April, 2017 was the biggest cluster (Color figure online)

Statistical data analysis shows that the sizes of the clusters versus their ranks fits Power Law for most months, especially the first two months of 2017. See Fig. 7. This trend indicates that the MDN is a scale-free network: a very small number of nodes have more persistent edges than others. The topological patterns help the analysts to pay attention to the largest clusters, rather than many, many smaller clusters. In our case, this would include the clusters after May. Besides, we found that during volatile cyber attack seasons, the Power Law effect becomes stronger in terms of the slopes of the curves.

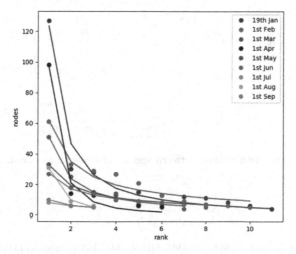

Fig. 7. The relationship between cluster sizes and rank fits the Power Law

6 Correlation of Events and Malware Clusters

Our dataset shows a correlation between major events and surge of malware distribution nodes. For example, the largest cluster formed after US Presidential Inauguration Day, between January 20 and February 13, 2017. Studies show the co-occurrence
of bonets on social media and political events, such as national elections, inaugurations,
and the controversial "Muslim Ban" [3]. After the election, the active bot accounts
continued and increased by a certain amount. After the Inauguration, the active bot
accounts increased even more. Our dataset only captured one of the significant events
in 2017. The causal relationship between botnets and events is to be further explored.
The number of nodes and malware can be fitted by:

$$Y = 9.027X + 125 \qquad (5)$$

The correlation coefficient between the number of nodes and malware is 0.60 (Fig. 8).
We detected the most popular single malware within our clusters by submitting the
domains to VirusTotal. Next, VirusTotal responded to us with all of the malware
downloaded from that domain with the last scanned date. We collected all of the
malware whose last scanned date was the same as our collection date of the domain.
The red nodes are those domains containing the single malware, and the other nodes
are domains that send or receive traffic between red nodes. The single malware appears
17 times in the top three biggest clusters.[1] The rest of the detected malware in the three
biggest clusters were discovered present on a server no more than two times with
several appearing only once. Seven malware events occurred twice and the remaining
102 malware appeared only once (Figs. 9, 10 and 11).

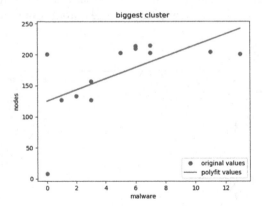

Fig. 8. The linear relation between species of malware and cluster size

---

[1] The SHA-256 of M is2eea543c86312c0fd361c31cba8774d2d6020c5ebcc1ce1a355482de74ed9863.

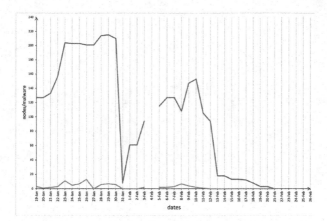

Fig. 9. The biggest cluster evolved over time in terms of size (nodes) and attributed malware. The red line is the number of malware in the cluster. (Color figure online)

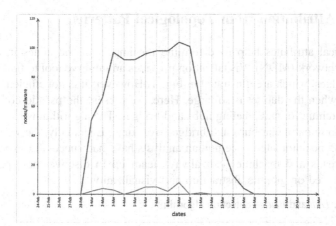

Fig. 10. The second big cluster evolved over time in terms of size (nodes) and attributed malware. The red line is the number of malware in the cluster. (Color figure online)

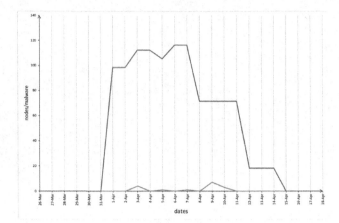

Fig. 11. The third big cluster evolved over time in terms of size (nodes) and attributed malware. The red line is the number of malware in the cluster. (Color figure online)

7 Cyber Attribution from Topological Patterns

The topological attributes help us determine the impact of the nodes in a malware distribution network (MDN). Visualization provides an intuitive tool to find the critical hubs and bridges, which are illustrated in Fig. 1. However, it is not efficient to identify those nodes when the dataset is so large. Here, we present the pseudo code for automatically searching for and labeling hubs and bridges. The algorithm is fast and can be used for tracking particular hubs and bridges over time. Eventually, the visual analytic process would be automated once human analysts have had successful experiences. In addition, humans and machines can always team up to discover new patterns and correlations based on graphic abstraction and visualization.

Algorithm 1. *Hub* and *Bridge* detection algorithm

Input:

The directed network, $G1$

The node set of the network, $M1$

The edge set of the network, $E1(M_s, M_d)$

Output:

Hub nodes, H_n

Bridge nodes, B_n

for $M1_1 \rightarrow M1_n$ **do**

    if $OutDegree(M1_i) > 0$ & $InDegree(M1_i) > 0$ **then**

    if $Degree(M1_i) > p$ **then**

        $M1_i \in H_n$

    end if

  end if

end for

Create new directed network G2, with nodes set M2

for $E1(M_a, M_d)_1 \rightarrow E1(M_a, M_d)_n$ **do**

  if $N_s \in H_n$ & $M_d \in H_n$ **then**

    $M_s \in M2$

    $M_d \in M2$

  end if

end for

for $M2_1 \rightarrow M2_n$ **do**

  if $OutDegree(M2_i) > 0$ & $InDegree(M2_i) > 0$ **then**

    if $Degree(M2_i) > q$ **then**

        $M2_i \in B_n$

    end if

  end if

end for

Figure 12 shows the infrastructural evolution of the malware distribution network between Jan. 19, 2017 and April 4, 2017. We found that there were several hubs in the biggest cluster, including *bit.ly, dlvr.it, smarturl.it, adf.ly, wp.me,* and *zip.net,* a bridge *bit.ly,* and a root node *brandnewbrand.br.* Amazingly, five out of six hubs are utility sites for shortening URL addresses: *bit.ly, adf.ly, smarturl.it,* and *wp.me.* Those sites redirect traffic to the malware host site.

Fig. 12. Dynamic graph of the infrastructure of the biggest cluster between Jan 19, 2017 and Feb 13, 2017

With the visualization and analytic model, we are able to track single Top Level Domain (TLD) nodes and reveal their "life cycle" in the malware distribution network, when the TLD address has been captured by both Google Safe Browsing (GSB) and VirusTotal (VT). Figure 12 shows the dynamics of the TLD *adf.ly* node and its inbound and outbound edges in the 8-months period. The plot shows that the node had persistent malware inbound and outbound traffic before January 19 through May 17. There are multiple recurrences during that period. The malware did not die out until May 17, 2017. It reached its peak between Feb 19 and March 19, in correlation with the cyber activities during that period.

We are also able to track a single malware from Jan 28 through March 9 based on the GSB and VT attributed dataset. Coincidentally, the single malware passed through the popular TLD address node adf.ly during Feb 6 and March 3. The multiple modality tracking enables us to cross-reference, discover new patterns, and ultimately to lead more accurate cyber attributions (Figs. 13 and 14).

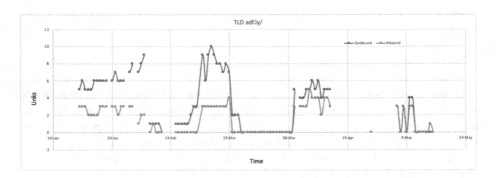

Fig. 13. The dynamics of a single TLD *adf.ly*

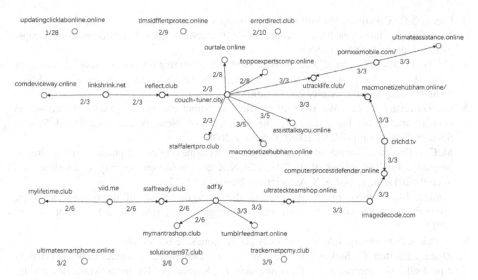

Fig. 14. The development of the single malware within clusters with time.

8   Conclusions

We developed a crawler to collect live malware distribution network data from publicly available sources including Google Safe Browser and VirusTotal. We then generated the graph with our visualization tool and performed malware attribution. We have discovered: 1) malware distribution networks form clusters; 2) those cluster sizes follow the Power Law; 3) there is a correlation between cluster size and the number of malware species in the cluster; 4) there is also a correlation between number of malware species and cyber events; and finally, 5) the infrastructure components such as bridges, hubs, and persistent links play significant roles in malware distribution dynamics.

Acknowledgement. The authors would like to thank VIS research assistants Sebastian Peryt, Pedro Pimentel, and Sihan Wang for participating in 3D model prototyping and data processing. This project is in part funded by Cyber-Security University Consortium of Northrop Grumman Corporation. The authors are grateful to the discussions with Drs. Neta Ezer, Justin King, and Paul Conoval.

References

1. Schiller, B., Deusser, C., Castrillon, J., Strufe, T.: Compile- and run-time approaches for the selection of efficient data structures for dynamic graph analysis. Appl. Network Sci. **1** (2016). Article number: 9 https://link.springer.com/article/10.1007/s41109-016-0011-2
2. DNA at GitHub. https://github.com/BenjaminSchiller/DNA

3. Carey, C.E.: Continued bot infiltration of Trump's Facebook Pages. Data for Democracy, 1 May 2017. https://medium.com/data-for-democracy/continued-bot-infiltration-of-trumps-facebook-pages-2df82ca86b5b
4. Gu, G., Perdisci, R., Zhang, J., Lee, W.: BotMiner: clustering analysis of network traffic for protocol- and structure-independent botnet detection. In: Proceedings of the 17th USENIX Security Symposium (Security 2008), (2008)
5. Gu, G., Zhang, J., Lee, W.: BotSniffer: detecting botnet command and control channels in network traffic. In: Proceedings of the 15th Annual Network and Distributed System Security Symposium (NDSS 2008), February 2008
6. McCoy, D., et al.: Pharmaleaks: understanding the business of online pharmaceutical affiliate programs. In: Proceedings of the 21st USENIX conference on Security symposium, ser. Security 2012, p. 1. USENIX Association, Berkeley (2012)
7. Karami, M., Damon, M.: Understanding the emerging threat of ddos-as-a-service. In: Proceedings of the USENIX Workshop on Large-Scale Exploits and Emergent Threats (2013)
8. Google safe browsing. https://developers.google.com/safe-browsing/
9. Zhang, J., Seifert, C., Stokes, J.W., Lee, W.: Arrow: Generating signatures to detect drive-by downloads. In: Srinivasan, S., Ramamritham, K., Kumar, A., Ravindra, M.P., Bertino, E., Kumar, R. (eds.) Proceedings of the 20th International Conference on World Wide Web, WWW 2011, Hyderabad, India, 28 March–1 April 2011. ACM (2011)
10. Rossow, C., Dietrich, C., Bos, H.: Large-scale analysis of malware downloaders. In: Flegel, U., Markatos, E., Robertson, W. (eds.) DIMVA 2012. LNCS, vol. 7591, pp. 42–61. Springer, Heidelberg (2013). https://doi.org/10.1007/978-3-642-37300-8_3
11. Caballero, J., Grier, C., Kreibich, C., Paxson, V.: Measuring pay-per-install: the commoditization of malware distribution. In: Proceedings of the 20th USENIX conference on Security, ser. SEC 2011. USENIX Association, Berkeley (2011)
12. Goncharov, M.: Traffic direction systems as malware distribution tools. Trend Micro, Technical report (2011)
13. Behfarshad, Z.: Survey of malware distribution networks. Electrical and Computer Engineering, University of British Columbia, Technical report (2012)
14. Provos, N., McNamee, D., Mavrommatis, P., Wang, K., Modadugu, N.: The ghost in the browser analysis of web-based malware. In: Proceedings of the first Conference on First Workshop on Hot Topics in Understanding Botnets, ser. HotBots 2007. USENIX Association, Berkeley (2007)
15. Provos, N., Mavrommatis, P., Rajab, M.A., Monrose, F.: All your iframes point to us. In: Proceedings of the 17th Conference on Security symposium, ser. SS 2008. USENIX Association, Berkeley (2008)
16. http://www.stachliu.com/2012/08/search-diggity-install/
17. http://virustotal.com
18. http://www.d3.org
19. Wigglesworth, V.B.: Insect Hormones, pp. 134–141. W.H. Freeman and Company (1970)
20. Cai, Y.: Instinctive Computing. Springer, London (2016). https://doi.org/10.1007/978-1-4471-7278-9
21. Bonabeau, E., Dorigo, M., Theraulaz, G.: Sawrm Intelligence: From Nature to Artificial Systems. Oxford University Press, Oxford (1999)
22. Cai, Y.: Ambient Diagnostics. CRC Press, Boca Raton (2014)
23. Jacobi, J.A., Benson, E.A., Linden, G.D.: Personalized recommendations of items represented within a database. US Patent. US 7113917 B2 (2006)

24. Peryt, S., Morales, J.A., Casey, W., Volkmann, A., Cai, Y.: Visualizing malware distribution network. In: IEEE Conference on Visualization for Security, Baltimore, October, 2016 (2016)
25. Rossi, R.A., Gallagher, B., Neville, J., Henderson, K.: Modeling dynamic behavior in large evolving graphs. In: Proceedings of the Sixth ACM International Conference on Web Search and Data Mining (WSDM 2013), pp. 667–676. ACM, New York (2013). http://dx.doi.org/10.1145/2433396.2433479
26. Yu, S., Gu, G., Barnawi, A., Guo, S., Stojmenovic, I.: Malware propagation in large-scale networks. IEEE Trans. Knowl. Data Eng. **27**, 170–179 (2015)

Applications of Data Assimilation Methods on a Coupled Dual Porosity Stokes Model

Xiukun Hu[1] and Craig C. Douglas[2(✉)]

[1] University of Wyoming Department of Mathematics and Statistics, Laramie, WY 82071-3036, USA
[2] University of Wyoming School of Energy Resources and Department of Mathematics and Statistics, Laramie, WY 82071-3036, USA
craig.c.douglas@gmail.com

Abstract. Porous media and conduit coupled systems are heavily used in a variety of areas such as groundwater system, petroleum extraction, and biochemical transport. A coupled dual porosity Stokes model has been proposed to simulate the fluid flow in a dual-porosity media and conduits coupled system. Data assimilation is the discipline that studies the combination of mathematical models and observations. It can improve the accuracy of mathematical models by incorporating data, but also brings challenges by increasing complexity and computational cost. In this paper, we study the application of data assimilation methods to the coupled dual porosity Stokes model. We give a brief introduction to the coupled model and examine the performance of different data assimilation methods on a finite element implementation of the coupled dual porosity Stokes system. We also study how observations on different variables of the system affect the data assimilation process.

Keywords: Data assimilation · Dual porosity · Stokes equation · Multiphysics

1 Introduction

Hou et al. [6] has proposed the Coupling of dual porosity flow with free flow as a replacement of the widely used Stokes-Darcy family. The proposed model has a better representation than the traditional Stokes Darcy model in modeling fractured porous media with large conduits. Potential applications of this model include petroleum extraction, hydrology, geothermal systems, and carbon sequestration. A finite element implementation of this model using FEniCS has been developed and studied by the authors [8]. Data assimilation is the discipline that studies the combination of mathematical models and observations. In this paper, we will apply data assimilation methods to the implementation of the coupled model to improve the accuracy of the model predictions [4,9].

Supported in part by National Science Foundation grant DMS-1722692.

V. V. Krzhizhanovskaya et al. (Eds.): ICCS 2020, LNCS 12142, pp. 72–85, 2020.
https://doi.org/10.1007/978-3-030-50433-5_6

In Sect. 2, we give an introduction to the mathematical model of the coupled dual porosity Stokes model proposed by Hou et al. [6]. In Sects. 3 and 4 we illustrate the applications of data assimilation methods on the coupled dual porosity Stokes model. We set up a data assimilation context from our model in Sect. 3. We present the numerical results based on synthetic data in Sect. 4. In Sect. 5 we draw conclusions and discuss future works.

2   A Coupled Dual Porosity Stokes Model

The dual porosity Stokes model proposed by Hou et al. [6] consists of a dual porosity porous subdomain and a conduit subdomain. An example is show in Fig. 1 where Ω_d represents the porous subdomain and Ω_c represents the conduit subdomain. Each subdomain has its own set of boundary conditions, represented by Γ_d and Γ_c respectively in the figure. The interface Γ_{cd} is the only place where the two subdomains communicate with each other.

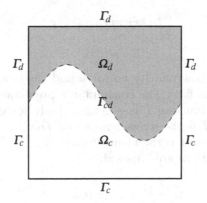

Fig. 1. A simplified coupled model in 2D.

Barenblatt et al. [2] first proposed the dual porosity model in 1960. Later in 1963, Warren and Root [16] studied the model thoroughly. In a dual porosity medium, two subsystems are assumed. One is the matrix subsystem, which has high porosity and low permeability, and the other is the microfracture subsystem, which has low porosity and high permeability. The dual porosity equations governing the dual porosity subdomain Ω_d in our coupled dual porosity Stokes model are

$$\phi_m C_{mt} \frac{\partial p_m}{\partial t} - \nabla \cdot \frac{k_m}{\mu} \nabla p_m = -Q, \tag{1}$$

$$\psi_f C_{ft} \frac{\partial p_f}{\partial t} - \nabla \cdot \frac{k_f}{\mu} \nabla p_f - Q + q_p. \tag{2}$$

The constant μ represents the dynamic viscosity. The constants k_m and k_f represent the intrinsic permeability, ϕ_m and ϕ_f the porosities, C_{mt} and C_{ft} the total compressibility, of the matrix and the microfracture subsystems respectively. The variables p_m and p_f are the flow pressure of the matrix and the microfracture subsystems respectively. The coefficient function q_p is the sink/source term. The term Q denotes the mass transfer rate per unit volume from the matrix subsystem to the microfracture subsystem and is defined as

$$Q = \frac{\sigma k_m}{\mu}(p_m - p_f), \tag{3}$$

where the parameter σ represents the characteristic of the fractured rock and is commonly known as the shape factor. Formulas for calculating σ can be found in Warren and Root [16] and Mora and Wattenbarger [12].

We assume the flow in the conduit domain is Stokes flow and thus describe it using the Stokes equation in (4) and (5). Note that the model can be extended to other free flow models such as the incompressible Navier-Stokes model, as proposed in [4].

$$\frac{\partial u}{\partial t} - \nabla \cdot \boldsymbol{T}(u, p) = f, \tag{4}$$

$$\nabla \cdot u = 0. \tag{5}$$

The two variables, the flow velocity vector u and the flow pressure p, together describe the state of the flow. The constant ν represents the kinematic viscosity. The vector valued function f is a general body force term. The operator $\boldsymbol{T}(u, p) := 2\nu \boldsymbol{D}(u) - p\boldsymbol{I}$ is the stress tensor and $\boldsymbol{D}(u) := \frac{1}{2}(\nabla u + \nabla^T u)$ is the deformation tensor, where $\boldsymbol{I}$ is the identity matrix.

Four interface conditions are imposed:

$$-\frac{k_m}{\mu}\nabla p_m \cdot (-n_{cd}) = 0, \tag{6}$$

$$u \cdot n_{cd} = -\frac{k_f}{\mu}\nabla p_f \cdot n_{cd}, \tag{7}$$

$$-n_{cd}^T \boldsymbol{T}(u, p)n_{cd} = \frac{p_f}{\rho}, \tag{8}$$

$$-\boldsymbol{P}_\tau(\boldsymbol{T}(uc, p)n_{cd}) = \frac{\alpha\nu\sqrt{N}}{\sqrt{\text{trace}(\boldsymbol{\Pi})}}\left(u + \frac{k_f}{\mu}\nabla p_f\right), \tag{9}$$

where n_{cd} is the unit normal vector of the interface Γ_{cd}, pointing toward Ω_d. The function $\boldsymbol{P}_\tau$ is the projection operator onto the local tangent plane of Γ_{cd}. The constant α is dimensionless and depends on the properties of the fluid and the permeable material. The constant ρ is the fluid density. The constant N is the space dimension. $\boldsymbol{\Pi} := k_f \boldsymbol{I}$ is the intrinsic permeability of the microfracture subsystem.

Equation (6) represents the no mass exchange condition between the matrix subsystem in Ω_d and the conduit. This is an assumption based on of the huge difference in permeabilities between the matrix and the microfracture subsystems. Equation (7) imposes conservation of mass exchange between the conduit and the microfracture subsystem on the interface. Equation (8) balances the two forces on the interface: the kinetic pressure in the microfracture subsystem and the normal component of the normal stress in the free flow. Equation (9) is the empirical Beavers-Joseph interface condition [3], which claims that the tangential component of the normal stress incurred by the free flow along the interface is proportional to the difference of the tangential component of flow velocities at two sides of the interface.

By introducing test function $[\psi_m, \psi_f, v^T, q]^T$, the coupled dual porosity Stokes PDE system defined by (1)–(9) has the variational form,

$$
\begin{aligned}
&\int_{\Omega_d} \left(\phi_m C_{mt} \frac{\partial p_m}{\partial t} \psi_m + \frac{k_m}{\mu} \nabla p_m \cdot \nabla \psi_m + \frac{\sigma k_m}{\mu} (p_m - p_f) \psi_m \right) d\Omega \\
&+ \int_{\Omega_d} \left(\phi_f C_{ft} \frac{\partial p_f}{\partial t} \psi_f + \frac{k_f}{\mu} \nabla p_f \cdot \nabla \psi_f + \frac{\sigma k_m}{\mu} (p_f - p_m) \psi_f \right) d\Omega \\
&+ \eta \int_{\Omega_c} \left(\frac{\partial u}{\partial t} \cdot v + 2\nu \, \boldsymbol{D}(u) : \boldsymbol{D}(v) - p \nabla \cdot v \right) d\Omega \\
&+ \eta \int_{\Gamma_{cd}} \left(\frac{1}{\rho} p_f v \cdot n_{cd} + \frac{\alpha \nu \sqrt{N}}{\sqrt{\text{trace}(\boldsymbol{\Pi})}} \boldsymbol{P}_\tau \left(u + \frac{k_f}{\mu} \nabla p_f \right) \cdot v \right) d\Gamma \\
&+ \eta \int_{\Omega_c} \nabla \cdot uqd\Omega - \int_{\Gamma_{cd}} u \cdot n_{cd} \psi_f d\Gamma \\
&= \eta \int_{\Omega_c} f \cdot vd\Omega + \int_{\Omega_d} q_p \psi_f.
\end{aligned}
\tag{10}
$$

A finite element implementation using the automated partial differential equation (PDE) solving platform FEniCS [1,11] has been developed by the authors [8]. The backward Euler time stepping scheme was used for time discretization.

3   A Data Assimilation Problem Based on the Coupled Model

In order to apply data assimilation methods to the coupled dual porosity Stokes model, we first convert the dual porosity Stokes model into a discrete dynamical system, and define the observations on it.

Following the finite element analysis with backward Euler scheme, at timestep t we solve the following equation system for the four variables in four finite functional spaces,

$$
\boldsymbol{A} \begin{bmatrix} p_m^{(t)} \\ p_f^{(t)} \\ u^{(t)} \\ p^{(t)} \end{bmatrix} = \boldsymbol{C} \begin{bmatrix} p_m^{(t-\Delta t)} \\ p_f^{(t-\Delta t)} \\ u^{(t-\Delta t)} \\ p^{(t-\Delta t)} \end{bmatrix} + \boldsymbol{b}.
$$

The matrix $\mathbf{A}$ is assembled from the bilinear form

$$
a\left(\begin{bmatrix} p_m^{(t)} \\ p_f^{(t)} \\ u^{(t)} \\ p^{(t)} \end{bmatrix}, \begin{bmatrix} \psi_m^{(t)} \\ \psi_f^{(t)} \\ v^{(t)} \\ q^{(t)} \end{bmatrix}\right)
$$

$$
= \int_{\Omega_d} \left(\phi_m C_{mt} \frac{p_m^{(t)}}{\Delta t} \psi_m + \frac{k_m}{\mu} \nabla p_m \cdot \nabla \psi_m + \frac{\sigma k_m}{\mu}(p_m - p_f)\psi_m \right) d\Omega
$$

$$
+ \int_{\Omega_d} \left(\phi_f C_{ft} \frac{p_f^{(t)}}{\Delta t} \psi_f + \frac{k_f}{\mu} \nabla p_f \cdot \nabla \psi_f + \frac{\sigma k_m}{\mu}(p_f - p_m)\psi_f \right) d\Omega
$$

$$
+ \eta \int_{\Omega_c} \left(\frac{u^{(t)}}{\Delta t} \cdot v + 2\nu \, \boldsymbol{D}(u) : \boldsymbol{D}(v) - p\nabla \cdot v \right) d\Omega
$$

$$
+ \eta \int_{\Gamma_{cd}} \left(\frac{1}{\rho} p_f v \cdot n_{cd} + \frac{\alpha\nu\sqrt{N}}{\sqrt{\mathrm{trace}(\boldsymbol{\Pi})}} \, \boldsymbol{P}_\tau \left(u + \frac{k_f}{\mu}\nabla p_f \right) \cdot v \right) d\Gamma
$$

$$
+ \eta \int_{\Omega_c} \nabla \cdot u q \, d\Omega - \int_{\Gamma_{cd}} u \cdot n_{cd} \psi_f d\Gamma.
$$

The vector $\mathbf{b}$ is assembled from the linear form

$$
L\left(\begin{bmatrix} p_m^{(t-\Delta t)} \\ p_f^{(t-\Delta t)} \\ u^{(t-\Delta t)} \\ p^{(t-\Delta t)} \end{bmatrix}\right) = \eta \int_{\Omega_c} f \cdot v \, d\Omega + \int_{\Omega_d} q_p \psi_f.
$$

The matrix $\mathbf{C}$ is assembled from the bilinear form

$$
c\left(\begin{bmatrix} p_m^{(t-\Delta t)} \\ p_f^{(t-\Delta t)} \\ u^{(t-\Delta t)} \\ p^{(t-\Delta t)} \end{bmatrix}, \begin{bmatrix} \psi_m^{(t)} \\ \psi_f^{(t)} \\ v^{(t)} \\ q^{(t)} \end{bmatrix}\right) = \int_{\Omega_d} \left(\phi_m C_{mt} \frac{p_m^{(t-\Delta t)}}{\Delta t} \psi_m + \phi_f C_{ft} \frac{p_f^{(t-\Delta t)}}{\Delta t} \psi_f \right) d\Omega
$$

$$
+ \eta \int_{\Omega_c} \frac{u^{(t-\Delta t)}}{\Delta t} \cdot v \, d\Omega,
$$

and thus has the form

$$
\mathbf{C} = \begin{bmatrix} \frac{\phi_m C_{mt}}{\Delta t} \boldsymbol{I}_{d_m} & & & \\ & \frac{\phi_f C_{ft}}{\Delta t} \boldsymbol{I}_{d_f} & & \\ & & \boldsymbol{I}_{d_u} & \\ & & & \boldsymbol{0}_{d_p} \end{bmatrix},
$$

where d_m, d_f, d_u, and d_p are the degrees of freedoms of p_m, p_f, u, and p, respectively,

If we let the state variable

$$v_t = \begin{bmatrix} p_m^{(t)} \\ p_f^{(t)} \\ u^{(t)} \\ p^{(t)} \end{bmatrix},$$

the dynamical system can be expressed as

$$v_{t+\Delta t} = \Psi(v_t) + \xi_t, \tag{11a}$$

$$\Psi(v_t) = \mathbf{A}^{-1}\mathbf{C}v_t + \mathbf{A}^{-1}\mathbf{b}, \tag{11b}$$

where $\xi_t \sim \mathcal{N}(\mathbf{0}, \Sigma)$ represents the model error. This dynamical system is linear. Note that the coefficient matrix $\mathbf{C}$ is singular as is $\mathbf{A}^{-1}\mathbf{C}$, the Jacobian of Ψ defined in (11a), (11b). Since some smoothing algorithms involve in inverting the Jacobian of the dynamical system, we need to avoid singularities.

In general we can use the singular value decomposition to get around with singularities. In our case, we let our state variable

$$v_t^* = \begin{bmatrix} p_m^{(t)} \\ p_f^{(t)} \\ u^{(t)} \end{bmatrix}.$$

The dynamical system becomes

$$v_{t+\Delta t}^* = \Psi^*(v_t^*) + \xi_t, \tag{12a}$$

$$\Psi^*(v_t^*) = (\mathbf{A}^{-1})^*\mathbf{C}^*v_t^* + (\mathbf{A}^{-1})^*\mathbf{b}^*, \tag{12b}$$

where $\mathbf{M}^*$ represents the matrix generated by removing the last d_p rows and columns from a matrix $\mathbf{M}$, and $\mathbf{b}^*$ is the vector from removing the last d_p components of a vector $\mathbf{b}$. In fact (12a), (12b) can also be formed from applying singular value decomposition to $\mathbf{A}^{-1}\mathbf{C}$ in (11a), (11b). Note that $p^{(t)}$ can still be calculated from $p_m^{(t-\Delta t)}$, $p_f^{(t-\Delta t)}$ and $u^{(t-\Delta t)}$, which in turn can be calculated from $p_m^{(t)}, p_f^{(t)}$ and $u^{(t)}$.

Similarly, the Dirichlet boundary conditions will also cause singularities as they do not depend on previous boundary values. We remove all Dirichlet boundary values from the state variable v_t using the same technique.

We base the dynamical model on a two dimensional dual porosity Stokes model shown in Fig. 2. Let $\Omega = [-0.5, 0.5] \times [0, 1]$ be a shifted unit square, $\Omega_c = \{(x, y) \in \Omega \mid x \leq 0\}$, and $\Omega_d = \{(x, y) \in \Omega \mid x \geq 0\}$. The interface is $\Gamma_{cd} = \{(x, y) \in \Omega \mid x = 0\}$. The domain is partitioned uniformly into $\frac{1}{16} \times \frac{1}{16}$ squares.

Dirichlet boundary conditions on Γ_c and Γ_d, initial conditions for all variables, and coefficients q_p and f are constructed such that

$$p_m = \cos(\pi t) \cos(x(-y+1))$$
$$p_f = \left((x^2 + y^2 - 2y + 2)\cos(\pi t) - 10\pi \sin(\pi t)\right)\cos(xy - x)$$
$$u = \begin{bmatrix} 2x\cos(\pi t) \\ 2x\cos(\pi t) - 2y\cos(\pi t) \end{bmatrix}$$
$$p = -10\pi\sin(\pi t) + (x^2 + 2x + y^2 - 2y + 6)\cos(\pi t)$$

is the solution to our problem.

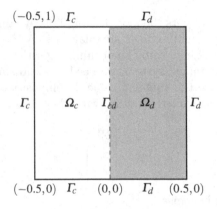

Fig. 2. The 2D example model with a shifted unit square domain $\Omega = [-0.5, 0.5] \times [0, 1]$, conduit subdomain $\Omega_c = \{(x, y) \in \Omega \mid x \le 0\}$ and dual porosity subdomain $\Omega_d = \{(x, y) \in \Omega \mid x \ge 0\}$.

Also, we let $\Delta t = 0.01$, $\xi_t \sim \mathcal{N}(\mathbf{0}, 5\boldsymbol{I})$, $v_0^* \sim \mathcal{N}(\mathbf{0}, 100\boldsymbol{I})$. The large variance of v_0^* indicates that we have little knowledge about the initial condition.

For the observations, we assume we have direct observations to every 4 components of v_t^* at time t:

$$y_t = h(v_t^*) + \eta_t, \tag{13a}$$
$$h(v_t^*) = \mathbf{H}v_t^*, \tag{13b}$$
$$\eta_t \sim \mathcal{N}(\mathbf{0}, 5\boldsymbol{I}), \tag{13c}$$

where

$$\mathbf{H} = \begin{bmatrix} 1 & 0 & 0 & 0 & 0 & 0 & 0 & 0 & 0 & \dots \\ 0 & 0 & 0 & 0 & 1 & 0 & 0 & 0 & 0 & \dots \\ 0 & 0 & 0 & 0 & 0 & 0 & 0 & 0 & 1 & \dots \\ & \dots & & & & & & & & \end{bmatrix}.$$

We observe every 0.01 time unit starting at $t = 0.01$. Equations (12a), (12b) and (13a), (13b), (13c) together defines the data assimilation problem we are solving. Data are generated synthetically.

4   Numerical Results

We run the model against the three dimensional variational method (3DVAR), the strong constraint four dimensional variational method (s4DVAR) with a time window with length 0.04, the extended Rauch-Tung-Striebel smoother (ExtRTS) [15], the extended Kalman Filter (ExtKF) [10], the ensemble Kalman Filter (EnKF) [5,7] with 100 particles, and ensemble Rauch-Tung-Striebel smoother (EnRTS) [13] with 100 particles. Note that since we have a linear data assimilation problem, the extended methods ExtRTS and ExtKF are just the Rauch-Tung-Striebel smoother (RTS) and the Kalman Filter (KF). We also use a baseline filtering method *Forward* that only uses the mathematical model $\boldsymbol{\Psi}$ and ignores all data. It starts at $v_0^* = \mathbf{0}$ and then applies $\boldsymbol{\Psi}^*$ to get an approximation for v_t^*. Since the model is linear, we expect an optimal solution by ExtKF for filtering and ExtRTS for smoothing.

All numerical experiments were run with the data assimilation package DAPPER [14] on the Teton computer cluster at the Advanced Research Computing Cluster (ARCC) at the University of Wyoming.

The results of filtering on the model with such observations are in Table 1 and the results of smoothing are in Table 2. Since we have a linear system with Gaussian errors, the Kalman Filter and the Kalman Smoother are expected to have optimal data assimilation solutions for filtering and smoothing, respectively, which is in accordance with our numerical results. We see that the Kalman Filter, the Kalman Smoother and 3DVAR are efficient in our small linear model while ensemble methods and s4DVAR are relatively slow.

Table 1. Average root mean square error for filtering (rmse_f) and elapsed time

	Forward	3DVAR	ExtKF	EnKF
rmse_f	0.4717	0.2824	0.2604	0.2651
elapsed time	5 s	1 s	5 s	56 s

Table 2. Average root mean square error for smoothing (rmse_s) and elapsed time

	s4DVAR	ExtRTS	EnRTS
rmse_s	0.3	0.1907	0.2033
elapsed time	121 s	32 s	62 s

The error of different data assimilation methods over time are shown in Figs. 3 and 4. Since Forward, 4DVAR, ExtKF, and EnKF all start with an initial guess $\tilde{v}_0 - \mathbf{0}$, they all have the same predictions at $t = 0.01$. This is why they all have the same error at $t = 0.01$ for forecasting as shown in Fig. 3. The predictions are made every 0.01 time units. ExtKF has a smaller forecasting error than all the other methods except for 3DVAR. Our 3DVAR implementation utilizes all true states to approximate the background covariance B_t. The exposure to the true states enables the 3DVAR implementation to surpass the theoretical optimal solution from the Kalman Filter. EnKF has a result very similar to that of ExtKF. In EnKF, the calculations of mean and variance of the states are approximated using the Monte Carlo method. Since the states follows a Gaussian process, the approximations converge to the truths as the number of particles increases. We can also see in Fig. 4 that by utilizing all observations, the smoothing error at $t = 0.01$ is reduced by half, comparing to the forecasting error in Fig. 3. Note that the Kalman Smoother ExtRTS achieves the best result at all time, and the ensemble Kalman Smoother EnRTS has a very similar result as ExtRTS, but consumes much more computation time as shown in Table 2.

Note that the baseline method *Forward* also has a decreasing error with respect to time. This is caused by the characteristics of our dynamical system. Because of the essential boundaries in our coupled model, solutions to the PDE system with different initial conditions all converge to each other as $t \to \infty$. This can also be explained by the linear dynamical system. Consider a linear dynamical system with $\boldsymbol{\Psi}(v_t) = M v_t$ where $\|M\| < 1$. Then $\boldsymbol{\Psi}^{(n)}(v_t) \to \boldsymbol{0}$ as $t \to 0$.

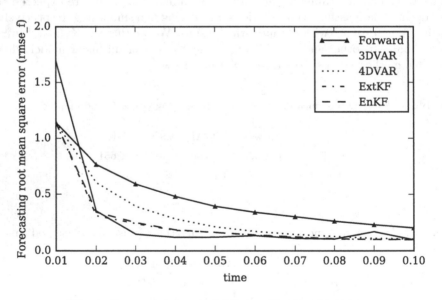

Fig. 3. Forecasting error of different filtering algorithms

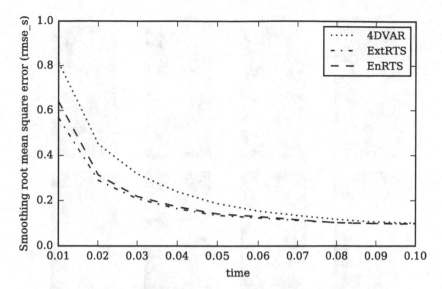

Fig. 4. Smoothing error of different smoothing algorithms

The results of smoothing at $t = 0.01$ from s4DVAR and ExtRTS are shown in Fig. 5. The results of filtering at $t = 0.02$ from 3DVAR and ExtKF are shown in Fig. 6. We can see from Fig. 5 that by using limited observations, 4DVAR and ExtRTS are able to recover the state close to true state. Also in Fig. 6, we see that by using only data at $t = 0.01$, 3DVAR and the Kalman Filter are able to predict a state at $t = 0.02$ that is much closer to true state comparing to the Forecast baseline method.

We also explore the importance of observations on different variables. With the same settings on the dynamical system, we apply the Kalman Filter (ExtKF) to observations on p_m, p_f, and u separately. We still observe from $t = 0.01$ and observe every 0.01 time unit, but on all grid points. The results are presented in Fig. 7.

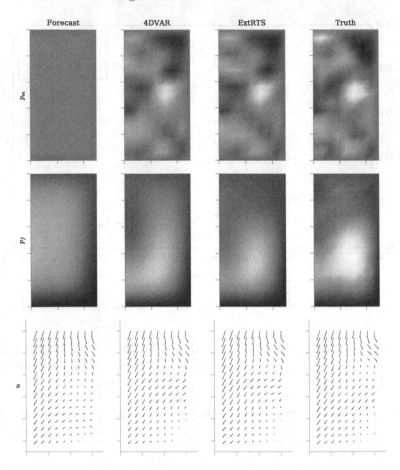

Fig. 5. Results of smoothers at time $t = 0.01$ in the 2D model.

We can see from Fig. 7 that the data on the flow pressure p_m in the matrix subsystem in the dual porosity subdomain provides most of the information while the other two variables provide little improvement over the *Forward* baseline method, which uses no observation at all. This behavior exists in all our test models with different boundary conditions, source terms and geometries. This phenomenon needs further investigation. Here we conclude that in our limited test cases, observations on p_m provide significant information about the true states while observations on p_f and u do not.

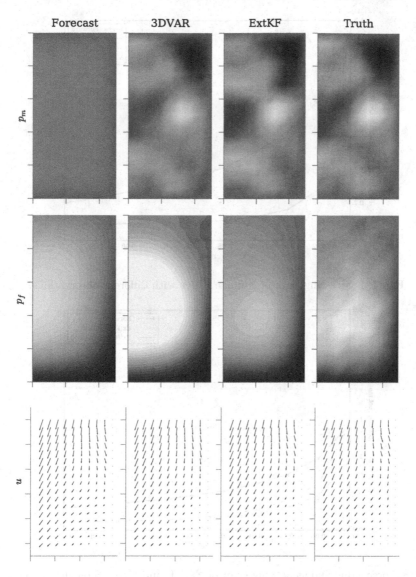

Fig. 6. Results of filters at time $t = 0.02$ in the 2D model.

Lastly we show the result of the Kalman Filter and the Kalman Smoother on a 3D coupled dual porosity Stokes model introduced in [8], with the mesh size $h = 1/8$ and observations on p_m only. All the other settings are the same as in the 2D model. The results in Fig. 8 validate the Kalman Filter and the Kalman Smoother on our 3D models and real world applications.

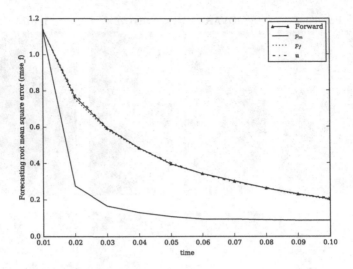

Fig. 7. Forecasting error of Kalman Filter with different observations.

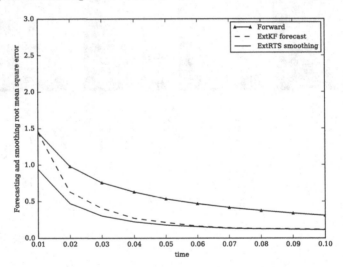

Fig. 8. Forecasting and smoothing error of ExtKF and ExtRTS on the 3D model.

5   Conclusions and Future Work

In this paper, we introduced the coupled dual porosity Stokes model. We set up a data assimilation problem based on the coupled model and applied different data assimilations to solve the problem. Due to the linearity of the coupled dual porosity Stokes model, the Kalman Filter and the Kalman Smoother achieve optimal solutions for filtering and smoothing, respectively, as expected. From our numerical experiments we have seen that observations of pressures in the matrix subsystem contain most of the useful information for data assimilation.

Future work includes exploring different data assimilation methods on the nonlinear coupled dual porosity Navier-Stokes model, applying data assimilation methods with experiment data and investigating the reason behind the uneven distribution of information in different variables.

References

1. Alnæs, M.S., et al.: The fenics project version 1.5. Arch. Numer. Softw. **3**, 9–23 (2015)
2. Barenblatt, G.I., Zheltov, I.P., Kochina, I.N.: Basic concepts in the theory of seepage of homogeneous liquids in fissured rocks [strata]. J. Appl. Math. Mech. **24**, 852–864 (1960)
3. Beavers, G.S., Joseph, D.D.: Boundary conditions at a naturally permeable wall. J. Fluid Mech. **30**, 197–207 (1967)
4. Douglas, C.C., Hu, X., Bai, B., He, X., Wei, M., Hou, J.: A data assimilation enabled model for coupling dual porosity flow with free flow. In: 2018 17th International Symposium on Distributed Computing and Applications for Business Engineering and Science (DCABES), pp. 304–307 (2018)
5. Evensen, G.: Sequential data assimilation with a nonlinear quasi-geostrophic model using Monte Carlo methods to forecast error statistics. J. Geophys. Res. **99**(C5), 10143–10162 (1994)
6. Hou, J., Qiu, M., He, X., Guo, C., Wei, M., Bai, B.: A dual-porosity-stokes model and finite element method for coupling dual-porosity flow and free flow. SIAM J. Sci. Comput. **38**, B710–B739 (2016)
7. Houtekamer, P.L., Mitchell, H.L.: Data assimilation using an ensemble kalman filter technique. Monthly Weather Rev. **126**, 796–811 (1998)
8. Hu, X., Douglas, C.C.: An implementation of a coupled dual-porosity-stokes model with FEniCS. In: Rodrigues, J.M.F., et al. (eds.) ICCS 2019. LNCS, vol. 11539, pp. 60–73. Springer, Cham (2019). https://doi.org/10.1007/978-3-030-22747-0_5
9. Hu, X., Douglas, C.C.: Performance and scalability analysis of a coupled dual porosity stokes model implemented with fenics. Japan J. Indust. Appl. Math. **36**, 1039–1054 (2019). https://doi.org/10.1007/s13160-019-00381-3
10. Kalman, R.E.: A new approach to linear filtering and prediction problems. J. Basic Eng. **82**, 35 (1960)
11. Logg, A., Mardal, K.A., Wells, G.N.: Automated Solution of Differential Equations by the Finite Element Method. LNCS, vol. 84. Springer, Heidelberg (2012). https://doi.org/10.1007/978-3-642-23099-8
12. Mora, C.A., Wattenbarger, R.A.: Analysis and verification of dual porosity and CBM shape factors. J. Can. Pet. Tech. **48**, 17–21 (2009)
13. Raanes, P.N.: On the ensemble Rauch-Tung-Striebel smoother and its equivalence to the ensemble Kalman smoother. Q. J. R. Meteorol. Soc. **142**, 1259–1264 (2016)
14. Raanes, P.N., et al.: Nansencenter/dapper: version 0.8, December 2018. https://doi.org/10.5281/zenodo.2029296
15. Rauch, H.E., Striebel, C.T., Tung, F.: Maximum likelihood estimates of linear dynamic systems. AIAA J. **3**, 1445–1450 (1965)
16. Warren, J.E., Root, P.J.: The behavior of naturally fractured reservoirs. SPE J. **3**, 245–255 (1963)

Spatiotemporal Filtering Pipeline for Efficient Social Networks Data Processing Algorithms

Ksenia Mukhina[✉], Alexander Visheratin, and Denis Nasonov

ITMO University, Saint Petersburg, Russia
mukhinaks@gmail.com, alexvish91@gmail.com, denis.nasonov@gmail.com

Abstract. One of the areas that gathers momentum is the investigation of location-based social networks (LBSNs) because the understanding of citizens' behavior on various scales can help to improve quality of living, enhance urban management, and advance the development of smart cities. But it is widely known that the performance of algorithms for data mining and analysis heavily relies on the quality of input data. The main aim of this paper is helping LBSN researchers to perform a preliminary step of data preprocessing and thus increase the efficiency of their algorithms. To do that we propose a spatiotemporal data processing pipeline that is general enough to fit most of the problems related to working with LBSNs. The proposed pipeline includes four main stages: an identification of suspicious profiles, a background extraction, a spatial context extraction, and a fake transitions detection. Efficiency of the pipeline is demonstrated on three practical applications using different LBSN: touristic itinerary generation using Facebook locations, sentiment analysis of an area with the help of Twitter and VK.com, and multiscale events detection from Instagram posts.

Keywords: Location-based social network · Data processing · Event detection · Sentiment analysis · Tourist path construction · Data filtering pipeline

1 Introduction

In today's world, the idea of studying cities and society through location-based social networks (LBSNs) became a standard for everyone who wants to get insights about people's behavior in a particular area in social, cultural, or political context [11]. Nevertheless, there are several issues concerning data from LBSNs in research. Firstly, social networks can use both explicit (i.e., coordinates) or implicit (i.e., place names or toponyms) geographic references [3]; it is a common practice to allow manual location selection and changing user's position. The Twitter application relies on GPS tracking, but user can correct the position using the list of nearby locations, which causes potential errors from both GPS and user sides [2]. Another popular source of geo-tagged data – Foursquare –

© Springer Nature Switzerland AG 2020
V. V. Krzhizhanovskaya et al. (Eds.): ICCS 2020, LNCS 12142, pp. 86–99, 2020.
https://doi.org/10.1007/978-3-030-50433-5_7

also relies on a combination of the GPS and manual locations selection and has the same problems as Twitter. Instagram provides a list of closely located points-of-interest [16], however, it is assumed that a person will type the title of the site manually and the system will advise the list of locations with a similar name. Although this functionality gives flexibility to users, there is a high chance that a person mistypes a title of the place or selects the wrong one. In Facebook, pages for places are created by the users [25], so all data including title of the place, address and coordinates may be inaccurate.

In addition to that, a user can put false data on purpose. The problem of detecting fake and compromised accounts became a big issue in the last five years [8,19]. Spammers misrepresent the real level of interest to a specific subject or degree of activity in some place to promote their services. Meanwhile, fake users spread unreliable or false information to influence people's opinion [8]. If we look into any popular LBSN, like Instagram or Twitter, location data contains a lot of errors [5]. Thus, all studies based on social networks as a data source face two significant issues: wrong location information stored in the service (wrong coordinates, incorrect titles, duplicates, etc.) and false information provided by users (to hide an actual position or to promote their content).

Thus, in this paper, we propose a set of methods for data processing designed to obtain a clean dataset representing the data from real users. We performed experimental evaluations to demonstrate how the filtering pipeline can improve the results generated by data processing algorithms.

2   Background

With more and more data available every minute and with a rise of methods and models based on extensive data processing [1,13], it was shown that the users' activity strongly correlates with human activities in the real world [21]. For solving problems related to LBSN analysis, it is becoming vital to reduce the noise in input data and preserve relevant features at the same time [10]. Thus, there is no doubt that such problem gathers more and more attention in the big data era. On the one side, data provided by social media is more abundant that standard georeferenced data since it contains several attributes (i.e., rating, comments, hashtags, popularity ranking, etc.) related to specific coordinates [3]. On the other side, the information provided by users of social networks can be false and even users may be fakes or bots. In 2013, Goodchild in [9] raised questions concerning the quality of geospatial data: despite that a hierarchical manual verification is the most reliable data verification method, it was stated that automatic methods could efficiently identify not only false but questionable data. In paper [15], the method for pre-processing was presented, and only 20% of initial dataset was kept after filtering and cleaning process.

One of the reasons for the emergence of fake geotags is a location spoofing. In [28], authors used the spatiotemporal cone to detect location spoofing on Twitter. It was shown that in the New York City, the majority of fake geotags are located in the downtown Manhattan, i.e., users tend to use popular places or locations in the city center as spoofing locations. The framework for the location spoofing detection was presented in [6]. Latent Dirichlet Allocation was used for the topic extraction. It was shown that message similarity for different users decreases with a distance increase. Next, the history of user check-ins is used for the probability of visit calculation using Bayes model.

The problem of fake users and bots identification become highly important in the last years since some bots are designed to distort the reality and even to manipulate society [8]. Thus, for scientific studies, it is essential to exclude such profiles from the datasets. In [27], authors observed tweets with specific hashtags to identify patterns of spammers' posts. It was shown that in terms of the age of an account, retweets, replies, or follower-to-friend ratio there is no significant difference between legitimate and spammer accounts. However, the combination of different features of the user profile and the content allowed to achieve a performance of 0.95 AUC [23]. It was also shown that the part of bots among active accounts varies between 9% and 15%. This work was later improved by including new features such as time zones and device metadata [26]. In contrast, other social networks do not actively share this information through a public API. In [12], available data from 12 social network sites were studied, and results showed that social networks usually provide information about likes, reposts, and contacts, and keep the data about deleted friends, dislikes, etc., private. Thus, advanced models with a high-level features are applicable only for Twitter and cannot be used for social networks in general.

More general methods for compromised accounts identification on Facebook and Twitter were presented in [22]. The friends ratio, URL ratio, message similarity, friend number, and other factors were used to identify spam accounts. Some of these features were successfully used in later works. For example, in [7], seven features were selected to identify a regular user from a suspicious Twitter account: mandatory – time, message source, language, and proximity – and optional – topics, links in the text, and user interactions. The model achieved a high value of precision with approximately 5% of false positives. In [20], Random Forest classifier was used for spammers identification on Twitter, which results in the accuracy of 92.1%. This study was focused on five types of spam accounts: sole spammers, pornographic spammers, promotional spammers, fake profiles, and compromised accounts. Nevertheless, these methods are user-centered, which means it is required to obtain full profile information for further analysis.

However, there is a common situation where a full user profile is not available for researches, for example, in spatial analysis tasks. For instance, in [17], authors studied the differences between public streaming API of Twitter and proprietary service Twitter Firehose. Even though public API was limited to 1% sample of data, it provided 90% of geotagged data, but only 5% of all sample contains spatial information. In contrast, Instagram users are on average 31 times more likely post data with geotag comparing to Twitter users [16]. Thus, LBSN data processing requires separate and more sophisticated methods that would be capable of identifying fake accounts considering incomplete data. In addition to that, modern methods do not consider cases when a regular user tags a false location for some reason, but it should be taken into account as well.

3   Pipeline Scheme

As it was discussed above, it is critical to use as clean data as possible for research. However, different tasks require different aspects of data to be taken into consideration. In this work, we focus on the main features of the LBSN data: space, time, and messages content. First of all, any LBSN contains data with geotags and timestamps, so the proposed data processing methods are applicable for any LBSN. Secondly, the logic and level of complexity of data cleaning depend on the study goals. For example, if some research is dedicated to studying daily activity patterns in a city, it is essential to exclude all data with wrong coordinates or timestamps. In contrast, if someone is interested in exploring the emotional representation of a specific place in social media, the exact timestamp might be irrelevant. In Fig. 1, elements of a pipeline are presented along with the output data from each stage. As stated in the scheme, we start from general methods for a large scale analysis, which require fewer computations and can be applied on the city scale or higher. Step by step, we eliminate accounts, places, and tags, which may mislead scientists and distort results.

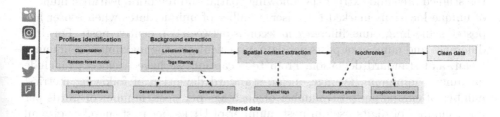

Fig. 1. Pipeline scheme

Suspicious Profiles Identification. First, we identify suspicious accounts. The possibility of direct contact with potential customers attracts not only global brands or local business but spammers, which try to behave like real persons and advertise their products at the same time. Since their goal differs from real people, their geotags often differ from the actual location, and they use tags or specific words for advertising of some service or product. Thus, it is important to exclude such accounts from further analysis. The main idea behind this method is to group users with the same spatial activity patterns. For the business profiles such as a store, gym, etc. one location will be prevalent among the others. Meanwhile, for real people, there will be some distribution in space. However, it is a common situation when people tag the city only but not a particular place, and depending on the city, coordinates of the post might be placed far from user's real location, and data will be lost among the others. Thus, on the first stage, we exclude profiles, who do not use geotags correctly, from the dataset.

We select users with more than ten posts with location to ensure that a person actively uses geotag functionality and commutes across the city. Users with less than ten posts do not provide enough data to correctly group profiles. In addition, they do not contribute sufficiently to the data [7]. Then, we calculate all distances between two consecutive locations for each user and group them by 1000 m, i.e., we count all distances that are less than 1 km, all distances between 1 and 2 km and so on. Distances larger than 50 km are united into one group. After that, we cluster users according to their spatial distribution. The cluster with a deficient level of spatial variations and with the vast majority of posts being in a single location represents business profiles and posts from these profiles can be excluded from the dataset.

At the next step, we use a Random Forest (RF) classifier to identify bots, business profiles, and compromised accounts – profiles, which do not represent real people and behave differently from them. It has been proven by many studies that a RF approach is efficient for bots and spam detection [23, 26]. Since we want to keep our methods as general as possible and to keep our pipeline applicable to any social media, we consider only text message, timestamp, and location as feature sources for our model. We use all data that a particular user has posted in the studied area and extract the following spatial and temporal features: number of unique locations marked by a user, number of unique dates when a user has posted something, time difference in seconds between consecutive posts. For time difference and number of posts per date, we calculated the maximum, minimum, mean, and standard deviation. From text caption we have decided to include maximum, minimum, average, mean, standard deviation of following metrics: number of emojis per post, number of hashtags per post, number of words per post, number of digits used in post, number of URLs per post, number of mail addresses per post, number of user mentions per post. In addition to that, we extracted money references, addresses, and phone numbers and included their maximum, minimum, average, mean, and standard deviation into the model. In addition, we added fraction of favourite tag in all user posts. Thus, we got 64 features in our model. As a result of this step, we obtain a list of accounts, which do not represent normal users.

City Background Extraction. The next stage is dedicated to the extraction of basic city information such as a list of typical tags for the whole city area and a set of general locations. General locations are places that represent large geographic areas and not specific places. For example, in the web version of Twitter user can only share the name of the city instead of particular coordinates. Some social media like Instagram or Foursquare are based on a list of locations instead of exact coordinates, and some titles in this list represent generic places such as streets or cities. Data from these places is useful in case of studying the whole area, but if someone is interested in studying actual temporal dynamics or spatial features, such data will distort the result. Also, it should be noted that even though throughout this paper we use the word 'city' to reference the particular geographic area, all stages are applicable on the different scales starting from city districts and metropolitan regions to states, countries, or continents.

Firstly, we extract names of administrative areas from Open Street Maps (OSM). After that, we calculate the difference between titles in social media data and data from OSM with the help of Damerau-Levenshtein distance. We consider a place to be general if the distance between its title and some item from the list of administrative objects is less than 2. These locations are excluded from the further analysis. For smaller scales such as streets or parks, there are no general locations.

Then, we analyze the distribution of tags mentions in the whole area. The term 'tag' denotes the important word in the text, which characterizes the whole message. Usually, in LBSN, tags are represented as hashtags. However, they can also be named entities, topics, or terms. In this work, we use hashtags as an example of tags, but this concept can be further extrapolated on tags of different types. The most popular hashtags are usually related to general location (e.g., #nyc, #moscow) or a popular type of content (#photo, #picsoftheday, #selfie) or action (#travel, #shopping, etc.). However, these tags cannot be used to study separate places and they are not relevant either to places or to events since they are actively used in the whole area. Nevertheless, scientists interested in studying human behavior in general can use this set of popular tags because it represents the most common patterns in the content. In this work, we consider tag as general if it was used in more than 1% of locations.

However, it is possible to exclude tags related to public holidays. We want to avoid such situations and keep tags, which have a large spatial distribution but narrow peak in terms of temporal distribution. Thus, we group all posts that mentioned a specific tag for the calendar year and compute their daily statistics. We then use the Gini index G to identify tags, which do not demonstrate constant behavior throughout the year. If $G \geq 0.8$ we consider tag as an event marker because it means that posts distribution have some peaks throughout the year. This pattern is common for national holidays or seasonal events such as sports games, etc. Thus, after the second stage, we obtain the dataset for further processing along with a list of common tags and general locations for the studying area.

Spatial Context Extraction. Using hashtags for events identification is a powerful strategy, however, there are situations where it might fail. The main problem is that people often use hashtags to indicate their location, type of activity, objects on photos and etc. Thus, it is important to exclude hashtags which are not related to the possible event. To do that, we grouped all hashtags by locations, thus we learn which tags are widely used throughout the city and which are place related. If some tag is highly popular in one place, it is highly likely that the tag describes this place. Excluding common place-related tags like #sea or #mall for each location, we keep only relevant tags for the following analysis. In other words, we get the list of tags which describe a normal state of particular places and their specific features. However, such tags cannot be indicators of events.

Fake Transitions Detection. The last stage of the pipeline is dedicated to suspicious posts identification. Sometimes, people cannot share their thoughts or photos immediately. It leads to situations where even normal users have a bunch of posts, which are not accurate in terms of location and timestamp. At this stage, we exclude posts that cannot represent the right combination of their coordinates and timestamps. This process is similar to the ideas for location spoofing detection – we search for transitions, which someone could not make in time. The standard approach for detection of fake transitions is to use space-time cones [28], but in this work, we suggest the improvement of this method – we use isochrones for fake transitions identification. In urban studies, isochrone is an area that can be reached from a specified point in equal time. Isochrone calculation is based on usage of real data about roads, that is why this method is more accurate than space-time cones. For isochrone calculation, we split the area into several zones depending on their distance from the observed point: pedestrian walking area (all locations in 5 km radius), car/public transport area (up to 300 km), train area (300–800 km) and flight area (further than 800 km). This distinction was to define a maximum speed for every traveling distance. The time required for a specific transition is calculated by the following formula:

$$t = \frac{1}{v} \sum_{i=1}^{N} s_i, \text{ where } v = \begin{cases} 5, & \text{if } S \leq 5 \\ 120, & 5 < S \leq 300 \\ 300, & 300 < S \leq 800 \\ 900, & S \geq 800 \end{cases} \tag{1}$$

where s_i is the length of the road segment and v is the maximum possible velocity depending on the inferred type of transport. The road data was extracted from OSM. It is important to note that on each stage of the pipeline, we get output data, which will be excluded, such as suspicious profiles, baseline tags, etc. However, this data can also be used, for example, for training novel models for fake accounts detection.

4 Experiments

4.1 Touristic Path Construction

The first experiment was designed to highlight the importance of general location extraction. To do that, we used the points-of-interest dataset for Moscow, Russia. The raw data was extracted from Facebook using the Places API and contained 40,473 places. The final dataset for Moscow contained 40,215 places, and 258 general sites were identified. However, it should be noted that among general locations, there were detected 'Russia' (8,984,048 visitors), 'Moscow', 'Russia' (7,193,235 visitors), 'Moscow Oblast' (280,128 visitors). For instance, the most popular non-general locations in Moscow are Sheremetyevo Airport and Red Square, with only 688,946 and 387,323 check-ins, respectively. The itinerary construction is based on solving the orienteering problem with functional profits (OPFP) with the help of the open-source framework FOPS [18]. In this approach, locations are scored by their popularity and by farness distance.

We used the following parameters for the Ant Colony Optimization algorithm: 1 ant per location and 100 iterations of the algorithm, as it was stated in the original article. The time budget was set to 5 h, the Red Square was selected as a starting point, and Vorobyovy Gory was used as a finish point since they two highly popular touristic places in the city center.

The resulting routes are presented in Fig. 2. Both routes contain extra places, including major parks in the city: Gorky park and Zaryadye park. However, there are several distinctions in these routes. The route based on the raw data contains four general places (Fig. 2, left) – 'Moscow', 'Moscow, 'Russia', 'Russia', and 'Khamovniki district', which do not correspond to actual places. Thus, 40% of locations in the route cannot be visited in real life. In contrast, in case of the clean data (Fig. 2, right), instead of general places algorithm was able to add real locations, such as Bolshoi Theatre and Central Children's Store on Lubyanka with the largest clock mechanism in the world and an observation deck with the view on Kremlin. Thus, the framework was able to construct a much better itinerary without any additional improvements in algorithms or methods.

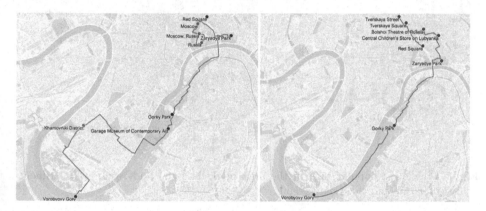

Fig. 2. Comparison of walking itineraries for original data (left) and cleaned dataset (right). Red dots related general locations, blue color indicates actual places (Color figure online)

4.2 Sentiment Analysis

To demonstrate the value of background analysis and typical hashtags extraction stages, we investigated the scenario of analysis of users' opinions in a geographical area via sentiment analysis. We used a combined dataset of Twitter and VK.com posts taken in Sochi, Russia, during 2016. Sochi is one of the largest and most popular Russian resorts. It was also the host of the Winter Olympics in 2014. Since Twitter and VK.com provide geospatial data with exact coordinates, we created a squared grid with a cell size equal to 350 m. We then kept only cells containing data (Fig. 3, right) – 986 cells in total. Each cell was considered as a separate location for the context extraction. The most popular tags in the area are presented in Fig. 3 (left). Tag '#sochi' was mentioned in 1/3 of cells (373 and 321 cells for Russian and English versions of the tag, respectively). The follow-up tags '#sochifornia' (used in 146 cells) and '#sea' (mentioned in 130 cells) were twice less popular. After that, we extracted typical tags for each cell. We considered a post to be relevant to the place if it contained at least one typical tag. Thus, we can be sure that posts represent the sentiment in that area.

Fig. 3. The most popular tags in the area of Sochi (left) and results of sentiment analysis for raw and clean data (right)

The sentiment analysis was executed in two stages. First, we prepare the text for polarity detection. To do that, we delete punctuation, split text in words, and normalized text with the help of [14]. In the second step, we used the Russian sentiment lexicon [4] to get the polarity of each word (1 indicates positive word and -1 negative word). The sentiment of the text is defined as 1 if a sum of polarities of all words more than zero and -1 if the sum is less than zero. The sentiment of the cell is defined as an average sentiment of all posts. On the Fig. 3, results of sentiment analysis are presented, cells with average sentiment less than 0.2 were marked as neutral. It can be noted from maps that after the filtering process, more cells have a higher level of sentiment. For Sochi city center, the number of posts with the sentiment $|s| \geq 0.75$ increased by 26.6%. It is also important that number of uncertain cell with sentiment rate $0.2 \leq |s| \leq 0.75$ decreased by 36.6% from 375 to 239 cells. Thus, we highlighted the strong positive and negative areas and decreased the number of uncertain areas by applying the context extraction stage of the proposed pipeline.

4.3   Event Detection

In this experiment, we applied the full pipeline on the Instagram data. New York City was used as a target city in the event detection approach [24] similar to this study the area was bounded by this latitude interval – [40.4745, 40.9445] and longitude varied in [−74.3060, −73.6790]. We collected the data from over 108,846 locations for a period of up to 7 years. The total number of posts extracted from the New York City area is 67,486,683.

In the first step, we try to exclude from the dataset all users who provide incorrect data, i.e. use several locations instead of the whole variety. We group users with the help of K-means clustering method. The appropriate number of clusters was obtained by calculating the distortion parameter. Deviant cluster contained 504,558 users out of 1,978,403. The shape of deviant clusters can be seen in Fig. 4. Suspicious profiles mostly post in the same location. Meanwhile, regular users have variety in terms of places.

Fig. 4. Obtained clusters' shape (top, red square indicates deviant clusters) and popular tags for Museum of Modern Art before and after filtering (bottom) (Color figure online)

After that, we trained our RF model using manually labelled data from both datasets. The training dataset contains 223 profiles with 136 ordinary users and 87 fake users; test data consists of 218 profiles including 146 normal profiles and 72 suspicious accounts. The model distinguishes a regular user from suspicious successfully. 122 normal user were detected correctly and 24 users were marked as suspicious. 63 suspicious users out of 72 were correctly identified. Thus, there were obtained 72% of precision and 88% of recall. Since the goal of this work is to get clean data as a result, we are interested in a high value of recall and precision is less critical. As a result, we obtained a list of 1,132,872 profiles which related to real people.

At the next step, we used only data from these users to extract background information about cities. 304 titles of general locations were derived for New York. These places were excluded from further analysis. After that, we extracted general hashtags; the example of popular tags in location before and after background tags extraction is presented on the Fig. 4. General tags contain mostly different term related to toponyms and universal themes such as beauty or life. Then, we performed the context extraction for locations. For each location typical hashtags were identified as 5% most frequent tags among users. We consider all posts from one user in the same location as one to avoid situations where someone tries to force their hashtag. We will use extracted lists to exclude typical tags from posts.

After that, we calculated isochrones for each normal users to exclude suspicious posts from data. In addition to that, locations with a high rate of suspicious posts (75% or higher part of posts in location was detected as suspicious) were excluded as well. There was 16 locations in New York City. The final dataset for New York consists of 103,977 locations. For event detection we performed the same experiment which was described in [24]. In the original approach the spike in activity in particular cell of the grid was consider as an event. To find these spikes in data, historical grids is created using retrospective data for a calendar year. Since we decrease amount of data significantly, we set threshold value to 12. We used data for 2017 to create grids, then we took two weeks from 2018 for the result evaluation: a week with a lot of events during 12–18 of March and an ordinary week with less massive events 19–25 February.

The results of the recall evaluation are presented in Table 1. As can be seen from the table on an active week, the recall increment was 14.9% and for non-active week recall value increase on 32.6%. It is also important to note that some events, which do not have specific coordinates, such as snowfall in March or Saint Patrick's day celebration, were detected in the less number of places. This leads to lesser number of events in total and more significant contribution to the false positive rate. Nevertheless, the largest and the most important events, such as nationwide protest '#Enough! National School Walkout' and North American International Toy Fair are still detected from the very beginning. In addition to that due to the altered structure of historical grids, we were able to discover new events such as a concert of Canadian R&B duo 'dvsn', 2018 Global Engagement Summit at UN Headquarters, etc. These events were covered with a low number of posts and stayed unnoticed during the original experiment. However, the usage of clean data helped to highlight small events which are essential for understanding the current situation in the city.

Table 1. Comparison of event detection results

Recall	Active week	Non-active week	Total
Before	81.8%	63.9%	73%
After	96.7%	96.5%	96.7%

5 Conclusion and Future Works

In this work, we presented a spatiotemporal filtering pipeline for data preprocessing. The main goal of this process is to exclude unreliable data in terms of space and time. The pipeline consists of four stages: during the first stage, suspicious user profiles are extracted from data with the help of K-means clustering and Random Forest classifier. On the next stage, we exclude the buzz words from the data and filter locations related to large areas such as islands or city districts. Then, we identify the context of a particular place expressed by unique tags. In the last step, we find suspicious posts using the isochrone method. Stages of the pipeline can be used separately and for different tasks. For instance, in the case of touristic walking itinerary construction, we used only general location extraction, and the walking itinerary was improved by replacing 40% of places. In the experiment dedicated to sentiment analysis, we used a context extraction method to keep posts that are related to the area where they were taken, and as a result, 36.2% of uncertain areas were identified either as neutral or as strongly positive or negative. In addition to that, for event detection, we performed all stages of the pipeline, and recall for event detection method increased by 23.7%.

Nevertheless, there are ways for further improvement of this pipeline. In Instagram, some famous places such as Times Square has several corresponding locations including versions in other languages. This issue can be addressed by using the same method from the general location identification stage. We can use distance to find places with a similar name. Currently, we do not address the repeating places in the data since it can be a retail chain, and some retail chains include over a hundred places all over the city. In some cases, it can be useful to interpret a chain store system as one place. However, if we want to preserve distinct places, more complex methods are required. Despite this, the applicability of the spatiotemporal pipeline was shown using the data from Facebook, Twitter, Instagram, and VK.com. Thus, the pipeline can be successfully used in various tasks relying on location-based social network data.

Acknowledgement. This research is financially supported by The Russian Science Foundation, Agreement #18-71-00149.

References

1. Biessmann, F., Salinas, D., Schelter, S., Schmidt, P., Lange, D.: "Deep" learning for missing value imputation in tables with non-numerical data. In: Proceedings of the 27th ACM International Conference on Information and Knowledge Management, CIKM 2018, pp. 2017–2025. ACM, New York (2018). https://doi.org/10.1145/3269206.3272005
2. Burton, S.H., Tanner, K.W., Giraud-Carrier, C.G., West, J.H., Barnes, M.D.: "Right time, right place" health communication on Twitter: value and accuracy of location information. J. Med. Inet. Res. **14**(6), e156 (2012). https://doi.org/10.2196/jmir.2121
3. Campagna, M.: Social media geographic information: why social is special when it goes spatial. In: European Handbook Crowdsourced Geographic Information, pp. 45–54 (2016). https://doi.org/10.5334/bax.d
4. Chen, Y., Skiena, S.: Building sentiment lexicons for all major languages. In: Proceedings of the 52nd Annual Meeting of the Association for Computational Linguistics (Short Papers), pp. 383–389 (2014)
5. Cvetojevic, S., Juhasz, L., Hochmair, H.: Positional accuracy of Twitter and Instagram Images in urban environments. GI_Forum **1**, 191–203 (2016). https://doi.org/10.1553/giscience2016_01_s191
6. Ding, C., et al.: A location spoofing detection method for social networks (short paper). In: Gao, H., Wang, X., Yin, Y., Iqbal, M. (eds.) CollaborateCom 2018. LNICST, vol. 268, pp. 138–150. Springer, Cham (2019). https://doi.org/10.1007/978-3-030-12981-1_9
7. Egele, M., Stringhini, G., Kruegel, C., Vigna, G.: COMPA: detecting compromised accounts on social networks (2013). https://doi.org/10.1.1.363.6606
8. Ferrara, E., Varol, O., Davis, C., Menczer, F., Flammini, A.: The rise of social bots. Commun. ACM **59**(7), 96–104 (2016). https://doi.org/10.1145/2818717
9. Goodchild, M.F.: The quality of big (geo)data. Dialogues Hum. Geogr. **3**(3), 280–284 (2013). https://doi.org/10.1177/2043820613513392
10. Silva, H., et al.: Urban computing leveraging location-based social network data: a survey. ACM Comput. Surv. **52**(1), 1–39 (2019)
11. Hochman, N., Manovich, L.: Zooming into an Instagram City: reading the local through social media. First Monday (2013). https://doi.org/10.5210/fm.v18i7.4711
12. John, N.A., Nissenbaum, A.: An agnotological analysis of APIs: or, disconnectivity and the ideological limits of our knowledge of social media. Inf. Soc. **35**(1), 1–12 (2019). https://doi.org/10.1080/01972243.2018.1542647
13. Kapoor, K.K., Tamilmani, K., Rana, N.P., Patil, P., Dwivedi, Y.K., Nerur, S.: Advances in social media research: past, present and future. Inf. Syst. Front. **20**(3), 531–558 (2017). https://doi.org/10.1007/s10796-017-9810-y
14. Korobov, M.: Morphological analyzer and generator for Russian and Ukrainian languages. In: Khachay, M.Y., Konstantinova, N., Panchenko, A., Ignatov, D.I., Labunets, V.G. (eds.) AIST 2015. CCIS, vol. 542, pp. 320–332. Springer, Cham (2015). https://doi.org/10.1007/978-3-319-26123-2_31
15. Lavanya, P.G., Kouser, K., Suresha, M.: Efficient pre-processing and feature selection for clustering of cancer tweets. In: Thampi, S.M., et al. (eds.) Intelligent Systems, Technologies and Applications. AISC, vol. 910, pp. 17–37. Springer, Singapore (2020). https://doi.org/10.1007/978-981-13-6095-4_2
16. Manikonda, L., Hu, Y., Kambhampati, S.: Analyzing user activities, demographics, social network structure and user-generated content on Instagram. arXiv preprint abs/1410.8 (2014)

17. Morstatter, F., Pfeffer, J., Liu, H., Carley, K.M.: Is the sample good enough? Comparing data from Twitter's streaming API with Twitter's Firehose. In: Proceedings of the 7th International Conference on Weblogs and Social Media, ICWSM 2013 (2013)
18. Mukhina, K.D., Visheratin, A.A., Nasonov, D.: Orienteering problem with functional profits for multi-source dynamic path construction. PLoS ONE **14**(4), e0213777 (2019). https://doi.org/10.1371/journal.pone.0213777
19. Shu, K., Sliva, A., Wang, S., Tang, J., Liu, H.: Fake news detection on social media: a data mining perspective. ACM SIGKDD Explor. Newsl. **19**(1), 22–36 (2017)
20. Singh, M., Bansal, D., Sofat, S.: Who is who on Twitter-Spammer, fake or compromised account? A tool to reveal true Identity in real-time. Cybern. Syst. **49**(1), 1–25 (2018). https://doi.org/10.1080/01969722.2017.1412866
21. Steiger, E., Westerholt, R., Resch, B., Zipf, A.: Twitter as an indicator for whereabouts of people? Correlating Twitter with UK census data. Comput. Environ. Urban Syst. **54**, 255–265 (2015). https://doi.org/10.1016/j.compenvurbsys.2015.09.007
22. Stringhini, G., Kruegel, C., Vigna, G.: Detecting spammers on social networks. In: Proceedings of the 26th Annual Computer Security Applications Conference, ACSAC 2010, pp. 1–9. ACM, New York (2010). https://doi.org/10.1145/1920261.1920263
23. Varol, O., Ferrara, E., Davis, C.A., Menczer, F., Flammini, A.: Online human-bot interactions: detection, estimation, and characterization. In: Proceedings of the 11th International Conference on Web and Social Media, ICWSM 2017, pp. 280–289 (2017)
24. Visheratin, A.A., Visheratina, A.K., Nasonov, D., Boukhanovsky, A.V.: Multiscale event detection using convolutional quadtrees and adaptive geogrids. In: 2nd ACM SIGSPATIAL Workshop on Analytics for Local Events and News, p. 10 (2018). https://doi.org/10.1145/3282866.3282867
25. Wilken, R.: Places nearby: Facebook as a location-based social media platform. New Media Soc. **16**(7), 1087–1103 (2014). https://doi.org/10.1177/1461444814543997
26. Yang, K.C., Varol, O., Davis, C.A., Ferrara, E., Flammini, A., Menczer, F.: Arming the public with artificial intelligence to counter social bots. Hum. Behav. Emerg. Technol. **1**(1), e115 (2019). https://doi.org/10.1002/hbe2.115
27. Yardi, S., Romero, D., Schoenebeck, G., boyd, D.: Detecting spam in a Twitter network. First Monday **15**(1) (2009). https://doi.org/10.5210/fm.v15i1.2793
28. Zhao, B., Sui, D.Z.: True lies in geospatial big data: detecting location spoofing in social media. Ann. GIS **23**(1), 1–14 (2017). https://doi.org/10.1080/19475683.2017.1280536

Normal Grouping Density Separation (NGDS): A Novel Object-Driven Indoor Point Cloud Partition Method

Jakub Walczak[1]([✉]) [iD], Grzegorz Andrzejczak[2] [iD], Rafał Scherer[3] [iD],
and Adam Wojciechowski[1] [iD]

[1] Institute of Information Technology, Lodz University of Technology, Łódź, Poland
{jakub.walczak,adam.wojciechowski}@p.lodz.pl
[2] Institute of Mathematics, Lodz University of Technology, Łódź, Poland
grzegorz.andrzejczak@p.lodzd.pl
[3] Częstochowa University of Technology, Częstochowa, Poland
rafal.scherer@pcz.pl

Abstract. Precise segmentation/partition is an essential part of many point cloud processing strategies. In the state-of-the-art methods, either the number of clusters or expected supervoxel resolution needs to be carefully selected before segmentation. This makes these processes semi-supervised. The proposed Normal Grouping- Density Separation (NGDS) strategy, relying on both grouping normal vectors into cardinal directions and density-based separation, produces clusters of better (according to use quality measures) quality than current state-of-the-art methods for widely applied object-annotated indoor benchmark dataset. The method reaches, on average, lower under-segmentation error than VCCS (by 45.9pp), Lin et al. (by 14.8pp), and SSP (by 26.2pp). Another metric - achievable segmentation accuracy - yields 92.1% across the tested dataset what is higher value than VCCS (by 14pp), Lin et al. (by 3.8pp), and SSP (by 10.3pp). The experiment carried out indicates superiority of the proposed method as a partition/segmentation algorithm - a process being usually a preprocessing stage of many object detection workflows.

Keywords: Space partition · Superpoints · Point cloud segmentation · NGDS

1 Introduction

Point clouds have recently become a powerful representation of the environment due to the inherent spatial cues that they possess. Depth information, provided either by a depth camera, multiple-view interpolation, or by a laser scanner, is clue information exploited for retrieval relationships among objects in a scene [7]. This is a reason for such wide and various application of point clouds [9,16,33,35]. Depth information can be also used in other fields, such as computer

© Springer Nature Switzerland AG 2020
V. V. Krzhizhanovskaya et al. (Eds.): ICCS 2020, LNCS 12142, pp. 100–114, 2020.
https://doi.org/10.1007/978-3-030-50433-5_8

graphics, from rasterization and ray tracing algorithms [28] known for decades to modern screen-space methods of accelerating calculations without loss of image quality based on the prior creation of depth information in individual image pixels [30].

One of the key aspects of a point cloud processing is object detection [12,17,26,37] and, in general, semantic analysis [8,29,36,38]. In many aspects of a point cloud processing, the strategy relying on either ordering points or excessive partition followed by the aggregation (top-bottom-top) is frequently applied [13]. The strategy, relies on excessive space/surfaces segmentation (top-bottom, in order to minimize under-segmentation) and aggregation (bottom-top, so that over-segmentation could be decreased, keeping under segmentation error low) [5,6,11] followed by features analysis [10,26] or deep learning pipeline [2,12,17,38]. Such an approach requires a process of over-segmentation (top-bottom) to be efficient, granular on objects' edges and corners, and, what is the most important, devoid of overlapping areas between semantically different objects. In this article, the novel, intuitive and high-quality method for over-segmentation, called Normal Grouping- Density Separation (NGDS), is presented. The novelty of the presented method relies on clever application of efficient grouping algorithm in order to detect primary plane directions in a point cloud followed by histogram- and density-based separation within points belonging to a primary direction. Unlike current state-of-the-art methods, the presented strategy does not require predefined, manually selected proper number of segments to produce expected and sufficiently granular over-segmentation result suitable in the context of object detection. Most of parameters are calculated based on provided point cloud characteristic and just a few are required to be set by a user.

2   Related Works

A popular strategy of a point cloud processing, in any form, either semantic segmentation, or object detection, is to reduce the size of a problem by means of over-segmentation algorithms [32] which group points - consistent according to some criterion - into clusters whose number is usually tremendously lower than the cardinality of a point cloud. This allows applying just a few groups instead of hundreds of thousands of points. Such an strategy lies behind the idea of compression - to use approximate larger regions by means of a small representative entity [18]. However, the problem of current over-segmentation algorithms is the fact that they focus on class-driven approach rather than object-driven one. As a result, such methods produce clusters of preserved boundaries between objects of different classes whereas objects within the same class are not partitioned properly. This disables benchmark methods to be successfully applied as over-segmentation strategy for object detection purposes.

One of first benchmark partition strategy - VCCS - was introduced in [22]. The method presented therein focuses on quasi-regular clusters called supervoxels. The researchers used 39 point features to make sure that points inside a single cluster/supervoxel are consistent (intra-cluster consistency). 33 of those 39 features are traits of Fast Point Feature Histogram, the others are geometrical coordinates and color in CIELab space. VCCS relies on efficient k-means clustering to group points to clusters of predefined seeds. Though pretty efficient, the method suffers some issues. The first problem is initialization of the algorithm itself. The seed points need to be carefully chosen and the number of those seeds (hence clusters) has to be selected before. The desired method solving the issue should somehow recognize the optimal number of resulting clusters in unsupervised manner, because improper choice can make a method pointless. Too low number of clusters results in insufficient segmentation - output clusters overlap many ground-truth clusters. Too high number, in turn, wreaks longer time processing and loss of the context. Furthermore, as indicated in [14], VCCS may segment borders inaccurately.

To overcome the issue related to borders and seed points selection, Lin et al. [14] proposed extension of VCCS. In these studies, suggested cost function, which optimization provides representative points, consists of two counteractive components: the first one ensuring a representative point approximates well a collection of points; and the second one - constraining the number of expected representative points to be as close to the predefined value as possible.

Having selected all representative points, optimization of cost function is continued by assigning non-representative points to those representative points for which dissimilarity distance is the lowest.

Though improved with respect to its predecessor, the method of Lin et al. still requires the number of resulting clusters to be selected in advance what cannot be reliably done in unsupervised partition process. Moreover, the resulting groups/clusters are similar to quasi-regular grid of VCCS what produces high over-segmentation error in regions where it is redundant. The clue is to design a method which maintains the proper balanced between both under- and over-segmentation keeping both of them low.

To reduce the problem of excessive partition keeping under-segmentation error low, Landrieu and Boussaha [11] proposed the SSP method mixing deep learning approach with analytical strategies. Applying PointNet-like neural network enables the authors to extract high-level object-oriented features, called embeddings. Such embeddings are calculated for each point in a data set based on its vicinity. Based on embeddings and spatial connectivity, Generalized Minimal Partition Problem is solved with the method ℓ_0 presented in [12]. The method yields good results in class-driven approach, however, taking into account single objects, the method is not reliable. It is caused by embeddings, which themselves cannot differentiate points belonging to different objects of the same class. Because of this, the method leaks intra-class separation what is crucial element of partition oriented to objects. The last drawback of SSP method is the fact that it may requires color information to produce reliable result.

To sum up, the methods of VCCS and Lin et al. inherently take into account spatial connectivity of points what is beneficial in terms of object-oriented separation of clusters, however, they perform many redundant subdivisions, which increase over-segmentation error. On the other hand, the SSP method avoid excessive partition at the cost of class-oriented partition rather than the object-one. In addition, all those methods require an expected number of clusters to be defined prior to computation what makes these methods difficult to be applied successfully in unsupervised segmentation. Hence, there is a need to develop a method which automatically splits points into geometrically coherent sets. And it turns out that relying only on normal vector may lead to sufficiently detailed space partition which quantizes well points of objects.

3   Methodology

At first, let the point cloud $\mathbf{P}$ be of the cardinality $\|\mathbf{P}\| = N$. Let the two clusterings $\mathcal{S}$ and $\mathcal{G}$ be also defined, where $\mathcal{S}$ of cardinality $|\mathcal{S}| = m$ consists of a set of m clusters: $\mathcal{S} = \{s_1, s_2, s_3, ..., s_m\}$ being the output of a method, and $\mathcal{G} = \{g_1, g_2, g_3, ..., g_n\}$, of cardinality $|\mathcal{G}| = n$, represents a set of n ground-truth (real) clusters (single objects in a scene: 1st chair, 2nd chair, 1st table, etc.). It is crucial to note that ground-truth clusters are single objects in a scene, while for algorithms' output clusters encompass usually subsets of objects and the goal of each partition method is to produce output clusters as alike to ground-truth ones as possible. Following the literature approaches for partition and segmentation validation [5,14,22], below quality measure were engaged.

3.1   Quality Measures

Under-Segmentation Error (UE). Also referred to as under-segmentation rate, indicates insufficient partition. In short, an output cluster overlaps more than one ground-truth cluster. Its value varies from 0 - if none of output clusters overlap more than one ground-truth cluster, and 1 if $|\mathcal{S}| = 1$ and $|\mathcal{G}| > 1$. In general, if $m < n$ then UE is in-between 0% and 100%. UE is expressed by (1). For visualization, see Fig. 1.

$$UE_{\mathcal{G}}(\mathcal{S}) = \frac{1}{m}[(\sum_{g_j \in \mathcal{G}} \sum_{s_i \in \mathcal{S}} < \frac{|s_i \cap g_j|}{|g_j|} > \epsilon <) - m] \tag{1}$$

where $< \cdot <$ is an Iverson bracket which takes 1 if inner condition is *True* and 0 otherwise, ϵ is a very small value (here it is 0.1%).

Weighted Under- and Over-Segmentation Error (wUE, wOE). The formula (1) relies on binary values (sum 1 if overlapping exceeds the threshold value ϵ. But, according to [27], the measure to express UE may be weighted with the intersection part- wUE (2). In similar manner wOE may be expressed- wOE (3). Their best values are 0% which means all points within a single cluster are associated with the only one ground-truth cluster, i.e. object and vice versa.

$$wUE_{\mathcal{G}}(\mathcal{S}) = \sum_{s_i \in \mathcal{S}} (|s_i| - \max_j |s_i \cap g_j|)/N \tag{2}$$

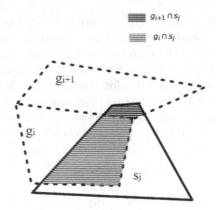

Fig. 1. Visualization of sample overlapping cases. Dashed lines represent ground- truth clusters whereas solid ones- the output cluster of a method. Black shaded region represents intersection lower than ϵ, so it is counted as 0 while red shaded region exceeds ϵ and is counted as 1. (Color figure online)

where N is the cardinality of a point cloud ($|\mathbf{P}| = \sum |g_j| = N$)

$$wOE_{\mathcal{G}}(\mathcal{S}) = \sum_{g_j \in \mathcal{G}} (|g_j| - \max_i |s_i \cap g_j|)/N \tag{3}$$

Harmonic Segmentation Error (HSE). Similarly to $F1$ score, which connected both precision and recall of classification in the form of harmonic mean, HSE may be defined as a single measure of error taking into account both weighted over- and under-segmentation errors (4).

$$HSE_{\mathcal{G}}(\mathcal{S}) = 2 \cdot \frac{wUE_{\mathcal{G}}(\mathcal{S}) \cdot wOE_{\mathcal{G}}(\mathcal{S})}{wUE_{\mathcal{G}}(\mathcal{S}) + wOE_{\mathcal{G}}(\mathcal{S})} \tag{4}$$

Achievable Segmentation Accuracy (ASA). ASA is one of quality metrics used by [15] to evaluate maximum possible accuracy in object detection task while applying proposed clusters as units. The best possible value it takes is 100%. Formally, this measure may be expressed by (5).

$$ASA_{\mathcal{G}}(\mathcal{S}) = \frac{\sum_{s_i \in \mathcal{S}} \max_j |s_i \cap g_j|}{\sum_{g_j \in \mathcal{G}} |g_j|} \tag{5}$$

Some literature studies made use of, so called, boundary recall and boundary precision [5,14]. However, indicating "boundary" points as done in [5,34] is, at least, questionable and ambiguous. Moreover, UE with low overlapping threshold ϵ and wUE directly point out if objects' borders are crossed or not. That is why boundary-based measures were skipped in the considerations presented in these studies.

3.2 Database

To make the method comparable, benchmark data sets need to be used. However, among all indoor databases widely used in studies, like: NYU RGBD v2 [21], ScanNet [4], or S3DIS [1] only the latter one distinguishes single objects. The others contained points labeled only by class what makes them useless in terms of verification of the partition method dedicated to object detection task. Therefore, S3DIS database was only selected as it is the only one available indoor database annotated by object.

S3DIS is one of the basic indoor benchmark dataset for semantic segmentation and object detection task. It was used, among others, in [5,11,12]. It contains 273 indoor-scene point sets of quite uniform densities with moderate scanning shadows present.

4   Proposed Method

In this paper, the novel partition method relying on spatial point connectivity and geometrical features - NGDS - is presented. The method consists of 6 stages:

a) normal vector estimation;
b) normal vector alignment;
c) primary directions detection;
d) level detection;
e) 2D density separation;
f) Lost points appending;

Normal Vector Estimation. Normal vector, yet trivial, is a kind of handcrafted point feature, contrary to traits learned with deep learning models [24]. Computation of a normal vector $\mathbf{n}_i$ of a point p_i is carried out by means of fitting a plane to the vicinity $\mathcal{N}$ of that point. In literature it is usually done with by eigendecomposition of the covariance matrix of neighbours' coordinates. This phase, taking into account efficient neighbours retrieving with kd-tree [3], is of time complexity: $O(n \cdot \log n) + O(|\mathcal{N}| \cdot n) + O(n) \equiv_{|\mathcal{N}|=const} O(n \cdot \log n)$

Normal Vector Alignment. Any method of calculating normal vector, which does not take into account constant reference point, cannot assure coherent orientation of normal vectors. This is the result of plane ambiguity (6).

$$\mathbf{n} \cdot p = -d \sim -\mathbf{n} \cdot p = d \tag{6}$$

Making normal vectors coherent across the same orientation simplifies to satisfying the condition (7) [23].

$$\mathbf{n}_i \cdot (v_p - p_i) > 0 \tag{7}$$

where v_p is a viewpoint, which in our studies is assumed to be in the mass center of a point cloud: $v_p = \bar{p} = \frac{1}{n} \sum_{i=1}^{n} p_i$.

In this way, parallel normal vectors get coherent orientation in symmetry with respect to v_p. This allows also distinguishing parallel planes on opposite sides of v_p (Fig. 2).

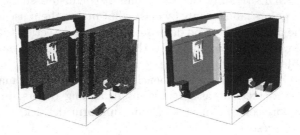

Fig. 2. Two parallel walls with associated normal vectors (brightness presents angle between a normal vector and camera optical axis). On the left- not aligned and on the right- aligned with respect to viewpoint

The stage of normal vector alignment is of linear time complexity- $O(n)$.

Primary Directions Detection (Normal Grouping). Having aligned point normal vectors, some primary directions may be identified taking a look at the distribution of normal vectors' orientations (Fig. 3a).

Such distribution seems to be dedicated for efficient k-means clustering algorithm. Though, relatively quick, it will be further accelerated if mini-batch based approach is used [25]. Though burdened with heuristics, mini-batch k-means clustering usually supplies results of sufficient accuracy with respect to optimum result in the sense of Maximum Likelihood estimation.

Widely known problem with k-means clustering concerns the proper selection of a number of resulting clusters k. In these studies, it is fixed. If normal vectors are oriented as in Fig. 3a - on unit sphere, then the allowed angle between normal vectors forms a spherical cap (Fig. 3b). Assuming expected angular tolerance to be $\Delta\theta$, the spherical cap surface associated to $\Delta\theta$ is expressed by (8).

$$P_c = 2 \cdot \pi \cdot (1 - \cos(2 \cdot \Delta\theta)) \tag{8}$$

Knowing that the unit sphere has the surface of $P_s = 4 \cdot \pi$, the expected number of clusters k may be calculated as $k = \lfloor P_s/P_c \rfloor$.

Time complexity of the mini-batch version of k-means algorithm is $O(n)$ assuming fixed number of maximum iterations and that the kd-tree for neighbours retrieving is already calculated.

Level Detection. The result of primary detection clustering provides groups of points representing a single primary direction (Fig. 4a).

Considering a single primary direction, let us define a group of points $\mathcal{D}_i \subseteq \mathbf{P}$ belonging to $i-$ th primary direction. As a one group, the points $\mathcal{D}_i$ should be

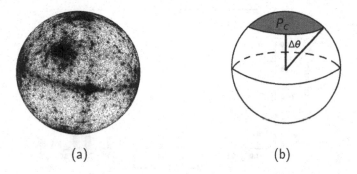

(a)   (b)

Fig. 3. (a) Unit sphere representing distribution of point normal vectors (b) spherical cap (in gray) of area P_c associated with angular tolerance $\Delta\theta$ on the unit sphere

described by a single normal vector, which may be calculated as a unit-length average normal vector $\hat{n}_i$ of the group. Since they are said to share a common normal vector $\hat{n}_i$, particular planar fragments may be easily distinguished by means of planes' constant factor (d, Eq. 6). For each point (and associated with it, plane fitted to its former vicinity $\mathcal{N}$), its constant factor is calculated according to the formula: $d = -p_j \cdot \hat{n}_i$ (assuming $p_j \in \mathcal{D}_i$). Based on those values, a histogram may be constructed. Its peaks would clearly indicate planar fragments hung at different levels (Fig. 5). As a result, sets of co-planar points are retrieved: $\mathcal{L}_1, \mathcal{L}_1, \mathcal{L}_3, ...$ (Fig. 4b).

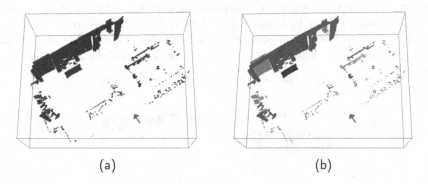

(a)   (b)

Fig. 4. (a) Groups of points around the single primary direction pointed by red arrow (b) points belonging to the same primary directions: $\mathcal{L}_1, \mathcal{L}_2, ...$, split according to associated constant factors (Color figure online)

In order to make extraction accurate, desired number of histogram bins should be carefully chosen. Intuitively, it is involved with a point cloud acquisition device tolerance, usually denoted as σ. It is involved with precision of points coordinates while noise is modelled with Gaussian distribution. The problem is,

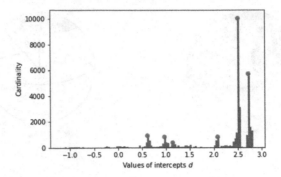

Fig. 5. Histogram of constant factors for point-planes of common primary direction

that sigma is usually not provided for indoor scans like S3DIS and it has to be somehow estimated. To do so, several random samples of all points are drawn and the distance (Euclidean or [31]) to their closest neighbours is calculated. The maximum value of this distance is saved. Across several samples, the mean maximum distance is calculated as the estimation of σ. The time complexity of this stage is linear- $O(n)$, since sigma approximation may be thought of to be of constant time complexity $O(1)$ and histogram construction takes $O(n)$ time for each primary direction.

2D Density-Based Clustering (Density Separation). Levels $\mathcal{L}_1, \mathcal{L}_3, \ldots$ detected in the previous stage contain points said to be co-planar. In real cases, especially when we bear in mind object detection task, co-planar points may quite often lead to insufficient partition is some areas. An example of such case is presented in Fig. 6, where two tops of two separate tables form a common group $\mathcal{L}_i$.

Fig. 6. Two tops of two separate tables detected as a single group $\mathcal{L}_i$

To avoid such issues, density-based HDBSCAN clustering [20] is carried out for points contained in each $\mathcal{L}_i$. Since points within $\mathcal{L}_i$ are co-planar, density separation may be reduced from 3D to 2D problem by means of Principal Component Analysis (9) so that computation time and occupied memory are both reduced. In this way, from a set $\mathcal{L}_i$ several subsets $\mathcal{L}_i = \{\mathcal{L}'_{i,1}, \mathcal{L}'_{i,2}, \mathcal{L}'_{i,3}, \ldots\}$ are extracted and the set of noise points $\mathcal{L}_{i,noise}$. (HDBSCAN is able to detect points deemed as noise in the context of local points density- those noise points are rejected and considered in the next stage.)

$$\mathcal{L}_i^{2D} = \mathcal{L}_i \times \mathbf{E} \tag{9}$$

where $\mathbf{E}$ is a column-wise matrix of eigenvectors

Though, HDBSCAN may yield sub-quadratic time complexity, it cannot achieve $O(n \cdot \log n)$, however, [19] suggests that it may approach log-linear asymptotic complexity for a number of data sets.

Lost Points Appending. During all previous stages, some points may be rejected due to rank-deficiency or density changes in HDBSCAN (noise points). To make the output point cloud conformed to the original one, these lost points need to be appended to the best matched clusters, to form $\mathcal{S}_1, \mathcal{S}_2, ...$ Assignment is done based on similarity function, defined like in [14] (10) taking as R the estimated values of σ ($R = \sigma$).

$$D(p_i, p_j) = 1 - abs(\mathbf{n}_i \cdot \mathbf{n}_j) + 0.4 \cdot \frac{|p_i - p_q|}{R} \tag{10}$$

where R is an assumed resolution of partition, $\mathbf{n}_i$ and $\mathbf{n}_j$ are normal vectors associated to the $i-$ th and $j-$th point, $|\cdot|$ states for a norm of a vector.

The proposed method NGDS was validated on the benchmark database for indoor scenes, namely S3DIS [1].

5   Experiments

To compare the proposed NGDS method with the state-of-the-art solutions, following experiments were conducted. For benchmark dataset - S3DIS [1] partition according to VCCS [22], Lin et al. [14], and [11] methods were conducted. Ground-truth clusters $\mathcal{G} = \{g_1, g_2, g_3, ..., g_n\}$ are single objects in a scene, provided by the dataset. Experiments for VCCS and for the method of Lin et al. were carried out for all point clouds contained in the aforementioned database. To compare the state-of-the-art solutions with the proposed one, publicly available implementation of those algorithms were employed:

- Point Cloud Library implementation of VCCS was used [23] in the experiments.
- The code for the method of Lin et al. is provided by the authors on the publicly available repository [1].
- The code for SSP method is accessible in the public repository managed by the author [2].

---

[1] https://github.com/yblin/Supervoxel-for-3D-point-clouds.
[2] https://github.com/loicland/superpoint_graph.

6   Results

The results for the proposed NGDS method juxtaposed with the current state-of-the-art methods [11,14,22] for benchmark object database S3DIS are presented below (Table 1). For quality measures of the proposed method associated standard deviations are given.

Table 1. Comparison of quality measures for the benchmark partition methods and the proposed method NGDS for S3DIS

Method	UE [%]	wUE [%]	wOE [%]	HSE [%]	ASA [%]
VCCS [22]	46.4	21.9	89.2	35.2	78.1
Lin et al. [14]	15.3	11.8	92.7	19.6	88.3
SSP [11]	26.7	18.2	35.9	24.2	81.8
Proposed method (NGDS)	0.5 ± 0.4	4.7 ± 3.2	48.7 ± 11.5	8.6	92.1 ± 3.9

There are exemplary results of three state-of-the-art methods and NGDS presented in the Fig. 7

The results presented in Table 1 show superiority of the NGDS method over state-of-the-art solutions in terms of all presented quality measures. They prove high quality of NGDS as a partition method dedicated to object detection task. Under-segmentation error for the proposed method is lower by 99%, 97%, and 98% for VCCS, Lin et al., and SSP respectively. This confirms that clusters created by NGDS method do not tend to cross **object** boundaries, even within the same class. Lower weighted under-segmentation error, in turn, proves that only 4.7% of points are mismatched. On the other hand, weighted over-segmentation error shows that less redundant subdivisions were done with respect to the method of Lin et al. and VCCS. In comparison to SSP, wOE is slightly higher. To infer an overall trade-off of under- and over-segmentation, HSE indicator was introduced. Undoubtedly, it attains the best (lowest) value for the proposed method - respectively lower by 26.6pp, 11pp, and 15.6pp than VCCS, Lin et al., and SSP. Also the average Achievable Segmentation Accuracy clearly points out that having applied NGDS method for the task of indoor object detection, the highest accuracy may be achieved among all four methods.

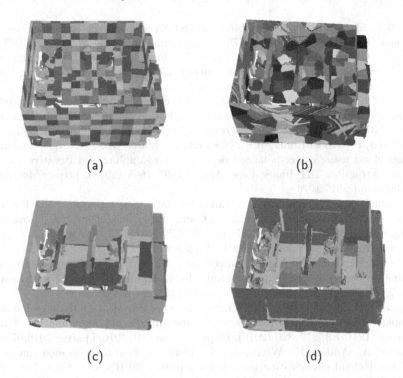

Fig. 7. Results for conferenceRoom_1 from Area_1 for (a) VCCS (b) Lin et al., (c) SSP, and (d) for NGDS

7 Conclusions

Based on performed evaluation, it may be clearly noted that the proposed method yields better results for indoor scenes than state-of-the-art partitioning algorithms. NGDS provides partition result less over-segmented than VCCS or the method of Lin et al. keeping under-segmentation ratio at the very low level (lower than competitive methods). The limitation of the method is the fact that in case of a point cloud of extremely uneven density, the over-segmentation ratio deteriorates significantly, keeping over-segmentation rate at similar, low level. Further research will focus on applying the proposed NGDS method for indoor object detection.

References

1. Armeni, I., Sax, A., Zamir, A.R., Savarese, S.: Joint 2D–3D-semantic data for indoor scene understanding. ArXiv e-prints, February 2017
2. Ben-Shabat, Y., Lindenbaum, M., Fischer, A.: Nesti-Net: normal estimation for unstructured 3d point clouds using convolutional neural networks (2018). http://arxiv.org/abs/1812.00709

3. Bentley, J.L.: Multidimensional binary search trees used for associative searching. Comm. ACM **18**(9), 509–517 (1975). https://doi.org/10.1145/361002.361007
4. Dai, A., Chang, A.X., Savva, M., Halber, M., Funkhouser, T., Nießner, M.: Scan-Net: richly-annotated 3D reconstructions of indoor scenes. In: Proceedings of the Computer Vision and Pattern Recognition (CVPR). IEEE (2017)
5. Dong, Z., Yang, B., Hu, P., Scherer, S.: An efficient global energy optimization approach for robust 3D plane segmentation of point clouds. ISPRS J. Phot. Rem. Sens. **137**, 112–133 (2018). https://doi.org/10.1016/j.isprsjprs.2018.01.013
6. El-Sayed, E., Abdel-Kader, R.F., Nashaat, H., Marei, M.: Plane detection in 3D point cloud using octree-balanced density down-sampling and iterative adaptive plane extraction. IET Image Proc. **12**(9), 1595–1605 (2018). https://doi.org/10.1049/iet-ipr.2017.1076
7. Forczmański, P., Nowosielski, A.: Multi-view data aggregation for behaviour analysis in video surveillance systems. In: Chmielewski, L.J., Datta, A., Kozera, R., Wojciechowski, K. (eds.) ICCVG 2016. LNCS, vol. 9972, pp. 462–473. Springer, Cham (2016). https://doi.org/10.1007/978-3-319-46418-3_41
8. Guerrero, P., Kleiman, Y., Ovsjanikov, M., Mitra, N.J.: PCPNet learning local shape properties from raw point clouds. In: Computer Graphics Forum, vol. 37, no. 2, pp. 75–85 (2018). https://doi.org/10.1111/cgf.13343
9. Kasaei, S.H., Tomé, A.M., Seabra Lopes, L., Oliveira, M.: GOOD: a global orthographic object descriptor for 3D object recognition and manipulation. Pattern Recogn. Lett. **83**, 312–320 (2016). https://doi.org/10.1016/j.patrec.2016.07.006
10. Kumar, A., Anders, K., Winiwarter, L., Höfle, B.: Feature relevance analysis for 3D point cloud classification using deep learning. ISPRS Ann. Phot. Rem. Sens. Spat. Inf. Sci. **4** (2019). https://doi.org/10.5194/isprs-annals-IV-2-W5-373-2019
11. Landrieu, L., Boussaha, M.: Supervized Segmentation with Graph-Structured Deep Metric Learning. IEEE Geosci. Remote Sens. Mag. (2019). http://arxiv.org/abs/1905.04014
12. Landrieu, L., Simonovsky, M.: Large-scale point cloud semantic segmentation with superpoint graphs. In: Proceedings of the IEEE CVPR Conference, pp. 4558–4567 (2018)
13. Lazarek, J., Pryczek, M.: A review on point cloud semantic segmentation methods. J. Appl. Comput. Sci. **26**(2), 99–106 (2018)
14. Lin, Y., Wang, C., Zhai, D., Li, W., Li, J.: Toward better boundary preserved supervoxel segmentation for 3D point clouds. ISPRS J. Phot. Rem. Sens. **143**, 39–47 (2018). https://doi.org/10.1016/j.isprsjprs.2018.05.004
15. Liu, M.Y., Tuzel, O., Ramalingam, S., Chellappa, R.: Entropy rate superpixel segmentation. In: CVPR 2011, pp. 2097–2104. IEEE (2011)
16. Lu, R., Brilakis, I., Middleton, C.R.: Detection of structural components in point clouds of existing RC bridges. Comput.-Aided Civil Infrastruct. Eng. **34**(3), 191–212 (2019). https://doi.org/10.1111/mice.12407
17. Ma, Y., Guo, Y., Lei, Y., Lu, M., Zhang, J.: 3DMAX-NET: a multi-scale spatial contextual network for 3D point cloud semantic segmentation. In: 2018 24th International Conference on Pattern Recognition (ICPR), pp. 1560–1566. IEEE (2018)
18. Maleika, W., Forczmański, P.: Adaptive modeling and compression of bathymetric data with variable density. IEEE J. Oceanic Eng. (2019). https://doi.org/10.1109/JOE.2019.2941120

19. McInnes, L., Healy, J.: Accelerated hierarchical density clustering. arXiv preprint arXiv:1705.07321 (2017)
20. McInnes, L., Healy, J., Astels, S.: HDBSCAN: hierarchical density based clustering. J. Open Source Softw. **2**(11), 205 (2017)
21. Silberman, N., Hoiem, D., Kohli, P., Fergus, R.: Indoor segmentation and support inference from RGBD images. In: Fitzgibbon, A., Lazebnik, S., Perona, P., Sato, Y., Schmid, C. (eds.) ECCV 2012. LNCS, vol. 7576, pp. 746–760. Springer, Heidelberg (2012). https://doi.org/10.1007/978-3-642-33715-4_54
22. Papon, J., Abramov, A., Schoeler, M., Worgotter, F.: Voxel cloud connectivity segmentation-supervoxels for point clouds. In: Proceedings of the IEEE Conference on Computer Vision and Pattern Recognition, pp. 2027–2034 (2013)
23. Rusu, R.B., Cousins, S.: 3D is here: point cloud library. Point Cloud Library. http://pointclouds.org/. Accessed 15 2017
24. Scherer, R.: Feature detection. In: Scherer, R. (ed.) Computer Vision Methods for Fast Image Classification and Retrieval. SCI, vol. 821, pp. 7–32. Springer, Cham (2020). https://doi.org/10.1007/978-3-030-12195-2_2
25. Sculley, D.: Web-scale k-means clustering. In: Proceedings of the 19th International Conference on WWW WWW 2010, pp. 1177–1178. ACM, New York (2010). https://doi.org/10.1145/1772690.1772862
26. Thomas, H., Goulette, F., Deschaud, J.E., Marcotegui, B.: Semantic classification of 3d point clouds with multiscale spherical neighborhoods. In: 2018 International Conference on 3D Vision (3DV), pp. 390–398. IEEE (2018)
27. Walczak, J., Wojciechowski, A.: Clustering quality measures for point cloud segmentation tasks. In: Chmielewski, L.J., Kozera, R., Orłowski, A., Wojciechowski, K., Bruckstein, A.M., Petkov, N. (eds.) ICCVG 2018. LNCS, vol. 11114, pp. 173–186. Springer, Cham (2018). https://doi.org/10.1007/978-3-030-00692-1_16
28. Walewski, P., Gałaj, T., Szajerman, D.: Heuristic based real-time hybrid rendering with the use of rasterization and ray tracing method. Open Phys. **17**(1), 527–544 (2019). https://doi.org/10.1515/phys-2019-0055
29. Wang, X., Liu, S., Shen, X., Shen, C., Jia, J.: Associatively segmenting instances and semantics in point clouds (2019)
30. Wawrzonowski, M., Szajerman, D.: Optimization of screen-space directional occlusion algorithms. Open Phys. **17**(1), 519–526 (2019). https://doi.org/10.1515/phys-2019-0054
31. Wróblewski, A., Andrzejczak, J.: Wave propagation time optimization for geodesic distances calculation using the heat method. Open Phys. **17**(1), 263–275 (2019)
32. Xie, Y., Tian, J., Zhu, X.X.: A review of point cloud semantic segmentation (2019). http://arxiv.org/abs/1908.08854
33. Xu, Y., Yao, W., Hoegner, L., Stilla, U.: Segmentation of building roofs from airborne LiDAR point clouds using robust voxel-based region growing. Remote Sens. Lett. **8**(11), 1062–1071 (2017). https://doi.org/10.1080/2150704X.2017.1349961
34. Yan, J., Shan, J., Jiang, W.: A global optimization approach to roof segmentation from airborne lidar point clouds. ISPRS J. Phot. Rem. Sens. **94**, 183–193 (2014). https://doi.org/10.1016/j.isprsjprs.2014.04.022
35. Yang, F., et al.: Automatic indoor reconstruction from point clouds in multi-room environments with curved walls. Sensors **19**(17), 3798 (2019). https://doi.org/10.3390/s19173798
36. Yin, Z., Liu, Z., Zhou, L., Zhang, F., Fu, K., Kong, X.: Superpixel based continuous conditional random field neural network for semantic segmentation. Neurocomputing **340**, 196–210 (2019). https://doi.org/10.1016/j.neucom.2019.01.016

37. Yousefhussien, M., Kelbe, D.J., Ientilucci, E.J., Salvaggio, C.: A multi-scale fully convolutional network for semantic labeling of 3D point clouds. ISPRS J. Phot. Rem. Sens. **143**, 191–204 (2018). https://doi.org/10.1016/j.isprsjprs.2018.03.018
38. Zhao, J., Liu, C., Zhang, B.: PLSTMNet: a new neural network for segmentation of point cloud. In: 2018 11th International Workshop on Human Friendly Robotics (HFR), pp. 42–47 (2019). https://doi.org/10.1109/hfr.2018.8633482

Machine Learning and Data Assimilation for Dynamical Systems

Learning Hidden States in a Chaotic System: A Physics-Informed Echo State Network Approach

Nguyen Anh Khoa Doan[1,2(✉)], Wolfgang Polifke[1], and Luca Magri[2,3]

[1] Department of Mechanical Engineering, Technical University of Munich, Garching, Germany
doan@tfd.mw.tum.de
[2] Institute for Advanced Study, Technical University of Munich, Garching, Germany
[3] Department of Engineering, University of Cambridge, Cambridge, UK

Abstract. We extend the Physics-Informed Echo State Network (PI-ESN) framework to reconstruct the evolution of an unmeasured state (hidden state) in a chaotic system. The PI-ESN is trained by using (i) data, which contains no information on the unmeasured state, and (ii) the physical equations of a prototypical chaotic dynamical system. Non-noisy and noisy datasets are considered. First, it is shown that the PI-ESN can accurately reconstruct the unmeasured state. Second, the reconstruction is shown to be robust with respect to noisy data, which means that the PI-ESN acts as a denoiser. This paper opens up new possibilities for leveraging the synergy between physical knowledge and machine learning to enhance the reconstruction and prediction of unmeasured states in chaotic dynamical systems.

Keywords: Echo state networks · Physics-Informed Echo State Networks · Chaotic dynamical systems · State reconstruction

1 Introduction

In experiments on physical systems, it is often difficult to measure all the physical states, whether it be because the instruments have a finite resolution, or because the measurement techniques have some limitations. Consequently, we are typically able to infer only a few states of the system from the measured observable quantities. The states that cannot be measured are *hidden*, that is, they may affect the system's evolution, but they cannot be straightforwardly

The authors acknowledge the support of the Technical University of Munich - Institute for Advanced Study, funded by the German Excellence Initiative and the European Union Seventh Framework Programme under grant agreement no. 291763. L.M. also acknowledges the Royal Academy of Engineering Research Fellowship Scheme.
L. Magri—(visiting) Institute for Advanced Study, Technical University of Munich, Germany.

V. V. Krzhizhanovskaya et al. (Eds.): ICCS 2020, LNCS 12142, pp. 117–123, 2020.
https://doi.org/10.1007/978-3-030-50433-5_9

measured. The accurate reconstruction of hidden states is crucial in many fields such as cardiac blood flow modelling [13], climate science [6], and fluid dynamics [2], to name only a few. For example, in fluid dynamics, measurements of the velocity field with particle image velocimetry may be limited to the in-plane two-dimensional velocity, although the three-dimensional velocity is the quantity of interest. The reconstruction of unmeasured quantities from experimental measurements has been the subject of recent studies, that used a variety of data assimilation and/or machine learning techniques. For example, spectral nudging, which combines data assimilation with physical equations, was used to infer temperature and rotation rate in 3D isotropic rotating turbulence [3]. Alternatively, [5] reconstructed the fine-scale features of an unsteady flow from large scale information by using a series of Convolutional Neural Networks. Using a similar approach, the reconstruction of the velocity from hydroxyl-radical planar laser induced fluorescence images in a turbulent flame was performed [1]. Another approach based on echo state networks has also been used for the reconstruction of time series of unmeasured states of chaotic systems [9]. While effective in reconstructing the unmeasured states, these approaches required training data with both the measured *and* unmeasured states. In this paper, we propose using physical knowledge to reconstruct hidden states in a chaotic system *without the need of any data of the unmeasured states during the training*. This is performed with the Physics-Informed Echo State Network (PI-ESN), which has been shown to accurately forecast chaotic systems [4]. The PI-ESN, and more generally *Physics-Informed* Machine Learning, relies on the physical knowledge of the system under study, in the form of its conservation equations, whose residuals are included in the loss function during the training of the machine learning framework [4,12]. These approaches, which combine physical knowledge and machine learning, have been shown to be efficient in improving the accuracy of neural networks [4,12]. Here the PI-ESN approach is applied to the Lorenz system, which is a prototypical chaotic system [8].

The paper is organized as follows. The problem statement and the methodology based on PI-ESN are detailed in Sect. 2. Then, results are presented and discussed in Sect. 3 and final comments are summarized in Sect. 4.

2   Methodology: Physics-Informed Echo State Network for Learning of Hidden States

We consider a dynamical system whose governing equations are:

$$\mathcal{F}(\boldsymbol{y}) \equiv \dot{\boldsymbol{y}} + \mathcal{N}(\boldsymbol{y}) = 0 \tag{1}$$

where $\mathcal{F}$ is a non-linear operator, $\dot{\ }$ is the time derivative and $\mathcal{N}$ is a nonlinear differential operator. Equation (1) represents a formal ordinary differential equation, which governs the dynamics of a nonlinear system. It is assumed that only a subset of the system states can be observed, which is denoted $\boldsymbol{z} \in \mathbb{R}^{N_z}$, while the hidden states are denoted $\boldsymbol{h} \in \mathbb{R}^{N_h}$. The full state vector is $\boldsymbol{y} \in \mathbb{R}^{N_y}$, which is the concatenation of z and h, i.e., $y = [z; h]$. The vectors' dimensions are related by $N_y = N_z + N_h$. The objective is to train a PI-ESN to reconstruct the hidden states, h. We assume that we have training data of the measured states $z(n)$ only, where $n = 0, 1, 2, \ldots, N_t - 1$ are the discrete time instants that span from 0 to $T = (N_t - 1)\Delta t$, where Δt is the sampling time. Thus, the specific goal for the PI-ESN is to reconstruct the hidden time series, $h(n)$, for the same time instants. To solve this problem, the PI-ESN of [4], which is based on the *data-only* ESN of [10], needs to be extended, as explained next.

The PI-ESN is composed of three main parts (Fig. 1): (i) an artificial high dimensional dynamical system, i.e., the reservoir, whose neurons' (or units') states at time n are represented by a vector, $x(n) \in \mathbb{R}^{N_x}$, representing the reservoir neuron activations; (ii) an input matrix, $W_{in} \in \mathbb{R}^{N_x \times (1 + N_u)}$, and (iii) an output matrix, $W_{out} \in \mathbb{R}^{N_y \times (N_x + N_u + 1)}$. The reservoir is coupled to the input signal, $u \in \mathbb{R}^{N_u}$, via W_{in}. A bias term is added to the input to excite the reservoir with a constant signal. The output of the PI-ESN, $\widehat{y}$, is a linear combination of the reservoir states, inputs and an additional bias:

$$\widehat{y}(n) = [\widehat{z}(n); \widehat{h}(n)] = W_{out}[x(n); u(n); 1] \tag{2}$$

where $[;]$ indicates a vertical concatenation and $\widehat{}$ denotes the predictions from the PI-ESN. The PI-ESN outputs both the measured states, $\widehat{z}$, and the hidden states, $\widehat{h}$ (Eq. (2)). The reservoir states evolve as:

$$x(n) = \tanh\left(W_{in}[u(n); 1] + W x(n - 1)\right) \tag{3}$$

where $W \in \mathbb{R}^{N_x \times N_x}$ is the recurrent weight matrix and the (element-wise) tanh function is the activation function for the reservoir neurons. Because we wish to predict a dynamical system, the input data for the PI-ESN corresponds to the measured system state at the previous time instant, $u(n) = z(n - 1)$, which is only a subset of the state vector. In the ESN approach [10], the input and recurrent matrices, W_{in} and W, are randomly initialized once and are not trained. Only W_{out} is trained. The sparse matrices W_{in} and W are constructed to satisfy the Echo State Property [10]. Following [11], W_{in} is generated such that each row of the matrix has only one randomly chosen nonzero element, which is independently taken from a uniform distribution in the interval $[-\sigma_{in}, \sigma_{in}]$. Matrix W is constructed with an average connectivity $\langle d \rangle$, and the non-zero elements are taken from a uniform distribution over the interval $[-1, 1]$. All the coefficients of W are then multiplied by a constant coefficient for the

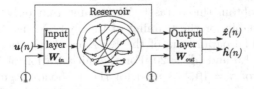

Fig. 1. Schematic of the ESN. ① indicates the bias.

largest absolute eigenvalue of $\boldsymbol{W}$, i.e. the spectral radius, to be equal to a value Λ, which is typically smaller than (or equal to) 1. To train the PI-ESN, hence $\boldsymbol{W}_{out}$, a combination of the data available and the physical knowledge of the system is used: the components of $\boldsymbol{W}_{out}$ are computed such that they minimize the sum of (i) the error between the PI-ESN prediction and the measured system states, E_d, and (ii) the physical residual, $\mathcal{F}(\widehat{\boldsymbol{y}}(n))$, on the prediction of the ESN, E_p:

$$E_{tot} = \underbrace{\frac{1}{N_t}\sum_{n=0}^{N_t-1}\frac{1}{N_z}\sum_{i=1}^{N_z}||\widehat{z}_i(n) - z_i(n)||^2}_{E_d} + \underbrace{\frac{1}{N_t}\sum_{n=0}^{N_t-1}\frac{1}{N_y}\sum_{i=1}^{N_y}||\mathcal{F}(\widehat{y}_i(n))||^2}_{E_p} \quad (4)$$

where $|| \cdot ||$ is the Euclidean norm. The training of the PI-ESN for the reconstruction of hidden states is initialized as follows. Matrix $\boldsymbol{W}_{out}$ is split into two partitions $\boldsymbol{W}_{z,out}$ and $\boldsymbol{W}_{h,out}$, i.e. $\boldsymbol{W}_{out} = [\boldsymbol{W}_{z,out}; \boldsymbol{W}_{h,out}]$, which are responsible for the prediction of the observed states, $\widehat{\boldsymbol{z}} = \boldsymbol{W}_{z,out}[\boldsymbol{x}(n); \boldsymbol{u}(n); 1]$, and the hidden states, $\widehat{\boldsymbol{h}} = \boldsymbol{W}_{h,out}[\boldsymbol{x}(n); \boldsymbol{u}(n); 1]$, respectively. $\boldsymbol{W}_{z,out}$ is initialized by Ridge regression of the data available for the measured states

$$\boldsymbol{W}_{z,out} = \boldsymbol{Z}\boldsymbol{X}^T\left(\boldsymbol{X}\boldsymbol{X}^T + \gamma\boldsymbol{I}\right)^{-1} \quad (5)$$

where $\boldsymbol{Z}$ and $\boldsymbol{X}$ are respectively the horizontal concatenation of the measured states, $\boldsymbol{z}(n)$, and associated ESN states, inputs signals and biases, $[\boldsymbol{x}(n); \boldsymbol{u}(n); 1]$ at the different time instants during training; γ is the Tikhonov regularization factor [10]; and $\boldsymbol{I}$ is the identity matrix. Matrix $\boldsymbol{W}_{h,out}$ is randomly initialized to provide an initial guess for the optimization of $\boldsymbol{W}_{out}$. The optimization process modifies the components of $\boldsymbol{W}_{out}$ to obtain the hidden states, while ensuring that the predictions on the hidden states satisfy the physical equations. The optimization is performed with a stochastic gradient method (the Adam-optimizer [7]) with a learning rate of 0.0001.

3   Results and Discussions

The approach described in Sect. 2 is tested for the reconstruction of the chaotic Lorenz system, which is described by [8]:

$$\dot{\phi}_1 = \sigma(\phi_2 - \phi_1), \quad \dot{\phi}_2 = \phi_1(\rho - \phi_3) - \phi_2, \quad \dot{\phi}_3 = \phi_1\phi_2 - \beta\phi_3 \quad (6)$$

where $\rho = 28$, $\sigma = 10$ and $\beta = 8/3$. The size of the training dataset is $N_t = 20000$ with a timestep between two time instants of $\Delta t = 0.01$. An explicit Euler scheme is used to obtain this dataset. We assume that only measurements of ϕ_1 and ϕ_2 are available for the training of the PI-ESN and the state ϕ_3 is to be reconstructed. The parameters of the reservoir of the PI-ESN are taken to be: $\sigma_{in} = 1.0$, $\Lambda = 1.0$ and $\langle d \rangle = 20$. For the initialization of $\boldsymbol{W}_{z,out}$ via Ridge regression, a value of $\gamma = 10^{-6}$ is used for the Tikhonov regularization. These values of the hyperparameters are taken from previous studies [9], who performed a grid search.

3.1 Reconstruction of Hidden States

In Fig. 2 where the time is normalized by the largest Lyapunov exponent, $\lambda_{max} = 0.934$, the reconstructed ϕ_3 time series is shown for the last 10% of the training data for PI-ESNs with reservoirs of 50 and 600 units. (The dominant Lyapunov exponent is the exponential divergence rate of two system trajectories, which are initially infinitesimally close to each other.) The small PI-ESN (50 units) can satisfactorily reconstruct the hidden state, ϕ_3. The accuracy slightly deteriorates when ϕ_3 has very large minima or maxima (e.g., $\lambda_{max}t = 202$). However, the large PI-ESN (600 units) shows an improved accuracy. The ability of the PI-ESN to reconstruct ϕ_3, which is not present in the training data, is a key-result. The reconstruction is enabled exclusively by the knowledge of the physical equation, which is constrained into the training of the PI-ESN. This constraint allows the PI-ESN to deduce the evolution of ϕ_3 from ϕ_1 and ϕ_2. Conversely, with neither the physical equation nor *training data for* ϕ_3, a data-only ESN cannot learn and reconstruct ϕ_3 because it has no information on it.

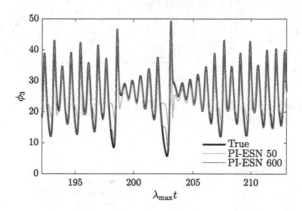

Fig. 2. Reconstruction of ϕ_3.

3.2 Effect of Noise

As the ultimate objective is to work with real-world experimental data, the effect of noise on the results is investigated. The training data for ϕ_1 and ϕ_2 are modified by adding Gaussian noise to the original signal to imitate additive measurements noise. Two Signal-to-Noise Ratios (SNRs) of 20 dB and 40 dB are considered. The results of the reconstructed ϕ_3 time series from the PI-ESN trained with the noisy training data are presented in Fig. 3. Despite the presence of noise in the training data, the PI-ESN well reconstructs the non-noisy ϕ_3 signal. This means that the physical constraints in Eq. (4) act as a physics-based smoother (or denoiser) of the noisy data. This can be appreciated also in the *prediction* of measured states. Figure 3b shows the prediction of state

ϕ_1: the non-noisy original data (full black line) and the prediction from the PI-ESN (dashed red line) overlap. This means that the PI-ESN provides a denoised prediction after training. Finally, Fig. 4 shows the root mean squared error of the reconstructed hidden state $\widehat{\phi_3}$, $RMSE = \sqrt{\frac{1}{N_t} \sum_{n=0}^{N_t-1} (\phi_3(n) - \widehat{\phi_3}(n))^2}$, for PI-ESNs of different reservoir sizes and noise levels, where $\phi_3(n)$ is the reference non-noisy data, which we wish to recover. For the non-noisy case, there is a large decrease in the RMSE when the PI-ESNs has 300 units (or more). With noise, the performance between the non-noisy and low-noise (SNR = 40 dB) cases are similar, whereas for a larger noise level (SNR = 20 dB), a larger reservoir is required to keep the RMSE small, as it may be expected. This suggests that the PI-ESN approach may be robust with respect to noise.

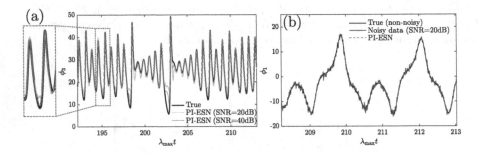

Fig. 3. (a) Reconstruction of ϕ_3 with PI-ESN of 600 units trained from noisy data (with zoomed inset). (b) Prediction of ϕ_1.

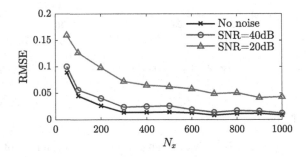

Fig. 4. RMSE of the reconstructed ϕ_3 time series in the training data.

4   Conclusions and Future Directions

We extend the Physics-Informed Echo State Network to reconstruct the hidden states in a chaotic dynamical system. The approach combines the knowledge of the system's physical equations and a small dataset. It is shown, on a prototypical chaotic system, that this method can (i) accurately reconstruct the hidden states; (ii) accurately reconstruct the states with training data contaminated by noise; and (iii) provide a physics-based smoothing of the noisy measured data. Compared to other reconstruction approaches, the proposed framework does not require any data of the hidden states during training. This has the potential to enable the reconstruction of unmeasured quantities in experiments of higher dimensional chaotic systems, such as fluids. This is being explored in on-going studies. Future work also aims at assessing the effect of imperfect physical knowledge on the reconstruction of the hidden states.

This paper opens up new possibilities for the reconstruction and prediction of unsteady dynamics from partial and noisy measurements.

References

1. Barwey, S., Hassanaly, M., Raman, V., Steinberg, A.: Using machine learning to construct velocity fields from OH-PLIF images. Combust. Sci. Technol. 1–24 (2019)
2. Brenner, M.P.: Perspective on machine learning for advancing fluid mechanics. Phys. Rev. Fluids 4(10), 100501 (2019)
3. Clark Di Leoni, P., Mazzino, A., Biferale, L.: Inferring flow parameters and turbulent configuration with physics-informed data assimilation and spectral nudging. Phys. Rev. Fluids 3, 104604 (2018)
4. Doan, N.A.K., Polifke, W., Magri, L.: Physics-informed echo state networks for chaotic systems forecasting. In: Rodrigues, J.M.F., et al. (eds.) ICCS 2019. LNCS, vol. 11539, pp. 192–198. Springer, Cham (2019). https://doi.org/10.1007/978-3-030-22747-0_15
5. Fukami, K., Fukagata, K., Taira, K.: Super-resolution reconstruction of turbulent flows with machine learning. J. Fluid Mech. 870, 106–120 (2019)
6. Kalnay, E.: Atmospheric Modeling, Data Assimilation, and Predictability. Cambridge University Press, Cambridge (2003)
7. Kingma, D.P., Ba, J.L.: Adam: a method for stochastic optimization. In: 3rd International Conference on Learning Representations, ICLR 2015 - Conference Track Proceedings, pp. 1–15 (2015)
8. Lorenz, E.N.: Deterministic nonperiodic flow. J. Atmos. Sci. 20(2), 130–141 (1963)
9. Lu, Z., Pathak, J., Hunt, B., Girvan, M., Brockett, R., Ott, E.: Reservoir observers: model-free inference of unmeasured variables in chaotic systems. Chaos 27(4), 041102 (2017)
10. Lukoševičius, M., Jaeger, H.: Reservoir computing approaches to recurrent neural network training. Comput. Sci. Rev. 3(3), 127–149 (2009)
11. Pathak, J., et al.: Hybrid forecasting of chaotic processes: using machine learning in conjunction with a knowledge-based model. Chaos 28(4), 041101 (2018)
12. Raissi, M., Perdikaris, P., Karniadakis, G.: Physics-informed neural networks: a deep learning framework for solving forward and inverse problems involving nonlinear partial differential equations. J. Comput. Phys. 378, 686–707 (2019)
13. Sankaran, S., Moghadam, M.E., Kahn, A.M., Tseng, A.M., Guccione, J.M.: Patient-specific multiscale modeling of blood flow for coronary artery bypass graft surgery. Ann. Biomed. Eng. 40(10), 2228–2242 (2012)

Learning Ergodic Averages in Chaotic Systems

Francisco Huhn[1] and Luca Magri[1,2(✉)]

[1] Department of Engineering, University of Cambridge, Cambridge, UK
lm547@cam.ac.uk
[2] Institute for Advanced Study, Technical University of Munich, Munich, Germany

Abstract. We propose a physics-informed machine learning method to predict the time average of a chaotic attractor. The method is based on the hybrid echo state network (hESN). We assume that the system is ergodic, so the time average is equal to the ergodic average. Compared to conventional echo state networks (ESN) (purely data-driven), the hESN uses additional information from an incomplete, or imperfect, physical model. We evaluate the performance of the hESN and compare it to that of an ESN. This approach is demonstrated on a chaotic time-delayed thermoacoustic system, where the inclusion of a physical model significantly improves the accuracy of the prediction, reducing the relative error from 48% to 1%. This improvement is obtained at the low extra cost of solving a small number of ordinary differential equations that contain physical information. This framework shows the potential of using machine learning techniques combined with prior physical knowledge to improve the prediction of time-averaged quantities in chaotic systems.

Keywords: Echo state networks · Hybrid echo state networks · Physics-informed echo state networks · Chaotic dynamical systems

1 Introduction

In the past decade, there has been a proliferation of machine learning techniques applied in various fields, from spam filtering [7] to self-driving cars [3], including the more recent physical applications in fluid dynamics [4,6]. However, a major hurdle in applying machine learning to complex physical systems, such as those in fluid dynamics, is the high cost of generating data for training [6].

L. Magri—Technical University of Munich, Germany (Visiting)

F.H. acknowledges the support from Fundação para a Ciência e Tecnologia under the Research Studentship no. SFRH/BD/134617/2017. L.M. acknowledges the support of the Technical University of Munich – Institute for Advanced Study, funded by the German Excellence Initiative and the European Union Seventh Framework Programme under grant agreement no. 291763.

V. V. Krzhizhanovskaya et al. (Eds.): ICCS 2020, LNCS 12142, pp. 124–132, 2020.
https://doi.org/10.1007/978-3-030-50433-5_10

Nevertheless, this can be mitigated by leveraging prior knowledge (e.g. physical laws). Physical knowledge can compensate for the small amount of training data. These approaches, called physics-informed machine learning, have been applied to various problems in fluid dynamics [4,6]. For example, [5,14] improve the predictability horizon of echo state networks by leveraging physical knowledge, which is enforced as a *hard* constraint in [5], without needing more data or neurons. In this study, we use a hybrid echo state network (hESN) [14], originally proposed to time-accurately forecast the evolution of chaotic dynamical systems, to predict the long-term time averaged quantities, i.e., the ergodic averages. This is motivated by recent research in optimization of chaotic multiphysics fluid dynamics problems with applications to thermoacoustic instabilities [8]. The hESN is based on reservoir computing [10], in particular, conventional Echo State Networks (ESNs). ESNs have shown to predict nonlinear and chaotic dynamics more accurately and for a longer time horizon than other deep learning algorithms [10]. However, we stress that the present study is not focused on the accurate prediction of the time evolution of the system, but rather of its ergodic averages, which are obtained by the time averaging of a long time series (we implicitly assume that the system is ergodic, thus, the infinite time average is equal to the ergodic average [2].). Here, the physical system under study is a prototypical time-delayed thermoacoustic system, whose chaotic dynamics have been analyzed and optimized in [8].

2   Echo State Networks

The ESN approach presented in [11] is used here. The ESN is given an input signal $u(n) \in \mathbb{R}^{N_u}$, from which it produces a prediction signal $\hat{y}(n) \in \mathbb{R}^{N_y}$ that should match the target signal $y(n) \in \mathbb{R}^{N_y}$, where n is the discrete time index. The ESN is composed of a reservoir, which can be represented as a directed weighted graph with N_x nodes, called neurons, whose state at time n is given by the vector $x(n) \in \mathbb{R}^{N_x}$. The reservoir is coupled to the input via an input-to-reservoir matrix, W_{in}, such that its state evolves according to

$$x(n) = \tanh(W_{\mathrm{in}}u(n) + Wx(n-1)), \tag{1}$$

where $W \in \mathbb{R}^{N_x} \times \mathbb{R}^{N_x}$ is the weighted adjacency matrix of the reservoir, i.e. W_{ij} is the weight of the edge from node j to node i, and the hyperbolic tangent is the activation function. Finally, the prediction is produced by a linear combination of the states of the neurons

$$\hat{y}(n) = W_{\mathrm{out}}x(n), \tag{2}$$

where $W_{\mathrm{out}} \in \mathbb{R}^{N_y} \times \mathbb{R}^{N_x}$. In this work, we are interested in dynamical system prediction. Thus, the target at time step n is the input at time step $n + 1$, i.e. $y(n) = u(n + 1)$ [14]. We wish to learn ergodic averages, given by

$$\langle \mathcal{J} \rangle = \lim_{T \to \infty} \frac{1}{T} \int_0^T \mathcal{J}(u(t)) \, dt, \tag{3}$$

where $\mathcal{J}$ is a cost functional, of a dynamical system governed by

$$\dot{u} = F(u), \tag{4}$$

where $u \in \mathbb{R}^{N_u}$ is the state vector and F is a nonlinear operator. The training data is obtained via numerical integration of Eq. (4), resulting in the time series $\{u(1), \ldots, u(N_t)\}$, where the different samples are taken at equally spaced time intervals Δt, and N_t is the length of the training data set. In the conventional ESN approach, W_{in} and W are generated once and fixed. Then, W is re-scaled to have the desired spectral radius, ρ, to ensure that the network satisfies the Echo State Property [9]. Only W_{out} is trained to minimize the mean-squared-error

$$E_d = \frac{1}{N_y} \sum_{i=1}^{N_y} \frac{1}{N_t} \sum_{n=1}^{N_t} (\hat{y}_i(n) - y_i(n))^2. \tag{5}$$

To avoid overfitting, we use ridge regularization, so the optimization problem is

$$\min_{W_{\text{out}}} E_d + \gamma \|W_{\text{out}}\|^2, \tag{6}$$

where γ is the regularization factor. Because the prediction $\hat{y}(n)$ is a linear combination of the reservoir state $x(n)$, the optimal W_{out} can be explicitly obtained with

$$W_{\text{out}} = Y X^T (X X^T + \gamma I)^{-1}, \tag{7}$$

where I is the identity matrix and Y and X are the column-concatenation of the various time instants of the output data, y, and corresponding reservoir states, x, respectively. After the optimal W_{out} is found, the ESN can be used to predict the time evolution of the system. This is done by looping back its output to its input, i.e. $u(n) = \hat{y}(n-1) = W_{\text{out}} x(n-1)$, which, on substitution into Eq. (1), results in

$$x(n) = \tanh(\widetilde{W} x(n-1)), \tag{8}$$

with $\widetilde{W} = W + W_{\text{in}} W_{\text{out}}$. Interestingly, Eq. (8) shows that if the reservoir follows an evolution of states $x(1), \ldots, x(N_p)$, where N_p is the number of prediction steps, then $-x(1), \ldots, -x(N_p)$ is also possible, because flipping the sign of x in Eq. (8) results in the same equation. This implies that either the attractor of the ESN (if any) is symmetric, i.e. if some x is in the ESN's attractor, then so is $-x$; or the ESN has a co-existing symmetric attractor. While this seemed not to have been an issue in short-term prediction, such as in [5], it does pose a problem in the long-term prediction of statistical quantities. This is because the ESN, in its present form, *can not* generate non-symmetric attractors. This symmetry needs to be broken to work with a general non-symmetric dynamical system. This can be done by including biases [10]. However, the addition of a bias can make the reservoir prone to saturation (results not shown), i.e. $x_i \to \pm 1$, and thus care needs to be taken in the choice of hyperparameters. In this paper, we break the symmetry by exploiting prior knowledge on the physics of the problem under investigation with a hybrid ESN.

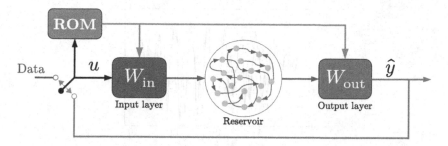

Fig. 1. Schematic of the hybrid echo state network. In training mode, the input of the network is the training data (switch is horizontal). In prediction mode, the input of the network is its output from the previous time step (switch is vertical).

3   Physics-Informed and Hybrid Echo State Network

The ESN's performance can be increased by incorporating physical knowledge during training [5] or during training and prediction [14]. This physical knowledge is usually present in the form of a reduced-order model (ROM) that can generate (imperfect) predictions. The authors of [5] introduced a physics-informed ESN (PI-ESN), which constrains the physics as a *hard* constraint with a physics loss term. The prediction is consistent with the physics, but the training requires nonlinear optimization. The authors of [14] introduced a hybrid echo state network (hESN), which incorporates incomplete physical knowledge by feeding the prediction of the physical model into the reservoir and into the output. This requires ridge regression. Here, we use an hESN (Fig. 1) because we are not interested in constraining the physics as a hard constraint for an accurate short-term prediction [5]. In the hESN, similarly to the conventional ESN, the input is fed to the reservoir via the input layer W_{in}, but also to a physical model, which is usually a set of ordinary differential equations that approximately describe the system that is to be predicted. In this work, that model is a reduced-order model (ROM) of the full system. The output of the ROM is then fed to the reservoir via the input layer and into the output of the network via the output layer.

4   Learning the Ergodic Average of an Energy

We use a prototypical time-delayed thermoacoustic system composed of a longitudinal acoustic cavity and a heat source modelled with a nonlinear time-delayed model [8,12,16], which has been used to optimize ergodic averages in [8] with a dynamical systems approach. The non-dimensional governing equations are

$$\partial_t u + \partial_x p = 0, \quad \partial_t p + \partial_x u + \zeta p - \dot{q}\delta(x - x_f) = 0, \tag{9}$$

where u, p, ζ and $\dot{q}$ are the non-dimensionalized acoustic velocity, pressure, damping and heat-release rate, respectively. δ is the Dirac delta. These equations

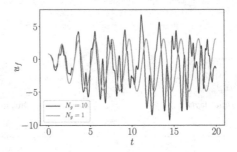

Fig. 2. Acoustic velocity at the flame location.

are discretized by using N_g Galerkin modes

$$u(x,t) = \sum_{j=1}^{N_g} \eta_j(t) \cos(j\pi x), \quad p(x,t) = -\sum_{j=1}^{N_g} \mu_j(t) \sin(j\pi x), \qquad (10)$$

which results in a system of $2N_g$ oscillators, which are nonlinearly coupled through the heat released by the heat source

$$\dot{\eta}_j - j\pi\mu_j = 0, \quad \dot{\mu}_j + j\pi\eta_j + \zeta_j\mu_j + 2\dot{q}\sin(j\pi x_f) = 0, \qquad (11)$$

where $x_f = 0.2$ is the heat source location and $\zeta_j = 0.1j + 0.06j^{1/2}$ is the modal damping [8]. The heat release rate, $\dot{q}$, is given by the modified King's law [8], $\dot{q}(t) = \beta[(1 + u(x_f, t - \tau))^{1/2} - 1]$, where β and τ are the heat release intensity parameter and the time delay, respectively. With the nomenclature of Sect. 2, $\boldsymbol{y}(n) = (\eta_1; \ldots; \eta_{N_g}; \mu_1; \ldots; \mu_{N_g})$. Using 10 Galerkin modes ($N_g = 10$), $\beta = 7.0$ and $\tau = 0.2$ results in a chaotic motion (Fig. 2), with the leading Lyapunov exponent being $\lambda_1 \approx 0.12$ [8]. (The leading Lyapunov exponent measures the rate of (exponential) separation of two close initial conditions, i.e. an initial separation $\|\boldsymbol{\delta u}_0\|$ grows asymptotically like $\|\boldsymbol{\delta u}_0\|e^{\lambda_1 t}$.) However, for the same choice of parameter values, the solution with $N_g = 1$ is a limit cycle (i.e. a periodic solution).

The echo state network is trained on data generated with $N_g = 10$, while the physical knowledge (ROM in Fig. 1) is generated with $N_g = 1$ only. We wish to predict the time average of the instantaneous acoustic energy,

$$E_{ac}(t) = \int_0^1 \frac{1}{2}(u^2 + p^2)\,dx, \qquad (12)$$

which is a relevant metric in the optimization of thermoacoustic systems [8]. The reservoir is composed of 100 units, a modest size, half of which receive their input from $\boldsymbol{u}$, while the other half receives it from the output of the ROM, $\hat{\boldsymbol{y}}_{\text{ROM}}$. The entries of $\boldsymbol{W}_{\text{in}}$ are randomly generated from the uniform distribution unif$(-\sigma_{\text{in}}, \sigma_{\text{in}})$, where $\sigma_{\text{in}} = 0.2$. The matrix $\boldsymbol{W}$ is highly sparse, with only 3% of non-zero entries from the uniform distribution unif$(-1, 1)$. Finally, $\boldsymbol{W}$ is scaled such that its spectral radius, ρ, is 0.1 and 0.3 for the ESN and the hESN, respectively. The time step is $\Delta t = 0.01$. The network is trained for $N_t = 5000$ units, which corresponds to 6 Lyapunov times, i.e. $6\lambda_1^{-1}$. The data is generated by integrating Eq. (11) in time with $N_g = 10$, resulting in $N_u = N_y = 20$. In the hESN, the ROM is obtained by integrating the same equations, but with $N_g = 1$ (one Galerkin mode only) unless otherwise stated. Ridge regression is performed with $\gamma = 10^{-7}$. The values of the hyperparameters are taken from the literature [5,14] and a grid search, which, while not the most efficient, is well suited when there are few hyperparameters, such as this work's ESN architecture.

On the one hand, Fig. 3a shows the instantaneous error of the first modes of the acoustic velocity and pressure $(\eta_1; \mu_1)$ for the ESN, hESN and ROM. None of these can accurately predict the instantaneous state of the system. On the other hand, Fig. 3b shows the error of the prediction of the average acoustic energy. Once again, the ROM alone does a poor job at predicting the statistics of the system, with an error of 50%. This should not come at a surprise since, as discussed previously, the ROM does not even produce a chaotic solution. The ESN, trained on data only, performs marginally better, with an error of 48%. In contrast, the hESN predicts the time-averaged acoustic energy satisfactorily, with an error of about 7%. This is remarkable, since both the ESN and the ROM do a poor job at predicting the average acoustic energy. However, when the ESN is combined with prior knowledge from the ROM, the prediction becomes significantly better. Moreover, while the hESN's error still decreases at the end of the prediction period, $t = 250$, which is 5 times the training data time, the ESN and the ROM stabilize much earlier, at a time similar to that of the training data. This result shows that complementing the ESN with a cheap physical model (only 10% the number of degrees of freedom of the full system) can greatly improve the accuracy of the predictions, with no need for more data or neurons. Figure 3c shows the relative error as a function of the number of Galerkin modes in the ROM, which is a proxy for the quality of the model. For each N_g, we take the median of 16 reservoir realizations. As expected, as the quality of the model increases, so does the quality of the prediction. This effect is most noticeable from $N_g = 1$ to 4, with the curve presenting diminishing returns. The downside of increasing N_g is obviously the increase in computational cost. At $N_g = 10$, the original system is recovered. However, the error does not tend exactly to 0 because $\boldsymbol{W}_{\text{out}}$ can not combine the ROM's output only (i.e. 0 entries for reservoir nodes) due to: i) the regularization factor in ridge regression that penalizes large entries; ii) numerical error. This graph further strengthens the point previously made that cheap physical models can greatly improve the prediction of physical systems with data techniques.

We stress that the optimal values of hyperparameters for a certain set of physical parameters, e.g. (β_1, τ_1), might not be optimal for a different set of physical parameters (β_2, τ_2). This should not be surprising, since different physical parameters will result in different attractors. For example, Fig. 4 shows that changing the physical parameters from $(\beta = 7.0, \tau = 0.2)$ to $(\beta = 6.0, \tau = 0.3)$ results in a change of type of attractor from chaotic to limit cycle. For the hESN to predict the limit cycle, the value of σ_{in} must change from 0.2 to 0.03 Thus,

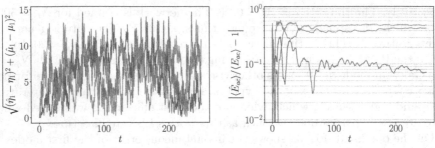

(a) Absolute error of $(\eta_1; \mu_1)$ prediction.   (b) Relative error of $\langle E_{ac} \rangle$ prediction.

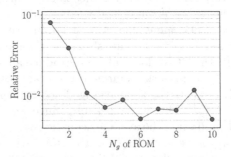

(c) Relative error vs number of Galerkin
modes of the ROM.

Fig. 3. Errors on the prediction from ESN (blue), hESN (red) and ROM (green). (Color figure online)

if the hESN (or any deep learning technique in general) is to be used to predict the dynamics of various physical configurations (e.g. the generation of a bifurcation diagram), then it should be coupled with a robust method for the automatic selection of optimal hyperparameters [1], with a promising candidate being Bayesian optimization [15, 17].

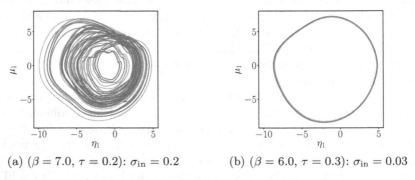

(a) $(\beta = 7.0, \tau = 0.2)$: $\sigma_{in} = 0.2$ (b) $(\beta = 6.0, \tau = 0.3)$: $\sigma_{in} = 0.03$

Fig. 4. Phase plot of system (black) and hESN (red) for two sets of (β, τ). (a) chaotic solution; (b) periodic solution. (Color figure online)

5 Conclusion and Future Directions

We propose the use of echo state networks informed with incomplete prior physical knowledge for the prediction of time averaged cost functionals in chaotic dynamical systems. We apply this to chaotic acoustic oscillations, which is relevant to aeronautical propulsion. The inclusion of physical knowledge comes at a low cost and significantly improves the performance of conventional echo state networks from a 48% error to 1%, without requiring additional data or neurons. This improvement is obtained at the low extra cost of solving a small number of ordinary differential equations that contain physical information. The ability of the proposed ESN can be exploited in the optimization of chaotic systems by accelerating computationally expensive shadowing methods [13]. For future work, (i) the performance of the hybrid echo state network should be compared against those of other physics-informed machine learning techniques; (ii) robust methods for hyperparameters' search should be coupled for a "hands-off" autonomous tool; and (iii) this technique is currently being applied to larger scale problems. In summary, the proposed framework is able to learn the ergodic average of a fluid dynamics system, which opens up new possibilities for the optimization of highly unsteady problems.

References

1. Bengio, Y.: Practical recommendations for gradient-based training of deep architectures. In: Montavon, G., Orr, G.B., Müller, K.-R. (eds.) Neural Networks: Tricks of the Trade. LNCS, vol. 7700, pp. 437–478. Springer, Heidelberg (2012). https://doi.org/10.1007/978-3-642-35289-8_26
2. Birkhoff, G.D.: Proof of the ergodic theorem. Proc. Natl. Acad. Sci. **17**(12), 656–660 (1931)
3. Bojarski, M., et al.: End to end learning for self-driving cars. arXiv e-prints, arXiv:1604.07316 (2016)
4. Brunton, S.L., Noack, B.R., Koumoutsakos, P.: Machine learning for fluid mechanics. Annu. Rev. Fluid Mech. **52**(1), 477–508 (2020)
5. Doan, N.A.K., Polifke, W., Magri, L.: Physics-informed echo state networks for chaotic systems forecasting. In: Rodrigues, J.M.F., Cardoso, P.J.S., Monteiro, J., Lam, R., Krzhizhanovskaya, V.V., Lees, M.H., Dongarra, J.J., Sloot, P.M.A. (eds.) ICCS 2019. LNCS, vol. 11539, pp. 192–198. Springer, Cham (2019). https://doi.org/10.1007/978-3-030-22747-0_15
6. Duraisamy, K., Iaccarino, G., Xiao, H.: Turbulence modeling in the age of data. Annu. Rev. Fluid Mech. **51**(1), 357–377 (2019)
7. Guzella, T.S., Caminhas, W.M.: A review of machine learning approaches to spam filtering. Expert Syst. Appl. **36**(7), 10206–10222 (2009)
8. Huhn, F., Magri, L.: Stability, sensitivity and optimisation of chaotic acoustic oscillations. J. Fluid Mech. **882**, A24 (2020)
9. Jaeger, H., Haas, H.: Harnessing nonlinearity: predicting chaotic systems and saving energy in wireless communication. Science **304**(5667), 78–80 (2004)
10. Lukoševičius, M., Jaeger, H.: Reservoir computing approaches to recurrent neural network training. Comput. Sci. Rev. **3**(3), 127–149 (2009)

11. Lukoševičius, M.: A practical guide to applying echo state networks. In: Montavon, G., Orr, G.B., Müller, K.-R. (eds.) Neural Networks: Tricks of the Trade. LNCS, vol. 7700, pp. 659–686. Springer, Heidelberg (2012). https://doi.org/10.1007/978-3-642-35289-8_36

12. Nair, V., Sujith, R.: A reduced-order model for the onset of combustion instability: physical mechanisms for intermittency and precursors. Proc. Combust. Inst. **35**(3), 3193–3200 (2015)

13. Ni, A., Wang, Q.: Sensitivity analysis on chaotic dynamical systems by non-intrusive least squares shadowing (NILSS). J. Comput. Phys. **347**, 56–77 (2017)

14. Pathak, J.: Hybrid forecasting of chaotic processes: using machine learning in conjunction with a knowledge-based model. Chaos: Interdisc. J. Nonlinear Sci. **28**(4), 41101 (2018)

15. Reinier Maat, J., Gianniotis, N., Protopapas, P.: Efficient optimization of echo state networks for time series datasets. arXiv e-prints, arXiv:1903.05071 (2019)

16. Traverso, T., Magri, L.: Data assimilation in a nonlinear time-delayed dynamical system with lagrangian optimization. In: Rodrigues, J.M.F., Cardoso, P.J.S., Monteiro, J., Lam, R., Krzhizhanovskaya, V.V., Lees, M.H., Dongarra, J.J., Sloot, P.M.A. (eds.) ICCS 2019. LNCS, vol. 11539, pp. 156–168. Springer, Cham (2019). https://doi.org/10.1007/978-3-030-22747-0_12

17. Yperman, J., Becker, T.: Bayesian optimization of hyper-parameters in reservoir computing. arXiv e-prints, arXiv:1611.05193 (2016)

Supermodeling: The Next Level of Abstraction in the Use of Data Assimilation

Marcin Sendera[1](✉), Gregory S. Duane[2,3], and Witold Dzwinel[4]

[1] Faculty of Mathematics and Computer Science, Jagiellonian University, Kraków, Poland
marcin.sendera@ii.uj.edu.pl
[2] University of Bergen, Bergen, Norway
[3] University of Colorado, Boulder, CO, USA
gregory.duane@colorado.edu
[4] AGH-UST, Kraków, Poland
dzwinel@agh.edu.pl

Abstract. Data assimilation (DA) is a key procedure that synchronizes a computer model with real observations. However, in the case of overparametrized complex systems modeling, the task of parameter-estimation through data assimilation can expand exponentially. It leads to unacceptable computational overhead, substantial inaccuracies in parameter matching, and wrong predictions. Here we define a *Supermodel* as a kind of ensembling scheme, which consists of a few *sub-models* representing various instances of the baseline model. The *sub-models* differ in parameter sets and are synchronized through couplings between the most sensitive dynamical variables. We demonstrate that after a short pretraining of the fully parametrized small *sub-model* ensemble, and then training a few latent parameters of the low-parameterized *Supermodel*, we can outperform in efficiency and accuracy the baseline model matched to data by a classical DA procedure.

Keywords: Data assimilation · Supermodeling · Dynamical systems

1 Introduction

Classical data assimilation (DA) procedure, which synchronizes a computer model with a real phenomenon through a set of observations, is an ill-posed inverse problem and suffers from the *curse of dimensionality* issue when used to estimate model parameters. That is, the time complexity of DA methods grows exponentially with the number of parameters and makes them helpless in the face of multiscale and sophisticated models such as models of climate&weather dynamics or tumor evolution (e.g. [12,13,21,30,37,38]). Our idea is to assimilate data to a hierarchically organized *Supermodel*[1] in which the

---

[1] See the *Chaos* Focus Issue introduced in [11] for the origin and history of supermodeling.

© Springer Nature Switzerland AG 2020
V. V. Krzhizhanovskaya et al. (Eds.): ICCS 2020, LNCS 12142, pp. 133–147, 2020.
https://doi.org/10.1007/978-3-030-50433-5_11

number of *trainable* metaparameters is much smaller than the number of *fixed* parameters in the *sub-models*, which themselves have to be trained in the usual schemes. We define the *Supermodel* as an ensemble of M imperfect *sub-models* μ, $\mu = 1 \ldots M$, synchronized with each other through d dynamic variables and coupled to reality by observed data. Each *sub-model* is described by a set of differential equations (ordinary ones or parabolic partial ones) for the state vectors $\mathbf{x}_\mu = (x_\mu^1, \ldots, x_\mu^i, \ldots, x_\mu^d)$, such that:

$$\dot{x}_\mu^i = f_\mu^i(\mathbf{x}_\mu) + \sum_{\nu \neq \mu} C_{\mu\nu}^i(x_\nu^i - x_\mu^i) + K^i(x_{GT}^i - x_\mu^i) \tag{1}$$

$$\mathbf{x}_s(t, \mathbf{C}) \equiv \frac{1}{M} \sum_\mu \mathbf{x}_\mu(t, \mathbf{C}), \tag{2}$$

where the coefficients $C_{\mu\nu}^i$ of tensor $\mathbf{C}$ are the coupling factors synchronizing the *sub-models*, K is a set of assimilation rates "attracting" the synchronized *Supermodel* to the *ground truth* (GT) observations x_{GT}, and $\mathbf{x}_s(.)$ is the *Supermodel* output calculated as the ensemble average.

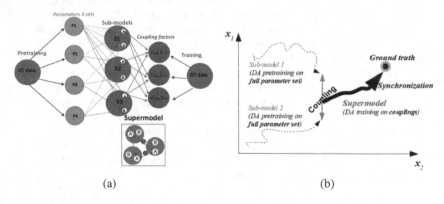

(a)   (b)

Fig. 1. (a) A *Supermodeling* scheme in which the *sub-models* are explicitly pre-trained and the inter-model couplings are trained without nudging the *sub-models* towards GT (as has been proposed also e.g., in [32]). We have assumed that the *sub-models* are coupled through only a single, the most sensitive, dynamical variable and the coupling factors $C_{\mu\nu}^i$ are matched to data by using a classical DA procedure ($K^i = 0$). (b) The concept of data adaptation by *Supermodeling*.

Unlike some previous applications of *Supermodeling* in climatology [29, 30], used for increasing the climate/weather forecast accuracy by relatively tight coupling ($\mathbf{C}$ is dense) of a few very complex and heterogenous climate models, we propose to explore *Supermodeling* from a somewhat different perspective. To this end, let us assume that the *Supermodel* is an ensemble of a few (here $M = 3$) homogeneous instances of the reference (baseline) model (see Fig. 1a). The *sub-models* are represented by pretrained (e.g., using a classical DA procedure) baseline models. This quick pretraining can be performed: (1) independently for M sub-models, each starting from different initial parameters or (2) exploiting M local minima of the loss function $F(\|\mathbf{x} - \mathbf{x_{GT}}\|)$ found during initial phases of a classical DA procedure for a single, initially parametrized sub-model. Though the second option is more elegant and efficient computationally, we have chosen the first one to assure a greater diversity of the sub-models. Let us also assume that the sub-models are coupled through only one - the most sensitive - dynamical variable i.e., $\mathbf{C}$ is sparse and $C^i_{\mu\nu} \neq 0$ only for $i = 1$ (see Fig. 1a). In addition, we refrain from attracting the Supermodel to GT via the assimilation rates K^i so we assume that $K^i = 0$ for $i = 1 \ldots d$ in Eq. 1. Instead, a classical DA algorithm (here ABC-SMC) will be employed directly for adaption of only $\mathbf{C}$ (latent parameters) to the GT data. Because of a small number of the coupling factors $\mathbf{C}$, we have expected that this training procedure will be very fast. We summarize our contribution as follows:

1. We propose a novel modeling methodology, which uses the Supermodeling scheme as a higher level of abstraction in the use of existing DA procedures. Our approach radically speeds up the process of model training. That is, just as DA estimates states and parameters by coupling the model to a "real" system, supermodeling allows a small set of different models to assimilate data from one another; only the inter-model coupling parameters need be estimated.
2. For better synchronization of the sub-models, we propose their fast pretraining by employing a classical DA scheme. In previous work [13], the arbitrary parametrization of the sub-models often caused their desynchronization.
3. Unlike in previous Supermodeling proofs-of-concept, a few $\mathbf{C}$ metaparameters can be quickly adapted to data by a classical DA method without coupling to truth. In the previous work (see, e.g. [13,37]), non-vanishing matrices $\mathbf{C}$ and K^i combined inter-model synchronization with a nudging scheme attracting the model to GT data.

In support of our modeling concept (see Fig. 1b), we present a case study: the process of parameter estimation in the Handy socio-economical model [25]. The model is a dynamical system that is an extended version of the predator-prey scheme. We have selected the Handy model due to its non-trivial behavior, reasonable computational complexity and relatively large number of parameters. On the basis of training data we try to predict the evolution of a "true" dynamical system. We compare the quality of the predictions for various time budgets for the classical ABC-SMC data assimilation method on the one hand, and the Supermodeling scheme on the other. Finally, we summarize and discuss the findings.

2   Classical Data Assimilation to the Handy Model

2.1   Handy Model

The Handy model is a substantial extension of the predator-prey system and is described by the time evolution of four dynamical variables: Commoners, Elites as well as *Nature* and *Wealth* (x_C, x_E, y, w). Their evolution is described by the following equations:

$$\begin{cases} \dot{x_C} = \beta_C x_C - \alpha_C x_C \\ \dot{x_E} = \beta_E x_E - \alpha_E x_E \\ \dot{y} = \gamma y(\lambda - y) - \delta x_C y \\ \dot{w} = \delta x_C y - C_C - C_E \end{cases}$$

$$\begin{cases} C_C = min(1, \frac{w}{w_{th}})s x_C \\ C_E = min(1, \frac{w}{w_{th}})\kappa s x_E \end{cases} \tag{3}$$

$$\begin{cases} \alpha_C = \alpha_m + max(0, 1 - \frac{C_C}{s x_C})(\alpha_M - \alpha_m) \\ \alpha_E = \alpha_m + max(0, 1 - \frac{C_E}{s x_E})(\alpha_M - \alpha_m) \end{cases}$$

$$w_{th} = \rho x_C + \kappa \rho x_E$$

In Table 1 we have compiled a glossary of parameters and variables, and their *ground truth* or initial values, respectively. In Fig. 2 we illustrate the typical evolution of the dynamical variables of the Handy model. The time evolution of the system is so variable and its parameters so sensitive that prediction of the model behavior is sufficiently difficult as to make data assimilation a non-trivial task.

Table 1. Parameters and initial values of dynamical variables of the Handy model.

Parameter	Description	Value
α_m	Normal (minimum) death rate	1.0×10^{-2}
α_M	Famine (maximum) death rate	7.0×10^{-2}
β_C	Commoners birth rate	3.0×10^{-2}
β_E	Elites birth rate	3.0×10^{-2}
s	Subsistence salary per capita	5.0×10^{-4}
ρ	Threshold wealth per capita	5.0×10^{-3}
γ	Regeneration rate of nature	1.0×10^{-2}
λ	Nature carrying capacity	1.0×10^{2}
κ	Inequality factor	1.0
δ	Depletion (production) factor	3.34
Variable	Description	Initial value
x_C	Commoners population	1.0×10^{2}
x_E	Elites population	2.9×10^{1}
y	Nature	1.0×10^{2}
w	Accumulated wealth	5.0×10^{1}

2.2 *Ground Truth* Data Generation

To further the testing of the supermodel concept, we generated artificial data, assuming that there exists a *ground-truth* "model" that simulates reality. Of course, because neither reality nor observations of reality can be accurately approximated by any mathematical model, we should somehow disturb both observations and the whole model as well. The comparison of the robustness of ABC and *Supermodeling* by using such a stochastic model would need many extensive tests. Nevertheless, conducting such research would make sense if the *Supermodeling* scheme outperforms a classical DA procedure for a much simpler *ground truth* model. Therefore, herein we have assumed that reality follows exactly a given baseline mathematical model with a rigid and "unknown" set of parameters. Our role is to guess them, having a limited number of observations, i.e., samples from this GT system evolution.

As presented in Fig. 2, the dynamical variables of the GT model evolve in a given time interval in a smooth but variable and non-trivial way. We consider here only one time interval (from $T_1 = 300$ up to $T_2 = 750$ timesteps) that was split into three subintervals of the same length ($A = [300, 450]$, $B = [450, 600]$, $C = [600, 750]$). The models (the baseline model, *sub-models* and *Supermodel*) will be trained on GT data sampled in the middle part B of the plot, and accuracies of predictions will be tested on A (backward forecasting), C (forward forecasting) and $A \cup C$ (overall) time intervals. We have decided to use both sparsely and densely sampled data, i.e., in each of the training subintervals we have generated "real" observations every $\Delta T_1 = 10$ or $\Delta T_2 = 3$ steps, respectively.

In the rest of this paper we present the results from the case study of data assimilation to the Handy model and arbitrarily selected fragments of its behavior (Fig. 2). We have tested our approach on other datasets, from which the same conclusions can be drawn. Some results and all numerical details can be found in the MSc thesis [31].

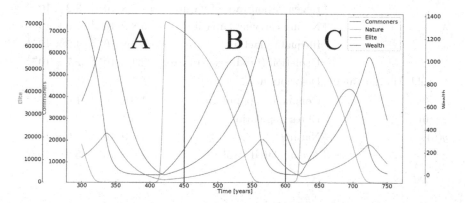

Fig. 2. The behaviour of the Handy model as the *ground-truth* model in selected time intervals. A fragment of the model time evolution is divided into three intervals (A, B, C), wherein the middle one B is used for generation of the training data while the remaining two for the test data.

2.3 Sensitivity Analysis

In many data assimilation tasks, knowledge of the most sensitive model parameters and dynamic variables, can help to give a faster and more precise search of the parameter space. This is particularly true if expert knowledge is unavailable. In the context of *Supermodeling*, the most sensitive dynamic variable has to be identified for use in synchronizing the sub-models. To determine the most significant dynamical variable, we performed Sobol Sensitivity Analysis (SA) [27,34]. Herein, we use the society quality measure:

$$Q = \frac{w}{x_C + x_E}, \tag{4}$$

to calculate the Sobol indices, where x_C, x_E are the populations of *Commoners* and *Elites* respectively, and w is the society's overall *Wealth*. We estimated that *Elites* is the most sensitive dynamical variable, also because it is closely connected with the most sensitive parameter, β_E, the Elites' birth rate (see Table 2). However, the SA procedure might be skipped if the most sensitive variable is already known, e.g., due to *a priori* possession of expert knowledge.

Table 2. The Sobol sensitivity indices S_1 and S_T for the parameters and dynamical variables of the Handy model (a greater value of the index means higher sensitvity).

Parameter	Description	S_1	S_T
α_m	Normal (minimum) death rate	2.4×10^{-2}	3.0×10^{-1}
α_M	Famine (maximum) death rate	1.6×10^{-3}	4.6×10^{-2}
β_C	Commoners birth rate	8.3×10^{-2}	4.6×10^{-1}
β_E	Elites birth rate	1.1×10^{-1}	4.8×10^{-1}
s	Subsistence salary per capita	3.3×10^{-3}	3.1×10^{-2}
ρ	Threshold wealth per capita	5.8×10^{-4}	2.1×10^{-3}
γ	Regeneration rate of nature	2.5×10^{-3}	3.2×10^{-2}
λ	Nature carrying capacity	5.8×10^{-3}	1.0×10^{-1}
κ	Inequality factor	5.9×10^{-2}	4.2×10^{-1}
δ	Depletion (production) factor	2.4×10^{-2}	2.7×10^{-1}
Variable (initial value)	Description	S_1	S_T
$x_C(0)$	Commoners population	1.9×10^{-3}	1.0×10^{-2}
$x_E(0)$	Elites population	4.5×10^{-3}	2.1×10^{-2}
$y(0)$	Nature	4.5×10^{-7}	5.9×10^{-9}
$w(0)$	Accumulated wealth	2.6×10^{-3}	3.2×10^{-4}

2.4 Approximate Bayesian Computation

Approximate Bayesian Computation (ABC) is not a single algorithm, but rather a very wide class of algorithms and methods that employ Bayesian inference for data assimilation purposes [7,14]. The main novelty of these methods is in their correct estimation of parameters even when the likelihoods are intractable [36]. In ABC algorithms, the functions of likelihood are not calculated, but the likelihood is approximated by the comparison of observed and simulated data [36].

Let us assume, that $\theta \in \mathbb{R}^n, n \geq 1$ is a vector of n parameters and $p(\theta)$ is a prior distribution. Then the goal of ABC approach is to approximate the posterior distribution $p(\theta|D)$ where D is the real data [1]. The posterior distribution is approximated in the following way:

$$p(\theta|D) \propto f(D|\theta)p(\theta), \tag{5}$$

where $f(D|\theta)$ is the function of likelihood of θ given the dataset D [35].

Among the variety of different approaches, one of the most useful is the ABC-SMC algorithm that uses the sequential Monte Carlo (M-C) method [7]. The major novelty, in comparison with previous methodologies (e.g., ABC-MCMC [24]), is the introduction of a set of *particles* $\theta^{(1)}, \ldots, \theta^{(S)}$ (parameter values sampled from a prior distribution $p(\theta)$), used to produce a sequence of intermediate distributions $p(\theta|d(D, \widetilde{D}) \leq \epsilon_i)$ (for $i = 1, \ldots, T - 1$) [36]. The *particles'* M-C propagation stops when a good representation of the target distribution $(p(\theta|d(D, \widetilde{D}) \leq \epsilon_T))$ is achieved. The set of error tolerance thresholds is chosen to be a decreasing sequence $\epsilon_1 > \cdots > \epsilon_T \geq 0$ that ensures the convergence of the intermediate probability distributions (of the parameters values) to the target ones. In the ABC-SMC algorithm, the parameter perturbation kernel can be simulated by the random walk procedure, with Gaussian or uniform functions [36]. Simultaneously, an adequately large set of *particles* will allow the Markov process to avoid low-probability regions and local minima in the parameter space.

2.5 ABC-SMC Training Results

For training the Handy model we use the ABC-SMC algorithm assuming that:

1. the number of *particles*, $S = 100$;
2. we fix the training time, t_{max};
3. we set intervals of possible values of parameters to be $\pm 10\%$ of exact (*ground truth*) ones (see Table 1);
4. the cost function is the root-mean-square error (RMSE).

Thus we assume that we know some approximate values of parameters. However, in the future we should also investigate the robustness of ABC-SMC and *Supermodeling* against prior selection of the values of the *sub-models'* parameters. In Table 3A, and Table 3B we present the training time (CPU time) and the RMSE errors of predictions for the Handy model for two pre-defined training error goals: RMSE = 50 and 100, respectively. The timings were measured for the layout presented in Fig. 2. We observe more than a ten-fold increase of the computational time when RMSE training precision goes from 100 to 50 (in dimensionless units) for sparsely sampled data. But for denser sampling, this increase is only two-fold.

We do not observe, in either case, any increase of the overall prediction quality with training precision. Meanwhile, one can notice signs of overfitting. A small increase is observed only for *forward prediction* (C). However, this improvement does not compensate the substantial decrease in *backward prediction* (A) quality. Summing up, both decreasing the training error and increasing the sampling frequency may lead to overfitting, so careful design is needed.

Table 3. The averaged CPU times and the prediction accuracies for the ABC-SMC training, to achieve given errors (RMSE = 50 and 100) with the *ground truth* training data. Results for sparser $\Delta T_1 = 10$ and denser $\Delta T_2 = 3$ sampled data.

Sampling	$\Delta T_1 = 10$				$\Delta T_2 = 3$			
Training error [RMSE]	Training time [s]	Overall prediction [RMSE]	Backward prediction [RMSE]	Forward prediction [RMSE]	Training time [s]	Overall prediction [RMSE]	Backward prediction [RMSE]	Forward prediction [RMSE]
50.0	1894.89	1387.58	1337.08	1369.54	248.51	1402.46	1303.10	1436.00
100.0	162.29	1303.83	1012.57	1456.40	111.74	1335.02	915.79	1559.29

3 *Supermodeling* the Handy System

3.1 Supermodeling by Data Assimilation Between Models

The *Supermodeling* approach is described in detail in the Introduction. Below we enumerate the main steps.

1. Create a small number M of instances (the *sub-models*) of the baseline model, initializing their parameters with a *rule-of-thumb* and/or using expert knowledge.
2. Pretrain every *sub-model* $\mu = 1, \ldots, M$ by using a classical DA procedure on the samples from Fig. 2B. New parameter sets will thus be generated for each *sub-model*.
3. Create the *Supermodel* by coupling the ODEs from Eqs. 3 through the most sensitive dynamical variable, as in Eq. 1, but with $K^i = 0$ and $C^i_{\mu\nu} = 0$ for $i \neq 1$ (Fig. 1a).
4. Train the coupling factors $C^i_{\mu\nu}$ of the *Supermodel* on the sampled data from Fig. 2B, according to the scheme sketched below, until either the RMS error relative to GT falls below a designated value or the elapsed training time reaches t_{max}.
5. The *Supermodel* trajectory is defined by averaging the sub-models states (Eq. 2).

3.2 Training Details

Unlike the classical DA training scheme described in Sect. 2.5, we fix not only the maximum time t_{max} but also the time needed for pretraining each of the *sub-models* t_{sub}. We pretrain $M = 3$ *sub-models* with ABC-SMC (one by one, each for a time t_{sub}) and couple them via the most sensitive variable x_E, to form the *Supermodel*. We also restrict the coupling coefficients to a fixed interval $[0, 0.5]$ (as in [12]). Furthermore, to speed-up the DA process, we divide the training data from Fig. 2B into five subintervals (*mini batches*) of the same length. Finally, we train the *Supermodel* with the ABC-SMC algorithm on the sequence of *mini batches* one after another for the estimated time $t_{sumo} = t_{max} - \overline{t_{sub}}$ (where $\overline{t_{sub}}$ is the mean time of pretraining the *sub-models*). Because the processes of pretraining the *sub-models* are independent, we have assumed that they are calculated in parallel. Then the normalized time for the Supermodel training will be equal to t_{max}.

We have performed the computations on the Prometheus supercomputer located in the ACK Cyfronet AGH UST, Krakow, Poland. We have used just one node, that consists of 8 CPUs (Intel Xeon E5-2680 v3, 2.5 GHz) with 12 cores each, giving 96 computational cores in total.

3.3 Results

Here we compare the *Supermodeling* scheme with the ABC-SMC DA algorithm with four different time budgets t_{max}: 14, 50, 100 and 250 s. Toward this end, we have constructed several *Supermodels*, each consisting of $M = 3$ differently initialized *sub-models*. Each *sub-model* was pretrained for a given short time period $t_{sub} < t_{max}$. We have selected several combinations, constructing four *Supermodels* which differ in the *sub-models'* pretraining time. We have repeated *Supermodel* training and testing procedure ten times for each pair (t_{max}, t_{sub}) and for various parameter initializations. Next, we have removed *zeroth* and *tenth* 10-quantiles from the results. The RMSE values on the test set (*backward prediction, forward prediction* and *overall prediction*) were averaged and the standard deviation was calculated. We present these averages for both sparsely ($\Delta T_1 = 10$) (see Table 4A) and densely ($\Delta T_2 = 3$) sampled datasets (see Table 4B).

As shown in Table 4A and Table 4B, the *forward prediction* RMSE error is a few times smaller for two-stage *Supermodeling* than for the classical parameter estimation with the ABC-SMC algorithm, for both sparse and denser datasets, and for all time regimes. Furthermore, with the ABC-SMC algorithm, longer learning appears to cause overfitting. It is important to mention that the ABC-SMC algorithm reaches the minimum RMSE after about 70 s of training. (The minimum is flat up to 120 s and afterwards RMSE grows due to overfitting.) Therefore, for Supermodel_70, composed of *sub-models* pretrained in 70 s, we obtain a radically lower RMSE, as compared to that for ABC-SMC, as total training time increases.

Turning attention to *backward prediction*, we note that although the *Supermodeling* approach is still convincingly better for overall prediction (except in one case) than the classical DA algorithm, the advantage for backward prediction is not so radical as for *forward prediction*. This bias can be clearly seen in Fig. 3 and Fig. 4, particularly, for the normalized RMSE plot. This behaviour is not seen with the ABC-SMC algorithm. It is the result of the specific training procedure we employed for the *Supermodeling* algorithm. The algorithm is

Table 4. RMS errors for the ABC algorithm and for the supermodel, for sparser (A) $\Delta T_1 = 10$, and denser (B) $\Delta T_2 = 3$ datasets. Supermodel_{X} is the *Supermodel* with the *sub-model* pretraining time t_{sub} set to X seconds. The better result for each case is shown in bold.

Sampling	$\Delta T_1 = 10$				$\Delta T_2 = 3$			
Time regime [s]	Method	Overall prediction [RMSE]	Backward prediction [RMSE]	Forward prediction [RMSE]	Method	Overall prediction [RMSE]	Backward prediction [RMSE]	Forward prediction [RMSE]
14 s	ABC-SMC	4586.38	4273.77	4879.00	ABC-SMC	5708.75	5181.07	6191.62
	Supermodel_10	**1897.73**	**2497.55**	**982.33**	Supermodel_12	**2453.95**	**3128.33**	**1502.42**
50 s	ABC-SMC	**1238.31**	**1213.21**	1262.92	ABC-SMC	2543.26	**2436.65**	2645.58
	Supermodel_35	1710.62	2350.98	**570.37**	Supermodel_20	**1959.12**	2652.73	**799.61**
100 s	ABC-SMC	1153.41	947.22	1327.96	ABC-SMC	1315.0	**930.0**	1510.0
	Supermodel_70	**665.46**	**762.62**	**551.45**	Supermodel_60	**855.24**	1098.18	**506.83**
250 s	ABC-SMC	1380.90	1511.72	1236.32	ABC-SMC	1410.4	1355.1	1401.0
	Supermodel_70	**466.33**	**548.09**	**366.78**	Supermodel_70	**600.08**	**795.76**	**294.87**

trained in five mini-batches starting from the left-hand side of the training interval (Fig. 2B). Consequently, the fitting accuracy is highest at the right-hand side of the B interval. At the last training point ($t = 600$) the standard deviation is equal to 0, while at the first point ($t = 450$) it is distinctly greater.

In summary, we conclude that the *Supermodeling* scheme results in predictions closer to the actual time series and with lower uncertainties, especially, for the *forward prediction* task. We have observed similar effects for other data, as presented in [31].

4   Discussion and Related Work

Classical data assimilation procedures were formulated on the basis of variational and Bayesian frameworks [2, 26]. The existing DA algorithms can be divided onto two main groups: (1) sequential-Monte-Carlo-based (e.g., [3]) and (2) Kalman-filter-based methods (e.g., [28])[2], which have formed the core of many other DA algorithms (e.g., [5]). Over the years, the majority of research in this direction was focused primarily on the improvement of the predictions' accuracy on tasks ranging from small-scale problems (e.g. [20]) to weather prediction [18]. Recently, more and more studies have attempted to speed-up data assimilation methods and to enable their use with extremely complex multi-scale models (e.g., [19, 26]).

The greatest challenge that arises with sequential Monte-Carlo-based methods (i.e., the ABC-SMC algorithm), is the requirement that a very large number of simulations need be performed, especially for the inverse problem of estimating parameters. That is, parameters can be adjoined to the model state and treated as variable quantities to be estimated - the second level of abstraction in the use of DA. But the number of required simulations increases exponentially with the number of model parameters (see e.g. [17]). To outperform the classical DA

---

[2] Kalman filtering is equivalent to the popular 4D-Var algorithm, for a perfect model.

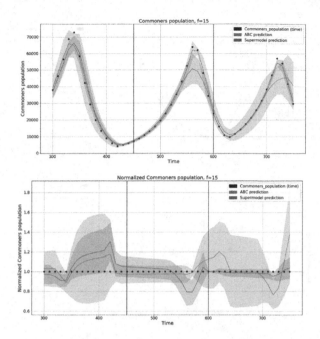

Fig. 3. The results for sparsely ($\Delta T_1 = 10$) sampled data and $t_{max} = 14$ s. Comparison between the value of *Commoners* predicted by the ABC-SMC method (grey), the *Supermodeling* (blue) and the *ground truth* (points), where lines are averaged predictions, while the boundaries of the shaded areas are at *mean ± one standard deviation* for each *ground truth* point: (**Top**) actual; (**Bottom**) normalized to the value of the *ground truth*. (Color figure online)

schemes, the current studies usually introduce either small algorithmic nuances (i.e. [6,15]) or algorithm implementations that support parallelization (i.e. [19]). For Kalman-filter-based data assimilation, the studies propose faster implementations of the algorithms [26] or hybridization with the ABC-SMC method (e.g. [8]). However, the aforementioned optimization approaches do not change the basic paradigms or improve DA performance radically.

In the era of deep learning, formal predictive models are often replaced (or supplemented) with faster data models for which the role of data assimilation in estimating parameters is played by the learning of black box (e.g., neural network) parameters. In general, learning a black box is a simpler procedure than data assimilation to a formal model. A very interesting data modeling concept, very competitive with formal models in the prediction of spatio-temporal patterns in chaotic systems, is that of Echo State Machines [16,22], particularly the Reservoir Computing (RC) approach [23]. No prior model based on physics or other knowledge is used.

In contrast, the *Supermodeling* paradigm, unlike the purely data-based RC and DA approaches, relies on the knowledge already encoded in formal models and on the partial synchronization of the chosen imprecise *sub-models* to

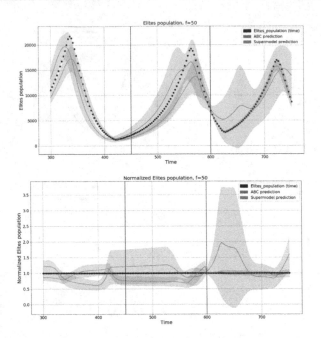

Fig. 4. The results for densely sampled data and $t_{max} = 14\,\text{s}$. Comparison between the value of *Elites* predicted by the ABC-SMC method (grey), the *Supermodeling* (green) and the *ground truth* (points), plotted as in Fig. 3 (Color figure online)

supplement the knowledge contained in any one sub-model. The original type of supermodel relied on synchronization of the *sub-models* by nudging them to one another, while simultaneously *nudging* them to the GT data [4]. The inter-model nudging effectively gives inter-model data assimilation, with nudging coefficients that can be estimated based on overall error relative to truth. Thus standard DA methods, having been employed first to estimate states, then to adjust a model itself by estimating its parameters, are now used to estimate inter-model couplings in a suite of models - an even higher level of abstraction in the application of DA [9,10]. This type of *Supermodeling* was successfully used for ensembling toy dynamical models [4,10] like *Lorenz systems* (Lorenz 63, Lorenz 84) and for combining simplified climate models (see e.g., [37]).

Recent results showed that the *Supermodeling* approach can also be applied in modeling complex dynamical biological processes such as tumor evolution. In [13] we demonstrated that in a *Supermodel* of melanoma the tumor evolution can be controlled by the *sub-models'* coupling factors **C**, producing a few qualitatively different tumor evolution patterns observed in reality. Recently, we have successfully assimilated *ground truth* data to the supermodel, using genetic algorithms [33]. However, due to computational complexity and the need for heavy High-Performance Computing, we are now implementing the more efficient procedure described in this paper.

5   Conclusions and Future Work

Herein we propose a novel metaprocedure for computational modeling, rooted in an extended use of data assimilation. It leads to a radical decrease in the number of free parameters, as compared to those in the source dynamical model, by ensembling a few imperfect *sub-models* - i.e., inaccurate and weak solutions of a classical DA-based pre-training scheme – within a single *Supermodel*. The case study demonstrates that due to the *sub-models'* synchronization, a small number of the *Supermodel* metaparameters can be estimated, based on assimilated observations, much faster than the full set of parameters in the overparametrized source model. Consequently, "effective parameter estimation" based on *Supermodeling* can produce more accurate predictions than those that could be obtained using traditional data assimilation methods to estimate a single model's parameters in reasonable time. It is crucial to mention that DA-based *Supermodeling* can be used with any given data assimilation procedure. The ABC-SMC algorithm was used here as the baseline classical DA method. *Supermodeling* plays only the role of the meta-framework dedicated to accelerating the modeling process.

We realize that our results can be treated as preliminary. A specific model was considered, and data assimilation was run on optimally selected working regimes and synthetic data. However, taking into account previous experience and more complicated phenomena simulated successfully by *Supermodeling*, one can expect that this procedure has wider prospects. Of course, there are still many unresolved issues, for example: how to generate efficiently the best *sub-models* and how many? How robust is the *Supermodel* against variations in noise, uncertainity and number of data samples? Herein we have assumed that the *sub-models* were generated in parallel because the pretraining of each can be performed independently. However, the total CPU time still increases proportionally with the number of *sub-models*. One can imagine that the *sub-models* could instead be generated by a single ABC-SMC procedure during the pre-training phase, by selecting more than one of the best solutions along the way. We plan to check this strategy in the very near future. We have taken as the *ground truth* the exact results from the reference (baseline) model. It would be worthwhile to check the quality of *Supermodel* predictions for disturbed data, which better simulate real observations. We are also considering a case study where the *sub-models* are simplfied versions of the baseline model (preliminary results can be found in [31]). This way, the differences between the *Supermodel* and the *ground-truth* simulator, could better reflect the differences between the computational model and reality. Summarizing, the application of *Supermodeling* can be an effective remedy to the *curse of dimensionality* problem, caused by model overparameterization.

Acknowledgments. W.D. and G.S.D. are thankful for support from the National Science Centre, Poland grant no. 2016/21/B/ST6/01539 and the funds assigned to AGH University of Science and Technology by the Polish Ministry of Science and Higher Education. G.S.D. was also supported by the ERC (Grant No. 648982). This research was supported in part by PLGrid Infrastructure.

References

1. Andrieu, C., De Freitas, N., Doucet, A., Jordan, M.I.: An introduction to MCMC for machine learning. Mach. Learn. **50**(1–2), 5–43 (2003)
2. Asch, M., Bocquet, M., Nodet, M.: Data Assimilation: Methods, Algorithms, and Applications, vol. 11. SIAM, New Delhi (2016)
3. Bain, A., Crisan, D.: Fundamentals of Stochastic Filtering, vol. 60. Springer, New York (2008). https://doi.org/10.1007/978-0-387-76896-0
4. Van den Berge, L., Selten, F., Wiegerinck, W., Duane, G.: A multi-model ensemble method that combines imperfect models through learning. Earth Syst. Dynam. Discuss **1**, 247–296 (2010)
5. Bergemann, K., Reich, S.: A mollified ensemble kalman filter. Q. J. R. Meteorol. Soc. **136**(651), 1636–1643 (2010)
6. Clarté, G., Robert, C.P., Ryder, R., Stoehr, J.: Component-wise approximate Bayesian computation via gibbs-like steps. arXiv preprint arXiv:1905.13599 (2019)
7. Del Moral, P., Doucet, A., Jasra, A.: An adaptive sequential monte carlo method for approximate bayesian computation. Stat. Comput. **22**(5), 1009–1020 (2012)
8. Drovandi, C., Everitt, R.G., Golightly, A., Prangle, D.: Ensemble MCMC: accelerating pseudo-marginal MCMC for state space models using the ensemble kalman filter. arXiv preprint arXiv:1906.02014 (2019)
9. Duane, G.: Data assimilation as artificial perception and supermodeling as artificial consciousness, pp. 209–222 (2012)
10. Duane, G.: Synchronicity from synchronized chaos. Entropy **17**(3), 1701–1733 (2015)
11. Duane, G., Grabow, C., Selten, F., Ghil, M.: Introduction to focus issue: synchronization in large networks and continuous media - data, models, and supermodels. Chaos: Interdisc. J. Nonlinear Sci. **27**(12), 126601 (2017)
12. Dzwinel, W., Kłusek, A., Paszyński, M.: A concept of a prognostic system for personalized anti-tumor therapy based on supermodeling. Procedia Comput. Sci. **108**, 1832–1841 (2017)
13. Dzwinel, W., Kłusek, A., Vasilyev, O.V.: Supermodeling in simulation of melanoma progression. Procedia Comput. Sci. **80**, 999–1010 (2016)
14. Fearnhead, P., Prangle, D.: Constructing summary statistics for approximate bayesian computation: semi-automatic approximate bayesian computation. J. R. Stat. Soc.: Ser. B (Stat. Methodol.) **74**(3), 419–474 (2012)
15. Ionides, E.L., Nguyen, D., Atchadé, Y., Stoev, S., King, A.A.: Inference for dynamic and latent variable models via iterated, perturbed bayes maps. Proc. Natl. Acad. Sci. **112**(3), 719–724 (2015)
16. Jaeger, H.: Echo state network. Scholarpedia **2**(9), 2330 (2007)
17. Järvenpää, M., Gutmann, M.U., Pleska, A., Vehtari, A., Marttinen, P., et al.: Efficient acquisition rules for model-based approximate bayesian computation. Bayesian Anal. **14**(2), 595–622 (2019)
18. Kalnay, E.: Atmospheric Modeling, Data Assimailation and Predictablity. Cambridge University Press, Cambridge (2003)
19. Klinger, E., Rickert, D., Hasenauer, J.: pyABC: distributed, likelihood-free inference. Bioinformatics **34**(20), 3591–3593 (2018)
20. Kypraios, T., Neal, P., Prangle, D.: A tutorial introduction to bayesian inference for stochastic epidemic models using approximate bayesian computation. Math. Biosci. **287**, 42–53 (2017)

21. Klusek, A., Łoś, M., Paszynski, M., Dzwinel, W.: Efficient model of tumor dynamics simulated in multi-GPU environment. Int. J. High Perform. Comput. Appl. **33**, 489–506 (2018). https://doi.org/10.1177/1094342018816772
22. Lukoševičius, M.: A practical guide to applying echo state networks. In: Montavon, G., Orr, G.B., Müller, K.-R. (eds.) Neural Networks: Tricks of the Trade. LNCS, vol. 7700, pp. 659–686. Springer, Heidelberg (2012). https://doi.org/10.1007/978-3-642-35289-8_36
23. Lukoševičius, M., Jaeger, H.: Reservoir computing approaches to recurrent neural network training. Comput. Sci. Rev. **3**(3), 127–149 (2009)
24. Marjoram, P., Molitor, J., Plagnol, V., Tavaré, S.: Markov chain monte carlo without likelihoods. Proc. Natl. Acad. Sci. **100**(26), 15324–15328 (2003)
25. Motesharrei, S., Rivas, J., Kalnay, E.: Human and nature dynamics (handy): modeling inequality and use of resources in the collapse or sustainability of societies. Ecol. Econ. **101**, 90–102 (2014)
26. Reich, S.: Data assimilation: the schrödinger perspective. Acta Numerica **28**, 635–711 (2019). https://doi.org/10.1017/S0962492919000011
27. Saltelli, A., Annoni, P., Azzini, I., Campolongo, F., Ratto, M., Tarantola, S.: Variance based sensitivity analysis of model output. Design and estimator for the total sensitivity index. Comput. Phys. Commun. **181**(2), 259–270 (2010)
28. Särkkä, S.: Bayesian Filtering and Smoothing, vol. 3. Cambridge University Press, Cambridge (2013)
29. Selten, F.M., Duane, G., Wiegerinck, W., Keenlyside, N., Kurths, J., Kocarev, L.: Supermodeling by combining imperfect models. Procedia Comput. Sci. **7**, 261–263 (2011)
30. Selten, F.M., Schevenhoven, F.J., Duane, G.S.: Simulating climate with a synchronization-based supermodel. Chaos: Interdisc. J. Nonlinear Sci. **27**(12), 126903 (2017)
31. Sendera, M.: Data adaptation in handy economy-ideology model. arXiv preprint arXiv:1904.04309 (2019)
32. Shen, M.L., Keenlyside, N., Bhatt, B., Duane, G.: Role of atmosphere-ocean interactions in super-modeling the tropical pacific climate. Chaos: Interdisc. J. Nonlinear Sci. **27**(12), 126704 (2017)
33. Siwik, L., Los, M., Klusek, A., Pingali, K., Dzwinel, W., Paszynski, M.: Supermodeling of tumor dynamics with parallel isogeometric analysis solver. arXiv preprint arXiv:1912.12836 (2019)
34. Sobol, I.M.: Global sensitivity indices for nonlinear mathematical models and their Monte Carlo estimates. Math. Comput. Simul. **55**(1–3), 271–280 (2001)
35. Sunnåker, M., Busetto, A.G., Numminen, E., Corander, J., Foll, M., Dessimoz, C.: Approximate bayesian computation. PLoS Comput. Biol. **9**(1), e1002803 (2013)
36. Toni, T., Stumpf, M.P.: Simulation-based model selection for dynamical systems in systems and population biology. Bioinformatics **26**(1), 104–110 (2009)
37. Wiegerinck, W., Selten, F.: Attractor learning in synchronized chaotic systems in the presence of unresolved scales. Chaos: Interdisc. J. Nonlinear Sci. **27**(12), 126901 (2017)
38. Wiegerinck, W., Burgers, W., Selten, F.: On the limit of large couplings and weighted averaged dynamics. In: Kocarev, L. (ed.) Consensus and Synchronization in Complex Networks. UCS, pp. 257–275. Springer, Heidelberg (2013)

A New Multilayer Network Construction via Tensor Learning

Giuseppe Brandi[1]([envelope]) and Tiziana Di Matteo[1,2,3]

[1] Department of Mathematics, King's College London, The Strand,
London WC2R 2LS, UK
giuseppe.brandi@kcl.ac.uk
[2] Department of Computer Science, University College London, Gower Street,
London WC1E 6BT, UK
[3] Complexity Science Hub Vienna, Josefstaedter Strasse 39, 1080 Vienna, Austria

Abstract. Multilayer networks proved to be suitable in extracting and providing dependency information of different complex systems. The construction of these networks is difficult and is mostly done with a static approach, neglecting time delayed interdependences. Tensors are objects that naturally represent multilayer networks and in this paper, we propose a new methodology based on Tucker tensor autoregression in order to build a multilayer network directly from data. This methodology captures within and between connections across layers and makes use of a filtering procedure to extract relevant information and improve visualization. We show the application of this methodology to different stationary fractionally differenced financial data. We argue that our result is useful to understand the dependencies across three different aspects of financial risk, namely market risk, liquidity risk, and volatility risk. Indeed, we show how the resulting visualization is a useful tool for risk managers depicting dependency asymmetries between different risk factors and accounting for delayed cross dependencies. The constructed multilayer network shows a strong interconnection between the volumes and prices layers across all the stocks considered while a lower number of interconnections between the uncertainty measures is identified.

Keywords: Tensor regression · Multidimensional data · Multilayer networks · Fractional differentiation

1 Introduction

Network structures are present in different fields of research. Multilayer networks represent a widely used tool for representing financial interconnections, both in industry and academia [1] and has been shown that the complex structure of the financial system plays a crucial role in the risk assessment [2,3]. A complex network is a collection of connected objects. These objects, such as stocks, banks or institutions, are called nodes and the connections between the nodes are called edges, which represent their dependency structure. Multilayer networks extend

© Springer Nature Switzerland AG 2020
V. V. Krzhizhanovskaya et al. (Eds.): ICCS 2020, LNCS 12142, pp. 148–154, 2020.
https://doi.org/10.1007/978-3-030-50433-5_12

the standard networks by assembling multiple networks 'layers' that are connected to each other via interlayer edges [4] and can be naturally represented by tensors [5]. The interlayer edges form the dependency structure between different layers and in the context of this paper, across different risk factors. However, two issues arise:

1 The construction of such networks is usually based on correlation matrices (or other symmetric dependence measures) calculated on financial asset returns. Unfortunately, such matrices being symmetric, hide possible asymmetries between stocks.
2 Multilayer networks are usually constructed via contemporaneous interconnections, neglecting the possible delayed cause-effect relationship between and within layers.

In this paper, we propose a method that relies on tensor autoregression which avoids these two issues. In particular, we use the tensor learning approach establish in [6] to estimate the tensor coefficients, which are the building blocks of the multilayer network of the intra and inter dependencies in the analyzed financial data. In particular, we tackle three different aspects of financial risk, i.e. market risk, liquidity risk, and future volatility risk. These three risk factors are represented by prices, volumes and two measures of expected future uncertainty, i.e. implied volatility at 10 days (IV10) and implied volatility at 30 days (IV30) of each stock. In order to have stationary data but retain the maximum amount of memory, we computed the fractional difference for each time series [7]. To improve visualization and to extract relevant information, the resulting multilayer is then filtered independently in each dimension with the recently proposed Polya filter [8]. The analysis shows a strong interconnection between the volumes and prices layers across all the stocks considered while a lower number of interconnection between the volatility at different maturity is identified. Furthermore, a clear financial connection between risk factors can be recognized from the multilayer visualization and can be a useful tool for risk assessment. The paper is structured as follows. Section 2 is devoted to the tensor autoregression. Section 3 shows the empirical application while Sect. 4 concludes.

2   Tensor Regression

Tensor regression can be formulated in different ways: the tensor structure is only in the response or the regression variable or it can be on both. The literature related to the first specification is ample [9,10] whilst the fully tensor variate regression received attention only recently from the statistics and machine learning communities employing different approaches [6,11]. The tensor regression we are going to use is the Tucker tensor regression proposed in [6]. The model is formulated making use of the contracted product, the higher order counterpart of matrix product [6] and can be expressed as:

$$\mathcal{Y} = \mathcal{A} + \langle \mathcal{X}, \mathcal{B} \rangle_{(\mathfrak{I}_{\mathcal{X}}; \mathfrak{I}_{\mathcal{B}})} + \mathcal{E} \qquad (1)$$

where $\mathcal{X} \in \mathbb{R}^{N \times I_1 \times \cdots \times I_N}$ is the regressor tensor, $\mathcal{Y} \in \mathbb{R}^{N \times J_1 \times \cdots \times J_M}$ is the response tensor, $\mathcal{E} \in \mathbb{R}^{N \times J_1 \times \cdots \times J_M}$ is the error tensor, $\mathcal{A} \in \mathbb{R}^{1 \times J_1 \times \cdots \times J_M}$ is the intercept tensor while the slope coefficient tensor, which represents the multi-layer network we are interested to learn, is $\mathcal{B} \in \mathbb{R}^{I_1 \times \cdots \times I_N \times J_1 \times \cdots \times J_M}$. Subscripts $\mathcal{I}_{\mathcal{X}}$ and $\mathcal{I}_{\mathcal{B}}$ are the modes over winch the product is carried out. In the context of this paper, $\mathcal{X}$ is a lagged version of $\mathcal{Y}$, hence $\mathcal{B}$ represents the multilinear inter-actions that the variables in $\mathcal{X}$ generate in $\mathcal{Y}$. These interactions are generally asymmetric and take into account lagged dependencies being $\mathcal{B}$ the mediator between two separate in time tensor datasets. Therefore, $\mathcal{B}$ represents a perfect candidate to use for building a multilayer network. However, the $\mathcal{B}$ coefficient is high dimensional. In order to resolve the issue, a Tucker structure is imposed on $\mathcal{B}$ such that it is possible to recover the original $\mathcal{B}$ with smaller objects.[1] One of the advantages of the Tucker structure is, contrarily to other tensor decompositions such as the PARAFAC, that it can handle dimension asymmetric tensors since each factor matrix does not need to have the same number of components.

2.1   Penalized Tensor Regression

Tensor regression is prone to over-fitting when intra-mode collinearity is present. In this case, a shrinkage estimator is necessary for a stable solution. In fact, the presence of collinearity between the variables of the dataset degrades the forecasting capabilities of the regression model. In this work, we use the Tikhonov regularization [12]. Known also as Ridge regularization, it rewrites the standard Least Squares problem as

$$\widehat{\mathcal{B}} = \operatorname*{arg\,min}_{Trk(\mathcal{B}) \leq \mathcal{R}_\bullet} \|\mathcal{Y} - \langle \mathcal{X}, \mathcal{B} \rangle_{(\mathcal{I}_{\mathcal{X}}; \mathcal{I}_{\mathcal{B}})}\|_F^2 + \lambda\|\mathcal{B}\|_F^2 \qquad (2)$$

where $\lambda > 0$ is the regularization parameter and $\|\|_F^2$ is the squared Frobenius norm. The greater the λ the stronger is the shrinkage effect on the parameters. However, high values of λ increase the bias of the tensor coefficient $\mathcal{B}$. Indeed, the shrinkage parameter is usually set via data driven procedures rather than input by the user. The Tikhonov regularization can be computationally very expensive for big data problem. To solve this issue, [13] proposed a decomposition of the Tikhonov regularization. The learning of the model parameters is a nonlinear optimization problem that can be solved by iterative algorithms such as the Alternating Least Squares (ALS) introduced by [14] for the Tucker decomposition. This methodology solves the optimization problem by dividing it into small least squares problems. Recently, [6] developed an ALS algorithm for the estimation of the tensor regression parameters with Tucker structure in both the penalized and unpenalized settings. For the technical derivation refer to [6].

---

[1] If the imposed Tucker rank is lower than the dimension of the tensor dataset, we have dimensionality reduction.

3   Empirical Application: Multilayer Network Estimation

In this section, we show the results of the construction of the multilayer network via the tensor regression proposed in Eq. 1.

3.1   Data and Fractional Differentiation

The dataset used in this paper is composed of stocks listed in the *Dow Jones* (DJ). These stocks time series are recorded on a daily basis from 01/03/1994 up to 20/11/2019, i.e. 6712 trading days. We use 26 over the 30 listed stocks as they are the ones for which the entire time series is available. For the purpose of our analysis, we use log-differenciated prices, volumes, implied volatility at 10 days (IV10) and implied volatility at 30 days (IV30). In particular, we use the fractional difference algorithm of [7] to balance stationarity and residual memory in the data. In fact, the original time series have the full amount of memory but they are non-stationary while integer log-differentiated data are stationary but have small residual memory due to the process of differentiation. In order to preserve the maximum amount of memory in the data, we use the fractional differentiation algorithm with different levels of fractional differentiation and then test for stationarity using the Augmented Dickey-Fuller test [15]. We find that all the data are stationary when the order of differentiation is $\alpha = 0.2$. This means that only a small amount of memory is lost in the process of differentiation.

3.2   Model Selection

The tensor regression presented in Eq. 1 has some parameters to be set, i.e. the Tucker rank and the shrinkage parameter λ for the penalized estimation of Eq. 2 as discussed in [6]. Regarding the Tucker rank, we used the full rank specification since we do not want to reduce the number of independent links. In fact, using a reduced rank would imply common factors to be mapped together, an undesirable feature for this application. Regarding the shrinkage parameter λ, we selected the value as follows. First, we split the data in a training set composed of 90% of the sample and in a test set with the remaining 10%. We then estimated the regression coefficients for different values of λ on the training set and then we computed the predicted R^2 on the test set. We used a grid of $\lambda = 0, 1, 5, 10, 20, 50$. and the predicted R^2 is maximized at $\lambda = 0$ (no shrinkage).

3.3   Results

In this section, we show the results of the analysis carried out with the data presented in Sect. 3.1. The multilayer network built via the estimated tensor autoregression coefficient $\mathcal{B}$ represents the interconnections between and within each layer. In particular $\mathcal{B}_{i,j,k,l}$ is the connection between stock i in layer j and stock k in layer l. It is important to notice that the estimated dependencies are in general not symmetric, i.e. $\mathcal{B}_{i,j,k,l} \neq \mathcal{B}_{k,j,i,l}$. However, the multilayer network constructed using $\mathcal{B}$ is fully connected. For this reason, a method for filtering those networks is necessary. Different methodologies are available for filtering information from complex networks [8,16]. In this paper, we use the Polya filter of [8] as it can handle directed weighted networks and it is both flexible and statistically driven. In fact, it employs a tuning parameter a that drives the strength of the filter and returns the p-values for the null hypotheses of random interactions. We filter every network independently (both intra and inter connections) using a parametrization such that 90% of the total links are removed.[2] In order to asses the dependency across the layers, we analyze two standard multilayer network measures, i.e. inter-layer assortativity and edge overlapping. A standard way to quantify inter-layer assortativity is to calculate Pearson's correlation coefficient over degree sequences of two layers and it represents a measure of association between layers. High positive (negative) values of such measure mean that the two risk factors act in the same (opposite) direction. Instead, overlapping edges are the links between pair of stocks present contemporaneously in two layers. High values of such measure mean that the stocks have common connections behaviour. As it can be possible to see from Fig. 1, prices and volatility have a huge portion of overlapping edges, still, these layers are disassortative as the correlation between the nodes sequence across the two layer is negative. This was an expected result since the negative relationship between prices and volatility is a stylized fact in finance. Not surprisingly, the two measures of volatility are highly assortative and have a huge fraction of overlapping edges.

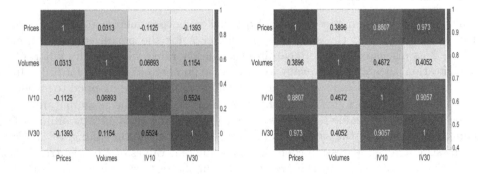

Fig. 1. Multilayer network assortativity matrix and edge overlapping matrix. Linear scale. Darker colour represents higher values.

Finally, we show in Fig. 2 the filtered multilayer network constructed via the tensor coefficient $\mathcal{B}$ estimated via the tensor autoregression of Eq. 1. As it can be possible to notice, the volumes layer has more interlayer connections rather than intralayer connections. Since each link represents the effect that one variable has on itself and other variables in the future, this means that stocks' liquidity

---

[2] Using hard thresholding the results are qualitatively equivalent.

risk mostly influences future prices and expected uncertainty. The two volatility networks have a relatively small number of interlayer connections despite being assortative. This could be due to the fact that volatility risk tends to increase or decrease through a specific maturity rather than across maturities. It is also possible to notice that more central stocks, depicted as bigger nodes in Fig. 2, have more connections but that this feature does not directly translate in a higher strength (depicted as darker colour of the nodes). This is a feature already emphasized in [3] for financial networks.

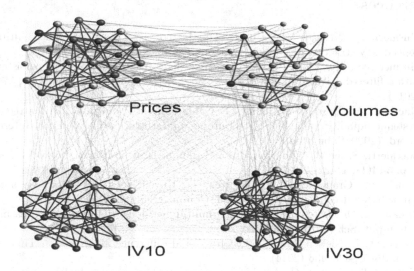

Fig. 2. Estimated multilayer network. Node colours: loglog scale; darker colour is associated to higher strength of the node. Node size: loglog scale; darker colour is associated to higher k-coreness score. Edge colour: uniform.

From a financial point of view, such graphical representation put together three different aspects of financial risk: market risk, liquidity risk (in terms of volumes exchanged) and forward looking uncertainty measures, which account for expected volatility risk. In fact, the stocks in the volumes layer are not strongly interconnected but produce a huge amount of risk propagation through prices and volatility. Understanding the dynamics of such multilayer network representation would be a useful tool for risk managers in order to understand risk balances and propose risk mitigation techniques.

4   Conclusions

In this paper, we proposed a methodology to build a multilayer network via the estimated coefficient of the Tucker tensor autoregression of [6]. This methodology, in combination with a filtering technique, has proven able to reproduce interconnections between different financial risk factors. These interconnections can be easily mapped to real financial mechanisms and can be a useful tool for monitoring risk as the topology within and between layers can be strongly affected in distressed periods. In order to preserve the maximum memory information in the data but requiring stationarity, we made use of fractional differentiation and found out that the variables analyzed are stationary with differentiation of order $\alpha = 0.2$. The model can be extendedto a dynamic framework in order to analyze the dependency structures under different market conditions.

References

1. Musmeci, N., Nicosia, V., Aste, T., Di Matteo, T., Latora, V.: The multiplex dependency structure of financial markets. Complexity **2017**, 1–13 (2017)
2. Musmeci, N., Aste, T., Di Matteo, T.: Risk diversification: a study of persistence with a filtered correlation-network approach. J. Netw. Theory Finan. **1**(1), 77–98 (2015)
3. Macchiati, V., Brandi, G., Cimini, G., Caldarelli, G., Paolotti, D., Di Matteo, T.: Systemic liquidity contagion in the European interbank market. J. Econ. Interact. Coord. (2020, Submitted to)
4. Boccaletti, S., et al.: The structure and dynamics of multilayer networks. Phys. Rep. **544**(1), 1–122 (2014)
5. Brandi, G., Gramatica, R., Di Matteo, T.: Unveil stock correlation via a new tensor-based decomposition method. J. Comput. Sci. (2020, Accepted in)
6. Brandi, G., Di Matteo., T.: Predicting multidimensional data via tensor learning. J. Comput. Sci. (2020, Submitted to)
7. Jensen, A.N., Nielsen, M.Ø.: A fast fractional difference algorithm. J. Time Ser. Anal. **35**(5), 428–436 (2014)
8. Marcaccioli, R., Livan, G.: A pólya urn approach to information filtering in complex networks. Nat. Commun. **10**(1), 1–10 (2019)
9. Zhou, H., Li, L., Zhu, H.: Tensor regression with applications in neuroimaging data analysis. J. Am. Stat. Assoc. **108**(502), 540–552 (2013)
10. Li, L., Zhang, X.: Parsimonious tensor response regression. J. Am. Stat. Assoc. **112**(519), 1131–1146 (2017)
11. Lock, E.F.: Tensor-on-tensor regression. J. Comput. Graph. Stat. **27**(3), 638–647 (2018)
12. Tikhonov, A.N.: On the stability of inverse problems. In: Doklady Akademii Nauk SSSR, vol. 39, pp. 195–198 (1943)
13. Arcucci, R., D'Amore, L., Carracciuolo, L., Scotti, G., Laccetti, G.: A decomposition of the tikhonov regularization functional oriented to exploit hybrid multilevel parallelism. Int. J. Parallel Prog. **45**(5), 1214–1235 (2017)
14. Kroonenberg, P.M., De Leeuw, J.: Principal component analysis of three-mode data by means of alternating least squares algorithms. Psychometrika **45**(1), 69–97 (1980)
15. Fuller, W.A.: Introduction to Statistical Time Series, vol. 428. Wiley, Hoboken (2009)
16. Aste, T., Di Matteo, T., Hyde, S.T.: Complex networks on hyperbolic surfaces. Phys. A: Stat. Mech. Appl. **346**(1–2), 20–26 (2005)

Neural Assimilation

Rossella Arcucci$^{(\boxtimes)}$ [iD], Lamya Moutiq, and Yi-Ke Guo

Data Science Institute, Imperial College London, London, UK
r.arcucci@imperial.ac.uk

Abstract. We introduce a new neural network for Data Assimilation (DA). DA is the approximation of the true state of some physical system at a given time obtained combining time-distributed observations with a dynamic model in an optimal way. The typical assimilation scheme is made up of two major steps: a *prediction* and a *correction* of the prediction by including information provided by observed data. This is the so called *prediction-correction* cycle. Classical methods for DA include Kalman filter (KF). KF can provide a rich information structure about the solution but it is often complex and time-consuming. In operational forecasting there is insufficient time to restart a run from the beginning with new data. Therefore, data assimilation should enable real-time utilization of data to improve predictions. This mandates the choice of an efficient data assimilation algorithm. Due to this necessity, we introduce, in this paper, the Neural Assimilation (NA), a coupled neural network made of two Recurrent Neural Networks trained on forecasting data and observed data respectively. We prove that the solution of NA is the same of KF. As NA is trained on both forecasting and observed data, after the phase of training NA is used for the *prediction* without the necessity of a correction given by the observations. This allows to avoid the *prediction-correction* cycle making the whole process very fast. Experimental results are provided and NA is tested to improve the prediction of oxygen diffusion across the Blood-Brain Barrier (BBB).

Keywords: Data Assimilation · Machine learning · Neural network

1 Introduction and Motivations

The current approach to forecasting modelling consists of simulating explicitly only the largest-scale phenomena, while taking into account the smaller-scale ones by means of "physical parameterisations". All numerical models introduce uncertainty through the selection of scales and parameters. Additionally, any computational methodology contributes to uncertainty due to discretization, finite precision and accumulation of round-off errors. Finally the ever growing size of the computational domains leads to increasing sources of uncertainties. Taking into account these uncertainties is essential for the acceptance of any numerical simulation. Numerical forecasting models often use Data Assimilation methods for the uncertainty quantification in the medium to long-term analysis.

© Springer Nature Switzerland AG 2020
V. V. Krzhizhanovskaya et al. (Eds.): ICCS 2020, LNCS 12142, pp. 155–168, 2020.
https://doi.org/10.1007/978-3-030-50433-5_13

Data Assimilation (DA) is the approximation of the true state of some physical system at a given time by combining time-distributed observations with a dynamic model in an optimal way. DA can be classically approached in two ways: as variational DA [16] and as filtering [5]. In both cases we seek an optimal solution. The most popular filtering approach for data assimilation is the Kalman Filter (KF) [15]. Statistically, KF seeks a solution with minimum variance. Variational methods seek a solution that minimizes a suitable cost function. In certain cases, the two approaches are identical and provide exactly the same solution [16]. However, the statistical approach, though often complex and time-consuming, can provide a richer information structure, i.e. an average and some characteristics of its variability (probability distribution). During the last 20 years hybrid approaches [11,18] have become very popular as they combine the two approaches into a single taking advantage of the relative rapidity and robustness of variational approaches, and at the same time, obtaining an accurate solution [2] thanks to the statistical approach. In this paper, in order to achieve the accuracy of the KF solution and reduce the execution time, we use Recurrent Neural Networks (RNN). Today the computational power of RNN is exploited for several application in different fields. Any non-linear dynamical system can be approximated to any accuracy by a Recurrent Neural Network, with no restrictions on the compactness of the state space, provided that the network has enough sigmoidal hidden units. This is what the Universal Approximation Theorem [12,20] claims. Only during the last few years, the DA community is starting to approach machine learning models to improve the efficiency of DA models. In [17], the authors combined Deep Learning and Data Assimilation to predict the production of gas from mature gas wells. They used a modified deep LSTM model as their prediction model in the EnKF framework for parameter estimation. Even if the *prediction* phase is speed up due to the introduction of Deep Learning, this only partially affects the whole *prediction-correction* cycle which is still time-consuming. In [9], the authors presented an approach for employing artificial neural networks (NNs) to emulate the local ensemble transform Kalman filter (LETKF) as a method of data assimilation. Even if the Feed Forward NN they implemented is able to emulate the DA process for the time window they fixed, when they need to assimilate observations in new time steps, it still needs the *prediction-correction* cycle and this affects the execution time which is just 90 times faster than the reference DA model. To further speed up the process, in [8] the authors combined the power of Neural Networks and High Performance Computing to assimilate meteorological data. These studies, alongside others discussed in conferences and still under publication, highlight the necessity to avoid the *prediction-correction* cycle by developing a Neural Network able to completely emulate the whole Data Assimilation process. In this context, we developed a Neural Assimilation (NA) as a Coupled Neural Network made of two RNNs. NA captures the features of a Data Assimilation process by interleaving the training of the two component RNNs on the forecasting data and the observed data. That is, the two component RNNs are trained on forecasting and observed data respectively with additional inputs provided by the interaction of these two. This NA network emulates the KF and runs much faster than the KF *prediction-correction* cycle for data assimilation. In this paper we develop the NA architecture and proved its equivalence to the KF. The equivalence between NA and KF is independent from the structure on the RNNs. In this paper we show results we obtained employing two Long short-term memory (LSTM) architectures for the two RNNs. Then we employ the NA model to a practical problem in predicting of oxygen (and drugs) diffusion across the Blood-Brain Barrier (BBB) [1] to justify its correctness and efficiency.

This paper is structured as follows. In Sect. 2 the Data Assimilation problem is described. The Neural Assimilation is introduced in Sect. 3, where we investigate the accuracy of the introduced method and we present a theorem demonstrating that the novel model is consistent with the KF result. Experimental results are provided in Sect. 4. Conclusions and future works are summarised in Sect. 5.

2 Data Assimilation

Data Assimilation (DA) is the approximation of the true state of some physical system at a given time by combining time-distributed observations $o(t)$ with a dynamic model $\dot{x} = \mathcal{M}(x, t)$ in an optimal way. DA can be classically approached in two ways: as variational DA [3] and as filtering. One of the best known tools for filtering approach is the Kalman filter (KF) [15]. We seek to estimate the state $x(t)$ of a discrete-time dynamic process that is governed by the linear difference equation

$$x(t) = M \ x(t-1) + w_t \tag{1}$$

with an observation $o(t)$:

$$o(t) = H \ x(t) + v_t \tag{2}$$

Note that M and H are discrete operators. The random vectors w_t and v_t represent the modeling and the observation errors respectively. They are assumed to be independent, white-noise processes with normal probability distributions

$$w_t \sim \mathcal{N}(0, B_t), \qquad v_t \sim \mathcal{N}(0, R_t) \tag{3}$$

where B_t and R_t are covariance matrices of the modeling and observation errors respectively. All these assumptions about unbiased and uncorrelated errors (in time and between each other) are not limiting, since extensions of the standard KF can be developed should any of these not be valid [5]. The KF problem can be summarised as follows: given a background estimate $x(t)$, of the system state at time t, what is the best analysis $z(t)$ based on the current available observation $o(t)$?

The typical assimilation scheme is made up of two major steps: a *prediction* step and a *correction* step. At time t we have the result of the previous forecast, $x(t)$ and the result of an ensemble of observations $o(t)$. Based on these two vectors, we perform an analysis that produces $z(t)$. We then use the evolution model to obtain a prediction of the state at time $t+1$. The result of the forecast at the *prediction* step is denoted with $x(t+1)$

$$x(t+1) = Mz(t), \tag{4}$$

$$B_{t+1} = M\left((1 - K_t H)B_t\right)M^T, \tag{5}$$

and becomes the background for the next *correction* time step:

$$K_{t+1} = B_{t+1}H^T(HB_{t+1}H^T + R_{t+1})^{-1}, \tag{6}$$

$$z(t+1) = x(t+1) + K_{t+1}\left(o(t+1) - Hx(t+1)\right), \tag{7}$$

We observe that, in case the observed data are defined in the same space of the state variable, the operator H_t in (2) is the identity matrix and the Eqs. (6)–(7) can be simplified becoming:

$$K_{t+1} = B_{t+1}(B_{t+1} + R_{t+1})^{-1}, \tag{8}$$

$$z(t+1) = x(t+1) + K_{t+1}\left(o(t+1) - x(t+1)\right), \tag{9}$$

Due to the high computational cost in updating the covariance matrices B_t by Eq. (5), it in operational DA, is often used to assume $B_t = B_{t+1}\ \forall t$. This assumption leads to a model which is also called Optimal Interpolation [16].

Statistically, KF seeks a solution with minimum variance. This approach, though often complex and time-consuming, can provide a rich information structure (often richer than information provided by variational DA), such as an average and some characteristics of its variability (probability distribution). In order to maintain the accuracy of the KF solution and reduce the execution time, we introduce, in the next section, a Neural Assimilation (NA) which is a network representing KF but much faster than a KF *prediction-correction* cycle.

3   Neural Assimilation

For a fixed time window $[t_0, t_1]$ and a fixed discretization time step Δt, let $x(t)$ still denote the forecasting result at each time step $t \in [t_0, t_1]$. Let $o(t)$ denotes an observation of the state value (Fig. 1). As it does not affect the generality of our study, we are assuming here the observed data defined in the same space of the state variable, i.e. the operator H_t in (2) is the identity matrix.

Given the data sets $\{x(t)\}_{t \in [t_0, t_1]}$ and $\{o(t)\}_{t \in [t_0, t_1]}$, the Neural Assimilation (NA) is a Coupled Neural Network (for temporal processing) as shown in Fig. 2, where:

- a first forecasting network NN_F is a Recurrent Neural Network trained on forecasting data $x(t)$ with an additional input provided by a second forecasting network NN_O trained on observed data $o(t)$;
- a second forecasting network NN_O is a Recurrent Neural Network trained on observed data $o(t)$ with an additional input provided by a first forecasting network NN_F.

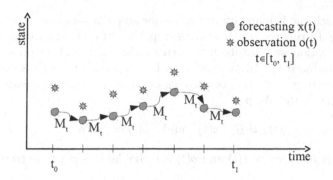

Fig. 1. Available data in the fixed time window.

A fundamental feature of each network is that it contains a feedback connection, so the activations can flow round in a loop. That enables the networks to do temporal processing and learn sequences with temporal prediction. The form of NA is a RNN with the previous set of hidden unit activations feeding back into the network along with the inputs.

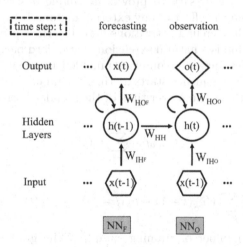

Fig. 2. Neural Assimilation

Note that the time t is discretized, with the activations updated at each time step. The time scale might correspond to any time step of size appropriate for the given problem. A delay unit given by the network NN_F needs to be introduced to hold activations in NN_O until they are processed at the next time step and vice versa. As for simple architectures and deterministic activation functions, learning will be achieved using similar gradient descent procedures to those leading to the back-propagation algorithm for feed forward networks.

The NA scheme is made up of two major steps: a *pre-processing* step and a *training* step. During the *pre-processing* step, the data set is normalized considering the information we have about the error estimations and the error covariance matrices introduced in (3). We consider, to normalise, the inverse of the error covariance matrices so that, data with big covariance/variance are assumed with a small weight [5, 16]. We pose

$$\bar{x}(t) = B_t^{-1}x(t) \quad \text{and} \quad \bar{o}(t) = R_t^{-1}o(t). \tag{10}$$

The computed vectors $\bar{x}(t)$ and $\bar{o}(t)$ are the data used in the *training* step:

$$\bar{o}(t) = f_{O_O}\left(W_{HO_O}h(t-1)\right) \tag{11}$$

$$h(t) = f_H\left(W_{IH}\bar{x}(t-1) + W_{HH}h(t-1)\right) \tag{12}$$

$$\bar{x}(t) = f_{O_F}\left(W_{HO_F}h(t)\right) \tag{13}$$

where the vectors $\bar{x}(t-1)$ are the inputs, the matrices W_{IH}, W_{HH}, W_{HO_F} and W_{HO_O} are the four connection weight matrices, and f_H, f_{O_F} and f_{O_O} are the hidden and outputs unit activation functions. The state of the dynamical system is a set of values that summarizes all the information about the past behaviour of the system that is necessary to provide a unique description of its future behaviour, apart from the effect of any external factors. In this case the state is defined by the set of hidden unit activations $h(t)$. The Back propagation Through Time for this algorithm is a natural extension of standard back propagation that performs gradient descent on a complete unfolded network ([21], Chapter 5 of [6]). If the NA training sequence starts at time t_0 and ends at time t_1, the total cost function is simply the sum over time of the standard error function $C(t)$ at each time-step:

$$C_{total} = \sum_{t=t_0}^{t_1} C(t) \tag{14}$$

where

$$C(t) = \frac{1}{2}\sum_{k=1}^{n}\left((\bar{o}_k(t-1) - h_k(t-1))^2 + (\bar{x}_k(t) - h_k(t))^2\right) \tag{15}$$

and n is the total number of training samples. The gradient descent weight updates have contributions from each time-step [19]:

$$\Delta w_{ij} = -\eta\frac{\partial C_{total}(t_0, t_1)}{\partial w_{ij}} = -\eta\sum_{t=t_0}^{t_1}\frac{\partial C(t)}{\partial w_{ij}} \tag{16}$$

where η is the learning rate [14]. The constituent partial derivatives $\frac{\partial C(t)}{\partial w_{ij}}$ have contributions from the multiple instances of each weight

$$w_{ij} \in \{W_{IH}, W_{HH}, W_{HO_O}, W_{HO_F}\}$$

and depend on the inputs and hidden unit activations at previous time steps. The errors now have to be back-propagated through time as well as through the network [23].

We prove that the output function $h(t)$ of the NA model corresponds to the solution of Kalman filter with fixed covariance matrices, i.e. in its Optimal Interpolation version [16]. The following result held.

Theorem 1. *Let $h(t)$ be the solution of NA given by Eqs. (10)–(16) and let $z(t)$ denote the solution of the KF algorithm as defined in (9). We have*

$$h(t) = z(t), \quad \forall t \in [t_0, t_1] \tag{17}$$

Proof: *Due to the definition of the L^2 norm, the loss function in (15) can be written as*

$$C(t) = \|\bar{o}(t-1) - h(t-1)\|_2^2 + \|\bar{x}(t) - h(t)\|_2^2 \tag{18}$$

then, from Eq. (1), and except for the numerical errors that will be introduced later as already included in the data sets, the (18) can be written as:

$$C(t) = \|\bar{o}(t-1) - h(t-1)\|_2^2 + \|M\,\bar{x}(t-1) - M\,h(t-1)\|_2^2 \tag{19}$$

From the properties of the L^2 norm, the (19) can be written as

$$C(t) = (\bar{o}(t-1) - h(t-1))^T(\bar{o}(t-1) - h(t-1))$$
$$+ (M\bar{x}(t-1) - Mh(t-1))^T(M\bar{x}(t-1) - Mh(t-1)). \tag{20}$$

To minimise this loss function, we compute the gradient

$$\nabla_{h(t-1)}C(t) = 2(\bar{o}(t-1) - h(t-1)) + 2M^T(M\bar{x}(t-1) - M\,h(t-1)) \tag{21}$$

where M^T denotes the Adjoint operator of the linear operator M [7] and we pose $\nabla_{h(t-1)}C(t) = 0$, then we have:

$$2h(t-1) = \bar{o}(t-1) + \bar{x}(t-1) \tag{22}$$

From the definition of $\bar{x}$ and $\bar{o}$ in (10), the (22) gives:

$$h(t-1)\,(B_{t-1} + R_{t-1}) = R_{t-1}x(t-1) + B_{t-1}o(t-1) \tag{23}$$

Then, adding and subtracting the quantity $B_{t-1}x(t-1)$ and merging the common factors, the (23) become

$$h(t-1)\,(B_{t-1} + R_{t-1}) = x(t-1)\,(B_{t-1} + R_{t-1}) + B_{t-1}\,(o(t-1) - x(t-1)) \tag{24}$$

Finally, posed $Q_{t-1} = B_{t-1}\,(B_{t-1} + R_{t-1})^{-1}$, the (24) gives:

$$h(t-1) = x(t-1) + Q_{t-1}\,(o(t-1) - x(t-1)) \tag{25}$$

which is the expression of the KF solution $z(t-1)$ in (9) for the time step $t-1$ and for the case of observed data defined in the same space of the state variable (i.e. $H = I$ and I is the identity matrix). Q_{t-1} is the Kalman gain matrix in (8).

The Eq. (25) in Theorem 1 represents a condition to assume that NA is consistent with KF.

In Sect. 4, we validate the results provided in this section. We also show that the employment of NA alleviates the computational cost making the running less expensive.

4   Experimental Results

In this section we provide experimental results that demonstrate the applicability and efficiency of NA. In our experiment, the NA is implemented by adopting Long short-term memory (LSTM) architecture for the two RNNs. The reason we use LSTMs is that they are suitable to contain information outside the normal flow of the recurrent network so it is easier to plug two networks together. Also, LSTMs allow to preserve the error that can be backpropagated through time and layers which is a very important point for discrete forecasting models. A description of the NA we implemented is provided in Fig. 3.

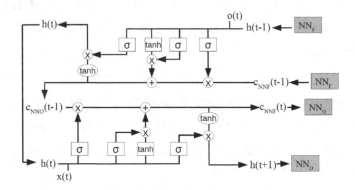

Fig. 3. Implementation of Neural Assimilation

The test case we consider is a numerical model to predict the oxygen diffusion across the Blood-Brain Barrier (BBB). Nevertheless the model can be used for any drugs by replacing the diffusion constant and the initial and boundary conditions [1]. The Blood-Brain Barrier protects the central nervous system, controls the entry of compounds into the brain by restricting access for blood borne compounds and facilitates access for nutrients. This protection makes it difficult to provide therapeutic compounds to brain cells when they are affected by brain diseases as Alzheimer, Autism [13]. The BBB is composed of endothelial cells connected by tight junctions. The main mechanisms allowing the transport of drugs across the membrane are passive transport, carrier-mediated transport, receptor-mediated transcytosis, and adsorption-mediated transcytosis [22]. The passive transport mechanism is the easiest method of drug transport for lipophilic and low molecular size molecules. It means a simple diffusion across any membrane without application of energy and carrier proteins. Opioids and steroids are examples of drugs which can be passively diffused [4]. Assuming that the main transport mechanism is through passive diffusion, the initial three-dimensional space problem can be reduced to a one-dimensional space problem. In fact, passive diffusion involves many simplifications as no reaction term, uniform movement in all directions and an overall diffusion constant. Therefore, a 1D partial differential equation (PDE) as (26) with one initial condition and two boundary conditions is an accurate model for this problem [22] where 0 corresponds to the location at which the blood meets the Blood-Brain Barrier and $L = 400\,nm$ is the real average thickness of the Blood-Brain Barrier.

$$\begin{cases} \frac{\partial x}{\partial t} = D\frac{\partial^2 x}{\partial y^2} \\ x(0, y) = x_{0,y} \\ x(t, 0) = x_{t,0} \\ x(t, L) = x_{t,L} \end{cases} \tag{26}$$

where $t \in [0, 10\,ms]$ (ms denotes microsecond) and $y \in [0, L]$. We consider that, at time 0 there is no oxygen, then $x_{0,y} = 0$. Moreover, for our boundary conditions, we consider that we have a constant concentration of oxygen in the bloodstream and that at the interface of the barrier and the brain tissue all oxygen will be consumed $x_{t,0} = 0.02945$ L/L blood and $x_{t,L} = 0$. We assume the diffusivity of oxygen through the Blood-Brain Barrier to be $3.24 * 10^{-5}$ cm^2/s [1].

Equation (26) is discretised by a second order central finite difference in space with $\Delta y = 8\,nm$ and a backward Euler method in time with $\Delta t = 0.1\,ms$:

$$-Fx_{i-1}^n + (1 + 2F)x_i^n - x_{i+1}^n = x_{i-1}^{n-1}$$

where $F = D\frac{\Delta t}{\Delta y^2}$, $i = 1, \ldots, 50$ and $n = 1, \ldots, 100$. As we know that it does not affect the generality of our study, in this paper we show results of NA using observed data $o(t)$ provided in [1] by the analytical solution of (26) for the oxygen diffusion. The model can be used for any drugs by replacing the diffusion constant and the initial and boundary conditions. Data sets for observed data can be found in http://cheminformatics.org/datasets/. The NA code and the pre-processed data can be downloaded using the link https://drive.google.com/drive/folders/1C_O-rk5wyqFsG5U-T7_vugBOddTPmOlY?usp=sharing.

The NA network has been trained using the 85% of the data and tested on the remaining 15%. Figure 4 shows the value of the Loos function for training and testing the forecasting network.

NA has been compiled as a sequential neural network with just one LSTM layer of 48 units using as loss function the mean squared error one and as optimiser the Adam one. Weights are automatically initialised by Keras using:

– Glorot uniform for the kernel weights matrix for the linear transformation of the inputs;
– Orthogonal for the linear transformation of the recurrent state.

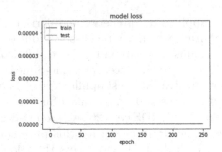

Fig. 4. Values of the loss function.

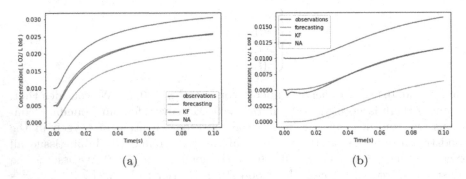

| (a) | (b) |

Fig. 5. Temporal evolution of the concentration at (a) $y = 12\,\text{nm}$ and (b) $y = 35\,\text{nm}$

Figure 5 shows the temporal evolution of the concentration at $y = 12\,\text{nm}$ (Fig. 5a) and $y = 35\,\text{nm}$ (Fig. 5b). The accuracy of the NA results is evaluated by the absolute error

$$e_{NA}(t, y) = |z(t, y) - h(t, y)| \tag{27}$$

and the mean squared error

$$MSE(h(t, y)) = \frac{\|z(t, y) - h(t, y)\|_{L^2}}{\|z(t, y)\|_{L^2}} \tag{28}$$

where $z(t, y)$ is the solution of KF performed at each time step. Table 1 shows values of absolute error computed every 10 time steps. We can see that the order of magnitude of the error is between $e{-}07$ and $e{-}04$. The corresponding values of mean squared error are $MSE(h(t, y)) = 1.31e{-}07$ for $y = 12\,\text{nm}$ and $MSE(h(t, y)) = 8.16e{-}08$ for $y = 35\,\text{nm}$ where $t \in [0, 0.10\,\text{ms}]$. Figure 6 shows the comparison of the KF result and the NA result for the temporal evolution of the concentration at each point of the BBB we are modelling. Values of execution time are provided in Table 2. The values are computed as mean of execution times from 100 runnings. We can observe that the time for *prediction* in NA is 1000 faster than the *prediction* with KF.

Table 1. Error computed every 10 time steps at (a) $y = 12\,\text{nm}$ and (b) $y = 35\,\text{nm}$

Time step t	$e_{NA}(t,y)$, $y = 12\,\text{nm}$	$e_{NA}(t,y)$, $y = 35\,\text{nm}$
0	0	0
10	7.05e−04	6.11e−04
20	4.17e−04	4.88e−04
30	4.29e−04	1.91e−04
40	1.52e−04	6.05e−07
50	2.51e−04	9.11e−05
60	3.40e−05	1.11e−04
70	4.13e−05	1.05e−04
80	4.72e−05	7.35e−05
90	1.11e−04	1.89e−05
100	1.60e−04	3.18e−05

Table 2. Execution time for 100 time steps and all the distances

	Executing time (s)
Neural Assimilation (training)	121.47
Neural Assimilation (prediction)	0.117
Kalman filter (prediction)	138

Finally, Table 3 shows the values of mean square forecasting error:

$$MSE^{F}(x(t,y)) = \frac{\|x(t,y) - o(t,y)\|_{L^2}}{\|o(t,y)\|_{L^2}} \tag{29}$$

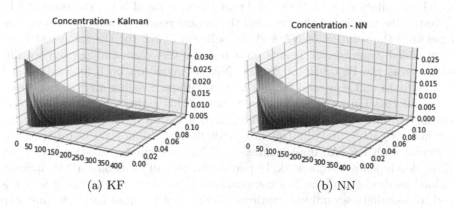

(a) KF   (b) NN

Fig. 6. Comparison between Data Assimilation (KF) and the Neural Network version for $t \in [0, 10\,\text{ms}]$ (*ms* denotes microsecond) and $y \in [0, 400\,\text{nm}]$.

and mean square assimilation error:

$$MSE^{NA}(h(t,y)) = \frac{\|h(t,y) - o(t,y)\|_{L^2}}{\|o(t,y)\|_{L^2}} \tag{30}$$

computed with respect observations $o(t,y)$. The values of the errors in the assimilation results present a reduction of approximately one order of magnitude with respect to the error in forecasting data.

Table 3. Mean square error forecasting error MSE^F and mean square assimilation error MSE^{NA} computed every 10 time steps at $y = 12\,\mathrm{nm}$

Time step t	$MSE^{NA}(h(t,y))$, $y = 12\,\mathrm{nm}$	$MSE^F(x(t,y))$, $y = 12\,\mathrm{nm}$
0	5.03e−03	1.00e−02
10	5.48e−03	1.00e−02
20	5.34e−03	9.99e−03
30	5.06e−03	9.95e−03
40	4.88e−03	1.00e−02
50	4.80e−03	1.00e−02
60	4.78e−03	1.00e−02
70	4.80e−03	1.00e−02
80	4.83e−03	1.00e−02
90	4.87e−03	1.00e−02
100	4.90e−03	1.00e−02

5   Conclusions and Future Works

We introduced a new neural network for Data Assimilation (DA) that we named Neural Assimilation (NA). We proved that the solution of NA is the same of KF. We tested the validity of the provided theoretical results showing values of misfit between the solution of NA and the solution of KF for the same test case. We provided experimental results on a realistic test case studying oxygen diffusion across the Blood-Brain Barrier. NA is trained on both forecasting and observed data and it is used for *predictions* without needing a correction given by the information provided by observations. This allows to avoid the *prediction-correction* cycle of a Kalman filter, and it makes the assimilation process very fast. We show that the time for *prediction* in NA is 1000 faster than the *prediction* with KF. An implementation of NA to emulate variational DA [10] will be developed as future work. In particular, we will focus on a 4D variational (4DVar) method [5]. 4DVar is a computational expensive method as it is developed to assimilate several observations (distributed in time) for each time step of the forecasting model. We will develop an extended version of NA able to assimilate set of distributed observations for each time step and, then, able to emulate 4DVar.

Acknowledgments. This work is supported by the EPSRC Centre for Mathematics of Precision Healthcare EP/N0145291/1 and the EP/T003189/1 Health assessment across biological length scales for personal pollution exposure and its mitigation (INHALE).

References

1. Agrawal, G., Bhullar, I., Madsen, J., Ng, R.: Modeling of oxygen diffusion across the blood-brain barrier for potential application in drug delivery (2013)
2. Arcucci, R., D'Amore, L., Pistoia, J., Toumi, R., Murli, A.: On the variational data assimilation problem solving and sensitivity analysis. J. Comput. Phys. **335**, 311–326 (2017)
3. Arcucci, R., Mottet, L., Pain, C., Guo, Y.K.: Optimal reduced space for variational data assimilation. J. Comput. Phys. **379**, 51–69 (2019)
4. Arya, M., Kumar, M.K.M., Sabitha, M., Menon, K.K., Nair, S.C.: Nanotechnology approaches for enhanced CNS delivery in treating Alzheimer's disease. J. Drug Deliv. Sci. Technol. **51**, 297–309 (2019)
5. Asch, M., Bocquet, M., Nodet, M.: Data Assimilation: Methods, Algorithms, and Applications, vol. 11. SIAM, Philadelphia (2016)
6. Bishop, C.M.: Pattern Recognition and Machine Learning. Springer, Boston (2006). https://doi.org/10.1007/978-1-4615-7566-5
7. Cacuci, D.G.: Sensitivity & Uncertainty Analysis, Volume 1: Theory. Chapman and Hall/CRC, Boca Raton (2003)
8. de Campos Velho, H., Stephany, S., Preto, A., Vijaykumar, N., Nowosad, A.: A neural network implementation for data assimilation using MPI. WIT Trans. Inf. Commun. Technol. **27** (2002)
9. Cintra, R.S., de Campos Velho, H.F.: Data assimilation by artificial neural networks for an atmospheric general circulation model. In: Advanced Applications for Artificial Neural Networks. IntechOpen (2018)
10. D'Amore, L., Arcucci, R., Marcellino, L., Murli, A.: A parallel three-dimensional variational data assimilation scheme. In: AIP Conference Proceedings, vol. 1389, pp. 1829–1831. American Institute of Physics (2011)
11. Desroziers, G., Camino, J.T., Berre, L.: 4DEnVar: link with 4D state formulation of variational assimilation and different possible implementations. Q. J. R. Meteorol. Soc. **140**(684), 2097–2110 (2014)
12. Funahashi, K., Nakamura, Y.: Approximation of dynamical systems by continuous time recurrent neural networks. Neural Netw. **6**(6), 801–806 (1993)
13. Gabathuler, R.: Approaches to transport therapeutic drugs across the blood-brain barrier to treat brain diseases. Neurobiol. Dis. **37**(1), 48–57 (2010)
14. Graves, A., Schmidhuber, J.: Framewise phoneme classification with bidirectional LSTM and other neural network architectures. Neural Netw. **18**(5–6), 602–610 (2005)
15. Kalman, R.E.: A new approach to linear filtering and prediction problems. J. Basic Eng. **82**(1), 35–45 (1960)
16. Kalnay, E.: Atmospheric Modeling, Data Assimilation and Predictability. Cambridge University Press, Cambridge (2003)
17. Loh, K., Omrani, P.S., van der Linden, R.: Deep learning and data assimilation for real-time production prediction in natural gas wells. arXiv preprint arXiv:1802.05141 (2018)

18. Lorenc, A.C., Bowler, N.E., Clayton, A.M., Pring, S.R., Fairbairn, D.: Comparison of hybrid-4DEnVar and hybrid-4DVar data assimilation methods for global NWP. Mon. Weather Rev. **143**(1), 212–229 (2015)
19. Salehinejad, H., Sankar, S., Barfett, J., Colak, E., Valaee, S.: Recent advances in recurrent neural networks. arXiv preprint arXiv:1801.01078 (2017)
20. Schäfer, A.M., Zimmermann, H.G.: Recurrent neural networks are universal approximators. In: Kollias, S.D., Stafylopatis, A., Duch, W., Oja, E. (eds.) ICANN 2006. LNCS, vol. 4131, pp. 632–640. Springer, Heidelberg (2006). https://doi.org/10.1007/11840817_66
21. Schuster, M.: On supervised learning from sequential data with applications for speech recognition. Daktaro disertacija, Nara Institute of Science and Technology (1999)
22. Simmons, J.M.: Effects of febuxostat on autistic behaviors and computational investigations of diffusion and pharmacokinetics. Ph.D. thesis, Virginia Tech (2019)
23. Werbos, P.J.: Generalization of backpropagation with application to a recurrent gas market model. Neural Netw. **1**(4), 339–356 (1988)

A Machine-Learning-Based Importance Sampling Method to Compute Rare Event Probabilities

Vishwas Rao[1(✉)], Romit Maulik[1], Emil Constantinescu[1], and Mihai Anitescu[1,2]

[1] Argonne National Laboratory, Lemont, IL, USA
{vhebbur,rmaulik,emconsta,anitescu}@anl.gov
[2] University of Chicago, Chicago, IL, USA

Abstract. We develop a novel computational method for evaluating the extreme excursion probabilities arising from random initialization of nonlinear dynamical systems. The method uses excursion probability theory to formulate a sequence of Bayesian inverse problems that, when solved, yields the biasing distribution. Solving multiple Bayesian inverse problems can be expensive; more so in higher dimensions. To alleviate the computational cost, we build machine-learning-based surrogates to solve the Bayesian inverse problems that give rise to the biasing distribution. This biasing distribution can then be used in an importance sampling procedure to estimate the extreme excursion probabilities.

Keywords: Machine learning · Rice's formula · Gaussian processes

1 Motivation

Characterizing high-impact rare and extreme events such as hurricanes, tornadoes, and cascading power failures are of great social and economic importance. Many of these natural phenomena and engineering systems can be modeled by using dynamical systems. The models representing these complex phenomena are approximate and have many sources of uncertainties. For example, the exact initial and boundary conditions or the external forcings that are necessary to fully define the underlying model might be unknown. Other parameters that are set based on experimental data may also be uncertain or only partially known. A probabilistic framework is generally used to formulate the problem of quantifying various uncertainties in these complex systems. By definition, the outcomes of interest that correspond to high-impact rare and extreme events reside in the tails of the probability distribution of the associated event space. Fully characterizing the tails requires resolving high-dimensional integrals over irregular domains. The most commonly used method to determine the probability

This material was based upon work supported by the U.S. Department of Energy, Office of Science, Office of Advanced Scientific Computing Research (ASCR) under Contract DE-AC02-06CH11347.

© Springer Nature Switzerland AG 2020
V. V. Krzhizhanovskaya et al. (Eds.): ICCS 2020, LNCS 12142, pp. 169–182, 2020.
https://doi.org/10.1007/978-3-030-50433-5_14

of rare and extreme events is Monte Carlo simulation (MCS). Computing rare-event probabilities via MCS involves generating several samples of the random variable and calculating the fraction of the samples that produce the outcome of interest. For small probabilities, however, this process is expensive. For example, consider an event whose probability is around 10^{-3} and the underlying numerical model for the calculation requires ten minutes per simulation. With MCS, estimating the probability of such an event to an accuracy of 10% will require two years of serial computation. Hence, alternative methods are needed that are computationally efficient.

Important examples of extreme events are rogue waves in the ocean [14], hurricanes, tornadoes [29], and power outages [3]. The motivation for this work comes from the rising concern surrounding transient security in the presence of uncertain initial conditions identified by North American Electric Reliability Corporation in connection with its long-term reliability assessment [32]. The problem can be mathematically formulated as a dynamical system with uncertain initial conditions. In this paper, the aim is to compute the extreme excursion probability: the probability that the transient due to a sudden malfunction exceeds preset safety limits. Typically, the target safe limit exceedance probabilities are in the range 10^{-4}–10^{-5}. We note that the same formulation is applicable in other applications such as data assimilation, which is used extensively for medium- to long-term weather forecasting. For example, one can potentially use the formulation in this paper to determine the likelihood of temperature levels at a location exceeding certain thresholds or the likelihood of precipitation levels exceeding safe levels in a certain area.

In [25], we presented an algorithm that uses ideas from excursion probability theory to evaluate the probability of extreme events [1]. In particular we used Rice's formula [27], which was developed to estimate the average number of upcrossings for a generic stochastic process. Rice's formula is given by

$$\mathbb{E}\left\{N_u^+(0,T)\right\} = \int_0^T \int_0^\infty y\varphi_t(u,y)\,\mathrm{d}y\,\mathrm{d}t\,, \tag{1}$$

where the left-hand side denotes the expected number of upcrossings of level u, y is the derivative of the stochastic process in a mean squared sense, and $\varphi_t(u,y)$ represents the joint probability distribution of the process and its derivative. In this paper, we build on our recent algorithm [25], which we used to construct an importance biasing distribution (IBD) to accelerate the computation of extreme event probabilities. A key step in the algorithm presented in [25] involves solving multiple Bayesian inverse problems, which can be expensive in high dimensions. Here, we propose to use machine-learning-based surrogates to obtain the inverse maps and hence alleviate the computational costs.

1.1   Mathematical Setup and Overview of the Method

The mathematical setup used in this paper consists of a nonlinear dynamical system that is excited by a Gaussian initial state and that results in a non-Gaussian stochastic process. We are interested in estimating the probability of the stochastic process exceeding a preset threshold. Moreover, we wish to estimate the probabilities when the underlying event of the process exceeding the threshold is a *rare event*. The rare events typically lie in the tails of the underlying event distribution. To characterize the tail of the resulting stochastic process, we use ideas from theory excursion probabilities [1]. Specifically, we use Rice's formula (1) to estimate the expected number of upcrossings of a stochastic process. For a description of the mathematically rigorous settings used for the rare event problem, we refer interested readers to [25, §2] and references therein.

Evaluating $\varphi_t(u, y)$, the joint probability distribution of the stochastic process and its derivative, is central to evaluating the integral in Rice's formula. However, $\varphi_t(u, y)$ is analytically computable only for Gaussian processes. Since our setup results in a non-Gaussian stochastic process, we linearize the nonlinear dynamical system variation around the trajectories starting at the mean of the initial state. We thus obtain a Gaussian approximation to the system trajectory distribution. In [25], we solve a sequence of Bayesian inverse problems to determine a biasing distribution to accelerate the convergence of the probability estimates. For high-dimensional problems, however, solving multiple Bayesian inverse problems can be expensive. In this work, we propose to replace multiple solutions to Bayesian inverse problems with machine-learning-based surrogates to alleviate the computational burden.

1.2   Organization

The rest of the paper is organized as follows. In Sect. 2 we review the existing literature for estimating rare event probabilities. In Sect. 3 we reformulate the problem of determining the IBD as a Bayesian inference problem, and in Sect. 4 we develop a machine-learning-based surrogate to approximate the solution to the Bayesian inference problem. In Sect. 5 we demonstrate this methodology on a simple nonlinear dynamical system excited by a Gaussian distribution. In Sect. 6 we present our conclusions and potential future research directions.

2   Existing Literature

2.1   Monte Carlo and Importance Sampling

Most of the existing methods to compute the probabilities of rare events use MCS directly or indirectly. The MCS approach was developed by Metropolis and his collaborators to solve problems in mathematical physics [22]. Since then, it has been used in a variety of applications [21,28]. When evaluating rare event probabilities, the MCS method basically counts the fraction of the random samples that cause the rare event. For a small probablilty P of the underlying event, the number of samples required to obtain an accuracy of $\epsilon \ll 1$ is $\mathcal{O}(\epsilon^{-2}P^{-1})$. Hence MCS becomes impractical for estimating rare event probabilities.

A popular sampling technique that is employed to compute rare event probablities is importance sampling (IS). IS is a variance reduction technique developed in the 1950s [17] to estimate the quantity of interest by constructing estimators that have smaller variance than MCS. In MCS, simulations from most of the samples do not result in the rare event and hence do not play a part in probability calculations. IS, instead, uses problem-specific information to construct an IBD; computing the rare event probability using the IBD requires fewer samples. Based on this idea, several techniques for constructing IBDs have been developed [8]. For a more detailed treatment of IS, we direct interested readers to [2,13]. One of the major challenges involved with importance sampling is the construction of an IBD that results in a low-variance estimator. We note that the approach may sometimes be inefficient for high-dimensional problems [19]. A more detailed description of MCS and IS in the context of rare events can be found in [25, §2] and references therein.

2.2   Nested Subset Methods

Other methods use the notion of conditional probability over a sequence of nested subsets of the probability space of interest. For example, one can start with the entire probability space and progressively shrink to the region that corresponds to the rare event. Furthermore, one can use the notion of conditional probability to factorize the event of interest as a product of conditional events. Subset simulation (SS) [4] and splitting methods [16] are ideas that use this idea. Several modifications and improvements have been proposed to both SS [6,9,10,18,33] and splitting methods [5,7]. Evaluating the conditional probabilities forms a major portion of the computational load. Compute the conditional probabilities for different nested subsets concurrently is nontrivial.

2.3   Other Approaches

Large deviation theory (LDT) is an efficient approach for estimating rare events in cases when the event space of interest is dominated by few elements such as rogue waves of a certain height. LDT also has been used to estimate the probabilities of extreme events in dynamical systems with random components [11,12]. A sequential sampling strategy has been used to compute extreme event statistics [23,24].

3   A Bayesian Inference Formulation to Construct IBD

Most of the work in this section is a review of our approach described in [25, §3]. Here we reformulate the problem of constructing an IBD as a sequence of Bayesian inverse problems. Consider the following dynamical system,

$$\mathbf{x}' = f(t, \mathbf{x}), \quad t = [0, T] \tag{2}$$
$$\mathbf{x}(0) = \mathbf{x}_0, \quad \mathbf{x}_0 \sim p, \quad \mathbf{x} \in \Omega,$$

where $\mathbf{c}$ is a canonical basis vector and $\mathbf{x}_0$, the initial state of the system, is uncertain and has a probability distribution p. The problem of interest is to estimate the probability that $\mathbf{c}^\top \mathbf{x}(t)$ exceeds the level u for $t \in [0, T]$. That is, we seek to estimate the following *excursion probability*,

$$P_T(u) := \mathbb{P}\left(\sup_{0 \leq t \leq T} \mathbf{c}^\top \mathbf{x}(t, \mathbf{x}_0) \geq u, \ \ t \in [0, T] \right), \tag{3}$$

where $\mathbf{x}(t, \mathbf{x}_0)$ represents the solution of the dynamical system (2) for a given initial condition $\mathbf{x}_0$. We note that

$$P_T(u) = \mu(\Omega(u)), \tag{4}$$

where μ is the respective measure transformation subject to (2) and $\Omega(u) \subset \Omega$ represents the *excursion set*

$$\Omega(u) := \left\{ \mathbf{x}_0 : \sup_{0 \leq t \leq T} \mathbf{c}^\top \mathbf{x}(t, \mathbf{x}_0) \geq u \right\}. \tag{5}$$

Hence, estimating $\Omega(u)$ will help us in estimating the excursion probability $P_T(u)$. In general, however, estimating the excursion set $\Omega(u)$ analytically is difficult. Rice's formula, (1) gives us insights about the excursion set and can be used to construct an approximation to the excursion set.

Recall that in Rice's formula (1), $\varphi_t(u, y)$ represents the joint probability density of $\mathbf{c}^\top \mathbf{x}$ and its derivative $\mathbf{c}^\top \mathbf{x}'$ for an excursion level u. The right-hand side of (1) can be interpreted as the summation of all times and slopes at which an excursion occurs. One can sample from $y \varphi_t(u, y)$ to obtain a slope-time pair (y_i, t_i) at which the sample paths of the stochastic process cause an excursion. Now consider the map $\mathcal{G} : \mathbb{R}^{d \times 1} \rightarrow \mathbb{R}^2$ that evaluates the vector $\begin{bmatrix} \mathbf{c}^\top \mathbf{x}(t) \\ \mathbf{c}^\top \mathbf{x}'(t) \end{bmatrix}$ based on the dynamical system (2), given an initial state $\mathbf{x}_0$ and a time t. By definition of the excursion set $\Omega(u)$, there exists an element $\mathbf{x}_i \in \Omega(u)$ that satisfies the following relationship,

$$\mathcal{G}(\mathbf{x}_i, t_i) = \begin{bmatrix} u + \varepsilon_i \\ y_i \end{bmatrix}, \tag{6}$$

where $\varepsilon > 0$. We can use this insight to construct an approximation of $\Omega(u)$ by constructing the preimages of multiple slope-time pairs. Observe that the problem of finding the preimage of a sample (y_i, t_i) is ill-posed since there could be multiple $\mathbf{x}_i$'s that map to $\begin{bmatrix} u + \varepsilon_i \\ y_i \end{bmatrix}$ at t_i via operator $\mathcal{G}$. We define the set

$$X_i := \left\{ \mathbf{x}_i \in \Omega : \mathcal{G}(\mathbf{x}_i, t_i) = \begin{bmatrix} u + \varepsilon_i \\ y_i \end{bmatrix} \right\}, \tag{7}$$

and an approximation $\widehat{\Omega}(u)$ to $\Omega(u)$ can be written as

$$\widehat{\Omega}(u) := \bigcup_{i=1}^{N} X_i. \tag{8}$$

Note that the approximation (8) improves as we increase N. For a discussion on the choice of ε_i, we refer interested readers to [25, §3.3].

The underlying computational framework to approximate $\widehat{\Omega}(u)$ consists of the following stages:

– Draw samples from unnormalized $y\varphi_t(u, y)$
– Find the preimages of these samples to approximate $\Omega(u)$.

We use MCMC to draw samples from unnormalized $y\varphi_t(u, y)$. We note that irrespective of the size of the dynamical system, $y\varphi_t(u, y)$ represents an unnormalized density in two dimensions; hence, using MCMC is an effective means , draw samples from it. Drawing samples from $y\varphi_t(u, y)$ requires evaluating it repeatedly, and in the following section we discuss the means to do so.

3.1   Evaluating $y\varphi_t(u, y)$

We note that $y\varphi_t(u, y)$ can be evaluated analytically only for special cases. Specifically, when $\varphi_t(u, y)$ is a Gaussian process, then the joint density function $y\varphi_t(u, y)$ is analytically computable. Consider the dynamical system described by (2). When p is Gaussian and f is linear, we have

$$\mathbf{x}' = A\mathbf{x}(t) + b, \quad \mathbf{x}(t_0) = \mathbf{x}_0, \quad \mathbf{x}_0 \sim \mathcal{N}(\overline{\mathbf{x}}_0, \Sigma). \tag{9}$$

Assuming A is invertible, $\mathbf{x}(t)$ can be written as

$$\mathbf{x}(t) = \exp(A(t - t_0))\,\mathbf{x}_0 - (I - \exp(A(t - t_0)))\,A^{-1}b, \tag{10}$$

where I represents an identity matrix of the appropriate size. Given that $\mathbf{x}_0$ is normally distributed, it follows that $\mathbf{x}(t)$ is a Gaussian process:

$$\mathbf{x}(t) \sim \mathcal{GP}\left(\overline{\mathbf{x}}, \mathrm{cov}_{\mathbf{x}}\right), \text{ where} \tag{11}$$
$$\overline{\mathbf{x}} = \exp(A(t - t_0))\overline{\mathbf{x}}_0 - (I - \exp(A(t - t_0)))\,A^{-1}b \text{ and}$$
$$\mathrm{cov}_{\mathbf{x}} = \exp(A(t - t_0))\Sigma\left(\exp(A(t - t_0))\right)^{\top}.$$

The joint probability density function (PDF) of $\mathbf{c}^{\top}\mathbf{x}(t)$ and $\mathbf{c}^{\top}\mathbf{x}'(t)$ is given by [26, equation 9.1]

$$\begin{bmatrix} \mathbf{c}^{\top}\mathbf{x} \\ \mathbf{c}^{\top}\mathbf{x}' \end{bmatrix} \sim \mathcal{GP}\left(\overline{\mathbf{x}}^{\varphi}, \begin{bmatrix} \mathbf{c}^{\top}\Phi\mathbf{c} & \mathbf{c}^{\top}\Phi A^{\top}\mathbf{c} \\ \mathbf{c}^{\top}A\Phi^{\top}\mathbf{c} & \mathbf{c}^{\top}A\Phi A^{\top}\mathbf{c} \end{bmatrix}\right), \tag{12}$$

where

$$\overline{\mathbf{x}}^{\varphi} := \begin{bmatrix} \mathbf{c}^{\top}\overline{\mathbf{x}} \\ \mathbf{c}^{\top}(A\overline{\mathbf{x}} + b) \end{bmatrix}$$

and

$$\Phi := \exp(A(t - t_0))\Sigma\left(\exp(A(t - t_0))\right)^{\top}.$$

We now can evaluate $y\varphi_t(u, y)$ for arbitrary values of u_i, y_i, and t_i as

$$y_i\varphi_{t_i}(u_i, y_i) = \frac{y_i}{2\pi |\Upsilon|} \exp\left(-\frac{1}{2}\left\|\begin{bmatrix}u_i \\ y_i\end{bmatrix} - \overline{\mathbf{x}}^\varphi\right\|_{\Upsilon^{-1}}^2\right), \tag{13}$$

where $\Upsilon := \begin{bmatrix} \mathbf{c}^\top \Phi \mathbf{c} & \mathbf{c}^\top \Phi A^\top \mathbf{c} \\ \mathbf{c}^\top A\Phi^\top \mathbf{c} & \mathbf{c}^\top A\Phi A^\top \mathbf{c} \end{bmatrix}$ and $|\Upsilon|$ denotes the determinant of Υ. Note that the right-hand side in (13) is dependent on t_i via Υ.

3.2   Notes for Nonlinear f

When f is nonlinear, $y\varphi_t(u, y)$ cannot be computed analytically—a key ingredient for our computational procedure. We approximate the nonlinear dynamics by linearizing f around the mean of the initial distribution. Assuming that the initial state of the system is normally distributed as described by Eq. (9), linearizing around the mean of the initial state gives

$$\mathbf{x}' \approx \mathbf{F} \cdot (\mathbf{x} - \overline{\mathbf{x}}_0) + f(\overline{\mathbf{x}}_0, 0), \tag{14}$$

where $\mathbf{F}$ represents the Jacobian of f at $t = 0$ and $\mathbf{x} = \overline{\mathbf{x}}_0$; this reduces the nonlinear dynamical system to a form that is similar to Eq. (9). Thus, we can now use Eqs. (11), (12), and (13) to approximate $y\varphi_t(u, y)$ for nonlinear f.

4   Machine-Learned Inverse Maps

In [25] we formulated the problem of determining preimages (7) as a Bayesian inverse problem. However, solving multiple Bayesian inverse problems can be expensive. Hence we approximated our IBD by using the solutions of a small number of Bayesian inverse problems. In this section we build a simple data-driven surrogate for approximating the preimages X_i described in Eq. (7). Using the surrogate, we can approximate the preimages of several $\mathbf{y}_i$'s obtained by sampling from $y\varphi_t(u, y)$. The surrogate developed here approximates the inverse of the map defined in Eq. (6). To that end, we wish to approximate the map

$$\mathcal{G}^{-1} : \mathbb{R}^2 \to \mathbb{R}^d, \tag{15}$$

where the input space corresponds to $(u + \varepsilon_i, y)|_{t_i}$ and the output lives in the domain of the state space (Ω here). This is equivalent to augmenting t_i as an additional input variable and building a surrogate that maps from $\mathbb{R}^3 \to \mathbb{R}^d$. We utilize a fully connected deep neural network to approximate this map. A one-layered neural network can be expressed as

$$\xi_j = F\left(\sum_{\ell=1}^{L} c_m^\ell x_\ell + \epsilon_m\right), \tag{16}$$

where F is a differentiable activation function that imparts nonlinearity to this transformation; L is the input dimension of an incoming signal; M is the number of hidden-layer neurons (in machine learning terminology); $c_m^\ell \in \mathbb{R}^{M \times L}$ are the weights of this map; $\epsilon_m \in \mathbb{R}^M$ are the biases; and $\xi_j \in \mathbb{R}^J$ is the nonlinear output of this map, which may be matched to targets available from data or "fed-forward" into future maps. Note that ξ_j is the postactivation value of each neuron in a hidden layer of J neurons. In practice, multiple compositions of this map may be used to obtain nonlinear function approximators, called deep neural networks, that are very expressive. For nonlinear activation, we utilize

$$F(\xi) = \max(\xi, 0), \tag{17}$$

for all its activation functions. In addition, we concatenate three such maps as shown in Eq. 16 to ultimately obtain an approximation for $\mathcal{G}^{-1}$. Two such submaps have J fixed at 256, and a final transformation utilizes $J = 3$. We note that the function F for the final transformation is the identity, as is common in machine learning algorithms. A schematic of this network architecture is shown in Fig. 1. The trainable parameters (c_m^ℓ and ϵ_m for each transformation) are optimized with the use of backpropagation [30], an adjoint calculation technique that obtains gradients of the loss function with respect to these parameters. A stochastic gradient optimization technique, ADAM, is used to update these parameters [20] with a learning rate of 0.001. Our loss function is given by the L_2-distance between the prediction of the network and the targets (i.e., the mean-squared error). Our network also incorporates a regularization strategy, called dropout [31], that randomly switches off certain units ξ_j (here we utilize a dropout probability of 0.1) in the forward propagation of the map (i.e., from $d \to 2$). Through this approach, memorization of data is avoided, while allowing for effective exploration of a complex nonconvex loss surface.

Our map is trained for 500 epochs with a batch size of 256; in other words, a weight update is performed after a loss is computed for 256 samples. Each epoch is completed when the losses from the entire data set are used for gradient update. During the network training, we set aside a random subset of the data for validation. Losses calculated from this data set are used only to monitor the learning of the framework for unseen data. These are plotted in Fig. 2, where one can see that both training and validation losses are reduced to an equal magnitude. Figure 3 also shows scatter plots for this validation data set where a good agreement between the true and predicted quantities can be seen. We may now use this map for approximating the IBD.

4.1   Using the Machine-Learned Inverse Map to Construct IBD

The following procedure is used to construct the IBD.

1. Obtain different realizations of the initial conditions of the dynamical system by sampling from the initial PDF p.
2. Use $\mathcal{G}$ to obtain the forward maps of these realizations.

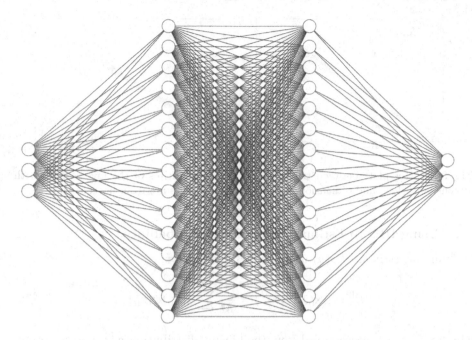

Fig. 1. Schematic of our neural network architecture. Note that the number of hidden layer units are not representative since this study utilizes 256 such units.

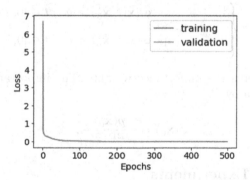

Fig. 2. Convergence of training for our network. Note how both training and validation losses diminish in magnitude concurrently.

3. Use the forward maps and the corresponding random realizations of the initial conditions to train the inverse map $\mathcal{G}^{-1}$.
4. We now apply this trained inverse map on samples generated from $y\varphi_t(u, y)$ to obtain the approximate preimages of samples $\mathbf{y}_i$.
5. Use a Gaussian approximation of these inverse maps is used as an IBD. Assume that this Gaussian approximation has PDF p^{IBD}.
6. Sample from the IBD, and use importance sampling to estimate the probabilities.

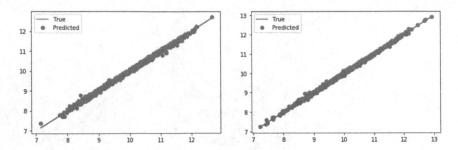

Fig. 3. Scatter plots between truth and predicted quantites of the inverse map with dimension 1 (left) and dimension 2 (right). These results are from unseen data.

4.2   Using IBD to Estimate Rare Event Probability

We can now estimate $P_T(u)$ using the IBD as follows:

$$P_T^{\mathrm{IS}}(u)(\widehat{\mathbf{x}}_0^1,\ldots,\widehat{\mathbf{x}}_0^M) = \frac{1}{M}\sum_{i=1}^{M}\mathbb{I}(\widehat{\mathbf{x}}_0^i)\psi(\widehat{\mathbf{x}}_0^i), \tag{18}$$

where $\widehat{\mathbf{x}}_0^1,\ldots,\widehat{\mathbf{x}}_0^M$ are sampled from the biasing distribution p^{IBD} and $\mathbb{I}(\widehat{\mathbf{x}}_0^i)$ represents the indicator function given by

$$\mathbb{I}(\widehat{\mathbf{x}}_0^i) = \begin{cases} 1, & \sup_{0\le t\le T}\mathbf{c}^\top\mathbf{x}(t,\widehat{\mathbf{x}}_0^i) \ge u, \quad t \in [0,T], \\ 0, & \sup_{0\le t\le T}\mathbf{c}^\top\mathbf{x}(t,\widehat{\mathbf{x}}_0^i) < u, \quad t \in [0,T]. \end{cases} \tag{19}$$

Also, $\psi(\widehat{\mathbf{x}}_0^i)$ represents the importance weights. The importance weight for an arbitrary $\widehat{\mathbf{x}}_0^i$ is given by

$$\psi(\widehat{\mathbf{x}}_0^i) = \frac{p(\widehat{\mathbf{x}}_0^i)}{p^{\mathrm{IBD}}(\widehat{\mathbf{x}}_0^i)}. \tag{20}$$

5   Numerical Experiments

We demonstrate the application of the procedure described in Sect. 3 and Sect. 4 for nonlinear dynamical systems excited by a Gaussian distribution. We use the Lotka-Volterra equations as a test problem. These equations, also known as the predator-prey equations, are a pair of first-order nonlinear differential equations and are used to describe the dynamics of biological systems in which two species interact, one as a predator and the other as a prey. The populations change through time according to the following pair of equations,

$$\frac{dx_1}{dt} = \alpha x_1 - \beta x_1 x_2, \tag{21}$$

$$\frac{dx_2}{dt} = \delta x_1 x_2 - \gamma x_2,$$

where x_1 is the number of prey, x_2 is the number of predators, and $\dfrac{dx_1}{dt}$ and $\dfrac{dx_2}{dt}$ represent the instantaneous growth rates of the two populations. We assume that the initial state of the system at time $t = 0$ is a random variable that is normally distributed:

$$\mathbf{x}(0) \sim \mathcal{N}\left(\begin{bmatrix} 10 \\ 10 \end{bmatrix}, 0.8 \times I_2\right),$$

and we are interested in estimating the probability of the event $P(\mathbf{c}^\mathsf{T}\mathbf{x} \geq u)$, where $\mathbf{c} = \begin{bmatrix} 0 \\ 1 \end{bmatrix}$, $t \in [0, 10]$, and $u = 17$. The first step of our solution procedure involves sampling from $y\varphi_t(u, y)$ to generate observations $\mathbf{y}_i$. We linearize the dynamical system about the mean of the distribution of $\mathbf{x}_0$ (Eq. (14)) and express $\varphi_t(u, y)$ as a function of t and y as described by Eq. (12). We compute $y\varphi_t(u, y)$ as shown in Eq. (13). We use the delayed rejection adaptive Metropolis (DRAM) Markov chain Monte Carlo (MCMC) method to generate samples from $y\varphi_t(u, y)$. (For more details about DRAM, see [15].) To minimize the effect of the initial guess on the posterior inference, we use a burn-in of 1,000 samples. Figure 4 shows the contours of $y\varphi_t(u, y)$ and samples drawn from it by using DRAM MCMC. In [25] we then solved the Bayesian inverse problem by using both MCMC and Laplace approximation at MAP to construct a distribution that approximately maps to likelihood constructed around $\mathbf{y}_i$. Here, we replace the solution to the Bayesian inverse problem with a machine-learned inverse map described in Sect. 4. Multiple samples generated from $y\varphi_t(u, y)$ can be used to construct the IBD, as described in Sect. 4.1, and the IBD can be used to estimate $P_T(u)$, as explained in Sect. 4.2. Figure 5 compares the results between the conventional MCS and Machine Learning based Importance Sampling (ML-based IS) methods. Note that we use an MCS estimate with 10 Million samples as a proxy for the true probabilities. The "true" probability is 3.28×10^{-5}. ML-based IS gives fairly good estimate even with small number of model evaluations. When the training dataset size is large enough, the improvements are dramatic. Notice that for a true probability of the order of 10^{-5} we obtain an estimate that has a relative error of less than 1%. Notice that our method gives the same (or better) accuracy as the MCS with hundred times lesser computational cost. The convergence with just 5000 training samples is acceptable and these results improve dramatically for 10000 and 20000 training samples. We believe the results could be even better when we use a Gaussian mixture to represent the IBD instead of a simple Gaussian approximation.

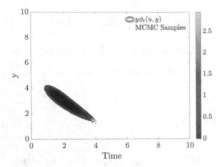
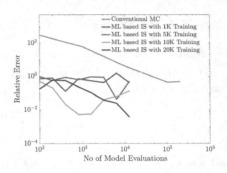

Fig. 4. Contours of $y\varphi_t(u, y)$ and samples drawn from it using DRAM MCMC.

Fig. 5. Comparison between conventional MCS and ML-based IS. We observe even with a small amount of training data, we obtain fairly accurate estimates; and as we increase the training data, the accuracy improves dramatically.

5.1   Computational Cost

In Fig. 5, we havent included the costs associated with generating training data, training costs, and cost for approximating the inverse map as these costs are almost negligible when compared to the overall costs. Note that generating 20000 samples is approximately equivalent to 400 model evaluations (this is because a single model evaluation can be used to generate the slope and state at 50 different times and each of them can be used as a training sample). The training of the ML framework, for this problem, required very little compute time. Each training was executed on an 8th-generation Intel Core-I7 machine with Python 3.6.8 and Tensorflow 1.14 and took less than 180 s for training 20000 samples (this is less than 50 model evaluations). Inference (for 20000 prediction points) costs were less than 2 s, on average.

6   Conclusions and Future Work

In this work we developed a ML-based IS to estimate rare event probabilities and we demostrated the algorithm on the prey-predator system. The method developed here builds on the approach in [25] and replaces the expensive Bayesian inference with a Machine learning based surrogate. This approach yields fairly accurate estimate of the probabilities and for a given accuracy requires atleast three orders of magnitude lesser computational effort than the traditional MCS. In future, we aim to test this algorithm for larger problems and also use an active learning based approach to pick the training samples. Scaling this algorithm to high dimensions (say $\mathcal{O}(1000)$) could be challenging and to address it, we will use state-of-the-art techniques developed by machine learning and deep learning community in the future.

References

1. Adler, R.J.: The geometry of random fields. SIAM (2010)
2. Asmussen, S., Glynn, P.W.: Stochastic Simulation: Algorithms and Analysis, vol. 57. Springer, New York (2007). https://doi.org/10.1007/978-0-387-69033-9
3. Atputharajah, A., Saha, T.K.: Power system blackouts-literature review. In: 2009 International Conference on Industrial and Information Systems (ICIIS), pp. 460–465. IEEE (2009)
4. Au, S.K., Beck, J.L.: Estimation of small failure probabilities in high dimensions by subset simulation. Probab. Eng. Mech. 16(4), 263–277 (2001)
5. Beck, J.L., Zuev, K.M.: Rare-event simulation. In: Ghanem, R., Higdon, D., Owhadi, H. (eds.) Handbook of Uncertainty Quantification, pp. 1075–1100. Springer, Cham (2017). https://doi.org/10.1007/978-3-319-12385-1_24
6. Bect, J., Li, L., Vazquez, E.: Bayesian subset simulation. SIAM/ASA J. Uncertain. Quan. 5(1), 762–786 (2017)
7. Botev, Z.I., Kroese, D.P.: Efficient monte carlo simulation via the generalized splitting method. Stat. Comput. 22(1), 1–16 (2012)
8. Bucklew, J.: Introduction to Rare Event Simulation. Springer, New York (2004). https://doi.org/10.1007/978-1-4757-4078-3
9. Ching, J., Beck, J.L., Au, S.: Hybrid subset simulation method for reliability estimation of dynamical systems subject to stochastic excitation. Probab. Eng. Mech. 20(3), 199–214 (2005)
10. Ching, J., Au, S.K., Beck, J.L.: Reliability estimation for dynamical systems subject to stochastic excitation using subset simulation with splitting. Comput. Methods Appl. Mech. Eng. 194(12–16), 1557–1579 (2005)
11. Dematteis, G., Grafke, T., Vanden-Eijnden, E.: Rogue waves and large deviations in deep sea. Proc. Natl. Acad. Sci. 115(5), 855–860 (2018)
12. Dematteis, G., Grafke, T., Vanden-Eijnden, E.: Extreme event quantification in dynamical systems with random components. SIAM/ASA J. Uncertain. Quan. 7(3), 1029–1059 (2019)
13. Dunn, W.L., Shultis, J.K.: Exploring Monte Carlo methods. Elsevier, Amsterdam (2011)
14. Dysthe, K., Krogstad, H.E., Müller, P.: Oceanic rogue waves. Annu. Rev. Fluid Mech. 40, 287–310 (2008)
15. Haario, H., Laine, M., Mira, A., Saksman, E.: Dram: efficient adaptive MCMC. Stat. Comput. 16(4), 339–354 (2006)
16. Kahn, H., Harris, T.E.: Estimation of particle transmission by random sampling. Natl. Bureau Stand. Appl. Math. Ser. 12, 27–30 (1951)
17. Kahn, H., Marshall, A.W.: Methods of reducing sample size in monte carlo computations. J. Oper. Res. Soc. Am. 1(5), 263–278 (1953)
18. Katafygiotis, L., Cheung, S.H.: A two-stage subset simulation-based approach for calculating the reliability of inelastic structural systems subjected to gaussian random excitations. Comput. Methods Appl. Mech. Eng. 194(12–16), 1581–1595 (2005)
19. Katafygiotis, L.S., Zuev, K.M.: Geometric insight into the challenges of solving high-dimensional reliability problems. Probab. Eng. Mech. 23(2–3), 208–218 (2008)
20. Kingma, D.P., Ba, J.: Adam: a method for stochastic optimization. arXiv preprint arXiv:1412.6980 (2014)
21. Liu, J.S.: Monte Carlo Strategies in Scientific Computing. Springer, New York (2004). https://doi.org/10.1007/978-0-387-76371-2

22. Metropolis, N., Ulam, S.: The monte carlo method. J. Am. Stat. Assoc. **44**(247), 335–341 (1949)
23. Mohamad, M.A., Sapsis, T.P.: A sequential sampling strategy for extreme event statistics in nonlinear dynamical systems. arXiv preprint arXiv:1804.07240 (2018)
24. Mohamad, M.A., Sapsis, T.P.: Sequential sampling strategy for extreme event statistics in nonlinear dynamical systems. Proc. Natl. Acad. Sci. **115**(44), 11138–11143 (2018)
25. Rao, V., Anitescu, M.: Efficient computation of extreme excursion probabilities for dynamical systems (2020)
26. Rasmussen, C., Williams, C.: Gaussian Processes for Machine Learning. Adaptive Computation and Machine Learning. MIT Press, Cambridge (2006)
27. Rice, S.O.: Mathematical analysis of random noise. Bell Labs Tech. J. **23**(3), 282–332 (1944)
28. Robert, C.P., Casella, G.: Monte Carlo Statistical Methods. Springer Texts in Statistics. Springer, New York (2005). https://doi.org/10.1007/978-1-4757-4145-2
29. Ross, T., Lott, N.: A climatology of 1980–2003 extreme weather and climate events. US Department of Commerece, National Ocanic and Atmospheric Administration, National Environmental Satellite Data and Information Service, National Climatic Data Center (2003)
30. Rumelhart, D.E., Hinton, G.E., Williams, R.J.: Learning representations by back-propagating errors. Nature **323**(6088), 533–536 (1986)
31. Srivastava, N., Hinton, G., Krizhevsky, A., Sutskever, I., Salakhutdinov, R.: Dropout: a simple way to prevent neural networks from overfitting. J. Mach. Learn. Res. **15**(1), 1929–1958 (2014)
32. "The North American Electricity Reliability Corporation": 2017 long-term reliability assessment (2017). https://www.nerc.com/pa/RAPA/ra/Reliability
33. Zuev, K.M., Beck, J.L., Au, S.K., Katafygiotis, L.S.: Bayesian post-processor and other enhancements of subset simulation for estimating failure probabilities in high dimensions. Comput. Struct. **92**, 283–296 (2012)

Node Classification in Complex Social Graphs via Knowledge-Graph Embeddings and Convolutional Neural Network

Bonaventure C. Molokwu[1]([mail]) [iD], Shaon Bhatta Shuvo[1] [iD], Narayan C. Kar[2] [iD], and Ziad Kobti[1] [iD]

[1] School of Computer Science, University of Windsor, 401 Sunset Avenue, Windsor, ON N9B-3P4, Canada
{molokwub,shuvos,kobti}@uwindsor.ca
[2] Centre for Hybrid Automotive Research and Green Energy (CHARGE), University of Windsor, 401 Sunset Avenue, Windsor, ON N9B-3P4, Canada
nkar@uwindsor.ca

Abstract. The interactions between humans and their environment, comprising living and non-living entities, can be studied via Social Network Analysis (SNA). Node classification, as well as community detection tasks, are still open research problems in SNA. Hence, SNA has become an interesting and appealing domain in Artificial Intelligence (AI) research. Immanent facts about social network structures can be effectively harnessed for training AI models in a bid to solve node classification and community detection problems in SNA. Hence, crucial aspects such as the individual attributes of spatial social actors, and the underlying patterns of relationship binding these social actors must be taken into consideration in the course of analyzing the social network. These factors determine the nature and dynamics of a given social network. In this paper, we have proposed a unique framework, Representation Learning via Knowledge-Graph Embeddings and ConvNet (RLVECN), for studying and extracting meaningful facts from social network structures to aid in node classification as well as community detection tasks. Our proposition utilizes an edge sampling approach for exploiting features of the social graph, via learning the context of each actor with respect to neighboring actors/nodes, with the goal of generating vector-space embedding per actor. Successively, these relatively low-dimensional vector embeddings are fed as input features to a downstream classifier for classification tasks about the social graph/network. Herein RLVECN has been trained, tested, and evaluated on real-world social networks.

Keywords: Node classification · Feature learning · Feature extraction · Dimensionality reduction · Semi-supervised learning

---

This research was supported by International Business Machines (IBM) and Compute Canada (SHARCNET).

V. V. Krzhizhanovskaya et al. (Eds.): ICCS 2020, LNCS 12142, pp. 183–198, 2020.
https://doi.org/10.1007/978-3-030-50433-5_15

1   Introduction and Related Literature

Humans inhabit in a planet comprised of several systems and ecosystems; and interaction is a natural phenomenon and characteristic obtainable in any given system or ecosystem. Thus, relationships between constituent entities in a given system/ecosystem is a strategy for survival and essentiality for the sustenance of a given system/ecosystem. Owing to the recent AI advances, these real-world complex systems and ecosystems can be effectively modelled as social network structures for analysis. Social network graphs [22] are intricate structures which present analytical challenges to Machine Learning (ML) and Deep Learning (DL) models because of their dynamic nature, complex links, and occasionally massive size. In this regard, we have proposed a new hybrid and DL-based model (RLVECN) based on an iterative learning approach [1] for solving (node) classification as well as clustering problems in SNA via an edge sampling strategy.

In SNA, the classification of nodes induces the formation of cluster(s). Consequently, clusters give rise to homophily in social networks. Basically, learning in RLVECN is induced via semi-supervised training. The architecture of RLVECN comprises two (2) distinct Representation Learning (RL) layers, viz: a Knowledge-Graph Embeddings (VE) layer and a Convolutional Neural Network (ConvNet) layer [13]. Both of these layers are trained by means of unsupervised learning. These layers are essentially feature-extraction and dimensionality-reduction layers where underlying knowledge and viable facts are automatically extracted from the social network structure [15]. The vector-embedding layer is responsible for projecting the feature representation of the social graph to a q-dimensional real-number space, $\mathbb{R}^q$. This is done by associating a real-number vector to every unique actor/node in the social network such that the cosine distance of any given tie/edge (a pair of actors) would capture a significant degree of correlation between the two associated actors. Furthermore, the ConvNet layer, which feeds on the vector-embedding layer, is responsible for further extraction of apparent features and/or representations from the social graph. Finally, a classification layer succeeds the RL layers; and it is trained by means of supervised learning. The classifier is based on a Neural Network (NN) architecture assembled using multiple and deep layers of stacked perceptrons (NN units) [6]. Every low-dimensional feature (X), extracted by the representation-learning layers, is mapped to a corresponding output label (Y); and these (X, Y) pairs are used to supervise the training of the classifier such that it can effectively and efficiently learn how to identify clusters and classify actors within a given social network structure. Hence, the novelty of our research contribution are as stated below:

(1) Proposition of a DL-based and hybrid model, RLVECN, which is aimed at solving node classification problems in SNA.
(2) Detailed benchmarking reports with respect to classic objective functions used for classification tasks.
(3) Comparative analysis, between RLVECN and state-of-the-art methodologies, against standard real-world social networks.

RLVECN is capable of learning the non-linear distributed features enmeshed in a social network [9]. We have evaluated RLVECN against an array of state-of-the-art models and RL methodologies which serve as our baselines, viz:

(i) DeepWalk: Online Learning of Social Representations [19].
(ii) GCN: Semi-Supervised Classification with Graph Convolutional Networks [11].
(iii) LINE: Large-scale Information Network Embedding [25].
(iv) Node2Vec: Scalable Feature Learning for Networks [8].
(v) SDNE: Structural Deep Network Embedding [26].

2   Proposed Methodology and Framework

2.1   Definition of Problem

Definition 1. *Social Network, SN: As expressed via Eq. 1 such that SN is a tuple comprising a set of actors/vertices, V; a set of ties/edges, E; a metadata function, f_V, which extends the definition of the vertices' set by mapping it to a given set of attributes, V'; and a metadata function, f_E, which extends the definition of the edges' set by mapping it to a given set of attributes, E'. Thus, a graph function, $G(V, E) \subset SN$*

$$SN = (V, E, f_V, f_E) \equiv (G, f_V, f_E)$$
$$V : |\{V\}| = M \qquad \text{set of actors/vertices with size, M}$$
$$E : E \subset \{U \times V\} \subset \{V \times V\} \text{ set of ties/edges between V} \qquad (1)$$
$$f_V : V \rightarrow V' \qquad \text{vertices' metadata function}$$
$$f_E : E \rightarrow E' \qquad \text{edges' metadata function}$$

Definition 2. *Knowledge Graph, KG: $(\mathbb{E}, \mathbb{R})$ is a set comprising entities, $\mathbb{E}$, and relations, $\mathbb{R}$, between the entities. Thus, a KG [24, 28] is defined via a set of triples, $t : (u, p, v)$, where $u, v \in \mathbb{E}$ and $p \in \mathbb{R}$. Also, a KG [27] can be modelled as a social network, SN, such that: $\mathbb{E} \rightarrow V$ and $\mathbb{R} \rightarrow E$ and $(\mathbb{E}, \mathbb{R}) \vdash f_V, f_E$.*

Definition 3. *Knowledge-Graph (Vector) Embeddings, X: The vector-space embeddings, X, generated by the embedding layer are based on a mapping function, f, expressed via Eq. 2. f projects the representation of the graph's actors to a q-dimensional real space, $\mathbb{R}^q$, such that the existent ties between any given pair of actors, (u_i, v_j), remain preserved via the homomorphism from V to X.*

$$f : V \rightarrow X \in \mathbb{R}^q$$
$$f : (u, p, v) \rightarrow X \in \mathbb{R}^q \quad \text{Knowledge-Graph Embeddings} \qquad (2)$$

Definition 4. *Node Classification: Considering, SN, comprising partially labelled actors (or vertices), $V_{lbl} \subset V : V_{lbl} \rightarrow Y_{lbl}$; and unlabelled vertices defined such that: $V_{ulb} = V - V_{lbl}$. A node-classification task aims at training a predictive function, $f : V \rightarrow Y$, that learns to predict the labels, Y, for all actors or vertices, $V \subset SN$, via knowledge harnessed from the mapping: $V_{lbl} \rightarrow Y_{lbl}$.*

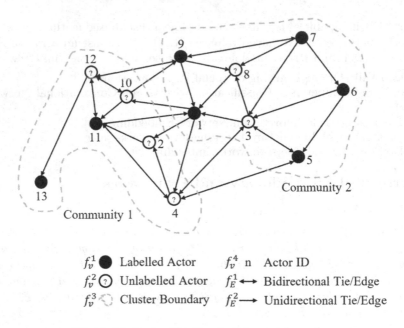

f_v^1 ● Labelled Actor f_v^4 n Actor ID
f_v^2 ? Unlabelled Actor f_E^1 ↔ Bidirectional Tie/Edge
f_v^3 ◌ Cluster Boundary f_E^2 → Unidirectional Tie/Edge

Fig. 1. Node classification task in social graphs

2.2   Proposed Methodology

Our proposition, RLVECN, is comprised of two (2) distinct Feature Learning (FL) layers, and one (1) classification layer (Fig. 2).

Representation Learning - Knowledge-Graph Embeddings Layer: Given a social network, SN, defined by a set of actors/vertices, $V : U \subset V \forall \{u_m, v_m\} \in V$, and $M : m \in M$ denotes the number of unique actors in SN. Additionally, let the ties/edges in SN be defined such that: $E \subset \{U \times V\}$; where $u_i \in V$ and $v_j \in V$ represent a source_vertex and a target_vertex in E, respectively.

The objective function of the vector-embedding layer aims at maximizing the average logarithmic probability of the source_vertex, u_i, being predicted as neighboring actor to the target_vertex, v_j, with respect to all training pairs, $\forall (u_i, v_j) \in E$. Formally, the function is expressed as in Eq. 3:

$$\mu = \frac{1}{M} \sum_{m=1}^{M} (\sum_{(u_i, v_j) \in E} log Pr(u_i | v_j)) \tag{3}$$

Consequently, in order to compute $Pr(u_i | v_j)$, we have to quantify the proximity of each target_vertex, v_j, with respect to its source_vertex, u_i. The vector-embedding model measures this adjacency/proximity as the cosine similarity between v_j and its corresponding u_i. Thus, the cosine distance is calculated as the dot product between the target_vertex and the source_vertex. Mathematically, $Pr(u_i|v_j)$ is computed via a softmax function as defined in Eq. 4:

$$Pr(u_i|v_j) = \frac{exp(u_i \cdot v_j)}{\sum_{m=1}^{M} exp(u_m \cdot v_j)} \qquad (4)$$

Hence, the objective function of our vector-embedding model with respect to the SN is as expressed by Eq. 5:

$$\sum_{(u_i,v_j)\in E} logPr(u_i|v_j) = \sum_{(u_i,v_j)\in E} log\frac{exp(u_i \cdot v_j)}{\sum_{m=1}^{M} exp(u_m \cdot v_j)} \qquad (5)$$

Representation Learning - ConvNet Layer: This layer comprises three (3) RL or FL operations, namely: convolution, non-linearity, and pooling operations. RLVECN utilizes a one-dimensional (1D) convolution layer [14] which is sandwiched between the vector-embedding and classification layers. Equation 6 expresses the 1D-convolution operation:

$$FeatureMap(F) = 1D\_InputMatrix(X) * Kernel(K)$$
$$f_i = (X * K)_i = (K * X)_i = \sum_{j=0}^{J-1} x_j \cdot k_{i-j} = \sum_{j=0}^{J-1} k_j \cdot x_{i-j} \qquad (6)$$

where f_i represents a cell/matrix position in the Feature Map; k_j denotes a cell position in the Kernel; and x_{i-j} denotes a cell/matrix position in the 1D-Input (data) matrix.

The non-linearity operation is a rectified linear unit (ReLU) function which introduces non-linearity after the convolution operation since real-world problems usually exist in non-linear form(s). As a result, the rectified feature/activation map is computed via: $r_i \in R = g(f_i \in F) = max(0, F)$.

The pooling operation is responsible for reducing the input width of each rectified activation map while retaining its vital properties. In this regard, the *Max Pooling* function is defined such that the resultant pooled (or downsampled) feature map is generated via: $p_i \in P = h(r_i \in R) = maxPool(R)$.

Classification - Multi-Layer Perceptron (MLP) Classifier Layer: This is the last layer of our proposed RLVECN's architecture, and it succeeds the RL layers. The pooled feature maps, generated by the Representation Learning layers, contain high-level features extracted from the constituent actors of the social network structure. Hence, the classification layer utilizes these extracted "high-level features" for identifying clusters, based on the respective classes, contained in the social graph. In this regard, a MLP [5] function is defined as a mathematical function, f_c, mapping some set of input values, P, to their respective output labels, Y. In other words, $Y = f_c(P, \Theta)$, and Θ denotes a set of parameters. The MLP [4] function learns the values of Θ that will result in the best decision, Y, approximation for the input set, P. The MLP classifier output is a probability distribution which indicates the likelihood of a feature representation belonging to a particular output class. Our MLP [10] classifier is modelled such that sequential layers of NN units are stacked against each other to form a Deep Neural Network (DNN) structure [3,16].

Node Classification Algorithm: Defined via Algorithm 1.

Algorithm 1. Proposed Algorithm for Node Classification

Input: $\{V, E, Y_{lbl}\} \equiv \{$Actors, Ties, Ground-Truth Labels$\}$
Output: $\{Y_{ulb}\} \equiv \{$Predicted Labels$\}$

Preprocessing:
$V_{lbl}, V_{ulb} \subset V = V_{lbl} \cup V_{ulb}$ // V_{lbl} : Labelled actors   // V_{ulb} : Unlabelled actors

$E : (u_i, v_j) \in \{U \times V\}$ // $(u_i, v_j) \equiv$ (source, target)
$E_{train} = E_t : u_i, v_j \in V_{lbl}$ // $|E_{train}| = \sum indegree(V_{lbl}) + \sum outdegree(V_{lbl})$
$E_{pred} = E_p : u_i, v_j \in V_{ulb}$

$f_c \leftarrow$ Initialize   // Construct classifier model

Training:
for $t \leftarrow 0$ **to** $|E_{train}|$ **do**
    $f : E_t \rightarrow [X \in \mathbb{R}^q]$ // Embedding operation
    $f_t \in F = (K * X)_t$ // Convolution operation
    $r_t \in R = g(F) = max(0, f_t)$
    $p_t \in P = h(R) = maxPool(r_t)$
    $f_c|\Theta : p_t \rightarrow Y_{lbl}$ // MLP classification operation
end for

return $Y_{ulb} = f_c(E_{pred}, \Theta)$

2.3   Proposed Architecture/Framework

Figure 2 illustrates the architecture of our proposition, RLVECN.

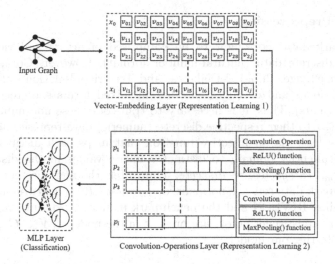

Fig. 2. Proposed system architecture

3   Data Sets and Materials

3.1   Data Sets

With regard to Table 1 herein, six (6) real-world benchmark social-graph data sets were employed for experimentation and evaluation, viz: Cora [20,23], CiteSeer [20,23], Facebook Page-Page webgraph [21], PubMed-Diabetes [17], Internet-Industry partnerships [2,12], and Terrorists-Relationship [29].

Table 1. Benchmark data sets

Data set	Classes → {label: 'description'}
Cora	$G(V, E) = G(2708, 5429)$
	{C1: 'Case_Based', C2: 'Genetic_Algorithms', C3: 'Neural_Networks', C4: 'Probabilistic_Methods', C5: 'Reinforcement_Learning', C6: 'Rule_Learning', C7: 'Theory'}
CiteSeer	$G(V, E) = G(3312, 4732)$
	{C1: 'Agents', C2: 'Artificial Intelligence', C3: 'Databases', C4: 'Information Retrieval', C5: 'Machine Learning', C6: 'Human-Computer Interaction'}
Facebook Page2Page	$G(V, E) = G(22470, 171002)$
	{C1: 'Companies', C2: 'Governmental Organizations', C3: 'Politicians', C4: 'Television Shows'}
PubMed Diabetes	$G(V, E) = G(19717, 44338)$
	{C1: 'Diabetes Mellitus - Experimental', C2: 'Diabetes Mellitus - Type 1', C3: 'Diabetes Mellitus - Type 2'}
Internet Industry	$G(V, E) = G(219, 631)$
	{C1: 'Content Sector', C2: 'Infrastructure Sector', C3: 'Commerce Sector'}
Terrorists Relation	$G(V, E) = G(851, 8592)$
	{C1: 'Content Sector', C2: 'Infrastructure Sector', C3: 'Commerce Sector'}

3.2   Data Preprocessing

All benchmark data sets ought to be comprised of actors and ties already encoded as discrete data (natural-number format). However, Cora, CiteSeer, Facebook-Page2Page, PubMed-Diabetes, and Terrorists-Relation data sets are made up of nodes and/or edges encoded in mixed formats (categorical and numerical formats). Thus, it is necessary to transcode these non-numeric (categorical) entities to their respective discrete (numeric) data representation, without semantic loss, via an injective function that maps each distinct entry in the categorical-entity domain to distinct numeric values in the discrete-data codomain, $f_m : categorical \rightarrow discrete$. Thereafter, the numeric representation of all benchmark data sets are normalized, $f_n : discrete \rightarrow continuous$, prior to training against RLVECN and the benchmark models. Also, only edgelist ties, $E \subset G$, whose constituent actors are present in the nodelist, $V \subset G$, were used for training/testing our model.

Table 2. Configuration of experimentation hyperparameters

Training Set: 80%	Test Set: 20%	Network Width: 640
Batch Size: 256	Optimizer: $AdaMax$	Network Depth: 6
Epochs: $1.8 * 10^2$	Activation: $ReLU$	Dropout: $4.0 * 10^{-1}$
Learning Rate: $1.0 * 10^{-3}$	Learning Decay: 0.0	Embed Dimension: 100

4   Experiment, Results, and Discussions

RLVECN's experimentation setup was tuned in accordance with the hyperparameters shown in Table 2. Our evaluations herein were recorded with reference to a range of objective functions. Thus, Categorical Cross Entropy was employed as the cost/loss function; while the fitness/utility was measured based on the following metrics: Precision (PC), Recall (RC), F-measure or F1-score (F1), Accuracy (AC), and Area Under the Receiver Operating Characteristic Curve (RO). Moreover, the objective functions have been computed against each benchmark data set with regard to the constituent classes (or categories) present in each data set. The Support (SP) represents the number of ground-truth samples per class/category for each data set.

   In a bid to avoid sample bias across-the-board, we have used exactly the same SP for all models inclusive of RLVECN model. However, since RLVECN is based on an edge-sampling technique; the SP recorded against RLVECN model represent the numbers of edges/ties used for computation as explained in Algorithm 1. With regard to the standard node-classification tasks herein, the performance of our RLVECN model during benchmarking against five(5) popular baselines (DeepWalk, GCN, LINE, Node2Vec, SDNE); and when evaluated against the validation/test samples for the benchmark data sets are as documented in Table 3, 4, 5, and 6 respectively. Consequently, Fig. 3 graphically shows the

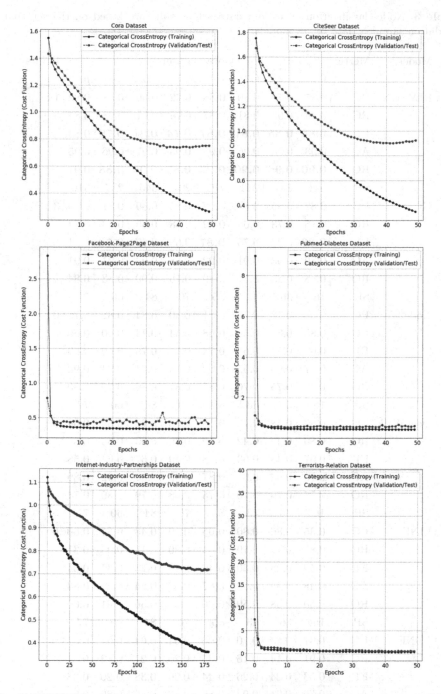

Fig. 3. RLVECN's learning curves during node-classification training against Cora, CiteSeer, Facebook-Page2Page, PubMed-Diabetes, Internet-Industry-Partnership, and Terrorists-Relation data sets (loss function *vs* training epochs)

Table 3. Node-classification over Cora data set. Results are based on the set apart validation sample - data set *vs* models.

Model	Metric	Cora Dataset								Points
		C1	C2	C3	C4	C5	C6	C7	μ	
RLVECN	PC	0.85	0.78	0.80	0.88	0.72	0.90	0.81	0.82	14
	RC	0.86	0.93	0.81	0.87	0.75	0.91	0.78	0.84	
	F1	**0.86**	**0.85**	**0.81**	**0.87**	**0.74**	**0.91**	**0.79**	0.83	
	AC	0.93	0.98	0.96	0.95	0.93	0.97	0.96	0.95	
	RO	**0.90**	**0.96**	**0.90**	**0.92**	**0.85**	**0.95**	**0.88**	0.91	
	SP	541	134	214	405	294	345	237	310	
GCN	PC	0.87	0.95	0.89	0.92	0.85	0.89	0.87	0.89	3
	RC	0.85	0.73	0.65	0.82	0.58	0.85	0.73	0.74	
	F1	**0.86**	0.83	0.75	**0.87**	0.69	0.87	**0.79**	0.81	
	AC	0.89	0.93	0.91	0.92	0.88	0.93	0.91	0.91	
	RO	0.88	0.83	0.80	0.89	0.75	0.90	0.83	0.84	
	SP	164	36	43	85	70	84	60	77	
Node2Vec	PC	0.58	0.78	0.72	0.81	0.80	0.84	0.82	0.76	0
	RC	0.85	0.50	0.53	0.68	0.64	0.74	0.60	0.65	
	F1	0.69	0.61	0.61	0.74	0.71	0.78	0.69	0.69	
	AC	0.77	0.96	0.95	0.92	0.93	0.94	0.94	0.92	
	RO	0.79	0.75	0.76	0.83	0.81	0.86	0.79	0.80	
	SP	164	36	43	85	70	84	60	77	
DeepWalk	PC	0.57	0.58	0.72	0.58	0.68	0.72	0.63	0.64	0
	RC	0.80	0.42	0.42	0.59	0.39	0.63	0.65	0.56	
	F1	0.67	0.48	0.53	0.58	0.49	0.67	0.64	0.58	
	AC	0.76	0.94	0.94	0.87	0.90	0.90	0.92	0.89	
	RO	0.77	0.70	0.70	0.75	0.68	0.79	0.80	0.74	
	SP	164	36	43	85	70	84	60	77	
LINE	PC	0.35	0.86	0.80	0.65	0.50	0.43	0.61	0.60	0
	RC	0.85	0.17	0.19	0.35	0.20	0.15	0.23	0.31	
	F1	0.50	0.28	0.30	0.46	0.29	0.23	0.34	0.34	
	AC	0.48	0.94	0.93	0.87	0.87	0.84	0.90	0.83	
	RO	0.59	0.58	0.59	0.66	0.59	0.56	0.61	0.60	
	SP	164	36	43	85	70	84	60	77	
SDNE	PC	0.37	0.83	0.70	0.60	0.54	0.64	0.64	0.62	0
	RC	0.91	0.14	0.16	0.35	0.20	0.27	0.12	0.31	
	F1	0.53	0.24	0.26	0.44	0.29	0.38	0.20	0.33	
	AC	0.50	0.94	0.93	0.86	0.87	0.86	0.89	0.84	
	RO	0.62	0.57	0.58	0.65	0.59	0.62	0.55	0.60	
	SP	164	36	43	85	70	84	60	77	

Table 4. Node-classification over CiteSeer data set. Results are based on the set apart test sample - data set *vs* models.

Model	Metric	CiteSeer Dataset							Points
		C1	C2	C3	C4	C5	C6	μ	
RLVECN	PC	0.76	0.81	0.78	0.43	0.88	0.60	0.71	12
	RC	0.84	0.83	0.79	0.60	0.79	0.65	0.75	
	F1	**0.80**	**0.82**	**0.79**	**0.50**	**0.83**	**0.63**	0.73	
	AC	0.93	0.88	0.92	0.93	0.96	0.89	0.92	
	RO	**0.90**	**0.87**	**0.87**	**0.78**	**0.89**	**0.79**	0.85	
	SP	304	609	377	107	225	275	316	
GCN	PC	0.80	0.78	0.86	0.95	0.91	0.75	0.84	2
	RC	0.76	0.76	0.73	0.08	0.67	0.54	0.59	
	F1	0.78	0.77	**0.79**	0.15	0.77	**0.63**	0.65	
	AC	0.88	0.87	0.88	0.91	0.89	0.83	0.88	
	RO	0.84	0.83	0.83	0.53	0.81	0.72	0.76	
	SP	119	134	140	50	102	118	111	
Node2Vec	PC	0.57	0.55	0.49	0.33	0.55	0.38	0.48	0
	RC	0.55	0.60	0.66	0.06	0.45	0.40	0.45	
	F1	0.56	0.58	0.56	0.10	0.50	0.39	0.45	
	AC	0.85	0.82	0.78	0.92	0.86	0.78	0.84	
	RO	0.73	0.74	0.74	0.53	0.69	0.63	0.68	
	SP	119	134	140	50	102	118	111	
DeepWalk	PC	0.46	0.53	0.43	0.43	0.47	0.33	0.44	0
	RC	0.51	0.54	0.57	0.06	0.41	0.32	0.40	
	F1	0.49	0.54	0.49	0.11	0.44	0.32	0.40	
	AC	0.81	0.81	0.75	0.92	0.84	0.76	0.82	
	RO	0.69	0.71	0.69	0.53	0.66	0.59	0.65	
	SP	119	134	140	50	102	118	111	
SDNE	PC	0.37	0.50	0.24	0.20	0.45	0.31	0.35	0
	RC	0.19	0.27	0.77	0.02	0.14	0.09	0.25	
	F1	0.25	0.35	0.36	0.04	0.21	0.14	0.23	
	AC	0.80	0.80	0.42	0.92	0.84	0.80	0.76	
	RO	0.56	0.60	0.55	0.51	0.55	0.52	0.55	
	SP	119	134	140	50	102	118	111	
LINE	PC	0.18	0.30	0.28	0.60	0.22	0.27	0.31	0
	RC	0.15	0.47	0.39	0.06	0.12	0.21	0.23	
	F1	0.16	0.36	0.32	0.11	0.15	0.24	0.22	
	AC	0.72	0.67	0.65	0.93	0.80	0.76	0.76	
	RO	0.50	0.59	0.56	0.53	0.52	0.55	0.54	
	SP	119	134	140	50	102	118	111	

learning-progress curves during the node-classification tasks using our proposed RLVECN model; and when training over the benchmark data sets. Hence, the dotted-black lines represent learning progress over the training set; and the dotted-blue lines represent learning progress over the test set.

Tables 3, 4, 5, and 6 have clearly tabulated our results as a multi-classification task over the benchmark data sets. Thus, for each class per data set, we have laid emphasis on the F1 (the weighted average of the PC and RC metrics) and

Table 5. Node-classification experiment over Facebook Page-Page webgraph and PubMed-Diabetes data sets. Results are based on the reserved test sample - data set *vs* models. N.B.: Mtc = Metric (fitness function); Pts = Points.

Model	Mtc	Facebook-Page2Page					Pts	PubMed-Diabetes				Pts
		C1	C2	C3	C4	μ		C1	C2	C3	μ	
RLVECN	PC	0.87	0.95	0.91	0.87	0.90	8	0.76	0.83	0.84	0.81	6
	RC	0.84	0.85	0.85	0.86	0.85		0.60	0.88	0.91	0.80	
	F1	**0.85**	**0.90**	**0.88**	**0.86**	0.87		**0.67**	**0.86**	**0.87**	0.80	
	AC	0.96	0.90	0.94	0.97	0.94		0.89	0.88	0.90	0.89	
	RO	**0.97**	**0.97**	**0.98**	**0.98**	0.98		**0.92**	**0.94**	**0.95**	0.94	
	SP	9989	33962	16214	6609	16694		3300	7715	7170	6062	
Node2Vec	PC	0.81	0.84	0.81	0.84	0.83	0	0.74	0.47	0.49	0.57	0
	RC	0.82	0.87	0.85	0.67	0.80		0.03	0.65	0.55	0.41	
	F1	0.81	0.85	0.83	0.74	0.81		0.05	0.55	0.52	0.37	
	AC	0.89	0.91	0.91	0.93	0.91		0.80	0.57	0.60	0.66	
	RO	0.87	0.90	0.89	0.82	0.87		0.51	0.58	0.59	0.56	
	SP	1299	1376	1154	665	1124		821	1575	1548	1315	
DeepWalk	PC	0.75	0.84	0.76	0.75	0.78	0	0.65	0.57	0.58	0.60	0
	RC	0.81	0.85	0.82	0.52	0.75		0.15	0.67	0.71	0.51	
	F1	0.78	0.84	0.79	0.62	0.76		0.24	0.62	0.63	0.50	
	AC	0.87	0.90	0.89	0.90	0.89		0.81	0.67	0.68	0.72	
	RO	0.85	0.89	0.87	0.75	0.84		0.56	0.67	0.69	0.64	
	SP	1299	1376	1154	665	1124		821	1575	1548	1315	
LINE	PC	0.53	0.66	0.72	0.66	0.64	0	0.48	0.42	0.44	0.45	0
	RC	0.72	0.71	0.59	0.29	0.58		0.05	0.60	0.46	0.37	
	F1	0.61	0.68	0.65	0.40	0.59		0.08	0.50	0.45	0.34	
	AC	0.73	0.80	0.83	0.87	0.81		0.79	0.51	0.56	0.62	
	RO	0.73	0.77	0.75	0.63	0.72		0.52	0.53	0.54	0.53	
	SP	1299	1376	1154	665	1124		821	1575	1548	1315	
SDNE	PC	0.49	0.80	0.70	0.65	0.66	0	0.65	0.43	0.74	0.61	0
	RC	0.90	0.63	0.50	0.19	0.56		0.05	0.96	0.17	0.39	
	F1	0.64	0.70	0.58	0.29	0.55		0.10	0.59	0.27	0.32	
	AC	0.70	0.84	0.82	0.86	0.81		0.80	0.48	0.65	0.64	
	RO	0.76	0.78	0.71	0.58	0.71		0.52	0.56	0.56	0.55	
	SP	1299	1376	1154	665	1124		821	1575	1548	1315	

Table 6. Node-classification experiment over Internet-Industry-Partnership and Terrorists-Relationship data sets. Results are based on the reserved validation sample - data set *vs* models. N.B.: Mtc = Metric (fitness function); Pts = Points.

Model	Mtc	Internet-Industry-Partnership				Pts	Terrorists-Relation Dataset					Pts
		C1	C2	C3	μ		C1	C2	C3	C4	μ	
RLVECN	PC	0.33	0.96	0.29	0.53	5	0.93	0.91	0.46	1.00	0.83	6
	RC	0.65	0.77	0.76	0.73		0.97	0.97	0.42	0.97	0.83	
	F1	**0.44**	**0.86**	0.42	0.57		**0.95**	**0.94**	0.44	**0.98**	0.83	
	AC	0.84	0.78	0.87	0.83		0.95	0.98	0.89	0.99	0.95	
	RO	**0.76**	**0.81**	**0.82**	0.80		**0.98**	**1.00**	0.85	**1.00**	0.96	
	SP	26	238	17	94		1706	491	319	561.	769	
GCN	PC	No experiment				0	0.94	0.74	0.67	0.96	0.83	5
	RC	due to the					0.90	0.95	0.60	1.00	0.86	
	F1	absence of					0.92	0.83	**0.63**	**0.98**	0.84	
	AC	vectorized feature					0.92	0.95	0.88	0.99	0.94	
	RO	set for this					**0.98**	0.99	**0.91**	**1.00**	0.97	
	SP	data set					92	21	30	27	43	
DeepWalk	PC	0.50	0.81	0.36	0.56	1	0.88	0.82	0.64	0.86	0.80	0
	RC	0.12	0.93	0.44	0.50		0.90	0.86	0.53	0.93	0.81	
	F1	0.20	**0.86**	0.40	0.49		0.89	0.84	0.58	0.89	0.80	
	AC	0.82	0.82	0.73	0.79		0.88	0.96	0.86	0.96	0.92	
	RO	0.55	0.79	0.62	0.65		0.88	0.92	0.73	0.95	0.87	
	SP	8	27	9	15		92	21	30	27	43	
Node2Vec	PC	0.00	0.68	0.75	0.48	1	0.86	0.82	0.60	0.86	0.79	0
	RC	0.00	1.00	0.33	0.44		0.88	0.86	0.50	0.93	0.79	
	F1	0.00	0.81	**0.46**	0.42		0.87	0.84	0.55	0.89	0.79	
	AC	0.82	0.70	0.84	0.79		0.86	0.96	0.85	0.96	0.91	
	RO	0.50	0.62	0.65	0.59		0.86	0.92	0.71	0.95	0.86	
	SP	8	27	9	15		92	21	30	27	43	
LINE	PC	0.00	0.61	0.00	0.20	0	0.82	0.82	0.58	0.92	0.79	0
	RC	0.00	1.00	0.00	0.33		0.92	0.86	0.37	0.85	0.75	
	F1	0.00	0.76	0.00	0.25		0.87	0.84	0.45	0.88	0.76	
	AC	0.82	0.61	0.80	0.74		0.85	0.96	0.84	0.96	0.90	
	RO	0.50	0.50	0.50	0.50		0.84	0.92	0.65	0.92	0.83	
	SP	8	27	9	15		92	21	30	27	43	
SDNE	PC	0.00	0.61	0.00	0.20	0	0.77	0.90	0.56	1.00	0.81	0
	RC	0.00	1.00	0.00	0.33		0.92	0.86	0.30	0.85	0.73	
	F1	0.00	0.76	0.00	0.25		0.84	0.88	0.39	0.92	0.76	
	AC	0.82	0.61	0.80	0.74		0.81	0.97	0.84	0.98	0.90	
	RO	0.50	0.50	0.50	0.50		0.80	0.92	0.62	0.93	0.82	
	SP	8	27	9	15		92	21	30	27	43	

RO. Therefore, we have highlighted the model which performed best (based on F1 and RO metrics) for each classification task using a **bold font**. Additionally, we have employed a point-based ranking standard to ascertain the fittest model for each node classification task. The model with the highest aggregate point signifies the fittest model for the specified task, and so on in a descending order of aggregate points. Accordingly, as can be seen from our tabular results, our proposed methodology (RLVECN) is at the top with the highest fitness points.

$L2$ regularization ($L2 = 0.04$) [7] and early stopping [18] were used herein as addon regularization techniques to overcome overfitting incurred during RLVECN's training. Hence, the application of early stopping with respect to RLVECN's training over the benchmark data sets were, viz: 50 epochs (Cora, CiteSeer, Facebook-Page2Page, PubMed-Diabetes, Terrorists-Relation) and 180 epochs (Internet-Industry-Partnership). We have used a mini-batch size of 256 for training and testing/validating because we want to ensure that sufficient patterns are extracted by RLVECN during training before its network weights are updated.

5   Limitations, Conclusion, Future Work, and Acknowledgements

The benchmark models evaluated herein were implemented using their default parameters. We were not able to evaluate GCN [11] against Facebook Page-Page webgraph, PubMed-Diabetes, and Internet-Industry-Partnership data sets; because each of these aforementioned data sets does not possess a vectorized feature set which is required by GCN model for input-data processing. Overall, RLVECN's remarkable performance with respect to our benchmarking results can be attributed to the following:

(1) The RL kernel of RLVECN is constituted of two (2) distinct layers of FL, viz: Knowledge-Graph Embeddings and ConvNet [13].
(2) The high-quality data preprocessing techniques employed herein with respect to the benchmark data sets. We ensured that all constituent actors of a given social graph were transcoded to their respective discrete data representations, without any loss in semantics, and normalized prior to training and/or testing.

In conclusion, we intend to expand our experimentation scope to include more social network data sets and benchmark models. This research was made possible by International Business Machines (IBM), SHARCNET and Compute Canada (www.computecanada.ca).

References

1. Aggarwal, C.C. (ed.): Social Network Data Analytics. Springer, Boston (2011). https://doi.org/10.1007/978-1-4419-8462-3

2. Batagelj, V., Doreian, P., Ferligoj, A., Kejzar, N. (eds.): Understanding Large Temporal Networks and Spatial Networks: Exploration, Pattern Searching. Visualization and Network Evolution. Wiley, Hoboken (2014)
3. Bengio, Y.: Learning deep architectures for AI. Found. Trends Mach. Learn. **2**, 1–113 (2009)
4. Deng, L.M., Yu, D.H.: Deep Learning: Methods and Applications. Foundations and trends in signal processing. Now Publishers (2014)
5. Goodfellow, I.G., Bengio, Y., Courville, A.C.: Deep learning. Nature **521**, 436–444 (2015)
6. Goodfellow, I.G., Bengio, Y., Courville, A.C. (eds.): Deep Learning. MIT Press, Cambridge (2017)
7. Gron, A. (ed.): Hands-On Machine Learning with Scikit-Learn and TensorFlow: Concepts, Tools, and Techniques to Build Intelligent Systems. O'Reilly Media Inc., Newton (2017)
8. Grover, A., Leskovec, J.: node2vec: scalable feature learning for networks. In: Proceedings of the 22nd ACM SIGKDD International Conference on Knowledge Discovery and Data Mining 2016, pp. 855–864 (2016)
9. Hinton, G.E.: Learning multiple layers of representation. TRENDS Cognit. Sci. **11**(10), 428–433 (2007)
10. Hinton, G.E., et al.: Deep neural networks for acoustic modeling in speech recognition. IEEE Sig. Process. Mag. **29**, 82–97 (2012)
11. Kipf, T.N., Welling, M.: Semi-supervised classification with graph convolutional networks. In: International Conference on Learning Representations (ICLR), abs/1609.02907 (2017)
12. Krebs, V.E.: Organizational adaptability quotient. In: IBM Global Services (2008)
13. Molokwu, B.C.: Event prediction in complex social graphs using one-dimensional convolutional neural network. In: Proceedings of the 28th International Joint Conference on Artificial Intelligence, IJCAI (2019)
14. Molokwu, B.C.: Event prediction in social graphs using 1-dimensional convolutional neural network. In: Canadian Conference on AI (2019)
15. Molokwu, B.C., Kobti, Z.: Event prediction in complex social graphs via feature learning of vertex embeddings. In: Gedeon, T., Wong, K.W., Lee, M. (eds.) Neural Information Processing, pp. 573–580. Springer, Cham (2019). https://doi.org/10.1007/978-1-4419-8462-3
16. Molokwu, B.C., Kobti, Z.: Spatial event prediction via multivariate time series analysis of neighboring social units using deep neural networks. In: 2019 International Joint Conference on Neural Networks (IJCNN), pp. 1–8 (2019)
17. Namata, G., London, B., Getoor, L., Huang, B.: Query-driven active surveying for collective classification. In: Proceedings of the Workshop on Mining and Learning with Graphs, MLG-2012 (2012)
18. Patterson, J., Gibson, A. (eds.): Deep Learning: A Practitioner's Approach. O'Reilly Media Inc., Newton (2017)
19. Perozzi, B., Al-Rfou', R., Skiena, S.: Deepwalk: online learning of social representations. In: Proceedings of the 20th ACM SIGKDD International Conference on Knowledge Discovery and Data Mining, abs/1403.6652 (2014)
20. Rossi, R.A., Ahmed, N.K.: The network data repository with interactive graph analytics and visualization. In: Proceedings of the Twenty-Ninth AAAI Conference on Artificial Intelligence (2015). http://networkrepository.com
21. Rozemberczki, B., Allen, C., Sarkar, R.: Multi-scale attributed node embedding (2019). arxiv: abs/1909.13021

22. Scott, J. (ed.): Social Network Analysis. SAGE Publications Ltd., Newbury Park (2017)
23. Sen, P., Namata, G., Bilgic, M., Getoor, L., Gallagher, B., Eliassi-Rad, T.: Collective classification in network data. AI Mag. **29**, 93–106 (2008)
24. Tabacof, P., Costabello, L.: Probability calibration for knowledge graph embedding models. In: International Conference on Learning Representations (ICLR), abs/1912.10000 (2020)
25. Tang, J., Qu, M., Wang, M., Zhang, M., Yan, J., Mei, Q.: Line: large-scale information network embedding. In: Proceedings of the 24th International Conference on World Wide Web (2015)
26. Wang, D., Cui, P., Zhu, W.: Structural deep network embedding. In: Proceedings of the 22nd ACM SIGKDD International Conference on Knowledge Discovery and Data Mining (2016)
27. Yang, S., Tian, J., Zhang, H., Yan, J., He, H., Jin, Y.: Transms: knowledge graph embedding for complex relations by multidirectional semantics. In: Proceedings of the 28th International Joint Conference on Artificial Intelligence, IJCAI (2019)
28. Zhang, Q., Sun, Z., Hu, W., Chen, M., Guo, L., Qu, Y.: Multi-view knowledge graph embedding for entity alignment. In: Proceedings of the 28th International Joint Conference on Artificial Intelligence, IJCAI. abs/1906.02390 (2019)
29. Zhao, B., Sen, P., Getoor, L.: Entity and relationship labeling in affiliation networks. In: Proceedings of the 23rd International Conference on Machine Learning (2006)

Accelerated Gaussian Convolution in a Data Assimilation Scenario

Pasquale De Luca[1]([✉]) [iD], Ardelio Galletti[2] [iD], Giulio Giunta[2] [iD],
and Livia Marcellino[2] [iD]

[1] Department of Computer Science, University of Salerno, Fisciano, Italy
p.deluca16@studenti.unisa.it
[2] Department of Science and Technology, University of Naples Parthenope,
Naples, Italy
{ardelio.galletti,giulio.giunta,livia.marcellino}@uniparthenope.it

Abstract. Machine Learning algorithms try to provide an adequate forecast for predicting and understanding a multitude of phenomena. However, due to the chaotic nature of real systems, it is very difficult to predict data: a small perturbation from initial state can generate serious errors. Data Assimilation is used to estimate the best initial state of a system in order to predict carefully the future states. Therefore, an accurate and fast Data Assimilation can be considered a fundamental step for the entire Machine Learning process. Here, we deal with the Gaussian convolution operation which is a central step of the Data Assimilation approach and, in general, in several data analysis procedures. In particular, we propose a parallel algorithm, based on the use of Recursive Filters to approximate the Gaussian convolution in a very fast way. Tests and experiments confirm the efficiency of the proposed implementation.

Keywords: Gaussian convolution · Recursive filters · Parallel algorithms · GPU

1 Introduction

Data Assimilation (DA) is a *prediction-correction* method for combining a physical model with observations. The Data Assimilation and the Machine Learning (ML) fields are closely related to each other. Machine Learning process is used to perform a specific task without using explicit instructions and it can be seen as a subset of the Artificial Intelligence (AI) field, because it creates new methods and applications for analyze and classify many natural phenomena (see for example [1]). In general, this process consists in two main phases: the *analysis phase* - some collected data are analyzed to detect patterns that help to create explicit features or parameters; the *training phase* - data parameter generated in the previous phase are used to create Machine Learning models.

However, the learning part of the training phase relies on a relevant training data-set, containing samples of spatio-temporal dependent structures. In many

© Springer Nature Switzerland AG 2020
V. V. Krzhizhanovskaya et al. (Eds.): ICCS 2020, LNCS 12142, pp. 199–211, 2020.
https://doi.org/10.1007/978-3-030-50433-5_16

fields, there is an absence of direct observations of the random variables, and therefore, learning techniques cannot be readily deployed.

Therefore, a correct training dataset is a basic need to get a right learning. This is because in order to perform a correct ML approach, often a classifier is used in the analyze phase. Each classifier is composed by a kernel which aims to correctly predict the classes by using a higher-dimension feature space to make data almost linearly separable. In order to compute a fair classification and accurate prediction a suitable method could be chosen [23].

The variational approach of the Data Assimilation process, characterized by a cost function minimization, is a good choice for classification. Numerically, this means to apply an iterative procedure using a covariance matrix defined by measuring the error between predictions and observed data. Here, we are interested in those numerical issues. In particular, since the error covariance matrix presents a Gaussian correlation structure, the Gaussian convolution process plays a key role in such a problem. Furthermore, it should be noted that, beyond its fundamental role in the Data Assimilation field, the convolution operation is always significant in the computational process of most big-data analysis problems. Hence, a correct Machine Learning process can use it as a basic step in the analysis phase. Moreover, because of the need to process large amount of data, parallel approaches and High Performance Computing (HPC) architectures, as multicore or Graphics Processing Units (GPUs), are mandatory [2–4]. In this direction, some recent papers deal with parallel data assimilation [5–7] but we just limit our attention to the basic step represented by a parallel implementation for the Gaussian Convolution. In particular, we propose an accelerated procedure to approximate the Gaussian convolution which is based on Recursive Filters (RFs). In fact, Gaussian RFs have been designed to provide an accurate and very efficient approximated Gaussian convolution [8–11]. Since the use of RFs is mainly suitable to overcome a large execution time, when there is a lot of data to analyze, many parallel implementations have been presented (see survey in [12]). Here, we propose a novel implementation that exploits the computational power of the GPUs which are very useful for solving numerical problems in several application fields [13,14].

More precisely, to manage big size input data, the parallelization strategy is based on a domain decomposition approach with overlapping, so that all possible interactions between forecasts and observations are included. In this way, this computational step becomes a very fast kernel specifically designed for exploiting the dynamic parallelism [15] approach available on the Compute Unified Device Architecture (CUDA) [16].

The paper is organized as follows. Section 2 recalls the variational Data Assimilation problem, and the use of the Recursive Filter to approximate the discrete Gaussian convolution. In Sect. 3, the underlying domain decomposition strategy and the GPU-CUDA parallel algorithm are provided. The experiments in Sect. 4 confirm the efficiency of the proposed implementation in terms of performance. Finally conclusions are drawn in Sect. 5.

2 Gaussian Convolutions in Data Assimilation

In this section, we show how the Gaussian convolution is involved in a Data Assimilation scenario. In particular, let us consider a three-dimensional variational data assimilation problem [17]: the objective is to give a best estimate of x, that is called the analysis or state vector, once a prior estimate vector x^b (background), usually provided by a numerical forecasting model, and a vector $y = \mathcal{H}(x) + \delta y$ of observations, related to the nonlinear model $\mathcal{H}$, are given. The unknown x solves the regularized constrained least-squared problem:

$$\min_x J(x) = \min_x \left[\|y - \mathcal{H}(x)\|^2 + \|x - x^b\|^2 \right], \tag{1}$$

where J denotes the objective function to minimize. Here, $\|x - x^b\|^2$ is a penalty term and $\|y - \mathcal{H}(x)\|^2$ is a quadratic data-fidelity term which compares measured data and solution obtained by the nonlinear model $\mathcal{H}$ [10]. In this scheme, the background error $\delta x = x^b - x$ and the observational error $\delta y = y - \mathcal{H}(x)$ are assumed to be random variables with zero mean and covariance matrices

$$\mathbf{B} = <\delta x, \delta x^T> \quad \text{and} \quad \mathbf{R} = <\delta y, \delta y^T>,$$

respectively. Following description in [9], let the matrix $\mathbf{H}$ be a first-order approximation of the Jacobian of $\mathcal{H}$ at x^b and denote by

$$d = y - \mathcal{H}(x^b)$$

the so-called *misfit*. Denoting by $\mathbf{V}$ the unique symmetric Gaussian matrix such that $\mathbf{V}^2 = \mathbf{B}$, and by introducing the variable $v = \mathbf{V}^{-1}\delta x$, the problem (1) can be proven to be equivalent to [9,18,19]:

$$\min_v \tilde{J}(v) = \min_v \frac{1}{2}(d - \mathbf{H}\mathbf{V}v)^T \mathbf{R}^{-1}(d - \mathbf{H}\mathbf{V}v) + \frac{1}{2}v^T v. \tag{2}$$

The minimization of the cost function $\tilde{J}(v)$ leads to the linear system:

$$(I + \mathbf{V}\Psi\mathbf{V})v = \mathbf{V}\mathbf{H}^T\mathbf{R}^{-1}d. \tag{3}$$

Since $I + \mathbf{V}\Psi\mathbf{V}$ is symmetric, the linear system (3) that can be handled by means of the CG method, whose basic operation is the matrix-vector multiplication:

$$(I + \mathbf{V}\Psi\mathbf{V})\rho = \rho + \mathbf{V}\Psi\mathbf{V}\rho.$$

Here, $\Psi = \mathbf{H}^T\mathbf{R}^{-1}\mathbf{H}$ is a diagonal matrix and ρ denotes the residual at the current step of the CG algorithm. More precisely, it turns out that such an operation involves three discrete Gaussian convolutions:

$$\mathbf{V}\rho, \qquad \mathbf{V}(\Psi\mathbf{V}\rho), \qquad \mathbf{V}(\mathbf{H}^T\mathbf{R}^{-1}d). \tag{4}$$

In conclusion, previous analysis shows that Gaussian convolution becomes a main kernel for Data Assimilation. From here comes the need to implement accurate and fast methods to perform it. In fact, in the described context, the matrix $\mathbf{V}$ is neither effectively used nor even assembled, and the matrix-vector multiplications in (4) are computed by introducing the so-called Gaussian Recursive Filters. It has been proved that these tools offer good accuracy and bring down the computational cost in time and space [20, 21].

In particular, in this work we just consider K-iterated first-order Gaussian RFs and follow the approach and notation used in [8]. Let:

$$s^{(0)} = \left\{ s_j^{(0)} \right\}_{j \in \mathbf{Z}} = \left(\dots, s_{-1}^{(0)}, s_0^{(0)}, s_1^{(0)}, \dots \right)$$

be an input signal and let g denote the Gaussian function with zero mean and standard deviation σ. The Gaussian filter is a filter whose response to the input $s^{(0)}$ is given by the discrete Gaussian convolution:

$$s_j^{(g)} = \left(g * s^{(0)} \right)_j = \sum_{t \in \mathbf{Z}} g_{j-t} s_t^{(0)}, \qquad \forall j \in \mathbf{Z}, \tag{5}$$

where $g_t \equiv g(t)$. A K-iterated first-order Gaussian recursive filter generates an output signal $s^{(K)}$, the so-called K-iterate approximation of $s^{(g)}$, whose entries solve the $2K$ recurrence relations:

$$p_j^{(k)} = \beta s_j^{(k-1)} + \alpha p_{j-1}^{(k)}, \qquad \forall j \in \mathbf{Z}, \tag{6}$$

$$s_j^{(k)} = \beta p_j^{(k)} + \alpha s_{j+1}^{(k)}, \qquad \forall j \in \mathbf{Z}. \tag{7}$$

$(k = 1, \dots, K)$ where values α and $\beta = 1 - \alpha$ are called *smoothing coefficients* and verify:

$$\alpha = 1 + E_\sigma - \sqrt{E_\sigma(E_\sigma + 2)}, \qquad \beta = \sqrt{E_\sigma(E_\sigma + 2)} - E_\sigma, \tag{8}$$

with $E_\sigma = K\sigma^{-2}$. It has been proved that as $K \to \infty$ the filter converges to the Gaussian filter [22]. If we consider a finite size input signal $s^{(0)}$ (i.e. with support in the grid $\{0, 1, 2, \dots, N - 1\}$) then the index j has to be used in increasing order in (6) and decreasing order in (7). Hence, relations (6) and (7) are suitable called advancing and backing filters, respectively [8]. We highlight that to prime the algorithm these filters requires to set values $p_0^{(k)}$ and $s_{N-1}^{(k)}$. This can be done using the boundary conditions [24]:

$$p_0^{(k)} = \frac{1}{1 + \alpha} s_0^{(k-1)}, \qquad s_{N-1}^{(k)} = \frac{1}{1 + \alpha} p_{N-1}^{(k)}$$

which are derived to simulate the effect of the neglected entries when using finite size input signals. Typically a well-known edge effect, i.e. a large perturbation error, can be seen on the boundary entries of the output. In [8], provided that the input support is in $[0, N - 1]$, this effect can be mitigated by increasing the input size including and putting artificial zero entries at the left and right boundaries of the input. Algorithm 1 describes a K-iterated first-order Gaussian RF straight implementation.

Algorithm 1. K-iterated first-order RF with boundary conditions

---

Input: $s^{(0)}$, σ, K

Output: $s^{(K)}$

1: set β, α as in (8); $M := 1/(1+\alpha)$

2: set $N := \texttt{size}(s^{(0)})$

3: for $k = 1, 2, \ldots, K$ % filter loop

4:      compute $p_0^{(k)} := M s_0^{(k-1)}$ % left end condition

5:      if $k = 1$ then

6:          $p_0^{(k)} := \beta s_0^{(k-1)}$

7:      end

8:      for $j = 1, \ldots, N-1$ % advancing filter

9:          $p_j^{(k)} := \beta s_j^{(k-1)} + \alpha p_{j-1}^{(k)}$

10:     endfor

11:     compute $s_{N-1}^{(k)} := M p_{N-1}^{(k)}$ % right end condition

12:     for $j = N-2, \ldots, 0$ % backing filter

13:         $s_j^{(k)} := \beta p_j^{(k)} + \alpha s_{j+1}^{(k)}$

14:     endfor

15: endfor

---

3   Parallel Approach and GPU Algorithm

In this section we give a description of our parallel algorithm, and the related strategy, to implement a fast and accurate version of the K-iterated first-order Gaussian RF. This approach exploits the main features of the GPU environment. The main idea relies on several macro steps in order to obtain a reliable and performing computation. The whole process can be partitioned in three steps. In the first phase, step 1, in order to perform a fair workload distribution, we use a Domain Decomposition (DD) approach with overlapping. More specifically, the strategy consists in splitting the input signal $s^{(0)}$ into t local blocks, one for each thread:

$$s_0^{(0),m}, \ s_1^{(0),m}, \ldots, \ s_{t-1}^{(0),m}. \tag{9}$$

Here, N denotes the problem size, while:

$$d = \left\lfloor \frac{N}{t} \right\rfloor \quad \text{and} \quad r = \text{mod}(N, t) \tag{10}$$

are the quotient and the remainder when dividing N by t, respectively. Moreover, the parameter m denotes the overlapping size. To be specific, each thread j loads in own local memory the block $s_j^{(0),m}$, whose size is $d + 2m$ or $d + 1 + 2m$ (depending on j). The entries of the j-th local block are formally defined using the subdivision:

$$
\left(s_j^{(0),m}\right)_i = \begin{cases} s_{jd+j+i-m}^{(0)}, & i = 0, \ldots, d+2m & \text{if } j < r \\ s_{jd+r+i-m}^{(0)}, & i = 0, \ldots, d+2m-1 & \text{if } j \geq r \end{cases} \tag{11}
$$

where the input signal entries are set to zero, when not available ($s_i^{(0)} = 0$ for $i < 0$ and $i \geq N$).

In other words, this partitioning consists in assigning to each thread a part of the signal, so that two consecutive threads have consecutive signal blocks and those blocks overlap on the edges by sharing exactly $2m$ entries. The reason of overlapping is because, to perform a good approximation of the convolution, block edge values need to use close values that lie in the neighboring blocks. We notice that by setting $m = 0$, i.e. by excluding the overlapping areas, could create possible perturbation errors and generate a bad accuracy close to the boundaries of the local output signals.

The **step** 2 deals with the approximated local Gaussian convolution for each block. More precisely, each thread **j** performs the K-iterated first-order Gaussian RF to $s_j^{(0),m}$, by applying Algorithm 1, and computes $s_j^{(K),m}$.

The last phase, **step** 3, is related to collect the local approximated results by loading them into a global output signal. Therefore, in order to remove the first and last m entries, a *resizing* operation is firstly performed for each local output. More in details, each thread **j** resizes the local computed signal $s_j^{(K),m}$, by removing its first and last m entries, and it generates the local output $s_j^{(K)}$. Finally, a gathering of local resized outputs into the global output signal is done.

A very important consequence of our strategy is that all previous steps, which are summarized in the following parallel algorithm, can be computed by all threads in a fully-parallel way.

Algorithm 2. Parallel K-iterated first-order Gaussian recursive filter based on domain decomposition with overlapping

---

Input: $s^{(0)}$, σ, m, K, **t**

Output: $s^{(K)}$

1: FOR ALL THREAD **j**
2: save in the private memory of thread the extended input signal $s_j^{(0),m}$ as described in step 1 (domain decomposition with overlapping)
3: apply Algorithm 1 to $s_j^{(0),m}$ with parameters σ, K as described in step 2, to obtain $s_j^{(K),m}$
4: resize $s_j^{(K),m}$ to recover $s_j^{(K)}$ and copy it in the shared memory in order to obtain the global output $s^{(K)}$ as described in step 3
5: ENDFOR ALL THREAD **j**

Now, we discuss how the Algorithm 2 is implemented in a CUDA environment. Firstly, input data are transferred to device global memory. Hence, in order to guarantee a reliable workload distribution, the described domain decomposition in **step 1** is performed. More in detail, we set for each thread the local size $n_{loc} = d$ or $n_{loc} = d + 1$, depending on the threads number **j** and the value r in (10). This confirms that if the input size value n is not divisible by **t**, according to (11) a suitable workload distribution is done. By also considering the overlapping entries, the block length becomes $n_{loc} + 2m$, and each thread can retrieve from the global array the required amount of data needed for its local computation.

Moreover, an adequate access to the global memory is performed by means of a suitable indexing, i.e. every thread loads data from global memory and stores them in its own local memory in order to perform each operation independently. Thanks to this operation, any overhead due to the contention and synchronization of the global memory is avoided. In the following, the overall GPU parallel algorithm is shown.

Algorithm 3. GPU parallel implementation

Input: N, input_data[], K
1: % set overlapping size value m
2: % compute local size value n_loc
3: % define the extended local size
4: length = 2m + n_loc
5: % define the index of each thread
6: index = threadIdx.x+(blockDim.x × blockIdx.x)
7: % define the local chunk interval
8: chunk_idx = (index × n_loc)+((index+1) × n_loc)
9: % parallel work: begin
10: **for** each thread **do**
11:     % compute the chunk overlapped from input
12:     x_local[index] = input_data[chunk_idx + length]
13:     % start the dynamic parallelism region, by setting the threads number using the iteration number K
14:     **for** each thread in dynamic region **do**
15:         % compute forward & backward filter
16:         x_local[index]
17:     **end for**
18:     %end the dynamic parallelism region
19:     % collect local results
20:     results[n_loc] = x_local
21: **end for**
22: % parallel work: end
    Output: results[ ]

Shortly, in Algorithm 3, starting from the input data size N, the iteration number K and the input signal vector **input_data**, which are loaded in the global device memory, the procedure returns the approximated Gaussian convolution by the signal vector `results` which is computed in a GPU-parallel way. More in detail, Algorithm 3 highlights several memory and computation strategies.

To be specific, first operations provide, lines 1–8 to set the local stacks for each thread by considering the padding pieces related to the overlapping value m. Hence, according to `step 1` each thread performs a preliminary check of the local chunk by means of the local index `chunk_idx`. Therefore, if the left and the right side of the input data are provided, these values are added, otherwise m values, set to zero, are inserted on the overlapped positions. In lines 9–18 the computation phase is performed and a *dynamic parallelism* approach [15] has been applied, when possible. CUDA allows us to exploit the dynamic parallelism which is an extension to the CUDA programming model by enabling a CUDA kernel to configure new thread grids in order to launch new kernels for reducing the computational time. The aim of dynamic parallelism in our implementation consists in to the assignment, by each thread corresponding to each input portion, therefore for every CUDA kernel, to K threads by scheduling each thread in order to perform the forward and the backward filter operations as described in Algorithm 1, in synchronous way. More in details, K different threads perform the operations on each element following a pipeline modality. The usage of dynamic parallelism is able to obtain very low execution times, despite the predictable start-up and end-up times. The lines 19–22 are related to gathering of the local results of each thread in the global output. The copy operation is designed according to avoid memory contention, so that it is memory-safe because each thread carries out only the `n_loc` central elements of own local result by removing the $2m$ boundary values. This property guarantees a strong memory consistency.

4   Experimental Results

In this section, several experimental results highlight and confirm the reliability and the efficiency of the proposed software. Following, the technical specifications where the GPU-parallel algorithm has been implemented, are shown:

- two CPU Intel Xeon with 6 cores, E5-2609v3, 1.9 GHz, 32 GB of RAM, 4 channels 51 Gb/s memory bandwidth
- two NVIDIA GeForce GTX TITAN X, 3072 CUDA cores, 1 GHz Core clock for core, 12 GB DDR5, 336 GBs as bandwidth.

Thanks to GPUs' computational power, our algorithm exploits the CUDA framework in order to take best advantage of parallel environment. Our approach relies on an ad-hoc memory strategy which provides to increase the size of local stack heap memory for each thread and for each thread blocks'. Exploiting this technique, when a large amount of input data will be loaded, the memory access time is reduced. Previous operations are executed by using the following CUDA routine: `cudaDeviceSetLimit`, by setting as first parameter `cudaLimitMallocHeapSize` and second `cudaLimitStackSize`; while as size, according to hardware architecture, the value $1024 \times 1024 \times 1024$ is fixed.

This trick allows us to allocate the dynamic memory, by using `malloc` system-call, directly on the device.

Therefore, in order to increase the performances an additional memory-based operation has been done. More precisely, this operation relies on L2 cache obtaining a gain of performance by varying dynamically the fetch granularity. More in details, after that each thread blocks computation is completed, we perform a dynamic fetch granularity by using the CUDA routine `cudaDeviceSetLimit` and setting as parameters: `cudaLimitMaxL2FetchGranularity` and `128*sizeof(int)`. The value 128 is related to hardware architecture that can be support this range of data loading. Applying this approach an appreciable increasing of performance has been obtained by exploiting the memory cache's property that can recover the most used data and instructions during the execution. Accordingly, due to the canonical operations of Recursive Filters during their execution, a reduced memory access time is obtained by increasing the fetch granularity.

In other words, in a classical execution each thread accesses to the global memory to retrieve the required data for the computation. In this case, according to the memory hierarchy and the L2 memory strategy, each thread accesses first in the cache, then in the local stack, and finally in the heap/global memory. With this procedure a considerable gain in terms of performance has been achieved. In the following tests we set $\sigma = 2$ and input signals randomly distributed (Gaussian or uniform). The choice $m = 2.5\sigma = 5$ guarantees a good accuracy level, as shown in [8, 20].

Test 1. Here, in order to highlight the performance gain, we set as input: $N = 10^5$, $m = 5$, $K = 10$ and the thread number $t = 100$. Averaged times, related to 10 executions, achieved are:

- 7.24 s, without increasing fetch granularity,
- 6.93 s, by varying dynamic fetch granularity.

The first test highlights a small time difference but, if we give a large dataset input, which requires a large execution time even on GPU, thanks to the granularity of the dynamic recovery a significant performance gain can be obtained. Thus, the dynamic operations are closely related to the size of the input, i.e. according to cache granularity size chosen as parameter into function `cudaDeviceSetLimit`, where in this case the maximum value is fixed to 128 bytes. This experiment provides a comparison among serial and GPU parallel execution times. More precisely, in Table 1 the execution times for both serial (CPU) and parallel version (GPU) by choosing different input sizes and the iteration numbers are shown. The input parameters are set as: `Blocks` $\times$ `Threads` $= 10 \times 100$ and $m = 5$.

Table 1. Execution times (in seconds), Blocks × Threads = 10 × 100, $m = 5$.

K	N	1×10^4	5×10^4	1×10^5
10	GPU	0.274	3.243	10.219
	CPU	5.892	129.790	403.871
50	GPU	0.483	3.592	11.883
	CPU	6.293	149.592	518.560
100	GPU	0.596	4.711	12.703
	CPU	8.431	172.197	650.195
500	GPU	0.778	5.719	13.177
	CPU	9.324	195.150	757.912
1000	GPU	0.984	6.135	14.297
	CPU	129.790	223.542	875.442

Test 2. This experiment confirms the reliability effects when choosing different CUDA thread configurations. Here, we emphasize the different input sizes given, while the iteration number and the overlapping value are set to $K = 500$ and $m = 5$, respectively. Indeed, reduction of execution times has been achieved by decreasing the Blocks number, this holds true for all given input sizes. This phenomenon is related to a good synchronization applied during the access to the global memory from each thread, which reduces the access time and consequently the overall execution time. These results are confirmed and verified also by choosing any possible CUDA configuration in the range 1000–3072 threads (3072 is the maximum threads number available for our hardware). Table 2 confirms the reliability of the parallelization strategy by highlighting the access time

Table 2. Execution times (in seconds), iteration number $K = 500$. $m = 5$.

CUDA configuration	N		
	1×10^4	5×10^4	1×10^5
10 × 100	0.77	5.72	13.17
4 × 250	0.59	5.01	11.88
2 × 512	0.41	4.07	10.91
1 × 1024	0.38	3.76	9.68
20 × 100	0.55	6.24	11.57
8 × 250	0.47	4.92	10.71
4 × 512	0.39	3.67	10.42
2 × 1024	0.31	3.10	8.96
30 × 100	0.21	3.24	6.58
12 × 250	0.17	3.12	5.70
6 × 512	0.14	3.02	4.99
3 × 1024	0.12	2.99	4.07

to the global memory. In particular, the results allow us to find the best CUDA thread configuration, `Blocks` × `Threads` = 3 × 1024, obtained in correspondence of the best execution times.

Test 3. This experiment is referred to the optimal CUDA configuration and aims to investigate the behaviour of the algorithm by varying both iteration number value K and the input size N. Figure 1 shows an appreciable gain of performance and, in particular, a sub-linear increase of execution time with respect to the problem size (which is linear in $N \times K$), typical for GPUs architectures.

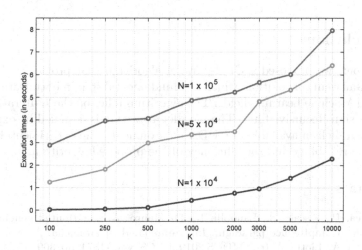

Fig. 1. Execution times by varying K and N (`Blocks` × `Threads` = 3 × 1024, $m = 5$)

Test 4. Here, we show a further improvement of the performance due to the use the power of dynamic parallelism approach. Table 3 exhibits the best execution times achieved by using the dynamic parallelism and choosing an ad-hoc, i.e. limited by our machine available resources, CUDA configuration. Comparison with Table 2 (first 4 lines) confirms the improvement for all data sizes. However, we underline that, because the hardware limits available, if we set a too large threads number, a big portion of them cannot work and, from a numerical point of view, the output result becomes completely unreliable. In other words, a fair CUDA configuration avoids a failed computation. For this reason, we have no results by using a greater number of threads. Finally, behaviour of results in Table 3 seems to suggest that an improving of performance should be obtained by exploiting a machine with higher computational resources.

Table 3. Execution times (in seconds) with dynamic parallelism, iteration number $K = 500$. $m = 5$.

CUDA configuration	N		
	1×10^4	5×10^4	1×10^5
10×100	0.28	4.64	10.09
4×250	0.21	3.86	9.71
2×512	0.19	3.60	8.15
1×1024	0.13	3.09	6.92

5   Conclusions

In this paper, we proposed a GPU-parallel algorithm that provides a fast and accurate Gaussian convolution, which is a fundamental step in both Data Assimilation and Machine Learning fields. The algorithm relies on the K-iterated first-order Gaussian Recursive filter. The parallel algorithm is designed by exploiting dynamic parallelism available in CUDA environment. The experimental results confirm the reliability and the efficiency of the proposed algorithm.

References

1. De Luca, P., Fiscale, S., Landolfi, L., Di Mauro, A.: Distributed genomic compression in MapReduce paradigm. In: Montella, R., Ciaramella, A., Fortino, G., Guerrieri, A., Liotta, A. (eds.) IDCS 2019. LNCS, vol. 11874, pp. 369–378. Springer, Cham (2019). https://doi.org/10.1007/978-3-030-34914-1_35
2. De Luca, P., Galletti, A., Giunta, G., Marcellino, L., Raei, M.: Performance analysis of a multicore implementation for solving a two-dimensional inverse anomalous diffusion problem. In: Sergeyev, Y.D., Kvasov, D.E. (eds.) NUMTA 2019. LNCS, vol. 11973, pp. 109–121. Springer, Cham (2020). https://doi.org/10.1007/978-3-030-39081-5_11
3. De Luca, P., Formisano, A.: Haptic data accelerated prediction via multicore implementation. In: Kohei, A. (ed.) Proceedings of the 2020 Computing Conference, CompCom. Advances in Intelligent Systems and Computing. Springer, Cham (2020)
4. Giunta, G., Montella, R., Mariani, P., Riccio, A.: Modeling and computational issues for air/water quality problems: a grid computing approach. Nuovo Cimento C Geophys. Space Phys. C **28**, 215 (2005)
5. Rao, V., Sandu, A.: A time-parallel approach to strong-constraint four-dimensional variational data assimilation. J. Comput. Phys. **313**, 583–593 (2016)
6. Bousserez, N., Guerrette, J.J., Henze, D.K.: Enhanced parallelization of the incremental 4D-Var data assimilation algorithm using the randomized incremental optimal technique (RIOT). Q. J. R. Meteorol. Soc. **146**, 1351–1371 (2020)
7. Fisher, M., Gürol, S.: Parallelization in the time dimension of four-dimensional variational data assimilation. Q. J. R. Meteorol. Soc. **143**(703), 1136–1147 (2017)
8. Galletti, A., Giunta, G.: Error analysis for the first-order Gaussian recursive filter operator. In: 2016 Federated Conference on Computer Science and Information Systems (FedCSIS), pp.673–378. IEEE (2016). APA

9. Cuomo, S., Galletti, A., Giunta, G., Marcellino, L.: Numerical effects of the Gaussian recursive filters in solving linear systems in the 3Dvar case study. Numer. Math. Theory Methods Appl., 520–540 (2017). https://doi.org/10.4208/nmtma.2017.m1528

10. D'Amore, L., Arcucci, R., Marcellino, L., Murli, A.: A parallel three-dimensional variational data assimilation scheme. In: AIP Conference Proceedings, vol. 1389, no. 1, pp. 1829–1831. American Institute of Physics, September 2011

11. De Luca, P., Galletti, A., Marcellino, L.: A Gaussian recursive filter parallel implementation with overlapping. In: 2019 15th International Conference on Signal-Image Technology & Internet-Based Systems (SITIS), Sorrento, Italy, pp. 633–640 (2019)

12. Chaurasia, G., Kelley, J.R., Paris, S., Drettakis, G., Durand, F.: Compiling high performance recursive filters. In: Proceedings of the 7th Conference on High-Performance Graphics, pp. 85–94 (2015)

13. De Luca, P., Galletti, A., Ghehsareh, H.R., Marcellino, L., Raei, M.: A GPU-CUDA framework for solving a two-dimensional inverse anomalous diffusion problem. In: Foster, I., Joubert, G.R., Kučera, L., Nagel, W.E., Peters, F. (eds.) Parallel Computing: Technology Trends, Advances in Parallel Computing, vol. 36, pp. 311–320. IOS Press, Amsterdam (2020). https://doi.org/10.3233/APC200056

14. Cuomo, S., Michele, P. D., Galletti, A., Marcellino, L.: A GPU-parallel algorithm for ECG signal denoising based on the NLM method. In: 2016 30th International Conference on Advanced Information Networking and Applications Workshops (WAINA), Crans-Montana, pp. 35–39 (2016)

15. Jones, S.: Introduction to dynamic parallelism. In: GPU Technology Conference Presentations, vol. 338, p. 2012, May 2012

16. https://docs.nvidia.com/cuda/cuda-c-programming-guide/index.html

17. Ghil, M., Malanotte-Rizzoli, P.: Data assimilation in meteorology and oceanography. In: Dmowska, R., Saltzman, B. (eds.) Advances in Geophysics, vol. 33, pp. 141–266. Elsevier, New York (1991)

18. Lorenc, A.C.: Development of an operational variational assimilation scheme. J. Meteorol. Soc. Jpn **75**, 339–346 (1997)

19. Hayden, C., Purser, R.J.: Recursive filter objective analysis of meteorological field: applications to NESDIS operational processing. J. Appl. Meteorol. **34**, 3–15 (1995)

20. Cuomo, S., Farina, R., Galletti, A., Marcellino, L.: An error estimate of Gaussian recursive filter in 3Dvar problem. In: 2014 Federated Conference on Computer Science and Information Systems, Warsaw, pp. 587–595 (2014). https://doi.org/10.15439/2014F279

21. Young, I.T., van Vliet, L.J.: Recursive implementation of the Gaussian filter. Signal Process. **44**, 139–151 (1995)

22. Wells, W.M.: Efficient synthesis of Gaussian filters by cascaded uniform filters. IEEE Trans. Pattern Anal. Mach. Intell. **2**, 234–239 (1986)

23. Gilbert, R.C., Richman, M.B., Trafalis, T.B., Leslie, L.M.: Machine learning methods for data assimilation. In: Computing Intelligence Architecturing Complex Engineering Systems, pp. 105–112 (2010)

24. Triggs, B., Sdika, M.: Boundary conditions for Young-van Vliet recursive filtering. IEEE Trans. Signal Process. **54**(6 I), 2365–2367 (2006)

On the Complementary Role of Data Assimilation and Machine Learning: An Example Derived from Air Quality Analysis

Richard Ménard[1]([✉])[iD], Jean-François Cossette[2][iD],
and Martin Deshaies-Jacques[2]

[1] Air Quality Research Division, Environment and Climate Change Canada,
Dorval, QC H9P 1J3, Canada
richard.menard@canada.ca
[2] Canadian Meteorological Center, Environment and Climate Change Canada,
Dorval, QC H9P 1J3, Canada

Abstract. We present a new formulation of the error covariances that derives from ensembles of model simulations, which captures terrain-dependent error correlations, without the prohibitive cost of ensemble Kalman filtering. Error variances are obtained from innovation variances empirically related to concentrations using a large data set. We use a k-fold cross-validation approach to estimate the remaining parameters. We note that by minimizing the cross-validation cost function, we obtain the optimal parameters for an optimal Kalman gain. Combined with the innovation variance consistent with the sum of observation and background error variances in observation space, yield a scheme that estimates the true error statistics, thus minimizing the true analysis error. Overall, this yield a new error statistics formulation and estimation out-performs the older optimum interpolation scheme using isotropic covariances with optimized covariance parameters. Yet, the analysis scheme is computationally comparable to optimum interpolation and can be used in real-time operational applications. These new error statistics comes as data-driven models, were we use validation techniques that are common to machine learning. We argue that the error statistics could benefit from a machine learning approach, while the air quality model and analysis scheme derives from physics and statistics.

Keywords: Air quality analysis · Cross-validation · Data driven model of error covariance

1 Introduction

Data assimilation was originally developed from a need to specify the initial conditions of numerical weather prediction models [1] that otherwise would have little predictive skill due to unstable dynamics of the atmosphere. Imperfectly known and incorrect assumptions (e.g. no model error covariance) on the (input) error statistics used for data assimilation, is not as critical for numerical weather prediction as opposed to other

© Her Majesty the Queen in Right of Canada 2020
V. V. Krzhizhanovskaya et al. (Eds.): ICCS 2020, LNCS 12142, pp. 212–224, 2020.
https://doi.org/10.1007/978-3-030-50433-5_17

applications, due to the presence of the unstable subspace in the forecast error. Indeed, it has been argued that by confining the forecast error corrections to the unstable and neutral subspace, four-dimensional variational data assimilation can better perform than without this confinement [2, 3]. However, not all application areas of data assimilation have an unstable subspace. Atmospheric chemistry models, for example, are known to be "slaved" by the meteorology [4, 5] and it is known that the precise knowledge of the chemical observation and model error covariances has a rather strong impact on the performance of the (chemical) data assimilation results [6, 7]. The estimation of correct error statistics is important for a truly optimal chemical data assimilation system. This has been recognized from the very beginning of Kalman filtering data assimilation using neutral or dissipative models (e.g. [8, 9]).

With dissipative models, the verification of the forecast as a measure of analysis accuracy has limited value. However, the true analysis error can be evaluated using cross-validation [10, 11]. In cross-validation we produce analyses with a subset of observations and verify the analysis with the remaining observations. There is no need to conduct a forecast. Since the optimality of analyses (measured against independent observations) depends on having the correct (input) error statistics, we may view the problem of estimating the true error statistics as an inverse problem of analyses (measured by cross-validation). But as with most inverse problems, the minimization of the verification error alone is not sufficient to determine the correct input error statistics [12]. Additional criteria are needed, such as the matching of the covariance of inno-vation with the sum of the background error covariance in observation space with the observation error covariance, called the innovation covariance consistency [13]. This is where (even elementary concepts) of machine learning can become useful.

Atmospheric models are based on the laws of physics, and in our case on chemical laws as well; they induce our prior knowledge of the system. On the other hand, the innovations or observation-minus-model residuals, are quantities that contain infor-mation that is unexplained by the model, and this is where machine learning can have a complementary role to data assimilation. Complementary roles of machine learning and data assimilation were also developed in a form of Kalman filtering (known as the Parametric Kalman filter) that requires closure on the form and parameters or error covariances [14].

At Environment and Climate Change Canada (ECCC) we have developed an operational surface analysis of atmospheric pollutants since 2002 [15] that provide a complete and more accurate representation of the air quality atmospheric chemistry, which has been used in several health impact studies (e.g. [16, 17]). Although the optimization of (isotropic) covariance model parameters improves the analysis [18], little is known about the more realistic and accurate covariance models suitable for chemical data assimilation [19, 20].

Since air quality models are computationally demanding (compared to numerical weather prediction models) the use of (ensemble) Kalman filtering data assimilation to obtain non-homogeneous non-isotropic error covariances is more in the realm of research than operational purposes. Recently at ECCC we have been developing a simple approach to generate non-homogeneous non-isotropic error correlations near the surface that does not involve rerunning the air quality model, by simply using pre-existing model simulations over a period of a few months. Note that chemical simulations are driven (slaved) by meteorological analyses. The error correlations that are obtained are not flow-dependent but are non-homogeneous non-isotropic and terrain-dependent, which account for a large fraction of the error correlation signal near the surface. We will presents some examples of these error correlations in Sect. 2.1. As with Kalman filtering localization of these ensembles is needed, and is obtained by minimizing the analysis error evaluated by cross-validation. This will be presented in Sect. 2.3. For error variances we use essentially an innovation-driven (data-driven) representation of the error variances as a function concentration, which is simple but somewhat appropriate for these fields. This will be discussed in Sect. 2.2. We show that this new analysis scheme is superior to the currently operational implementation using optimum interpolation with homogeneous isotropic correlation models [21]. We realize that the method we are using, is similar (yet much simpler) to some of the methods used in (simple) machine learning. We argue that a data-driven approach to model error covariances, and more sophisticated machine learning algorithms [22] could potentially improve those representations and be beneficial for truly optimizing an assimilation system.

2   Description of the Method

2.1   Model Representors

In our current operational version of the analysis of surface pollutants, we use homogeneous isotropic correlation models [21] with tuning of covariance parameters [15, 18] to optimize the system. In this new version, we obtain anisotropic error correlations from an ensemble of pre-existing air quality simulations (i.e. forecasts with no chemical data assimilation). The ensemble is in fact climatological, where the air quality simulations are collected over a period of two months. For each hour of the day, we thus have an ensemble of about 60 realizations over that time period that captures non-homogenous and non-isotropic correlations. As in the ensemble Kalman filter, there is a need for localization to avoid spurious correlations at large distances that infiltrate the analysis and thus significantly reduce its optimality. The idea behind using an ensemble of pre-existing model forecasts is that those error correlations will be able to capture stationary and terrain-dependent structures such as those induced by the proximity of water surfaces, mountain ranges, valleys, chemical sources, and so on, since these features are always present. This method does not capture the day-to-day variability (or flow dependence) of the error correlations like in an ensemble Kalman filter, but may capture the effect of prevailing winds for example.

In the example presented below, we computed the spatial correlations for each observation sites using a time series of the Canadian operational air quality model (GEM-MACH) output of PM2.5 at 21 UTC over a two-month period (July-August 2016) and applied a compact support correlation function (the 5th-order piecewise rational function of Gaspari and Cohn [23]) as a method for localization.

Examples of spatial correlation of PM2.5 are presented in Fig. 1, with respect to an observation site in Toronto (Canada) and in Los Angeles (USA). We note that the correlation around Toronto (upper panel) is more or less oval, while strong anisotropic structures extending along the coast and influenced by the presence of water and terrain are depicted with the correlation about Los Angeles (lower panel). Correlations at Winnipeg in the Canadian prairies over a flat terrain are nearly circular or isotropic (result not shown). Likewise, the spatial correlation with respect to Surrey (in the neighborhood of Vancouver, Canada) shows a minimum over the Coast Mountain range and over the Rockies with a maximum in between over the central interior plateau (results not shown). In general, the spatial correlation structures shows the presence of mountain ranges, valleys, and extended water surfaces. These correlation structures were obtained after localization using a Schur product with a compact support correlation function. The length-scale of the compact correlation model is estimated by cross-validation (see Sect. 2.4).

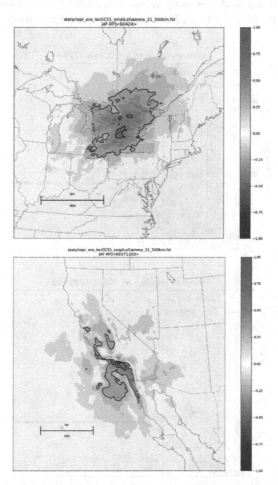

Fig. 1. Error correlation obtained for a time series of the air quality model GEM-MACH simulations valid at 21 UTC over a period of 2 months (July-August 2016), after localization using the compact support correlation function. Upper panel is the correlation with respect to Toronto (Canada) and lower panel with respect to Los Angeles (USA) stations. The solid black line depicts the correlation contour of 0.3.

2.2 Innovation Scatter Plot

The innovation (i.e. the observation-minus-model residual) variance has been plotted against the mean concentration for each station at a given local time (21 UTC) using a 2-months' time series to generate the statistics for each station. The variance of the innovation gives the sum of observation and model (i.e. background) error variances, but its partition into observation and model errors is yet unknown. Furthermore, the observation error is not simply the instrument error but should also include; errors due to the mismatch in scales being sampled in a single observation vs that of the model grid box, subgrid scale variability that may be captured by the observation but not by the model, and missing modeling processes, etc., that collectively we call representativeness error. The observation error is thus not well known and needs to be estimated. This will be done in the Sect. 2.4 using cross-validation and assuming that observation error variance is simply a fraction of the variance of the innovation.

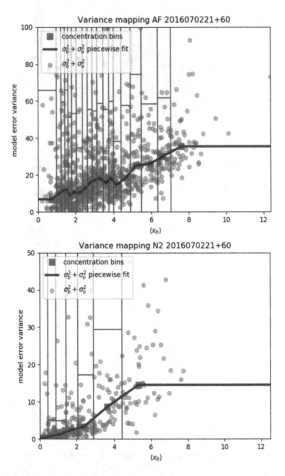

Fig. 2. Variance of the innovation as a function of the mean concentration, station by station. Upper panel is for $PM_{2.5}$ and lower panel for NO_2 (same time period as in Fig. 1). The red squares represent the median in each bin, and the horizontal line in each bin determines the third inter quartile of the distribution in a bin. (Color figure online)

The behavior as a function of concentration is different for $PM_{2.5}$ than for NO_2. We note that for NO_2 the innovation variance goes to zero when the concentration goes to zero, which is not the case for $PM_{2.5}$. The $PM_{2.5}$ is nearly linear, while NO_2 (especially at night - result not shown) has a shape of a quadratic for low concentrations and saturate at higher concentrations.

The fitting of the innovation variance as a function of concentration (as in Fig. 2) is important for two reasons. First, as it will become relevant in the following subsection, we need to have the sum of observation and model (background) error variances match the innovation variance - a property known as the innovation variance consistency. Secondly, after obtaining the portion of the innovation variance due to the model (background) error variance, we can then apply the relationship between background error variance with concentrations, to determine the background error variance at each model surface grid point (not only at the observation locations).

2.3 Estimation of the True Error Statistics in Observation Space

Under the assumptions of uncorrelated observation and background errors, it has been established that two necessary and sufficient conditions to estimate the true observation and background error covariances are [13]. One, is that the Kalman gain in observation space (i.e. **HK** where **H** is the observation operator and **K** the Kalman gain) is optimal in the sense that the true analysis error (in observation space) is minimum. The second condition is that the innovation covariance matches the sum of the background error covariance in observation space plus the observation error covariance, i.e. $\mathbf{HBH}^T + \mathbf{R}$ where **B** is the background error covariance and **R** the observation error covariance.

The fit of the innovation variance presented in Fig. 2 assures by construction that the sum of error variances $\sigma_o^2 + \sigma_b^2$ is consistent with innovation variance (not the innovation covariance), thus at least partly fulfilling the second condition above.

The first condition about the optimal Kalman gain is obtained by using cross-validation [10, 11]. As a way to illustrate this, we use a geometric interpretation where random variables y_1, y_2 are represented in a Hilbert space whose inner product is defined as the covariance between the two random variables.

$$\langle y_1, y_2 \rangle := \mathrm{E}\left[(y_1 - \mathrm{E}(y_1)) (y_2 - \mathrm{E}(y_2)) \right], \tag{1}$$

where E is the mathematical expectation. In this framework, random variables form an Hilbert space. For example, uncorrelated random variables are represented as perpendicular "vectors", and the error variance as the norm squared of that "vector" (see [11] and references therein).

Figure 3 illustrates geometrically the cross-validation. Let the active observation y^o be illustrated as O in the Fig. 3, the prior or background y^b (illustrated as B), the analysis y^a (illustrated as A), and the independent (or passive) observation y^c (illustrated as O_c). The origin T corresponds to the truth. Note that although the illustration is made for a (scalar) random variable, the same principle holds for random vectors for each components. We assume that the background error is uncorrelated with observations errors, and that observations are spatially uncorrelated horizontally. Then, the background error, the active and the passive observation errors are uncorrelated with one another, so the three axes; ε^o for the active observation error, ε^b for the background error, and ε^o_c for the passive observation error are orthogonal. The plane defined by axes ε^o and ε^b is the space where the analysis takes place, and is called the analysis plane. Since we define the analysis to be linear and unbiased, the analysis is a linear combination of the observation and background. Thus, in this illustration the analysis A lies on the line (B,O). The thick lines in Fig. 4 represent the norm of the associated error. For example, the thick line along the ε^o axis depict the (active) observation standard deviation σ_o, and similarly for the other axes and other random variables. Since the active observation error is uncorrelated with the background error, the triangle $\triangle$ OTB is a right triangle, and by Pythagoras theorem we have, $\overline{(y^o - y^b)^2} := \langle (O - B), (O - B) \rangle = \sigma^2_o + \sigma^2_b$. This is the usual statement that the innovation variance is the sum of background and observation error variances. The analysis is optimal $\hat{A}$ when the analysis error $\| \varepsilon^a \|^2 = \sigma^2_a$ is minimum, that is when the point A is closest to T. In this case the line $(T, \hat{A})$ is perpendicular to line (O,B). Since passive observations are never collocated with active observations, they have errors uncorrelated to active observations and background errors, and thus ε^o_c lies perpendicular to the analysis plane. So when the point A is closest to T it is also closest to O_c. Since the distance (O_c,A) squared corresponds to the variance of $(O_c - A)_{CV}$ evaluated in cross-validation, we conclude that when Var $(O_c - A)_{CV}$ is minimum, the analysis is optimal (i.e. minimal error), $A = \hat{A}$ [11].

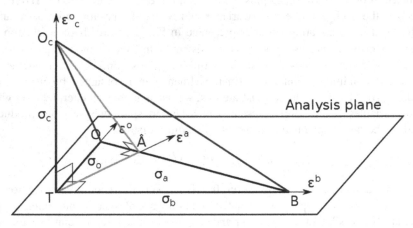

Fig. 3. Geometric representation (i.e. Hilbert space) of a scalar analysis with cross-validation (reproduced from [24] Figure 37.2 b)).

2.4   Optimization by Cross-Validation

An evaluation by cross-validation is performed to evaluate the quality of the analysis. The k-fold cross-validation methodology consists of separating the data set of observations into k equal size spatially random-distributed subsets and using each subset to evaluate an analysis while using the remaining observations in the k-1 subsets to compute the analysis. Here we use $k = 3$ [16, 17]. The separation into spatially random-distributed subsets has been achieved by selecting one station over three in an ordered station ID number list. An illustration of the selection method for the PM2.5 surface monitoring stations is depicted in Fig. 4, below. Another method which makes this selection rigorous is the application Hilbert curves [25].

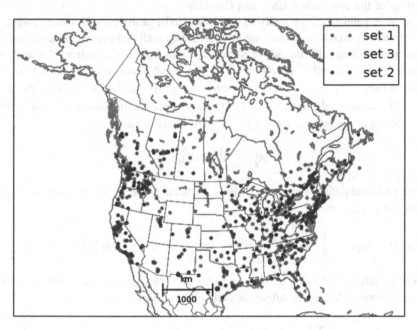

Fig. 4. Spatial distribution of three subsets of PM$_{2.5}$ observation sites (reproduced from [10] Figure 1)

Let $O_j(n; t)$ be the concentration measured at a station j at the local hour t, on day n. First, we calculate the verification statistics for each station at a given local time using an ensemble of days $\{n = 1, \ldots, N_{days}\}$. The average at a station j at the local time t, is simply

$$\bar{O}_j(t) = \frac{1}{N_{days}} \sum_{n=1}^{N_{days}} O_j(n; t). \tag{2}$$

The verification statistics are often defined over a region. For example, the mean concentration variance over an ensemble i, i.e. $\{O^i\}$, of stations, for a total number of N_s stations, is defined as

$$\text{var}\left(O^i(t)\right) = \frac{1}{N_s} \sum_{j \in \{O^i\}} \left(\frac{1}{N_{days} - 1} \sum_{n=1}^{N_{days}} \left(O_j(n; t) - \bar{O}_j(t)\right)^2 \right). \tag{3}$$

In general the ensemble $\{O^i\}$ (or simply denoted by O^i) can either be an:

- ensemble over all stations in the whole domain
- ensemble over a region (or subdomain)
- ensemble over all passive stations (i.e. stations not used in the analysis)

or any variants or combination thereof. In this document, there is only a single domain consisting of the continental USA and Canada.

For cross-validation, the analyses are evaluated against passive observations, i.e. observations not used to construct the analyses. We recall that passive observation sites are never collocated with the active stations (stations used to construct the analyses). Specifically, the ensemble of observations (for each local time) is split into three disjoint subsets O^1, O^2, O^3, and we denote the cross-validation analysis by $A^{[1]} = A^{[1]}(O^2, O^3)$ as an analysis that uses O^2, O^3 and excludes the subset O^1. The interpolated analysis at the passive station $j \in O^1$ will be denoted by

$$A_j^{[1]} = A_j^{[1]}(O^2, O^3). \tag{4}$$

The cross-validation variance statistics are then given by the average over the 3 subsets $i = 1, 2, 3$, i.e.

$$\mathrm{var}(O - A)_{CV} = \frac{1}{3} \left\{ \mathrm{var}\left(O^1 - A^{[1]}\right) + \mathrm{var}\left(O^2 - A^{[2]}\right) + \mathrm{var}\left(O^3 - A^{[3]}\right) \right\}, \tag{5}$$

where the statistics of each passive subset i are calculated by an average over all passive stations in the given subset. Specifically we have,

$$
\begin{aligned}
\mathrm{var}\left(O^i - A^{[i]}\right) &= \frac{1}{N_s} \sum_{j \in O^i} \mathrm{var}\left(O_j^i - A_j^{[i]}\right) \\
&= \frac{1}{N_s} \sum_{j \in O^i} \left\{ \frac{1}{N_{days} - 1} \sum_{n=1}^{N_{days}} \left[\left(O_j^i(n) - A_j^{[i]}(n)\right) - \overline{\left(O_j^i(n) - A_j^{[i]}(n)\right)} \right]^2 \right\} .
\end{aligned}
\tag{6}
$$

Note that a cross-validation statistic is evaluated for each local time and we have omitted the variable t, to keep the notation simple.

In our context where we enforce innovation variance consistency through the innovation variance fitting (see Sect. 2.2), we are left with only 2 parameters to estimate: 1) the ratio of observation error variance to background error variance σ_o^2 / σ_b^2, and 2) the compact support correlation length-scale L_s. These parameters are estimated by minimizing $\mathrm{var}(O - A)_{CV}$ and thus result in the end in a nearly optimal analysis.

In Fig. 5, we plotted $\mathrm{var}(O - A)_{CV}$ for different values of σ_o^2 / σ_b^2 and L_s. We find a single minimum of the fit of the analysis to independent (or passive observations) for an error variance ratio of 1.5 and for a compact support correlation length of 300 km, used to localized the raw model correlations.

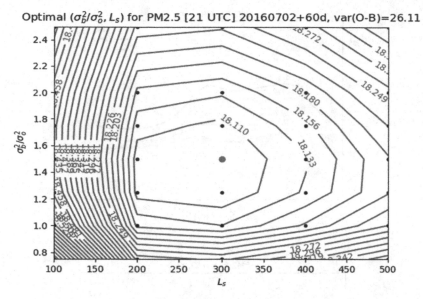

Fig. 5. Contours of the cross-validation of independent observation-minus-analysis (at the passive observation sites), as a function of compact support correlation length L_s and the ratio of model error variance to observation error variance.

3 Comparison with the Operational Analysis

Using these optimal parameter values and the modeling based in the innovation variance as function of concentration and with model output statistics to construct the spatial correlation structures, we evaluate the new analysis (i.e. Av2). We then compare it against the old analysis scheme (i.e. Av1) which is using homogenous isotropic correlation functions and χ^2-optimized error statistics [15]. The result evaluated by cross-validation is presented in the Fig. 6 below.

We observe a sensitivity to the error statistics used to generate the analysis, with superior analyses when the modeling is based on our methodology using the data rather than using some specified isotropic models.

It is by letting the data itself (model output and observations) provide the modeling elements of the observation and background error covariances that we arrive at an improved analysis. Thus, we thus argue that data-driven modeling of the observation and background error covariances plays a complementary role to data assimilation, resulting in a nearly optimal system.

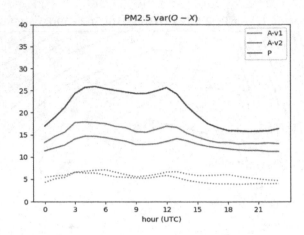

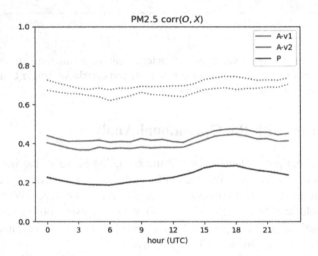

Fig. 6. Verification of the PM2.5 analysis against passive observations using cross-validation for Av1 (old scheme) and Av2 (new scheme). The solid line (green and red) uses independent observations, while the dotted lines are the statistics using the same observations as those used to construct the analysis. The solid blue line represent the verification of the model (i.e. no analysis). The upper panel displays the variance and lower panel the correlation. (Color figure online)

References

1. Daley, R.: Atmospheric Data Analysis. Cambridge University Press, New York (1991). 455 p.
2. Trevisan, A., D'Isidoro, M., Talagrand, O.: Four-dimensional variational assimilation in the unstable subspace and the optimal subspace dimension. Q. J. Roy. Meteorol. Soc. **136**, 487–496 (2010). https://doi.org/10.1002/qj.571

3. Grudzien, C., Carrassi, A., Bocquet, M.: Asymptotic forecast uncertainty and the unstable subspace in presence of additive model error. SIAM/ASA J. Uncertainty Quantification **6**(4), 1335–1363 (2018). https://doi.org/10.1137/17M114073X

4. Lahoz, W., Errera, Q.: Constituent Assimilation. In: Lahoz, W., Khattatov, B., Menard, R. (eds.) Data Assimilation, pp. 449–490. Springer, Heidelberg (2010). https://doi.org/10.1007/978-3-540-74703-1_18

5. Ménard, R., et al.: Coupled stratospheric chemistry-meteorology data assimilation. Part I: Physical background and coupled modeling aspects. Atmosphere **11**, 150 (2020). https://doi.org/10.3390/atmos11020150

6. Errera, Q., Ménard, R.: Technical Note: Spectral representation of spatial correlations in variational assimilation with grid point models and application to the Belgian Assimilation System for Chemical Observations (BASCOE). Atmos. Chem. Phys. **12**, 10015–10031 (2012). https://doi.org/10.5194/acp-12-10015-2012

7. Ménard, R., et al.: Coupled stratospheric chemistry-meteorology data assimilation. Part II: Weak and strong coupling. Atmosphere **10**(12), 798 (2019). https://doi.org/10.3390/atmos10120798

8. Daley, R.: The lagged innovation covariance: a performance diagnostic for atmospheric data assimilation. Mon. Wea. Rev. **120**, 178–196 (1992).

9. Daley, R.: The effect of serially correlated observation and model error on atmospheric data assimilation. Mon. Wea. Rev. **120**, 164–177 (1992).

10. Ménard, R., Deshaies-Jacques, M.: Evaluation of analysis by cross-validation. Part I: Using verification metrics. Atmosphere **9**(3) (2018). https://doi.org/10.3390/atmos9030086

11. Ménard, R., Deshaies-Jacques, M.: Evaluation of analysis by cross-validation. Part II: Diagnostic and optimization of analysis error covariance. Atmosphere **9**(2), 70 (2018). https://doi.org/10.3390/atmos9020070

12. Talagrand, O.: A posteriori verification of analysis and assimilation algorithms. In: Proceedings of the ECMWF Workshop on Diagnosis of Data Assimilation Systems, 2–4 November 1999, pp. 17–28. Reading, UK (1999)

13. Ménard, R.: Error covariance estimation methods based on analysis residuals: theoretical foundation and convergence properties derived from simplified observation networks. Q. J. Roy. Meteorol. Soc. **142**, 257–273 (2016). https://doi.org/10.1002/qj.2650. http://onlinelibrary.wiley.com/doi/10.1002/qj.2650/full

14. Pannekoucke, O., Fablet, R.: PDE-NetGen 1.0: from symbolic PDE representations of physical processes to trainable neural network representations. Geoscientific Model Development Discussion (2020). https://doi.org/10.5194/gmd-2020-35

15. Robichaud, A., Ménard, R.: Multi-year objective analysis of warm season ground-level ozone and PM2.5 over North-America using real-time observations and Canadian operational air quality models. Atmos. Chem. Phys. **14**, 1769–1800 (2014). https://doi.org/10.5194/acp-14-1769-201

16. Crouze, D.L., et al.: Ambient PM2.5, O3, and NO2 exposures and association with mortality over 16 years of follow-up in the Canadian Census Health and Environment Cohort (CanCHEC). Environ. Health Perspect. **123,** 1180–1186. https://doi.org/10.1289/ehp.1409276. Accessed 2 Nov 2015

17. To, T., et al.: Early life exposure to air pollution and incidence of childhood asthma, allergic rhinitis and eczema: Eur. Respir. J. pii, 1900913 (2019). https://doi.org/10.1183/13993003.00913-2019

18. Ménard, R., Deshaies-Jacques, M., Gasset, N.: A comparison of correlation-length estimation methods for the objective analysis of surface pollutants at Environment and Climate Change Canada. J. Air Waste Manag. Assoc. **66**(9), 874–895 (2016). https://doi.org/10.1080/10962247.2016.1177620

19. Constantinescu, E.M., Chai, T., Sandu, A., Carmichael, G.R.: Autoregressive models of background errors for chemical data assimilation. J. Geophys. Res. **112**, D12309 (2007). https://doi.org/10.1029/2006JD008103

20. Singh, K., Jardak, M., Sandu, A., Bowman, K., Lee, M., Jones, D.: Construction of non-diagonal background error covariance matrices for global chemical data assimilation. Geosci. Model Dev. **4**, 299–316 (2011). www.geosci-model-dev.net/4/299/2011, https://doi.org/10.5194/gmd-4-299-2011

21. Ménard, R., Robichaud, A.: The chemistry-forecast system at the Meteorological Service of Canada. In: The ECMWF Seminar Proceedings on Global Earth-System Monitoring, Reading, UK, 5–9 September 2005, pp. 297–308 (2005)

22. Rasmussen, C.E., Williams, C.K.: Gaussian Processes for Machine Learning. MIT Press, Cambridge (2006). 248 p.

23. Gaspari, G., Cohn, S.E.: Construction of correlation functions in two and three dimensions. Q. J. Roy. Meteorol. Soc. **125**, 723–757 (1999)

24. Ménard, R., Deshaies-Jacques, M.: Evaluation of air quality maps using cross-validation: Metrics, diagnostics and optimization. In: Mensink, C., Gong, W., Hakami, A. (eds.) Air Pollution Modelling and Its Application XXVI, pp. 237–242. Springer Proceedings in Complexity (2020). https://doi.org/10.1007/978-3-030-22055-6_37

25. De Pondeca, M.S.F.V., Park, S.-Y., Purser, J., DiMego, G.: Applications of Hilbert curves to the selection of subsets of spatially inhomogeneous observational data for cross-validation and to the construction of super-observations. Preprints, AGU Fall Meeting, San Francisco, CA, Amer. Geophys. Union, A31A-0868 (2006)

Recursive Updates of Wildfire Perimeters Using Barrier Points and Ensemble Kalman Filtering

Abhishek Subramanian[1], Li Tan[1], Raymond A. de Callafon[1(✉)],
Daniel Crawl[2], and Ilkay Altintas[2]

[1] Department of Mechanical and Aerospace Engineering,
University of California San Diego, La Jolla, CA, USA
{absubram,ltan,callafon}@eng.ucsd.edu
[2] San Diego Supercomputer Center, University of California San Diego,
La Jolla, CA, USA
{crawl,altintas}@sdsc.edu

Abstract. This paper shows how the wildfire simulation tool FAR-
SITE is augmented with data assimilation capabilities that exploit the
notion of barrier points and a constraint-point ensemble Kalman filtering to update wildfire perimeter predictions. Based on observations of
the actual fire perimeter, stationary points on the fire perimeter are
identified as barrier points and combined with a recursive update of the
initial fire perimeter. It is shown that the combination of barrier point
identification and using the barrier points as constraints in the ensemble
Kalman filter gives a significant improvement in the forward prediction
of the fire perimeter. The results are illustrated on the use case of the
2016 Sandfire that burned in the Angeles National Forest, east of the
Santa Clarita Valley in Los Angeles County, California.

Keywords: Wildfire · Barrier points · Ensembles · Ensemble Kalman
filter · FARSITE

1 Introduction

The ability to reliably predict fire perimeter propagation during a wildfire event
has a large potential in resource allocation and fire fighting planning to help
save lives and valuable infrastructure. Studying wildfire dynamics is done by
collecting data [3] and combining wildfire simulation tools with experimental
data to assimilate or adjust the widfire simulation [11–14]. Previous work on
data assimilation using the FARSITE [7] wildfire simulation tool combined with
ensemble Kalman filtering can be found in [5,16]. Next to large body of work that
use some form of Kalman filtering, [2,4,8,10] there are alternative approaches

This work was partly funded by NSF 1331615 under CI, Information Technology
Research and SEES Hazards programs.

that use Genetic Algorithms to determine the best set of input parameters to match the measurements [1]. The power of a data driven approach is also confirmed by [17] illustrating improvements to wildfire prediction on data obtained from physical experiments.

Fire perimeters that may be obtained periodically during a wildfire event may be well-suited for periodic or recursive updates of the initial conditions (e.g. initial fire perimeter) and the relevant parameters that govern the wildfire dynamics. For a Kalman filter-based approach it is essential that wildfire perimeter measurements are quantified with a measurement accuracy to find the optimal trade-off in adjusting initial conditions and wildfire parameters. To this extend, the work by [14] uses thermal-infrared imaging to measure the true fire perimeter on a controlled fire experiment done on a (4 m × 4 m) patch of land and [17] used ForeFire/Meso-NH simulations produced by [6] as observations. Unfortunately, such methods cannot be employed for time-sensitive and large scale wildfires, where perimeters are obtained with aerial measurements and computations of future wildfire perimeters must be done in near real-time.

To improve the quality of wildfire prediction and to speed up computations, one piece of critical information is often neglected in wildfire data assimilation: stationary points at which a (part of the) fire perimeter remains at the same locations between periodic updates. Clearly, those points can be characterized with a relatively high accuracy and do not require computational updates. In terms of data assimilation, such stationary points can be viewed as constrains in a constrained Ensemble Kalman Filtering [15] formulation. Fortunately, the FARSITE [7] wildfire simulation tool has the notion of barrier points to account for stationary fire perimeters. Identifying such stationary or barrier points on the fire perimeter and combining this information with ensemble Kalman filtering is the main contribution of this paper.

The results presented are based on the work in [5, 16] to fully use the information of barrier points in the prediction and update steps of the data assimilation tools for FARSITE. The approach presented in this paper performs recursive data assimilation to estimate the true values of fuel dependent adjustment factors along with wind speed and direction that influence fire spread rate by including them in the state updates. Estimation of these input parameters along with the identification of barrier points further improve the periodic prediction of fire perimeters. The data assimilation tools are tested on actual wildfire perimeter data that was obtained for the 2016 Sandfire that burned in the Angeles National Forest, east of the Santa Clarita Valley in Los Angeles County, California.

2   Contour and Stationary Points

2.1   Fourier Analysis

To introduce the concept of stationary points, we first formalize the approximation of a fire perimeter as a n-polygon described by a ordered sequence of n piece-wise linear line segments parametrized in Eastern e_j and Northern n_j

coordinate pairs (e_j, n_j), $j = 0, 1, \ldots, n-1$. To simplify notation, we may represent the n coordinates of the n-polygon as a complex number $p_j = e_j + i \cdot n_j$ for which we can define a complex Discrete Fourier Transform (DFT)

$$u_l = \sum_{j=0}^{n-1} p_j e^{-il\frac{2\pi}{n}j} \tag{1}$$

and represent the fire perimeter p_j, $k = 0, 1, \ldots, n-1$ by the complex sequence Fourier series u_l, $l = 0, 1, \ldots, n-1$. Since a fire perimeter is always a closed polygon, e.g. p_1 is connected to p_n via a linear line segment, the parametrization of the n-polygon should be independent of the starting point p_1 and the (anti)clock wise rotation of the sequence p_j in the complex plane. A shift in the starting point or rotation can be easily represented in the Fourier series $\bar{u}_l$ by an additional phase shift of u_l and given by $\bar{u}_l = u_l e^{-i\phi}$, where the rotation angle ϕ is determined by integer shift in the starting point and the binary choice on the anti-clockwise or clockwise rotation of p_j [9]. The Fourier series representation u_l of the n-polygon approximation of a fire perimeter allows fire perimeters at subsequent time steps k and $k+1$ to be compared as parts of the fire perimeter might be stationary. A non-moving or stationary part of the fire perimeter may be due to the presence of a nonburnable surface fuel or explicit fire fighting efforts in which part of the surface fuel has been removed or extinguished. Such information must be taken into account to improve the prediction of fire perimeter progression over time.

To identify stationary parts of the fire perimeter, we consider fire perimeters represented by the n-polygon $p_j(k)$, $j = 0, 1, \ldots, n-1$ at time step k and a m-polygon $p_j(k+1)$, $j = 0, 1, \ldots, m-1$ at a subsequent time step $k+1$. As $n \neq m$ and the starting point $p_1(k) \neq p_1(k+1)$, the simple check of $p_j(k) = p_j(k+1)$ will not suffice in determining the stationary points $p_j(k)$ of the fire perimeter. Instead, we first consider the Least Squares minimization

$$\bar{\phi} = \arg \min_{\phi} \sum_{l=0}^{d-1} |u_l(k) - u_l(k+1)e^{-i\phi}|^2 \tag{2}$$

where $d = \min(m, n)$ and $u_l(k)$, $u_l(k+1)$ are given by the DFT in (1). The optimization in (2) recomputes the optimal starting point and rotation of the Fourier transform of the m-polynomial $p_j(k+1)$ by evaluating the difference between $s = \min(m, n)$ Fourier coefficients. The end result is a set of Fourier coefficients $\bar{u}_l(k+1) = u_j(k+1)e^{-i\bar{\phi}}$ for which the inverse DFT will lead to a re-oriented m-polygon $\bar{p}_j(k+1)$, $j = 0, 1, \ldots, m-1$ of the fire perimeter at time step $k+1$ that can be compared with the fire perimeter $p_j(k)$, $k = 0, 1, \ldots, n-1$ at time step k. Stationary points are now defined as the set of points $p_j(k) \in \mathcal{P}_k$ on the fire perimeter $p_j(k)$ at time step k for which

$$\mathcal{P}_k : |p_j(k) - \bar{p}_j(k+1)| \leq \varepsilon \text{ for } k = t, t+1, \ldots, t+t_{stat}-1, \; t = 0, 1, \ldots, d-t_{stat} \tag{3}$$

where $t_{stat} > 1$ ensures no single points for which $|p_j(k) - \bar{p}_j(k+1)| \leq \varepsilon$ are iden-
tified as stationary points. Only a sequence of t_{stat} points on the fire perimeter
$p_j(k)$ at time step k and $\bar{p}_j(k+1)$ at time step $k+1$ must lie within a distance
of ε to qualify as stationary points.

2.2   Illustration and Boundary Points

The above proposed identification of stationary points is illustrated in Fig. 1,
that shows fire perimeter measurements for two consecutive (time step $k = 0$
and time step $k = 1$) for the use case of the 2016 Sandfire that burned in the
Angeles National Forest, east of the Santa Clarita Valley in Los Angeles County,
California. It can be seen that only the lower portion of the fire has propagated
from time step $k = 0$ to $k = 1$, indicating a large set of stationary points in the
progression of ther wildfire.

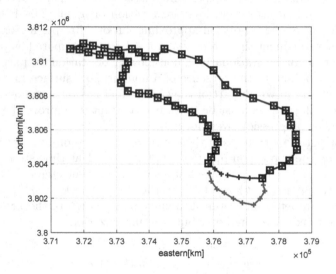

Fig. 1. Polygon approximation of fire perimeter measurement at time step k = 0 (blue)
compared with the fire perimeter at the subsequent time step $k = 1$ (red) for the 2016
Sandfire. Black squares are the identified stationary points of the fire perimeter at time
step $k = 0$. (Color figure online)

With the re-orientation of the fire perimeter points found by the optimization in (2), the location of stationary points identified by the black squares in
Fig. 1 become apparent. It should be noted that stationary points along the fire
perimeter might persist only for certain amount of time, as the fire could eventually progress. For example, the stationary points shown in Fig. 1 are only valid
for the fire perimeter at time step $k = 0$. To identify the stationary points for the
fire perimeter at $k = 1$, measurement at time step $k = 2$ are required. Clearly,
the information on the set of stationary points is important in predicting the

spread of wildfires accurately. The wildfire simulation tool FARSITE [7] has the notion of a barrier perimeter defined by barrier points to account for the identified stationary points in the fire perimeter by temporarily defining surface fuels as non-burnable.

3   Data Assimilation

3.1   Ensemble Forward Simulation

The FARSITE wild fire simulation tool takes in n real valued eastern- and northern-coordinates of a fire perimeter $p_j(k)$ at time step k to simulate a fire perimeter $p_j(k+1)$ at time step $k+1$. Next to the initial perimeter $p_j(k)$ specified in a shape file, FARSITE also uses information on surface fuels, topography, wind speed, wind direction and fuel adjustment factors, collectively combined in the environmental parameter θ_k at time step k, to adjust the fire perimeter prediction [16]. In addition, FARSITE can account for a set of barrier points $b_j(k) \in \mathcal{P}_k$ as defined in (3) to approximate the stationary points of the fire perimeter. Using only real valued calculations, FARSITE can be viewed as a non-linear mapping

$$X(k+1) = f(X(k), \theta_k, B_k) \tag{4}$$

where $X(k)$ is a vector of the eastern- and northern coordinates of the wild fire perimeter $p_j(k)$ and B_k is a vector of the eastern- and northern coordinates of the barrier points $b_j(k)$.

It should be noted that the non-linear map $f(\cdot)$ is not known analytically and both sensitivity or uncertainty of the inputs $X(k)$, θ_k and B_k can be evaluated numerically via ensemble averaging. For that purpose, random samples (ensembles) are chosen from a probability description of the initial fire perimeter $X(k)$ and possibly the environmental parameters θ_k. In this paper only uncertainty on the initial fire perimeter $X(k)$ is considered in the form of a covariance on the vector $X(k)$, whereas the environmental parameters θ_k and the barrier points B_k are assumed to be fixed. The later is a reasonable assumption as barrier points are defined as stationary points, whereas variability in θ_k can be considered as a possible improvement for the ensemble forward simulation presented in this paper.

The approach of ensemble forward simulation is similar to presented earlier in [16], but with the important addition of the vector of barrier points B_k. For the initialization of the covariance matrix P_k on the vector $X(k)$ for fire perimeter points, the proximity of the eastern- and northern-coordinates to its neighboring points is used. The proximity measure for the covariance σ_{e_j,n_j} for each point (e_j, n_j) on the n-polygon of the fire perimeter is defined by $\sigma_{e_j,n_j} = h(e_{j-1}, y_{j-1}, e_j, n_j, e_{j+1}, n_{j+1})$, where $h(\cdot)$ is a function that computes the measure of closeness of each point to its neighboring points on the fire perimeter. The value of σ_{e_j,e_j} is inversely proportional to the distance of (e_j, n_j) with its neighboring points. Using the environmental parameters θ_k, barrier points B_k and N ensembles $X^i(k)$ taken from a normal distribution determined by the mean value $X(k)$ and the covariance σ_{e_j,n_j}, each ensemble member $X^i(k)$ at time step k is advanced through the forward model

$$X^i(k+1) = f(X^i(k), \theta_k, B_k), \quad i = 1, 2, \ldots, N,$$

where B_k is determined *recursively* from a measurement of a fire perimeter $X(k+1)$ at time step $k+1$ by the optimization in (2) and the definition of the stationary points in (3) using the mean value $X(k)$.

With the ensemble forward simulation, N predicted fire perimeters $X^i(k+1)$ or $p_j^i(k+1)$ for $i = 1, 2, \ldots, N$ and $j = 0, 1, \ldots, n_i$ have become available. For ensemble averaging, a mean value and a covariance must be estimated, but typically two different fire perimeter boundaries $X^i(k+1)$ and $X^j(k+1)$ will not have the same number n_i of polygon points, starting point of orientation. To address this issue, first all N predicted fire perimeters $p_j^i(k+1)$, $i = 1, 2, \ldots, N$ are interpolated to the same number of (maximum) points $n_{max} = \max_i(n_i)$. Subsequently, an optimization similar to (2) is performed to align the starting point and orientation of all the ensembles. Denoting $u_l^1(k+1)$ as the DFT transform of the first (primary) ensemble $p_j^1(k+1)$ for $j, l = 0, 1, \ldots, n_{max}$, the other ensembles are aligned using the minimization

$$\bar{\phi}^i = \arg\min_\phi \sum_{l=0}^{n_{max}-1} |u_l^1(k+1) - u_l^i(k+1)e^{-i\phi}|^2, \tag{5}$$

for the index $i = 2, \ldots, N$, while i in the exponent $e^{-i\phi}$ still denotes the complex number $i^2 = -1$. The $N-1$ optimizations in (5) leads to a re-oriented set of ensembles represented by the vector $\bar{X}^i(k+1)$ for $i = 2, \ldots, N$ and found by the inverse DFT of $\bar{p}_j^i(k+1) = u_l^i(k+1)e^{-i\phi^i}$. The end result of this process is set of properly aligned N ensembles $X^1(k+1)$ and $\bar{X}^i(k+1)$, $i = 2, \ldots, n$ for which a mean $X^\mu(k+1) \in \mathcal{R}^{n_{max}}$ and a covariance $P^\mu(k+1) \in \mathcal{R}^{n_{max}, n_{max}}$ can be computed.

3.2   Ensemble Kalman Filter Update

In the ensemble Kalman filter update, a measurement of a fire perimeter obtained at time step $k+1$ is consolidated with the prediction of the mean $X^\mu(k+1) \in \mathcal{R}^{n_{max}}$ and the covariance $P^\mu(k+1) \in \mathcal{R}^{n_{max}, n_{max}}$ obtained from the ensemble forward simulation described above. For the optimal consolidation of the measurement and the prediction, we assume that the fire perimeter obtained at time step $k+1$ is described by its mean $Y(k+1)$ and a covariance $P^y(k+1)$. It is worth noting that the covariance $P^y(k+1)$ may be determined by either the inherent limited accuracy in obtaining the measurement $Y(k+1)$, the relative distance between the points on the perimeter $Y(k)$ or estimated by computing a two-dimensional variance from multiple measurements [5].

As the points of the measured fire perimeter $Y(k+1) \in \mathcal{R}^m$ and typically $m \neq n_{max}$, the different size of the observation and prediction is handled by linear interpolation of $Y(k+1)$ and $P^y(k+1)$ to n_{max} points. Subsequently, the following steps are done for each ensemble pair contained in $\tilde{X}$ and $\tilde{Y}$.

1. Find the closest point on $\tilde{X}^e$ to each point in $\tilde{Y}^e$ and pair them up, store the pair for which magnitude of distance is minimum and discard others, i.e. capture i that satisfies,

$$\min_i(\min_j |y_i^e - x_j^e|) \quad \forall i = 1, ..., m/m_i$$

$$j = \{1, 2, ..., n_{max}\}/m_j$$

along with i store its corresponding j in m_i and m_j respectively
2. Repeat the step above until each point in $\tilde{Y}^e$ is paired with a unique point in $\tilde{X}^e$.
3. This pairing scheme is used to construct the C matrix that is required to perform the Kalman update step to get the updated perimeter using the Kalman gain K via

$$X_{updated}^e = \tilde{X}^e + K[\tilde{Y}^e - C\tilde{X}^e] \tag{6}$$

The steps above are repeated until all ensembles have been exhausted. In this manner we perform data assimilation to improve our prediction by optimally combining results from a forward model (with errors in input) and measurement (with errors). Incorporation of stationary point information in the form of the barrier points B_k is done in the forward model itself to improve the ensemble forward simulation.

4 Results for the 2016 Sand Fire

4.1 Illustration for Single Step Data Assimilation

For the illustration of the data assimilation with barrier points, ensemble forward simulations and actual measurements of the 2016 Sandfire in Los Angeles County, California are used at time steps k separated by 2 h intervals, as shown in Fig. 2.

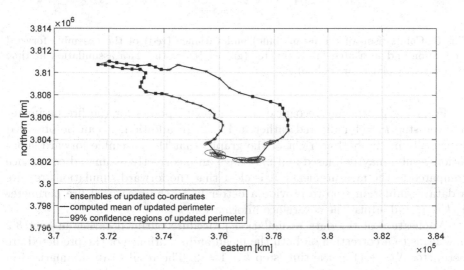

Fig. 2. Initial fire perimeter at time step $k = 1$.

In this case, the input fire perimeter given to FARSITE at step $k = 1$ shown in Fig. 2 is obtained from ensembles of the last fire perimeter depicted earlier in Fig. 1. Note that the variability along the fire perimeter at step $k = 1$ is very small and virtually non-existent in certain parts due to the stationary points identified from the previous time step at $k = 0$.

The stationary points for the fire perimeter at time step $k = 1$ are indicated by the black squares in Fig. 2. The stationary points are found by the procedure outlined earlier in Sect. 2 using the interpolated fire perimeter $X(k+1)$ obtained from a measurement of the actual fire perimeter $Y(k+1)$ at time step $k+1 = 2$. With the identified stationary points, barrier perimeters are defined in FARSITE and the ensemble forward simulation now lead to a predicted fire perimeter characterized by a mean $X^\mu(k + 1) \in \mathcal{R}^{n_{max}}$ and a covariance $P^\mu(k + 1) \in \mathcal{R}^{n_{max},n_{max}}$ as described earlier in Sect. 3.1. A comparison of the measurement of the actual fire perimeter $Y(k + 1)$ and the predicted mean fire perimeter $X^\mu(k + 1)$ with the covariance at each point is summarized in Fig. 3.

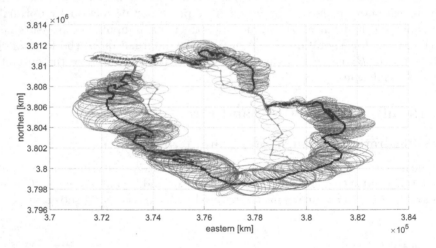

Fig. 3. Comparison of the mean (blue) and variance (red) of the ensemble forward simulation and measured fire perimeter (green) *before* the data assimilation at time step $k + 1$. (Color figure online)

From Fig. 3 one may recognize the stationary points for the fire perimeter at time step $k = 1$ indicated earlier in Fig. 2. In addition, it can be observed (especially in the bottom right of the graph) that the ensemble forward simulation results may be biased and may have an incorrectly estimated covariance compared to the measurements. It is clear that the forward simulations require a data assimilation step to provide a better fit to the measured fire perimeter $Y(k + 1)$ and adjust the covariance information.

Application of the data assimilation procedure outlined earlier in Sect. 3.2 now leads to a correction on both the mean and covariance of the predicted fire perimeter $X(k + 1)$ at the time step $k + 1 = 2$. The results are summarized in

Fig. 4 where it can be observed that the ensemble Kalman filter adjusted forward simulation now provides a much better fit to the measured fire perimeter at the next time step $k + 1$.

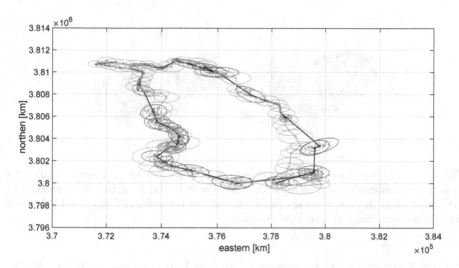

Fig. 4. Comparison of mean value (black) and variance (red) of Kalman filtered updated fire perimeter and measured fire perimeter (green) *after* the data assimilation at step $k + 1$. (Color figure online)

4.2   Definition of Fire Coverage Error

For the evaluation of the performance and improvement of wildfire data assimilation it is important to carefully characterize the error between two fire perimeters. Typically, the error is computed using the euclidean distance between points in a predicted fire perimeter and points along the measured fire perimeter [17]. This method is not suitable in large scale fires as the number of points along a fire perimeter can be vastly different in the predicted and the measured fire perimeter. Instead, we define the error using lower and upper bounds on the surface of the overlapping area of the fire perimeter, taking into accounts the uncertainty or variability of the fire perimeters.

Figure 5 has a visual illustration of how the error and its lower and upper bounds are computed. All points along the fire perimeter are associated with a value of uncertainty for the predicted fire perimeter and the measured fire perimeter. The uncertainty is represented in the form of an ellipse, which represents the confidence interval of where that particular point could lie. For the computation of the lower and upper bounds, we compute the area A_p covered by the predicted fire perimeter and the area A_m covered by the measured fire perimeter. Based on these areas, the mean value of the fire coverage error A_e is computed via

$$A_e = A_p \cup A_m - 2(A_p \cap A_m) \tag{7}$$

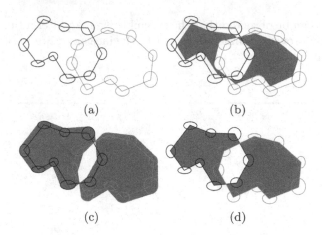

(a)   (b)

(c)   (d)

Fig. 5. (a) Predicted fire perimeter (black) and measured fire perimeter (red), grey area in (b): minimum error in area coverage (lower bound on error uncertainty), (c): maximum error in area coverage (upper bound on error uncertainty) and (d): mean error in area coverage. (Color figure online)

whereas lower and upper bounds are created by taking into account the variability or uncertainty on each fire perimeter.

4.3   Performance Comparison of Data Assimilation with Barrier Points

The single step data assimilation summarized earlier in Fig. 4 illustrates an improvement in the fire coverage error. The question remains whether the identification of stationary points and used as barrier points in FARSITE indeed improves the fire coverage error, in addition to improvements achieved by standard ensemble Kalman filter based data assimilation techniques. To illustrate the improvement of the fire coverage error, the subsequent steps of ensemble forward simulation and data assimilation for several time step $k = 0, 1, \ldots, 4$ is performed *with* and *without* the identification of stationary points.

To summarize the improvement in performance, the mean value of the fire coverage error A_e as defined in (7) along with the upper and lower bounds due to the variability on the fire perimeters is computer for the different time steps. The results are summarized in Fig. 6 and the improvement in performance measured in fire coverage error is evident from the graph. Both uncertainty and magnitude of the fire coverage error have been reduced when data assimilation is performed, but results are further improved when stationary points are identified and used as barrier points in the ensemble forward simulations. Especially the upper bound on the fire coverage error remains at acceptable levels during several data assimilation steps when using the identified stationary points on the fire perimeter.

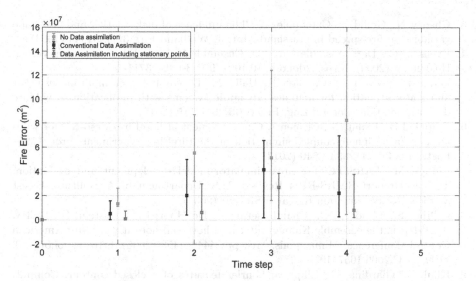

Fig. 6. Mean value, lower bounds and upper bounds on the fire coverage error A_e as defined in (7) for: no data assimilation (green), ensemble Kalman filtering *without* stationary points (black) and ensemble Kalman filtering *with* identification of stationary points (magenta). (Color figure online)

5 Conclusions

This papers shows the importance of data assimilation combined with the identification of stationary points in correcting the prediction of the spread of wildfires at a large scale. Stationary points are identified by comparison of subsequent fire perimeters and used in FARSITE to define barrier points to limit fire propagation. Based on the observations of the 2016 Sandfire, it is shown that the combined use of data assimilation and barrier points can significantly improve the fire coverage error. It is worth noting that an Ensemble Kalman Filter (EKF) approach is used to perform data assimilation, where a Gaussian representation of the fire perimeter vector is assumed. However, in certain situations it would be more suitable to implement a particle filter to take into account non-Gaussian distributions of fire perimeter uncertainty for future research directions.

References

1. Artés, T., Cencerrado, A., Cortés, A., Margalef, T.: Relieving the effects of uncertainty in forest fire spread prediction by hybrid MPI-OpenMP parallel strategies. Procedia Comput. Sci. **18**, 2278–2287 (2013)
2. Mandel, J., Beezley, J.D.: An ensemble Kalman-particle predictor-corrector filter for non-Gaussian data assimilation. In: Allen, G., Nabrzyski, J., Seidel, E., van Albada, G.D., Dongarra, J., Sloot, P.M.A. (eds.) ICCS 2009. LNCS, vol. 5545, pp. 470–478. Springer, Heidelberg (2009). https://doi.org/10.1007/978-3-642-01973-9_53

3. Cheney, N., Gould, J., Catchpole, W.: The influence of fuel, weather and fire shape variables on fire-spread in grasslands. Int. J. Wildland Fire **3**(1), 31–44 (1993)
4. Evensen, G.: Data Assimilation: The Ensemble Kalman Filter. Springer-Verlag, Heidelberg (2009). https://doi.org/10.1007/978-3-642-03711-5
5. Fang, M., Srivas, T., de Callafon, R., Haile, M.: Ensemble-based simultaneous input and state estimation for nonlinear dynamic systems with application to wildfire data assimilation. Control Eng. Pract. **63**, 104–115 (2017)
6. Filippi, J.B., Pialat, X., Clements, C.: Assessment of ForeFire/Meso-NH for wildland fire/atmosphere coupled simulation of the FireFlux experiment. Proc. Combust. Inst. **34**(2), 2633–2640 (2013)
7. Finney, M.: FARSITE: fire area simulator-model development and evaluation. Technical report RMRS-RP-4 Revised, U.S. Department of Agriculture, Forest Service, Rocky Mountain Research Station (2004)
8. Gillijns, S., Mendoza, O., Chandrasekar, J., De Moor, B., Bernstein, D., Ridley, A.: What is the ensemble Kalman filter and how well does it work? In: American Control Conference, Minneapolis, MN, pp. 4448–4453 (2006). https://doi.org/10.1109/ACC.2006.1657419
9. Khul, F., Giardina, C.: Elliptical Fourier features of a closed contour. Comput. Graphics Image Process. **18**, 236–258 (1982)
10. Mandel, J., Beezley, J., Cobb, L., Krishnamurthy, A.: Data driven computing by the morphing fast Fourier transform ensemble Kalman filter in epidemic spread simulations. Procedia Comput. Sci. **1**, 1221–1229 (2010)
11. Mandel, J., et al.: Towards a dynamic data driven application system for wildfire simulation. In: Sunderam, V.S., van Albada, G.D., Sloot, P.M.A., Dongarra, J.J. (eds.) ICCS 2005. LNCS, vol. 3515, pp. 632–639. Springer, Heidelberg (2005). https://doi.org/10.1007/11428848_82
12. Mandel, J., et al.: A note on dynamic data driven wildfire modeling. In: Bubak, M., van Albada, G.D., Sloot, P.M.A., Dongarra, J. (eds.) ICCS 2004. LNCS, vol. 3038, pp. 725–731. Springer, Heidelberg (2004). https://doi.org/10.1007/978-3-540-24688-6_94
13. Oriol, R., Miguel, V., Elsa, P., Planas, E.: Data-driven fire spread simulator: validation in Vall-llobrega's fire. Front. Mech. Eng. **5**, 1–11 (2019)
14. Rochoux, M., Ricci, S., Lucor, D., Cuenot, B., Trouvé, A., Bart, J.: Towards predictive simulations of wildfire spread using a reduced-cost ensemble Kalman filter based on polynomial chaos approximations. In: Proceedings of the Summer Program, Center for Turbulence Research (2012)
15. Simon, D.: Kalman filtering with state constraints: a survey of linear and nonlinear algorithms. IET Control Theory Appl. **4**(8), 1303–1318 (2010)
16. Srivas, T., Artés, T., de Callafon, R., Altintas, I.: Wildfire spread prediction and assimilation for FARSITE using ensemble Kalman filtering. Procedia Comput. Sci. **80**, 897–908 (2016)
17. Zhang, C., Rochoux, M., Tang, W., Gollner, M., Filippi, J.B., Trouvé, A.: Evaluation of a data-driven wildland fire spread forecast model with spatially-distributed parameter estimation in simulations of the FireFlux I field-scale experiment. Fire Saf. J. **91**, 758–767 (2017)

Meshfree Methods in Computational Sciences

Finding Points of Importance for Radial Basis Function Approximation of Large Scattered Data

Vaclav Skala[1]($\boxtimes$) (iD), Samsul Ariffin Abdul Karim[1,2] (iD),
and Martin Cervenka[1] (iD)

[1] Department of Computer Science and Engineering,
Faculty of Applied Sciences, University of West Bohemia, Univerzitni 8,
301 00 Plzen, Czech Republic
skala@kiv.zcu.cz, samsul_ariffin@utp.edu.my

[2] Fundamental and Applied Sciences Department and Centre for Smart Grid
Energy Research (CSMER), Institute of Autonomous System,
Universiti Teknologi PETRONAS, Bandar Seri Iskandar, 32610 Seri Iskandar,
Perak DR, Malaysia

Abstract. Interpolation and approximation methods are used in many fields such as in engineering as well as other disciplines for various scientific discoveries. If the data domain is formed by scattered data, approximation methods may become very complicated as well as time-consuming. Usually, the given data is tessellated by some method, not necessarily the Delaunay triangulation, to produce triangular or tetrahedral meshes. After that approximation methods can be used to produce the surface. However, it is difficult to ensure the continuity and smoothness of the final interpolant along with all adjacent triangles. In this contribution, a meshless approach is proposed by using radial basis functions (RBFs). It is applicable to explicit functions of two variables and it is suitable for all types of scattered data in general. The key point for the RBF approximation is finding the important points that give a good approximation with high precision to the scattered data. Since the compactly supported RBFs (CSRBF) has limited influence in numerical computation, large data sets can be processed efficiently as well as very fast via some efficient algorithm. The main advantage of the RBF is, that it leads to a solution of a system of linear equations (SLE) $Ax = b$. Thus any efficient method solves the systems of linear equations that can be used. In this study is we propose a new method of determining the importance points on the scattered data that produces a very good reconstructed surface with higher accuracy while maintaining the smoothness of the surface.

Keywords: Meshless methods · Radial Basis Functions · Approximation

© Springer Nature Switzerland AG 2020
V. V. Krzhizhanovskaya et al. (Eds.): ICCS 2020, LNCS 12142, pp. 239–250, 2020.
https://doi.org/10.1007/978-3-030-50433-5_19

1 Introduction

Interpolation and approximation techniques are used in the solution of many engineering problems. However, the interpolation of unorganized scattered data is still a severe problem. In the one dimensional case, i.e., curves represented as $y = f(x)$, it is possible to order points according to the x-coordinate. However, in a higher dimensionality this is not possible. Therefore, the standard approaches are based on the tessellation of the domain in x, y or x, y, z spaces using, e.g. Delaunay triangulation [7], etc. This approach is applicable for static data and t-varying data, if data in the time domain are "framed", i.e. given for specific time samples. It also leads to an increase of dimensionality, i.e. from triangulation in E^2 to triangulation in E^3 or from triangulation in E^3 to triangulation in E^4, etc. It results in significant increase of the triangulation complexity and complexity of a triangulation algorithm implementation. This is a significant factor influencing computation in the case of large data sets and large range data sets, i.e. when x, y, z values are spanned over several magnitudes.

On the contrary, meshless interpolations based on Radial Basis Functions (RBF) offer several significant advantages, namely:

- RBF interpolation is applicable generally to d-dimensional problems and does not require tessellation of the definition domain
- RBF interpolation and approximation is especially convenient for scattered data interpolation, including interpolation of scattered data in time as well
- RBF interpolation is smooth by a definition
- RBF interpolation can be applied for interpolation of scalar fields and vector fields as well, which can be used for scalar and vector fields visualization
- If the Compactly Supported RBFs (CSRBF) are used, sparse matrix data structures can be used which decreases memory requirements significantly.

However, there are some weak points of RBF application in real problems solution:

- there is a real problem for large data sets with robustness and reliability of the RBF application due to high conditionality of the matrix A of the system of linear equations, which is to be solved
- numerical stability and representation is to be applied over a large span of x, y, z values, i.e. if values are spanned over several magnitudes
- problems with memory management as the memory requirements are of $O(N^2)$ complexity, where N is a number of points in which values are given
- the computational complexity of a solution of the linear system, which is $O(N^3)$, resp. $O(kN^2)$, where k is a number of iteration if the iterative method are used, but k is relatively high, in general.
- Problems with unexpected behavior at geometrical borders

Many contributions are solving some issues of the RBF interpolation and approximation available. Numerical tests are mostly made using some standard testing functions and restricted domain span, mostly taking interval $\langle 0, 1 \rangle$ or similar. However, in many physically based applications, the span of the domain is higher, usually over several magnitudes and large data sets need to be processed. Also large data sets are to be processed.

As the meshless techniques are easily scalable to higher dimensions and can handle spatial scattered data and spatial-temporal data as well, they can be used in many engineering and economical computations, etc. Polygonal representations (tessellated domains) are used in computer graphics and visualization as a surface representation and for surface rendering. In time-varying objects, a surface is represented as a triangular mesh with constant connectivity.

On the other hand, all polygonal based techniques, in the case of scattered data, require tessellations, e.g. Delaunay triangulation with $O(N^{\lfloor d/2+1 \rfloor})$ computational complexity for N points in d-dimensional space or another tessellation method. However, the complexity of tessellation algorithms implementation grows significantly with dimensionality and severe problems with robustness might be expected, as well.

In the case of data visualization smooth interpolation or approximation on unstructured meshes is required, e.g. on triangular or tetrahedral meshes, when physical phenomena are associated with points, in general. This is quite a difficult task especially if the smoothness of interpolation is needed. However, it is a natural requirement in physically-based problems.

2 Meshless Interpolation

Meshless (meshfree) methods are based on the idea of Radial Basis Function (RBF) interpolation [1, 2, 22, 23], which is not separable. RBF based techniques are easily scalable to d-dimensional space and do not require tessellation of the geometric domain and offer smooth interpolation naturally. In general, meshless techniques lead to a solution of a linear system equations (LS) [4, 5] with a full or sparse matrix.

Generally, meshless methods for scattered data can be split into two main groups in computer graphics and visualization:

- "implicit" – $F(x) = 0$, i.e. $F(x, y, z) = 0$ used in the case of a surface representation in E^3, e.g. surface reconstruction resulting into an implicit function representation. This problem is originated from the implicit function modeling [15] approach,
- "explicit" – $F(x) = h$ used in interpolation or approximation resulting in a functional representation, e.g. a height map in E^2, i.e. $h = F(x, y)$.

where: x is a point represented generally in d-dimensional space, e.g. in the case of 2-dimensional case $x = [x, y]^T$ and h is a scalar value or a vector value.

The RBF interpolation is based on computing of the distance of two points in the d–dimensional space and it is defined by a function:

$$f(x) = \sum_{j=1}^{M} \lambda_j \varphi(\|x - x_j\|) = \sum_{j=1}^{M} \lambda_j \varphi(r_j) \tag{1}$$

where: $r_j = \|x - x_j\|_2 \underset{=}{\text{def}} \sqrt{(x - x_j)^2 + (y - y_j)^2}$ (in 2-dimensional case) and λ_j are weights to be computed. Due to some stability issues, usually a polynomial $P_k(x)$ of a degree k is added [6]. It means that for the given data set $\{\langle x_i, h_i \rangle\}_1^M$, where h_i are associated values to be interpolated and x_i are domain coordinates, we obtain a linear system of equations:

$$h_i = f(x_i) =$$
$$\sum_{j=1}^{M} \lambda_j \, \varphi(\|x_i - x_j\|) + P_k(x_i) \qquad i = 1, \ldots, M \qquad x = [x, y : 1]^T \tag{2}$$

For a practical use, a polynomial of the 1st degree is used, i.e. linear polynomial $P_1(x) = a^T x$ in many applications. Therefore, the interpolation function has the form:

$$f(x_i) = \sum_{j=1}^{M} \lambda_j \, \varphi(\|x_i - x_j\|) + a^T x_i \quad h_i = f(x_i) \qquad i = 1, \ldots, M$$
$$= \sum_{j=1}^{M} \lambda_j \, \varphi_{i,j} + a^T x_i \tag{3}$$

and additional conditions are to be applied:

$$\sum_{j=1}^{M} \lambda_i x_i = 0 \quad \text{i.e.} \quad \sum_{j=1}^{M} \lambda_i x_i = 0 \quad \sum_{j=1}^{M} \lambda_i y_i = 0 \quad \sum_{j=1}^{M} \lambda_i = 0 \tag{4}$$

It can be seen that for the d-dimensional case a system of $(M+d+1)$ linear system has to be solved, where M is a number of points in the dataset and d is the dimensionality of data. For $d = 2$ vectors x_i and a are in the form $x_i = [x_i, y_i, 1]^T$ and $a = [a_x, a_y, a_0]^T$, we can write:

$$\begin{bmatrix} \varphi_{1,1} & \cdots & \varphi_{1,M} & x_1 & y_1 & 1 \\ \vdots & \ddots & \vdots & \vdots & \vdots & \vdots \\ \varphi_{M,1} & \cdots & \varphi_{M,M} & x_M & y_M & 1 \\ x_1 & \cdots & x_M & 0 & 0 & 0 \\ y_1 & \cdots & y_M & 0 & 0 & 0 \\ 1 & \cdots & 1 & 0 & 0 & 0 \end{bmatrix} \begin{bmatrix} \lambda_1 \\ \vdots \\ \lambda_M \\ a_x \\ a_y \\ a_0 \end{bmatrix} = \begin{bmatrix} h_1 \\ \vdots \\ h_M \\ 0 \\ 0 \\ 0 \end{bmatrix} \tag{5}$$

This can be rewritten in the matrix form as:

$$\begin{bmatrix} B & P \\ P^T & 0 \end{bmatrix} \begin{bmatrix} \lambda \\ a \end{bmatrix} = \begin{bmatrix} f \\ 0 \end{bmatrix} \qquad Ax = b \quad a^T x_i = a_x x_i + a_y y_i + a_0 \tag{6}$$

For the two-dimensional case and M points given a system of $(M+3)$ linear equations has to be solved. If "global" functions, e.g. $\varphi(r) = r^2 lg\, r$, are used, then the matrix B is "full", if "local" functions CSRBFs are used, the matrix B can be sparse.

The RBF interpolation was originally introduced by Hardy as the multiquadric method in 1971 [5], which was called Radial Basis Function (RBF) method. Since then many different RFB interpolation schemes have been developed with some specific properties, e.g. 4 uses $\varphi(r) = r^2 lg\, r$, which is called Thin-Plate Spline (TPS), a function $\varphi(r) = e^{-(\in r)^2}$ was proposed in [23]. However, the shape parameter $\in$ might leads to an ill-conditioned system of linear equations [26].

The CSRBFs were introduced as:

$$\varphi(r) = \begin{cases} (1-r)^q P(r), & 0 \leq r \leq 1 \\ 0, & r > 1 \end{cases} \tag{7}$$

where: $P(r)$ is a polynomial function and q is a parameter. Theoretical problems with numerical stability were solved in [4]. In the case of global functions, the linear system of equations is becoming ill conditioned and problems with convergence can be expected. On the other hand, if the CSRBFs are taken, the matrix A is becoming relatively sparse, i.e. computation of the linear system will be faster, but we need to carefully select the scaling factor α (which can be "tricky") and the final function might tend to be "blobby" shaped, see Table 1 and Fig. 1.

Table 1. Typical examples of "local" functions – CSRBF ("+" means – value zero out of $\langle 0, 1 \rangle$)

ID	Function	ID	Function
1	$(1-r)_+$	6	$(1-r)_+^6 (35r^2 + 18r + 3)$
2	$(1-r)_+^3 (3r+1)$	7	$(1-r)_+^8 (32r^3 + 25r^2 + 8r + 3)$
3	$(1-r)_+^5 (8r^2 + 5r + 1)$	8	$(1-r)_+^3$
4	$(1-r)_+^2$	9	$(1-r)_+^3 (5r+1)$
5	$(1-r)_+^4 (4r+1)$	10	$(1-r)_+^7 (16r^2 + 7r + 1)$

The compactly supported RBFs are defined for the "normalized" interval $r \in 0, 1$, but for the practical use a scaling is used, i.e. the value r is multiplied by shape parameter α, where $\alpha > 0$.

Meshless techniques are primarily based on the approaches mentioned above. They are used in engineering problem solutions, nowadays, e.g. partial differential equations, surface modeling, surface reconstruction of scanned objects [13, 14], reconstruction of corrupted images [21], etc. More generally, meshless object representation is based on specific interpolation or approximation techniques [1, 6, 23].

The resulting matrix A tends to be large and ill-conditioned. Therefore, some specific numerical methods have to be taken to increase the robustness of a solution, like preconditioning methods or parallel computing on GPU [9, 10], etc. In addition, subdivision or hierarchical methods are used to decrease the sizes of computations and increase robustness [15, 16, 27].

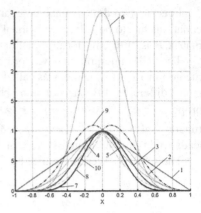

Fig. 1. Properties of CSRBFs

It should be noted, that the *computational complexity* of meshless methods actually covers the complexity of tessellation itself and interpolation and approximation methods. This results in problems with large data set processing, i.e. numerical stability and memory requirements, etc.

If global RBF functions are considered, the RBF matrix is full and in the case of 10^6 of points, the RBF matrix is of the size approx. $10^6 \times 10^6$! On the other hand, if CSRBF used, the relevant matrix is sparse and computational and memory requirements are decreased significantly using special data structures [8, 10, 20, 27].

In the case of physical phenomena visualization, data received by simulation, computation or obtained by experiments usually are oversampled in some areas and also numerically more or less precise. It seems possible to apply approximation methods to decrease computational complexity significantly by adding virtual points in the place of interest and use analogy of the least square method modified for the RBF case [3, 12, 17, 25].

Due to the CSRBF representation the space of data can be subdivided, interpolation, resp. the approximation can be split to independent parts and computed more or less independently [20]. This process can be also parallelized and if appropriate computational architecture is used, e.g. GPU, etc. it will lead to faster computation as well. The approach was experimentally verified for scalar and vector data used in the visualization of physical phenomena.

3   Points of Importance

Algorithms developed recently were based on different specific properties of "global" RBFs or "local" compactly supported RBFs (CS-RBFs) and application areas expected, e.g. for interpolation, approximation, solution of partial differential equations, etc., expecting "reasonable" density of points. However, there are still some important problems to be analyzed and hopefully solved, especially:

- What is an acceptable compromise between the precision of approximation and compression ratio, i.e. reduction of points, if applicable?
- What is the optimal constant shape parameter, if does exist and how to estimate it efficiently [26]?

- What are optimal shape parameters α for every single $\varphi(r, \alpha)$ [24, 26]?
- What is the robustness and stability of the RBF for large data and large range span of data with regard to shape parameters [16, 17]?

In this contribution, we will analyze a specific problem related to the first question.

Let us consider given points of a curve (samples of a signal), described by explicit function $y = f(x)$. According to the Nyquist-Shannon theorem, the sampling frequency should be at least double the frequency of the highest frequency of the original signal. The idea is, how "points of importance", i.e. points of inflection and extrema can be

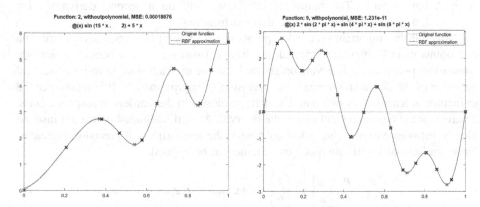

Fig. 2. Testing functions and resulting approximation based on the points of importance (red points are extrema, black points are additional points of importance) (Color figure online)

used for smooth precise curve approximation.

Let us consider sampled curves in Fig. 2, i.e. a signal without noise (the blue points are values at the borders, red are maxima, the black are inflection and added points. It can be seen that the reconstruction based on radial basis functions (RBF) has to pass:

- points at the interval borders
- points at extremes, maxima and minima
- some other important points, like points of inflection etc., and perhaps some additional points of the given data to improve signal reconstruction.

However, there several factors to be considered as well, namely:

- extensibility from 2 D to 3 D for explicit functions of two variables, i.e. $z = f(x, y)$ and hopefully to higher dimension robustness of computation as given discrete data are given.

For extrema finding, the first derivative $f'(x)$ is to be replaced by a standard discrete scheme. At the left, resp. right margin, forward, resp. the backward difference is to be used. Inside of the interval, the central difference scheme is recommended, as it also "filters" high frequencies. The simple scheme for the second derivative estimation is shown, too. It can be seen, that this is easily extensible for the 3 D case as well.

$$f'(x) \approx \frac{f(x_{i+1})-f(x_i)}{x_{i+1}-x_i} \qquad f'(x) \approx \frac{f(x_i)-f(x_{i-1})}{x_i-x_{i-1}}$$
$$f''(x) \approx \frac{(x_{i+1})-f(x_{i-1})}{2(x_{i+1}-x_{i-1})} \qquad f''(x) \approx \frac{f(x_{i+1})-2f(x_i)+f(x_{i-1})}{(x_{i+1}-x_i)(x_i-x_{i-1})} \qquad (8)$$

So far, a finding of extrema is a simple task, now. However, due to the discrete data, the extrema is detected by

$$\text{sign}(f(x_{i+1}) - f(x_i)) \neq \text{sign}(f(x_i) - f(x_{i-1})) \qquad (9)$$

as we need to detect the change of the sign, only. This increases the robustness of computation as well. The points of inflections rely on a second derivative, i.e. $f''(x) = 0$; a similar condition can be derived from (8).

Now, all the important points, i.e. points at the interval borders, maxima, minima and points of inflection, are detected and found. However, it is necessary to include some more points at the interval borders (at least one on each side) to respect the local behavior of the curve and increase the precision of approximation. It is recommended to include at least one or two points which are closest to the borders to respect a curve behavior at the beginning and end of the interval. Also, if additional points are inserted ideally between extreme and inflection points, the approximation precision increases. Now, the standard RBF interpolation scheme can be applied.

$$\begin{bmatrix} B & P \\ P^T & 0 \end{bmatrix} \begin{bmatrix} \lambda \\ a \end{bmatrix} = \begin{bmatrix} f \\ 0 \end{bmatrix} \qquad Ax = b \quad a^T x_i = a_x x_i + a_0 \qquad (10)$$

where: B represents the RBF submatrix, λ the weights of RBFs, P represents points for the polynomial a represents coefficients of the polynomial, f given function values.

It should be noted, that in the case of scattered data, neighbors for each point are to be found, before the estimation of the derivative is made. In the 2 D case, ordering is possible, in the 3 D case computation is to be made on neighbors found. If the regular sampling in each dimension (along the axis) is given, computation simplifies significantly.

It is necessary to note that the curve reconstruction is at the Nyquist-Shannon theorem boundary and probably limits of the compression were obtained with very low relative error, which is less than 0.1%. However, we have many more points available and if a higher precision is needed, the approximation based on Least Square Error (LSE) computational scheme with Lagrange multipliers might be used [11]. The RBF methods usually lead to an ill-conditioned system of linear equations [26]. In the case of approximation, it can be partially improved by geometry algebra in projective space [18, 19] approach.

4 Experimental Results

The presented approach was tested on several testing functions used for evaluation of errors, stability, robustness of computation, see Table 2:

Table 2. Examples of testing functions

ID	Function	ID	Function
1	$y = \sin(15x^2 + 5x)$	2	$y = \cos(20x)/2 + 5x$
3	$y = 50(0.4\sin(15x^2) + 5x)$	4	$y = \sin(8\pi x)$
5	$y = \sin(6\pi x^2)$	6	$y = \sin(25x + 0.1)/(25x + 0.1)$
7	$y = 2\sin(2\pi x) + \sin(4\pi x)$	8	$y = 2\sin(2\pi x) + \sin(4\pi x) + \sin(8\pi x)$
9	$y = 2\sin(\pi(2x - 1)) + \sin(3\pi(2x - 1/2))$	10	$y = 2\sin(\pi(1 - 2x)) + \sin(3\pi(2x - 1/2))$
11	$y = 2\sin(\pi(2x - 1)) + \sin(3\pi(2x - 1/2)) - x$	12	$y = 2\sin\left(2\pi x - \frac{\pi}{2}\right) + \sin(3\pi(2x - 1/2))$
13	$y = \operatorname{atan}(10x - 5)^3 + \operatorname{atan}(10x - 8)^3/2$	14	$y = (4.88x - 1.88) * \sin(4.88x - 1.88)^2 + 1$
15	$y = \exp(10x - 6) * \sin(5x - 2)^3 + (3x - 1)^3$	16	$y = \tanh(9x + 1/2)/9$

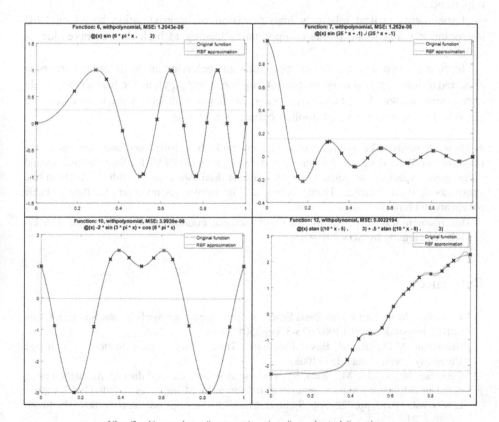

Fig. 3. Examples of approximation for selected functions.

The experiments have also proven, that for large data and data with a large span of data a polynomial $P_k(x)$ should be $P_k(x) = a_0$, i.e. $k = 0$, see [16, 17].

Selected results of the approximation of some functions are presented at Fig. 3. It can be seen, that the proposed approximation based actually on RBF interpolation scheme using points of importance offers good precision of approximation a with good compression ratio. The functions were sampled in 200 points approx. and 10–20 points are actually used for the proposed approximation method.

5   Conclusion

This contribution briefly describes a method for efficient RBF approximation of large scattered data based on finding points of importance. This leads to a simple RBF based approximation of data with relatively low error with high compression. The precision of approximation can be increased significantly by covering some additional points. The approach is easily extensible to the 3D case, especially if data are ordered. However, if data are scattered, the neighbor points must be evaluated to find points of importance.

Experiments proved relatively high precision of approximation based on RFB interpolation using found points of importance leading to high data compression as well.

In future, deep analysis of an approximation behavior at the interval borders is expected as it is a critical issue for the 3D case, i.e. $z = f(x, y)$, as the first already made experiments shown. Also, the discrete points of curves of inflection are to be taken into account, i.e. discrete points of implicit curves $F(x, y) = 0$.

Acknowledgments. The authors would like to thank their colleagues and students at the University of West Bohemia and Universiti Teknologi PETRONAS for their discussions and suggestions; especially to Michal Smolik, Zuzana Majdisova and Jakub Vasta from the University of West Bohemia. Thanks belong also to anonymous reviewers for their valuable comments and hints provided.

This research was supported by the Czech Science Foundation (GACR) project GA 17-05534S and partially by SGS 2019-016.

References

1. Biancolini, M.E.: Fast Radial Basis Functions for Engineering Applications. Springer, Cham (2017). https://doi.org/10.1007/978-3-319-75011-8
2. Buhmann, M.D.: Radial Basis Functions: Theory and Implementations. Cambridge University Press, Cambridge (2008)
3. Cervenka, M., Smolik, M., Skala, V.: A new strategy for scattered data approximation using radial basis functions respecting points of inflection. In: Misra, S., et al. (eds.) ICCSA 2019. LNCS, vol. 11619, pp. 322–336. Springer, Cham (2019). https://doi.org/10.1007/978-3-030-24289-3_24

4. Duchon, J.: Splines minimizing rotation-invariant semi-norms in Sobolev space. In: Schempp, W., Zeller, K. (eds.) Constructive Theory of Functions of Several Variables. LNCS, vol. 571. Springer, Heidelberg (1997). https://doi.org/10.1007/BFb0086566
5. Hardy, L.R.: Multiquadric equation of topography and other irregular surfaces. J. Geophys. Res. **76**(8), 1905–1915 (1971)
6. Fasshauer, G.E.: Meshfree Approximation Methods with MATLAB. World Scientific Publishing, Singapore (2007)
7. Karim, S.A.A., Saaban, A., Skala, V.: Range-restricted interpolation using rational bi-cubic spline functions with 12 parameters. **7**, 104992–105006 (2019). SSN: 2169-3536. https://doi.org/10.1109/access.2019.2931454
8. Majdisova, Z., Skala, V.: A new radial basis function approximation with reproduction. In: CGVCVIP 2016, Portugal, pp. 215–222 (2016). ISBN 978-989-8533-52-4
9. Majdisova, Z., Skala, V.: Radial basis function approximations: comparison and applications. Appl. Math. Model. **51**, 728–743 (2017). https://doi.org/10.1016/j.apm.2017.07.033
10. Majdisova, Z., Skala, V.: Big geo data surface approximation using radial basis functions: a comparative study. Comput. Geosci. **109**, 51–58 (2017). https://doi.org/10.1016/j.cageo.2017.08.007
11. Majdisova, Z., Skala, V., Smolik, M.: Determination of stationary points and their bindings in dataset using RBF methods. In: Silhavy, R., Silhavy, P., Prokopova, Z. (eds.) CoMeSySo 2018. AISC, vol. 859, pp. 213–224. Springer, Cham (2019). https://doi.org/10.1007/978-3-030-00211-4_20
12. Majdisova, Z., Skala, V., Smolik, M.: Determination of reference points and variable shape parameter for RBF approximation. Integr. Comput.-Aided Eng. **27**(1), 1–15 (2020). https://doi.org/10.3233/ICA-190610. ISSN 1069-2509
13. Pan, R., Skala, V.: A two level approach to implicit modeling with compactly supported radial basis functions. Eng. Comput. **27**(3), 299–307 (2011). https://doi.org/10.1007/s00366-010-0199-1. ISSN 0177-0667
14. Pan, R., Skala, V.: Surface reconstruction with higher-order smoothness. Vis. Comput. **28**(2), 155–162 (2012). https://doi.org/10.1007/s00371-011-0604-9. ISSN 0178-2789
15. Ohtake, Y., Belyaev, A., Seidel, H.-P.: A multi-scale approach to 3D scattered data interpolation with compactly supported basis functions. In: Shape Modeling, pp. 153–161. IEEE, Washington (2003). https://doi.org/10.1109/smi.2003.1199611
16. Skala, V.: RBF interpolation with CSRBF of large data sets, ICCS 2017. Procedia Comput. Sci. **108**, 2433–2437 (2017). https://doi.org/10.1016/j.procs.2017.05.081
17. Skala, V.: RBF interpolation and approximation of large span data sets. In: MCSI 2017 – Corfu, pp. 212–218. IEEE (2018). https://doi.org/10.1109/mcsi.2017.44
18. Skala, V., Karim, S.A.A., Kadir, E.A.: Scientific computing and computer graphics with GPU: application of projective geometry and principle of duality. Int. J. Math. Comput. Sci. **15**(3), 769–777 (2020). ISSN 1814-0432
19. Skala, V.: High dimensional and large span data least square error: numerical stability and conditionality. Int. J. Appl. Phys. Math. **7**(3), 148–156 (2017). https://doi.org/10.17706/ijapm.2017.7.3.148-156. ISSN 2010-362X
20. Smolik, M., Skala, V.: Large scattered data interpolation with radial basis functions and space subdivision. Integr. Comput.-Aided Eng. **25**(1), 49–62 (2018). https://doi.org/10.3233/ica-170556
21. Uhlir, K., Skala, V.: Reconstruction of damaged images using radial basis functions. In: EUSIPCO 2005 Conference Proceedings, Turkey (2005). ISBN 975-00188-0-X
22. Wenland, H.: Scattered Data Approximation. Cambridge University Press (2010). http://doi.org/10.1017/CBO9780511617539

23. Wright, G.B.: Radial basis function interpolation: numerical and analytical developments. Ph.D. thesis, University of Colorado, Boulder (2003)
24. Skala, V., Karim, S.A.A., Zabran, M.: Radial basis function approximation optimal shape parameters estimation: preliminary experimental results. In: ICCS 2020 Conference (2020)
25. Vasta, J., Skala, V., Smolik, M., Cervenka, M.: Modified radial basis functions approximation respecting data local features. In: Informatics 2019, IEEE Proceedings, Poprad, Slovakia, pp. 445–449 (2019). ISBN 978-1-7281-3178-8
26. Cervenka, M., Skala, V.: Conditionality analysis of the radial basis function matrix. In: International Conference on Computational Science and Applications ICCSA (2020)
27. Smolik, M., Skala, V.: Efficient speed-up of radial basis functions approximation and interpolation formula evaluation. In: International Conference on Computational Science and Applications ICCSA (2020)

High Accuracy Terrain Reconstruction from Point Clouds Using Implicit Deformable Model

Jules Morel[1][(✉)], Alexandra Bac[2], and Takashi Kanai[1]

[1] Kanai Laboratory, Graduate School of Arts and Sciences,
The University of Tokyo, Tokyo, Japan
jules.morel@ifpindia.org
[2] Laboratoire d'Informatique et des Systèmes, Aix Marseille University,
Marseille, France

Abstract. Few previous works have studied the modeling of forest ground surfaces from LiDAR point clouds using implicit functions. [10] is a pioneer in this area. However, by design this approach proposes over-smoothed surfaces, in particular in highly occluded areas, limiting its ability to reconstruct fine-grained terrain surfaces. This paper presents a method designed to finely approximate ground surfaces by relying on deep learning to separate vegetation from potential ground points, filling holes by blending multiple local approximations through the partition of unity principle, then improving the accuracy of the reconstructed surfaces by pushing the surface towards the data points through an iterative convection model.

Keywords: Implicit surface · Deformable model · Deep learning

1 Introduction

Digital terrain model (DTM) extraction is an important issue in the field of LiDAR remote sensing of Earth surface. Indeed, further data processing procedures (segmentation, surface reconstruction, digital volume computation) usually rely on a prior DTM computation to focus on features of interest (buildings, roads or trees for instance). In the last two decades, many filtering algorithms have been proposed to solve this problem using Airborne LiDAR sensors (ALS) data.

However, most previous works address DTM extraction for ALS data, which largely differs from terrestrial LiDAR sensors (TLS) data. Unlike ALS point clouds, TLS ones provide very dense sampling rates at the ground level, describing the micro-topography around the sensor. However, the presence of vegetation and the terrain topography itself generate strong occlusions causing large data gaps at the ground level, and a risk of integrating objects above the ground within the DTM. Additionally, the scanning resolution of TLS devices depends,

© Springer Nature Switzerland AG 2020
V. V. Krzhizhanovskaya et al. (Eds.): ICCS 2020, LNCS 12142, pp. 251–265, 2020.
https://doi.org/10.1007/978-3-030-50433-5_20

by nature, on their distance to the sensor, resulting in spatial variations in point density. Therefore, DTM extraction for TLS data requires dedicated approaches, in particular for 3D samples acquired in forest environments. [10] is a pioneering effort designed to reconstruct detailed DTMs from TLS data under forest canopies using implicit function. The basic idea of this work is (1) to approximate locally the ground surface through adaptive scale refinement based on a quad-tree division of the scene, (2) then to simultaneously filter vegetation and correct approximations based on the points distribution in each quad-tree cell, and (3) to blend the local approximations into a global implicit model. In the present paper, we build upon this previous work and propose several innovations in order to reconstruct high quality terrain models from laser scan point clouds.

2   Overview of the Method

Based on the previous work [10], our method relies on adaptive scale refinement through a quad-tree division of the scene, local approximations in the quad-tree cells, and estimation of a global model through the blending of local approximations, to compute the first approximation of the ground surface. The method presented in this paper offers several improvements of the previous work to enhance the accuracy of the reconstructed terrain model: First, we relegate the filtering out of vegetation points to a deep network based on PointNet++ [14]. Second, we attribute weights to the data points according to their local density in order to obtain a more robust local approximation. Third, we refine the blending of local approximations to make it compliant with the partition of unity concept with compactly supported radial basis functions (CSRBF). Finally, we introduce an original deformable model, based on a convection model under a moving least squares field and a proper functional basis. We thus push the implicit ground surface towards data points to obtain centimeter accuracy. The complete sequence of computational steps is shown in Fig. 1.

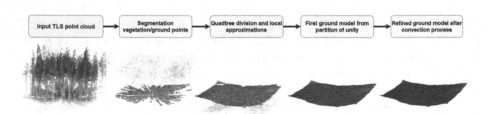

Fig. 1. Overview of method. The segmented ground points are colored according to their weights coming from the local density.

3   Filtering of Vegetation

3.1   Selection of Ground Points Using Deep Learning

In the estimation of ground surface from LiDAR point clouds, it is common to first project the point cloud into a fine regular 2D (x,y) grid, of resolution res_{MIN}, and further select the points of minimum elevation in each cell. The resulting point cloud is denoted P_{min}. In order to ease the segmentation, we enriched each point of P_{min} by geometric descriptors encoding the local linearity and planarity. In practice, we consider for each points its Cartesian coordinates plus the averaged eigenvalues of the local covariance matrix of the 3D distribution of the K_{PCA} neighbors, K_{PCA} being defined by the user. In order to feed these enriched point clouds into a convolutional architecture, we define a partition of space allowing splitting of an unordered 3D point cloud into overlapping fixed size regions, each of them encoding the 3D local points distribution. In practice, we divide the point cloud P_{min} into a regular 2D (x,y) grid, then distribute collocation points to the centroid of every occupied voxel. Then, we form each batch $\mathscr{B}_i$ by considering the 1024 nearest neighbors of each collocation point in the (x,y) plane. These batches are designed to feed and train a deep learning model; this model eventually predicts a label for each point of each batch. As batches largely overlap, each input point belongs to several batches and thus receive multiple predictions, one for every batch it belongs to. In the last step, we define a voting process to extract a final segmentation from the multiple votes per point. The resulting set of points denoted by $P = \{\mathbf{p}_i\}_{i=1...N} \subset \mathbb{R}^3$ serves as raw data for the proposed algorithm.

3.2   Weighting of Points According to the Local Density

A common practice in surface reconstruction from in-homogeneous point clouds consists of applying a weight to each point according to the local density around $\mathbf{p}_i$. In order to take into account density irregularities due to overlapping scans and the confidence measures attributed to the points, the proposed method by [11] scales down the point influence in high density areas. We were inspired by this idea but implement the opposite scaling: in forest 3D scans, while considering terrain modeling, clusters of points tend to describe accurately ground topology whereas isolated points appear less reliable and their influence needs to be reduced. To do so, we assigned a weight d_i at each point $\mathbf{p}_i$ given by

$$d_i = 1 - \frac{1}{max_d} \sum_{\mathbf{p}_j \in \mathscr{P}_j^K} \|\mathbf{p}_i - \mathbf{p}_j\| \tag{1}$$

where P_j^K is the K-neighborhood of point $\mathbf{p}_i$ (from our experiments $K = 20$ appears as a good choice), max_d is defined as:

$$max_d = \max \left[\sum_{\mathbf{p}_j \in \mathscr{P}_j^K} \|\mathbf{p}_i - \mathbf{p}_j\|, \forall\ \mathbf{p}_i \in P \right] \tag{2}$$

4 First Surface Estimation by Blending Local Quadrics Approximations

4.1 Quadtree Division of the Plane

Initially, the minimum bounding rectangle of filtered minimum points is inserted as the root of the quadtree. Then, we iteratively subdivide each cell into four leaves. The process stops when a leaf presents a side smaller than $Size_{min}$, the minimal size allowed defined by the user, or when n_{cell} the number of points in the cell becomes inferior to the number of parameters of the local quadric patches (i.e. $n_{cell} < 6$) described in the following Sect. 4.2.

4.2 Local Approximations

Taking into account a quadtree that divides the data points, for each of its leaves $\mathscr{O}_i$, we denote by $\mathbf{c}_i$ the center of $\mathscr{O}_i$.

In each cell $\mathscr{O}_i$, an approximate tangent plane is computed using least square fitting. Let (u, v, w) denote a local orthonormal coordinate system such that the direction z_l is orthogonal to the fitting plane. We approximate $\mathscr{O}_i$ by a local implicit quadric function $g_i(u, v, w) = w - h_i(u, v)$ where, in the local coordinate system, h_i is a quadratic parametric surface of the following form: $h(u, v) = A \cdot u^2 + B \cdot u \cdot v + C \cdot v^2 + D \cdot u + E \cdot v + F$. In each leaf $\mathscr{O}_i$, considering $\mathbf{p}_j = (u_j, v_j, w_j) \in \mathscr{P}$, coefficients A, B, C, D, E, F are determined by the weighted least square minimization of:

$$\sum_{\mathbf{p}_j \in \mathscr{P}} d_i \cdot (w_j - h(u_j, v_j))^2 \cdot \Phi_{\sigma^i}(\|\mathbf{p}_j - \mathbf{c}_i\|) \tag{3}$$

where d_i is the weight defined in Sect. 3.2 and $\Phi_{\sigma^i}(\|x - c_i\|) = \phi(\frac{\|x - c_i\|}{\sigma^i})$ and $\phi(r) = (1 - r)_+^4 (1 + 4r)$ is a compactly supported Wendland's RBF function [16]. The function ϕ is $\mathscr{C}^2$ and radial on $\mathbb{R}^3$. The parameter σ^i controls the influence of the local approximation of the leaf $\mathscr{O}_i$. In our case, in order to always cover the bounding box of the points contained in the leaf $\mathscr{O}_i$ by a ball of radius σ^i, we chose $\sigma^i = a_i \times \sqrt{3} \times 0.75$, where a_i is the longest side of the leaf $\mathscr{O}_i$.

4.3 Global Approximation by Blending Local Approximations

After estimating the local approximations of the data points for each leafs $\mathscr{O}_i$, the global implicit model by blending together local patches by means of Wendland's CSRBF. The global implicit model f is computed as:

$$f(\mathbf{x}) = \sum_{\mathbf{c}_i \in C} g_i(\mathbf{x}) \cdot \frac{\Phi_{\sigma^i}(\|\mathbf{x} - \mathbf{c}_i\|)}{\sum_{\mathbf{c}_j \in C} \Phi_{\sigma^j}(\|\mathbf{x} - \mathbf{c}_j\|)} \tag{4}$$

where $g_i(\mathbf{x})$ is the local approximation in the leaf $\mathscr{O}_i$ defined in the previous Section.

More precisely we use compactly supported Wendland's RBF functions [16] as a partition of unity to merge the local implicit functions (issued from parametric patches) and compute a global implicit model.

5 Improvement of Surface Model with a Convection Model

Because occlusion phenomenon forces large quadtree cells, and thus coarser local approximations, the first surface estimation described in Sect. 4 can be further improved by being pushed closer to the data point. Inspired by the level-sets formulations [12,15] in the field of image processing, our method builds particularly on the work of Gelas [6] and describes the evolution of the surface driven by a time-dependent partial differential equation (PDE) where the so-called velocity term reflects the 3D point cloud features. The level-set PDE is then solved through a collocation method using CSRBF.

5.1 From Blended Quadrics Model to CSRBF Model

In order to both express the refined implicit surface and discretize the numerical problem, we create a new base of functions. This basis is built according to the following process: collocation centers are regularly distributed on a 3D grid, of user-defined resolution r, around the zero-levelset of the implicit function 4. In order to limit the number of collocations while allowing the surface to move in the vicinity of data points but, we decimate the clouds of collocation by removing each center further than $2 \times r$ from a data point or further than r from the zero-levelset. Considering $\mathscr{Q}$ the set of collocation, the basis of function is then defined as the set of translates at each remaining collocation $\mathbf{o} \in \mathscr{Q}$ of Wendland's CSRBF $\Phi_{r_o}(\|\mathbf{x} - \mathbf{o}\|) = \phi(\frac{\|\mathbf{x}-\mathbf{o}\|}{r_o})$: where $r_o = r \times \sqrt{3} \times 0.75$.

First, we approximate the implicit function f defined in Eq. 4 in the new basis $Span\{\Phi_{r_o}\}$ as $g(\mathbf{x}) = \sum_{\mathbf{o} \in \mathscr{Q}} \alpha_{\mathbf{o}} \cdot \Phi_{r_o}(\|\mathbf{x} - \mathbf{o}\|)$.

To retrieve $\{\alpha_{\mathbf{o}}\}_{\mathbf{o} \in \mathscr{Q}}$, we consider the system:

$$
\begin{vmatrix} \ddots & & \\ & \mathscr{A}_{\mathbf{o},\mathbf{o}'} & \\ & & \ddots \end{vmatrix} \begin{vmatrix} \vdots \\ \alpha_{\mathbf{o}} \\ \vdots \end{vmatrix} = \begin{vmatrix} \vdots \\ f(\mathbf{o}') \\ \vdots \end{vmatrix} \tag{5}
$$

where o and o' are two collocations, $\mathscr{A}_{\mathbf{o},\mathbf{o}'} = \Phi_{r_o}(\|\mathbf{o}' - \mathbf{o}\|)$ and the implicit function f describing the blended quadrics model is expressed by Eq. 4. Thanks to the properties of Wendland's CSRBF Φ_{r_o}, the matrix of the $\mathscr{A}_{\mathbf{o},\mathbf{o}'}$ elements is sparse, self-adjoint and positive definite. Thus, the system is inverted by using the LDL^T Cholesky decomposition implemented by Eigen library.

5.2 Solving Convection Evolution Equation Using CSRBF Collocation

We define now the velocity vector $\mathscr{V}(\mathbf{p}, t), \mathbf{p} \in \mathbb{R}^3, t \in \mathbb{R}$, a function reflecting the geometrical properties of the interface according to the data, and quantifying the local deformations over time. The velocity is actually computed at each collocation $\mathbf{o} \in \mathscr{O}$ by minimization of an energy function inspired by the snake approaches introduced by Kass [8]. Indeed, we define the energy function $\forall \mathbf{q} \in \mathbb{R}^3, \xi(\mathbf{q}) = \gamma \cdot \xi_{data}(\mathbf{q}) + (1 - \gamma) \cdot \xi_{surface}(\mathbf{q})$, where $\xi_{data}(\mathbf{q})$ gives rise to sample points force and $\xi_{surface}(\mathbf{q})$ gives rise to external constraint forces. A weighting scalar parameter γ, defined by the user, adjusts the influence of both terms.

At each collocation $\mathbf{o} \in \mathscr{O}$, we consider a local plane of normal $\mathbf{n}$ fitting the data points. For a given point $\mathbf{q} \in \mathbb{R}^3$, $\mathbf{p}$ is its projection on the this plane, and t the distance between $\mathbf{q}$ and the plane; so we have $\mathbf{p} = \mathbf{q} - t \times \mathbf{n}$. Figure 2 summarizes this setup.

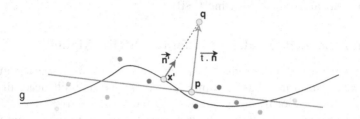

Fig. 2. The weight on an data point $\mathbf{p}_i \in P$, denoted here by different shades of gray, is a function of the distance from $\mathbf{p}_i$ to $\mathbf{q}$.

Thus, ξ_{data} and $\xi_{surface}$ are expressed at the point $\mathbf{q}$ as functions of t and $\mathbf{n}$:

$$\xi_{data}(t, \mathbf{n}) = \sum_{\mathbf{p}_i \in P} \langle \mathbf{p}_i - \mathbf{q} + t \times \mathbf{n}, \frac{\mathbf{n}}{\|\mathbf{n}\|} \rangle^2 \quad \cdot \theta(\|\mathbf{p}_i - \mathbf{q} + t \times \mathbf{n}\|)$$

$$\xi_{surface}(t, \mathbf{n}) = \langle \mathbf{x}' - \mathbf{q} + t \times \mathbf{n}, \mathbf{n}' \rangle^2 \tag{6}$$

where $\mathbf{n}'$ is the gradient of the implicit function defining the continuous tubular model, $\mathbf{x}'$ is the projection of $\mathbf{p}$ on the zero level-set of this function, and θ is a compactly supported Wendland's RBF.

For clarity, $\xi_{surface}$ quantifies the distance between the continuous tubular model and MLS model [1] locally approximating the data. The gradient of ξ is then expressed as $\nabla \xi = \gamma \cdot \nabla \xi_{data} + (1 - \gamma) \cdot \nabla \xi_{surface}$. Following Eqs. 7 and 8 detail the computation of both data and surface gradient terms, respectively ξ_{data} and $\xi_{surface}$.

$$\nabla \xi_{data}(t, \mathbf{n}) = \sum_{\mathbf{p}_i} 2 \cdot \langle \mathbf{p}_i - \mathbf{q} + t \times \mathbf{n}, \frac{\mathbf{n}}{\|\mathbf{n}\|} \rangle \cdot \nabla_{\mathscr{G}}(t, \mathbf{n}) \cdot \theta(\|\mathbf{p}_i - \mathbf{q} + t \times \mathbf{n}\|)$$

$$+ \langle \mathbf{p}_i - \mathbf{q} + t \times \mathbf{n}, \frac{\mathbf{n}}{\|\mathbf{n}\|} \rangle^2 \cdot \mathscr{M}^T_{(t, \mathbf{n})} \cdot \nabla \theta(\|\mathbf{p}_i - \mathbf{q} + t \times \mathbf{n}\|) \tag{7}$$

$$
\text{where}
\begin{cases}
\nabla_{\mathscr{G}}(t, \mathbf{n}) = \left[\dfrac{\dfrac{1}{\|\mathbf{n}\|} \cdot (I_3 - \dfrac{\mathbf{n} \cdot \mathbf{n}^T}{\|\mathbf{n}\|^2}) \cdot (\mathbf{p}_i - \mathbf{q}) + t \cdot \dfrac{\mathbf{n}}{\|\mathbf{n}\|}}{\|\mathbf{n}\|} \right] \\[4ex]
\mathscr{M}_{(t,\mathbf{n})} = \left[I_3 \cdot t \begin{array}{c} n_x \\ n_y \\ n_z \end{array} \right] \\[4ex]
\nabla\theta(\|\mathbf{r}\|) = (-20 \cdot \|\mathbf{r}\|) \cdot (1 - \|\mathbf{r}\|)^3 \cdot \dfrac{\mathbf{r}}{\|\mathbf{r}\|}
\end{cases}
$$

The second gradient term is computed as:

$$
\nabla\xi_{surface}(t, \mathbf{n}) = 2 \cdot \mathscr{M}_{(t,\mathbf{n})}^T \cdot \langle \mathbf{x}' - \mathbf{q} + t \times \mathbf{n}, \mathbf{n}' \rangle \cdot \mathbf{n}' \tag{8}
$$

We retrieve $(t_{min}, \mathbf{n}_{min})$, minimizing $\xi(t, \mathbf{n})$ with the Fletcher-Reeves [5] conjugate gradient algorithm available in the GSL library. The minimizer is initiated at the point $\mathbf{q} \in \mathbb{R}^3$ to $(t_0, \mathbf{n}_0) = (\overrightarrow{\mathbf{q} - \mathbf{x}'} \cdot \mathbf{n}', \mathbf{n}')$. From this minimization process computed at each collocation $\mathbf{o}$ arises a deformation vector defined as $\mathbf{v_o} = -t_{min} \times \mathbf{n}_{min}$.

To push the surface to the data points according to the deformation vector field, we propose to follow a convection model as proposed by Osher [12]:

$$
\frac{dg(\mathbf{p}, t)}{dt} = \mathbf{v_p} \circ \nabla g(\mathbf{p}, t) \tag{9}
$$

where $\mathbf{v_p}$ is the deformation vector expressed at $\mathbf{p}$, and $\circ$ is the element-wise product. To solve Eq. 9, we assume that space and time are separable. We decompose $g(\mathbf{p}, t) = \boldsymbol{\alpha}(t) \cdot \boldsymbol{\Phi}(\mathbf{p})$ where $\boldsymbol{\Phi}(\mathbf{p})$ is made of the basis functions values evaluated at the point $\mathbf{p}$ and $\boldsymbol{\alpha}(t)$ composed of the respective α values. Equation 9 thus becomes the ordinary differential equation of evolution:

$$
\frac{d\boldsymbol{\alpha}(t)}{dt} \cdot \boldsymbol{\Phi}(\mathbf{p}) = \mathbf{v_p} \circ [\boldsymbol{\alpha}(t) \cdot \nabla\boldsymbol{\Phi}(\mathbf{p})] \tag{10}
$$

After the application of Euler's method, Eq. 10 becomes:

$$
\boldsymbol{\alpha}(t + \tau) = \boldsymbol{\alpha}(t) - \tau \cdot \boldsymbol{\Phi}(\mathbf{p})^{-1} \cdot \mathscr{H}(t, \mathbf{p}) \tag{11}
$$

where τ is the time step and $\mathscr{H}(t, \mathbf{p}) = \mathbf{v_p} \circ [\boldsymbol{\alpha}(t) \cdot \nabla\boldsymbol{\Phi}(\mathbf{p})]$. The evolution of the CSRBF coefficients is finally given by Eq. 11. From this set of CSRBF, we compute an implicit function and finally extract its zero level-set to produce the terrain surface model.

5.3   Re-normalization of Implicit Function

In order to prevent the apparition of new zero level components far away from the initial surface, periodically reshaping the implicit function is a common strategy. In order to bound the function g defined in Sect. 5.1, hence its gradient ∇g, we bound the expansion of its coefficients $\boldsymbol{\alpha}_{o_c}$. In practice, inspired by [6], we bound $\|\boldsymbol{\alpha}(t)\|_\infty$ if $\|\boldsymbol{\alpha}(t)\|_\infty > \beta$, where β is a positive constant. The evolution equation becomes

$$
\begin{cases}
\boldsymbol{\alpha}(t+\tau) = \widetilde{\boldsymbol{\alpha}}(t) - \tau \cdot \boldsymbol{\Phi}(\mathbf{p})^{-1} \cdot \mathscr{H}(t, \mathbf{p}) \\
\widetilde{\boldsymbol{\alpha}}(t+\tau) = \frac{\beta}{\|\boldsymbol{\alpha}(t+\tau)\|_\infty} \cdot \boldsymbol{\alpha}(t+\tau), & \text{if } \|\boldsymbol{\alpha}(t+\tau)\|_\infty > \beta \\
\widetilde{\boldsymbol{\alpha}}(t+\tau) = \boldsymbol{\alpha}(t+\tau), & \text{otherwise}
\end{cases}
\tag{12}
$$

5.4 Complexity

Using an octree data structure for the collocations layout, as advised by Wendland [16], the matrix $\boldsymbol{\Phi}(\mathbf{p})$ computation is $\mathscr{O}(N log N)$ and the implicit function evaluation g is $\mathscr{O}(log N)$. According to Botsch [2], the inversion of $\boldsymbol{\Phi}(\mathbf{p})$ through the Cholesky factorization is $\mathscr{O}(nzf)$, where nzf is the number of nonzero factors, which depends on the CSRBF center position and on the CSRBF support size. The cost of the re-normalization described in Sect. 5.3 is $\mathscr{O}(N)$.

6 Experiment and Validation

In order to train our deep learning model, labeled point clouds of forest plot mock-ups are required. To generate such data-sets, we simulate 3D point clouds from artificial 3D terrains and trees models, as described in the following Sect. 6.1. As the accuracy of the terrain surface reconstructed by our method depends first on the efficiency of our deep network segmentation, we first measured its ability to segment vegetation points from ground points on local minima of simulated scenes. Section 6.3 describes the quantitative and qualitative results of terrain reconstruction on simulated and real TLS scans. Moreover, we evaluate the improvement brought by the deformable model.

6.1 Training Data Generation

Fig. 3. Models of forest $32\,\mathrm{m} \times 32\,\mathrm{m}$ plot: left side, mesh model; center and right side, side and top view of a simulated point cloud.

To build a training data-set whose point clouds resemble the real TLS forest scans, described later in Sect. 6.2, we designed realistic tree mesh models of pines, spruces and birches using SpeedTree [7]. This software is providing nowa-days high quality 3D renderings for the gaming and film industries as well as architectural visualization projects. Ground surfaces meshes were produced using hybrid multi-fractal terrain method [4] implemented within Meshlab [3]. Then, we associated ground and trees models to build 3D models of complete forest plots, and used a LiDAR simulator based on PBRT [13] to generate point clouds. In order to simulate the point distribution close to the ground, which is scattered and complex due to leaves and very small vegetation, we applied a Gaussian noise (std $= 10$ cm) on points originating from ray hitting the ground surface. Figure 3 shows an example of a meshed forest plot mock-up along with two views of the resulting simulated point clouds. As the goal of our deep network is to learn how to handle different patterns of trees occlusions, we artificially increased the training data by repeating the simulation process for six virtual scan positions evenly distributed on a 4 m circle centered on the plot. Each simulated point cloud has around 7 million points.

6.2   Real LiDAR Scans

The real TLS scans used in the validation are the ones used in the benchmark [9]. They were acquired in a southern boreal forest of Finland. Each of the originating plot, of a fixed size of 32 m × 32 m, was selected from varying forest-stand conditions representing different developing stages with a range of species, growth stages, and management activities. The plots were divided into three categories (two plots per category) based on the complexity of their structure (from the point-of-view of a TLS survey): easy (plot 1 and 2), medium (plot 3 and 4) and difficult (plot 5 and 6). This data-set comes with a reference ground model, fully checked by the operator, composed of 3D points laying over a 2D grid of 20 cm resolution. Each plot was scanned from five positions: one scan at the plot center and four scans at the four quadrant directions.

6.3   Experiments and Discussions

As pointed out earlier, the performance of our method relies first on its abil-ity to filter out vegetation points based on the deep learning segmentation. In order to assess this performance, we first compute the mean kappa indicator of the classification of all the simulated scenes, for different values of res_{MIN} the resolution of the 2D grid used in the extraction of the minimum points P_{min} and K_{PCA} the number of neighbors used in the local PCA analysis described in Sect. 3.1. Through sensitivity analysis, we obtained the best average kappa (0.977) on the simulated data-set for $res_{MIN} = 10$ cm and $K_{PCA} = 128$. We use those parameter values for the rest of the validation. Using a pre-trained model with these parameters, we analyzed the distance between the input data of our method and the reconstructed surfaces it produces.

Simulated Data. In each step of the validation, we checked the robustness of our method by asserting that the reconstructed surfaces were made of a single continuous patch without holes covering the full extent of the input data. Then, we measured its accuracy by computing the distance between the input data and the surface model. To do so, we defined a distance distribution by computing the euclidean distance in between each vertices of the resulting surfaces to the closest vertex of the reference surfaces. The Table 1 presents the segmentation result and the distance distribution for one scan of each artificial forest scan.

Table 1. Distance between reconstructed terrain surface and reference: (Top) Results of segmentation of the simulated point cloud. The vegetation points that are filtered out are represented in green. (Middle) The euclidean distance to the reconstructed surface represented as a color gradient at each vertex of the reference surface. (Bottom) The distribution of distances represented as box-plots (without outliers).

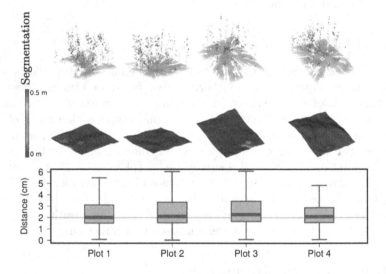

For every artificial plot, our model efficiently filters out the vegetation points. Moreover, the reconstructed surfaces are placed at a mean distance of 2 cm from the terrain model produced with Meshlab, except in large occlusions where ground points are missing and where the reconstructed surface is approximated from the surroundings.

Real Data. While dealing with real TLS scans, we reconstructed, for each plot, the terrain surface from the single central scan and also from a point cloud resulting of the fusion of all five available scans per plot. This allows evaluation of the method performance for various point density in different forest environments. Having access to the reference terrain model, we computed the distance in between surfaces, vertex to vertex. In the case of real data, we first extract the ground points using our deep learning network, then compute the distance between the vertices of the reconstructed surface and the ground points packaged with the LiDAR scans. The results are shown in Table 2.

Finally, we analyzed the evolution of terrain model the iteration of deformation, by measuring its distance distribution to the segmented ground points. The results are presented in Tables 3 and 4, for single and multiple scans respectively.

Table 2. Mean Hausdorff distance (cm) between reconstructed surface from single/multiple scan(s) and reference ground model.

Plot	1	2	3	4	5	6
Single - Mean Distance (cm)	6.04	5.82	22.23	26.89	12.85	21.76
Multiple - Mean Distance (cm)	2.58	3.14	5.93	8.27	2.88	7.17

As pointed out in the Table 2, the efficiency of our method to reconstruct a terrain model close to the reference depends on the terrain complexity and on the nature of the scans: for the simpler forest plots (1 and 2) single scans, our method produces accurate terrain models that are positioned 6 cm from the reference. Due to the missing points in the occlusion that expands with complexity, this distance increases for plots 3, 4, 5, and 6. However, in the case of multiple scans, the occlusion phenomenon lessens, which makes available anchor points allowing sharper terrain surfaces to be produced. In such cases, with our method, the mean distances are around 3 cm on the simpler plot, and from 3 to 8 cm on a more complex topology.

In Tables 3 and 4, we analyzed the contribution of the deformation produced by the convection in the final accuracy of our terrain surfaces. For every plot, the mean of the distance distribution and its standard deviation decreases for both single and multiple scans. The stronger drops occurs at the first iteration, the following ones remaining light in comparison. Except for plot 3 and 4 single scans, the reconstructed surface sets at 2 cm on average from the segmented ground points after five iterations of convection.

The single scans of plots 3 and 4 highlight the limitation of our method: while the mean distance from the segmented ground points remained at 2 to 3 cm, the distance distribution presents extreme values above 20 cm. This is due to the failure of our deep learning model to correctly filter out vegetation points. Those two particular plots, while being scans from a single point presents problematic 3D points patterns, which vanish if scanned from multiple points of view. The forest mock-ups are crucial; they need to take proper account of the actual reality. Indeed, they are used to produce simulated data training and determine the quality of the segmentation, which is the first step of our method and condition the reconstruction of the overall scene.

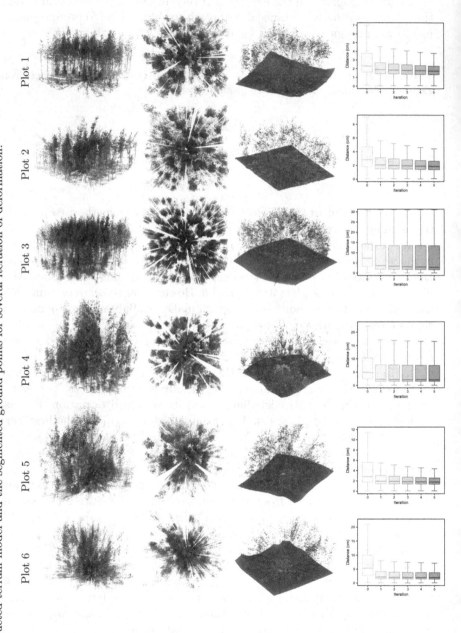

Table 3. Estimation of forest plots ground surface based on TLS single scans. The two first columns show the side and top view of the point cloud. For visualization purpose, the point clouds have been colorized according to the point elevation and the local point density. The third column present the segmentation of the minimum points. The last column show the evolution of the distance between the reconstructed terrain model and the segmented ground points for several iteration of deformation.

Table 4. Estimation of forest plots ground surface based on TLS multiple scans. The two first columns show the side and top view of the point cloud. For visualization purpose, the point clouds have been colorized according to the point elevation and the local point density. The third column present the segmentation of the minimum points. The last column show the evolution of the distance between the reconstructed terrain model and the segmented ground points for several iteration of deformation.

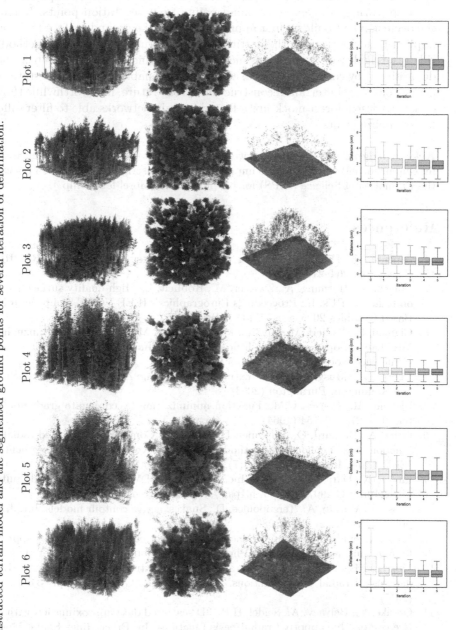

7   Conclusion

In this work, we propose an efficient method designed to recover terrain surface model from 3D sample data acquired in forest environments. Our approach relies on deep learning to separate ground points from vegetation points. It handles the occlusion and builds a first approximation of the ground surface by blending implicit quadrics through the partition of unity principle. Then our method gets rid of the rigidity of the previous model by projecting it in a CSRBF basis before deforming it by convection. These contributions enable us to achieve state of the art performance in terrain reconstruction. In the future, it is worthwhile thinking of how to design forest mock-ups adapted to train networks able to filter different forest environments.

Acknowledgments. The authors would like to thank the reviewers for their thoughtful comments and efforts towards improving our manuscript and the Japan Society for the Promotion of Science (JSPS) for providing Jules Morel fellowship.

References

1. Amenta, N., Kil, Y.J.: Defining point-set surfaces. In: ACM SIGGRAPH 2004 Papers, pp. 264–270 (2004)
2. Botsch, M., Hornung, A., Zwicker, M., Kobbelt, L.: High-quality surface splatting on today's GPUs. In: Proceedings Eurographics - IEEE VGTC Symposium Point-Based Graphics 2005, pp. 17–141. IEEE (2005)
3. Cignoni, P., Callieri, M., Corsini, M., Dellepiane, M., Ganovelli, F., Ranzuglia, G.: MeshLab: an open-source mesh processing tool. In: Eurographics Italian Chapter Conference, vol. 2008, pp. 129–136 (2008)
4. Ebert, D.S., Musgrave, F.K.: Texturing & Modeling: A Procedural Approach. Morgan Kaufmann, Burlington (2003)
5. Fletcher, R., Reeves, C.M.: Function minimization by conjugate gradients. Comput. J. **7**(2), 149–154 (1964)
6. Gelas, A., Bernard, O., Friboulet, D., Prost, R.: Compactly supported radial basis functions based collocation method for level-set evolution in image segmentation. IEEE Trans. Image Process. **16**(7), 1873–1887 (2007)
7. Interactive Data Visualization, Inc.: SpeedTree. IDV, 5446 Sunset Blvd. Suite 201 Lexington, SC 29072 (2017). https://store.speedtree.com
8. Kass, M., Witkin, A., Terzopoulos, D.: Snakes: active contour models. Int. J. Comput. Vis. **1**(4), 321–331 (1988)
9. Liang, X., et al.: International benchmarking of terrestrial laser scanning approaches for forest inventories. ISPRS J. Photogramm. Remote Sens. **144**, 137–179 (2018)
10. Morel, J., Bac, A., Véga, C.: Terrain model reconstruction from terrestrial LiDAR data using radial basis functions. IEEE Comput. Graphics Appl. **37**(5), 72–84 (2017)
11. Ohtake, Y., Belyaev, A., Seidel, H.P.: 3D scattered data approximation with adaptive compactly supported radial basis functions. In: Proceedings Shape Modeling Applications 2004, pp. 31–39. IEEE (2004)
12. Osher, S., Fedkiw, R.: Level Set Methods and Dynamic Implicit Surfaces, vol. 153. Springer, Heidelberg (2006)

13. Pharr, M., Jakob, W., Humphreys, G.: Physically Based Rendering: From Theory to Implementation. Morgan Kaufmann, Burlington (2016)
14. Qi, C.R., Yi, L., Su, H., Guibas, L.J.: PointNet++: deep hierarchical feature learning on point sets in a metric space. In: Advances in Neural Information Processing Systems, pp. 5099–5108 (2017)
15. Tsai, R., Osher, S., et al.: Review article: level set methods and their applications in image science. Commun. Math. Sci. 1(4), 1–20 (2003)
16. Wendland, H.: Scattered Data Approximation. Cambridge University Press, Cambridge (2005)

Solving Non-linear Elasticity Problems by a WLS High Order Continuation

Oussama Elmhaia[✉], Youssef Belaasilia, Bouazza Braikat,
and Noureddine Damil

Laboratoire d'Ingénierie et Matériaux, Faculté des Sciences Ben M'Sik,
Hassan II University of Casablanca, Casablanca, Morocco
oussama.elmhaia@gmail.com, belaasilia@gmail.com, b.braikat@gmail.com,
noureddine.damil@gmail.com

Abstract. In this paper, a high order mesh-free continuation for non-linear elasticity problems is presented. This proposal consists to introduce the Weighted Least Squares (WLS) in a High Order Continuation (HOC). The WLS has been employed to create shape functions using a local support domain. The HOC permits to transform the nonlinear problems in a succession of linear problems of the same tangent matrix. A strong formulation of the problem is adopted to avoid the numerical integration and mesh generation. In this work, a numerical study has been conducted in nonlinear elasticity problems in order to study the behaviour and stability of the proposed approach. Several examples are investigated numerically in order to demonstrate the robustness and efficiency of the proposed approach. This proposed approach has shown its efficiency in management of complex geometries and irregular nodal distributions with respect to other approaches.

Keywords: Nonlinear elasticity · High Order Continuation · Weighted Least Squares · Strong formulation · Irregular nodal distributions

1 Introduction

In the recent years, many authors have shown the interest in developing mesh-free methods to overcome some difficulties. These methods provide the possibility to avoid mesh generation and connectivity between the nodes which require a considerable computation time. Nevertheless, not all mesh-free methods have this advantage. They are divided in two categories. The first one concerns the mesh-free methods based on weak formulation that still require a background mesh like the Diffuse Element Method (DEM) [1], Element Free Galerking (EFG) [2] and Point Interpolation Method (PIM) [3] etc. The second category of mesh-free methods is based on strong formulation. This category don't need a background mesh and the numerical integration [4,5].

The Weighted Least Squares (WLS) approximation is wildly used in data fitting [6]. Wallstedt et al. [7] took advantage of the concepts of weighted least

© Springer Nature Switzerland AG 2020
V. V. Krzhizhanovskaya et al. (Eds.): ICCS 2020, LNCS 12142, pp. 266–279, 2020.
https://doi.org/10.1007/978-3-030-50433-5_21

squares surface estimation and implicit surface definition to more precisely define the region of integration for solid mechanic's simulations that are performed within a PIC framework. Onate et al. [8] presented an approach termed generically the 'Finite Point Method' based on a Weighted Least Square Interpolation of point data and point collocation for evaluating the approximation integrals. Baeza et al. [9] have presented a technique based on the application of a variant of the Lagrange extrapolation through the computation of weights capable of detecting regions with discontinuities. Recently several authors developed a high order continuation using the Moving Least Squares approximation [10] as a discretization method under a strong formulation to solve various non linear problems [11–19]. Other authors used a high order continuation using Radial Point Interpolation method under a strong formulation [20].

In this paper, a new numerical approach is proposed. This approach combines a high order continuation [21] and the Weighted Least Squared (WLS) as a discretization method. Thanks to the development of the main variables into Taylor series, non-linear elastic problem is transformed into a succession of linear differential equations with the same tangent operator followed of a continuation technique is used to obtain the whole solution. In this work, the proposed approach has been referred to it as HOC-WLS. The performance of this proposed approach is illustrated on examples of non-linear elasticity problems. A comparison has been held between the proposed approach, the High Order Continuation coupled with Moving Least Squares (HOC-MLS) and the High Order Continuation coupled with Finite Element Method (HOC-FEM).

2   Description of Weighted Least Squares (WLS) and Moving Least Squares (MLS) Methods

In this section, we present in detail the description of the approximations by the Weighted Least Squares and by the Mobile Least Squares methods.

2.1   Weighted Least Squares Approximation

The Weighted Least Squares (WLS) approximation is highly used for data fitting [6]. The approximation $u^h(x)$ of a field $u(x)$ at a point x by WLS in an influence domain Ω is given by:

$$u^h(x) = \sum_{i=1}^{m} p_i(x)a_i = <p(x)>\{a\} \tag{1}$$

where $<p(x)> = <p_1(x), p_2(x), \cdots, p_m(x)>$ is a vector of monomial basis functions, with m is the number of monomials and $^t\{a\} = <a_1, a_2, \cdots, a_m>$ is a vector of the unknown constant coefficients. The most used vector of monomial basis functions is defined as follows:

$$\begin{cases} m = 3 \quad p = <1, x, y> \\[2mm] m = 6 \quad p = <1, x, y, xy, x^2, y^2> \\[2mm] m = 10 \; p = <1, x, y, xy, x^2, y^2, x^2y, xy^2, x^3, y^3> \\[2mm] m = 15 \; p = <1, x, y, xy, x^2, y^2, x^2y, xy^2, x^3, y^3, x^3y, xy^3, x^4, y^4> \end{cases} \quad (2)$$

To determine the vector of constant coefficients $\{a\}$ in Eq. (1), we select n nodes in the local support domain for the approximation. We can solve the Eq. (1) for all nodes in the support domain using a Weighted Least Square (WLS) method by minimizing the weighted discrete L_2 norm:

$$J = \sum_{i=1}^{n} w_i \left(u^h(x) - u(x_i) \right)^2 \quad (3)$$

where w_i is the weight function associated to the node i and n is the number of nodes in the support domain. We note that weight functions plays an important role in constructing the WLS shape function. We should note that the weights used here are considered constant i.e they do not depend on x. They can be computed from any weight function with the bell shape [6].

For an arbitrary point x, the vector $\{a\}$ is chosen to minimize the weighted residual. The stationary condition of the quadrature form J is given by:

$$\frac{\partial J}{\partial \{a\}} = 0 \quad (4)$$

After the calculation, the stationarity condition (4) can take the following form:

$$[A]\{a\} = [B]\{U\} \quad (5)$$

where the matrices $[A]$ and $[B]$ are defined by:

$$[A] = {}^T[P_m][w][P_m]; \quad [B] = {}^T[P_m][w] \quad (6)$$

where $[P_m]$ is a matrix of order $n \times n$ evaluated at all nodes of influence domain, $[w]$ is the diagonal matrix constructed from the weight constants. The vector $\{U\}$ collects all nodal unknowns. The resolution of Eq. (5) gives the following expression of the vector $\{a\}$:

$$\{a\} = [A]^{-1}[B]\{U\} \quad (7)$$

Using Eq. (7), the local approximation $u^h(x)$ of Eq. (1) is written as follows:

$$u^h(x) = <\phi(x)> \{U\} \quad (8)$$

where the vector of shape functions $<\phi(x)>$ and its derivatives are defined as follows:

$$\begin{cases} <\phi(x)> \; = <\phi_1(x), \phi_2(x), \cdots, \phi_n(x)> \qquad = <p(x)>[A]^{-1}[B] \\[2mm] <\phi(x)_{,i}> \; = <\phi_{1,i}(x), \phi_{2,i}(x), \cdots, \phi_{n,i}(x)> \quad = <p_{,i}(x)>[A]^{-1}[B] \\[2mm] <\phi(x)_{,ij}> = <\phi_{1,ij}(x), \phi_{2,ij}(x), \cdots, \phi_{n,ij}(x)> = <p_{,ij}(x)>[A]^{-1}[B] \end{cases} \quad (9)$$

where the symbol $(\bullet)_{,i}$ denotes the derivative with respect to the ith coordinate and the symbol $(\bullet)_{,ij}$ denotes the derivative with respect to ith and to jth coordinates. It should be noted that the vector $\{a\}$ is constant and the derivatives of the matrices $[A]$ and $[B]$ are null. This leads to a considerable reduction in computation time. Under a strong formulation, the shape function can be derived twice or more and we notice when the matrix $[A]$ is differentiable and it's derivatives increase the instability of the solution. If the matrix $[A]$ is constant, the degree of instability of the solution decreases. This effect has been investigated in the numerical examples.

The shape functions so constructed do not have the Kronecker delta function property, which can cause difficulties in imposing the boundary conditions, [6]. In this case, the collocation methods are necessary [6] to overcome this difficulty. Note also that the WLS shape functions are compatible only in the influence domain rather than in the global domain. This is not a problem when the WLS shape functions are used in the local weak-form methods or collocation methods. But, care needs to be taken when it is used in a global weak-form formulation for which the Moving Least Squares method is chosen of a smart manner [6]. Fortunately in this paper we are using the WLS under a strong formulation (mesh-free collocation methods) and a local support domain.

2.2   Moving Least Squares Approximation

The Moving Least Squares (MLS) method is an efficient numerical method that is classed as a meshless approach that have a highly accurate approximation. Due to the method flexibility, numerous authors have used it to solve a large number of problems [6,10]. For the implementation of this method, we use the same procedure as presented in Sect. 2.1. In this case, the approximation $u^h(x)$ is defined as follows:

$$u^h(x) = \sum_{j=1}^{m} p_j(x)a_j(x) = <p(x)> \{a(x)\} \tag{10}$$

In the same way as before, the weighted residual is constructed and minimized with respect to the vector $\{a(x)\}$ which depends on the point x in this case. In addition, in this case the weighted function depends on the variable x. In the same manner as in Sect. 2.1, the vector $\{a(x)\}$ is the solution of the following system:

$$[A(x)]\{a(x)\} = [B(x)]\{U\} \tag{11}$$

where the matrices $[A(x)]$ and $[B(x)]$ are defined by:

$$[A(x)] = {}^T[P_m][W_I(x)][P_m]; \quad [B(x)] = {}^T[P_m][W(x)] \tag{12}$$

and $\{U\}$ is the vector that collects the nodal unknowns of the influence domain. We should note that the matrix $[A(x)]$ is not always invertible and depends on the nodal distribution inside the support domain. Assuming that $[A(x)]$ is invertible, the vector $\{a(x)\}$ is given by:

$$\{a(x)\} = [A(x)]^{-1}[B(x)]\{U\} \tag{13}$$

Substituting the Eq. (13) in Eq. (10) leads to:

$$u^h(x) = <\phi(x)> \{U\} \tag{14}$$

In this case, the vector of shape functions $<\phi(x)>$ and its derivatives are defined as follows:

$$
\begin{cases}
<\phi(x)> \quad = <p(x)>[A(x)]^{-1}[B(x)] \\[2mm]
<\phi_{,i}(x)> \ = <p_{,i}(x)>[A(x)]^{-1}[B(x)] + <p(x)>[A_{,i}(x)]^{-1}[B(x)] \\[2mm]
\qquad\qquad + <p(x)>[A(x)]^{-1}[B_{,i}(x)] \\[2mm]
<\phi_{,ij}(x)> = <p_{,ij}(x)>[A(x)]^{-1}[B(x)] + <p_{,i}(x)>[A_{,j}(x)]^{-1}[B(x)] \\[2mm]
\qquad\qquad + <p_{,i}(x)>[A(x)]^{-1}[B_{,j}(x)] + <p_{,j}(x)>[A_{,i}(x)]^{-1}[B(x)] \\[2mm]
\qquad\qquad + <p(x)>[A_{,ij}(x)]^{-1}[B(x)] + <p(x)>[A_{,i}(x)]^{-1}[B_{,j}(x)] \\[2mm]
\qquad\qquad + <p_{,j}(x)>[A(x)]^{-1}[B_{,i}(x)] + <p(x)>[A_{,j}(x)]^{-1}[B_{,i}(x)] \\[2mm]
\qquad\qquad + <p(x)>[A(x)]^{-1}[B_{,ij}(x)]
\end{cases} \tag{15}
$$

Properly constructed MLS shape functions are compatible and consistent with the order of polynomials included in the formulation. These shape functions do not have the Kronecker delta function property because they are not interpolation functions.

3   Strong Form Formulation of Two-Dimensional Nonlinear Elastic Problems

We consider a two dimensional elastic problem described by the equilibrium equations and boundary conditions. Let Ω be the domain occupied by a two dimensional solid, $\partial\Omega_u$ and $\partial\Omega_f$ are complementary portions of $\partial\Omega$ in which the Dirichlet and Newman conditions are applied. The equilibrium of this solid is governed by the following matrix equations:

$$
\begin{cases}
[div]\{T\} = 0 & \text{in } \Omega \\[2mm]
[N]\{T\} \ = \lambda\{f\} & \text{in } \partial\Omega_f \\[2mm]
\{u\} \qquad = \lambda\{u_d\} & \text{in } \partial\Omega_u
\end{cases} \tag{16}
$$

where the matrices $[div]$ and $[N]$ are given by:

$$
[div] = \begin{bmatrix} \frac{\partial}{\partial x} & 0 & \frac{\partial}{\partial y} & 0 \\ 0 & \frac{\partial}{\partial y} & 0 & \frac{\partial}{\partial x} \end{bmatrix}; \qquad
[N] = \begin{bmatrix} N_x & 0 & N_y & 0 \\ 0 & N_y & 0 & N_x \end{bmatrix} \tag{17}
$$

$\{T\}$ is the vector containing all the components of the first Piola-Kirchhoff stress tensor, N_x and N_y are the components of the normal vector outward the boundary $\partial\Omega_f$, λ is a control parameter, $\{u_d\}$ is the imposed displacement on Ω_u and $\{f\}$ is the stress vector applied on Ω_f. The relation between the first and second Piola-Kirchhoff stress tensors and the constitutive law are represented by the following equations:

$$\begin{cases} \{T\} = ([III] + [B(\theta)])\{S\} \\[2mm] \{S\} = [D]\{\gamma\} \\[2mm] \{\gamma\} = ([H] + \frac{1}{2}[A(\theta)])\{\theta\} \\[2mm] {}^{T}\{\theta\} = <\frac{\partial u}{\partial x}, \frac{\partial u}{\partial y}, \frac{\partial v}{\partial x}, \frac{\partial v}{\partial y}> \end{cases} \tag{18}$$

The matrices $[D]$, $[H]$, $[A(\theta)]$, $[III]$ and $[B(\theta)]$ are defined as follows [20]:

$$[D] = \frac{E'}{1-\nu'^2}\begin{bmatrix} 1 & \nu' & 0 \\ \nu' & 1 & 0 \\ 0 & 0 & \frac{(1-\nu')}{2} \end{bmatrix}; \quad [H] = \begin{bmatrix} 1 & 0 & 0 & 0 \\ 0 & 0 & 1 & 1 \\ 0 & 1 & 0 & 0 \end{bmatrix}; \quad [III] = \begin{bmatrix} 1 & 0 & 0 \\ 0 & 1 & 0 \\ 0 & 0 & 1 \\ 0 & 0 & 1 \end{bmatrix}$$

$$[A(\theta)] = \begin{bmatrix} \frac{\partial u}{\partial x} & 0 & \frac{\partial v}{\partial x} & 0 \\ 0 & \frac{\partial u}{\partial y} & 0 & \frac{\partial v}{\partial y} \\ \frac{\partial u}{\partial y} & \frac{\partial u}{\partial x} & \frac{\partial v}{\partial y} & \frac{\partial v}{\partial x} \end{bmatrix}; \quad [B(\theta)] = \begin{bmatrix} \frac{\partial u}{\partial x} & 0 & \frac{\partial u}{\partial y} \\ 0 & \frac{\partial v}{\partial y} & \frac{\partial v}{\partial x} \\ 0 & \frac{\partial u}{\partial y} & \frac{\partial u}{\partial x} \\ \frac{\partial v}{\partial x} & 0 & \frac{\partial u}{\partial y} \end{bmatrix}$$

$$\tag{19}$$

where u and v describe respectively the horizontal and vertical displacements, $[D]$ is the elasticity matrix and $\{\gamma\}$ is the strain vector. For a plane stress $E' = E$ and $\nu' = \nu$ and for a plane strain $E' = \frac{E}{(1-\nu^2)}$ and $\nu' = \frac{\nu}{1-\nu^2}$ with E is the Young modulus and ν is the Poisson ratio.

Finally, we have obtained a nonlinear problem that requires a resolution numerical method. For this effect, we have proposed a high order continuation coupled with the Weighted Least Squares method HOC-WLS and with Moving Least Squares method HOC-MLS to solve this nonlinear problem.

4   Resolution Strategy

The solution of the problem governed by Eqs. (16) and (18) is sought under the Taylor series expansion truncated at order P with respect to the path parameter "r" [21,22]:

$$\begin{cases} \{T\} = \{T_0\} + \sum_{i=1}^{P} r^i \{T_i\} \\[2mm] \{S\} = \{S_0\} + \sum_{i=1}^{P} r^i \{S_i\} \\[2mm] \{\gamma\} = \{\gamma_0\} + \sum_{i=1}^{P} r^i \{\gamma_i\} \\[2mm] \{u\} = \{u_0\} + \sum_{i=1}^{P} r^i \{u_i\} \\[2mm] \lambda \quad = \lambda_0 \quad + \sum_{i=1}^{P} r^i \lambda_i \end{cases} \qquad (20)$$

where the terms with the subscript 0 represent the initial solutions of the starting point of each solution branch and the terms with the subscript i are the solutions at each order i. Injecting the development (20) in Eqs. (16) and (18), we have obtained a succession of linear problems which have the same tangent operator. This succession of linear problems is written as follows: **Problem at order $i = 1$**

$$\begin{cases} [div]\{T_1\} = \{0\} & in \ \Omega \\[2mm] [N]\{T_1\} \ = \lambda_1 \{f\} & on \ \partial\Omega_f \\[2mm] \{T_1\} \quad = ([III] + [B_0])\{S_1\} + [B_1]\{S_0\} & in \ \Omega \\[2mm] \{S_1\} \quad = [D]\{\gamma_1\} & in \ \Omega \\[2mm] \{\gamma_1\} \quad = ([H] + [A_0])\{\theta_1\} & in \ \Omega \\[2mm] \{u_1\} \quad = \lambda_1 \{u_d\} & on \ \partial\Omega_u \end{cases} \qquad (21)$$

Problem at order $2 \leq i \leq P$

$$\begin{cases} [div]\{T_i\} = \{0\} & in \ \Omega \\[2mm] [N]\{T_i\} \ = \lambda_i \{f\} & on \ \partial\Omega_f \\[2mm] \{T_i\} \quad = ([III] + [B_0])\{S_i\} + \{T_i^{nl}\} & in \ \Omega \\[2mm] \{S_i\} \quad = [D]\{\gamma_i\} & in \ \Omega \\[2mm] \{\gamma_i\} \quad = ([H] + [A(\theta_0)])\{\theta_i\} + \{\gamma_i^{nl}\} & in \ \Omega \\[2mm] \{u_i\} \quad = \lambda_i \{u_d\} & on \ \partial\Omega_u \end{cases} \qquad (22)$$

where the terms with the superscript nl are found using the solutions at the previous orders. The approximation of the unknown $\{u_i\}$ using WIS approximation is given by:

$$\{u_i\} = [\phi(x)]\{U_n\} \qquad (23)$$

where n is the number of nodes inside the support domain, $[\phi(x)]$ is the matrix of the shape functions and $\{U_n\}$ is the vector collecting the nodal unknown displacements in the support domain. Taking into consideration the approximation of the unknown in (23) and after substitution and assembly technique, we get the discrete problems of Eqs. (21) and (22) completed by the pilotage equation in the following compact form:

$$\text{Problem at order } i = 1 : \begin{cases} [K_T]\{U_1\} = \lambda_1\{F\} \\ <U_1>\{U_1\} + \lambda_1^2 = 1 \end{cases}$$

$$\text{Problem at order } 2 \leq i \leq P : \begin{cases} [K_T]\{U_i\} = \lambda_i\{F\} + \{F_i^{nl}\} \\ <U_i>\{U_1\} + \lambda_i\lambda_1 = 0 \end{cases} \tag{24}$$

where $[K_T]$ is the tangent stiffness matrix evaluated at the initial solution ($\{T_0\}$, $\{S_0\}$, $\{U_0\}$, $\{\gamma_0\}$), $\{F_i^{nl}\}$ is a right hand side that depends on the previous orders and $\{U_i\}$ is the vector of unknown nodal displacements at each order i. The Taylor series expansion of Eq. (20) has a validity radius "r_{max}" given by:

$$r_{max} = \left(\varepsilon \frac{\| \{U_1\} \|}{\| \{U_P\} \|} \right)^{\frac{1}{P-1}} \tag{25}$$

where ε is a given tolerance parameter.

5   Numerical Examples and Discussions

In this section, the robustness and efficiency of the proposed approach has been putting to the test. To compute the shape functions in each examples, the truncated Gauss-type distribution is used as a weight function.

$$\begin{cases} w(q) = \frac{e^{-(\frac{q}{c})^2} - e^{-(\frac{1}{c})^2}}{1 - e^{-(\frac{1}{c})^2}} & q \leq 1 \\ w(q) = 0 & q > 1 \end{cases} \tag{26}$$

the influence domain for all nodes is a circle with a varying radius $r = h \times d_{max}$, $q = d_i/r$, $d_i = \| x - x_i \|$, h is controlled by the user and d_{max} is the distance between two successive nodes (see Fig. 1). In the case of a nodes irregular distribution inside a varying influence domain, we measure d_{max} only for the points inside this domain.

The obtained solutions of each problem will be verified by computing the residual logarithmic norm $log_{10}(\| Res \|)$ of problem (16) via inserting the obtained solution of Eq. (20).

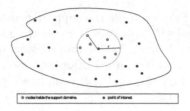

Fig. 1. Influence domain of a point

5.1   Two Dimensional Elastic Plate in Bending

To investigate the convergence of the two approaches, geometrical nonlinearity is considered in this example. A two dimensional plate is clamped on the left hand side and subjected to loading λF on the right hand side, with $F = 1\,\mathrm{N}$ (Fig. 2). The mechanical and geometrical characteristics of the plate are: length $L = 100\,\mathrm{mm}$, height $l = 10\,\mathrm{mm}$, Young's modulus $E = 210\,\mathrm{GPa}$ and Poison's ratio $\nu = 0.3$. The comparison has been executed with same control parameters of the both approaches for various values of parameter h, the truncation order $P = 15$, the tolerance parameter $\varepsilon = 10^{-6}$, a shape parameter $c = 0.2$ and fixed nodal distribution 605 and for HOC-FEM approach as the reference solution, we take 825 $T6$ elements. In Table 1, we present the validity range "r_{max}" and the solution quality measured by "$log_{10}(\| \, Res \, \|)$" obtained by HOC-WLS approach for different numbers of monomial basis and parameter h. In this table, we remark that the proposed approach converges from $h = 2.3$ for the quadratic basis while for cubic and quartic basis it converges from $h = 2.8$. When this approach converges, the validity range is almost constant. In Table 2, we present the validity range "r_{max}" and the solution quality "$log_{10}(\|Res\|)$" obtained by HOC-MLS approach for different numbers of monomial basis and parameter h. In this table, we remark that the HOC-MLS approach converges for the values $2.3 \leq h \leq 3$ for the quadratic basis while for cubic basis it converges only for $h = 3$. The HOC-MLS approach diverges always for quartic basis. When this approach converges, the validity range is almost constant.

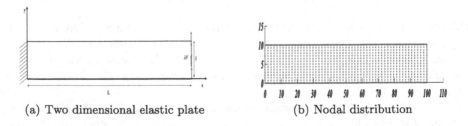

(a) Two dimensional elastic plate   (b) Nodal distribution

Fig. 2. Two dimensional elastic plate and nodal distribution

Table 1. Effects of parameters of HOC-WLS on the validity range and the solution quality

Basis type	Quadratic basis		Cubic basis		Quartic basis	
h	r_{max}	$log_{10}(\|\|Res\|\|)$	r_{max}	$log_{10}(\|\|Res\|\|)$	r_{max}	$log_{10}(\|\|Res\|\|)$
2	637.54	Diverge	976.57	Diverge	0.91	Diverge
2.3	678.63	−3.95	685.12	Diverge	211.10	Diverge
2.5	638.82	−4.59	583.56	Diverge	634.11	Diverge
2.8	639.61	−4.67	648.32	−4.58	410.65	−2.87
3	640.36	−4.68	678.69	−3.39	537.70	−3.01
3.5	643.27	−4.66	620.88	−4.29	456.87	−5.13
4	648.77	−4.54	573.36	−4.75	683.45	−3.35
4.5	658.52	−4.26	649.59	−3.69	641.42	−4.45

Table 2. Effects of parameters of HOC-MLS on the validity range and the solution quality

Basis type	Quadratic basis		Cubic basis		Quartic basis	
h	r_{max}	$log_{10}(\|\|Res\|\|)$	r_{max}	$log_{10}(\|\|Res\|\|)$	r_{max}	$log_{10}(\|\|Res\|\|)$
2	774.78	Diverge	188.81	Diverge	37.04	Diverge
2.3	667.08	−3.79	31.62	Diverge	22.17	Diverge
2.5	637.67	−4.45	124.08	Diverge	580.49	Diverge
2.8	642.63	−4.49	255.19	Diverge	336.27	Diverge
3	642.98	−4.44	588.97	−4.61	319.49	Diverge
3.5	195.19	Diverge	595.89	Diverge	634.96	Diverge
4	80.16	−Diverge	381.12	Diverge	183.50	Diverge
4.5	635.83	Diverge	349.34	−2.11	205.55	Diverge

In Fig. 3, we plot the load-displacement curve obtained by HOC-WLS, HOC-MLS and HOC-FEM algorithms. The horizontal displacement u and the vertical displacement v are plotted at the interest point. The HOC-WLS and HOC-MLS algorithms are in good agreement with the reference solution obtained by HOC-MEF algorithm. However, the HOC-WLSM algorithm shows a better solution quality than that of HOC-MLS algorithm.

5.2 Two Dimensional Plate Under Traction Load

We consider a two dimensional plate with a length $L = 100\,\text{mm}$ and a height $l = 10\,\text{mm}$. The structure is clamped at the left hand side $x = 0$ and subjected to a loading at the right hand side $x = 100$ (see Fig. 4a). The mechanical properties are: Young's modulus $E = 210\,\text{GPa}$ and Poison's ratio $\nu = 0.3$. We adopt the

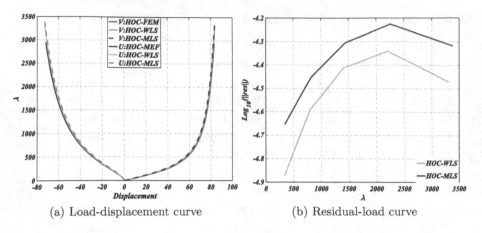

(a) Load-displacement curve   (b) Residual-load curve

Fig. 3. Load-displacement and residual-load curves at the registration point ($x = 100, y = 10$) for a quadratic basis and $h = 2.5$

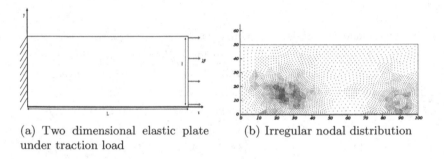

(a) Two dimensional elastic plate under traction load   (b) Irregular nodal distribution

Fig. 4. Two dimensional elastic plate under traction load and irregular nodal distribution

following numerical parameters: a truncation order $P = 15$, a tolerance parameter $\varepsilon = 10^{-8}$ and a shape parameter $c = 0.2$. The domain occupied by the plate is replaced by 583 nodes irregularly distributed (see Fig. 4b). For the reference solution, 3825 $T6$ elements are used.

The Fig. 5 represents the load-displacements curves at the interest point ($x = 100, y = 50$) obtained the both mesh-free approaches and by the reference algorithm. The horizontal displacement u and the vertical displacement v are plotted at the interest point. Even when using an irregular nodal distribution, the load-displacement curve computed by the HOC-WLS is in a good agreement with the reference solution. Whereas the load-displacement curve obtained by HOC-MLS algorithm could not reach the same load as the proposed approach with the same number of steps. This shows that the MLS approximation does not give the same solution when the nodes are distributed irregularly.

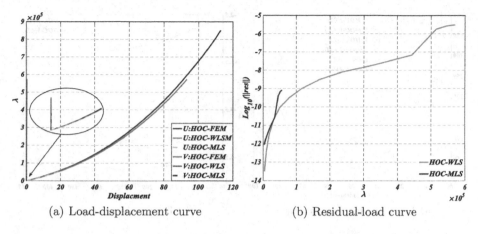

(a) Load-displacement curve   (b) Residual-load curve

Fig. 5. Load-displacement and residual-load curves at the registration point ($x = 100, y = 50$)

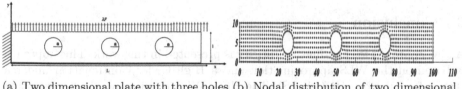

(a) Two dimensional plate with three holes (b) Nodal distribution of two dimensional
plate with three holes

Fig. 6. Two dimensional plate with three holes and nodal distribution

5.3 Two Dimensional Plate with Three Holes

In this numerical study, the effect of complex geometry on the proposed approach is tested. Two dimensional plate with three holes is considered which is clamped on the left hand side $x = 0$ and subjected to a load λF on the top hand side $y = l$ (see Fig. 6a). The mechanical and geometrical properties are: Young's modulus $E = 210\,\mathrm{GPa}$, Poison's ratio $\nu = 0.3$, Length $L = 100\,\mathrm{mm}$, height $l = 10\,\mathrm{mm}$ and radius of each hole is $a = 3\,\mathrm{mm}$. We adopt the following numerical parameters: a truncation order $P = 15$, tolerance parameter $\varepsilon = 10^{-8}$ and a shape parameter $c = 0.2$. The domain occupied by the plate is replaced by 1030 nodes (see Fig. 6b). For the reference solution a 1784 $T6$ elements are used. In Fig. 7, the load-displacement curve is plotted at the interest point ($x = 100, y = 5$).The horizontal displacement u and the vertical displacement v are plotted at the interest point. The obtained solution is compared to HOC-MLS approach and the reference solution. From Fig. 7a, the obtained solution is in a good agreement with the reference solution, while the one obtained by HOC-MLS diverges. In Fig. 7b, the proposed approach shows a good solution quality than the HOC-MLS approach.

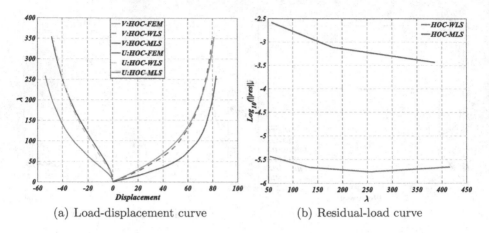

(a) Load-displacement curve   (b) Residual-load curve

Fig. 7. Load-displacement and residual-load curves at the interest point ($x = 100$, $y = 5$)

6   Conclusion

A new approach for non-linear elasticity is presented in this work. This approach is based on the WLS approximation and a High Order Continuation under a strong formulation (HOC-WLS). The HOC-WLS requires no mesh generation and it avoids numerical integration which is computationally expensive. The proposed approach is tested and verified in various numerical examples of $2D$ elasticity with different nodal distributions and geometries. These tests were done also by the HOC-MLS approach which is another mesh-free approach of the same category as the proposed one and HOC-FEM that represents a reference solution. These numerical tests are done to find the best numerical parameters for the proposed approach. Both mesh-free methods show a good quality results when the nodal distribution is regular. However, the HOC-WLS approach shows a better convergence and efficiency than HOC-MLS regarding irregular nodal distributions and complex geometries.

References

1. Nayroles, B., Touzot, G., Villon, P.: Generalizing the finite element method: diffuse approximation and diffuse elements. Comput. Mech. **10**(5), 307–318 (1992)
2. Belytschko, T., Lu, Y.Y., Gu, L.: Element-free Galerkin methods. Int. J. Numer. Meth. Eng. **37**(2), 229–256 (1994)
3. Liu, G.R., Gu, Y.: A point interpolation method for two-dimensional solids. Int. J. Numer. Meth. Eng. **50**(4), 937–951 (2001)
4. Kim, D.W., Kim, Y.: Point collocation methods using the fast moving least-square reproducing kernel approximation. Int. J. Numer. Methods Eng. **56**(10), 1445–1464 (2003). ISO 690
5. Zong, Z.: A complex variable boundary collocation method for plane elastic problems. Comput. Mech. **31**(3–4), 284–292 (2003)

6. Liu, G.R., Gu, Y.T.: An Introduction to Meshfree Methods and Their Programming. Springer, Heidelberg (2005). https://doi.org/10.1007/1-4020-3468-7
7. Wallstedt, P.C., Guilkey, J.E.: A weighted least squares particle-in-cell method for solid mechanics. Int. J. Numer. Meth. Eng. **85**(13), 1687–1704 (2011)
8. Onate, E., Idelsohn, S., Zienkiewicz, O.C., Taylor, R.L.: A finite point method in computational mechanics. Applications to convective transport and fluid flow. Int. J. Numer. Methods Eng. **39**(22), 3839–3866 (1996)
9. Baeza, A., Mulet, P., Zorío, D.: High order weighted extrapolation for boundary conditions for finite difference methods on complex domains with Cartesian meshes. J. Sci. Comput. **69**(1), 170–200 (2016)
10. Lancaster, P., Salkauskas, K.: Surfaces generated by moving least squares methods. Math. Comput. **37**(155), 141–158 (1981)
11. Timesli, A., Braikat, B., Lahmam, H., Zahrouni, H.: An implicit algorithm based on continuous moving least square to simulate material mixing in friction stir welding process. Model. Simul. Eng. **2013**, 22 (2013)
12. Timesli, A., Braikat, B., Lahmam, H., Zahrouni, H.: A new algorithm based on moving least square method to simulate material mixing in friction stir welding. Eng. Anal. Bound. Elem. **50**, 372–380 (2015)
13. Mesmoudi, S., Timesli, A., Braikat, B., Lahmam, H., Zahrouni, H.: A 2D mechanical-thermal coupled model to simulate material mixing observed in friction stir welding process. Eng. Comput. **33**(4), 885–895 (2017)
14. Timesli, A., Braikat, B., Jamal, M., Damil, N.: Prediction of the critical buckling load of multi-walled carbon nanotubes under axial compression. C. R. Mécanique **345**(2), 158–168 (2017)
15. Belaasilia, Y., Timesli, A., Braikat, B., Jamal, M.: A numerical mesh-free model for elasto-plastic contact problems. Eng. Anal. Bound. Elem. **82**, 68–78 (2017)
16. Belaasilia, Y., Braikat, B., Jamal, M.: High order mesh-free method for frictional contact. Eng. Anal. Bound. Elem. **94**, 103–112 (2018)
17. Mesmoudi, S., Braikat, B., Lahmam, H., Zahrouni, H.: Three-dimensional numerical simulation of material mixing observed in FSW using a mesh-free approach. Eng. Comput. **36**(1), 13–27 (2018). https://doi.org/10.1007/s00366-018-0683-6
18. Fouaidi, M., Hamdaoui, A., Jamal, M., Braikat, B.: A high order mesh-free method for buckling and post-buckling analysis of shells. Eng. Anal. Bound. Elem. **99**, 89–99 (2019)
19. Rammane, M., Mesmoudi, S., Tri, A., Braikat, B., Damil, N.: Solving the incompressible fluid flows by a high order mesh-free approach. Int. J. Numer. Methods Fluids **92**(5), 422–435 (2019)
20. Askour, O., Mesmoudi, S., Braikat, B.: On the use of Radial Point Interpolation Method (RPIM) in a high order continuation for the resolution of the geometrically nonlinear elasticity problems. Eng. Anal. Bound. Elem. **110**, 69–79 (2020)
21. Cochelin, B., Damil, N., Potier-Ferry, M.: The asymptotic-numerical method: an efficient perturbation technique for nonlinear structural mechanics. Rev. européenne des éléments finis **3**(2), 281–297 (1994)
22. Mottaqui, H., Braikat, B., Damil, N.: Discussion about parameterization in the asymptotic numerical method: application to nonlinear elastic shells. Comput. Methods Appl. Mech. Eng. **199**(25–28), 1701–1709 (2010)

Advanced Radial Basis Functions Mesh Morphing for High Fidelity Fluid-Structure Interaction with Known Movement of the Walls: Simulation of an Aortic Valve

Leonardo Geronzi[1,2,3], Emanuele Gasparotti[1,2], Katia Capellini[1,2], Ubaldo Cella[4] , Corrado Groth[4], Stefano Porziani[4], Andrea Chiappa[4], Simona Celi[1(✉)] , and Marco Evangelos Biancolini[4]

[1] BioCardioLab, Bioengineering Unit, Fondazione Toscana "G. Monasterio", Heart Hospital, Massa, Italy
s.celi@ftgm.it
[2] Department of Information Engineering, University of Pisa, Pisa, Italy
[3] RBF Morph srl, Rome, Italy
[4] Department of Enterprise Engineering, University of Rome Tor Vergata, Rome, Italy

Abstract. High fidelity Fluid-Structure Interaction (FSI) can be tackled by means of non-linear Finite Element Models (FEM) suitable to capture large deflections of structural parts interacting with fluids and by means of detailed Computational Fluid Dynamics (CFD). High fidelity is gained thanks to the spatial resolution of the computational grids and a key enabler to have a proper exchange of information between the structural solver and the fluid one is the management of the interfaces. A class of applications consists in problems where the complex movement of the walls is known in advance or can be computed by FEM and has to be transferred to the CFD solver. The aforementioned approach, known also as one-way FSI, requires effective methods for the time marching adaption of the computation grid of the CFD model. A versatile and well established approach consists in a continuum update of the mesh that is regenerated so to fit the evolution of the moving walls. In this study, an innovative method based on Radial Basis Functions (RBF) mesh morphing is proposed, allowing to keep the same mesh topology suitable for a continuum update of the shape. A set of key configurations are exactly guaranteed whilst time interpolation is adopted between frames. The new framework is detailed and then demonstrated, adopting as a reference the established approach based on remeshing, for the study of a Polymeric-Prosthetic Heart Valve (P-PHV).

Keywords: Morphing · RBF · Multi-physics · RBF Morph · FSI · Fluid-Structure Interaction · Polymeric aortic valve · Aortic valve

© Springer Nature Switzerland AG 2020
V. V. Krzhizhanovskaya et al. (Eds.): ICCS 2020, LNCS 12142, pp. 280–293, 2020.
https://doi.org/10.1007/978-3-030-50433-5_22

1 Introduction

In all the engineering areas of development, multiphysics analysis appears highly difficult to be carried out because of the interactions between more than one physics involved. During the achievement and the analysis of coupled systems, the way in which the shape of the object influences its performances is required to be carefully taken into account. In the biomedical engineering field for example, the design and the evaluation of the behaviour of prosthetic valves, stent-grafts and ventricular assist devices are related to both the structural and fluid mechanics physics [1–3]. Generally speaking, numerical meshes need to be created for the specific kind of analysis and, in a multi-physics context, two or more meshes have to be realised. In this environment, each geometric change is applied to all the numerical models involved in the analysis: such update has to be performed in a rapid way and as easy as possible. This task, usually carried out through remeshing methods, may be also obtained faster with the use of mesh morphing techniques. This approach allows the changing of the shape of a meshed surface so that the topology is preserved while nodal positions are updated [4,5]: modifications are applied on a baseline grid by moving the surface nodes and propagating displacements inside the surrounding volume mesh nodes. Concerning biomedical applications, morphing methods have been applied in both bone and cardiovascular fields. Recently, in [6], an interactive sculpting and RBF mesh morphing approach has been proposed to address geometry modifications using a force-feedback device, while in [7] RBF have been applied to improve cranioplasty applications. In the cardiovascular field, morphing approaches were employed in [8] for the registration procedures of the cardiac muscle and to model an aorta aneurysm carrying a one-way FSI [9–11]. In literature RBF have been extensively employed to tackle FSI problems [12], using the modal method for both static [13,14] and transient simulations [15–17], as well as the two-way approach [18,19] with RBF-based mapping methods [20]. In this work the state of the art regarding biomedical one-way FSI applications, as shown in [11], is furtherly improved for a transient simulation by taking into account the non-linear deformations of the wetted surfaces during motion. An ad-hoc workflow was developed to transfer and to update, using an RBF-based morphing technique, the CFD mesh, incrementally adjusting the geometry according to the non-linear evolution predicted by the FEM solver. To demonstrate the effectiveness of the proposed approach, it was applied to a tailor-made Polymeric-Prosthetic Heart Valve; these devices [21] proved to significantly reduce blood coagulation problems, maintaining excellent properties in term of strength, efficient function and long-term durability [22,23]. This work is arranged as follows: at first, an introduction on FSI coupling is given, comparing remeshing and morphing workflows. The proposed procedure is implemented in the following paragraph, in which the incremental approach is shown on the P-PHV case. Results are finally discussed and compared to those obtained by remeshing.

2   FSI Coupling: Known-Imposed Motion of the Walls

One of the most delicate processes in a Fluid-Structure Interaction analysis concerns the management of the fluid-solid interfaces; at the boundary surfaces between these two domains, solution data is shared between the fluid solver and the structural one. A useful and employed kind of FSI analysis concerns all the simulations in which a moving body interacts with an internal or external surrounding fluid in a unidirectional way. This one-way coupling does not guarantee energy conservation at the fluid-solid interface but holds the benefit of lower computational time in comparison to the bidirectional one, in which a continuous data exchange between Computational Structural Mechanics (CSM) and CFD solvers is required. In this work two Arbitrary Lagrangian Eulerian (ALE) methods [24–26] for moving meshes are employed: the first based on remeshing algorithms, called in this paper "standard" approach and the novel one using RBF mesh morphing procedures.

2.1   FSI Analysis Based on Remeshing

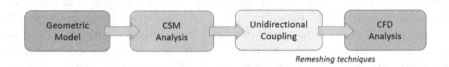

Fig. 1. Fluid-Structure Interaction analysis flowchart using remeshing tools.

In this first kind of analysis, as shown in Fig. 1, a specific component is responsible to transfer the deformation of the CSM mesh to the grid of the CFD solver. In fact, one of the most applied strategies to manage body meshes in FSI applications is simply to move the grid for as long as possible and, when the quality of the mesh becomes critical, to enable the remeshing tools updating the low-quality cells generated by high displacements in the fluid domain [27]. Larger are the displacements of the grid, wider are the distortions of the cells; remeshing method agglomerates cells that violate the initially defined Skewness [28] or size criteria and remeshes them. If the new cells or faces satisfy the Skewness criterion, the mesh is locally updated with the new cells, interpolating the solution from the old cells. Otherwise, the new cells are discarded and another remeshing step is required. Obviously, when small displacements are involved and just few remeshing steps are necessary, this approach may be considered efficient. However, for larger displacements, the number of remeshing steps increases to avoid the presence of negative and invalid elements; as a result, the simulation is slowed down because a new mesh with different numbers and positions of nodes and cells has to be generated more frequently.

2.2   A Novel FSI Workflow Using RBF Mesh Morphing

To overcome the problems related to remeshing, a procedure based on mesh morphing to adapt the shape according to a target one [29] is here described. Among the morphing methods available in literature, RBF are well known for their interpolation quality also on very large meshes [4, 30]. RBF allow to interpolate everywhere in the space a scalar functions known at discrete points, called Source points (Sp). By interpolating three scalar values it is possible, solving a linear system of order equal to the number of Sp employed [4], to describe a displacement of the Sp in the three directions in space. The interpolation function is defined as follows:

$$s(x) = \sum_{i=1}^{N} \gamma_i \varphi \left(\| x - x_{s_i} \| \right) + h(x) \tag{1}$$

where x is a generic position in the space, x_{s_i} the Sp position, $s(\cdot)$ the scalar function which represents a transformation $\mathbb{R}^n \to \mathbb{R}$, $\varphi\,(\cdot)$ the radial function of order m, γ_i the weight and $h(x)$ a polynomial term with degree $m-1$ added to improve the fit assuring uniqueness of the problem and polynomial precision.

The unknowns of the system, namely the polynomial coefficients and the weights γ_i of the radial functions, are retrieved by imposing the passage of the function on the given values and an orthogonality condition on the polinomials. If the RBF is conditionally positive definite, it can be demonstrated that a unique interpolant exists and in 3D, if the order is equal or less than 2, a linear polynomial in the form $h(x) = \beta_1 + \beta_2 x + \beta_3 y + \beta_4 z$ can be used. The linear problem can be also written in matrix form:

$$\begin{bmatrix} \mathbf{M} & \mathbf{P} \\ \mathbf{P}^T & \mathbf{0} \end{bmatrix} \begin{Bmatrix} \gamma \\ \beta \end{Bmatrix} = \begin{Bmatrix} \mathbf{g} \\ \mathbf{0} \end{Bmatrix} \tag{2}$$

in which $\mathbf{M}$ is the interpolation matrix containing all the distances between RBF centres $\mathbf{M}_{ij} = \varphi\,(\| x_i - x_j \|)$, $\mathbf{P}$ the matrix containing the polynomial terms that has for each row j the form $\mathbf{P}_j = [1\ x_{1j}\ x_{2j}\ ...\ x_{nj}]$ and $\mathbf{g}$ the known values at Sp. The new nodal positions, if interpolating the displacements, can be retrieved for each node as:

$$x_{node_{new}} = x_{node} + \begin{bmatrix} s_x(x_{node}) \\ s_y(x_{node}) \\ s_z(x_{node}) \end{bmatrix} \tag{3}$$

Being the problem solved pointwise, the approach is meshless and able to manage every kind element (tetrahedral, hexahedral, polyhedral and others), both for surface and volume mesh smoothing ensuring the preservation of their topology.

In general, a morphing operation can introduce a reduction of the mesh quality but a good morpher has to minimize this effect and to maximize the possible shape modifications. If mesh quality is well preserved, morphing has a clear benefit over remeshing because it avoids introducing noise [4]. This procedure

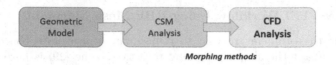

Morphing methods

Fig. 2. Novel FSI workflow based on morphing techniques.

provides for the sampling of the displacements of the structural model. In this way, the surface nodes of the FE model are used as Sp to apply RBF mesh morphing in the space. The detailed workflow shown in Fig. 2, in addition to the realization of the geometric model, provides for three main steps:

- CSM settings
- RBF Morph application
- CFD set-up

CSM Settings. Given that in this kind of FSI analysis the deformations of the fluid grid are imposed by the motion of the walls of the CSM domain, a structural simulation has to be conducted to obtain the deformed positions in the space during time. During the structural simulation, Source points (S_p) and Target points (T_p), that are the surface nodal positions of the grid in this case, are recursively extracted and stored in files. These files are structured as a list of nodes and the related displacements to obtain the nodes in the following sampled configuration.

RBF Morph Application. To apply the shape deformation without remeshing, the morpher RBF Morph is in this study employed [31]. To delimit the morphing action, an encapsulation technique has to be implemented, defining in this way sub-domains or parts of the fluid domain within which the morpher action is applied [32]. According to [33], S_p effectively used are a random subgroup which includes a percentage close to 2% of the total extracted surface nodes of the structural model. A Radial Basis Function is selected and, when the settings of the procedure are completed, morphing solutions are calculated, each per morphing transitions, from the starting configuration to the final position of the model. Every transition, a new deformed mesh is generated and the solution files are stored.

CFD Set-Up. Once all the morphed meshes are obtained, the CFD analysis is carried out every time step on a new deformed mesh, by using a specific script in Scheme Programming Language. This script is recalled before the starting of the CFD simulation and it allows the automatization of the MultiSol tool of RBF Morph software in order to guarantee the synchronous implementation of the deformed shape in fluid environment. The nodal displacement inside two subsequent morphed positions is simply handled inside the Scheme script by means of a parameter called amplification $A(t)$ which varies over time from 0 to A_0 (usually equal to $+1$ or -1). Because of the samplings available at different times, an accurate and effective solution can be obviously achieved, handling all the morphing deformations separately and linearly modulating the amplification within each of the sampled grids: calling i the index that concerns all the mesh conformations $P_0...P_n$, the amplification $A_i(t)$ inside each time step can be evaluated as:

$$A_i(t) = \begin{cases} 0, & \text{if } t \leq t_i \\ \frac{t-t_i}{t_{i+1}-t_i}, & \text{if } t_i < t < t_{i+1} \\ 1, & \text{if } t \geq t_{i+1} \end{cases} \tag{4}$$

In this way, the fluid grid is able to follow the displacements of the CSM mesh and solution can be quickly calculated every time on a new deformed grid.

3 Implementation of the Workflows to Study the Opening Phase of a Polymeric Aortic Valve

In this paper, both the presented FSI solution approaches are applied to study the opening phase of a P-PHV. This kind of analysis is conducted on a prosthetic valve to study its haemodynamic behaviour during the opening phase in which the leaflets or cusps [34,35] are pushed open to allow the ejection of the blood flow and to investigate the effects of the solid domain (P-PHV) towards the fluid domain (blood flow).

The design of a prosthetic aortic valve is realised using the software Space-Claim, following the geometry proposed in [36], and then imported inside ANSYS Workbench (v193). Figure 3 depicts the CAD model of the polymeric valve here investigated and the fluid domain of the blood in which also the Valsalva sinuses are represented [37]. This fluid domain is obtained by a boolean intersection between the solid domain of the valve and a vessel made up of an inlet and outlet tube.

The computational grid of this structural model is made up of 301256 tetrahedral elements and it is obtained inside ANSYS Meshing. The material chosen to model the valve is isotropic linear elastic with a Young's modulus of 3 MPa, and a Poisson's ratio of 0.4. The mesh of the fluid domain is obtained inside Ansa (v20.0.1) and made up of 1564792 tetra-hexaedral elements with a maximum starting Skewness (Sk) of 0.694. A velocity inlet and a systolic aortic pressure outlet varying over time boundary conditions have been provided. The flow model selected is Viscous-Laminar and the fluid is considered as incompressible and Newtonian with a viscosity of 4 cP.

3.1 Study of the Valve Using the Remeshing Workflow

The workflow of the first approach based on remeshing and implemented inside ANSYS Workbench is proposed in Fig. 4. In this case, the displacements of a

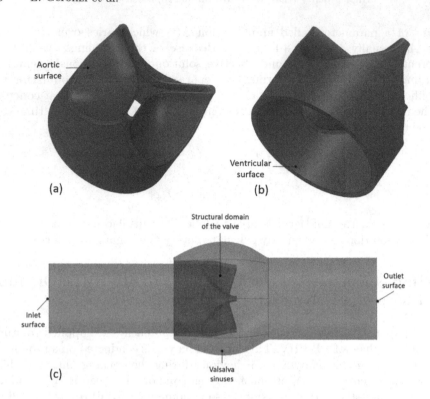

Fig. 3. CAD model of the Polymeric-Prosthetic Heart Valve. Indications of aortic side (a) and ventricular side (b). Blood domain with the hollow volume of the valve (c).

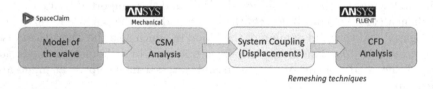

Fig. 4. Flowchart of the FSI analysis implemented inside ANSYS Workbench to study the behaviour of a P-PHV using Remeshing Method.

transient CSM in ANSYS Mechanical (v193) are transferred to the fluid domain inside ANSYS Fluent (v193) by means of System Coupling. The body of the prosthetic device is fixed in its bottom surface, i.e. the circular ring in the ground. Concerning the loads, a physiological time-varying pressure is uniformly applied to the ventricular portion of the valve to simulate the opening pressure.

Fluent was used as CFD solver, activating the Dynamic Mesh to deform and regenerate the computational grid [38]: the Spring-Laplace based Smoothing Method and the Remeshing Method, two tools to solve the analysis of deformed domains due to boundary movement over time [39], are applied.

Inside the System Coupling component, all the surfaces of the valve in contact with the blood are set as fluid-solid interfaces and a time step of 1e-5 s is selected.

3.2   Study of the Valve Using the Morphing Approach

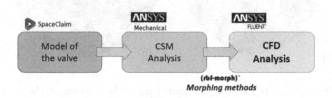

Fig. 5. Flowchart with the application of morphing.

The novel workflow, based on a RBF mesh morphing, also in this case developed into the ANSYS Workbench environment is shown in Fig. 5 which depicts the overall incremental procedure; in particular, the System Coupling component, fundamental in the previously shown approach, here is not employed.

Results obtained with the use of RBF Morph software are compared to those achieved using Fluent remeshing tools. They are reported in terms of pressure map, velocity streamlines and computational time.

Structural Simulations. The same CSM analysis implemented in the remeshing approach is conducted. A specific Application Customization Toolkit (ACT) Extension, i.e. a method to achieve custom applications inside ANSYS Mechanical, is developed to collect Sp and Tp according to a sampling time chosen by the user: in this case, ten different surface nodal positions ($P_0...P_9$) are collected.

RBF Morph Set-Up. The RBF Morph Fluent (v193) Add-On is adopted: the RBF are used to step by step project the S_p to the T_p and interpolate the displacements of the volume nodes inside the blood domain over time. RBF Morph tools may impose a blood domain adaptation in the CFD analysis following the deformed shapes of the valvular domain, extracted from the previous CSM simulation. In the case of the valve, to control the morphing action in the space, no encapsulation volumes [32] are generated but a fixed Surf Set (with null displacement during the movement of the leaflets) on the wall of the Sinotubular Junction (STJ) [40] is preferred (Fig. 6). In this way, since the distortion of the mesh is extremely high, such deformations can be distributed also in the other zones of the fluid domain and not only in proximity to the valve.

To prevent mesh quality reduction, in this work the bi-harmonic kernel $\varphi(r) = r$ is used: it is the RBF that guarantees to keep the mesh quality degradation at minimum [41]. The nine solutions are calculated using the Solve Panel of the software, each per morphing action. Two different strategies are evaluated:

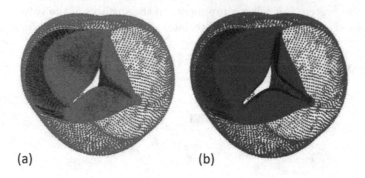

Fig. 6. First morphing step of the opening: Sp on the surface of the valve and fixed Surf Set on the wall of the Valsalva sinuses (a), Tp on the surface of the prosthetic device to obtain a first movement keeping stationary the points on the STJ surface (b).

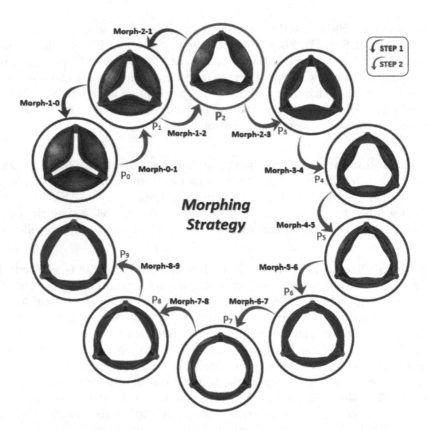

Fig. 7. Backwards/Forwards strategy: step 1 represents a backwards morphing procedure from P_2 (partially open position) to P_0 (initial position) while step 2 allows to reach position P_9 (completely open) starting from position P_0. This strategy allows to prevent excessive mesh quality degradation.

- *Forwards*: from P_0 (initial position) directly to P_9 (totally open position).
- *Backwards/Forwards* (Fig. 7): extracting a new already deformed valve in position P_2 of which a new mesh is achieved, from P_2 going back to P_0 (backwards - step 1) and then from P_0 moving on to P_9 (forwards - step 2).

The first strategy produced an excessive mesh distortion with negative cells volume in particular at the centre of the leaflets. No mesh problems are detected with the adopted second approach.

Fluent Setting. The same time step of the standard approach (1e-5 s) is selected. The flow is initialized with the valve in position P_0 and the Scheme file with the amplification of each of the nine different transitions is recalled inside Fluent. Thanks to this workflow, the FSI simulation can be conducted only by means of this CFD analysis in which the mesh is updated every time step.

4   Results and Discussions

The first important parameter to check in the implementation of both the approaches is the Skewness: no negative cells are detected and maximum Sk values are 0.858 and 0.961 respectively for the procedure implemented through remeshing and morphing. No convergence issues due to mesh degradation are recorded.

In this work, only pressure and velocity results of the 3D-model at a specific and significant time step are shown using the CFD-Post software.

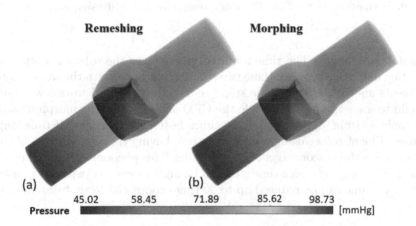

Fig. 8. Volume rendering of the pressure at $t = 7$ ms, when the valve is for the first time open: remeshing approach (a), morphing approach (b).

Pressure Map - A 3D-pressure rendering at the time in which the valve is for the first time open (7 ms) is represented in Fig. 8; as it can be observed, no significant differences in terms of pressure values are recorded between the two approaches: the mean pressure are 69.64 mmHg and 70.31 mmHg respectively for the simulation based on remeshing and the analysis using mesh morphing (difference of less than 1%).

Velocity Streamlines - Streamlines of both the solution methods are reported in Fig. 9. It is possible to observe how in this 3D-representation the local maximum values are placed in the same zones. Maximum local fluid velocities are reached right at $t = 7$ ms because the fast movement of the leaflets pushes away the blood: maximum value for the FSI implemented with the novel approach is 3.011 m/s, about 5.58% less than the value detected in the analysis based on remeshing, that is 3.189 m/s.

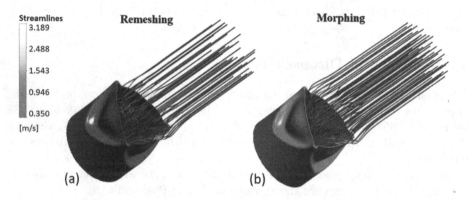

Fig. 9. Streamlines at $t = 7$ ms: FSI using remeshing (a), FSI using morphing (b).

Computational Time - The time required to calculate the solution is the most interesting difference between these two solution methods: with the same number of elements and to parity of time step (1e-5 s), the standard workflow requires 6283 min to solve the problem while the CFD analysis with the morphing application only 396 min, approximately 16 times faster. Furthermore, if time step is increased, Fluent solver using remeshing starts having troubles and the obvious consequence is the no-convergence of the model. This phenomenon does not happen with morphing by which time step can be highly increased without problems and solution time can be reduced up to 60 times compared to that of remeshing.

5   Conclusions

In this work, a fast high fidelity workflow for a multi-physic Fluid-Structure Interaction analysis exploiting ANSYS Workbench and RBF Morph is presented. In particular, this powerful system has proved to be useful and accurate in all those applications in which the movement of the fluid domain is strongly controlled by the deformation of the solid domain. The possibility to replicate the non-linear analysis in which the motion of a solid model is transferred to a fluid domain carefully handling the interfaces is demonstrated. In a specific application case, the attempt to transfer aortic valve opening kinematics to the CFD analysis through a mesh morphing technique results to be successful and the effectiveness of the FSI simulation implemented through mesh morphing to reproduce the fluid dynamic field of the aortic valve is here shown and validated. Morphing has turned out to be very consistent in comparison to remeshing. Further research is needed to asses in more detail the differences between these two distinct approaches. Testing the novel workflow on a FSI analysis about the study of the opening of a polymeric aortic valve, it has allowed to reduce the simulation time up to 16 times in comparison to that required by remeshing methods using the same time step. Until now, numerical FSI simulations have been used in a restricted manner precisely because of the long time required to solve them, as in the case of heart valve computational analyses. Dealing with clinical trials, this approach could ensure a considerable time saving with a decrease of the final mesh quality which results acceptable for the numerical solver here adopted. This new workflow has definitely broken down this limit, granting in this case the possibility to test several models of P-PHV in a lot of flow conditions. In addition, a parametric design of P-PHV combined with this method could finally make a strong contribution to the patient-specific aortic valve replacement.

Acknowledgment. The research has received funding from the European Union's Horizon 2020 research and innovation programme under the Marie Skłodowska-Curie grant agreement No. 859836 and has been partially supported by RBF Morph®.

References

1. Marom, G.: Numerical methods for fluid-structure interaction models of aortic valves. Arch. Comput. Methods Eng. **22**(4), 595–620 (2015)
2. Roy, D., Kauffmann, C., Delorme, S., Lerouge, S., Cloutier, G., Soulez, G.: A literature review of the numerical analysis of abdominal aortic aneurysms treated with endovascular stent grafts. Comput. Math. Methods Med. **2012**, Article ID 820389, 16 p. (2012). https://doi.org/10.1155/2012/820389
3. Avrahami, I., Rosenfeld, M., Raz, S., Einav, S.: Numerical model of flow in a sac-type ventricular assist device. Artif. Organs **30**(7), 529–538 (2006)
4. Biancolini, M.E.: Fast Radial Basis Functions for Engineering Applications, 1st edn. Springer, New York (2017). https://doi.org/10.1007/978-3-319-75011-8
5. Staten, M.L., Owen, S.J., Shontz, S.M., Salinger, A.G., Coffey, T.S.: A comparison of mesh morphing methods for 3D shape optimization. In: Quadros, W.R. (ed.) Proceedings of the 20th International Meshing Roundtable, pp. 293–311. Springer, Heidelberg (2011). https://doi.org/10.1007/978-3-642-24734-7_16
6. Biancolini, M.E., Valentini, P.P.: Virtual human bone modelling by interactive sculpting, mesh morphing and force-feedback. Int. J. Interact. Des. Manuf. (IJI-DeM) **12**(4), 1223–1234 (2018). https://doi.org/10.1007/s12008-018-0487-3

7. Atasoy, F., Sen, B., Nar, F., Bozkurt, I.: Improvement of radial basis function interpolation performance on cranial implant design. Int. J. Adv. Comput. Sci. Appl. **8**(8), 83–88 (2017)
8. This, A., et al.: One mesh to rule them all: registration-based personalized cardiac flow simulations. In: Pop, M., Wright, G.A. (eds.) FIMH 2017. LNCS, vol. 10263, pp. 441–449. Springer, Cham (2017). https://doi.org/10.1007/978-3-319-59448-4_42
9. Capellini, K., et al.: An image-based and RBF mesh morphing CFD simulation for parametric aTAA hemodynamics. In: Vairo, G. (ed.) Proceedings VII Meeting Italian Chapter of the European Society of Biomechanics (ESB-ITA 2017) (2017)
10. Capellini, K., et al.: Computational fluid dynamic study for aTAA hemodynamics: an integrated image-based and radial basis functions mesh morphing approach. J. Biomech. Eng. **140**(11), 111007 (2018)
11. Capellini, K., et al.: A coupled CFD and RBF mesh morphing technique as surrogate for one-way FSI study. In: Proceedings VIII Meeting Italian Chapter of the European Society of Biomechanics, (ESB-ITA 2018) (2018)
12. Groth, C., Cella, U., Costa, E., Biancolini, M.E.: Fast high fidelity CFD/CSM fluid structure interaction using RBF mesh morphing and modal superposition method. Aircr. Eng. Aerosp. Technol. **91**, 893–904 (2019)
13. Biancolini, M.E., Cella, U., Groth, C., Genta, M.: Static aeroelastic analysis of an aircraft wind-tunnel model by means of modal RBF mesh updating. J. Aerosp. Eng. **29**(6), 04016061 (2016)
14. Groth, C., Biancolini, M.E., Costa, E., Cella, U.: Validation of high fidelity computational methods for aeronautical FSI analyses. In: Biancolini, M.E., Cella, U. (eds.) Flexible Engineering Toward Green Aircraft. LNACM, vol. 92, pp. 29–48. Springer, Cham (2020). https://doi.org/10.1007/978-3-030-36514-1_3
15. Van Zuijlen, A.H., de Aukje, B., Bijl, H.: Higher-order time integration through smooth mesh deformation for 3D fluid-structure interaction simulations. J. Comput. Phys. **224**(1), 414–430 (2007)
16. Di Domenico, N., et al.: Fluid structure interaction analysis: vortex shedding induced vibrations. Procedia Struct. Integrity **8**, 422–432 (2018)
17. Costa, E., Groth, C., Lavedrine, J., Caridi, D., Dupain, G., Biancolini, M.E.: Unsteady FSI analysis of a square array of tubes in water crossflow. In: Biancolini, M.E., Cella, U. (eds.) Flexible Engineering Toward Green Aircraft. LNACM, vol. 92, pp. 129–152. Springer, Cham (2020). https://doi.org/10.1007/978-3-030-36514-1_8
18. Cella, U., Marco, E.B.: Aeroelastic analysis of aircraft wind-tunnel model coupling structural and fluid dynamic codes. J. Aircr. **49**(2), 407–414 (2012)
19. Keye, S.: Fluid-structure coupled analysis of a transport aircraft and flight-test validation. J. Aircr. **48**(2), 381–390 (2011)
20. Biancolini, M.E., Chiappa, A., Giorgetti, F., Groth, C., Cella, U., Salvini, P.: A balanced load mapping method based on radial basis functions and fuzzy sets. Int. J. Numer. Meth. Eng. **115**(12), 1411–1429 (2018)
21. Ghosh, R.P., et al.: Comparative fluid-structure interaction analysis of polymeric transcatheter and surgical aortic valves' hemodynamics and structural mechanics. J. Biomech. Eng. **140**(12), 121002 (2018)
22. Bezuidenhout, D., Williams, D.F., Zilla, P.: Polymeric heart valves for surgical implantation, catheter-based technologies and heart assist devices. Biomaterials **36**, 6–25 (2015)

23. Ghanbari, H., Viatge, H., Kidane, A.G., Burriesci, G., Tavakoli, M., Seifalian, A.M.: Polymeric heart valves: new materials, emerging hopes. Trends Biotechnol. **27**(6), 359–367 (2009)
24. Hirt, C.W., Amsden, A.A., Cook, J.L.: An arbitrary Lagrangian-Eulerian computing method for all flow speeds. J. Comput. Phys. **14**(3), 227–253 (1974)
25. Ferziger, J.H., Perić, M.: Computational Methods for Fluid Dynamics, vol. 3. Springer, Heidelberg (2002). https://doi.org/10.1007/978-3-642-56026-2
26. Zhang, Q., Hisada, T.: Analysis of fluid-structure interaction problems with structural buckling and large domain changes by ALE finite element method. Comput. Methods Appl. Mech. Eng. **190**(48), 6341–6357 (2001)
27. Baum, J., Luo, H., Loehner, R.: A new ALE adaptive unstructured methodology for the simulation of moving bodies. In 32nd Aerospace Sciences Meeting and Exhibit, pp. 1–14 (1994)
28. Braaten, M., Shyy, W.: A study of recirculating flow computation using body-fitted coordinates: consistency aspects and mesh skewness. Numer. Heat Trans. Part A: Appl. **9**(5), 559–574 (1986)
29. Profir, M.M.: Mesh morphing techniques in CFD. In: International Student Conference on Pure and Applied Mathematics, pp. 195–208 (2011)
30. Yu, H., Xie, T., Paszczyñski, S., Wilamowski, B.M.: Advantages of radial basis function networks for dynamic system design. IEEE Trans. Industr. Electron. **58**(12), 5438–5450 (2011)
31. Biancolini, M.E.: Mesh morphing and smoothing by means of radial basis functions (RBF): a practical example using fluent and RBF morph. In: Handbook of Research on Computational Science and Engineering: Theory and Practice, pp. 347–380. IGI Global (2012)
32. RBF Morph for FLUENT: User's Guide. Release V1.93 (2019). Accessed 2019
33. RBF Morph: Modelling Guidelines and Best Practices Guide. Release V1.93. Accessed 2019
34. Anderson, R.H.: Clinical anatomy of the aortic root. Heart **84**(6), 670–673 (2000)
35. Iaizzo, P.A.: Handbook of Cardiac Anatomy, Physiology, and Devices, 2nd edn. Springer, Heidelberg (2009). https://doi.org/10.1007/978-1-60327-372-5
36. Jiang, H., Campbell, G., Boughner, D., Wan, W.K., Quantz, M.: Design and manufacture of a polyvinyl alcohol (PVA) cryogel tri-leaflet heart valve prosthesis. Med. Eng. Phys. **26**(4), 269–277 (2004)
37. Reid, K.: The anatomy of the sinus of Valsalva. Thorax **25**(1), 79–85 (1970)
38. 3.1 Dynamic Mesh Update Methods. Release V12.0. Accessed Jan 2009
39. Si, H., Fuxiang, Y., Jing, G.: Numerical simulation of 3D unsteady flow in centrifugal pump by dynamic mesh technique. Procedia Eng. **61**, 270–275 (2013)
40. Maselli, D., et al.: Sinotubular junction size affects aortic root geometry and aortic valve function in the aortic valve reimplantation procedure: an in vitro study using the Valsalva graft. Ann. Thorac. Surg. **84**(4), 1214–1218 (2007)
41. Helenbrook, B.T.: Mesh deformation using the biharmonic operator. Int. J. Numer. Meth. Eng. **56**(7), 1007–1021 (2003)

Radial Basis Functions Mesh Morphing

A Comparison Between the Bi-harmonic Spline and the Wendland C2 Radial Function

Marco Evangelos Biancolini[1]([✉])[iD], Andrea Chiappa[1][iD],
Ubaldo Cella[1][iD], Emiliano Costa[2][iD], Corrado Groth[1][iD],
and Stefano Porziani[1]

[1] University of Rome "Tor Vergata", Via Politecnico 1, 00133 Rome, Italy
biancolini@ing.uniroma2.it
[2] RINA Consulting SPA, Viale Cesare Pavese 305, 00144 Rome, Italy

Abstract. Radial basis functions (RBFs) based mesh morphing allows to adapt the shape of a computational grid onto a new one by updating the position of all its nodes. Usually nodes on surfaces are used as sources to define the interpolation field that is propagated into the volume mesh by the RBF. The method comes with two distinctive advantages that makes it very flexible: it is mesh independent and it allows a node wise precision. There are however two major drawbacks: large data set management and excessive distortion of the morphed mesh that may occur. Two radial kernels are widely adopted to overtake such issues: the bi-harmonic spline (BHS) and the Wendland C2 (WC2). The BHS minimizes the mesh distortion but it is computational intense as a dense linear system has to be solved whilst the WC2 leads to a sparse system easier to solve but which can lack in smoothness. In this paper we compare these two radial kernels with a specific focus on mesh distortion. A detailed insight about RBF fields resulting from BHS and WC2 is first provided by inspecting the intensity and the distribution of the strain for a very simple shape: a square plate with a central circular hole. An aeronautical example, the ice formation onto the leading edge of a wing, is then exposed adopting an industrial software implementation based on the state of the art of RBF solvers.

Keywords: CAE · Mesh morphing · Radial basis functions

1 Introduction

Radial basis functions (RBFs) are a powerful mathematical tool introduced by Hardy [1] in the late sixties for the interpolation of scattered data in the field of surveying and mapping. A review of the multiquadric (MQ) approach was published by Hardy himself [2] twenty years later; here the author explains that the MQ is bi-harmonic for 3D problems. Since their inception, RBF where adopted in many different fields and their mathematical framework developed [3] by exploring a variety of RBF kernels; among them the compact supported ones were introduced by Wendland [4].

© Springer Nature Switzerland AG 2020
V. V. Krzhizhanovskaya et al. (Eds.): ICCS 2020, LNCS 12142, pp. 294–308, 2020.
https://doi.org/10.1007/978-3-030-50433-5_23

An RBF interpolation at a generic point x (see Eq. 1) consists in the weighted summation of inter-distance interactions of the point x with a set of given points, defined sources x_{si}, where the inter-distance interaction consists in the Euclidean distance transformed by the radial function $\varphi(r)$. The weights γ_i are computed so that the RBF gets at the sources the known input values to be interpolated.

$$s(x) = \sum_{i=1}^{N} \gamma_i \varphi(\|x - x_{si}\|) \tag{1}$$

RBF are nowadays adopted in several engineering applications [5] and are well accepted as one of the most powerful and versatile mathematical approach to manage mesh morphing. The RBF allows creating a scalar field that interpolates a function defined on a set of source points: in the case of 3d mesh morphing a vector field is defined by individually interpolating the three components of the displacement known at source points. The RBF vector field is a point function independent from the mesh itself, all the points in the space receiving the field are called targets. A typical mesh morphing problem that can be faced using RBF consists in a three-dimensional mesh to be adapted according to a known displacement of the surface; surface nodes are extracted from the surface mesh as sources and all the nodes of the volume mesh receive the morphing field as targets.

Computer Aided Engineering (CAE) is more and more demanding for advanced methods capable to generate and adapt computational grids for multi-physics models. High fidelity models are widely employed, for instance, in computational fluid dynamics (CFD) and computational structural mechanics (CSM). The size of the grids daily adopted for CFD according to industrial best practices can be comprised of many millions of cells and, in some situation, close to one billion [6]; structured meshes of hexahedrons, hybrid meshes of tetrahedrons with prisms layers at the wall and meshes of Cartesian polyhedrons with inflation of prisms at surfaces are common adopted options for CFD. For CSM applications the mesh size is about one-two order lower than CFD ranging from hundred thousand up to some million of nodes for the most complex cases [7]; parabolic tetrahedrons are in this case widely adopted and the extra complexity of mid-side nodes management is added. In view of mesh morphing we have to consider that to the complexity of the mesh typology (i.e. linear/parabolic, tetra/hexa/poly) we have to add special cases for management of interfaces: moving/rotating parts are common in CFD, contact connections are common in CSM.

Mesh morphing for CAE applications is required as a companion/replacement of meshing when multiple shape variations of the same component/system are required. This is typical of design optimization which requires the automatic update of the CAE grid onto new design configurations to be explored: the baseline mesh is updated onto the new configuration instead of generating a new mesh onto the updated geometry [8, 9]. Morphing becomes very useful for optimization where the new shape is not known in advance but predicted by automatic sculpting methods as the adjoint based [10, 11] and the biological growth method [12]. Mesh morphing can furthermore help the solver itself as a tool to support the shape evolution (ice/snow deposition, erosion) [27], to enable fluid structure interaction (1-way, 2-way, structural modes embedding) [13, 14] and to move the CAE model onto "as built" configurations surveyed. Mesh morphing is also a key enabler toward the creation of reduced order models (ROM) that are adopted for the creation of digital twins [15].

Whatever is the need for mesh morphing, there are common requirements to fulfill to set-up an effective mesh morphing approach. Among these there are the ability to:

- manage different element typology (linear/parabolic, tetra/hexa/poly);
- go across interfaces (contact/rotating parts) and across partitions (distributed meshes for parallel calculations);
- adapt the mesh with the minimum distortion (stretching/compression);
- preserve the aspect ratio of cells at the boundary so that the ability to capture stress raisers/boundary layers is preserved;
- adapt the mesh in a reasonable time even for very large models.

RBF mesh morphing satisfactorily fits above mentioned requirements because it comes with two distinctive advantages that makes it very flexible: it is mesh independent and it allows a node wise precision. Its meshless nature is due to the fact that once the RBF field is defined, the mesh deformation becomes a point function; nodal positions are updated regardless the attached mesh elements, and interfaces (contact/partitions boundaries) are implicitly preserved as the same field is received on coincident locations. The meshless nature is a distinctive feature of another widely adopted mesh morphing method: the free form deformation (FFD) [16]. One of the major drawbacks of FFD is that a point-wise precision cannot be achieved. Mesh based methods, as for instance the use of an auxiliary FEM solution [17, 18], allow the user to have pointwise control; despite this benefit their mesh based nature makes complex the management of arbitrary elements typology and interfaces.

Considering the great advantages of RBF, it is clear why a large research effort has been invested over the last decades toward their effective implementation. There are however two major drawbacks to be handled: numerical complexity and deformed mesh quality.

RBFs require the solution of large linear systems whose size is equal to the number of source points. The source points count grows fast especially if a node-wise control is needed because large portions of the nodes on the surface mesh are used as sources. Most of the implementations that are considered for research purposes exploits direct linear solvers because a great flexibility is possible and different radial function kernels can be seamlessly adopted. Unfortunately the numerical complexity scales up in this case with a cubic law so the maximum size of the RBF problem (number of sources) is limited to about 10 000 points. To overtake such limit there are different strategies: replace the original cloud with a smaller one (point decimation) fine enough to guarantee the desired precision [19], adopt iterative [20] solvers, use of the partition of unity [21] and use of fast multipole expansion [22]. The best "recipe" for a fast and effective RBF solver is still an open issue and the acceleration extent is strictly related to the adopted kernel.

The quality of the morphed mesh is of paramount importance and the effectiveness of deformation depends on the radial kernel adopted. The distortion (stretching/compression) of the mesh should be limited as much as possible because the validity of the numerical solution is maintained only within a certain range (which is solver related) of the element/cell[1] quality (skeweness, aspect ratio, ...). Keeping a good quality usually is not enough; a proper spacing of the cells close to the surfaces is required, as example, for CFD meshes in order to preserve the capability to correctly solve the boundary layer. The same spacing is expected in the deformed mesh. This means that while the shape of the surface is changed, the deformation orthogonal to the surface itself has to be minimum so that the boundary layers are solved keeping a similar overall thickness and spacing.

The aim of this paper is to provide useful insights on two specific kernels that are widely adopted for mesh morphing: the bi-harmonic spline (BHS) and the Wendland C2 (WC2). The BHS minimizes the mesh distortion and it is known to have the feature of "smoothest interpolant" for the three-dimensional case; it is computationally expensive because of the full support of the radial function and so a dense linear system has to be solved. The WC2 is compactly supported and so a sparse system has to be solved; however it may lack in smoothness and the behavior close to the surfaces is strongly dependent on the radius of the support.

An overview of the studied problem is given in the first section (this introduction), a refresh of the math of RBF, including first derivatives calculation, is provided in the second section; the third section demonstrates how various RBFs perform for a square plate with a central circular hole, the fourth section deals with an aeronautical example faced with an industrial software and, finally, the fifth section wraps the study with the concluding remarks.

2 Radial Basis Functions Mesh Morphing

2.1 Radial Basis Functions Background

The RBF interpolation of a generic scalar field can be adopted to represent a variety of quantities. A generic component of a displacement field can be interpolated as

$$\delta(\mathbf{x}) = \sum_{i=1}^{N} \gamma_i^\delta \varphi(\|\mathbf{x} - \mathbf{x}_{si}\|) \tag{2}$$

In this study we consider the case where the quantity to be interpolated δ_s is given at the source point locations

$$\delta(\mathbf{x}_{si}) = \delta_{si}, 1 \leq i \leq N \tag{3}$$

The unknown coefficient vector γ^δ can be computed by solving the linear system

$$M\gamma^\delta = \delta_s \tag{4}$$

---

[1] In this paper we use both "element" and "cell" terms for the same entity because in CSM the finite element method is the standard and the "element" term is commonly adopted whilst in CFD the finite volume method is the standard and the "cell" term is commonly adopted.

Where the interpolation matrix M is

$$M_{ij} = \varphi(\|x_{si} - x_{sj}\|), 1 \leq i \leq N, 1 \leq j \leq N \tag{5}$$

Table 1. Radial functions for RBF interpolation

Full supported radial functions	
Linear spline (BHS)	$\varphi(r) = r$
Cubic spline (THS)	$\varphi(r) = r^3$
Compact supported radial functions	
Wendland C0 (WC0)	$\varphi(r) = \left(1 - \frac{r}{R_{sup}}\right)^2, r \leq R_{sup}$
Wendland C2 (WC2)	$\varphi(r) = \left(1 - \frac{r}{R_{sup}}\right)^4 \left(4\frac{r}{R_{sup}} + 1\right), r \leq R_{sup}$
Wendland C4 (WC4)	$\varphi(r) = \left(1 - \frac{r}{R_{sup}}\right)^6 \left(\frac{35}{3}\frac{r^2}{R_{sup}^2} + 6\frac{r}{R_{sup}} + 1\right), r \leq R_{sup}$

Different radial functions can be adopted. In this study we focus on the splines and on the Wendland, summarized in Table 1. In the case of three-dimensional spaces the interpolated function can be rewritten as

$$\delta(x) = \sum_{i=1}^{N} \gamma_i^\delta \varphi\left(\sqrt{(x - x_{si_x})^2 + (y - x_{si_y})^2 + (z - x_{si_z})^2}\right) \tag{6}$$

And can be differentiated, for instance, with respect to x

$$\frac{\partial \delta(x)}{\partial x} = \sum_{i=1}^{N} \gamma_i^\delta \frac{d\varphi(r)}{dr} \frac{1}{2\sqrt{(x - x_{si_x})^2 + (y - x_{si_y})^2 + (z - x_{si_z})^2}} 2(x - x_{si_x}) \tag{7}$$

The full gradient of the interpolated function can be computed accordingly

$$\nabla \delta(x) = \begin{pmatrix} \frac{\partial \delta(x)}{\partial x} \\ \frac{\partial \delta(x)}{\partial y} \\ \frac{\partial \delta(x)}{\partial z} \end{pmatrix} = \sum_{i=1}^{N} \gamma_i^\delta \frac{\frac{\partial \varphi(\|x - x_{si}\|)}{\partial r}}{\|x - x_{si}\|} \begin{pmatrix} x - x_{si_x} \\ y - x_{si_y} \\ z - x_{si_z} \end{pmatrix} \tag{8}$$

2.2   RBF Mesh Morphing

When RBFs are used for mesh morphing, the three components of a displacement field (Eq. 9), that is typically assigned at a cloud of control points here defined as RBF centers or source points, are interpolated in the space and used to update the nodal positions of the mesh nodes to be morphed.

$$\begin{cases} u(\boldsymbol{x}) = \sum_{i=1}^{N} \gamma_i^u \varphi(\|\boldsymbol{x} - \boldsymbol{x}_{si}\|) \\ v(\boldsymbol{x}) = \sum_{i=1}^{N} \gamma_i^v \varphi(\|\boldsymbol{x} - \boldsymbol{x}_{si}\|) \\ w(\boldsymbol{x}) = \sum_{i=1}^{N} \gamma_i^w \varphi(\|\boldsymbol{x} - \boldsymbol{x}_{si}\|) \end{cases} \qquad (9)$$

The field is applied to process all the nodal positions to be updated.

$$\boldsymbol{x}_{node\_new} = \boldsymbol{x}_{node} + \begin{bmatrix} u(\boldsymbol{x}_{node}) \\ v(\boldsymbol{x}_{node}) \\ w(\boldsymbol{x}_{node}) \end{bmatrix} \qquad (10)$$

The local deformation due to the morphing field can be inspected by computing the derivatives of the three components of displacement thus obtaining the strains as follows

$$\varepsilon_x = \frac{\partial u}{\partial x} \; \varepsilon_y = \frac{\partial v}{\partial y} \; \varepsilon_z = \frac{\partial w}{\partial z} \; \varepsilon_{xy} = \frac{\partial u}{\partial y} + \frac{\partial v}{\partial x} \; \varepsilon_{yz} = \frac{\partial v}{\partial z} + \frac{\partial w}{\partial y} \; \varepsilon_{xz} = \frac{\partial u}{\partial z} + \frac{\partial w}{\partial x} \qquad (11)$$

The definition of the RBF problem, i.e. the arrangement of the cloud of RBF sources and their input values, is the key enabler for RBF based mesh morphing and, considering its meshless nature and the great flexibility offered by the RBF mathematics, there are a variety of options to impose a desired morphing action. An overview of RBF mesh morphing strategies is given in [23, 24], while a deeper presentation about the use of RBF mesh morphing in industrial applications can be found in [25]. Among such a variety of mesh morphing paradigms the simpler one, which is the easiest to be automated, consists in the usage of all the nodes (or a subset) onto the surfaces as sources [26]. For sake of simplicity the study herein presented is based on such approach. The example of Fig. 1 demonstrates how the mesh of a notched bar can be updated controlling the radius of the notch. In this case the new CAD representation allows the user to define the positions of the RBF sources. An uniform spacing is imposed to all curves of the original geometry. The deformed positions of sources are then computed onto the updated CAD by keeping the same parametric distribution along the curve. The RBF field produced by source points is then used to update all the nodes of the mesh.

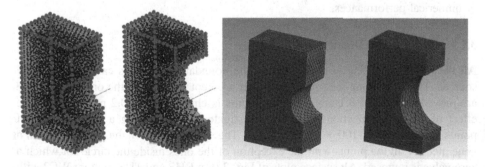

Fig. 1. The notched bar is reshaped adopting RBF mesh morphing. The nodes on all the curves are updated onto the new geometrical model keeping the same parametric spacing along the curves.

3 A Square Plate with a Circular Hole

The first application faced to understand how the different radial functions are capable to handle mesh morphing consists in a simple geometry: the square plate (1.0 m side) with a circular hole (0.2 m radius) represented in Fig. 2.

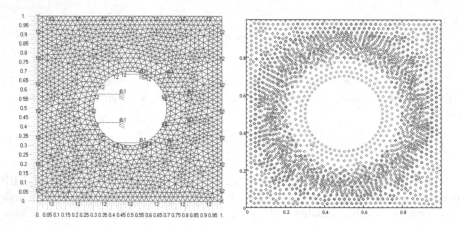

Fig. 2. Square plate with a circular hole: the nodes on the curves are used to define the RBF sources, the nodes of the mesh are the morphing targets. FEM mesh on the left, FEM nodes in original and morphed position on the right.

The implementation of this RBF mesh morphing demonstrator is divided in two parts. The mesh morphing set-up is performed using a pre-processor for FEA applications, the NX Femap in this case. The mesh is exported in a readable ASCII format (Nastran data deck in this example) so that the full mesh (nodes and elements) together with boundary conditions (constrained nodes with prescribed displacements) can be translated. The second part is a Mathcad application that implements basic RBF according to [5] and provides a fast bench to play around with the math and with obtainable morphed mesh. This implementation is good for investigate the method thanks to great flexibility provided, but it is not intended to be used for the assessment of numerical performances.

3.1 Changing the Diameter of the Hole

All the nodes belonging to the curves on the boundary are in this case used as RBF sources. The ones along the square perimeter are kept fixed while the ones on the hole are moved radially of 0.1 m so that the radius is changed from 0.2 m to 0.3 m. The effect of the radial functions is firstly investigated by generating a uniform map of points in the square (100 × 100 samples). The surface plot of the ε_y strain is then generated for all the points with the exception of the ones inside the circle for which a zero value is imposed. All surface plots of Fig. 3 (top BHS and THS, bottom WC2 with support radius 0.2 m and 0.1 m) have the same scale (−2, 1). We can clearly notice that, as expected, the maximum strain is 0.5 for all the situations: it is the hoop strain imposed at the circle and it occurs along the x direction. As far as the minimum strain, this is related to the compression of the cell required to accommodate morphing, and it has a peak along the y direction (pure radial on the circle). The minimum value represents the severity of element compression due to morphing. It clearly grows moving form BHS to THS and becomes more and more severe with WC2, especially if a small support radius is selected.

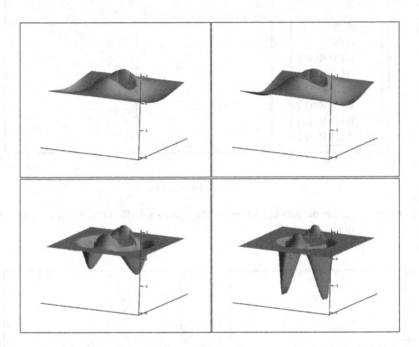

Fig. 3. Map of the ε_y strain for the BHS (top left), THS (top right) and WC2 with support radius 0.2 m (bottom left) and 0.1 m (bottom right)

A better insight can be gained by plotting the ε_y strain along the vertical segment that starts from the midpoint of the bottom side of the square and ends at the intersection with the circle (Fig. 4). In this comparison also the WC0 and WC4 functions are included. The value of the ε_y strain at the intersection with the circle for BHS, THS is respectively: 0.04, 0.109; its values with support radius 0.1 m for WC0, WC2, WC4 are respectively: 0.276, −0.242, −0.244; its values with support radius 0.2 m for WC0, WC2, WC4 are respectively: −0.093, −0.235, −0.238.

The ability of the RBF to succesfully adapt the mesh is investigated by plotting the minimum quality of the triangles (i.e. the ratio between the inner circle and the outer circle for each triangle) as a function of the deformation intensity (Fig. 5). The chart for BHS and THS is very similar so only BHS is plotted. The WC2 with radius 0.2 m is at the limit whilst the case of WC2 with radius 0.1 m (Fig. 6 right) produces negative quality at an half of the intensity.

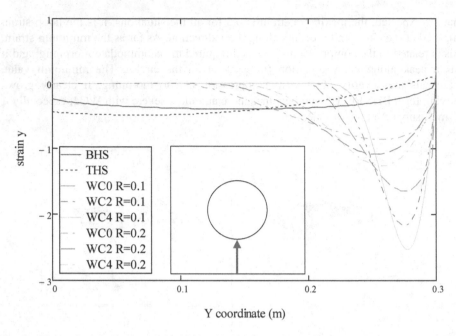

Fig. 4. Chart of ε_y along the path highlighted in the figure for BHS, THS, WC0, WC2 and WC4 (support radii 0.1 m and 0.2 m).

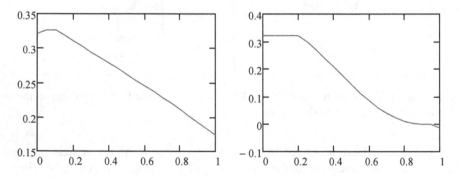

Fig. 5. Chart of minimum mesh quality as a function of the deformation intensity for the BHS (left) and for the WC2 with support radius 0.2 m (right)

To better understand the effect of the support radius for the WC2 a larger value of 0.4 m has been considered. Figure 6 (right) shows that in this case a valid mesh can be achieved for the full interval. The case of 0.1 m is plotted as well to show how the quality is strictly related to the support radius. It is worth to notice that such large support radius allows to add the long distance interactions; the benefit of generating sparse system will be in this case reduced or lost.

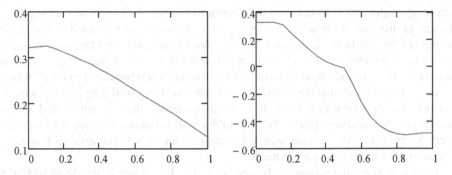

Fig. 6. Chart of minimum mesh quality as a function of the deformation intensity for the WC2 with support radius 0.4 m (left) and for the WC2 with support radius 0.1 m (right)

4   Industrial Example

The application shown in this section was tackled adopting an industrial software implementation: the RBF Morph module for the ANSYS Fluent CFD solver [25]. RBF Morph allows to set-up the RBF mesh morphing problems with a variety of methods including the simple ones relevant for this study. The software features a fast RBF solver that comes with specific acceleration for the BHS. A library of Wendland functions is available as well.

4.1   Ice Profile on an Aircraft Wing Airfoil

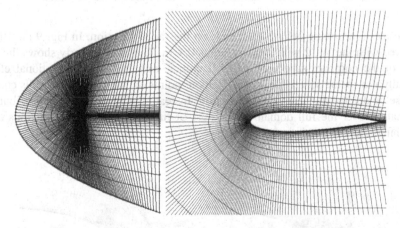

Fig. 7. Lateral view of the CFD mesh (left) with a detail of the mesh around the airfoil profile (right). Mesh vertexes at the walls are used as RBF sources (red points). (Color figure online)

The study presented in this section refers to the problem studied in [27, 28]. Mesh morphing is adopted to simulate the growth of ice; advanced CFD simulations coupled with icing models allow to compute the distribution of ice thickness onto the surfaces. The ice profile has an aerodynamic impact which cannot be neglected if the shape change is relevant. In the icing workflow the mesh is adapted onto the new "as iced" shape after a certain number of iterations. The CFD model of the 2D geometry is meshed in 3D as a single layer of hexahedrons; the mesh is represented in Fig. 7 where a detail of the mesh around the airfoil profile is shown. The chord length of the wing is 1 m and the thickness is 0.12 m. It is worth to notice that, as anticipated in the introduction, the spacing of the cells close to the wall is defined to solve the boundary layer. A key feature of the mesh adapted onto new shapes is to preserve such spacing similar to the one of the baseline mesh.

The mesh morphing set-up can be appreciated in Fig. 7 and in the detail of Fig. 8 where the shape of the ice profile is shown.

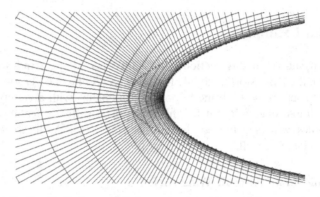

Fig. 8. Detail of the controlled sources that define the ice nose profile.

The mesh morphing behavior depends on the radial function. In Fig. 9 the BHS is compared with the WC2 with support radius 0.1 m. The figure clearly shows the local nature of the compact supported RBF. It comes with lighter computational efforts, especially if a sparse solver is adopted. Remote points on the inlet outlet walls could in this case be deleted as there is no interaction. Instead, the BHS allows to distribute the deformation over the full domain at the cost of full long distance interactions and a dense interpolation matrix.

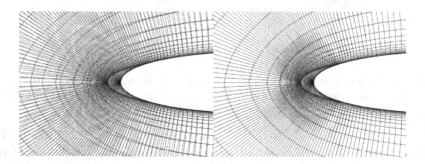

Fig. 9. Morphed mesh around the ice nose for the BHS (left) and the WC2 with support radius 0.1 m (right).

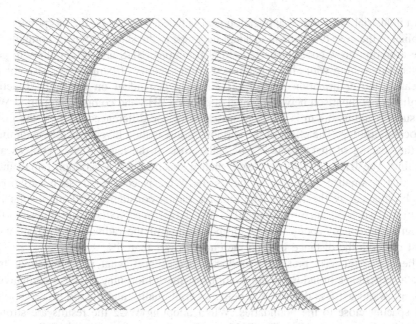

Fig. 10. Detail of the morphed mesh for the BHS (top left) and the WC2 with support radius 0.5 m (top right), 0.3 m (bottom left) and 0.1 m (bottom right).

The effect of the radial function can be visually inspected in Fig. 10 where a detail of the deformed mesh is represented in overlay with the baseline one. To have a quantitative metric of mesh distortion close to the wall, the distance from the surface of the fifth layer of cells is measured along the horizontal direction starting from the nose of the leading edge. Such distance on the baseline mesh is 21.7 mm and, after morphing, it becomes 22.3 mm for the BHS, 28.9 mm for the WC2 with radius 0.5 mm, 27.5 mm for the radius 0.3 mm, 17.5 mm for the radius 0.1 mm. Adopting a radius of 0.15 mm the distance becomes 22.7 mm.

5 Conclusions

In this paper a brief introduction about RBF mesh morphing is given with a special focus on the selection of the radial function. Among the many available options two of the most adopted ones have been considered in detail: the BHS and the WC2. The first example examined is a very simple mesh constituted of triangular elements representing a square plate with a circular hole. The deformation fields are in this case examined in details for BHS, WC2 with various support radii and for THS, WC0, WC4 as well. The study clearly shows that the radial function has an important effect on how deformations are distributed and on their peaks. A successful morphing is obtained with BHS and THS whilst the success of the WC2 depends from the radius. If too small values are adopted the morphed mesh results to have reversed cells. With increased radii (0.1 m, 0.2 m, 0.4 m) the filling ratio of the matrix increases (4.0%, 7.9%, 23.7%) together with the linear system solving effort.

An industrial aeronautical application is also faced. In this case the morphing action is applied to the nose of an airfoil to adapt its shape to the shape of the ice computed by an ice accretion grow code. Inspected radial functions (BHS and WC2) allow to get a successful morphing with all the examined combinations of parameters. Also for this application the deformation intensity is strongly dependent on the radial function. A qualitative comparison between global supported BHS and compact supported WC2 with support radius 0.1 m is firstly given showing the differences between the global and local morphing. The effect of the support radius is then investigated by examining how the spacing of first five layers of cells close to the leading edge evolves after morphing. The BHS allows to preserve such a spacing with a variation (small increment) that is smaller than 3%. As far as the WC2 is concerned, such variation is strongly affected by the radius. The smaller radius considered (0.1 m) results in a reduction of such distance of about 20%. The larger one produces an opposite effect with an increment of about 33%. A proper tuning of the radius (0.15 m) allows to almost preserve the distance with an increment lower than 5%.

The numerical tests presented in this paper support what is well known from practical experience in industrial applications of mesh morphing. BHS, if compared to WC2, allows to get less distortion, a better overall quality and a good preservation of mesh spacing close to the boundaries. When adopting WC2 the results are strongly affected by the support radius. The BHS, on the other hand, produces a full populated linear system and requires far source points to limit the morphing field not needed with the WC2 that, thanks to the compact support, produces a decaying morphing field.

References

1. Hardy, R.L.: Multiquadric equations of topography and other irregular surfaces. J. Geophys. Res. **76**(8), 1905–1915 (1971)
2. Hardy, R.L.: Theory and applications of the multiquadric-biharmonic method, 20 years of discovery, 1968–1988. Comput. Math Appl. **19**(8/9), 163–208 (1990)
3. Buhmann, M.D.: Radial Basis Functions: Theory and Implementation. Cambridge University Press, New York (2003)
4. Wendland, H.: Piecewise polynomial, positive definite and compactly supported radial basis functions of minimal degree. Adv. Comput. Math. **4**(1), 389–396 (1995). ISSN 1019-7168
5. Biancolini, M.E.: Fast Radial Basis Functions for Engineering Applications. Springer, Heidelberg (2017). https://doi.org/10.1007/978-3-319-75011-8
6. Cella, U., Biancolini, M.E., Wade, A.: RBF morph and ANSYS contribution to GMGW-2 (case 3 – OPAM-1). In: 2nd AIAA Geometry and Mesh Generation Workshop, San Diego, California (USA), 5–6 January 2019 (2019)
7. Groth, C., et al.: Structural validation of a realistic wing structure: the RIBES test article. Procedia Struct. Integrity **12**, 448–456 (2018)
8. Biancolini, M.E., Costa, E., Cella, U., Groth, C., Veble, G., Andrejašič, M.: Glider fuselage-wing junction optimization using CFD and RBF mesh morphing. AEAT **88**(6), 740–752 (2016)

9. Biancolini, M.E., Del Bene, C., Larsson, T., Groth, C.: Evaluation of go-kart aerodynamic efficiency using CFD, RBF mesh morphing and lap time simulation. Int. J. Aerodyn. **5**(3–4), 146–171 (2016)
10. Groth, C., Chiappa, A., Biancolini, M.E.: Shape optimization using structural adjoint and RBF mesh morphing. Procedia Struct. Integrity **8**, 379–389 (2018)
11. Papoutsis-Kiachagias, E.M., Porziani, S., Groth, C., Biancolini, M.E., Costa, E., Giannakoglou, K.C.: Aerodynamic optimization of car shapes using the continuous adjoint method and an RBF morpher. In: Minisci, E., Vasile, M., Periaux, J., Gauger, Nicolas R., Giannakoglou, Kyriakos C., Quagliarella, D. (eds.) Advances in Evolutionary and Deterministic Methods for Design, Optimization and Control in Engineering and Sciences. CMAS, vol. 48, pp. 173–187. Springer, Cham (2019). https://doi.org/10.1007/978-3-319-89988-6_11
12. Porziani, S., Groth, C., Biancolini, M.: Automatic shape optimization of structural components with manufacturing constraints. Procedia Struct. Integrity **12**, 416–428 (2018)
13. Biancolini, M., Cella, U., Groth, C., Genta, M.: Static aeroelastic analysis of an aircraft wind-tunnel model by means of modal RBF mesh updating. J. Aerosp. Eng. **29**(6), 04016061 (2016)
14. Groth, C., Cella, U., Costa, E., Biancolini, M.E.: Fast high fidelity CFD/CSM fluid structure interaction using RBF mesh morphing and modal superposition method. Aircr. Eng. Aerosp. Technol. **91**, 893–904 (2019)
15. Groth, C., et al.: The medical digital twin assisted by reduced order models and mesh morphing (2019)
16. Sederberg, T.W., Parry, S.R.: Free-form deformation of solid geometric models. In: Evans, D.C., Athay, R.J. (eds.) Proceedings of the 13th Annual Conference on Computer Graphics and Interactive Techniques (SIGGRAPH 1986), New York, NY, pp. 151–160 (1986)
17. Masud, A., Bhanabhagvanwala, M., Khurram, R.A.: An adaptive mesh rezoning scheme for moving boundary flows and fluid–structure interaction. Comput. Fluids **36**(1), 77–91 (2007)
18. Biancolini, M.E., Brutti, C., Pezzuti, E.: Shape optimisation for structural design by means of finite elements method. In: ADM, vol. 12 (2001)
19. Rendall, T.C.S., Allen, C.B.: Reduced surface point selection options for efficient mesh deformation using radial basis functions. J. Comput. Phys. **229**(8), 2810–2820 (2010)
20. Faul, A.C., Goodsell, G., Powell, M.J.D.: A Krylov subspace algorithm for multiquadric interpolation in many dimensions. IMA J. Numer. Anal. **25**, 1–24 (2005)
21. Wendland, H.: Fast evaluation of radial basis functions: methods based on partition of unity. In: Chui, C.K., Schumaker, L.L., Stöckler, J. (eds.) Approximation Theory X: Wavelets, Splines, and Applications, pp. 473–483. Vanderbilt University Press (2002)
22. Beatson, R.K., Powell, M.J.D., Tan, A.M.: Fast evaluation of polyharmonic splines in three dimensions. IMA J. Numer. Anal. **27**, 427–450 (2007)
23. Cella, U., Groth, C., Biancolini, M.E.: Geometric parameterization strategies for shape optimization using RBF mesh morphing. Advances on Mechanics, Design Engineering and Manufacturing. LNME, pp. 537–545. Springer, Cham (2017). https://doi.org/10.1007/978-3-319-45781-9_54
24. Botsch, M., Kobbelt, L.: Real-time shape editing using radial basis functions. In: Marks, J., Alexa, M. (eds.) Computer Graphics. Eurographics, vol. 24, no. 3 (2005)
25. Biancolini, M.E.: Mesh morphing and smoothing by means of radial basis functions (RBF): a practical example using fluent and RBF morph. In: Handbook of Research on Computational Science and Engineering: Theory and Practice, p. 34 (2012)
26. Sieger, D., Menzel, S., Botsch, M.: RBF morphing techniques for simulation-based design optimization. Eng. Comput. **30**(2), 161–174 (2014)

27. Biancolini, M.E., Groth, C.: An efficient approach to simulating ice accretion on 2D and 3D airfoils. In: Advanced Aero Concepts, Design and Operations (2014)
28. Groth, C., Costa, E., Biancolini, M.E.: RBF-based mesh morphing approach to perform icing simulations in the aviation sector. Aircr. Eng. Aerosp. Technol. **91**(4), 620–633 (2019)

Radial Basis Function Approximation Optimal Shape Parameters Estimation

Vaclav Skala[1][(✉)] , Samsul Ariffin Abdul Karim[2] ,
and Marek Zabran[1]

[1] Department of Computer Science and Engineering,
Faculty of Applied Sciences, University of West Bohemia,
Univerzitni 8, CZ 301 00 Plzen, Czech Republic
skala@kiv.zcu.cz
[2] Fundamental and Applied Sciences Department and Centre for Smart Grid
Energy Research (CSMER), Institute of Autonomous System, Universiti
Teknologi PETRONAS, Bandar Seri Iskandar, 32610 Seri Iskandar
Perak DR, Malaysia
samsul_ariffin@utp.edu.my

Abstract. Radial basis functions (RBF) are widely used in many areas especially for interpolation and approximation of scattered data, solution of ordinary and partial differential equations, etc. The RBF methods belong to meshless methods, which do not require tessellation of the data domain, i.e. using Delaunay triangulation, in general. The RBF meshless methods are independent of a dimensionality of the problem solved and they mostly lead to a solution of a linear system of equations. Generally, the approximation is formed using the principle of unity as a sum of weighed RBFs. These two classes of RBFs: global and local, mostly having a shape parameter determining the RBF behavior. In this contribution, we present preliminary results of the estimation of a vector of "optimal" shape parameters, which are different for each RBF used in the final formula for RBF approximation. The preliminary experimental results proved, that there are many local optima and if an iteration process is to be used, no guaranteed global optima are obtained. Therefore, an iterative process, e.g. used in partial differential equation solutions, might find a local optimum, which can be far from the global optima.

Keywords: Approximation · Radial basis function · RBF · Shape parameters · Optimal variable shape parameters

1 Introduction

Interpolation and approximation of acquired data is required in many areas. Usually a data domain is tessellated, i.e. meshed by triangles or tetrahedrons, using Delaunay tessellation [7]. However, there are some complications using this approach: Delaunay tessellation is not acceptable for some computational methods (other criteria for the shape of elements are required) and for computational and memory requirements in the case of higher dimensions. Also, if used for physical phenomena interpolation, there are additional problems with smoothness physical phenomena interpolation over

© Springer Nature Switzerland AG 2020
V. V. Krzhizhanovskaya et al. (Eds.): ICCS 2020, LNCS 12142, pp. 309–317, 2020.
https://doi.org/10.1007/978-3-030-50433-5_24

resulting triangles or tetrahedrons. In the case of approximation, i.e. when the input data are to be reduced with regard to the values given, standard mesh reduction methods are complex as they are primarily designed for surface representation.

Meshless methods based on radial basis function (RBF) are based on the principle partition of unity, in general [2, 8, 28]. They are used for computations [1], e.g. partial differential equations (PDE) [6, 30, 31], interpolation and approximation of given data [16, 17], which lead to implicit representation [9, 18, 19] or explicit representation [10–12], impainting removal [25, 26, 29] vector data approximation (fluids, etc.) [24]. In the following, the RBFs for explicit representation will be explored more deeply from the shape parameter behavior and its estimation.

The RBFs interpolation is defined as [6, 20]:

$$f(x) = \sum_{j=1}^{M} c_j \varphi(r_j) \quad r_j = \|x - x_j\| \tag{1}$$

where c_j are coefficients and $\varphi(r_j)$ is a RBF kernel, and $x \in R^d$ is an independent variable in d-dimensional space, in general. If scalar values are given in N points, i.e. $h_i = f(x_i)$, $i = 1, \ldots, N$, then a system of linear equations is obtained

$$h_i = f(x_i) = \sum_{j=1}^{M} c_j \varphi(r_j) \quad i = 1, \ldots, N \tag{2}$$

This leads to a system of linear equations $Ax = b$. If $N = M$ and all points are distinct, then we have an interpolation scheme, if $N > M$ than the approximation scheme is obtained, as the system of equations is overdetermined [6, 10, 13, 20, 21, 27].

However, in our case we use the interpolation scheme, i.e. $N = M$, and used only points of extrema, inflection points and some additional points, using the RBF interpolation scheme, to study estimation of a suboptimal selection of weights and shape parameters of the RBF interpolation. In this case, we obtain a system of linear equations:

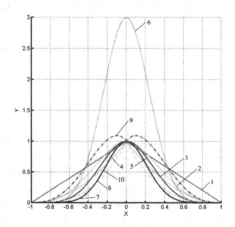

Fig. 1. CS RBF functions

$$
\begin{bmatrix} \varphi_{1,1} & \cdots & \varphi_{1,M} \\ \vdots & \ddots & \vdots \\ \varphi_{M,1} & \cdots & \varphi_{M,M} \end{bmatrix} \begin{bmatrix} c_1 \\ \vdots \\ c_M \end{bmatrix} = \begin{bmatrix} h_1 \\ \vdots \\ h_M \end{bmatrix} \tag{3}
$$

The RBF interpolation was originally introduced as a multiquadric method by Hardy [5] in 1971. Since then, many different RFB interpolation schemes have been developed with some specific properties [3, 4].

The RBF kernel functions can be split to two main groups (just some examples):

- Global functions:
  - Gauss - $\varphi(r,) = e^{-(\varepsilon r)^2}$,
  - Multiquadric (MQ) - $\varphi(r) = \sqrt{1 - (\varepsilon r)^2}$,
  - Inverse multiquadric (IMQ) - $\varphi(r) = 1/\sqrt{1 - (\varepsilon r)^2}$,
  - MQ-LG - $\varphi(r) = \frac{1}{9}(4c^2 + r^2)\sqrt{r^2 + c^2} - \frac{c^3}{3}\ln(c + \sqrt{r^2 + c^2})$ [31]
  - Thin Spline (TPS)
    - $\varphi(r) = r^k, k = 1, 3, 5, ..,$
    - $\varphi(r) = r^k \ln r, k = 2, 4, 6, ..$ etc.
- Local - Compactly Supported RBF (CSRBF), e.g.

$$
\varphi(r) = (1 - r/\varepsilon)^4_+ (1 + 4r/\varepsilon))\, 0 \leq r \leq 1
$$

"+" means zero if a function argument is out of the interval.
Some of those are presented at Table 1 and Fig. 1.
New RBF functions were recently introduced by Menandro [15].

Table 1. Typical examples of "local" functions – CSRBF (" + " means – value zero out of $\langle 0, 1 \rangle$)

ID	Function	ID	Function
1	$(1 - r)_+$	6	$(1 - r)^6_+ (35r^2 + 18r + 3)$
2	$(1 - r)^3_+ (3r + 1)$	7	$(1 - r)^8_+ (32r^3 + 25r^2 + 8r + 3)$
3	$(1 - r)^5_+ (8r^2 + 5r + 1)$	8	$(1 - r)^3_+$
4	$(1 - r)^2_+$	9	$(1 - r)^3_+ (5r + 1)$
5	$(1 - r)^4_+ (4r + 1)$	10	$(1 - r)^7_+ (16r^2 + 7r + 1)$

They generally depend on a shape parameter, which must be carefully set up. Unfortunately, global functions lead to ill-conditioned systems, while CSRBF causes "blobby" behavior [23].

In this contribution, a slightly different problem is solved. Consider given points of an explicit curve; we aim to approximate it, strictly complying with the following requirements for the approximated curve and we want to approximate it having strict the following requirements for the approximated curve:

- it has to pass through all points of extreme and also points of inflection
- it keeps the value at the data interval border

Usually, the approximation case is solved using least square error (LSE) methods. The LSE application leads to good results in general, however, the LSE cannot guarantee the above-stated requirements. On the other hand, signal theory says that the sampling frequency should be two times higher than the highest frequency in the data. Also, different parts of a signal might have different high frequencies.

Therefore, the signal reconstruction should follow the local properties of the signal, if it would respect only the global properties, the compression ratio of the approximated signal would be lower and the sampling theorem is fulfilled. However, in practical experiments error behavior at the interval borders is to be improved by adding two additional points close to the border.

It means, that formally standard RBF interpolation scheme is obtained in which only points of extrema and inflection, points at the interval borders and two additional points are taken into account. Additional points only slightly improve interpolation precision.

However, the majority of RBFs depends on the "magic" constant – a shape parameter, which has a significant influence to the robustness, stability and precision of computation. Usually, some standard estimation formulas are used for the constant shape parameter, i.e. all RBFs have the same shape parameter. In the following, the case, when each kernel RBF has a non-constant parameter is described.

2 Determination of RBF Shape Parameters

Signal reconstruction using radial basis functions has to respect the basic requirements stated above. The sampling frequency should be locally two times higher than the highest frequency as the sampling theorem says. Therefore, if the values at the points of extrema and points of inflections are respected together with the values at the interval border, the sampling theory is fulfilled [14].

The question of choosing the shape parameters is not considered, yet. The global constant shape parameter, i.e. for all RBFs in the RBF approximation is to be used, the minimization process can be used

$$\varepsilon = argmin\left\{ \left\| h_i - \sum_{j=1}^{M} c_j\varphi\left(r_{ij}, \varepsilon_j\right) \right\| \right\} \quad r_{ij} = \left\| x_i - x_j \right\| \tag{4}$$

where $\varepsilon_j = \varepsilon_k$ for all j, k, i.e. all shape parameters are the same.

However, if the shape parameters can differ from each other, the approximation will be more precise with fewer reference points, which also speed-up the RBF function evaluation. The question is, whether there is a unique optimum. Therefore, the "Monte-Carlo" approach was taken and the minimization process was initiated for different starting vector (uniformly generated using Halton's distribution) of shape parameters using Gauss function.

$$
\varepsilon = argmin \left\{ \left\| h_i - \sum_{j=1}^{M} c_j \varphi \left(r_{ij}, \varepsilon_j \right)^2 \right\| \right\} \quad
\begin{aligned}
r_{ij} &= \left\| x_i - x_j \right\| \\
\varepsilon &= [\varepsilon_1, \ldots, \varepsilon_M]^T
\end{aligned}
\tag{5}
$$

It can be seen, that finding optimum shape parameter vector is computationally expensive. In the following, only one representative example from experiments made is presented.

3 Experimental Results

Several explicit functions $y = f(x)$ have been used to understand the behavior of the optimal shape parameter vector determination (Table 2).

Table 2. Examples of testing functions.

ID	Function	ID	Function
1	$y = \sin(15x^2 + 5x)$	2	$y = \cos(20x)/2 + 5x$
3	$y = 50(0.4\sin(15x^2) + 5x)$	4	$y = \sin(8\pi x)$
5	$y = \sin(6\pi x^2)$	6	$y = \sin(25x + 0.1)/(25x + 0.1)$
7	$y = 2\sin(2\pi x) + \sin(4\pi x)$	8	$y = 2\sin(2\pi x) + \sin(4\pi x) + \sin(8\pi x)$
9	$y = 2\sin(\pi(2x - 1)) + \sin(3\pi(2x - 1/2))$	10	$y = 2\sin(\pi(1 - 2x)) + \sin(3\pi(2x - 1/2))$
11	$y = 2\sin(\pi(2x - 1)) + \sin(3\pi(2x - 1/2)) - x$	12	$y = 2\sin\left(2\pi x - \frac{\pi}{2}\right) + \sin(3\pi(2x - 1/2))$
13	$y = \text{atan}(10x - 5)^3 + \text{atan}(10x - 8)^3/2$	14	$y = (4.88x - 1.88)*\sin(4.88x - 1.88)^2 + 1$
15	$y = \exp(10x - 6)*\sin(5x - 2)^3 + (3x - 1)^3$	16	$y = \tanh(9x + 1/2)/9$

The experiments were implemented in MATLAB. The experiments proved, that there are several different shape parameters vectors, which give a local minimum of approximation error. In all cases, the error of the function values was less than 10^{-5}, with a minimum number of points.

In the following, just two examples of RBF approximation are presented, additional examples can be found at [32]. Figure 2 presents two functions and their RBF approximation using found points of importance. Figure 3a presents shape parameters for each local optima found; the blue one is for a starting vector with an optimal global shape constant for all RBFs. Figure 3b presents computed weights of the RBF approximation.

The radial distances in the graphs are transformed monotonically, but nonlinearly to obtain "visually reasonable" graphs as:

$$\varepsilon = 1 - e^{-2(\varepsilon - \varepsilon_{min})/(\varepsilon_{max} - \varepsilon_{min})} \qquad\qquad c = 1 - e^{-2(c - c_{min})/(c_{max} - c_{min})}$$

Transformation for the shape parameters   Transformation for the RBF weights
(all the values of c were taken as $abs(c)$)

It can be seen, that the shape parameters for each local optimum changes significantly, while the computed weights are similar. The experiments made on several explicit functions proved a hypothesis, that there are several local optima and for all of those a similar behavior was identified.

However, it leads to a serious question, how the RBFs use in the solution of ordinary and partial differential equations, in approximation and interpolation, etc. is a computationally reliable method, as results depend on "good" choice of the vector of shape parameters.

Detailed test results for several testing functions can be found at [32] http://wscg.zcu.cz/RBF-shape/contents-new.htm.

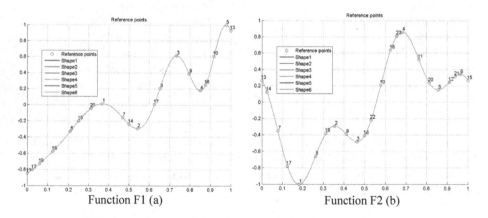

Function F1 (a)   Function F2 (b)

Fig. 2. Two examples of an approximated functions with approximation $error < 10^{-5}$

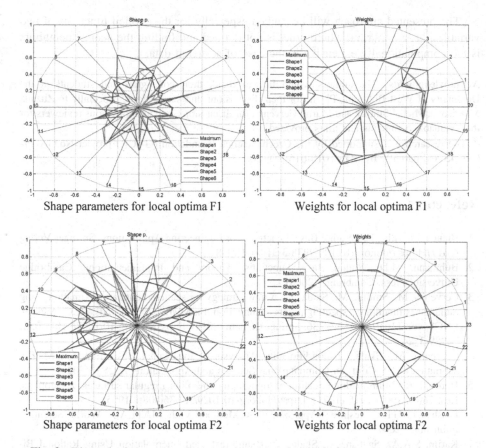

Shape parameters for local optima F1   Weights for local optima F1

Shape parameters for local optima F2   Weights for local optima F2

Fig. 3. Diagram of shape parameters vectors (a) and weights for local optima found (b)

4 Conclusion

In this contribution, we shortly described preliminary experimental results in finding optimal shape vector parameters, i.e. when each RBF has a different shape parameter, which leads to the higher precision of approximation of the given data. The experiments were made on different explicit curves to prove basic properties of the approach. The experiments proved that there are several local optima for the shape vector parameters, which leads to different precision of the final approximation. However, it should be stated, that the approximation was made with relatively high compression and the error was lower than 10^{-5}, which is in many cases acceptable. The given approach approximates data having different local frequencies will be explored in future together with an extension to explicit functions of two variables. The presented approach and results obtained should also have an influence to solution of partial differential equations (PDE), as the precision of a solution depends on good shape parameter selection.

The presented approach will be studied especially for the explicit functions of two variables, i.e. $z = f(x, y)$, where order is not defined and finding the nearest neighbors is too computationally expensive in the case of scattered data.

Acknowledgments. The authors would like to thank their colleagues and students at the University of West Bohemia, Plzen, for their discussions and suggestions, especially to Zuzana Majdisova, Michal Smolik and Martin Cervenka for discussions, and to anonymous reviewers for their valuable comments and hints provided.

The research was supported by the Czech Science Foundation (GACR) project No.GA 17-05534S and partially by SGS 2019-016 projects.

References

1. Biancolini, M.E.: Fast Radial Basis Functions for Engineering Applications. Springer Verlag (2017). https://doi.org/10.1007/978-3-319-75011-8
2. Buhmann, M.D.: Radial Basis Functions: Theory and Implementations. Cambridge University Press, Cambridge (2008)
3. Cervenka, M., Smolik, M., Skala, V.: A new strategy for scattered data approximation using radial basis functions respecting points of inflection. In: Misra, S., et al. (eds.) ICCSA 2019. LNCS, vol. 11619, pp. 322–336. Springer, Cham (2019). https://doi.org/10.1007/978-3-030-24289-3_24
4. Cervenka, M., Skala,V.: Conditionality Analysis of the Radial Basis Function Matrix, submitted to Int. Conf. on Computational Science and Applications ICCSA (2020)
5. Hardy, L.R.: Multiquadric equation of topography and other irregular surfaces. J. Geophys. Res. **76**(8), 1905–1915 (1971). https://doi.org/10.1029/JB076i008p01905
6. Fasshauer, G.E.: Meshfree Approximation Methods with MATLAB. World Scientific Publishing (2007)
7. Karim, S.A.A., Saaban, A., Skala, V.: Range-restricted Interpolation Using Rational Bi-Cubic Spline Functions with 12 Parameters, vol. 7, pp. 104992–105006. IEEE (2019). https://doi.org/10.1109/access.2019.2931454, ISSN:2169-3536
8. Lazzaro, D., Montefusco, L.B.: Radial Basis functions for multivariate interpolation of large data sets. J. Comput. Appl. Math. **140**, 521–536 (2002). https://doi.org/10.1016/S0377-0427(01)00485-X
9. Macedo, I., Gois, J.P., Velho, L.: Hermite interpolation of implicit surfaces with radial basis functions. Comput. Graph. Forum **30**(1), 27–42 (2011). https://doi.org/10.1109/SIBGRAPI.2009.11
10. Majdisova, Z., Skala, V.: A new radial basis function approximation with reproduction. In: CGVCVIP 2016, Portugal, pp. 215–222 (2016). ISBN 978-989-8533-52-4
11. Majdisova, Z., Skala, V.: Big Geo data surface approximation using radial basis functions: a comparative study. Comput. Geosci. **109**, 51–58 (2017). https://doi.org/10.1016/j.cageo.2017.08.007
12. Majdisova, Z., Skala, V.: Radial basis function approximations: comparison and applications. Appl. Math. Model. **51**, 728–743 (2017). https://doi.org/10.1016/j.apm.2017.07.033
13. Majdisova, Z., Skala, V., Smolik, M.: Determination of stationary points and their bindings in dataset using RBF Methods. In: Silhavy, R., Silhavy, P., Prokopova, Z. (eds.) CoMeSySo 2018. AISC, vol. 859, pp. 213–224. Springer, Cham (2019). https://doi.org/10.1007/978-3-030-00211-4_20

14. Majdisova, Z., Skala, V., Smolik, M.: Determination of reference points and variable shape parameter for RBF approximation. Integr. Compute. Aided Eng. **27**(1), 1–15, https://doi.org/10.3233/ica-190610, ISSN 1069-2509, IOS Press (2020)

15. Menandro, F.C.M.: Two new classes of compactly supported radial basis functions for approximation of discrete and continuous data. Eng. Rep. **1**, e12028 (2019). https://doi.org/10.1002/eng2.12028

16. Pan, R., Skala, V.: A two level approach to implicit modeling with compactly supported radial basis functions. Eng. Comput. **27**(3), 299–307 (2011). https://doi.org/10.1007/s00366-010-0199-1, ISSN 0177-0667, Springer

17. Pan, R., Skala, V.: Surface Reconstruction with higher-order smoothness. Vis. Comput. **28**(2), 155–162 (2012). https://doi.org/10.1007/s00371-011-0604-9, ISSN 0178-2789

18. Ohtake, Y., Belyaev, A., Seidel, H.-P.: A multi-scale approach to 3d scattered data interpolation with compactly supported basis functions. Shape modeling, IEEE, Washington, pp 153–161 (2003). https://doi.org/10.1109/smi.2003.1199611

19. Savchenko, V., Pasco, A., Kunev, O., Kunii, T.L.: Function representation of solids reconstructed from scattered surface points & contours. Comput. Graph. Forum **14**(4), 181–188 (1995). https://doi.org/10.1111/1467-8659.1440181

20. Skala, V.: RBF interpolation with CSRBF of large data sets, ICCS 2017. Procedia Comput. Sci. **108**, 2433–2437 (2017). https://doi.org/10.1016/j.procs.2017.05.081

21. Skala, V.: RBF interpolation and approximation of large span data sets. In: MCSI 2017 – Corfu, pp. 212–218. IEEE (2018). https://doi.org/10.1109/mcsi.2017.44

22. Skala, V., Karim, S.A.A., Kadir, E.A.: Scientific Computing and Computer Graphics with GPU: application of projective geometry and principle of duality. Int. J. Math. Comput. Sci. **15**(3), 769–777 (2020). ISSN 1814-0432

23. Skala, V.: High dimensional and large span data least square error: numerical stability and conditionality. Int. J. Appl. Phys. Math. **7**(3), 148–156 (2017). https://doi.org/10.17706/ijapm.2017.7.3.148-156, ISSN 2010-362X, IAP, California, USA

24. Smolik, M., Skala, V.: Large scattered data interpolation with radial basis functions and space subdivision. Integr. Comput. Aided Eng. **25**(1), 49–62 (2018). https://doi.org/10.3233/ICA-170556

25. Smolik, M., Skala, V., Majdisova, Z.: Vector field radial basis function approximation. Adv. Eng. Softw. **123**(1), 117–129 (2018). https://doi.org/10.1016/j.advengsoft.2018.06.013

26. Uhlir, K., Skala, V.: Reconstruction of damaged images using radial basis functions. In: EUSIPCO 2005 Conference Proceedings, Turkey (2005). ISBN 975-00188-0-X

27. Vasta, J., Skala, V., Smolik, M., Cervenka, M.: Modified radial basis functions approximation respecting data local features. In: Informatics 2019, IEEE proceedings, Poprad, Slovakia, pp. 445–449 (2019). ISBN 978-1-7281-3178-8

28. Scattered Data Approximation. Cambridge University Press (2010). https://doi.org/10.1017/cbo9780511617539

29. Wright, G.B.: Radial Basis Function Interpolation: Numerical and Analytical Developments. University of Colorado, Boulder. Ph.D. thesis (2003)

30. Zhang, X., Song, K.Z., Lu, M.W., Liu, X.: Meshless methods based on collocation with radial basis functions. Comput. Mech. **26**(4), 333–343 (2000). https://doi.org/10.1007/s004660000181

31. Zheng, H., Yao, G., Kuo, L.H., Li, X.: On the selection of good shape parameter of the localized method of approximated particular solution. Adv. Appl. Math. Mech. **10**, 896–911 (2018). https://doi.org/10.4208/aamm.oa-2017-0167, ISSN 2075-1354

32. Detailed results of test available at http://wscg.zcu.cz/RBF-shape/contents-new.htm

Multiscale Modelling and Simulation

Projective Integration for Moment Models of the BGK Equation

Julian Koellermeier[✉] and Giovanni Samaey

Department of Computer Science, NUMA, KU Leuven,
Celestijnenlaan 200A, 3001 Leuven, Belgium
julian.koellermeier@kuleuven.be

Abstract. In this paper, we apply the projective integration method
to moment models of the Boltzmann-BGK equation and investigate the
numerical properties of the resulting scheme. Projective integration is an
explicit, asymptotic-preserving scheme that is tailored to problems with
a large spectral gap between fast and slow eigenvalues of the model.
A spectral analysis of the moment model shows a clear spectral gap
and reveals the multi-scale nature of the model. The new scheme over-
comes the severe time step constraint of standard explicit schemes like
the forward Euler scheme by performing a number of inner iterations and
then extrapolating the solution forward in time. The projective integration scheme is non-intrusive and yields fast and accurate solutions, as
demonstrated using a 1D shock tube test case. These observations open
up many possibilities for further use of the scheme for high-resolution
discretizations and different collision models.

Keywords: Kinetic theory · Moment model · Asymptotic preserving

1 Introduction

The Boltzmann BGK equation is widely used to model applications in kinetic
theory, such as rarefied gases [15,19]. It is a high-dimensional equation for the
evolution of the particle density function that describes the distribution of particles in position-velocity phase space. An efficient discretization of the BGK
equation is necessary due to its high-dimensionality. A simple mesh-based discretization in the velocity dimensions, the so-called Discrete Velocity Method
(DVM) typically requires a large number of variables and thus leads to a large
system of equations [1,15]. Another approach is the discretization by means
of moments [19,23]. Moments are integrals of some quantities of interest with
respect to the distribution functions over the velocity dimensions. These moment
models lead to smaller PDE systems and allow for more physical insight.

Both for the moment models and the DVM, the resulting PDE models are
in balance law form and typically contain a stiff right-hand side governed by a

---

J. Koellermeier—Supported by Research Foundation - Flanders (FWO), grant no.
0880.212.840.

V. V. Krzhizhanovskaya et al. (Eds.): ICCS 2020, LNCS 12142, pp. 321–333, 2020.
https://doi.org/10.1007/978-3-030-50433-5_25

smallness parameter ϵ. For small values of ϵ, which correspond to flow conditions close to the continuum regime, suitable time-stepping schemes need to be applied to overcome the severe time step constraint posed by the stiff right-hand side.

Implicit time stepping methods can in principle be used with a very large time step, but they are costly because they require the solution of a large non-linear system of equations in each time step. As they use the information of all spatial grid points for the update of each single unknown, this method may not reflect the properties of the underlying BGK equation, which is a hyperbolic transport problem describing a directed transport of information in our application.

Explicit time stepping methods on the other hand are fast and straight-forward to apply but typically suffer from a severe time step constraint when the stiff right-hand side function features multiple scales. The stability condition then poses a more severe restriction to the time step size than the typical CFL constraint, which might otherwise be sufficient to acquire a desired level of accuracy.

Several methods have been developed to overcome these problems. For example, IMEX Runge-Kutta methods split the right hand side in a stiff and a non-stiff part and apply the respective explicit and implicit solvers to each of the parts to allow for an efficient solution [5,16]. However, they need to be tailored to the specific form of the right-hand side. In the context of discrete velocity methods, we refer to [22] for an efficient method to evaluate the full Boltzmann operator and [20,21] which both make use of a splitting into a relaxation and a transport stages for near-equilibrium flows.

In this paper we will use projective integration, a fully explicit, potentially high-order scheme that is asymptotic preserving [14]. The asymptotic preserving property means that the time step constraint of the projective integration scheme does not depend on the stiffness parameter ϵ in the limit where ϵ tends to zero [8]. Instead, a standard CFL-type time step constraint will be sufficient to allow for accurate and fast solutions. The scheme is furthermore non-intrusive and can be used with different collision models and spatial discretizations.

In this paper we present the first application of a first order scheme using projective integration for moment models. We thus combine the efficient discretization in velocity space using moment models with the asymptotic preserving discretization in time using projective integration. This is a first step towards high-order schemes in future work.

The rest of this paper is organized as follows: in Sect. 2 we present a fully non-linear, hyperbolic moment model and a linearized version thereof. In Sect. 3 we analyse the spectrum of the linearized, semi-discrete system to determine the parameters of the projective integration scheme in the following Sect. 4. Simulation results for a shock tube test case are presented in Sect. 5. The paper ends with a short conclusion.

2   Moment Models for the Boltzmann BGK Equation

The Boltzmann BGK equation is used to model rarefied gases. It describes the evolution of the particles' mass density distribution function $f(t, x, c)$, where

$x, c \in \mathbb{R}^d$ denote the physical space and the velocity space, respectively. For the rest of this paper we consider the 1D case $(d = 1)$ so that the kinetic equation reads

$$\frac{\partial}{\partial t} f(t, x, c) + c \frac{\partial}{\partial x} f(t, x, c) = \frac{1}{\epsilon} (f_M - f),\tag{1}$$

The right-hand side operator models collisions. Several collision models are possible, but we use the simplified BGK model [2], which describes a relaxation with relaxation time $\epsilon \in \mathbb{R}_+$ towards the local Maxwellian $f_M(t, x, c)$ given by

$$f_M(t, x, c) = \frac{\rho(t, x)}{\sqrt{2\pi\theta(t, x)}} \exp\left(-\frac{(c - u(t, x))^2}{2\theta(t, x)}\right).\tag{2}$$

The macroscopic quantities density $\rho(t, x)$, bulk velocity $u(t, x)$ and temperature $\theta(t, x)$ are so-called *moments* of the distribution function $f(t, x, c)$ in velocity space and they are computed via integration over velocity space

$$\rho(t, x) = \int_{\mathbb{R}} f(t, x, c)\, dc,\tag{3}$$

$$\rho(t, x) u(t, x) = \int_{\mathbb{R}} c f(t, x, c)\, dc,\tag{4}$$

$$\rho(t, x) \theta(t, x) = \int_{\mathbb{R}} |c - u|^2 f(t, x, c)\, dc.\tag{5}$$

For vanishing ϵ, Eq. (1) only allows solutions that are in the kernel of the right-hand side collision operator. These solutions are given by the standard compressible Euler equations for ρ, u and θ. The dynamics of the kinetic equation is governed by the relaxation speed ϵ with propagation speeds proportional to $\frac{1}{\epsilon}$, as we will see in Sect. 5, whereas the macroscopic quantities might evolve on a much slower time scale. We are interested in numerical schemes that allow for an efficient solution in the limit $\epsilon \to 0$.

2.1 Hyperbolic Quadrature-Based Moment Equations (QBME)

Equation(1) is difficult to solve because of the additional microscopic velocity dimension. A direct discretization of the microscopic velocity variable, the so-called Discrete Velocity Method (DVM) is possible, but leads to a large system of equations [1,15]. A more efficient choice of variables is the discretization via moments leading to a smaller set of moment equations [23].

The distribution function f is therefore expanded in a series of basis functions multiplied with basis coefficients around some equilibrium state. In this paper the expansion will be performed around local equilibrium (2) first, before we consider a linearized version in the next section. In [7] the expansion uses a series of Hermite basis functions

$$f(t, x, c) = \sum_{\alpha=0}^{M} f_\alpha(t, x) \mathcal{H}_\alpha^{[u(t,x), \theta(t,x)]}(c),\tag{6}$$

for coefficients (also called moments) $f_\alpha(t, x)$, $\alpha \in [0, M]$, and weighted Hermite basis functions $\mathcal{H}_\alpha^{[u,\theta]}$

$$\mathcal{H}_\alpha^{[u,\theta]}(c) = (-1)^\alpha \frac{d^\alpha}{dc^\alpha} \omega^{[u,\theta]}(c), \quad \alpha \geq 0, \quad \omega^{[u,\theta]}(c) = \frac{1}{\sqrt{2\pi\theta}} \exp\left(-\frac{(c-u)^2}{2\theta}\right). \tag{7}$$

We note that the coefficients encode information about how much the distribution function deviates from the equilibrium distribution function, which is given by the Maxwellian in Eq. (2). The more coefficients are used, the better the deviations can be represented in the weighted Hermite basis. It is the goal of a moment model to use relatively few variables in comparison to a standard discretization of the velocity space while still allowing for high model accuracy. It can be attributed to the fact that the coefficients are a more clever choice of variables than normal point values of the distribution function.

The macroscopic variables as defined in Eq. (3) result in the following additional constraints that can be easily applied

$$f_0 = \rho, \, f_1 = f_2 = 0. \tag{8}$$

Substituting expansion (6) into the kinetic equation (1) a system of equations can be derived by either matching coefficients of the basis functions [6] or by multiplying with a set of test functions (also called testing) [10]. The result is an explicit set of PDEs in space and time for the unknown coefficients f_α of the expansion and the additional macroscopic variables. This set of PDEs is called the moment model.

The whole set of variables reads $\boldsymbol{w}_M = (\rho, u, \theta, f_3, f_4, \ldots, f_M)^T \in \mathbb{R}^{M+1}$. Using this definition the moment system can be given by

$$\frac{\partial \boldsymbol{w}_M}{\partial t} + \mathbf{A}(\boldsymbol{w}_M) \frac{\partial \boldsymbol{w}_M}{\partial x} = -\frac{1}{\epsilon} \boldsymbol{S}(\boldsymbol{w}_M), \tag{9}$$

with system matrix $\mathbf{A}(\boldsymbol{w}_M) \in \mathbb{R}^{(M+1)\times(M+1)}$ depending on the variables. The vector $\boldsymbol{S}(\boldsymbol{w}_M) \in \mathbb{R}^{M+1}$ results from the corresponding discretization of the right-hand side collision operator [12].

Following the derivation in [6], the so-called Grad model's system matrix $\mathbf{A}$ reads

$$\mathbf{A}(\boldsymbol{w}_M) = \begin{pmatrix} u & \rho & & & & & \\ \frac{\theta}{\rho} & u & 1 & & & & \\ & 2\theta & u & \frac{6}{\rho} & & & \\ & 4f_3 & \frac{\rho\theta}{2} & u & 4 & & \\ -\frac{\theta f_3}{\rho} & 5f_4 & \frac{3f_3}{2} & \theta & u & 5 & \\ \vdots & \vdots & \vdots & & \vdots & \ddots & \ddots & \ddots \\ -\frac{\theta f_{M-2}}{\rho} & Mf_{M-1} & \frac{(M-2)f_{M-2}+\theta f_{M-4}}{2} & -\frac{3f_{M-3}}{\rho} & & \theta & u & M \\ -\frac{\theta f_{M-1}}{\rho} & (M+1)f_M & \frac{(M-1)f_{M-1}+\theta f_{M-3}}{2} & -\frac{3f_{M-2}}{\rho} & & & \theta & u \end{pmatrix}, \tag{10}$$

where all other entries in the system matrix $\mathbf{A}$ are set to zero.

The right-hand side S is given by

$$\mathbf{S}(\boldsymbol{w}_M) = (0, 0, 0, f_3, f_4, \ldots, f_M). \tag{11}$$

Unfortunately, Grad's system (9) is not hyperbolic, as shown in [3,4]. Numerical simulations using this model might suffer from instabilities and break down due to nonphysical values [3,12]. New hyperbolic models have recently been developed that overcome these problems and make an efficient solution of Eq. (1) possible. In this paper, we will use the Quadrature-Based Moment Equations (QBME), first developed in [10]. The model was derived using a different framework in [6] and was extended to the multi-dimensional case in [11]. It requires only a small modification of the Grad model in (10) to ensure hyperbolicity while still preserving the conservation laws and the most important physical properties of the model. For more information we refer to [9].

In comparison to Grad's model, the QBME model adds regularization terms in the last and the second to last equation

$$\mathbf{A}_{\mathrm{QBME}}(M, 3) = \frac{(M-2) f_{M-2} + \theta f_{M-4}}{2} - \frac{M(M+1)}{2} \frac{f_M}{\theta}, \tag{12}$$

$$\mathbf{A}_{\mathrm{QBME}}(M+1, 3) = -f_{M-1} + \frac{f_{M-3}\theta}{2}, \tag{13}$$

$$\mathbf{A}_{\mathrm{QBME}}(M+1, 4) = -\frac{3 f_{M-2}}{\rho} + \frac{3(M+1) f_M}{\rho\theta}. \tag{14}$$

Due to the additional terms, the two changed equations can no longer be written in conservative form. However, the effect of the non-conservative form on the solution was investigated in [3,12] and it was found that stable and accurate non-equilibrium solutions can be computed despite the non-conservative form. In addition, we expect no problems for the equilibrium case $\epsilon \to 0$, as the non-conservative terms then vanish.

2.2   Hermite Spectral Method (HSM)

The expansion (6) is performed around local equilibrium (ρ, u, θ), which leads to a non-linear model. A simpler, linear model can be derived when considering the expansion around a Gaussian distribution function instead. The expansion then reads

$$f(t, x, c) = \sum_{\alpha=0}^{M} f_\alpha(t, x) \mathcal{H}_\alpha(c), \tag{15}$$

where the weighted Hermite basis functions $\mathcal{H}_\alpha$ are defined as

$$\mathcal{H}_\alpha(c) = \frac{1}{\sqrt{2\pi}} \exp\left(-\frac{c^2}{2}\right) He_\alpha(c) \cdot \frac{1}{\sqrt{2^\alpha \alpha!}}, \tag{16}$$

where He_α is the standard Hermite polynomial of degree α and the last factor is chosen for normalization of the basis functions.

Similar to the non-linear model, the following constraints hold for the linear model

$$f_0 = \rho, \quad f_1 = \rho u, \quad f_2 = \frac{1}{\sqrt{2}} \left(\rho\theta + \rho u^2 - \rho \right). \tag{17}$$

The linear moment model is then derived in the same way as the non-linear model, by testing Eq. (1), and can be written as

$$\frac{\partial \boldsymbol{w}_M}{\partial t} + \mathbf{A} \frac{\partial \boldsymbol{w}_M}{\partial x} = -\frac{1}{\epsilon} \boldsymbol{S} \left(\boldsymbol{w}_M \right), \tag{18}$$

using a constant system matrix $\mathbf{A} \in \mathbb{R}^{(M+1) \times (M+1)}$ given by

$$\mathbf{A} = \begin{pmatrix} & 1 & & & \\ 1 & & \sqrt{2} & & \\ & \sqrt{2} & & \ddots & \\ & & \ddots & & \sqrt{M} \\ & & & \sqrt{M} & \end{pmatrix}. \tag{19}$$

The right-hand side vector $\boldsymbol{S} \left(\boldsymbol{w}_M \right) \in \mathbb{R}^{M+1}$ is given by

$$\boldsymbol{S}_\alpha = \int_{\mathbb{R}} f(t, x, c) \psi_\alpha(c) \, dc, \quad \text{for } \psi_\alpha(c) = He_\alpha(c) \cdot \frac{1}{\sqrt{2^\alpha \alpha!}}, \tag{20}$$

using the ansatz from (15) and can be computed analytically beforehand. We omit the details of the derivation here for conciseness.

3   Reference Splitting Scheme

In the case of the simple BGK collision operator, a first-order time splitting scheme can be easily implemented. In a splitting scheme the computation of a single time step with step size $\Delta t = t^{n+1} - t^n$ is formally split into two separate steps that are performed each after the other as follows

$$\begin{aligned} 1. \quad & \frac{\partial \boldsymbol{w}_M}{\partial t} + \mathbf{A} \frac{\partial \boldsymbol{w}_M}{\partial x} = \mathbf{0}, \\ 2. \quad & \frac{\partial \boldsymbol{w}_M}{\partial t} = \mathbf{S} \left(\boldsymbol{w}_M \right), \end{aligned} \tag{21}$$

so that the first step solves only the hyperbolic transport part of the PDE and the second step solves only the relaxation with the right-hand side collision term.

The second step can be solved exactly for the simple BGK model, both in the linear and in the non-linear case. We refer to [9] for more details. However, a splitting scheme highly depends on the specific form of the right-hand side operator and can become difficult for different (and more realistic) collision operators. Furthermore, a splitting scheme is not easily extendable to higher-order accuracy, which is a significant disadvantage if high-order solutions are necessary. In this paper we will thus use the splitting scheme only as a reference method to compare it to our new projective integration scheme for moment models. For this first application we will use a standard first-order splitting (see [12] for more details) to compare it with the first-order projective integration method.

4   Projective Integration Scheme

Projective Integration (PI) is an asymptotic preserving time stepping scheme consisting of an inner integrator and an extrapolation step [14]. The first order PI scheme uses the standard forward Euler method as inner integrator, but higher order schemes using Runge-Kutta methods have been derived, see [13]. For the description of the scheme we use that the semi-discrete version of the model equation after spatial discretization is given by

$$\frac{\partial \boldsymbol{w}_M}{\partial t} = D_t\left(\boldsymbol{w}_M\right), \quad D_t\left(\boldsymbol{w}_M\right) = -D_x\left(\boldsymbol{w}_M\right) + \frac{1}{\epsilon}S\left(\boldsymbol{w}_M\right), \qquad (22)$$

where the term D_x is the result of the spatial discretization and the second term represents the collision operator. Note that this form does not rely on any special form of the collision operator and it can also be used for models other than the BGK model.

As inner integrator we use the explicit forward Euler scheme with time step size δt for $K + 1$ steps

$$\boldsymbol{w}_M^{n,k+1} = \boldsymbol{w}_M^{n,k} + \delta t D_t\left(\boldsymbol{w}_M^{n,k}\right), \quad k = 0, 1, \dots, K. \qquad (23)$$

After the $K + 1$ inner steps a discrete derivative using the last two values is obtained and used in an outer step to compute the value at the new time step $\boldsymbol{w}_M^{n+1}$ via extrapolation in time

$$\boldsymbol{w}_M^{n+1} = \boldsymbol{w}_M^{n,K+1} + (\Delta t - (K+1)\delta t)\frac{\boldsymbol{w}_M^{n,K+1} - \boldsymbol{w}_M^{n,k+1}}{\delta t}. \qquad (24)$$

5   Spectral Analysis

The dynamic behavior of the semi-discrete system in Eq. (22) is governed by the eigenvalue spectrum of the right-hand side function $D_t\left(\boldsymbol{w}_M\right)$. In [14] a spectral analysis was performed for the similar DVM model and a spectral gap could be shown analytically. Here, we do something similar numerically for the moment model. For the linear moment model from Eq. (18) using $M = 5$ and the UPRICE spatial discretization for 400 spatial discretization points, Fig. 1 shows the spectrum of the right-hand side operator.

The spectrum shows a clear spectral gap, ideally suited for projective integration. The spectral gap increases with smaller ϵ, such that the time step constraint for a standard explicit Euler method becomes more and more severe. However, projective integration can be used to overcome that time step constraint.

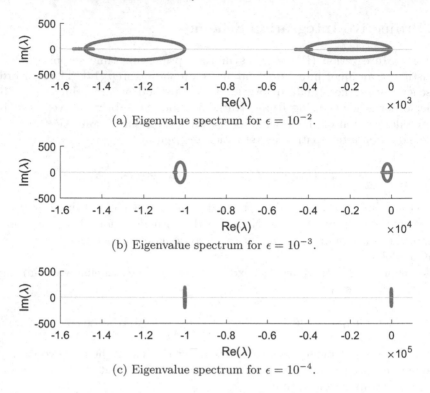

(a) Eigenvalue spectrum for $\epsilon = 10^{-2}$.

(b) Eigenvalue spectrum for $\epsilon = 10^{-3}$.

(c) Eigenvalue spectrum for $\epsilon = 10^{-4}$.

Fig. 1. The eigenvalue spectra for $\epsilon = 10^{-2}, 10^{-3}, 10^{-4}$ clearly show an increasing spectral gap, which is ideally suited for the application of projective integration.

Following the stability criterion of the projective integration scheme [14] and the eigenvalue spectrum of the moment model, we can choose the inner time step size as $\delta t = \epsilon$. Furthermore, we choose $K = 2$ so that three inner time steps are performed before the extrapolation step.

6   Computational Speedup

In this paper projective integration is used to speed up simulations of moment models close to equilibrium, where the stiffness of the model equation would normally require an extremely small time step size. Here we want to derive an estimate of the speedup factor when using projective integration in comparison to a standard time stepping method.

A standard time stepping method needs to resolve the fast eigenvalues, the time step size is thus $\Delta t \sim \epsilon$. This method then needs $\frac{1}{\Delta t} \sim \frac{1}{\epsilon}$ time steps to compute the solution over a unit time interval.

The projective integration method only does $K + 1$ small inner time steps and then extrapolates over the remainder of a standard CFL time step size

$\Delta t \sim \Delta x$. The projective integration method then needs $\frac{K+1}{\Delta t} \sim \frac{K+1}{\Delta x}$ time steps to compute the solution over a unit time interval.

Neglecting the computational overhead of the extrapolation step, the speedup S can be computed by the ratio of the respective number of time steps that need to be performed. For the projective integration method, we get

$$S = \frac{time\ steps\ standard\ method}{time\ steps\ projective\ integration} = \frac{\Delta x}{(K+1)\cdot \epsilon}. \tag{25}$$

In the test cases below, we used $\Delta x = 0.001$, $K = 2$ and up to $\epsilon = 10^{-5}$. In this case, the speedup can thus be estimated as $S = \frac{100}{3}$. This is mainly due to the larger time step of the projective integration method, which uses an outer time step size $\Delta t \approx 0.001$ versus the standard method, which uses the finer $\Delta t = 10^{-5}$.

We note that the speedup will be significantly higher for smaller values of ϵ or coarser spatial discretization, i.e. larger Δx. This will be possible when using higher-order spatial discretization.

7   Simulation Results

7.1   Shock-Tube Test Case Setup

For the numerical tests in this paper we consider a 1D shock tube test case, which is a standard benchmark problem in rarefied gases, see [3,12]. The shock tube features a strong shock wave propagating forward. Close to the shock, the solution will be in non-equilibrium if the relaxation time ϵ is large. However, for small relaxation time, the solution will quickly relax to the equilibrium Maxwellian and in the limit it can be derived easily by the well known Euler equations. It is important for any model, to correctly predict the limit of vanishing relaxation time, as large parts of any simulation will typically feature equilibrium conditions. This is exactly the region in which the kinetic Boltzmann-BGK equation becomes stiff and is difficult to solve. We are thus interested in a speedup of moment models for simulations close to equilibrium like in this test case.

At $t = 0$, the gas is in exact equilibrium, whereas the density, velocity, and temperature are given by

$$(\rho, u, \theta) = \begin{cases} (7,0,1) & \text{if } x < 0 \\ (1,0,1) & \text{if } x > 0 \end{cases}, \tag{26}$$

modeling a jump in density at the discontinuity at $x = 0$.

The computational domain is $[-2,2]$. The simulations run until $t_{END} = 0.3$ and $\Delta t = 0.0001$ corresponds to a CFL number of approximately 0.45 on a spatial grid discretized with 4000 cells as used in [12]. The first order, path conservative UPRICE method from [12] is used for the spatial discretization, see [17,18]. The scheme needs the eigenvalues of the system matrix, but not the eigenvectors. It is therefore computationally more expensive for the non-linear model from Sect. 2.1, as the eigenvalues change throughout the simulation. The linear model from Sect. 2.2 has a constant system matrix and thus only requires one eigenvalue computation in the beginning of the simulation. Note that higher-order spatial discretization methods are available (for examples see [9], but we will focus on the first order scheme for the first application of the new projective integration method.

According to the spectral analysis in Sect. 5, we will use $\delta t = \epsilon$ for the inner time step size use $K = 2$ for the number of inner time steps.

7.2 Hermite Spectral Method

First we verify that the projective integration method results in the same solution as the reference method, which uses the splitting scheme. For both schemes we use the same HSM model (18) with $M = 10$ on a grid using 4000 cells and outer time step size $\Delta t = 0.0001$, so that the only cause of difference could be the time stepping method. Figure 2 shows the results for both $\epsilon = 10^{-2}$ and $\epsilon = 10^{-5}$. There is no visual difference between the reference method and the projective integration method for both values of ϵ.

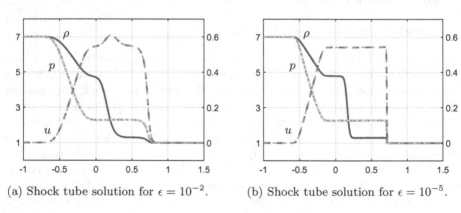

(a) Shock tube solution for $\epsilon = 10^{-2}$.   (b) Shock tube solution for $\epsilon = 10^{-5}$.

Fig. 2. Reference method (grey) and projective integration (colored) yield same accuracy. Left axis is for ρ and $p = \rho\theta$, right axis is for u.

With the help of the projective integration method we can thus get a reliable solution for smaller and smaller values of ϵ. In Fig. 3, we compare the different solutions and can see that the limit approaches the shock structure of a standard Euler equations model. This has been shown for the discrete velocity model in [14]. It is important to note that the runtime of the new projective integration scheme for the HSM moment model does not depend on ϵ and the scheme is thus asymptotic preserving in terms of computational cost. A standard explicit forward Euler scheme would lead to an increasing number of time steps for smaller values of ϵ to ensure stability.

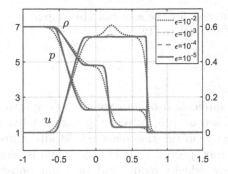

Fig. 3. The projective integration solution of HSM model approaches the Euler shock structure for vanishing ϵ. The runtime of the projective integration scheme does not depend on ϵ.

7.3 Hyperbolic Moment Equations

The projective integration method can readily be extended towards using the non-linear QBME model in (9), as it requires no special treatment of the right-hand side collision operator. In this case the non-linearity of the system matrix is not a problem, as the system is close to equilibrium. We choose the standard five moment case $M = 4$ [3,9] and use the same settings as for the linear model from before to obtain stability from the previous linear stability analysis. The solutions in Fig. 4 show that the QBME model gives the same solution as the HSM model for this range of the parameter ϵ despite the simplicity of the linear model. This is due to the small value of ϵ that leads to solutions very close to equilibrium with no significant rarefaction effects. Due to that, the linear model yields accurate solutions and can be used equivalently for the more complex, non-linear model. We note that this is not the case for large values of ϵ, but this

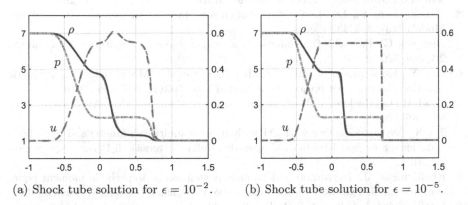

(a) Shock tube solution for $\epsilon = 10^{-2}$.   (b) Shock tube solution for $\epsilon = 10^{-5}$.

Fig. 4. HSM (grey) shows the same accuracy as QBME model (colored) for both $\epsilon = 10^{-2}$ and $\epsilon = 10^{-5}$.

is not the regime of interest for the scope of this paper. For simulations further in the kinetic regime, we refer to [12].

8   Conclusion

We showed the first application of an explicit, asymptotic preserving scheme for moment models based on projective integration. A linear stability analysis showed that the projective integration scheme is ideally suited for the eigenvalue structure due to the clear spectral gap. The spectrum is essentially the same as in the case of a discrete velocity model and the same parameters can be used. The projective integration scheme was validated in comparison to a splitting scheme that is specifically tailored to the type of the collision operator. However, the projective integration scheme is non-intrusive and does not depend on the implementation of the right-hand side collision operator. It can thus be used in further tests and applications using a more advanced collision operator. We have consistently extended the projective integration scheme to the non-linear moment model and achieved fast and accurate solutions. This allows for further applications of the scheme to full high-order simulations using the non-linear model.

References

1. Baranger, C., Claudel, J., Herouard, N., Mieussens, L.: Locally refined discrete velocity grids for deterministic rarefied flow simulations. AIP Conf. Proc. **1501**(1), 389–396 (2012)
2. Bhatnagar, P.L., Gross, E.P., Krook, M.: A model for collision processes in gases. 1. Small amplitude processes in charged and neutral one-component systems. Phys. Rev. **94**, 511–525 (1954)
3. Cai, Z., Fan, Y., Li, R.: Globally hyperbolic regularization of Grad's moment system in one dimensional space. Commun. Math. Sci. **11**(2), 547–571 (2013)
4. Cai, Z., Fan, Y., Li, R.: On hyperbolicity of 13-moment system. Kinet. Relat. Models **7**(3), 415–432 (2014)
5. Dimarco, G., Pareschi, L.: Numerical methods for kinetic equations. Acta Numer. **23**, 369–520 (2014)
6. Fan, Y., Koellermeier, J., Li, J., Li, R., Torrilhon, M.: Model reduction of kinetic equations by operator projection. J. Stat. Phys. **162**(2), 457–486 (2016)
7. Grad, H.: On the kinetic theory of rarefied gases. Commun. Pure Appl. Math. **2**(4), 331–407 (1949)
8. Jin, S.: Asymptotic preserving (AP) schemes for multiscale kinetic and hyperbolic equations: a review. Rivista di Matematica della Università di Parma. New Series, 2 (2010)
9. Koellermeier, J.: Derivation and numerical solution of hyperbolic moment equations for rarefied gas flows. Dissertation, RWTH Aachen University, Aachen (2017)
10. Koellermeier, J., Schaerer, R.P., Torrilhon, M.: A framework for hyperbolic approximation of kinetic equations using quadrature-based projection methods. Kinet. Relat. Models **7**(3), 531–549 (2014)

11. Koellermeier, J., Torrilhon, M.: Hyperbolic moment equations using quadrature-based projection methods. AIP Conf. Proc. **1628**(1), 626–633 (2014)
12. Koellermeier, J., Torrilhon, M.: Numerical study of partially conservative moment equations in kinetic theory. Commun. Comput. Phys. **21**(4), 981–1011 (2017)
13. Lafitte, P., Lejon, A., Samaey, G.: A high-order asymptotic-preserving scheme for kinetic equations using projective integration. SIAM J. Numer. Anal. **54**(1), 1–33 (2016)
14. Melis, W., Rey, T., Samaey, G.: Projective Integration for nonlinear BGK kinetic equations. In: Cancès, C., Omnes, P. (eds.) FVCA 2017. SPMS, vol. 200, pp. 145–153. Springer, Cham (2017). https://doi.org/10.1007/978-3-319-57394-6_16
15. Mieussens, L.: Discrete velocity model and implicit scheme for the BGK equation of rarefied gas dynamics. Math. Models Methods Appl. Sci. **10**(08), 1121–1149 (2000)
16. Pareschi, L., Russo, G.: Implicit-explicit runge-kutta schemes and applications to hyperbolic systems with relaxation. J. Scientific Comput. **25**, 129–155 (2005)
17. Stecca, G., Siviglia, A., Toro, E.F.: Upwind-biased FORCE schemes with applications to free-surface shallow flows. J. Comput. Phys. **229**(18), 6362–6380 (2010)
18. Stecca, G., Siviglia, A., Toro, E.F.: A finite volume upwind-biased centred scheme for hyperbolic systems of conservation laws: application to shallow water equations. Commun. Comput. Phys. **12**(4), 1183–1214 (2012)
19. Struchtrup, H.: Macroscopic Transport Equations for Rarefied Gas Flows: Approximation Methods in Kinetic Theory. Interaction of Mechanics and Mathematics. Springer, Heidelberg (2006). https://doi.org/10.1007/3-540-32386-4
20. Tcheremissine, F.G.: Solution of the boltzmann equation in stiff regime. In: Freistühler, H., Warnecke, G. (eds.) Hyperbolic Problems: Theory, Numerics, Applications, pp. 883–890. Birkhäuser Basel, Basel (2001)
21. Tcheremissine, F.G.: Solution to the Boltzmann kinetic equation for high-speed flows. Comput. Math. Math. Phys. **46**, 315–329 (2006)
22. Tcheremissine, F.G.: Conservative evaluation of boltzmann collision integral in discrete ordinates approximation. Comput. Math. Appl. **35**(1), 215–221 (1998)
23. Torrilhon, M.: Modeling nonequilibrium gas flow based on moment equations. Annu. Rev. Fluid Mech. **48**(1), 429–458 (2016)

Open Boundary Modeling in Molecular Dynamics with Machine Learning

Philipp Neumann$^{(\boxtimes)}$ (iD) and Niklas Wittmer

Helmut-Schmidt-Universität, Holstenhofweg 85, 22043 Hamburg, Germany
philipp.neumann@hsu-hh.de
https://www.hsu-hh.de/hpc

Abstract. Molecular-continuum flow simulations combine molecular dynamics (MD) and computational fluid dynamics for multiscale considerations. A specific challenge in these simulations arises due to the "open MD boundaries" at the molecular-continuum interface: particles close to these boundaries do not feel any forces from outside which results in unphysical behavior and incorrect thermodynamic pressures. In this contribution, we apply neural networks to generate approximate boundary forces that reduce these artefacts. We train our neural network with force-distance pair values from periodic MD simulations and use this network to later predict boundary force contributions in non-periodic MD systems. We study different training strategies in terms of MD sampling and training for various thermodynamic state points and report on accuracy of the arising MD system. We further discuss computational efficiency of our approach in comparison to existing boundary force models.

Keywords: Open boundary · Machine learning ·
Molecular-continuum · Boundary forcing · Molecular dynamics

1 Introduction

1.1 Molecular Dynamics

Molecular dynamics (MD) enables investigations of fluids, suspensions and materials at the nanoscale. For this purpose, the considered system is modeled in terms of molecules or atoms that are characterized through positions $\mathbf{x}_i$ and velocities $\mathbf{v}_i$, as well as through forces $\mathbf{F}_i$, with the latter typically arising from pairwise inter-molecular interactions in terms of pair potentials [13]. The interplay of these variables is described by Newton's equations of motion

$$
\begin{aligned}
\frac{d\mathbf{x}_i}{dt} &= \mathbf{v}_i, \\
\frac{d\mathbf{v}_i}{dt} &= \frac{1}{m_i}\mathbf{F}_i.
\end{aligned}
\tag{1}
$$

P. Neumann and N. Wittmer thank the Scientific Computing Group, University of Hamburg, for providing computational resources.

V. V. Krzhizhanovskaya et al. (Eds.): ICCS 2020, LNCS 12142, pp. 334–347, 2020.
https://doi.org/10.1007/978-3-030-50433-5_26

We will restrict considerations in the following to short-range single-site Lennard-Jones NVT systems, that is two spherical particles i, j interact via forces

$$\mathbf{F}_{ij} = \frac{48\epsilon}{\sigma^2} \left[\left(\frac{\sigma}{r_{ij}} \right)^{14} - \frac{1}{2} \left(\frac{\sigma}{r_{ij}} \right)^{8} \right] \mathbf{r}_{ij} \qquad (2)$$

with $\mathbf{r}_{ij} := \mathbf{x}_i - \mathbf{x}_j$, $r_{ij} := \|\mathbf{r}_{ij}\|$, as long as the particles lie within a cut-off distance $r_{ij} \leq r_c$; the total force on a particle arises as $\mathbf{F}_i = \sum_{i \neq j} \mathbf{F}_{ij}$. The parameters ϵ, σ are material-dependent parameters. Temperature is controlled via a thermostat. Despite its simplicity, this model is used in a great variety of applications and implemented in basically all popular MD packages; we made use of the package LAMMPS for all our tests [12].

1.2 The Challenge: Modeling Open Boundaries

MD for fluid systems is often used in combination with periodic boundary conditions. This approach naturally extends the molecular interaction potential across the actual boundaries of the considered system. However, *modeling open boundaries for non-periodic domains*—which are relevant for actual flow scenarios in engineering applications or, in particular, in multiscale flow simulation such as molecular-continuum coupling [3,9,14]—poses a grand challenge, especially for dense particle systems: as no particle interactions exist between near-boundary particles and the fluid domain beyond the boundary, particles would be pushed out of the domain. This results in invalid thermodynamic conditions and pressure distributions as well as in unphysical particle fluxes. Consequently, *an open boundary force model* is required to counteract this behavior and ties back into the MD system via an additional forcing term $\mathbf{F}_i^{ext}$, i.e. $\mathbf{F}_i = \sum_{i \neq j} \mathbf{F}_{ij} + \mathbf{F}_i^{ext}$ for particles close to the open boundary.

1.3 Open Boundary Force Modeling: State-of-the-Art

The goal of an open boundary force model is typically (i) to impose the correct average hydrodynamic pressure on the MD system, (ii) to take into account the molecular information of the system, (iii) to yield the correct molecular structure close to the boundary, i.e. the model shall add as little physical perturbations to the particle system as possible, and (iv) to perform at acceptable computational cost, that is particle updates close to the boundary must not be inhibitively more expensive than particle updates in the inner part of the computational domain.

The first developed open boundary force models were based on analytical formulae, incorporating the pressure, particle density and, potentially, weighting functions to take into account the distance of a particle from the boundary [3,4,11]. These approaches, however, lack molecular information (ii) and resulted in rather severe density perturbations in proximity of the open boundary (iii). A significant reduction of perturbations could be achieved through the use of radial distribution functions (RDFs) [14], which impose the average hydrodynamic pressure (i) and—through the RDFs—incorporate molecular information (ii). Although accurate density profiles near the boundary were obtained for several particle systems, stronger oscillations were observed in case of very dense particle systems (iii). Besides, the creation of the actual force model from the measured RDFs required an additional interpolation/integration step. An extension of this approach to multi-site molecules was presented in [10].

A purely density-driven approach was presented in [7]: the particle density is measured close to the boundary, and its trend is captured over time via noise filters and gradient approximations. Based on this trend, the boundary force is adopted to yield a flat density profile at the boundary. While this algorithm also naturally extends towards multi-site molecules [6], easily extends towards non-stationary flows (i.e. adopting boundary forces to varying flow velocities) and can be automatically employed for arbitrary MD systems, it misses molecular information (ii); to the authors' knowledge, no molecular structure investigations have been provided for this method (iii), except for molecular orientation measurements in case of multi-site molecules [6]. Besides, an amplification factor for the force relaxation needs to be prescribed which is not necessarily known a priori.

Finally, a parameter space exploration was carried out in [15,16] for single-site Lennard-Jones systems and a regression formula for a boundary force in dependence of a particle's distance from the open boundary was derived over a wide range of temperature and density values. While this formula is highly valuable for many systems and provides accurate forcing (i), (ii), no molecular investigations were reported so far (iii). Besides, entire parameter space explorations are computationally very expensive.

All derived methods were found to perform at acceptable cost (iv) in molecular-continuum simulations with open boundaries, with some limitations of the approach presented in [10].

1.4   Outline and Objective

In the following, we describe a methodology that uses neural networks to estimate open boundary forces from existing MD data (ii) via regression. We demonstrate that the neural network-based approach

- provides accurate open boundary force predictions that are in good agreement with and partly outperform the model by Zhou et al. [15,16] (i),
- retains the molecular structure rather well, even in very close proximity of the open boundary (iii); this is also shown for the Zhou model in this context,
- performs at acceptable computational cost (iv).

MD data is very noisy and there are several variants available to improve neural networks for these cases. For the sake of simplicity and facilitated applicability of our method, we make use of standard neural network formulations and we further implemented the algorithm using the freely available open source software TensorFlow.

We describe the design and discuss the choice of our ML-based approach in Sect. 2. The sensitivity of our ML-based approach with regard to sampling methodology and used parameters as well as of the Zhou model are studied in Sect. 3.1; more in-depth comparison of the molecular structure close to the open boundary for the Zhou model and the best ML-based approximation is given in Sect. 3.2. Estimates on sampling numbers and training epochs are provided in Sect. 3.3. To underpin the generality of the method, various MD state points are examined in Sect. 3.4, followed by a discussion of run times of the open boundary force-augmented MD systems in Sect. 3.5. We close with a summary and give an outlook to future work in Sect. 4.

2 Machine-Learning Approach to Boundary Forcing

We use TensorFlow [1, 2] to develop a neural network model for nonlinear regression of boundary forces in non-periodic boundary conditions. Our initial model contains one hidden layer with five neurons. This number results from a reasonable estimate to approximate the average force and its gradient sufficiently well, given, for example, the interpolating representation described by Zhou et al. [15, 16]. The commonly used sigmoid function tanh [8] serves as activation function within the hidden layer whereas on the output layer a linear activation function is used.

For optimization of the parameters, we apply the *ADAM* [5] optimizer in combination with the mean-squared-error (*MSE*) function. ADAM is an extension of the standard stochastic gradient descent which claims to be computationally efficient, especially in terms of memory usage; MSE is a default choice for the error function. Due to rotational symmetry of single-site particles, we use solely the distance of particles from the open boundary as input feature and the boundary force perpendicular to the respective boundary as output feature. Both input and output features are obtained from periodic MD simulations in which we assume a virtual wall inside the domain: we determine corresponding distances of the particles from the virtual wall as well as forces that act onto the particles from beyond the virtual wall. The network thus only contains one input and one output neuron. Input data are normalized to lie within the unit interval $[0, 1]$. Since MD systems are typically equilibrated initially in periodic settings, the equilibration phase can be immediately used to generate these values for both training and validation of the network.

The learning rate was set so 0.001 in all of our scenarios which is the default for the ADAM optimizer, and the network was trained with a number of epochs between 1000 and 30,000.

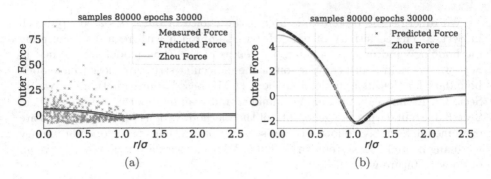

Fig. 1. Predicted boundary forces compared to Zhou force. (a) Force predictions and actual, measured samples. (b) Close-up of predicted force profiles.

3 Results

For our initial tests, we set up a short-range single-site Lennard-Jones-based system with reduced parameters $\sigma = \epsilon = m = k_B = 1$, where k_B denotes the Boltzmann constant. The Lennard-Jones potential has a cut-off distance $r_c = 2.5\sigma$. We simulate a box of size $30 \times 30 \times 30$ with a mass density $\rho = 0.81$, resulting in $21,952$ particles homogeneously distributed within the simulation box. Temperature was set to 1.1. The time step size Δt was set to 0.002.

MD equilibration was performed over 20,000 time steps employing periodic boundary conditions on all sides of the simulation box. This results in a random distribution of the molecules with a fluctuating density profile with the same mean value ρ; every corresponding MD simulation with open boundaries should feature the same mean value and, optimally, no or only small deviations from it. We used this state to generate training samples for the neural network from a subsequent period of 110,000 time steps. Afterwards, the left and the right boundary were changed into reflecting boundaries to hinder particles from escaping. Besides, the force model was activated.

The system was then equilibrated for another 5,000 time steps. From this point on, each simulation was run over a period of 20,000 time steps. Sampling of all quantities reported in this work was performed within this last part of the simulations.

3.1 Sensitivity of Machine Learning-Based Algorithm

For first investigations, we generated 200,000 samples from the periodic simulation. As a base setup for training, we randomly chose 80,000 of these samples (we refer to this strategy as regular sampling in the following) for training over 30,000 epochs.

In Fig. 1(a), the resulting boundary forcing estimate is displayed together with the Zhou profile, and actual force contributions used to train the model.

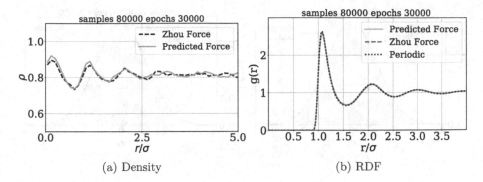

Fig. 2. (a) Density profile of a simulation using a model trained with 80,000 samples, compared to the density profile of a simulation using Zhou forcing. (b) RDF of the same simulation computed over a strip close to the boundary with thickness r_c. Dashed: Density and RDF as result of applying the Zhou force at the boundaries. Continuous: Resulting quantities when using our predicted force model in the simulation. Dotted: Periodic RDF

Both our ML model and Zhou's method perform a regression upon the Lennard-Jones force model; Fig. 1(b) shows a close-up of the arising force profiles. Apparently, the trained ML model diverges from the Zhou profile in close proximity of the boundary. The minimum of the force profile is approximately the same for both models, yet it is slightly shifted towards the right in the ML approach.

The remaining parameters are the same as in Sect. 2. The following analyses have been computed by sampling periodic and, especially, open boundary simulations over 20,000 timesteps.

Figure 2(a) shows the density profile of our simulation using this setup in comparison to a simulation using Zhou boundary forcing. The quality of the neural network's force computation reaches basically the same accuracy as the Zhou method: the maximum density deviation of our ML-based simulation is 12.9% compared to 11.4% in the Zhou model. Note that our sampling is carried out in small bins of size 0.1 to capture fine-scale structures in the different quantities in very close proximity of the boundary, which explains the actual visibility of these deviations. This is done for a detailed comparison of the methods; very good accuracy was already shown for the Zhou model in [15]. In a distance of one cut-off region (2.5σ), the density profiles of periodic and open boundary simulation are basically indistinguishable.

The RDF in Fig. 2(b) has been computed over the boundary region within a boundary strip of thickness r_c. Due to non-periodicity of our domain, we scaled the distribution by a volumetric factor per particle depending on the distance from the open boundary to account for the correspondingly missing particle pairs across the open boundary. The RDF of our simulated model near the boundary exhibits the same contour as the RDF obtained from a fully periodic and a Zhou-based open boundary simulation. We further checked the distribution of x-velocities (i.e., the velocity component of the molecules perpendicular to the

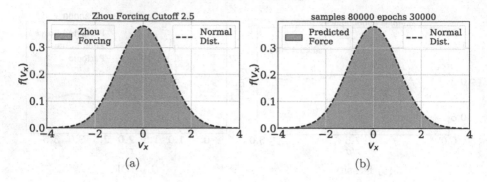

Fig. 3. Distribution of x-velocities for particles using Zhou model and ML-based model within a boundary distance r_c. Gray: Distribution resulting from a simulation using (a) the Zhou boundary force and (b) our predicted boundary force

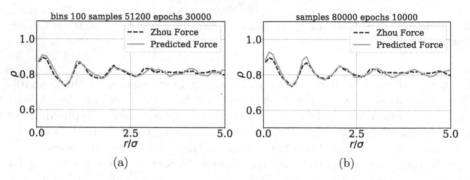

Fig. 4. (a) Density profile of a simulation trained with binned sampling approach. (b) Density profile for run with regular sampling using 10,000 epochs for training

open boundary) within the same boundary strip. Figure 3 demonstrates that the expected normal distribution is captured well.

In order to increase the efficiency of the training, we reduced the number of samples needed to train the model. For this purpose, we organized the boundary region in bins and evenly distributed the training samples over these bins.

Figure 4(a) shows the results for training with binned samples, using 100 bins with 512 samples in each bin to discretize the boundary strip r_c. The resulting density exhibits only slight fluctuations similar to the Zhou model and the original ML-based regular sampling model, i.e. the density deviates by less than 10.8%; yet, through the binning and corresponding homogeneous distribution of samples, we require 36% less samples than in the regular sampling case.

We further investigated the influence of the number of epochs used in the training. A resulting density profile for a reduced number of 10,000 epochs is shown in Fig. 4(b). The deviations are higher, amounting to approx. 16% in this case.

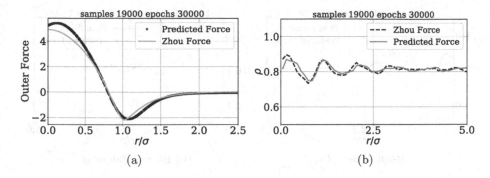

Fig. 5. Force and density profiles of best case scenario using 19,000 samples over 30,000 epochs

Using the ML-based approach, we were able to even outperform the Zhou model in terms of particle density approximation. One of the results is shown in Figs. 5(a) and (b). This run used 19,000 randomly selected samples and training was carried out over 30,000 epochs. The force estimate behaves similarly compared to the force estimate from Fig. 1(b) but the gradient is smoother within the range $[0, 0.5]$ and there seems to be little more emphasis on attracting forces.

The density in this run deviates by 8.4%, which is ca. 3% less than the deviations in the Zhou-based forcing scenario.

We further studied the influence of the size of the ML-based model's hidden layer, e.g. changing from five to three hidden neurons. Two of the resulting density profiles are shown in Fig. 6. Figure 6(a) shows the best run using regular sampling during training, Fig. 6(b) shows the best result using binned sampling. Using binning results in a density deviation of 9.8%, whereas the regular sampling method reaches 11.5%. As we were not able to produce better results in most configurations that could compare to the other models, we decided to stay with the initial five neurons. We further experimented with different learning rates. This approach did not significantly improve the quality of the trained models, but could possibly be employed for optimizing the performance of the training phase.

3.2   Best Case Properties

In the previous section we have demonstrated the general applicability of our approach to a Lennard-Jones fluid. Next, we examine the best run so far more closely and compare its results with those of the Zhou simulation.

Figure 7 shows the RDF profiles sampled from the simulation in significantly thinner boundary strips. Both RDFs from Zhou and ML-based model slightly underestimate the maximum peak of the expected RDF (obtained from the periodic case).

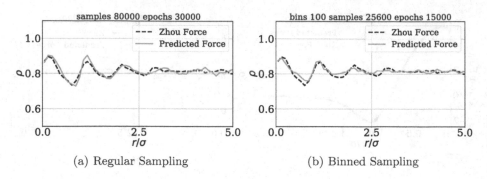

(a) Regular Sampling   (b) Binned Sampling

Fig. 6. Two examples from running simulations using models which were trained with three instead of five neurons

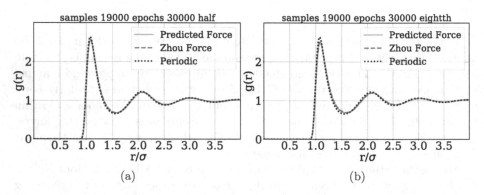

(a)   (b)

Fig. 7. RDF profile sampled over boundary strip of thickness (a) 1.25σ and (b) 0.3125σ

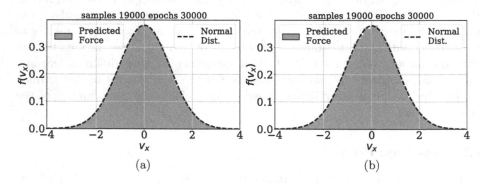

(a)   (b)

Fig. 8. Distribution of x-velocities sampled over boundary strip of thickness (a) 1.25σ and (b) 0.3125σ

We sampled the distribution of x-velocities in the same boundary strips, cf. Fig. 8. While the estimation in the strip of thickness 1.25σ resembles the normal distribution, (very) slight fluctuations are visible for the 0.3125σ-thick strip.

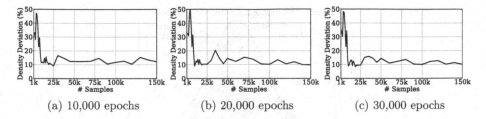

(a) 10,000 epochs   (b) 20,000 epochs   (c) 30,000 epochs

Fig. 9. Maximum density deviations for varying sample size using different numbers of epochs

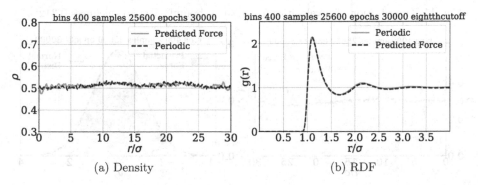

(a) Density   (b) RDF

Fig. 10. Density and RDF for setup $\rho = 0.5$ using boundary force prediction

3.3 Estimating the Sampling Properties

We further aimed at an estimate for how many samples are needed to obtain accurate force profiles for the subsequent simulation. We therefore ran tests with training sample sizes ranging from 1000 to 150,000 samples. We ran theses tests with 10,000, 20,000, and 30,000 epochs and extracted the maximum density deviation per simulation.

The results of these experiments are shown in Fig. 9. Obviously, less than 10,000 samples is not enough to generate a valid model.

3.4 Varying State Points

We further validated our method in simulations of varying densities. For this purpose, we set up two more simulations, using density values $\rho = 0.5$ and $\rho = 0.3$. In both cases, temperature was set to $T = 1.1$. Simulation time was scaled up linearly according to the base setup of $\rho = 0.81$ to provide comparable sampling quality and, thus, to enable comparison with the results reported in prior sections. We did not compare the resulting models to the Zhou model, since this model's underlying parameter space exploration does not capture these state points.

Figure 10(a) displays the resulting density profile of one simulation for the case of $\rho = 0.5$. The periodic density profile deviates from the expected density

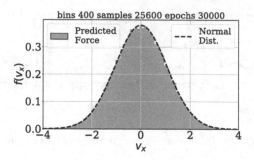

Fig. 11. Distribution of x-velocities for $\rho = 0.5$, boundary strip thickness 0.3125σ

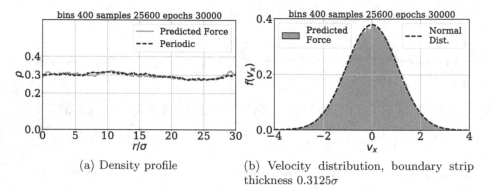

(a) Density profile

(b) Velocity distribution, boundary strip thickness 0.3125σ

Fig. 12. Density and distribution of x-velocities for $\rho = 0.3$ using boundary force prediction

by 6.5%, the ML profile deviates by 5.9%. Close to the boundaries, we can still observe slightly higher fluctuations within the ML profile. Figure 10(b) shows the RDF estimate from the boundary strip of thickness 0.3125σ. The estimated RDF fits the expected radial distribution nearly perfectly.

The estimated velocity distribution is shown in Fig. 11. Similar to the default case $\rho = 0.81$, the normal distribution is basically perfectly matched.

We made similar observations for the test case with density $\rho = 0.3$. In this case, the density in the periodic simulation deviates by 6.7% whereas the shown ML model deviates by 8.2% (cf. Fig. 12(a)). The estimated velocity distribution is again captured well near the boundary, (cf. Fig. 12(b)).

3.5 Run Time Considerations

Performance tests were conducted on a small cluster at the Scientific Computing Group, University of Hamburg. The utilized nodes contain two CPUs of type Intel Xeon X5650 with six cores each. The size of the memory per node is 12 GB. All tests were performed in sequential mode.

Setup	Training Time(s)
80,000 Samples 30,000 Epochs	3160
100 Bins 12 800 Samples 15,000 Epochs	364

(a)

Setup	Run time(s)	Steps/s
Periodic	589.05	42.44
Reflecting	563.33	44.36
Zhou	579.04	43.18
ML, 5 neurons	617.06	40.48
ML, 3 Neurons	612.8	40.78

(b)

Fig. 13. (a) Training times of different sampling setups. (b) Run times of different setups, running periodic simulations or open boundary simulations with the Zhou or ML-based models

Regarding the training step of our method, the number of samples plays a considerable role. The higher the size of our sample set, the longer the training takes to complete. Furthermore, the complexity of the model affects its run time during training as well as during a simulation.

Figure 13(a) shows the times needed for training of some of the ML models considered in this work. Using the binning method, the training completes ca. nine times faster. Due to the number of samples, not only the run time of one epoch is lower, but also the overall number could be reduced while retaining the quality of the density profile or even improving it.

Figure 13(b) contains the measured run times of the different setups. The Zhou model is slightly faster than a purely periodic simulation. The ML-based runs are not as efficient as Zhou and yield higher run times than the periodic version. The force computation during a simulation employing the ML model is conducted in bulk mode. That is, at each time step the force and distance values of each particle in the boundary regions are collected and passed at once to the neural network. As Tensorflow is optimized for parallel computation, there is quite some overhead expected when the tool is used in sequential mode.

4   Conclusion and Outlook

We have introduced a novel ML-based model to predict open boundary forcing. The model is built from molecular data from a prior equilibration and thus reflects the molecular structure of the fluid well. It further yields accurate forcings at acceptable computational performance, which has been demonstrated in various tests, including a detailed comparison with the parameter space exploration-based Zhou model. It shall be remarked that relying on a prior equilibration is only necessary for the startup phase: in case of molecular systems, that change dynamically over time, the ML model could be adjusted "on-the-fly" by simply using more molecular data from the inner part of the MD domain where valid MD information should be available. This should in principal work well, since our analysis showed that potential perturbations in the molecular systems are very marginal and have only been observed in close proximity of the open boundary.

Yet, enough samples need to be found in this case in a sufficiently short time frame. This appears a promising route that shall be investigated in the future.

We have discussed various aspects of parametrization of our ML-based model and also provided estimates on how many samples and training epochs are required to generate valid open boundary force models, cf. Sect. 3.3. Yet, improved explainability of the arising network weights in relation to, e.g., the impact of thermal fluctuations or the RDFs of the fluid would be desirable.

Further research to optimize the overall neural network and improve its outputs, e.g. by optimizing the learning rate as briefly discussed in Sect. 3.1 would be desirable. Employing ensemble strategies is another possibility to improve the quality of ML-based boundary forcing. Both aspects were, however, beyond the scope of this work.

The actual power of machine learning lies in the prediction of highly complex systems, e.g. deducing information from higher-dimensional inputs. With the method presented in this contribution delivering very good results, an extension towards multi-site molecules or dynamic systems with locally varying densities as well as further computational performance improvements would therefore be logical next steps in our analysis and are, partly, work in progress. Since the ML-based model can potentially be used in conjunction with arbitrary MD solvers and is not restricted to LAMMPS that has been used in our studies, we further plan to incorporate the model into MaMiCo, a coupling tool for hybrid molecular-continuum simulations [9]. This would thus provide the functionality and make it re-usable for different combinations of MD solvers and molecular-continuum coupling algorithms in the future.

In terms of performance, we observed, besides the slight overheads of the ML approach compared to periodic simulations, that modifications of the learning rate could reduce the number of epochs needed for training in some of our test scenarios. Optimization of the learning rate would, thus, be also helpful in our case, similar to many other ML systems.

References

1. TensorFlow. https://www.tensorflow.org/
2. Abadi, M., et al.: TensorFlow: large-scale machine learning on heterogeneous distributed systems (2016). arXiv:1603.04467 [cs], http://arxiv.org/abs/1603.04467
3. Delgado-Buscalioni, R., Coveney, P.: Continuum-particle hybrid coupling for mass, momentum, and energy transfers in unsteady flow. Phys. Rev. E **67**, 046704 (2003)
4. Flekkøy, E., Wagner, G., Feder, J.: Hybrid model for combined particle and continuum dynamics. EPL **52**, 271–276 (2000)
5. Kingma, D., Ba, J.: Adam: a method for stochastic optimization (2017). arXiv:1412.6980 [cs], http://arxiv.org/abs/1412.6980
6. Kotsalis, E.M., Koumoutsakos, P.: A control algorithm for multiscale simulations of liquid water. In: Bubak, M., van Albada, G.D., Dongarra, J., Sloot, P.M.A. (eds.) ICCS 2008. LNCS, vol. 5102, pp. 234–241. Springer, Heidelberg (2008). https://doi.org/10.1007/978-3-540-69387-1_26
7. Kotsalis, E., Walther, J., Koumoutsakos, P.: Control of density fluctuations in atomistic-continuum simulations of dense liquids. Phys. Rev. E **76**, 16709 (2007)

8. Murphy, K.: Machine Learning: A Probabilistic Perspective. Adaptive Computation and Machine Learning Series. MIT Press, Cambridge (2012)

9. Neumann, P., Bian, X.: MaMiCo: transient multi-instance molecular-continuum flow simulation on supercomputers. Comput. Phys. Commun. **220**, 390–402 (2017)

10. Neumann, P., Eckhardt, W., Bungartz, H.J.: A radial distribution function-based open boundary force model for multi-centered molecules. Int. J. Mod. Phys. C **25**, 1450008 (2014)

11. O'Connell, S., Thompson, P.: Molecular dynamics-continuum hybrid computations: a tool for studying complex fluid flow. Phys. Rev. E **52**, R5792–R5795 (1995)

12. Plimpton, S.: Fast parallel algorithms for short-range molecular dynamics. J. Comput. Phys. **117**, 1–19 (1995)

13. Rapaport, D.: The Art of Molecular Dynamics Simulation, 2nd edn. Cambridge University Press, Cambridge (2004)

14. Werder, T., Walther, J., Koumoutsakos, P.: Hybrid atomistic-continuum method for the simulation of dense fluid flows. J. Comput. Phys. **205**, 373–390 (2005)

15. Zhou, W., Luan, H., He, Y., Sun, J., Tao, W.: A study on boundary force model used in multiscale simulations with non-periodic boundary condition. Microfluid. Nanofluid. **16**(3), 587–595 (2013). https://doi.org/10.1007/s10404-013-1251-4

16. Zhou, W., Luan, H., He, Y., Sun, J., Tao, W.: Erratum to: a study on boundary force model used in multiscale simulations with non-periodic boundary condition. Microfluid. Nanofluid. **20**(6), 93 (2016). https://doi.org/10.1007/s10404-016-1756-8

Microtubule Biomechanics and the Effect of Degradation of Elastic Moduli

Sundeep Singh[1] and Roderick Melnik[1,2]($\boxtimes$)

[1] MS2Discovery Interdisciplinary Research Institute, Wilfrid Laurier University, 75 University Avenue West, Waterloo, ON N2L 3C5, Canada
rmelnik@wlu.ca
[2] BCAM - Basque Center for Applied Mathematics, Alameda de Mazarredo 14, 48009 Bilbao, Spain

Abstract. The present study aims at quantifying the effect of mechanical degradation of microtubules on their electro-elastic response. A three-dimensional continuum-based hollow cylindrical domain of a microtubule has been considered in this work. A fully coupled electro-mechanical model has been developed for conducting the comparative analysis considering three different cases, viz., no degradation, 50% degradation and 90% degradation of elastic modulus of the microtubule. The microtubule has been subjected to dynamic forces adopted from the commonly used loading-unloading conditions in nanoindentation experiments. The results show that the degradation of microtubules significantly influences their electro-elastic response when subjected to externally applied forces. The transient response of the model in terms of induced displacement, electric potential and volumetric strain has also been analyzed for different magnitudes of mechanical degradation. The modelling study presented here represents a more accurate electro-mechanical model compared to the classical mechanical model for quantifying the effects of mechanical transductions on microtubules biomechanics.

Keywords: Coupled biological problems · Microtubules · Piezoelectricity · Electro-mechanical model · Finite element analysis

1 Introduction

Microtubules are long protein polymers involved in a wide range of cellular activities e.g., maintaining cell stiffness/shape, intracellular transport, regulation of cell morphology and cell mechanics along with facilitation of other physiological processes [1–3]. Microtubules are the stiffest element in the cytoskeleton of eukaryotic cells that helps them to withstand both static and dynamic loads. Several experimental and numerical studies have been reported in the past for understanding the mechanical properties of microtubules [1, 2, 4–7]. Owing to the previously reported studies in literature, the mechanical characteristics and the basics of microtubules biomechanics are now understood reasonably well [5, 6]. However, most of the reported numerical studies on the mechanical characterization of microtubules consider only the coupling between applied stress and induced strain under the application of externally applied forces. It has also been

© Springer Nature Switzerland AG 2020
V. V. Krzhizhanovskaya et al. (Eds.): ICCS 2020, LNCS 12142, pp. 348–358, 2020.
https://doi.org/10.1007/978-3-030-50433-5_27

theoretically proved that microtubules possess piezoelectric properties [8], a two-way linear coupling that results in the conversion of mechanical deformation into an applied electric field and vice-versa. The experimental basis of piezoelectric properties of microtubules is yet to come and is mainly limited due to associated intricacies in the intracellular probing of microtubules at nanoscales [9]. Importantly, such electro-mechanical integration will result in fostering many exciting applications of the bio-compatible piezoelectric materials, paving a way to a new age in the field of medicine [10]. Potential applications of this promising research area in the biomedical field include minimally invasive sensors, regenerative medicine, eco-friendly energy harvesters, drug delivery, etc. [10, 11].

In what follows, a three-dimensional coupled electro-mechanical model of a microtubule has been developed including the piezoelectric effect. The main novelty of this work is in a quantitative evaluation of the effect of degradation of mechanical properties of microtubules on the electro-elastic response, based on the proposed electro-mechanical model. The motivation for studying this effect lies in the fact that the cytoskeleton of the biological cell is dynamic and the cytoskeleton mechanics can be disrupted owing to aging or pathological disease. Moreover, cytoskeleton can also be degraded, e.g., due to the application of various anti-cancer drugs used for thera-peutic treatments of cancer cells [12]. Furthermore, previously reported *in vitro* studies indicate that the elastic modulus of cancer cells is significantly lower as compared to the healthy cells [12, 13]. Thus, in the present contribution, we focus on quantifying the effect of degradation of the microtubule, one of the most important cytoskeletal structures, on the electro-elastic response of coupled electro-mechanical model sub-jected to externally applied forces. Importantly, the microtubule degradation is simu-lated by altering the mechanical properties of the microtubule, keeping the structural features intact.

2   Coupled Electro-Mechanical Model of Microtubule

The computational domain of a microtubule modeled as an elastic hollow cylinder is presented in Sect. 2.1. The governing equations and mathematical framework adopted in the present numerical study are given in Sect. 2.2. The details of the numerical setup and necessary boundary conditions used for solving the proposed problem are provided in Sect. 2.3.

2.1   Geometrical Assumptions and Microtubule Computational Domain

A microtubule comprising of 13 protofilaments has been modelled as a hollow cylinder having inner and outer diameters of 15 nm and 23 nm, respectively [14]. The total length of the microtubule has been considered to be 320 nm [5, 15]. Figure 1(a) presents the three-dimensional continuum-based model of a microtubule considered in the present numerical study. The material properties considered in this study can be found in Table 1 [12, 16]. Motivated by [12], mechanical degradation of microtubules have been modelled by reducing their elastic moduli (see Table 1) by 50% and 90%.

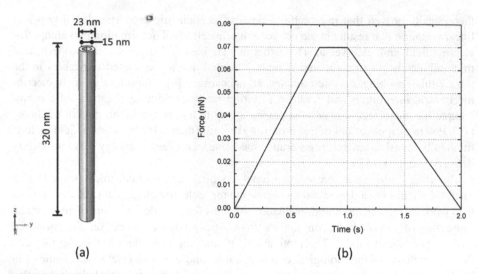

(a)   (b)

Fig. 1. (a) A schematic representation of the continuum-based three-dimensional model of a microtubule considered in this study and (b) Illustration of the force-time load function used in the present analysis.

2.2   Coupled Electro-Mechanical Model of Microtubules

The basic relationships for a linearly coupled electrical and mechanical fields are given by [17, 18]:

$$\sigma_{ij} = c_{ijkl}\varepsilon_{kl} - e_{ijk}E_k, \tag{1}$$

$$D_j = e_{jkl}\varepsilon_{kl} + \kappa_{jk}E_k, \tag{2}$$

where σ_{ij} and ε_{ij} are the components of elastic stress and strain tensors, respectively, D_j are the components of electric flux density vector, E_k are the components of electric field vector, c_{ijkl} are the elastic coefficients, e_{ijk} are the piezoelectric coefficients and κ_{ij} is the dielectric permittivity coefficients, where subscript $i, j = 1, 2, 3$ and $k, l = 1, 2, 3, 4, 5, 6$. Further, the strain field is related to the displacement vector field by the Cauchy relationship, given by

$$\varepsilon_{ij} = \frac{1}{2}\left(u_{i,j} + u_{j,i}\right). \tag{3}$$

The relationship between the electric field (E) and the electric potential (ϕ) is given by

$$E = -\nabla \cdot \phi. \tag{4}$$

The constitutive equations (Eqs. (1–4)) are further subjected to the equilibrium condition and Gauss's law with the assumption of vanishing body forces and vanishing free charges:

$$\rho \frac{\partial^2 u}{\partial t^2} = \nabla \cdot \sigma, \tag{5}$$

$$\nabla \cdot D = 0. \tag{6}$$

The elastic coefficients of microtubule obtained from its Young's (elastic) modulus (E_m) and Poisson's ratio (v) are given by

$$c = \begin{bmatrix} \lambda + 2\mu & \lambda & \lambda & 0 & 0 & 0 \\ \lambda & \lambda + 2\mu & \lambda & 0 & 0 & 0 \\ \lambda & \lambda & \lambda + 2\mu & 0 & 0 & 0 \\ 0 & 0 & 0 & \mu & 0 & 0 \\ 0 & 0 & 0 & 0 & \mu & 0 \\ 0 & 0 & 0 & 0 & 0 & \mu \end{bmatrix}, \tag{7}$$

where λ and μ are the Lame's constants:

$$\lambda = \frac{vE_m}{(1+v)(1-2v)}, \quad \mu = \frac{E_m}{2(1+v)}. \tag{8}$$

The present numerical study considers the piezoelectric coefficients of microtubules to be similar to that of collagen and have been adapted from [16]. Using Voigt's notation, piezoelectric strain coefficients are given as [16]

$$d_{ij} = \begin{bmatrix} 0 & 0 & 0 & d_{14} & d_{15} & 0 \\ 0 & 0 & 0 & d_{15} & -d_{14} & 0 \\ d_{31} & d_{31} & d_{33} & 0 & 0 & 0 \end{bmatrix}, \tag{9}$$

where i subscript represents the direction of electric field displacement of the piezoelectric tensor and subscript j represents the associated mechanical deformation. In the present numerical study, the constitutive relations are expressed in the stress-charge form (Eqs. (1–2)), accordingly, the piezoelectric coefficients expressed in the strain-charge form (Eq. (9)) have been converted into the stress-charge form utilizing the following relation

$$e_{ijk} = c_{jklm}d_{ilm}, \tag{10}$$

where e_{ijk} are the piezoelectric stress coefficients, c_{jklm} are the components of the elastic tensor given in Eq. (7) and d_{ilm} are the piezoelectric strain coefficients given in Eq. (9).

Table 1. Mechanical, piezoelectric and dielectric properties of microtubules considered in the present numerical study.

Parameter	Value
Density (ρ)	1000 kg/m^3
Elastic modulus (E_m)	1.9 GPa
Poisson's ratio (v)	0.3
Relative permittivity (κ)	40
Piezoelectric coefficients	
d_{14}	−12 pC/N
d_{15}	6.21 pC/N
d_{31}	−4.84 pC/N
d_{33}	0.89 pC/N

2.3 Numerical Setup and Boundary Conditions

A finite-element method (FEM) has been used for solving the coupled electro-mechanical model of microtubule presented in Fig. 1(a). A trapezoidal load function (with a loading period of 1 s), as shown in Fig. 1(b), has been used to apply the compressive force at the top surface of the microtubule. This applied force is similar to the commonly used loading-unloading paths used in nanoindentation experiments for evaluating cell biomechanics [12]. The bottom part of the microtubule has been approximated by fixed and electrically grounded boundary conditions. The initial displacement and electric potential within the computational domain have been assumed to be 0 m and 0 V, respectively. The transient electro-elastic response of the coupled model has been quantified for different values of elastic modulus of the microtubule considered in the present study. The implementation has been completed in the COMSOL Multiphysics software [19]. It's built-in mesh generator has been used for meshing the computational domain of microtubule with heterogeneous tetrahedral mesh elements obtained after conducting a mesh convergence analysis. All the transient three-dimensional simulations have been conducted on a Dell T7400 workstation with Quad-core 2.0 GHz Intel® Xeon® processors.

3 Results and Discussion

As an initial test, the developed three-dimensional electro-mechanical model's fidelity and integrity have been evaluated by comparing the previously reported results of zinc oxide nanowires [17] to those obtained from the current model utilizing similar geometrical details, boundary conditions and electro-mechanical parameters. Table 2 presents the comparative analysis of the maximum piezoelectric potential and displacements reported in [17] to those obtained from the current model under the application of a compressive load of 100 nN and 85 nN at the top surface of the nanowire with a fixed and electrically grounded bottom. As presented in Table 2, values of both the maximum piezoelectric potential and displacements obtained from the present model and reported in [17] are in good agreement with each other, lending confidence in the developed coupled electro-mechanical model and its extensions to microtubules, inspired by the earlier modelling of nanowire.

Table 2. Comparison of the maximum piezoelectric potential and displacements reported in the previous nanowire study [17] to that computed as a result of the present study.

Force (nN)	Maximum piezoelectric potential (V)		Maximum displacement (nm)	
	Previous study [17]	Present study	Previous study [17]	Present study
100	0.48	0.48	0.03	0.03
85	0.4	0.41	Not given	0.02

The main motive of the present numerical study was to quantify the effect of mechanical degradation of microtubules on the electro-elastic response of the coupled electro-mechanical model under the application of dynamic loads. A compressive force (whose profile has been presented in Fig. 1(b)) has been applied at the top surface of the microtubule with fixed bottom and consequently, the electro-elastic response of the microtubule has been predicted for three conditions, viz., an undegraded microtubule, a 50% degraded microtubule and a 90% degraded microtubule. Figure 2 presents the total displacement distribution at the end of 1 s of the loading period for different cases considered in the present numerical study. As depicted in Fig. 2, the maximum magnitude of the displacement at the end of the loading period (or under the application of maximum applied force) has been found to be 0.04 nm, 0.07 nm and 0.37 nm for the undegraded, 50% degraded and 90% degraded microtubules, respectively. Thus, there prevail significant variations in the induced displacements among different considered cases of the microtubule with the larger magnitudes of displacement actually induced in the highly degraded microtubule. Further, for all three cases, the maximum displacement occurs at the top surface where the external force was applied and it decreases in magnitude moving away from the top surface. It is noteworthy to mention that the concentric circles above the microtubules in Fig. 2 (and also Fig. 3) indicates the top of the undeformed microtubule.

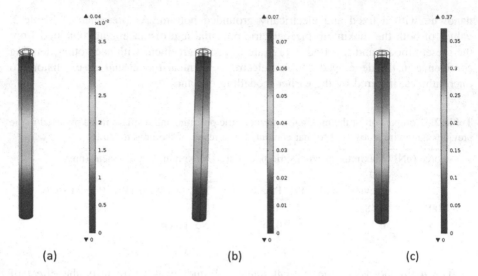

Fig. 2. (Color online). Total displacement distribution (in nm) at the end of the loading period for: (a) undegraded microtubule, (b) 50% degraded microtubule, and (c) 90% degraded microtubule.

The electric potential distribution predicted from the electro-mechanical model of microtubule under the application of dynamic load has been presented in Fig. 3 for different cases considered in this study. In Fig. 3, the maximum absolute value of electric potential has been obtained for the highly degraded microtubule, which can be attributed to a higher magnitude of the displacement induced for the microtubule with 90% degraded elastic modulus. Further, the blue side in Fig. 3 denotes the negative potential and the red side denotes the electrically ground condition. Moreover, there prevail a very negligible variation in the electric potential distribution among the undegraded and 50% degraded microtubules due to relatively smaller differences in their induced displacements.

The force-displacement curve for the three levels of mechanical degradation of the microtubules has been presented in Fig. 4. As mentioned earlier, a trapezoidal function based loading-unloading path has been used for applying force at the top surface of microtubule with a fixed bottom. It is evident from Fig. 4 that there prevail noticeable differences in the three curves representing three cases considered in the present study. The slope of the force-displacement curve is higher for the undegraded microtubule and decreases with an increase in the percentage of degradation of elastic modulus of the microtubule. In other words, the force-displacement response of microtubule subjected to externally applied load is significantly softer for the highly degraded microtubule. For example, for a displacement of about 25 pm, the corresponding forces are about 0.05 nN, 0.02 nN and 0.005 nN, respectively. Thus, the mechanical degradation of microtubule dramatically affects the force-displacement response of microtubules.

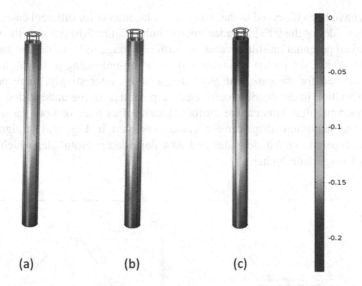

Fig. 3.   (Color online). Electric potential distribution (in mV) at the end of the loading period for: (a) undegraded microtubule, (b) 50% degraded microtubule, and (c) 90% degraded microtubule.

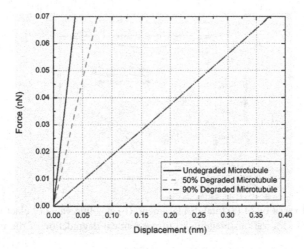

Fig. 4.   Force-displacement response of the three-dimensional coupled electro-mechanical model of microtubule subjected to loading and unloading path for various magnitudes of mechanical degradation.

Figure 5 presents the temporal variation of the total displacement, electric potential and volumetric strain for various magnitudes of the mechanical degradation of microtubule at a point on the top surface where the external force has been applied. As can be seen from Fig. 5(a), the displacement profile follows the same loading-unloading profile of the applied force presented in Fig. 1(b). Furthermore, significant differences have been observed in the induced displacements for different cases and it is on the higher side for the 90% degraded microtubule. Figure 5(b) presents the variation of the electrical potential (absolute value) for different magnitudes of degradation of the microtubule. Again, the profile is similar to the loading-unloading path with maximum potential obtained for the case with 90% degradation. Interestingly, there prevails a negligible variation in the distribution of electric potential for the undegraded and 50% degraded microtubules. This can be attributed to the fact that for both cases the difference in the maximum displacement value presented in Fig. 5(a) is significantly smaller as compared to the undegraded and 90% degraded microtubules which is about one order of magnitude higher.

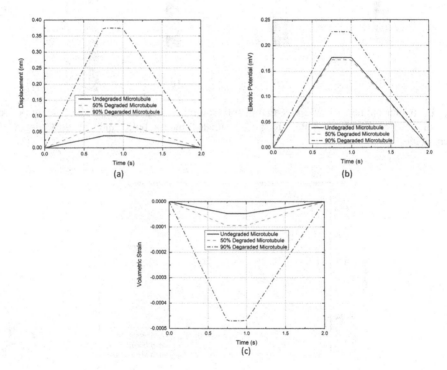

Fig. 5. Temporal variation of (a) total displacement, (b) absolute value of electric potential, and (c) volumetric strain for various magnitudes of mechanical degradation of the microtubule.

The variation of volumetric strain with respect to time for different cases of microtubule degradation has been presented in Fig. 5(c). Again, the maximum volumetric strain has been induced for the case with a 90% degraded microtubule. Thus, the results predicted from the FEM simulations of microtubules, considering our developed fully coupled electro-mechanical model, confirm that the mechanical degradation of microtubules has a pronounced effect on the electro-elastic response under dynamic loadings. The proposed model can be further extended to a more complex situation [20] by considering the complete biological cell embedded with intricate cytoskeleton and organelles for extracting more critical information of cell biomechanics subjected to different loading conditions at the cellular and sub-cellular levels. One of the major limitations of the proposed model is the consideration that microtubules behave as a linearly elastic and isotropic material, which has been motivated by [9, 12]. Future studies will be conducted to quantify the electro-elastic response of microtubules considering visco-elastic and anisotropic models.

4 Conclusion

A finite-element-based three-dimensional coupled electro-mechanical model of microtubules has been developed. A comparative analysis has been conducted for evaluating the effects of degradation of elastic moduli of microtubules on their electro-elastic response under the application of dynamic load. The obtained results demonstrate that the mechanical degradation of microtubules can significantly affect their electro-elastic response. It has been found that the 90% degraded microtubule results in one order of magnitude higher displacement and around 35% higher electric potential generation as compared to the undegraded microtubule. Accordingly, it becomes very important to account for such changes in the mechanical characteristics for more accurately quantifying the cellular biomechanics under the influence of static and dynamic forces. It is expected that the proposed model can be further extended for different applications in the quest for our better understanding of the complex behavior of cells biomechanics in general and microtubules in particular, along with designing electro-mechanical devices for medical applications.

Acknowledgments. Authors are grateful to the NSERC and the CRC Program for their support. RM is also acknowledging the support of the BERC 2018-2021 program and Spanish Ministry of Science, Innovation and Universities through the Agencia Estatal de Investigacion (AEI) BCAM Severo Ochoa excellence accreditation SEV-2017-0718 and the Basque Government fund AI in BCAM EXP. 2019/00432. Authors are also grateful to Prof. Jack Tuszynski as well as to Dr. Jagdish Krishnaswamy for useful information, valuable suggestions, and the number of important references.

References

1. Li, S., Wang, C., Nithiarasu, P.: Effects of the cross-linkers on the buckling of microtubules in cells. J. Biomech. **72**, 167–172 (2018)
2. Li, S., Wang, C., Nithiarasu, P.: Simulations on an undamped electromechanical vibration of microtubules in cytosol. Appl. Phys. Lett. **114**(25), 253702 (2019)
3. Melnik, R.V.N., Wei, X., Moreno-Hagelsieb, G.: Nonlinear dynamics of cell cycles with stochastic mathematical models. J. Biol. Syst. **17**(3), 425–460 (2009)
4. Kučera, O., Havelka, D., Cifra, M.: Vibrations of microtubules: physics that has not met biology yet. Wave Motion **72**, 13–22 (2017)
5. Havelka, D., Deriu, M.A., Cifra, M., Kučera, O.: Deformation pattern in vibrating microtubule: Structural mechanics study based on an atomistic approach. Scientific Rep. **7** (1), 4227 (2017)

6. Liew, K.M., Xiang, P., Zhang, L.W.: Mechanical properties and characteristics of microtubules: a review. Compos. Struct. **123**, 98–108 (2015)
7. Marracino, P., et al.: Tubulin response to intense nanosecond-scale electric field in molecular dynamics simulation. Scientific Rep. **9**(1), 1–14 (2019)
8. Tuszynski, J.A., Kurzynski, M.: Introduction to Molecular Biophysics. CRC Press LLC, Boca Raton (2003)
9. Kushagra, A.: Thermal fluctuation induced piezoelectric effect in cytoskeletal microtubules: Model for energy harvesting and their intracellular communication. J. Biomed. Sci. Eng. **8** (08), 511 (2015)
10. Chorsi, M.T., et al.: Piezoelectric biomaterials for sensors and actuators. Adv. Mater. **31**(1), 1802084 (2019)
11. Chae, I., Jeong, C.K., Ounaies, Z., Kim, S.H.: Review on electromechanical coupling properties of biomaterials. ACS Appl. Bio Mater. **1**(4), 936–953 (2018)
12. Katti, D.R., Katti, K.S.: Cancer cell mechanics with altered cytoskeletal behavior and substrate effects: a 3D finite element modeling study. J. Mech. Behav. Biomed. Mater. **76**, 125–134 (2017)
13. Suresh, S.: Biomechanics and biophysics of cancer cells. Acta Mater. **55**(12), 3989–4014 (2007)
14. Thackston, K.A., Deheyn, D.D., Sievenpiper, D.F.: Simulation of electric fields generated from microtubule vibrations. Phys. Rev. E **100**(2), 022410 (2019)
15. Jin, M.Z., Ru, C.Q.: Localized buckling of a microtubule surrounded by randomly distributed cross linkers. Phys. Rev. E **88**(1), 012701 (2013)
16. Denning, D., et al.: Piezoelectric tensor of collagen fibrils determined at the nanoscale. ACS Biomater. Sci. Eng. **3**(6), 929–935 (2017)
17. Hao, H., Jenkins, K., Huang, X., Xu, Y., Huang, J., Yang, R.: Piezoelectric potential in single-crystalline ZnO nanohelices based on finite element analysis. Nanomaterials **7**(12), 430 (2017)
18. Krishnaswamy, J.A., Buroni, F.C., Garcia-Sanchez, F., Melnik, R., Rodriguez-Tembleque, L., Saez, A.: Improving the performance of lead-free piezoelectric composites by using polycrystalline inclusions and tuning the dielectric matrix environment. Smart Mater. Struct. **28**, 075032 (2019)
19. COMSOL Multiphysics® v. 5.2. COMSOL AB, Stockholm, Sweden. www.comsol.com
20. Singh, S., Krishnaswamy, J.A., Melnik, R.: Biological cells and coupled electro-mechanical effects: the role of organelles, microtubules, and nonlocal contributions, J. Mech. Behav. Biomed. Mater. (2020). https://doi.org/10.1016/j.jmbbm.2020.103859

Formation of Morphogenetic Patterns in Cellular Automata

Manan'Iarivo Rasolonjanahary and Bakhtier Vasiev$^{(\boxtimes)}$ (iD)

Department of Mathematical Sciences, University of Liverpool, Liverpool, UK
bnvasiev@liv.ac.uk

Abstract. One of the most important problems in contemporary science, and especially in biology, is to reveal mechanisms of pattern formation. On the level of biological tissues, patterns form due to interactions between cells. These interactions can be long-range if mediated by diffusive molecules or short-range when associated with cell-to-cell contact sites. Mathematical studies of long-range interactions involve models based on differential equations while short-range interactions are modelled using discrete type models. In this paper, we use cellular automata (CA) technique to study formation of patterns due to short-range interactions. Namely, we use von Neumann cellular automata represented by a finite set of lattices whose states evolve according to transition rules. Lattices can be considered as representing biological cells (which, in the simplest case, can only be in one of the two different states) while the transition rules define changes in their states due to the cell-to-cell contact interactions. In this model, we identify rules resulting in the formation of stationary periodic patterns. In our analysis, we distinguish rules which do not destroy preset patterns and those which cause pattern formation from random initial conditions. Also, we check whether the forming patterns are resistant to noise and analyse the time frame for their formation. Transition rules which allow formation of stationary periodic patterns are then discussed in terms of pattern formation in biology.

Keywords: Pattern formation · Mathematical modelling · Cellular automata

1 Introduction

Biological pattern formation is a complex phenomenon which is studied experimentally in a number of model organisms [1] and theoretically by means of various mathematical techniques [2, 3]. One of the model organisms commonly used in biological studies of pattern formation is the fly (Drosophila) embryo [4]. Patterning in the fly embryo takes place in two directions: along the head-to-tail (antero-posterior) axis where the pattern occurs as a repeated structure formed by segments and along the dorso-ventral axis where patterning results into formation of internal morphological structures. It is known that the formation of segments is preconditioned by the formation of spatially periodic patterns of gene expressions which takes place at early stages of embryonic development [5]. There are many other examples of formation of periodic patterns in biology including formation of stripes on skin of animals (zebra) and fish.

© Springer Nature Switzerland AG 2020
V. V. Krzhizhanovskaya et al. (Eds.): ICCS 2020, LNCS 12142, pp. 359–373, 2020.
https://doi.org/10.1007/978-3-030-50433-5_28

In this work, we focus on formation of periodic patterns which allow discrete representation. Discrete models often complement continuous models by giving description of patterning processes on different spatial or time scales. For example, the continuous (reaction-diffusion) model of anterio-posterior patterning in fly embryo presented in [6] allows to conclude that the modelled units (nuclei) quickly reach their stable equilibrium states and there are only four such states in normally evolving embryo. This, in turn, allows to simplify the model by reducing its quantitative details and to describe the considered patterning as evolution of a chain of interacting nuclei on discrete level. Another example is given by the continuous model of pigmentation patterning on a skin of growing reptilian reported in [7]. Due to slow diffusion between skin scales this model also reduces to a discrete model for the interactions between differently pigmented neighbouring scales, although the pigmentation of each scale is described by differential equations. Generally, discrete models, as compared to continuous models, often offer a simple explanation for basic properties (such as spatial scaling) of morphogenetic patterns.

Since we are primarily interested in formation of periodic patterns, we can reformulate the problem and ask the general question: what kind of local interactions can result in formation of a stationary periodic pattern? In our study, we use von Neumann's cellular automata (CA) to address this question. This model has been invented, more than fifty years ago, by Stanislaw Ulam and John von Neumann [8, 9]. The model is represented by a collection of cells forming a regular grid which evolves over discrete time steps according to a set of rules based on the state of the cells [10]. A detailed description of Neumann's CA can be found in a number of sources [11, 12].

Extensive study of various modifications (involving different numbers of dimensions, allowed states per site and neighbours affecting transitions) of this model have been performed by Wolfram [12–14]. He has found that patterns forming in cellular automata fall into one of the following four classes: 1. Homogenous, 2. Periodic structures, 3. Chaotic structures and 4. Complex. Correspondingly, all automata also fall into four classes, although the class in which the given automata falls depends on the imposed initial conditions.

In this work, we will explore rules resulting in the formation of periodic patterns in the CA where each cell can only be in two distinct states. This is an extension of the research reported in [12, 13] in two ways: firstly, we consider all 256 rules (while the classification made by Wolfram was restricted by 32 so called "legal" rules) and, secondly, we are interested in periodic patterns forming in a domain of finite size, and therefore the found rules appear to fall into classes 2 and 4 identified by Wolfram. Although this model is very simple, the results obtained allow conclusions to be drawn on the mechanisms of periodic pattern formation in biology, for example the anterio-posterior patterning in fly embryo.

2   Chain of Logical Elements

The simplest version of von Neumann's cellular automata was designed by Wolfram [12, 13] and called "elementary cellular automaton" (ECA). It consists of a regular lattice of cells which form a chain and can only be in two states. The position of a cell in the chain will be denoted by the symbol i, the total number of cells - by n and the state of cell by s_i ($s_i = 0$ or $s_i = 1$). The state s_i^{t+1} of a cell i at time $t + 1$ is determined by the states of cells $i - 1$, i and $i + 1$ at time t according to a transition rule, defined as $s_i^{t+1} = f\left(s_{i-1}^t, s_i^t, s_{i+1}^t\right)$. A cell interacting with its two closest neighbours leads to consideration of $2^3 = 8$ possible configurations, which are 000, 001, 010, 011, 100, 101, 110 and 111. For each of these 8 configurations, the resulting states can be represented as: $0s_00$, $0s_11$, $0s_20$, $0s_31$, $1s_40$, $1s_51$, $1s_60$ and $1s_71$. The complete set of middle elements of all resulting states forms a binary number $s_7s_6s_5s_4s_3s_2s_1s_0$ which varies between 0 and 255. In other words, the configurations above allow $2^8 = 256$ possible rules or cellular automata. The binary number, translated into decimal, is taken to be the rule number.

Different rules applied to different initial conditions can result into formation of various nontrivial spatio-temporal patterns including oscillations and propagating waves. Among these 256 rules, a few symmetry groups can be distinguished so that the rules belonging to one group result in the formation of similar patterns. These groups contain up to four rules and their symmetry-based relationships termed complement, mirror image and mirror complement [14]. The definitions of these relationships are based on certain features of the binary representation of rules, namely:

The complement of a given rule is a rule whose binary representation is obtained from the binary representation of the given rule where (a) the digits are taken in the reverse order and (b) each digit is interchanged with the opposite digit.

The mirror image of a given rule is the transition rule such that if the given rule contains an elementary rule $f(b, c, d) = a$ then the mirror image contains an elementary rule $f(d, c, b) = a$, that is, the same output is allocated to the rule with mirrored states of neighbours.

The mirror complement is a complement to a mirror image.

Our study is performed using the one-dimensional CA described above. Note that this model can also be seen as a chain of logical elements. It is represented by a one-dimensional regular lattice of elements which can be in one of two states. The state s_i of the element i, can be either "false" (commonly denoted by "0" and illustrated in black colour) or "true" (denoted by "1" and drawn in white). The state of each element changes with each time increment (except for the border cells which remain unaltered) according to the applied rules.

The goal of this study is to find out what rules can cause the formation of stationary periodic patterns in a chain of finite size where the states of border elements are fixed. We also check whether the periodic patterns are stable, i.e. noise sensitive. To do so, we introduce a noise by changing the state of each element randomly with a certain (small) probability. In application to biology, each logical element corresponds to a biological cell and the cell's state reflects the expression of certain genes. The ECA represents a two-state model and can only be used for study of patterns formed by expression of a single gene along cells forming a chain.

3 Results

Our study of periodic patterns forming in the ECA has been performed in a few steps. First, we have focused on two-periodic patterns and identified the rules which (1) do not destroy a preset two-periodic pattern, (2) allow recovery of two-periodic patterns perturbed by noise and (3) allow formation of two-periodic patterns from random initial conditions. Then, we have repeated the above study for the case of three- and more-periodic patterns.

3.1 Conservation of Preset Two-Periodic Patterns in ECA

Our first set of simulations aimed to find all transition rules which do not destroy pre-existing periodic patterns. These simulations have been started with preset two-periodic patterns (1 white + 1 black) like the one shown in Fig. 1 but with more (usually 60 or more) elements. Starting with two-periodic initial conditions, we have found that 64 rules out of 256 (25%) do not destroy it. Their values, in decimal form, are as follows:
    4, 5, 6, 7, 12, 13, 14, 15, 20, 21, 22, 23, 28, 29, 30, 31, 68, 69, 70, 71, 76, 77, 78, 79, 84, 85, 86, 87, 92, 93, 94, 95, 132, 133, 134, 135, 140, 141, 142, 143, 148, 149, 150, 151, 156, 157, 158, 159, 196, 197, 198, 199, 204, 205, 206, 207, 212, 213, 214, 215, 220, 221, 222 and 223.

Fig. 1. Two-periodic pattern in the ECA composed by 20 cells. The binary expression for all these rules has the form: $xx0xx1xx$, where x can be either 0 or 1. That is, the pattern consisting of a repetition of black and white remains the same when the transition rule leaves the configurations 3 and 6 unaltered: $(010) \rightarrow (010)$ and $(101) \rightarrow (101)$.

The second set of simulations was aimed at finding all rules which are such that, starting with periodic initial conditions (see Fig. 1) and applying a noise, the perturbed pattern is able to recover. The noise level of 0.1% has been applied meaning that the state of each cell can be altered with probability of 0.001 (i.e. one out of 1000 cells is altered each time step). Simulations show that 33 out of the above 64 rules (52%) would allow resistance to the perturbation caused by the noise. These rules are given by the following numbers:
    6, 13, 14, 15, 20, 28, 30, 69, 70, 77, 78, 79, 84, 85, 86, 92, 93, 134, 135, 141, 143, 148, 149, 156, 157, 158, 159, 197, 198, 199, 213, 214 and 215.

The recovery of a perturbed pattern can happen in two ways: locally, when the perturbation disappears without affecting surrounding cells, or globally, when the perturbation propagates along the medium to one of its borders and then vanishes. Also, for both scenarios there are exceptional rules (28, 70, 78, 92, 141, 156, 157, 197, 198 and 199) such that if the perturbed cell is located next to the border cell then it cannot recover (unless it is hit by another perturbation).

3.2   Local Recovery of Perturbed Two-Periodic Patterns

The following rules allow the local extinction of perturbations:

13, 28, 69, 70, 77, 78, 79, 92, 93, 141, 156, 157, 197, 198 and 199.

For the analysis of the recovery processes, we point out that if the probability of perturbation is small, then the probability that two neighbouring cells are perturbed simultaneously is much smaller and, therefore, can be neglected. With this in mind, let us consider the recovery of the periodic pattern in a chain described, for example, by the rule 77, whose binary number is 01001101. Consider the periodic sequence ... 010101... and the scenario when, for example, the element in the 4th position changes to **0** due to the perturbation caused by noise. Then, the above sequence will become ... 010001.... We want to know how this sequence will evolve over time and how many time steps it will take for the pattern to recover. The changed element can only affect its two closest neighbours, i.e. the zeros in 3rd and 5th positions. The 3rd zero is in the middle of 100 and will not change following the transition (100)→(100). The perturbed cell is in the middle of **0**00 and therefore will change to **1** following the transition (000)→(010). The 5th zero is in the middle of **0**01 and will remain unaltered following the transition (001)→(001). Thus, we conclude that it takes one time step for the pattern to restore. The above analysis is valid for all perturbations changing any "1" to "0" inside the periodic chain and therefore it will take one time step for the pattern to recover from all these perturbations. Similarly, one can show that perturbations changing any "0" to "1" inside the chain will also vanish in one time step. From here, we conclude that under rule 77 it takes one time step for two-periodic pattern to recover from any single perturbation. Since rule 77 coincides with its own complement, mirror image and mirror complement (rule 77 is unique in this respect), the above result cannot be extended to any other rule.

There are rules which allow the local recovery of perturbed patterns in more than one time step. Let us, for example, consider the rule 13 (00001101 in binary representation) applied to the preset periodic pattern (i.e. ...0101010...) in a medium of arbitrary (finite) size. It is easy to show that if the noise changes any of the 1s to 0 in the periodic sequence it will take only one time step to recover. On the other hand, it takes two time steps to restore when the noise changes any of 0s to 1. To illustrate this, let us consider the finite periodic sequence ...01010101.... where the noise changes the digit 0 at the 3rd position into 1, so that the sequence becomes ...01110101.... The 1 in the second position will stay unaltered following the transition: (011)→(011). The 1 in the third position will change to 0 following the transition (111)→(101). The 1 in the fourth position will change to 0 following the transition (110)→(100). So, after the first step we get the sequence ...01000101.... This is identical to the case (considered above) when the state of the (4th) cell in periodic pattern is changed from "1" to "0". Its recovery takes only one time step and thus it takes two time steps for the pattern to recover when, owing to the noise, any of 0s change to 1. The stated result is exactly the same for the rule 69, which is the mirror image of the rule 13, and it is slightly amended for the rules 79 and 93, which are complements of rules 13 and 69 respectively. Namely, in the cases of rules 79 and 93 it takes one time step to recover when any "0" is changed to "1" in the preset periodic pattern and two time steps when any "1" is changed to "0".

Table 1. Summary of recovery dynamics for rules allowing local recovery of perturbed two-periodic pattern. The notation "$a{\rightarrow}b$ (n TS)" is used to state that the perturbed by noise cell recovers from state "a" to its original state "b" in n time steps. The exceptional cases, when the recovery doesn't take place, (and when the noise hits the cell next to the border cell) are described in the extra (last) line. For example, the notation "(...000)" indicates that the rule doesn't allow recovery in the case when the perturbed chain ends with three 0s.

Rule	Complement	Mirror image	Mirror Complement
77 $1{\rightarrow}0$ (1 TS) $0{\rightarrow}1$ (1 TS)			
13 $1{\rightarrow}0$ (1 TS) $0{\rightarrow}1$ (2 TS)	79 $1{\rightarrow}0$ (2 TS) $0{\rightarrow}1$ (1 TS)	69 $1{\rightarrow}0$ (1 TS) $0{\rightarrow}1$ (2 TS)	93 $1{\rightarrow}0$ (2 TS) $0{\rightarrow}1$ (1 TS)
70 $1{\rightarrow}0$ (2 TS) $0{\rightarrow}1$ (3 TS) (...000)	157 $1{\rightarrow}0$ (3 TS) $0{\rightarrow}1$ (2 TS) (...111)	28 $1{\rightarrow}0$ (2 TS) $0{\rightarrow}1$ (3 TS) (000...)	199 $1{\rightarrow}0$ (3 TS) $0{\rightarrow}1$ (2 TS) (111...)
78 $1{\rightarrow}0$ (3 TS) $0{\rightarrow}1$ (1 TS) (...000)	141 $1{\rightarrow}0$ (1 TS) $0{\rightarrow}1$ (3 TS) (...111)	92 $1{\rightarrow}0$ (3 TS) $0{\rightarrow}1$ (1 TS) (000...)	197 $1{\rightarrow}0$ (1 TS) $0{\rightarrow}1$ (3 TS) (111...)
156 $1{\rightarrow}0$ (2 TS) $0{\rightarrow}1$ (2 TS) (...111) (000...)	198 $1{\rightarrow}0$ (2 TS) $0{\rightarrow}1$ (2 TS) (...000) (111...)		156 Same as in the first column.

The process of recovery of a perturbed periodic pattern takes even longer for some other rules. For example, in the case of rule 70 whose binary representation is 01000100, it takes two time steps for restoration when the noise changes (almost) any "1" to "0" in the preset periodic pattern and three time steps after changing any "0" to "1". There is an exceptional case for this rule when the perturbed pattern does not recover. Namely if, in the medium with a preset periodic pattern such that the states of the rightmost cells are given as (010), the noise strikes the second cell from the right border (resulting into ...101000) then the pattern does not recover. This is because rule 70 involves the transition (100)→(100), which allows for the 3rd cell from the right border remaining unaltered, and (000)→(000), for which the state of the second (perturbed) cell remains unaltered. The only way for the periodic pattern to recover is that the noise strikes the same cell once again. The results for rule 70 can, with small modifications, be extended to rules 157, 28 and 199 which are respectively complement, mirror image and mirror complement of rule 70. The properties of all rules which allow local recovery of perturbations on preset two-periodic patterns are given in Table 1.

3.3   Recovery of Two-Periodic Patterns by Means of Propagating Waves

In this section, we consider a set of rules which also allow the recovery of perturbed periodic patterns but in a different manner: namely, the perturbation is driven to one of the two edges of the chain where it commonly disappears. In the most common scenario, the perturbation moves like a wave with a constant speed and the time required for the perturbation to get to the chain edge is proportional to the initial distance from the perturbed cell to the corresponding edge of the chain. However, for some rules the waves of perturbation which form are not regular, so that their speed and the size of perturbed area change over time. The rules exhibiting propagating waves of perturbation are:

6, 14, 15, 20, 30, 84, 85, 86, 134, 135, 143, 148, 149, 158, 159, 213, 214 and 215.

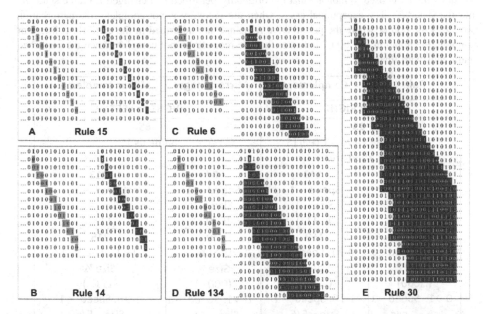

Fig. 2. Illustrations of propagating perturbation waves in recovering patterns. Each row shows a fragment (perturbed) of initially periodic pattern. Rows (top to bottom) show the states of the fragment after each consecutive time step. Perturbed cells are highlighted in **green/blue** to distinguish between the initial perturbations from state "1" to state "0" or from "0" to "1" respectively. **A:** Each perturbed cell recovers in one time step. **B:** Each perturbed cell recovers in two time steps. **C: Green case** - perturbed "0" recovers in one time step and perturbed "1" – in two. **Blue case** - perturbed "0" recovers in five and perturbed "1" in six time steps. **D: Green case** – same as on Panel C. **Blue case** - perturbed "0" recovers in nine and perturbed "1" in ten time steps. **E:** Recovery is slow and completely irregular. (Color figure online)

The perturbation waves propagate from LEFT to RIGHT in the case of rules 6, 14, 15, 30, 134, 135, 143, 158 and 159 and from RIGHT to LEFT for the remaining rules. The most important question about the propagating waves of perturbations is at what speed they propagate and, consequently, how long it takes for the perturbation to reach the chain's border where it usually vanishes. Our study shows that for rules 15 and 85 (which form a symmetry group, see Table 2), the perturbed cell recovers in one time step and at the same time the perturbation shifts one cell (to the right/left in the case of rule 15/85 respectively). Thus, the size of the wave is 1 cell and the speed of the perturbation wave is 1 cell/time-step (see Fig. 2A) and, therefore, it takes no more than n time steps (where n is a size of the chain) for the perturbations to reach the chain edge and disappear.

Table 2. Summary on the recovery dynamics for rules allowing formation of propagating perturbation waves. The direction, speed and size of the waves formed are given. If the properties of perturbation waves depend on the form of original perturbation, they are stated separately: the notation "$a{\to}b$" is used to indicate that the originally perturbed cell changed from the state "a" into the state "b". The exceptional cases, when the recovery is not complete (the perturbation wave hits the border where the perturbation remains), are described in the extra (last) line. For example, the notation "(…000)" indicates that the recovery is not complete when the perturbed chain ends with three 0s.

Rule	Complement	Mirror image	Mirror complement
LEFT to RIGHT		**RIGHT to LEFT**	
15 Speed of the wave – 1 cell/TS Size of perturbed area – 1 cell		**85** Same as 15.	
14 Speed – 1 Size – 2 (…000)	**143** Speed – 1 Size – 2 (…111)	**84** Speed – 1 Size – 2 (000…)	**213** Speed – 1 Size – 2 (111…)
6 1→0: Speed – 1 Size ½ (…000) 0→1: Speed – 1 Size – 5/6 (…000)	**159** 0→1: Speed – 1 Size ½ (…111) 1→0: Speed – 1 Size – 5/6 (…111)	**20** 1→0: Speed – 1 Size ½ (000…) 0→1: Speed – 1 Size – 5/6 (000…)	**215** 0→1: Speed – 1 Size ½ (111…) 1→0: Speed – 1 Size – 5/6 (111…)
134 1→0 (same as 6) 0→1: Speed – 1 Size – 9/10 (…000)	**158** 0→1 (same as 159) 1→0: Speed – 1 Size – 9/10 (…111)	**148** 1→0 (same as 20) 0→1: Speed – 1 Size – 9/10 (000…)	**214** 0→1 (same as 215) 1→0: Speed – 1 Size – 9/10 (111…)
30 Same as 149.	**135** Same as 149.	**86** Same as 149.	**149** Speed of expansion 1. Recovery is slow and irregular. Size is unlimited.

Properties of the perturbation waves forming under rules 14, 143, 84 and 213 (form a symmetry group, see Table 2) are similar to what was observed for rules 15 and 85. However, now it takes two time steps for each perturbed cell to recover (see Fig. 2B). The recovery (that is, the contracting border of the perturbed area) as well as the perturbation (that is, the expanding border of the perturbed area) shifts one cell per time-step and the number of perturbed cells (i.e. the size of the wave) is 2 cells. When the perturbation reaches the edge of the chain, it doesn't always disappear: for example, in the case of rule 14 the state of the second rightmost cell remains perturbed (at "0") if the state of rightmost cell in the initial periodic pattern is 0 (the chain in the stationary state has '000' on its right border). Similar exceptions occur for the other three rules in the symmetry group (see Table 2).

The recovery of the periodic pattern under rule 6 is more complicated and takes place in one of two scenarios (Fig. 2C). The first scenario takes place when the originally perturbed cell was in the state "1" and the perturbation changes its state to "0". In this case, the perturbation expands with the speed of 1 cell/time-step and the recovery takes place in a way that two cells recover simultaneously every second time step so that the size of the wave alternates between 1 and 2 cells. This is because it takes two time steps to recover from "0" to "1" and one time step to recover from "1" to "0". However, if the originally perturbed cell was in state "0" then the recovery is much slower. After the initial transition process, the perturbation wave stabilizes, so that each perturbed "0" recovers in five time steps while each "1" recovers in six. The speed of the perturbation expansion is still 1 cell per time-step, but its size alternates between 5 and 6 cells. In similarity to the case of rule 15 the perturbation after reaching the edge of the chain, doesn't always disappear but the second leftmost cell remains perturbed (at "0") if the state of leftmost cell in the initial periodic pattern is 0 (the chain in the stationary state ends with "000" on the right). Similar scenarios are observed for the three other rules (159, 20 and 215) in the symmetry group (see Table 2).

The recovery of the periodic pattern under rule 134 is identical to that under rule 6 when the initially perturbed cell was in state "1" (Fig. 2D). A difference is observed when the state of the initially perturbed cell changes from "0" to "1". After the initial transition process when the perturbation wave stabilizes, the recovery turns to be even slower: each perturbed "0" recovers in 9 time steps while each "1" – in 10. The speed of the wave is still 1 cell per time-step but its size alternates between 9 and 10 cells. This observation naturally extends to rules 158, 148 and 214 forming the symmetry group with rule 134.

The recovery of the periodic pattern under rules 30, 135, 86 and 149 (forming a symmetry group) is significantly different from what we have seen so far (see Fig. 2E). The perturbed area under these rules still expands with the speed of 1 cell/time-step (i.e. under rule 30, the right border of the disturbed area shifts 1 cell/time-step) but the recovery is much slower: the contracting border (i.e. the left border under rule 30) of the perturbed area is slow, is affected by oscillations and its motion is completely irregular. As a result, the perturbed area quickly expands until it reaches the border of the chain, i.e. its size is only limited by the size of unrecovered area in the chain (i.e. the area to the right of the contracting border under rule 30). Although the contracting border moves much more slowly than the expanding one, it can be shown that its speed can't be less than 2 cells per 9 time-steps. The properties of propagating perturbation waves under all the rules allowing their formation are summarized in Table 2.

3.4 Formation of Two-Periodic Patterns from Random Initial Condition

In the last two sections, we dealt with the rules which allow the recovery of preset periodic patterns perturbed by a rare and random noise. In this section, we shall focus on a subset of those rules which not only allow the recovery of perturbed periodic pattern but also the generation of periodic patterns from any initial conditions. These are the following six rules: 15, 30, 85, 86, 135 and 149.

These six rules form two symmetry groups: (15, 85) and (30, 86, 135 and 149) (see Table 2). The process of periodic pattern formation is significantly different for these two groups: under rules 15 and 85 periodic patterns are generated considerably faster than under the rules forming the second group.

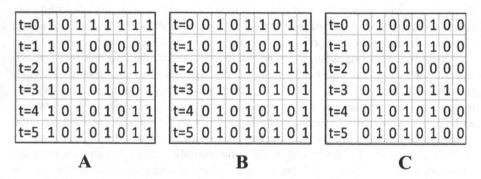

A   B   C

Fig. 3. Formation of periodic patterns from random initial conditions under rule 15.

Three examples of short simulations illustrating the formation of stationary periodic patterns under the rule 15 are shown in Fig. 3. One can see that periodic patterns form in the course of time, first arising at the right border and then expanding to the left with a speed of 1 cell/time-step. For rules 15 and 85, one can formulate propositions like the one below:

Proposition for Rule 15: *Starting from any initial conditions, the state of the chain evolves towards two-periodic pattern (010101... or 101010...) developing from the LEFT to the RIGHT with a speed of at least 1 cell/time-step. The final stationary pattern may end with "11" or "00" on its RIGHT edge.*

Proof of this proposition is omitted. Proposition for rule 85 differs only in the directionality of the process: under the rule 85 the periodic pattern evolves from right to left. The formation of patterns under the other four rules (30, 86, 135 and 149) is considerably different and in a line with the information provided in Table 2. For example, in the case of rule 30 the formation of a periodic pattern starts at the right border and develops to the left. The speed at which the periodic pattern expands is irregular and alters from 4 cells per time-step to 2 cells per 9 time-steps. Furthermore, depending on the initial conditions, the final pattern may remain non-stationary and exhibit oscillations at the right border of the chain. The oscillating spot can contain one element, $(\ldots0100)\leftrightarrow(\ldots0110)$, two elements $(\ldots1001)\leftrightarrow(\ldots1111)$ or even three elements $(\ldots11111) \rightarrow (\ldots10001) \rightarrow (\ldots11011) \rightarrow (\ldots10011) \rightarrow (\ldots11111)$. The periodicity of oscillations is 2 for the first and second cases and 4 for the last case. The properties of patterns forming under rules 86, 135 and 149 are similar to those for rule 30.

3.5   Three- and More- Periodic Patterns in ECA

So far, we have focused on rules allowing existence/formation of two-periodic patterns. In this section, we address the same questions with respect to three-periodic patterns. Two types of three-periodic patterns can be set in ECA: two black cells followed by a white cell or two white cells followed by a black (see Fig. 4).

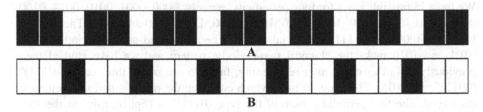

Fig. 4. Three-periodic patterns in ECA composed of 20 cells. A: Repetitions of 2 black and 1 white cells. B: Repetitions of 2 white and 1 black cells.

Consider the periodic pattern formed by the repetition of one white and two black cells. There are 32 rules which do not destroy preset periodic pattern of this type. These rules are:

4, 5, 12, 13, 36, 37, 44, 45, 68, 69, 76, 77, 100, 101, 108, 109, 132, 133, 140, 141, 164, 165, 172, 173, 196, 197, 204, 205, 228, 229, 236 and 237.

All these numbers have binary expressions in the form of $xxx0x10x$ where x is either 0 or 1. That is, the pattern consisting of a repetition of 2 blacks and 1 white is conserved under the rules keeping the configurations 2, 3 and 5 unaltered: $(001) \rightarrow (001)$, $(010) \rightarrow (010)$ and $(100) \rightarrow (100)$.

Similarly, there are 32 rules which conserve a periodic pattern containing one black and two white cells. The binary expressions of these rules have the form of $x10x1xxx$. That is, the pattern formed by repetitions of 2 whites and 1 black is conserved under the rules keeping the configurations 4, 6 and 7 unaltered: $(011) \rightarrow (011)$, $(101) \rightarrow (101)$ and $(110) \rightarrow (110)$. One can notice that there are four rules (76, 77, 204 and 205) which belong to both groups and allow the existence of both types of three-periodic patterns.

Our next task is to identify rules which allow recovery of three-periodic pattern perturbed by a random and rare noise. Our simulations show that there are only two such rules: rule 133, which allows recovery of three-periodic patterns formed by two blacks and one white, and its complement rule 94, which allows recovery of the second type of three-periodic pattern. Note that the mirror image of rule 133 is itself and its mirror complement is rule 94. Let's consider the recovery of perturbed three-periodic pattern under rule 133 (whose binary representation is 10000101). Consider the fragment of periodic pattern ...001001001... and assume that the noise strikes it at either 2nd or 3rd or 4th position. The perturbed sequence becomes either ...011001001... or ... 000001001... or ...001101001... respectively. Simple analysis shows that it takes 3, 2 or 3 time steps respectively to recover for each of the cases considered. Thus, the recovery under the rule 133 is local. All results obtained for rule 133 are naturally extended to rule 94.

We found no rules which allow the recovery of three-periodic patterns by means of propagating waves. Furthermore, there are no rules which allow the formation of three-periodic patterns from random initial conditions.

It is easy to show that there are three types of strictly four-periodic patterns which can form in ECA. Indeed, the four-periodic patterns in a generic form can be represented as $(a_1a_2a_3a_4)^n$ where elements a_i can be either 0 or 1 and the chain has $4n$ cells. We have 16 possible cases for four-periodicity, namely: 0000, 0001, 0010, 0011, 0100, 0101, 0110, 0111, 1000, 1001, 1010, 1011, 1100, 1101, 1110 and 1111. The first and last cases (0000 and 1111) are trivial and can be considered as one-periodic. Cases 0101 and 1010 make the classical two-periodic pattern and we have studied them previously. All other cases, after re-indexing, fall into one of the three cases: $(0111)^n$, $(0011)^n$ or $(0001)^n$. There are 16 rules which can keep the preset 4-periodic pattern of each type. The four-periodic pattern of the type $(0111)^n$ is kept by rules in the form $110x1xxx$ (where x is either 0 or 1); patterns of the type $(0011)^n$ - in the form $x1x01x0x$, and patterns of the type $(0001)^n$ - in the form $xxx0x100$. None of these rules allow noise resistance or generation of a strictly four-periodic pattern from random initial conditions.

In similarity to the four-periodic patterns, the patterns of higher periodicity can exist under some rules but they are never noise resistant or can they be generated from random initial conditions. The number of rules keeping preset periodic patterns gets smaller with an increase of the periodicity. Obviously, the pre-set pattern of any periodicity exists under rule 204: this is an identity rule and doesn't impose any change to any pattern.

4 Discussion

The study presented here was motivated by the problem of biological pattern formation governed by local (cell-to-cell) signals. As our primary interest was about segmentation in fly embryos, we have focused on the formation of periodic patterns. Furthermore, since the segmentation takes place along the embryonic anterio-posterior axis and is essentially a one-dimensional process, we have modelled it using a one-dimensional chain of logical elements represented by ECA. In the framework of this model, we have identified all rules allowing the existence or formation of periodic patterns. At the same time, our study was mainly focused on properties of ECA and, we believe, that it contributes to an extensive research [15] performed by a large community of scientists in this field.

In the presented study, we have explored whether formation of morphogenetic patterns in developing tissue can be modelled using ECA. The two-state model used in this study represents a chain of locally interacting cells, where cells in state "1" express some particular gene while in state "0" - don't. Interactions of logical elements in ECA can be considered as representing contact (membrane-to-membrane) interactions between cells in biological tissues. The modelled interactions can be seen as regulating the differentiation of cells. For example, the two-periodic stationary pattern forming in the model represents a chain of cells where each second cell expresses a certain gene. We have identified sets of interaction rules which allow existence, recovery from noise and formation from random initial conditions of various periodic patterns in CA. Particularly, we have found that 64 (out of 256) rules allow existence of two-periodic patterns, 33 of them allow recovery of two-periodic patterns from rare and random noise and only six rules (15, 85, 30, 86, 135 and 149) allow formation of two-periodic patterns from any initial conditions. Furthermore, although there are many rules allowing the existence of three- and more- periodic patterns, only two rules allow the recovery of three-periodic patterns from random and rare noise and none allow the recovery of four- and more- periodic patterns or the formation of three- and more-periodic patterns from random initial conditions.

It is known that the pattern associated with the expression of segmentation genes in the fly embryo is four-periodic. As this pattern forms in a few steps and under the influence of maternal, gap and pair-rule genes, one could consider this pattern as preset four-periodic which is allowed by a number of rules. However, it is also known that the periodic pattern formed by segmentation genes undergoes certain correction in position [16] and this wouldn't be allowed by any of transition rules in two-state model.

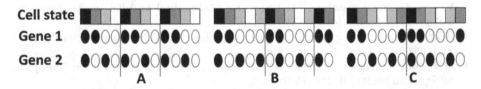

Fig. 5. Periodic patterns in a four-state model. Four- (A), five- (B) and six- (C) periodic patterns with illustration of how each state corresponds to different expressions of two genes. Cell state is indicated in black if both genes are expressed; in dark grey - Gene 1 is expressed and Gene 2 is not; in light grey – Gene 1 is not expressed and Gene 2 is expressed; white – both genes are not expressed.

Four-periodicity can easily be obtained if the model allows at least three distinct states for cells. While the three-state model doesn't have direct biological implementation, the four-state model seems to have better perspective as it can be viewed as modelling cells whose differentiation is associated with expression of pair of genes. The four-periodic pattern can correspond to the alternation of expressions of two genes as illustrated in Fig. 5A. The four-state model is considerably more sophisticated compared with the two-state model and can be used to reproduce the formation four- and even more- periodic patterns (Fig. 5). While the number of possible transition rules in the two-state model is relatively small (256), this number is extremely large for the four-state model (4^{4^3}, that is, every transition rule is a combination of $64 = 4^3$ elementary rules, while each elementary rule has four forms). This number makes the analysis of 4-state models challenging. Our preliminary simulations show that the formation of periodic stationary patterns in four-state model is extremely sensitive to the initial conditions, i.e. these patterns form only when very special initial conditions are met. This may explain the multi- (four-) level of segmentation in the fly embryo. The model can account for interactions between segment polarity genes with specific initial conditions set by the above three levels of patterning.

Mathematical study of the segmentation in fly embryo performed in [6] and based on gene network modelling has indicated that all nuclei, during normal development, can be only in four distinct states. This points to the direction of further extension of the presented work: our extensive analysis of ECA can be considered as a step towards similar analysis of four-state CA. Also, we note that more sophisticated CA (two-dimensional and including randomization of transition rules) have recently been used to successfully model the dynamics of skin patterning in growing lizard [7].

Acknowledgements. This work was funded by BBSRC grant BB/K002430/1. Authors are grateful to Dr Igor Potapov for fruitful discussions.

References

1. Wolpert, L.: Principles of Development. Oxford University Press, Oxford (2002)
2. Murray, J.D. (ed.): Mathematical Biology. IAM, vol. 17. Springer, New York (2002). https://doi.org/10.1007/b98868
3. Vasieva, O., Rasolonjanahary, M., Vasiev, B.: Mathematical modelling in developmental biology. Reproduction **145**(6), 175–184 (2013)
4. Hake, S., Wilt, F.: Principles of Developmental Biology. W. W. Norton and Company, New York (2003)
5. St Johnston, D., Nusslein-Volhard, C.: The origin of pattern and polarity in the Drosophila embryo. Cell **68**(2), 201–219 (1992)
6. Albert, R., Othmer, H.G.: The topology of the regulatory interactions predicts the expression pattern of the segment polarity genes in Drosophila melanogaster. J. Theor. Biol. **223**(1), 1–18 (2003)
7. Manukyan, L., et al.: A living mesoscopic cellular automaton made of skin scales. Nature **544**(7649), 173–176 (2017)
8. von Neumann, J.: The general and logical theory of automata. In: Hixon Symposium. Pasadena, California, USA (1948)
9. von Neumann, J. and Burks, A.W.: The Theory of Self-Reproducing Automata. Univ. Illinois Press (1966)
10. Wainer, G., et al.: Applying cellular automata and DEVS methodologies to digital games: a survey. Simul. Gaming **41**(6), 796–823 (2010)
11. Yang, X.-S., Young, Y.: Cellular automata, PDEs, and Pattern Formation. arXiv.1003.1983 (2010)
12. Wolfram, S.,: A new kind of science. Wolfram Media Inc, 1197 (2002)
13. Wolfram, S.: Statistical mechanics of cellular automata. Rev. Mod. Phys. **55**(3), 601–644 (1983)

14. Wolfram, S.: Theory and Applications of Cellular Automata. World Scientific, Singapore (1986)
15. Gravner, J., Griffeath, D.: Robust periodic solutions and evolution from seeds in one-dimensional edge cellular automata. Theoret. Comput. Sci. **466**, 64–86 (2012)
16. Jaeger, J., et al.: Dynamic control of positional information in the early Drosophila embryo. Nature **430**(6997), 368–371 (2004)

Multilevel Monte Carlo with Improved Correlation for Kinetic Equations in the Diffusive Scaling

Emil Løvbak[(✉)][iD], Bert Mortier[iD], Giovanni Samaey[iD],
and Stefan Vandewalle[iD]

Department of Computer Science, KU Leuven, Leuven, Belgium
{emil.loevbak,bert.mortier,giovanni.samaey,stefan.vandewalle}@kuleuven.be

Abstract. In many applications, it is necessary to compute the time-dependent distribution of an ensemble of particles subject to transport and collision phenomena. Kinetic equations are PDEs that model such particles in a position-velocity phase space. In the low collisional regime, explicit particle-based Monte Carlo methods simulate these high dimensional equations efficiently, but, as the collision rate increases, these methods suffer from severe time-step constraints. In the high collision regime, asymptotic-preserving particle schemes are able to produce stable results. However, this stability comes at the cost of a bias in the computed results. In earlier work, the multilevel Monte Carlo method was used to reduce this bias by combining simulations with large and small time steps. This multilevel scheme, however, still has large variances when correlating fine and coarse simulations, which leads to sub-optimal multilevel performance. In this work, we present an improved correlation approach that decreases the variance when bridging the gap from large time steps to time steps of the order of magnitude of the collision rate. We further demonstrate that this reduced variance results in a sharply reduced simulation cost at the expense of a small bias.

Keywords: Multilevel Monte Carlo · Kinetic equations · Diffusive scaling · Particle methods

1 Introduction

In many application domains, one encounters kinetic equations that model particle behavior in a high-dimensional position-velocity phase space. One such motivating application for this work is modelling neutral transport in plasma

We thank the anonymous reviewers for their feedback on improving the quality of this paper. This work was funded by the Research Foundation - Flanders (FWO) under fellowship numbers 1SB1919N and 1189919N. The computational resources and services used in this work were provided by the VSC (Flemish Supercomputer Center), funded by the Research Foundation - Flanders (FWO) and the Flemish Government - department EWI.

© Springer Nature Switzerland AG 2020
V. V. Krzhizhanovskaya et al. (Eds.): ICCS 2020, LNCS 12142, pp. 374–388, 2020.
https://doi.org/10.1007/978-3-030-50433-5_29

simulations of Tokamak fusion reactors. Simulation codes such as EIRENE [19] and DEGAS2 [20] simulate these models with particle-based Monte Carlo methods. Other approaches such as finite-volume methods or discrete Galerkin typically either suffer from dimension dependent computational costs, or additional errors due to the use of lower-dimensional approximations.

Often, one is interested in low-dimensional moments of the particle distribution, computed as averages over velocity space. In high-collisional regimes, such those in the plasma edge simulations, the time-scale at which these quantities of interest change is typically much slower than that of the particle dynamics. This presence of multiple time-scales results in stiffness: a naive simulation requires small time steps due to the fast dynamics, but long time horizons to capture the evolution of the macroscopic moments. The exact nature of the macroscopic behavior depends on the problem scaling, which can be hyperbolic or diffusive [4].

We consider a diffusively-scaled, spatially homogeneous kinetic equation modelling a distribution of particles $f(x, v, t)$ as a function of space $x \in \mathcal{D}_x \subset \mathbb{R}^d$, velocity $v \in \mathcal{D}_v \subset \mathbb{R}^d$ and time $t \in \mathbb{R}^+$. The scaling is captured by the dimensionless parameter ε. Simultaneously dividing the collision rate and multiplying the simulation time-scale with ε produces the model equation

$$\partial_t f(x, v, t) + \frac{v}{\varepsilon} \nabla_x f(x, v, t) = \frac{1}{\varepsilon^2} (\mathcal{M}(v)\rho(x, t) - f(x, v, t)), \qquad (1)$$

with $\rho(x, t) = \int_{\mathcal{D}_v} f(x, v, t) \, dv$ the particle density.

The right-hand side of (1) consists of the BGK operator [1], which linearly drives the velocity to a steady-state distribution $\mathcal{M}(v)$ with characteristic velocity $\tilde{v}$.

Individual particles in the distribution $f(x, v, t)$ follow a velocity-jump process, i.e., they travel in a straight line with a fixed velocity for an exponentially distributed time interval $\mathcal{E}(1/\varepsilon^2)$, at which point a collision is simulated by resampling their velocity from the distribution $\mathcal{M}(v)$. Taking the diffusion limit $\varepsilon \to 0$ in (1) drives the collision rate to infinity, while simultaneously increasing the velocity. In this limit, it can be shown that the particle density resulting from (1) converges to the diffusion equation [11],

$$\partial_t \rho(x, t) = \nabla_x^2 \rho(x, t). \qquad (2)$$

Equation (1) can be simulated using a wide selection of methods, which broadly fall into two categories. Deterministic methods solve (1) for $f(x, v, t)$ on a grid over the position-velocity phase space (using, for instance, finite differences or finite volumes). This approach quickly becomes computationally infeasible when the dimension grows, since a grid must be formed over the $2d$-dimensional domain, $\mathcal{D}_x \times \mathcal{D}_v$. Stochastic methods, on the other hand, perform simulations of individual particle trajectories, with each trajectory sampling the probability distribution $f(x, v, t)$. These methods do not suffer from the dimensionality of the phase space, but introduce a statistical error in the computed solution. When using explicit time steps, both approaches become prohibitively expensive for small values of ε due to the time step restriction.

To avoid the issues when ε becomes small, one can use asymptotic-preserving schemes [9]. Such methods preserve the macroscopic limit equation, in our case given by (2), as ε tends to zero, while avoiding time step constraints. Many such methods have been developed in literature for deterministic methods in the diffusive limit. For more information, we refer to a recent review paper [4] and the references therein. In the particle setting, only a few asymptotic-preserving methods exist, mostly in the hyperbolic scaling [3,6,7,16–18]. In the diffusive scaling, we are only aware of three works [2,5,14]. We use the approach taken in [5], which uses operator splitting to produce an unconditionally stable fixed time step particle method. This stability comes at the cost of an extra bias in the model, proportional to the size of the time step.

In [13] the multilevel Monte Carlo method [8] was applied to the scheme presented in [5]. Multilevel Monte Carlo methods first compute an initial estimate, using a large number of samples with a large time step (low cost per sample). This estimate has low variance, but is expected to have a large bias. Next, the bias of the initial estimate is reduced by performing corrections using a hierarchy of simulations with decreasing time step sizes. Under correct conditions, far fewer samples with small time steps are needed, compared to a direct Monte Carlo simulation with the smallest time step. This approach results in a reduced computational cost for a given accuracy. A core component of developing a multilevel Monte Carlo method is the development of a numerical approach to generate samples with different time step sizes which are correlated, i.e., they approximate the same continuous particle trajectory. This paper presents a strategy for improving the correlation developed in [13] and thus further reducing the computational cost of particle based simulations of (1).

The remainder of this paper is structured as follows. In Sect. 2 we present a naive particle based Monte Carlo scheme for simulating the kinetic equation (1) and show why this approach fails. In Sect. 3, we show how combining the asymptotic-preserving scheme from [5] and the multilevel Monte Carlo method combats these issues. In Sect. 4, we show the flaws of the existing method and introduce our improved correlation. In Sect. 5, we present experimental results using the new correlation, demonstrating computational speed-up. In Sect. 6, we summarize our results and list further challenges.

2   Monte Carlo Simulation Near the Diffusive Limit

For the sake of exposition, we limit this work to one spatial dimension, i.e., $d = 1$ but, our approach can be applied straightforwardly in more dimensions. Equation (1) now becomes

$$\partial_t f(x,v,t) + \frac{v}{\varepsilon}\partial_x f(x,v,t) = \frac{1}{\varepsilon^2}(\mathcal{M}(v)\rho(x,t) - f(x,v,t)). \tag{3}$$

We also restrict this paper to simulations with two discrete velocities, i.e., $\mathcal{M}(v) \equiv \frac{1}{2}(\delta_{v,-1} + \delta_{v,1})$, with δ the Kronecker delta function.

We focus on computing a quantity of interest $Y(t^*)$, which takes the form of an integral over a function $F(x, v)$ of the particle position $X(t)$ and velocity $V(t)$ at time $t = t^*$, with respect to the measure $f(x, v, t) \, dx \, dv$, i.e.,

$$Y(t^*) = \mathbb{E}\left[F\left(X(t^*), V(t^*)\right)\right] = \int_{\mathcal{D}_v} \int_{\mathcal{D}_x} F(x, v) f(x, v, t^*) \, dx \, dv.$$

Equation (3) can be simulated using a particle scheme with a fixed time step size Δt. Each particle p has a state $\left(X_{p,\Delta t}^n, V_{p,\Delta t}^n\right)$ in the position-velocity phase space at each time step n, with $X_{p,\Delta t}^n \approx X_p(n\Delta t)$ and $V_{p,\Delta t}^n \approx V_p(n\Delta t)$. We represent the distribution $f(x, v, t)$ of particles at time $n\Delta t$ by an ensemble of P particles, with indices $p \in \{1, \ldots, P\}$,

$$\left\{\left(X_{p,\Delta t}^n, V_{p,\Delta t}^n\right)\right\}_{p=1}^P. \tag{4}$$

A classical Monte Carlo estimator $\hat{Y}(t^*)$ for $Y(t^*)$ averages over such an ensemble

$$\hat{Y}(t^*) = \frac{1}{P} \sum_{p=1}^P F\left(X_{p,\Delta t}^N, V_{p,\Delta t}^N\right), \quad t^* = N\Delta t. \tag{5}$$

To generate an ensemble (4) we perform forward time stepping based on operator splitting, which is first order in the time step Δt [15]. For (3), operator splitting results in two actions for each time step:

1. **Transport step.** Each particle's position is updated based on its velocity

$$X_{p,\Delta t}^{n+1} = X_{p,\Delta t}^n + \Delta t V_{p,\Delta t}^n. \tag{6}$$

2. **Collision step.** Between transport steps, each particle's velocity is either left unchanged (no collision) or re-sampled from $\mathcal{M}(v)$ (collision), i.e.,

$$V_{p,\Delta t}^{n+1} = \begin{cases} V_{p,\Delta t}^{n,\prime} \sim \mathcal{M}(v), & \text{with collision probability } p_{c,\Delta t} = \Delta t / \varepsilon^2, \\ V_{p,\Delta t}^n, & \text{otherwise.} \end{cases} \tag{7}$$

Scheme (6)–(7) has a severe time step restriction $\Delta t = \mathcal{O}(\varepsilon^2)$ when approaching the limit $\varepsilon \to 0$. This time step restriction results in unacceptably high simulation costs, despite the well-defined limit (2) [5]. Given that the variance of the estimator $\hat{Y}(t^*)$ scales with $\frac{1}{P}$, a high individual simulation cost will severely constrain the precision with which we can approximate $Y(t^*)$.

3   Multilevel Monte Carlo Using AP Schemes

In the previous section, we demonstrated that a time step $\Delta t = \mathcal{O}(\varepsilon^2)$ is needed to resolve the collision dynamics of the kinetic equation (3). As ε becomes small, this causes a naive Monte Carlo estimator to be computationally infeasible. In this section, we first present an asymptotic-preserving Monte Carlo scheme which maintains stability for large time steps by reducing the contribution of collision dynamics for larger time steps. We will then present a multilevel Monte Carlo estimator, using this scheme, to combine simulations with different time step sizes, reducing the overall computational cost of the estimator $\hat{Y}(t^*)$.

3.1   Asymptotic-Preserving Monte Carlo Scheme

We use the asymptotic-preserving scheme from [5] as an alternative to (6)–(7). This asymptotic-preserving scheme has no time step constraints, so ε and Δt can be chosen independently. The scheme simulates a modified version of (3):

$$\partial_t f + \frac{\varepsilon v}{\varepsilon^2 + \Delta t}\partial_x f = \frac{\Delta t}{\varepsilon^2 + \Delta t}\partial_{xx}f + \frac{1}{\varepsilon^2 + \Delta t}(\mathcal{M}(v)\rho - f). \tag{8}$$

In (8), we have omitted the space, velocity and time dependency of $f(x, v, t)$ and $\rho(x, t)$, for conciseness. Note that the coefficients of (8) now explicitly contain the simulation time step Δt.

Equation (8) has the following properties [5]:

1. For $\varepsilon \to 0$, (8) converges to the diffusion equation (2).
2. For $\Delta t \to 0$, (8) converges to the original kinetic equation (3) $\mathcal{O}(\Delta t)$.
3. For $\Delta t \to \infty$, (8) converges to the diffusion equation (2) with a rate $\mathcal{O}\left(\frac{1}{\Delta t}\right)$.

The first property states that (8) has the same asymptotic limit in ε as (3). The second and third properties provide us with an intuitive understanding of (8). We can interpret the modified equation as a combination of the diffusion equation (2) and the original kinetic equation (3), where the two contributions are weighted in function of Δt. For large time steps the diffusion equation dominates over the kinetic equation. At the particle level, the diffusion equation corresponds with Brownian motion, which has no time constraints, hence the stability of (8).

Particle trajectories are now simulated as follows:

1. **Transport-diffusion step.** The position of the particle is updated based on its velocity and a Brownian increment

$$X_{p,\Delta t}^{n+1} = X_{p,\Delta t}^n + V_{p,\Delta t}^n \Delta t + \sqrt{2\Delta t}\sqrt{D_{\Delta t}}\xi_p^n, \tag{9}$$

in which we generate $\xi_p^n \sim \mathcal{N}(0, 1)$ and introduce a Δt-dependent diffusion coefficient $D_{\Delta t} = \frac{\Delta t}{\varepsilon^2 + \Delta t}$ and velocity $V_{p,\Delta t}^n \sim \mathcal{M}_{\Delta t}(v)$, where the time step dependent distribution $\mathcal{M}_{\Delta t}(v)$ with characteristic velocity $\tilde{v}_{\Delta t} = \frac{\varepsilon}{\varepsilon^2 + \Delta t}$ can be decomposed as

$$\mathcal{M}_{\Delta t}(v) = \tilde{v}_{\Delta t}\mathcal{M}(v). \tag{10}$$

2. **Collision step.** During collisions, each particle's velocity is updated as:

$$V_{p,\Delta t}^{n+1} = \begin{cases} V_{p,\Delta t}^{n,\prime} \sim \mathcal{M}_{\Delta t}(v), & \text{with probability} p_{c,\Delta t} = \frac{\Delta t}{\varepsilon^2 + \Delta t}, \\ V_{p,\Delta t}^n, & \text{otherwise.} \end{cases} \tag{11}$$

The scheme (9)–(11) stably generates trajectories for large Δt. However, these trajectories are biased with $\mathcal{O}(\Delta t)$ compared to those generated by (6)–(7).

3.2   Multilevel Monte Carlo

The key idea behind the multilevel Monte Carlo method (MLMC) [8] is to combine simulations with different time step sizes to simulate many trajectories (low variance), while also simulating accurate trajectories (low simulation bias).

First, a coarse Monte Carlo estimator $\hat{Y}_0(t^*)$ with a time step Δt_0 is used

$$\hat{Y}_0(t^*) = \frac{1}{P_0} \sum_{p=1}^{P_0} F\left(X_{p,\Delta t_0}^{N_0}, V_{p,\Delta t_0}^{N_0}\right), \quad t^* = N_0 \Delta t_0. \tag{12}$$

Estimator (12) is based on a large number of trajectories P_0, but has a low computational cost as few time steps N_0 are required to reach time t^*.

Next, $\hat{Y}_0(t^*)$ is refined upon by a sequence of L difference estimators at levels $\ell = 1, \ldots, L$. Each difference estimator uses an ensemble of P_ℓ particle pairs

$$\hat{Y}_\ell(t^*) = \frac{1}{P_\ell} \sum_{p=1}^{P_\ell} \left(F\left(X_{p,\Delta t_\ell}^{N_\ell}, V_{p,\Delta t_\ell}^{N_\ell}\right) - F\left(X_{p,\Delta t_{\ell-1}}^{N_{\ell-1}}, V_{p,\Delta t_{\ell-1}}^{N_{\ell-1}}\right) \right), \quad t^* = N_\ell \Delta t_\ell. \tag{13}$$

Each correlated particle pair consists a particle with a fine time step Δt_ℓ and a particle with a coarse time step $\Delta t_{\ell-1} = M \Delta t_\ell$, with M a positive integer. The particles in each pair undergo correlated simulations, which intuitively can be understood as making both particles follow the same qualitative trajectory for two different simulation accuracies. One can interpret the difference estimator (13) as using the fine simulation to estimate the bias in the coarse simulation.

Given a sequence of levels $\ell \in \{0, \ldots, L\}$, with decreasing step sizes, and the corresponding estimators given by (12)–(13), the multilevel Monte Carlo estimator for the quantity of interest $Y(t^*)$ is computed by the telescopic sum

$$\hat{Y}(t^*) = \sum_{\ell=0}^{L} \hat{Y}_\ell(t^*). \tag{14}$$

It is clear that the expected value of the estimator (14) is the same as that of (5), with the finest time step $\Delta t = \Delta t_L$. Given a sufficiently fast reduction in the number of simulated (pairs of) trajectories P_ℓ as ℓ increases, it can be shown that the multilevel Monte Carlo estimator requires a lower computational cost than a classical Monte Carlo estimator to achieve the same mean square error.

3.3   Term-by-term Correlation Approach

The differences in (13) will only have low variance if the simulated paths up to $X_{p,\Delta t_\ell}^{N_\ell}$ and $X_{\Delta t_{p,\ell-1}}^{N_{\ell-1}}$, with time steps related by $\Delta t_{\ell-1} = M \Delta t_\ell$ are correlated. To discuss these correlated pairs of trajectories, we define a sub-step index $m \in \{1, \ldots, M\}$, i.e., $X_{p,\Delta t_\ell}^{n,m} \approx X_p(n\Delta t_{\ell-1} + m\Delta t_\ell) \equiv X_p((nM+m)\Delta t_\ell)$. In order to span a time interval of size $\Delta t_{\ell-1}$, the coarse simulation requires a single time step of size $\Delta t_{\ell-1}$ while the fine simulation requires M time steps of size Δt_{ℓ}:

$$\begin{cases} X_{p,\Delta t_{\ell-1}}^{n+1} = X_{p,\Delta t_{\ell-1}}^n + \Delta t_{\ell-1} V_{p,\Delta t_{\ell-1}}^n + \sqrt{2\Delta t_{\ell-1}}\sqrt{D_{\Delta t_{\ell-1}}}\xi_{p,\ell-1}^n \\ X_{p,\Delta t_{\ell}}^{n+1,0} = X_{p,\Delta t_{\ell}}^{n,0} + \displaystyle\sum_{m=1}^{M}\left(\Delta t_{\ell} V_{p,\Delta t_{\ell}}^{n,m} + \sqrt{2\Delta t_{\ell}}\sqrt{D_{\Delta t_{\ell}}}\xi_{p,\ell}^{n,m}\right) \end{cases},$$

with $\xi_{p,\ell-1}^n, \xi_{p,\ell}^{n,m} \sim \mathcal{N}(0,1)$, $V_{p,\Delta t_{\ell-1}}^n \sim \mathcal{M}_{\Delta t_{\ell-1}}(v)$ and $V_{p,\Delta t_{\ell}}^{n,m} \sim \mathcal{M}_{\Delta t_{\ell}}(v)$.

There are two sources of stochastic behavior in the asymptotic-preserving scheme (9)–(11). On the one hand, a new Brownian increment ξ_p^n is generated at each time step. On the other hand, each time step has a finite collision probability. Collisions produce a new particle velocity $V_{p,\Delta t}^{n+1}$ for the next time step.

To correlate these simulations we first perform M independent time steps of the fine simulation. We then combine the M sets of random numbers in the fine path into one set of random numbers which can be used in a single time step for the coarse simulation. If the fine simulation random numbers are combined in such a way that the resulting coarse simulation random numbers are distributed as if they had been generated independently, then the coarse simulation statistics will be preserved, while also being dependent on the fine simulation.

In [13], an approach was developed to independently correlate Brownian increments and collision phenomena, which we briefly describe here. For a full derivation, demonstrative figures and an analysis, we refer to [12,13].

Correlating Brownian Increments. In each fine sub-step $m \in \{1,\dots,M\}$ a Brownian increment $\xi_{p,\ell}^{n,m} \sim \mathcal{N}(0,1)$ is generated. The sum of these M increments is distributed as $\mathcal{N}(0,M)$. This means that a Brownian increment for the corresponding coarse simulation step can be generated as

$$\xi_{p,\ell-1}^n = \frac{1}{\sqrt{M}}\sum_{m=1}^{M}\xi_{p,\ell}^{n,m} \sim \mathcal{N}(0,1). \tag{15}$$

Correlating Collisions. Implementing (11) requires simulating a collision in a fine sub-step $m \in \{1,\dots,M\}$, with the collision probability $p_{c,\Delta t_{\ell}}$. In practice this is achieved by drawing a uniformly distributed random number $u_{p,\ell}^{n,m} \sim \mathcal{U}([0,1])$ and performing a collision if this number is larger than the probability $p_{nc,\Delta t_{\ell}} = 1 - p_{c,\Delta t_{\ell}}$ that no collision has occurred in the time step

$$u_{p,\ell}^{n,m} \geq p_{nc,\Delta t_{\ell}} = \frac{\varepsilon^2}{\varepsilon^2 + \Delta t_{\ell}}. \tag{16}$$

At least one collision has taken place in the fine simulation if the largest of the generated $u_{p,\ell}^{n,m}$, $m \in \{1,\dots,M\}$, satisfies (16), i.e.,

$$u_{p,\ell}^{n,\max} = \max_m u_{p,\ell}^{n,m} \geq p_{nc,\Delta t_{\ell}}. \tag{17}$$

When (17) is satisfied, the probability of a collision taking place in the correlated coarse simulation should be large. This is the case if we compare

$$u_{p,\ell-1}^n = \left(u_{p,\ell}^{n,\mathrm{max}}\right)^M \sim \mathcal{U}([0,1]), \tag{18}$$

with $p_{nc,\Delta t_{\ell-1}}$ in the coarse simulation.

If a collision is performed in the simulation at level $\ell-1$ using (18), then the coarse simulation velocity should be correlated with that of the fine simulation, at the end of the time interval. We consider the decomposition of $\mathcal{M}(v)$ given in (10). In those of the M fine sub-steps that contain a collision, we draw a time step independent velocity $\bar{v}_{p,\ell}^{n,m}$ to generate $V_{p,\ell}^{n,m}$. Given that the $\bar{v}_{p,\ell}^{n,m}$ are i.i.d., we can select one freely to use as $\bar{v}_{p,\ell-1}^n$. To maximize the correlation of the velocities at the end of the time interval, we take the last generated $\bar{v}_{p,\ell}^{n,m}$, i.e.,

$$\bar{v}_{p,\ell-1}^n = \bar{v}_{p,\ell}^{n,i}, \quad i = \arg\max_{1\le m\le M}\left(m\Big|u_{p,\ell}^{n,m} \ge p_{nc,\Delta t_\ell}\right). \tag{19}$$

4   Improved Coupling of Particle Trajectories

4.1   Limitations of Term-by-Term Correlation

In [12], it was demonstrated that a multilevel Monte Carlo scheme, based on the correlation in Sect. 3.3 achieves an asymptotic speed-up $\mathcal{O}\left(E^{-2}\log^2(E)\right)$ in the root mean square error bound E for the following sequence of levels:

1. At level 0, generate an initial estimate of $\hat{Y}(t^*)$ by simulating with $\Delta t_0 = t^*$.
2. At level 1, perform correlated simulations to t^* using $\Delta t_0 = t^*$ and $\Delta t_1 = \varepsilon^2$.
3. Continue to generate a geometric sequence of levels $\Delta t_l = \varepsilon^2 M^{1-l}$ for $l > 1$ until an acceptably low bias has been achieved.

This result only holds when $\Delta t_\ell \ll \varepsilon^2$, however, meaning that the scheme only significantly reduces computational cost when calculating very accurate results.

To demonstrate this phenomenon, we present a multilevel simulation with a two-speed velocity distribution using the scheme from Sect. 3.3 to estimate the squared particle displacement $F(x,v) = x^2$. The results for $\varepsilon = 0.1$, a RMSE $E = 0.1$, $f(x,v,0) \equiv \delta_{x,0}\frac{1}{2}(\delta_{v,-1} + \delta_{v,1})$, and $t^* = 0.5$ are shown in Table 1. In this table, we list various statistics of the samples $\hat{F}_\ell$, sample differences $\hat{F}_\ell - \hat{F}_{\ell-1}$ and resulting estimators $\hat{Y}_\ell$ as well as the computational cost. We define $F_{-1} \equiv 0$ and define the sample cost C_ℓ relative to a simulation with $\Delta t = \varepsilon^2$.

Looking at the values of $\mathbb{V}[\hat{F}_\ell - \hat{F}_{\ell-1}]$ in Table 1, it is clear that the variance at level 1 is almost as large as that at level 0. Comparing the values of $\mathbb{V}[\hat{F}_\ell - \hat{F}_{\ell-1}]$ with those of $\mathbb{E}[\hat{F}_\ell - \hat{F}_{\ell-1}]$, we see, contrastingly, that the expected value of the estimator at level 1 is an order of magnitude smaller than that at level 0. This indicates that the lack of decreasing variance from level 0 to level 1 is due to insufficient correlation between the trajectories used in the estimator at level 1. As a result a large number of samples P_1 are needed with cost C_1, making the total cost $P_1 C_1$ of level 1 much higher than that of level 0.

Table 1. Computing $\hat{Y}(t^*)$ for $\varepsilon = 0.1$ and $E = 0.1$. Listed quantities are the fine time step size Δt_ℓ, number of samples P_ℓ, expected value $\mathbb{E}$ and variance $\mathbb{V}$ of the differences of simulations $\hat{F}_\ell - \hat{F}_{\ell-1}$, estimator variance $\mathbb{V}[\hat{Y}_\ell]$, sample cost C_ℓ and level cost $P_\ell C_\ell$.

ℓ	Δt_ℓ	P_ℓ	$\mathbb{V}[\hat{F}_\ell]$	$\mathbb{E}[\hat{F}_\ell - \hat{F}_{\ell-1}]$	$\mathbb{V}[\hat{F}_\ell - \hat{F}_{\ell-1}]$	$\mathbb{V}[\hat{Y}_\ell]$	C_ℓ	$P_\ell C_\ell$
0	5.00×10^{-1}	17 716	2.02	1.01×10^{0}	2.02×10^{0}	1.14×10^{-4}	0.02	354
1	1.00×10^{-2}	2 157	1.43	-1.17×10^{-1}	1.48×10^{0}	6.85×10^{-4}	1.02	2 200
2	5.00×10^{-3}	602	1.49	-1.36×10^{-2}	3.71×10^{-1}	6.16×10^{-4}	3	1 806
3	2.50×10^{-3}	489	1.65	3.55×10^{-2}	4.10×10^{-1}	8.39×10^{-4}	6	2 934
4	1.25×10^{-3}	345	1.44	-3.94×10^{-2}	4.43×10^{-1}	1.29×10^{-3}	12	4 140
5	6.25×10^{-4}	185	1.61	3.88×10^{-2}	1.84×10^{-1}	9.94×10^{-4}	24	4 440
6	3.13×10^{-4}	50	2.87	-5.41×10^{-2}	1.31×10^{-1}	2.62×10^{-3}	48	2 400
Σ				8.60×10^{-1}		7.16×10^{-3}		18 274

The reason for this poor correlation can be found in the intuitive interpretation given to Eq. (8) in Sect. 3.1. For large time steps, the diffusion term becomes dominant over the transport and collision effects, whereas the inverse is true for smaller time steps. At level 1, the coarse diffusive simulation is being generated from a simulation with a $\Delta t = \varepsilon^2$, which contains a significant transport component. The correlation approach in Sect. 3.3 correlates the Brownian increments independently from the collisions and resulting velocity changes. Hence the transport-collision effects from the fine simulation, which form a significant part of the particle behavior, are ignored in the coarse simulation.

4.2   Improved Correlation Approach

In this section, we present an improved correlation approach for particles simulating the scheme (9)–(11). This scheme will let the velocities generated in the fine simulation $\bar{v}_{p,1}^{n,m}$ influence the Brownian increments $\xi_{p,0}^{n,m}$ in the correlated coarse simulation, avoiding the correlation issues described in Sect. 4.1.

We start from [10], which shows how one can simulate Brownian motion in the weak sense, using an approximate weak Euler scheme. The lower order moments of the approximate Brownian increments in this scheme must be bounded by

$$\left| \mathbb{E}\left[\xi_{p,\ell-1}^{n} \right] \right| + \left| \mathbb{E}\left[\left(\xi_{p,\ell-1}^{n} \right)^{3} \right] \right| + \left| \mathbb{E}\left[\left(\xi_{p,\ell-1}^{n} \right)^{2} \right] - 1 \right| \leq K \Delta t_{\ell-1}, \qquad (20)$$

for some constant K. If this condition holds, the same weak convergence as classical Euler-Maruyama is achieved. The new, improved, correlation generates $\xi_{p,0}^{n}$ as a weighted sum of a contribution $\xi_{p,\ell-1,W}^{n}$ from the fine simulation diffusion and a contribution $\xi_{p,\ell-1,T}^{n}$ from the fine simulation re-scaled velocities $\bar{v}_{p,\ell}^{n,m}$

$$\xi_{p,\ell-1}^{n} = \sqrt{\theta_\ell} \, \xi_{p,\ell-1,W}^{n} + \sqrt{1 - \theta_\ell} \, \xi_{p,\ell-1,T}^{n}, \qquad (21)$$

with $\theta_\ell \in [0,1]$. If both $\xi_{p,\ell-1,W}^{n}$ and $\xi_{p,\ell-1,T}^{n}$ have mean zero and variance one, the same holds for $\xi_{p,\ell-1}^{n}$. This means that $K = 0$ in (20). The correlation of the transport increments (18)–(19) is left unaltered, to maintain the correlation of the simulation velocities. We now discuss each value in (21):

Diffusive Contribution $\xi^n_{p,\ell-1,W}$. The coupling of Brownian increments is done in a way identical to (15) from the correlation approach in Sect. 3.3:

$$\xi^n_{p,\ell-1,W} = \frac{1}{\sqrt{M}} \sum_{m=1}^{M} \xi^{n,m}_{p,\ell}. \tag{22}$$

Transport Contribution $\xi^n_{p,\ell-1,T}$. We can generate a value $\xi^n_{p,\ell-1,T}$ with expected value zero and unit variance from the $\bar{v}^{n,m}_{p,\ell}$ as

$$\xi^n_{p,\ell-1,T} = \left(\mathbb{V} \left[\sum_{m=1}^{M} \bar{v}^{n,m}_{p,\ell} \right] \right)^{-\frac{1}{2}} \sum_{m=1}^{M} \bar{v}^{n,m}_{p,\ell}.$$

As subsequent $\bar{v}^{n,m}_{p,\ell}$ are correlated, computing this variance is slightly more involved than in the case of $\xi^n_{p,\ell-1,W}$, but still straightforward

$$\mathbb{V} \left[\sum_{m=1}^{M} \bar{v}^{n,m}_{p,\ell} \right] = \sum_{m=1}^{M} \mathbb{V} \left[\bar{v}^{n,m}_{p,\ell} \right] + 2 \sum_{m=1}^{M-1} \sum_{m'=m+1}^{M} \mathrm{Cov} \left(\bar{v}^{n,m}_{p,\ell}, \bar{v}^{n,m'}_{p,\ell} \right)$$

$$= M + 2 \sum_{\Delta m=1}^{M} (M - \Delta m) \left(p_{nc,\Delta t_\ell} \right)^{\Delta m}$$

$$= M + 2 \frac{p_{nc,\Delta t_\ell} \left((p_{nc,\Delta t_\ell})^M - M(p_{nc,\Delta t_\ell} - 1) - 1 \right)}{(p_{nc,\Delta t_\ell} - 1)^2}$$

$$= M + 2 \frac{p_{nc,\Delta t_\ell} \left((p_{nc,\Delta t_\ell})^M + M p_{c,\Delta t_\ell} - 1 \right)}{(p_{c,\Delta t_\ell})^2}, \tag{23}$$

where $p_{c,\Delta t_\ell}$ is the collision probability as defined in (11) and $p_{nc,\Delta t_\ell} = 1 - p_{c,\Delta t_\ell}$. We want to improve the correlation of level 1 where a large time step Δt_0 is correlated with $\Delta t_1 = \varepsilon^2$. In that case $p_{nc,\Delta t_1} = p_{c,\Delta t_1} = \frac{1}{2}$ so (23) simplifies to

$$\mathbb{V} \left[\sum_{m=1}^{M} \bar{v}^{n,m}_{p,1} \right] = 3M + 2^{2-M} - 4 \approx 3M - 4, \tag{24}$$

where the approximation quickly becomes more accurate for larger values of M.

Contribution Weight θ_ℓ. What remains is to choose the weight between the two contributions from the simulation with Δt_ℓ to the diffusion in the simulation with $\Delta t_{\ell-1}$. We aim to pick θ_ℓ so that (22) has an equally large contribution when simulating over a time interval of size $\Delta t_{\ell-1}$ at both level ℓ and $\ell - 1$.

At level ℓ the variance contribution of (22) relative to the variance of the total simulation spanning $\Delta t_{\ell-1}$ with M steps is given by

$$\frac{\mathbb{V}\left[\sum_{m=1}^{M}\sqrt{2\Delta t_{\ell}}\sqrt{D_{\ell}}\xi_{p,\ell}^{n,m}\right]}{\mathbb{V}\left[\sum_{m=1}^{M}\sqrt{2\Delta t_{\ell}}\sqrt{D_{\ell}}\xi_{p,\ell}^{n,m}+\Delta t_{\ell}\tilde{v}_{\ell}\bar{v}_{p,\ell}^{n,m}\right]}. \tag{25}$$

The contribution of (22) when spanning the same time interval of size $\Delta t_{\ell-1}$ with a single step is given by

$$\frac{\mathbb{V}\left[\sqrt{2\Delta t_{\ell-1}}\sqrt{D_{\ell-1}}\sqrt{\theta_{\ell}}\xi_{p,\ell-1,W}^{n}\right]}{\mathbb{V}\left[\sqrt{2\Delta t_{\ell-1}}\sqrt{D_{\ell-1}}\xi_{p,\ell-1}^{n}+\Delta t_{\ell-1}\tilde{v}_{\ell-1}\bar{v}_{p,\ell-1}^{n}\right]}. \tag{26}$$

Equating (25) and (26) and solving for θ_{ℓ} gives

$$\theta_{\ell}=\frac{\mathbb{V}\left[\sum_{m=1}^{M}\sqrt{2\Delta t_{\ell}}\sqrt{D_{\ell}}\xi_{p,\ell}^{n,m}\right]\mathbb{V}\left[\sqrt{2\Delta t_{\ell-1}}\sqrt{D_{\ell-1}}\xi_{p,\ell-1}^{n}+\Delta t_{\ell-1}\tilde{v}_{\ell-1}\bar{v}_{p,\ell-1}^{n}\right]}{\mathbb{V}\left[\sqrt{2\Delta t_{\ell-1}}\sqrt{D_{\ell-1}}\xi_{p,\ell-1,W}^{n}\right]\mathbb{V}\left[\sum_{m=1}^{M}\sqrt{2\Delta t_{\ell}}\sqrt{D_{\ell}}\xi_{p,\ell}^{n,m}+\Delta t_{\ell}\tilde{v}_{\ell}\bar{v}_{p,\ell}^{n,m}\right]}.$$

Given (23) and the unit variance of $\xi_{p,\ell-1,W}^{n},\xi_{p,\ell-1}^{n},\xi_{p,\ell}^{n,m}$ and $\bar{v}_{p,\ell-1}^{n}$, we get

$$\theta_{\ell}=\frac{D_{\ell}\left(2\Delta t_{\ell-1}D_{\ell-1}+\Delta t_{\ell-1}^{2}\tilde{v}_{\ell-1}^{2}\right)}{D_{\ell-1}\left(2M\Delta t_{\ell}D_{\ell}+\Delta t_{\ell}^{2}\tilde{v}_{\ell}^{2}\left(M+2\frac{p_{nc,\Delta t_{\ell}}(p_{nc,\Delta t_{\ell}}^{M}+Mp_{c,\Delta t_{\ell}}-1)}{p_{c,\Delta t_{\ell}}^{2}}\right)\right)}. \tag{27}$$

At level 1, a very large time step Δt_{0} is correlated with $\Delta t_{1}=\varepsilon^{2}$. Here, we can use approximations $D_{0}\approx1,\tilde{v}_{0}\approx\frac{\varepsilon}{\Delta t_{0}},p_{nc,\Delta t_{0}}\approx\frac{\varepsilon^{2}}{\Delta t_{0}}$ and (24) to simplify (27) to

$$\theta_{1}\approx\frac{4M+1}{7M-4}.$$

It is important to note that (21) does not produce true Brownian increments, in contrast to the independent simulation at level 0. On the one hand, the distribution of (21) is not Gaussian. On the other hand, subsequent coarse time steps will be correlated. Both issues become more pronounced for higher levels and for that reason we only apply this approach in level 1. In the next section, numerical experiments asses whether this mismatch produces biased results.

5   Experimental Results

5.1   Application to a Low-Precision Case

We re-do the computations from Sect. 4.1, using the new correlation (21) at level 1. Comparing Tables 1 and 2, we see that the variance $\mathbb{V}[F_{1}-F_{0}]$ has reduced by a factor 30. The reduced variance reduces the cost of the full simulation by a factor 3.9. Given the low tolerance $E=0.1$, it cannot be observed whether the improved correlation produces biased results without more accurate simulations.

Table 2. Computing $\hat{Y}(t^*)$ for $\varepsilon = 0.1$ and $E = 0.1$ with improved correlation. Quantities as listed in Table 1.

ℓ	Δt_ℓ	P_ℓ	$\mathbb{V}[\hat{F}_\ell]$	$\mathbb{E}[\hat{F}_\ell - \hat{F}_{\ell-1}]$	$\mathbb{V}[\hat{F}_\ell - \hat{F}_{\ell-1}]$	$\mathbb{V}[\hat{Y}_\ell]$	C_ℓ	$P_\ell C_\ell$
0	5.00×10^{-1}	8 123	1.88	9.76×10^{-1}	1.88×10^0	2.31×10^{-4}	0.02	162
1	1.00×10^{-2}	189	1.74	-1.44×10^{-1}	4.89×10^{-2}	2.59×10^{-4}	1.02	193
2	5.00×10^{-3}	315	1.76	-1.74×10^{-2}	4.30×10^{-1}	1.36×10^{-3}	3	945
3	2.50×10^{-3}	152	1.28	-5.32×10^{-2}	2.71×10^{-1}	1.78×10^{-3}	6	912
4	1.25×10^{-3}	102	3.16	6.78×10^{-2}	1.58×10^{-1}	1.55×10^{-3}	12	1 224
5	6.25×10^{-4}	50	1.08	3.61×10^{-2}	1.76×10^{-1}	3.52×10^{-3}	24	1 200
$\sum$				8.65×10^{-1}		8.70×10^{-3}		4 636

5.2 Application to a High-Precision Case

In Table 3, we repeat the simulations in Tables 1 and 2 for $E = 0.01$. Despite the variance decrease at level 1, both simulation costs are of the same order of magnitude. This is due to the high cost of levels $\ell > 1$, reducing the impact of the improvement at level 1. At first sight, this seems a major setback, however the penultimate column of Table 3 shows that $\mathbb{V}[\hat{F}_\ell - \hat{F}_{\ell-1}] > \mathbb{V}[F_1 - F_0]$ for many of these levels. This indicates that we can skip levels by correlating Δt_0 with $\Delta t_1 \ll \varepsilon^2$ at level 1. In Table 4, we do this with $\Delta t_1 = \frac{\varepsilon^2}{128}$. For a fair comparison, we also apply the same sequence of levels to the term-by-term correlation.

Table 3. Computing $\hat{Y}(t^*)$ for $\varepsilon = 0.1$ and $E = 0.01$. Quantities as listed in Table 1.

ℓ	Δt_ℓ	Term-by-term correlation			Improved correlation		
		$\mathbb{E}[\hat{F}_\ell - \hat{F}_{\ell-1}]$	$\mathbb{V}[\hat{F}_\ell - \hat{F}_{\ell-1}]$	$P_\ell C_\ell$	$\mathbb{E}[\hat{F}_\ell - \hat{F}_{\ell-1}]$	$\mathbb{V}[\hat{F}_\ell - \hat{F}_{\ell-1}]$	$P_\ell C_\ell$
0	5.0×10^{-1}	9.91×10^{-1}	1.96×10^0	6.84×10^4	9.89×10^{-1}	1.96×10^0	6.87×10^4
1	1.0×10^{-2}	-1.23×10^{-1}	1.42×10^0	4.14×10^5	-1.36×10^{-1}	5.25×10^{-2}	7.99×10^4
2	5.0×10^{-3}	8.34×10^{-3}	4.36×10^{-1}	3.92×10^5	1.06×10^{-2}	4.38×10^{-1}	3.98×10^5
3	2.5×10^{-3}	2.57×10^{-2}	4.02×10^{-1}	5.34×10^5	3.02×10^{-2}	4.03×10^{-1}	5.42×10^5
4	1.3×10^{-3}	3.05×10^{-2}	3.12×10^{-1}	6.68×10^5	2.65×10^{-2}	3.04×10^{-1}	6.61×10^5
5	6.3×10^{-4}	2.29×10^{-2}	1.96×10^{-1}	7.47×10^5	2.49×10^{-2}	1.91×10^{-1}	7.42×10^5
6	3.1×10^{-4}	8.53×10^{-3}	1.15×10^{-1}	8.01×10^5	9.44×10^{-3}	1.10×10^{-1}	7.92×10^5
7	1.6×10^{-4}	9.38×10^{-3}	5.21×10^{-2}	7.69×10^5	4.06×10^{-3}	5.98×10^{-2}	8.22×10^5
8	7.8×10^{-5}	1.88×10^{-3}	2.35×10^{-2}	7.38×10^5	2.64×10^{-3}	3.12×10^{-2}	8.67×10^5
9	3.9×10^{-5}	3.07×10^{-4}	1.23×10^{-2}	8.35×10^5	5.54×10^{-3}	1.86×10^{-2}	1.06×10^6
10	2.0×10^{-5}	3.54×10^{-3}	1.74×10^{-2}	7.68×10^5	1.88×10^{-3}	1.08×10^{-2}	7.68×10^5
$\sum$		9.79×10^{-1}		6.73×10^6	9.69×10^{-1}		6.80×10^6

Comparing Tables 3 and 4, shows that leaving out levels reduces the computational cost when using the improved correlation at level 1, while the term-by-term correlation becomes more expensive. The simulation with term-by-term correlation in Table 4 requires a factor 2 less computation than that with improved correlation in Table 3.

As a reference for the bias, we note that a simulation with $E = 0.001$ in [12], computed an estimate of 9.80×10^{-1} for $\hat{Y}(t^*)$. Tables 3 and 4 indicate that the results using the term-by-term correlation are consistently within the set tolerance $E = 0.01$. Those using the new correlation are consistently too small. Twenty repetitions of the simulation with term-by-term correlation in Table 3 and that with improved correlation in Table 4 estimates a bias of 2.31×10^{-2} with standard deviation 1.14×10^{-2}. This shows that the improved correlation comes with the cost of an additional bias.

Table 4. Computing $\hat{Y}(t^*)$ for $\varepsilon = 0.1$ and $E = 0.01$ with skipped levels. Quantities as listed in Table 1.

ℓ	Δt_ℓ	Term-by-term correlation			Improved correlation		
		$\mathbb{E}[\hat{F}_\ell - \hat{F}_{\ell-1}]$	$\mathbb{V}[\hat{F}_\ell - \hat{F}_{\ell-1}]$	$P_\ell C_\ell$	$\mathbb{E}[\hat{F}_\ell - \hat{F}_{\ell-1}]$	$\mathbb{V}[\hat{F}_\ell - \hat{F}_{\ell-1}]$	$P_\ell C_\ell$
0	5.0×10^{-1}	9.91×10^{-1}	1.96×10^0	9.88×10^5	9.89×10^{-1}	1.95×10^0	3.70×10^4
1	7.8×10^{-5}	-4.93×10^{-3}	3.76×10^0	1.09×10^7	-3.12×10^{-2}	3.76×10^{-2}	4.06×10^5
2	3.9×10^{-5}	1.55×10^{-4}	1.57×10^{-2}	1.41×10^6	-5.25×10^{-3}	1.42×10^{-2}	4.28×10^5
3	2.0×10^{-5}	3.01×10^{-3}	3.16×10^{-3}	7.68×10^5	2.68×10^{-4}	2.58×10^{-2}	8.80×10^5
4	9.8×10^{-6}	-	-	-	-1.25×10^{-3}	6.75×10^{-3}	1.54×10^6
$\sum$		9.89×10^{-1}		1.32×10^7	9.51×10^{-1}		3.29×10^6

6   Conclusion

We presented a new, improved particle based multilevel Monte Carlo scheme for simulating the Boltzmann-BGK equation in diffusive regimes. This scheme improves upon earlier work by introducing a correlation between the kinetic transport in the fine simulation and the diffusion in the coarse simulation. The approach reduces the difference estimator variance at level 1 by an order of magnitude. In turn, the reduced variance significantly decreases the computation required for low accuracy simulations. For more accurate simulations, the reduced variance allows for several computationally expensive levels to be skipped, resulting in a greatly improved computational efficiency. A caveat is the inter time-step correlation and non-normal distribution of the coarse random numbers, resulting in biased results. Resolving these inconsistencies is the topic of future work, which will aid the extension of the improved correlation to all levels. It also remains to be seen whether the observed biases have a significant effect in practical applications, which typically contain, among others, modelling errors.

References

1. Bhatnagar, P.L., Gross, E.P., Krook, M.: A model for collision processes in gases. I. Small amplitude processes in charged and neutral one-component systems. Phys. Rev. **94**(3), 511–525 (1954). https://doi.org/10.1103/PhysRev.94.511
2. Crestetto, A., Crouseilles, N., Lemou, M.: A particle micro-macro decomposition based numerical scheme for collisional kinetic equations in the diffusion scaling. Commun. Math. Sci. **16**(4), 887–911 (2018). https://doi.org/10.4310/CMS.2018.v16.n4.a1
3. Degond, P., Dimarco, G., Pareschi, L.: The moment-guided Monte Carlo method. Int. J. Numer. Methods Fluids **67**(2), 189–213 (2011). https://doi.org/10.1002/fld.2345
4. Dimarco, G., Pareschi, L.: Numerical methods for kinetic equations. Acta Numer. **23**, 369–520 (2014). https://doi.org/10.1017/S0962492914000063
5. Dimarco, G., Pareschi, L., Samaey, G.: Asymptotic-preserving Monte Carlo methods for transport equations in the diffusive limit. SIAM J. Sci. Comput. **40**(1), A504–A528 (2018). https://doi.org/10.1137/17M1140741
6. Dimarco, G., Pareschi, L.: Hybrid multiscale methods II. Kinetic equations. Multiscale Model. Simul. **6**(4), 1169–1197 (2008). https://doi.org/10.1137/070680916
7. Dimarco, G., Pareschi, L.: Fluid solver independent hybrid methods for multiscale kinetic equations. SIAM J. Sci. Comput. **32**(2), 603–634 (2010). https://doi.org/10.1137/080730585
8. Giles, M.B.: Multilevel Monte Carlo path simulation. Oper. Res. **56**(3), 607–617 (2008). https://doi.org/10.1287/opre.1070.0496
9. Jin, S., Pareschi, L., Toscani, G.: Uniformly accurate diffusive relaxation schemes for multiscale transport equations. SIAM J. Numer. Anal. **38**(3), 913–936 (2000). https://doi.org/10.1137/S0036142998347978
10. Kloeden, P.E., Platen, E.: Weak taylor approximations. Numerical Solution of Stochastic Differential Equations, vol. 23, pp. 457–484. Springer, Heidelberg (1992). https://doi.org/10.1007/978-3-662-12616-5_14
11. Lapeyre, B., Pardoux, É., Sentis, R., Craig, A.W., Craig, F.: Introduction to Monte Carlo Methods for Transport and Diffusion Equations, vol. 6. Oxford University Press, Oxford (2003)
12. Løvbak, E., Samaey, G., Vandewalle, S.: A multilevel Monte Carlo method for asymptotic-preserving particle schemes. arXiv (jul 2019)
13. Løvbak, E., Samaey, G., Vandewalle, S.: A Multilevel Monte Carlo asymptotic-preserving particle method for kinetic equations in the diffusion limit. In: Tuffin, B., L'Ecuyer, P. (eds.) MCQMC 2018. SPMS, vol. 324, pp. 383–402. Springer, Cham (2020). https://doi.org/10.1007/978-3-030-43465-6_19
14. Mortier, B., Baelmans, M., Samaey, G.: Kinetic-diffusion asymptotic-preserving Monte Carlo algorithms for plasma edge neutral simulation. Contrib. Plasma Phys. e201900134 (2019). https://doi.org/10.1002/ctpp.201900134
15. Pareschi, L.: Hybrid multiscale methods for hyperbolic and kinetic problems. ESAIM Proc. **15**, 87–120 (2005). https://doi.org/10.1051/proc:2005024
16. Pareschi, L., Caflisch, R.E.: An implicit Monte Carlo method for rarefied gas dynamics. J. Comput. Phys. **154**(1), 90–116 (1999). https://doi.org/10.1006/jcph.1999.6301
17. Pareschi, L., Russo, G.: An introduction to Monte Carlo method for the Boltzmann equation. ESAIM Proc. **10**, 35–75 (2001). https://doi.org/10.1051/proc:2001004

18. Pareschi, L., Trazzi, S.: Numerical solution of the Boltzmann equation by time relaxed Monte Carlo (TRMC) methods. Int. J. Numer. Methods Fluids **48**(9), 947–983 (2005). https://doi.org/10.1002/fld.969
19. Reiter, D., Baelmans, M., Börner, P.: The EIRENE and B2-EIRENE codes. Fusion Sci. Technol. **47**(2), 172–186 (2005). https://doi.org/10.13182/FST47-172
20. Stotler, D., Karney, C.: Neutral gas transport modeling with DEGAS 2. Contrib. Plasma Phys. **34**(2–3), 392–397 (1994). https://doi.org/10.1002/ctpp.2150340246

Development and Application of the Statistically Similar Representative Volume Element for Numerical Modelling of Multiphase Materials

Łukasz Rauch$^{(\boxtimes)}$ (iD), Krzysztof Bzowski(iD), Danuta Szeliga(iD), and Maciej Pietrzyk(iD)

AGH University of Science and Technology,
al. Mickiewicza 30, 30-059 Kraków, Poland
`lrauch@agh.edu.pl`

Abstract. Modern aerospace, automotive and construction industries rely on materials with non-homogeneous properties like composites or multiphase structures. Such materials offer a lot of advantages, but they also require application of advanced numerical models of exploitation condition, which are of high importance for designers, architects and engineers. However, computational cost is one of the most important problems in this approach, being very high and sometimes unacceptable. In this paper we propose approach based on Statistically Similar Representative Volume Element (SSRVE), which is generated by combination of isogeometric analysis and optimization methods. The proposed solution significantly decreases computational cost of complex multiscale simulations and simultaneously maintains high reliability of solvers. At first, the motivation of the work is described in introduction, which is followed by general idea of the SSRVE as a modelling technique. Afterwards, examples of generated SSRVEs based on two different cases are given and passed further to numerical simulations of exploitation conditions. The results obtained from these calculations are used in the model predicting gradients of material properties, which are crucial results for discussion on uniqueness of the proposed solution. Additionally, some aspects of computational cost reduction are discussed, as well.

Keywords: Multiscale modelling · SSRVE · Multiphase materials · Gradients of properties

1 Introduction

In recent years it was observed that heterogeneous materials benefit from the best features due to the mix of phases they are made of. Taking an advantage of heterogeneity is the main strengthening mechanism for multiphase steels, which are developed today. On the other hand, steep gradients of properties between various phases in these steels may lead to low local fracture resistance [1, 2]. Thus, searching for a compromise between strengthening by multiphase microstructure and tendency to local

© Springer Nature Switzerland AG 2020
V. V. Krzhizhanovskaya et al. (Eds.): ICCS 2020, LNCS 12142, pp. 389–402, 2020.
https://doi.org/10.1007/978-3-030-50433-5_30

fracture due to steep gradients of properties became an important field of research in materials engineering [3]. This research requires multiscale material models, which explicitly account for the microstructure. Thus, a detailed description of the microstructure features of the Complex Phase (CP) steels is required to investigate the correlation between the multiphase structure and exploitation properties. Smoothing of gradients of properties can be reached by control of such phenomena as segregation of elements, evolution of the morphology of hard phases, evolution of dislocation populations in the soft phase and precipitation. Modelling of these phenomena requires advanced models. Large spectrum of material models with various complexity and various predictive capabilities are available now [4]. However, mean field models, which predict only average parameters of the microstructure, are not applicable. Multiscale models, which use representative volume element (RVE) have to be used. Since computing costs are an important criterion in the model selection, designers intensively search for methods of model reduction, see discussion in [4]. Application of the statistically similar representative volume element (SSRVE), which was proposed in [5], is one of the possibilities.

The basic idea of the SSRVE is to replace a RVE with an arbitrary complex inclusion morphology by a periodic one composed of optimal unit cells [5]. To reach this goal, the parameters describing fraction of different phases and their geometrical characteristics are determined for a considered microstructure. Following this, optimization methods are used to design morphology of hard inclusions in the SSRVE. Authors have already applied SSRVE to modelling deformation of Dual Phase (DP) steels with the microstructure composed of hard martensite islands in soft ferrite matrix [6]. In publication [7] four shape coefficients were considered to describe multiphase microstructure. Seven more parameters, which are used in the image analysis, were added and described in the paper [6]. Since a large number of parameters can be used to describe multiphase microstructure, uniqueness of the solution is always questionable.

The objectives of the present paper were formulated with the above comments in mind. Computational aspects of the SSRVE design for the microstructures were discussed and accuracy as well as uniqueness of the solution were evaluated. Beyond this, as it has been mentioned above, complex phase steels microstructures, which allow to obtain smoother gradients of properties, are of particular interest now. Thus, an attempt to design SSRVE for multiphase microstructures composed of various constituents and phases was the main objective of the work.

2   Statistically Similar Representative Volume Element

2.1   General Idea

In the micro-macro modelling approach an RVE representing the underlying microscale domain is usually attached at each Gauss point of the macroscopic solution [8]. The constitutive law describing material behaviour in the macroscale is obtained by averaging the first Piola-Kirchoff stresses with respect to the RVE. The theoretical basis of the micro-macro modelling is well described in the scientific literature (e.g. [5]) and it is not repeated here. In the SSRVE application to DP steel, already mentioned in introduction [6], the focus was on development of the simplest SSRVE, which allows to decrease the computing costs and will make micro-macro modelling approach more efficient. The basic idea was to replace an RVE with an arbitrary complex inclusion morphology by a periodic one composed of optimal unit cells, as shown in Fig. 1.

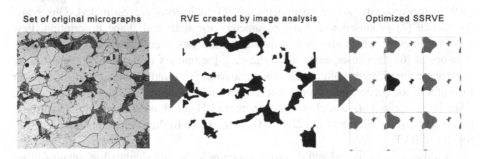

Fig. 1. Illustration of the basic concept of the SSRVE, a) RVE, b) periodically arranged SSRVE.

This idea is applied in the present work to the analysis of the more sophisticated steels microstructures composed of more than two phases like CP or TRIP steels with Transformation Induced Plasticity effect. Both of these materials are difficult to analyse and model, while various phases are characterized by different properties and morphologies. This fact highly influences computational complexity of both creation and application of SSRVE models for such materials in practice.

2.2 Design of the SSRVE

The process of SSRVE creation consists of the following steps (Fig. 2):

- image analysis aiming at conversion of set of original micrographs from optical microscope to RVE containing separated phases and grains inside of them,
- qualitative and quantitative analysis leading to obtain information on shape coefficients based on grains/phases shapes, which allows to apply sensitivity analysis and indicate the most important coefficients characterizing microstructure,
- construction of a cost function for optimization procedures, implementation of a proper optimization procedure and selection of the most suitable results.

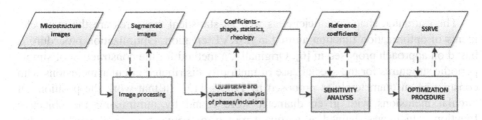

Fig. 2. Procedure of SSRVE creation

The procedure of SSRVE creation starts with an analysis of original micrographs, which aims at creation of binary (segmented) images with separated phases or inclusions. The algorithms for processing of various micrographs are presented in details in [9]. Then, in the case of 2D SSRVE, the shape coefficients of inclusions in original images are estimated directly from the segmented pictures. In the case of 3D procedure, the reconstruction of 3D microstructure on the basis of 2D images has to be performed [10]. Afterwards, the shape coefficients of 3D inclusions are estimated. Ohser and Muecklich [7] proposed four basic parameters for statistical shape description. Few more parameters, which are used in the image analysis, were added and described by Authors in [6]. In consequence, the following parameters describing shape of inclusions were considered: volume fraction, area/volume, roundness, ellipsoid fit, contour to center ratio, border index, mean curvature, total curvature, Malinowska coefficient, Blair-Bliss coefficient, Danielsson coefficient and Haralick coefficient. Not all of these parameters can be adapted from 2D to 3D. Thus, some of the coefficients are used only for 2D SSRVE.

Additionally, statistical and rheological coefficients are calculated to obtain full set of reference coefficients, which describe all aspects of material properties and which are used further in optimization procedure. Brandts et al. [11] introduced the higher order statistical measures for microstructures: n-point probability functions, spectral density and lineal-path function. The latter parameter is crucial in description of anisotropic microstructures, while it describes the probability that a complete line segment $\mathbf{a} = \overrightarrow{\mathbf{a_1 a_2}}$ is located in specific direction in the same phase, where $\mathbf{a_1} = \{x_1, y_1\}$ and $\mathbf{a_2} = \{x_2, y_2\}$ are coordinates of the ends of the line segment. Lu and Torquato [12] gave a general mathematical description of this measure for multi-phase anisotropic materials. Simplified approach, which is applicable to DP microstructures represented in form digital images composed of a set of pixels, can be defined by the following equation:

$$\zeta_{LP}(m,k) = \frac{1}{N_x N_y} \sum_{p=1}^{N_x} \sum_{q=1}^{N_y} \chi\left(\overrightarrow{\mathbf{a_1 a_2}}\right) \tag{1}$$

where m and k are the length of vector $\overrightarrow{\mathbf{a_1 a_2}}$ in x and y directions respectively, N_x, N_y are the dimensions of micrograph in pixels, and $\chi^{(i)}\left(\overrightarrow{\mathbf{a_1 a_2}}\right)$ is modified indicator function defined for phase $D^{(i)}$ as:

$$\chi^{(i)}\left(\overrightarrow{\mathbf{a_1 a_2}}\right) = \begin{cases} 1 & \text{if } \overrightarrow{\mathbf{a_1 a_2}} \in D^{(i)} \\ 0 & \text{otherwise} \end{cases} \tag{2}$$

The estimated shape coefficients as well as statistical measures are the main elements of optimization function aiming at SSRVE creation. Optimization procedure is based on approach proposed in [6]. Originally, a method for the construction of simple periodic structures for the special case of randomly distributed circular inclusions with constant equal diameters was proposed by Povirk [13]. In that work the positions of circular inclusions with given diameter were found by minimizing the objective function, which was defined as a square root error between spectral density of the periodic RVE and non-periodic real microstructure. In our work this function was adapted to the following form:

$$\Phi = \sqrt{\sum_{i=1}^{n}\left[w_i\left(\frac{\varsigma_i - \varsigma_{iSSRVE}}{\varsigma_i}\right)\right]} \qquad (3)$$

where: w_i – normalized weights, n – number of coefficients, ζ_i – i^{th} reference coefficient obtained from original microstructure, ζ_{iSSRVE} – i^{th} coefficient obtained from SSRVE. The coefficients include all mentioned parameters describing shape, rheology and statistics. The current implementation of optimization procedure is based on genetic algorithm (GA), where chromosome is composed of m elements representing coordinates of control points determining SSRVE shape. These points are connected with spline functions forming smooth shape of SSRVE inclusion. Calculations of the objective function are performed iteratively for each proposition of new SSRVE shape. The optimization loop is preceded by sensitivity analysis (SA), which allows to determine the most influential parameters of the optimization or to determine the weights used in the objective function (3).

2.3   Optimization Procedure

On the basis of results obtained from sensitivity analysis the reference coefficients, their values and weights were established. A set of these parameters was used further in optimization procedure as ς_i in (3). Multi-iterative genetic algorithm [14] was applied as an optimization procedure. The procedure is composed of the following steps:

- Generation of initial population – random generation of n specimens containing information about coordinates of control points and their weights. Size of the specimen depends strongly on a number of control points describing inclusions in the SSRVE. At the beginning of calculations the control points form m random shapes, where m is a number of inclusions.
- Estimation of the objective function value – the procedure calculates values of ς_{iSSRVE} on the basis of shapes of inclusions in subsequently generated SSRVEs. Shapes of these inclusions are described by Non-uniform Relational B-Splines (NURBS). These parametric curves are controlled by basic interpolation functions and a set of mentioned control points. The weight of each control point influences the position of curve near the particular control point. The shapes of inclusions influence also rheological properties and statistical description of microstructure. Rheological model of SSRVE is determined by processing of this element in virtual uniaxial compression, tensile and shear deformation tests. Obtained stress-strain relations allow to calculate equivalent tensile stress, which describes material rheology. The main statistical measure is lineal-path function, calculated directly from pixels or voxels values of SSRVE.
- Stop conditions – two fundamental conditions are implemented i.e. number of iterations and mean square error between expected and actual objective function.
- Application of genetic operators – crossing and mutation operators were implemented in the presented approach. The former operator is responsible for exchange of random number of genes between two specimens (Fig. 3).

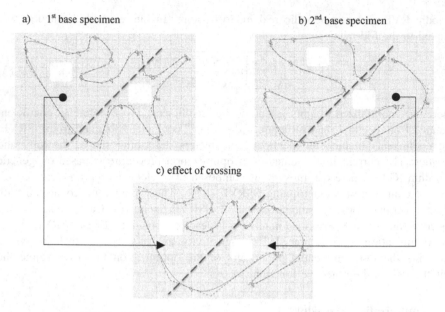

Fig. 3. Illustration of crossing operator for 2D specimens.

The first of mutation operators (Fig. 4) changes positions of control points regarding centre of gravity of the shape. The operator determines random set of control points, which will be mutated, and then the new position is calculated also on the basis of randomized operations. The second mutation operator changes the positions of control points in X or Y axis direction by using random values of coordinates in the vector of translation.

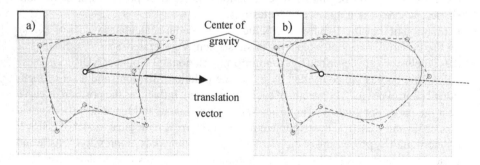

Fig. 4. Illustration of mutation operator for 2D specimens.

- Generation of a new population – the specimens obtained after operations of crossing and mutation are included in the new population. In each subsequent iteration, each specimen is validated. On the basis of validation the worst specimens are removed from population, some of the best specimens go further without changes and the rest is passed again to crossing and mutation operators.

In the case of 3D calculations shape of inclusion is composed of 2D layers, which are used in reconstruction algorithm mentioned at the beginning of this section. Thus, the specimen in 3D contains additional Z coordinates, which influence all the algorithms inside optimization procedure. Therefore, the crossing operator is based on exchange of a set of whole layers between two specimens instead of a set of single control points. The first mutation operator behaves similarly by translating of whole layers in Z axis direction, while the second mutation operator chooses two layers randomly and replaces them inside one SSRVE.

3   SSRVE for Complex Phase Steel

Verification of the SSRVE concept for more than two phases inside material microstructure was performed on micrographs obtained for CP steels. This group of steels is a part of Advanced High Strength Steels (AHSSs), widely used by modern industry and characterized by elevated mechanical properties [15]. High strength of CP steels is gained through extremely fine grain size and microstructure containing small amounts of martensite, pearlite and retained austenite embedded in a ferrite-bainite matrix. Therefore, from numerical simulations point of view such material is very difficult to be modelled, especially meshing of computational domain and coverage of larger scale with full field micro model. Example of CP micrograph is presented in Fig. 5.

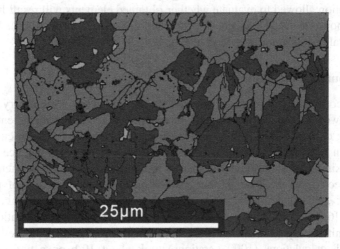

Fig. 5. Microstructure of CP steel composed of bainite (1st phase in blue), martensite (2nd phase in green), ferrite matrix (yellow), small inclusions of retained austenite (light yellow). (Color figure online)

A set of fundamental input data is composed of series of images, like in Fig. 5, obtained from optical microscope or EBSD, which marks various phases and inclusions in different colours. Such set of images is passed to image analysis procedures to divide original micrographs into subsets dedicated to different phases (Fig. 6).

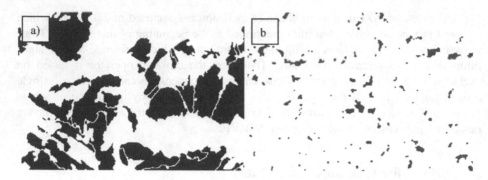

Fig. 6. Separated phases of microstructure: (a) bainite – 40% of volume fraction, (b) martensite – 2%.

3.1 SSRVE Generation

For validation purposes and SSRVE generation two CP steels were analysed with different volume fraction of phases: I – 40% of bainite and 7% of martensite, II – 30% of bainite and 15% of martensite. Both steels contained 50% of ferrite. Optimization procedure responsible for generation of SSRVEs was performed several times for both cases assuming random initial generation of SSRVE in the first iteration. The set of images for each steel contained six micrographs taken in various places in material. This assumption allowed to evaluate whether obtained elements will result in the same material response as far as gradient of properties is considered (Sect. 4.2). The results of SSRVE calculations are presented in Fig. 7.

3.2 Computing Cost Analysis

Computing cost mostly depends on selected shape coefficients. Preliminary sensitivity analysis allowed for the identification of representative geometric coefficients for each phase. In the case of bainite grains: volume fraction of phases ξ_1, mean curvature ξ_8, ratio between maximum distance in inclusion phase and its circumference Lp2, ratio between maximum and minimum distance in inclusion phase Lp3, compactness factor R_c were selected. In the case of martensite grains: volume fraction of phases ξ_1, ellipsoid fit ξ_5, mean curvature ξ_8 were used. The details of each coefficient are broadly described in [6, 16]. Satisfactory results were obtained after about 300 iteration for each attempt. Minimal improvements in the objective functions were observed in further analysis. Full calculations (1000 iterations) took about 10 h on a typical, 8 cores, desktop computer, but a satisfactory threshold was reached after just 2 h. Examples of the changes of the Root Mean Square Error (RMSE) in the optimization procedure are shown in Fig. 8.

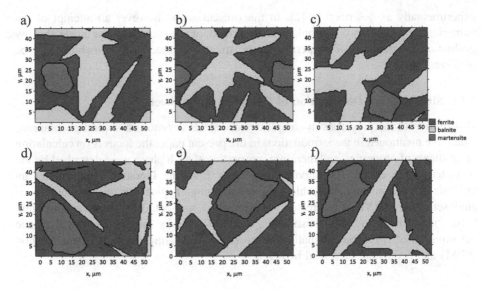

Fig. 7. Three selected results of SSRVE generation for CP steel variant I (a-c) and variant II (d-f)

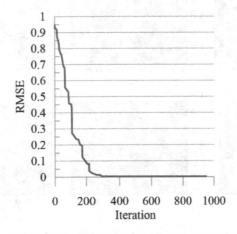

Fig. 8. Root Mean Square Error minimization during SSRVE generation procedure.

4 Solution Uniqueness Analysis

As it has been mentioned in the introduction, depending on their microstructure multiphase steels show differing damage mechanisms. In general global formability and local formability is distinguished [2]. The former is an ability of a material to undergo plastic deformation without formation of a localized neck respectively to distribute strains uniformly. The latter is an ability of a material to undergo plastic deformation in a local area without fracture. These two features can be characterized experimentally as described in [2]. In the present work, however an attempt of the numerical approach to this problem is described. An assumption was made that by evaluation of gradients of properties in the microstructure, local formability can be predicted.

4.1   Simulation of Deformation of Complex Phase Steels

Distribution of the properties in complex phase steels depends on several phenomena, which are mentioned in the introduction. In the present paper the focus is on calculation of gradients of properties, and therefore, properties of each phase were described by the work-hardening curve. The curves were taken from the literature [17] without an analysis of phenomena responsible for the hardening. Typical CP microstructure was analysed and SSRVE was created, as described in Sect. 3.1. To reveal different behaviour of hard and soft constituents, that SSRVE was subjected to small plastic deformation of 5%. The calculations were performed using Finite Element Method (FEM) and results are presented in Fig. 9.

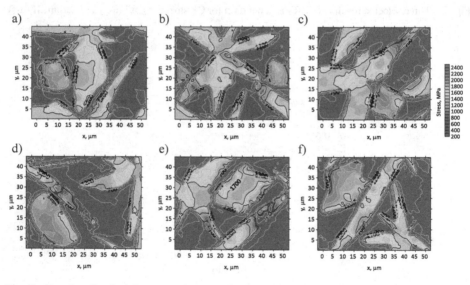

Fig. 9. Results of uniaxial compression tests for SSRVE with distribution of the flow stress for variant I (a-c) and variant II (d-f).

4.2   Gradients of Properties

As it was shown in [3], gradients of mechanical properties in the microstructure can be considered as a measure of the local formability. The local flow stress of the material was selected as a measure of these properties in the present work. It was assumed that description of the gradients can be based on the local and global sensitivity analysis (SA) methods [18]. Considered physical phenomenon during material manufacturing or processing can be described as a non-stationary system $\mathbf{y}(t, \mathbf{p}, \mathbf{x})$, where $\mathbf{p}$ is a vector of process parameters including model coefficients, **x** is a position vector in the computation domain, t is time. To predict gradient distributions in the whole material volume, the first order sensitivity indices were applied and expressed through a Taylor series expansion with respect to a position vector **x**. Thus, locally, sensitivity matrix can be computed as:

$$S = \{s_{ij}\} = \{\partial y_i / \partial x_j\} \qquad (4)$$

where y_i - model outputs, x_i - components of a position vector **x**.

On the other hand, the screening sensitivity analysis method, called Morris Design (MD) [15], allows to estimate a global measure of the sensitivity. In this algorithm elementary effects (ξ_i) are defined:

$$\xi_j(\bar{\mathbf{x}}) := \frac{\mathbf{y}(\bar{x}_1, \ldots, \bar{x}_{j-1}, \bar{x}_j + \Delta_j, \bar{x}_{j+1}, \ldots, \bar{x}_k) - \mathbf{y}(\bar{\mathbf{x}})}{\Delta_j} \qquad (5)$$

where: **y** - model output, $\bar{\mathbf{x}} \in \Omega \subset P^k - k$ - dimensional vector of model inputs x_j, Δ_j - an increment of the model input $\bar{x}_j$.

Sampling $\bar{x}_j$ in the space Ω gives a finite distribution of elementary effects ξ_j calculated for the jth component of an input vector $\bar{\mathbf{x}}$. Based on this distribution expected value μ_j for j^{th} model input can be estimated through the classic estimators for independent random samples.

For estimation material gradient properties let $\bar{\mathbf{x}}$ from MD algorithm be a position vector **x** in the equation describing a process. Thus, the elementary effects (5), for $\Delta_j \to 0$, correspond to derivatives in the sensitivity matrix (4). In the present work only one model output was considered ($i = 1$). The matrix **S** is a vector calculated for one model input $\bar{\mathbf{x}}$ and for gradient estimation $\|S\|$ was taken. The distribution of $\|S\|$ was obtained by sampling $\bar{\mathbf{x}}$ over the whole computation domain. Next, based on the MD method assumptions, two sensitivity indices were estimated: expected values μ of all $\|S\|$ as a global measure of properties gradient and $\mu_{0.05}$ as expected values of the gradient properties for 5% of maximum values. The last measure was introduced to evaluate maximum gradients in the microstructure. However, since a single maximum value of gradient could be a result of a numerical error, a mean of 5% of maximum values was calculated.

One of the main assumptions of the gradients estimation algorithm is that the data for which gradients are estimated may be either experimental or derived from numerical simulations. In general, the data is defined as a points cloud form. The efficient management of this cloud required the introduction of a specific data structure for a flexible search of single points and their neighbours. For this, a quadtree structure [19] was developed and adopted for gradients estimation requirements.

To approximate value v for point of coordinates (x_1, x_2), based on N points of known values v_i, $i = 1, \ldots, N$, Shepard's method and the inverse distance weighting were applied [20]:

$$v(x_1, x_2) = \begin{cases} v_i & \exists i : x_1 = x_1^i \wedge x_2 = x_2^i \\ \dfrac{\sum_{i=1}^{N} w_i(x_1, x_2) v_i}{\sum_{i=1}^{N} w_i(x_1, x_2)} & otherwise \end{cases}$$

$$w_i(x_1, x_2) = \left(\sqrt{(x_1 - x_1^i)^2 + (x_2 - x_2^i)^2} \right)^{-2}$$

(6)

Gradients of the flow stress in the SSRVE were calculated in an analytical way for the approximation given by Eq. (6) and using the finite difference method. Finally, expected values μ and $\mu_{0.5}$ were computed as global measures of gradients of properties. The gradient distributions for two variants of SSRVE are presented in Fig. 10 and the quantities of μ were as follows: 55.2 MPa/µm, 53.4 MPa/µm and 502.5 MPa/µm for variant I and plots (a-c), and 62.6 MPa/µm, 36.8 MPa/µm, 110.4 MPa/µm for variant II and plots (d-f), respectively. For the measure $\mu_{0.5}$ the results were: 412.0 MPa/µm, 277.6 MPa/µm and 17508.7 MPa/µm for variant I (plots (a-c) Fig. 10), and 524.8 MPa/µm, 179.7 MPa/µm, 1802.9 MPa/µm for variant II (plots (d-f) Fig. 10).

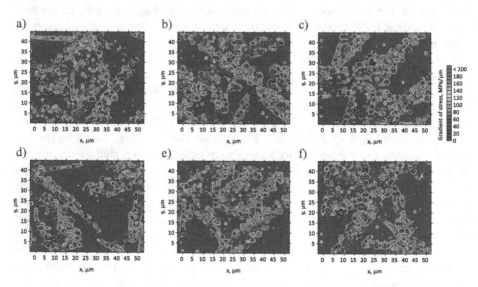

Fig. 10. Gradient distributions for SSRVE: variant I (a-c), variant II (d-f).

4.3   Discussion of Results

SSRVE elements were generated with the goal function based on the selected geometric coefficients and a proper description of a microstructure morphology was obtained. Optimization based on the genetic algorithms was efficient and satisfactory results were obtained after about 300 iteration for each attempt. On the other hand, although the geometrical effect was correct, subsequent runs of the optimization procedure yielded the properties of material, which were not unique and different values of gradients were obtained for the same phase composition. In the present work the SSRVEs were selected and analysis was performed to highlight those differences.

Analysis of the results of global measures of material gradient properties showed that the differences are significant and they can reach an order of magnitude in extreme examples. Keeping in mind the results of calculations performed so far, a suggestion was made that the goal function for SSRVE generation cannot be based solely on the geometrical parameters and it should include also the data of maximum strains or stresses for each of the material phase. Further investigations over these problems will be carried out.

5 Conclusions

Methodology of design of the SSRVE for complex phase steel composed of three phases and application of this SSRVE to evaluation of gradients of properties is described in the paper. The following conclusions were drawn:

- Application of the SSRVE allowed significant decrease of the computing costs comparing to classical RVE solution in terms of element dimension. Following this, it is possible to apply more SSRV elements in a multiscale simulation than when RV elements are with the same computation cost.
- Good convergence of the optimization was observed during SSRVE generation. Satisfactory results were obtained after about 300 iteration for each attempt.
- Design of the SSRVE based on the geometrical features only does not guarantee uniqueness of the solution. Different gradients were obtained for one phase composition and several optimization runs. It leads to a conclusion that additional parameters based on maximum strains and stresses in various phases should be introduced in the objective function. This will be subject of the future works.
- The methodology based on the sensitivity analysis proved to be efficient for evaluation of gradients of materials properties when they are defined as a cloud of points.

Acknowledgement. Financial assistance of the NCN, project no. 2016/23/G/ST5/04059, is acknowledged.

References

1. Karelova, A., Krempaszky, C., Werner, E., Tsipouridis, P., Hebesberger, T., Pichler, A.: Hole expansion of dual-phase and complex-phase AHS steels – effect of edge conditions. Steel Res. Int. **80**, 71–77 (2009). https://doi.org/10.2374/SRI08SP110
2. Heibel, S., Dettinger, T., Nester, W., Clausmeyer, T., Tekkaya, A.E.: Damage mechanisms and mechanical properties of high-strength multiphase steels. Materials **11**, 761 (2018). https://doi.org/10.3390/ma11050761

3. Szeliga, D., Chang, Y., Bleck, W., Pietrzyk, M.: Evaluation of using distribution functions for mean field modelling of multiphase steels. Procedia Manuf. **27**, 72–77 (2019). https://doi.org/10.1016/j.promfg.2018.12.046

4. Pietrzyk, M., Madej, Ł., Rauch, Ł., Szeliga, D.: Computational Materials Engineering: Achieving High Accuracy and Efficiency in Metals Processing Simulations. Elsevier, Amsterdam (2015)

5. Schroeder, J., Balzani, D., Brands, S.: Approximation of random microstructures by periodic statistically similar representative volume elements based on lineal-path functions. Arch. Appl. Mech. **81**, 975–997 (2011). https://doi.org/10.1007/s00419-010-0462-3

6. Rauch, Ł., Pernach, M., Bzowski, K., Pietrzyk, M.: On application of shape coefficients to creation of the statistically similar representative element of DP steels. Comput. Methods Mater. Sci. **11**, 531–541 (2011)

7. Ohser, J., Muecklich, F.: Statistical Analysis of Microstructure in Materials Science. Wiley, New York (2000)

8. Allix, O.: Multiscale strategy for solving industrial problems. Comput. Methods Appl. Sci. **6**, 107–126 (2006). https://doi.org/10.1007/1-4020-5370-3_6

9. Rauch, Ł., Madej, Ł.: Application of the automatic image processing in modelling of the deformation mechanisms based on the digital representation of microstructure. Int. J. Multiscale Comput. Eng. **8**, 1–14 (2010). https://doi.org/10.1615/IntJMultCompEng.v8.i3.90

10. Rauch, Ł., Imiołek, K.: Reconstruction of 3D material microstructure of one phase steels. Rudy i Metale Nieżelazne **58**, 726–730 (2013)

11. Brands, S., Schroeder, J., Balzani, D.: Statistically similar reconstruction of dual-phase steel microstructures for engineering applications. In: Computer Methods in Mechanics, CD-ROM, Warsaw (2011)

12. Lu, B.L., Torquato, S.: Lineal-path function for random heterogeneous materials. J. Appl. Mech. **36**, 922–929 (1992). https://doi.org/10.1103/physreva.45.922

13. Povirk, G.L.: Incorporation of microstructural information into models of two-phase materials. Acta Metall. **43**, 3199–3206 (1995). https://doi.org/10.1016/0956-7151(94)00487-3

14. Kus, W., Burczynski, T.: Parallel evolutionary optimization in multiscale problems. Comput. Methods Material Sci. **9**, 347–351 (2009)

15. Schmitt, J.H., Iung, T.: New developments of advanced high-strength steels for automotive applications. C R Phys. **19**, 641–656 (2018). https://doi.org/10.1016/j.crhy.2018.11.004

16. Matusiewicz, P., Czarski, A., Adrian, H.: Estimation of materials microstructure parameters using computer program SigmaScan Pro. Metall. Foundry Eng. **33**, 33–40 (2007). https://doi.org/10.7494/mafe.2007.33.1.33

17. Wang, X., Zurob, H.S., Xu, G., Ye, Q., Bouazis, O., Embury, D.: Influence of microstructural length scale on the strength and annealing behavior of pearlite, bainite and martensite. Metall. Mater. Trans. A **44A**, 1454–1461 (2013). https://doi.org/10.1007/s11661-012-1501-1

18. Szeliga, D.: Identification problems in metal forming. AGH University of Science and Technology, Kraków, A comprehensive study (2013)

19. Veenadevi, S.V., Ananth, A.G.: Fractal image compression using quadtree decomposition and Huffman coding. Signal Image Process. Int. J. **3**, 207–212 (2012). https://doi.org/10.5121/sipij.2012.3215

20. Shepard, D.: A two-dimensional interpolation function for irregularly-spaced data. In: 23rd Association for Computing Machinery, New York, pp. 517–524 (1968)

A Heterogeneous Multi-scale Model for Blood Flow

Benjamin Czaja[1]([✉]) [iD], Gábor Závodszky[1,2] [iD], and Alfons Hoekstra[1] [iD]

[1] Computational Science Lab, Informatics Institute, University of Amsterdam,
Amsterdam, Netherlands
B.E.Czaja@uva.nl
[2] Department of Hydrodynamic Systems, Budapest University of Technology
and Economics, Budapest, Hungary

Abstract. This research focuses on developing a heterogeneous multi-scale model (HMM) for blood flow. Two separate scales are considered in this study, a Macro-scale, which models whole blood as a continuous fluid and tracks the transport of hematocrit profiles through an advection diffusion solver. And a Micro-scale, which computes directly local diffusion coefficients and viscosities using cell resolved simulations. The coupling between these two scales also includes the use of a surrogate model, which saved local viscosity and diffusion coefficients from previously simulated local hematocrit and shear rate combinations. As the HMM model progresses fewer micro models will be spawned. This is accomplished through the surrogate by interpolating from previously computed viscosities and diffusion coefficients. The benefit of using the HMM method for blood flow is that it, along with resolving the rheology of whole blood, can be extended with other types computational models to model physiological processes like thrombus formation.

Keywords: Multi-scale · Blood flow · Rheology

1 Introduction

The primary function of the human red blood cell (RBC) is to deliver oxygen to the tissues of the body. From a physiological perspective there are multiple scales at which a red blood cell interacts with the human body. On the largest scales is a healthy human, typical size ≈ 1–$2\,\mathrm{m}$, which is maintained though the constant supply of oxygen via the oxygen diffusion from the red blood which occurs on scales of $\mu\mathrm{m}$. The physiological processes that maintains a healthy oxygenated human being is a connection of multiple heterogeneous systems that exists and operate on multiple spatial scales. Mutli-scale computational models are being developed in order to understand how these individual processes are coupled together to maintain a working healthy human being [17,18].

The CompBioMed2 project grant agreement No. 823712.

From a rheological perspective whole blood, which travels the through the cardiovascular system, is a multi-scale process in itself. On the smallest scales (μm), are the deformations of a RBC and the corresponding interactions with the suspending blood plasma. Because of the cell nature of whole blood, typically dominated by the deformable RBCs, blood flow on scales below 300 μm exhibits non-Newtonian Behavior. Such a hallmark of this behavior is the Fåhraeus-Lindqvist effect [13], which is the decrease in the relative apparent viscosity of whole blood that has been contributed by the lubrication layer on the vessel wall provided by the existence of a red blood cell-free layer (CFL) [14]. Migration of RBCs away from the vessel wall has been since observed in multiple in vitro studies, and has been shown to create two phases over the cross section of a tube; a central region of mainly RBCs, and an on average a cell depleted region, the CFL [10]. The shear thinning behavior of whole blood however is not only limited to the Fåhraeus-Lindqvist effect and may not only occur on the micro-vascular scale. There are also prominent shear thinning effects of whole blood and has been found to be affected by many things; such as the break up of rouleaux structures [6], RBC alignment, and RBC stretching in flow [11,16]. These phenomena have given rise to the fact that whole blood viscosity is also dependent on shear rate [9].

In order to properly resolve the non-Newtonian phenomena, cell resolved blood flow models are required [5,14,27], as they account for the mechanical deformations of the red blood cells and the subsequent influence on the suspension rheology. The transport of other blood cells types on these scales are also influenced by the presence of red blood cells such as the margination of platelets [28] and white blood cells [15] to the vessel wall. Currently numerical models that adequately resolve the transport and rheological properties of whole blood have large amounts of computational overhead and require high performance computing [3,23].

Blood flow on scales larger than 300 μm is consistently modeled as a continuous fluid, as it is computationally more convenient because models no longer include the individual cell dynamics [20,24,25], which greatly reduces the computational overhead. Continuous models either assume whole blood as a Newtonian fluid on larger scales or use a non-Newtonian blood viscosity model to approximate the departure of whole from the Newtonian description. Non-Newtonian models describe the change in blood viscosity with a dependency either on shear rate like a power law fluid [22] or Carreau-Yasuda [2], or depends on yield stress like the Casson model [19]. Since such models do not include the dynamics of the cells they may over estimate the transport behaviors which are a result of cell-cell collisions within whole blood suspensions. This may lead to an invalid description of particle diffusivities within whole blood.

In order to capture both the non-Newtonian viscosity change of whole blood along with the proper treatment of the transport of suspended blood cells a multi-scale model must be developed in order to correctly account for both processes on all scales of the cardiovascular system. Because systems of the body are heterogeneous across multiple scales present in the body we chose to develop a heterogeneous multi-scale model (HMM) for blood flow. Though this research is focused primarily on the rheology and transport properties of blood, HMM is a more general method for modeling multi scale problems and can be useful for future development as it can incorporate model beyond strictly fluid dynamic problems.

2   Heterogeneous Multi-scale Model

The Heterogeneous Multi-scale Method [1] is a modeling technique used to numerically solve multi-scale problems by coupling multiple sub-models together that each solve a component separately, but by combining each separate sub-model together an overall macro model emerges. HMM relies on efficient coupling between macro and micro models [12]. A macro scale model can be either limited or too computational expensive to numerical solve an entire problem on its own; therefore micro models are employed to resolve each component of the problem separately and return the result to the macro model. Heterogeneous here suggests the problem is multi-physics in nature [8]. Scale separation is a common exploitation in HMM as many numerical problems are difficult to capture solely with one single model.

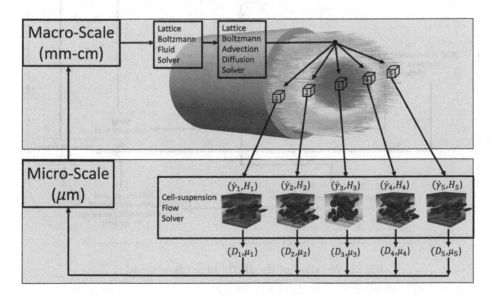

Fig. 1. Schematic highlighting the physical scales and types of numerical models employed in the HMM model.

A benefit of such an HMM model applied to blood flow is that on the largest scales a continuous blood flow solver will be informed by a micro scale cell resolved blood flow solver, resolving the cell nature of whole blood by keeping computational overhead in mind. The heterogeneity of the models allows for extension to be added, such as the diffusion of chemicals. A schematic highlighting the different numerical models and physical scales present in the HMM model is shown in Fig. 1, and a flow chart of the operations through one time-step of this model is highlighted in Fig. 2.

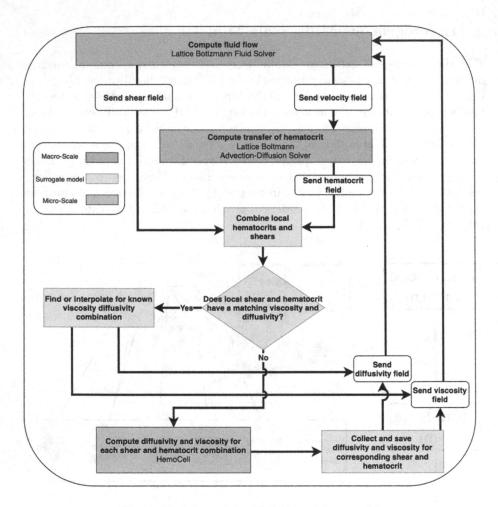

Fig. 2. Workflow of the HMM blood flow model.

2.1   Macro Model

On the largest scales in this model (Macro-scale) blood flow is modelled as a continuous fluid using the lattice Boltzmann method (LBM) [7]. Fluid will be driven by an external body force through a straight vessel geometry. After a time-step of the LBM fluid solver the velocity field across the macroscopic domain will be passed as input to an LBM advection-diffusion solver, which is also on the macro scale and will compute the transport of RBCs (hematocrit) given the underlying fluid flow. After this step the Hematocrit field from the advection diffusion solver and the velocity field (shear rate) from the fluid solver will be sent to a surrogate model. Here the surrogate model it will determine, via interpolation, whether a viscosity and diffusion coefficient is known for the shear rate and hematocrit combination. If there is no such parameter, either viscosity or diffusion coefficient, then a cell resolved blood flow model is spawned which will compute the diffusion coefficient and viscosity directly. The macro scale model will wait for either the surrogate model or the micro-scale model to return viscosities and diffusion coefficients for the entire domain. The viscosity field will be return to the fluid solver and the diffusion field will be returned to the advection diffusion solver. The time-step will then be incremented and the process will begin again.

2.2   Coupling Between Micro and Macro

To avoid duplicating shear rate and local hematocrit micro-scale simulations for each iteration of the HMM, we aim to build a surrogate model. This process will decrease the required number of micro-scale models requested, which in turn will decrease computation time. Hematocrit fields and shear rate fields will be passed down from the macro scale. First local hematocrit values and shear rate values will be combined, then the surrogate will query a database if there already exists a known viscosity and diffusion coefficient for the combination of hematocrit and shear rate. The surrogate will also conduct interpolations (based on a Gaussian process) for similar parameters. If the is a known viscosity and diffusion coefficients then it will return them to the macro scale. If there are missing viscosities and/or diffusion coefficients then a micro model will be spawned for each unknown shear rate and hematocrit combination. A heterogeneous multi-scale computing framework has recently been developed to handle the scheduling and sub-model synchronization of such an application in order to efficiently use computational resources [4].

2.3   Micro Model

The micro scale is modelled with the cell resolved blood flow model HemoCell, in which plasma is modelled by the LBM, the mechanical model of the RBCs are described by a discrete element method and are couple to the plasma via the immersed boundary method [26,27]. From each local hematocrit and shear rate combination on the macro scale, a cell resolved micro scale will simulate perfect sheared environments using Lees-Edwards boundary conditions [21]. A single simulation will be spawned for each unknown diffusivity and shear rate pair, this will likely result in multiple spawns if there are multiple unknown input parameters. The resulting diffusion coefficients and viscosities will be computed from each cell resolved simulation. After each model has finished the results will be stored into the surrogate model and passed up towards the macro scale. As the HMM model progresses through time however a smaller amount of micro model will be spawned due to the surrogate model.

3 Discussion

The immediate focus of this work is to implement the HMM model for blood flow in a 3D straight vessel. Validation to known blood quantities such as the Fåhraeus-Lindqvist effect is needed. A correct physical implementation of this model will allow the determination of the minimum spatial size required from the cell resolved micro models, in order to preserve accurate viscosity and diffusivity measurements that will maximize computational efficiency. The determination of the minimum spatial resolution of local viscosity and diffusivity is also required on the macro scale in order to accurately and efficiently model blood flow.

The novelty of developing an HMM model for blood flow is that it can be extended to include models that are not limited to only model fluid dynamics. Given the heterogeneous nature of this model, computational models which include the reaction and perfusion of chemical species in the human microbiome, as well as the models for the smooth muscle cells, for example, can also be included to contribute to a more complete HMM model for the physiological human.

References

1. Abdulle, A., Weinan, E., Engquist, B., Vanden-Eijnden, E.: The heterogeneous multiscale method. Acta Numer. **21**, 1–87 (2012). https://doi.org/10.1017/S0962492912000025
2. Abraham, F., Behr, M., Heinkenschloss, M.: Shape optimization in steady blood flow: a numerical study of non-newtonian effects. Comput. Methods Biomech. Biomed. Eng. **8**(2), 127–137 (2005)
3. Alowayyed, S., Závodszky, G., Azizi, V., Hoekstra, A.G.: Load balancing of parallel cell-based blood flow simulations. J. Comput. Sci. **24**, 1–7 (2018). https://doi.org/10.1016/j.jocs.2017.11.008
4. Alowayyed, S.A., Vassaux, M., Czaja, B., Coveney, P.V., Hoekstra, A.G.: Towards heterogeneous multi-scale computing on large scale parallel supercomputers. Supercomputing Front. Innov. **6**(4), 20–43 (2020)
5. Bagchi, P.: Mesoscale simulation of blood flow in small vessels. Biophys. J. **92**(6), 1858–1877 (2007)
6. Barshtein, G., Wajnblum, D., Yedgar, S.: Kinetics of linear rouleaux formation studied by visual monitoring of red cell dynamic organization. Biophys. J. **78**(5), 2470–2474 (2000)
7. Chen, S., Doolen, G.D.: Lattice boltzmann method for fluid flows. Annu. Rev. Fluid Mech. **30**(1), 329–364 (1998)
8. Cheng, L.T., Weinan, E.: The heterogeneous multi-scale method for interface dynamics. Contempo. Math. **330**, 43–54 (2003)
9. Chien, S., Usami, S., Taylor, H.M., Lundberg, J.L., Gregersen, M.I.: Effects of hematocrit and plasma proteins on human blood rheology at low shear rates. J. Appl. Physiol. **21**(1), 81–87 (1966)

10. Cokelet, G.R., Goldsmith, H.L.: Decreased hydrodynamic resistance in the two-phase flow of blood through small vertical tubes at low flow rates. Circ. Res. **68**(1), 1–17 (1991)
11. Dintenfass, L.: Internal viscosity of the red cell and a blood viscosity equation. Nature **219**(5157), 956–958 (1968)
12. Weinan, E., Engquist, B.: The heterognous multiscale methods. Commun. Math. Sci. **1**(1), 87–132 (2003). https://projecteuclid.org:443/euclid.cms/1118150402
13. Fåhræus, R., Lindqvist, T.: The viscosity of the blood in narrow capillary tubes. Am. J. Physiol. Legacy Content **96**(3), 562–568 (1931)
14. Fedosov, D.A., Caswell, B., Popel, A.S., Karniadakis, G.E.: Blood flow and cell-free layer in microvessels. Microcirculation **17**(8), 615–628 (2010)
15. Fedosov, D.A., Fornleitner, J., Gompper, G.: Margination of white blood cells in microcapillary flow. Phys. Rev. Lett. **108**(2), 028104 (2012)
16. Fischer, T.M., Stohr-Lissen, M., Schmid-Schonbein, H.: The red cell as a fluid droplet: tank tread-like motion of the human erythrocyte membrane in shear flow. Science **202**(4370), 894–896 (1978)
17. Hoekstra, A.G., et al.: Towards the virtual artery: a multiscale model for vascular physiology at the physics-chemistry-biology interface. Philos. Trans. Royal Soc. A Math. Phys. Eng. Sci. **374**(2080), 20160146 (2016)
18. Hoekstra, A.G., van Bavel, E., Siebes, M., Gijsen, F., Geris, L.: Virtual physiological human 2016: translating the virtual physiological human to the clinic (2018)
19. Jung, J., et al.: Reference intervals for whole blood viscosity using the analytical performance-evaluated scanning capillary tube viscometer. Clin. Biochem. **47**(6), 489–493 (2014)
20. Latt, J.: Palabos, parallel lattice boltzmann solver (2009)
21. Lorenz, E., Hoekstra, A.G., Caiazzo, A.: Lees-edwards boundary conditions for lattice boltzmann suspension simulations. Phys. Rev. E **79**(3), 036706 (2009)
22. Nadeem, S., Akbar, N.S., Hendi, A.A., Hayat, T.: Power law fluid model for blood flow through a tapered artery with a stenosis. Appl. Math. Comput. **217**(17), 7108–7116 (2011)
23. Tarksalooyeh, V.A., Závodszky, G., Hoekstra, A.G.: Optimizing parallel performance of the cell based blood flow simulation software hemocell. In: Rodrigues, J., et al. (eds.) ICCS 2019. LNCS, vol. 11538, pp. 537–547. Springer, Cham (2019). https://doi.org/10.1007/978-3-030-22744-9_42
24. Updegrove, A., Wilson, N.M., Merkow, J., Lan, H., Marsden, A.L., Shadden, S.C.: Simvascular: an open source pipeline for cardiovascular simulation. Ann. Biomed. Eng. **45**(3), 525–541 (2017)
25. Závodszky, G., Paál, G.: Validation of a lattice boltzmann method implementation for a 3D transient fluid flow in an intracranial aneurysm geometry. Int. J. Heat Fluid Flow **44**, 276–283 (2013)
26. Zavodszky, G., van Rooij, B., Azizi, V., Alowayyed, S., Hoekstra, A.: Hemocell: a high-performance microscopic cellular library. Procedia Comput. Sci. **108**, 159–165 (2017)
27. Závodszky, G., van Rooij, B., Azizi, V., Hoekstra, A.: Cellular level in-silico modeling of blood rheology with an improved material model for red blood cells. Front. Physiol. **8**, 563 (2017)
28. Závodszky, G., van Rooij, B., Czaja, B., Azizi, V., de Kanter, D., Hoekstra, A.G.: Red blood cell and platelet diffusivity and margination in the presence of cross-stream gradients in blood flows. Phys. Fluids **31**(3), 031903 (2019)

Towards Accurate Simulation of Global Challenges on Data Centers Infrastructures via Coupling of Models and Data Sources

Sergiy Gogolenko[1]([✉]), Derek Groen[2,3], Diana Suleimenova[2], Imran Mahmood[2], Marcin Lawenda[4], F. Javier Nieto de Santos[5], John Hanley[6], Milana Vučković[6], Mark Kröll[7], Bernhard Geiger[7], Robert Elsässer[8], and Dennis Hoppe[1]

[1] High Performance Computing Center Stuttgart, Stuttgart, Germany
{gogolenko,hoppe}@hlrs.de
[2] Department of Computer Science, Brunel University London, Uxbridge, UK
[3] Centre for Computational Science, University College London, London, UK
[4] Poznan Supercomputing and Networking Center, Poznan, Poland
[5] Atos Research and Innovation (ARI), ATOS, Bilbao, Spain
[6] European Centre for Medium-Range Weather Forecasts, Reading, UK
[7] Know Center GmbH, Graz, Austria
[8] Department of Computer Sciences, University of Salzburg, Salzburg, Austria

Abstract. Accurate digital twinning of the global challenges (GC) leads to computationally expensive coupled simulations. These simulations bring together not only different models, but also various sources of massive static and streaming data sets. In this paper, we explore ways to bridge the gap between traditional high performance computing (HPC) and data-centric computation in order to provide efficient technological solutions for accurate policy-making in the domain of GC. GC simulations in HPC environments give rise to a number of technical challenges related to coupling. Being intended to reflect current and upcoming situation for policy-making, GC simulations extensively use recent streaming data coming from external data sources, which requires changing traditional HPC systems operation. Another common challenge stems from the necessity to couple simulations and exchange data across data centers in GC scenarios. By introducing a generalized GC simulation workflow, this paper shows commonality of the technical challenges for various GC and reflects on the approaches to tackle these technical challenges in the HiDALGO project.

Keywords: Global systems science · Global challenges · Coupling · Multiscale modelling · HPC · HPDA · Workflow · Data management · Streaming data · Cloud Data-as-a-Service

The authors would like to thank Dr. M. Gienger and Prof. Z. Horváth from the University of Györ for sharing their invaluable thoughts which influenced our project. This research has received funding from the European Union's Horizon 2020 research and innovation programme under Grant Agreement no. 824115 (HiDALGO).

V. V. Krzhizhanovskaya et al. (Eds.): ICCS 2020, LNCS 12142, pp. 410–424, 2020.
https://doi.org/10.1007/978-3-030-50433-5_32

1   Introduction

In recent years, there has been an increasing interest in evidence-based computer-aided approaches to address global challenges (GC). It has led to a rapid formation and methodological development of the domain of computational global systems science (GSS) which strives to provide "scientific evidence to support policy-making, public action, and civic society" [10]. Accurate digital twinning of the GC inevitably results in computationally expensive coupled simulations comprised of multiple models for diverse social and physical phenomena, often multiscale by design. These simulations assemble not only different models, but also various sources of massive static and streaming data sets. In effect, accurate numerical treatment of the GC requires combining up-to-date theoretical developments in GSS with high performance computing (HPC) and high performance data analytics (HPDA) technologies.

The HiDALGO project – HPC and Big Data Technologies for Global Challenges – aims to bridge the gap between traditional HPC and data-centric computation with a view to provide efficient technological solutions for accurate policy-making in the domain of GC. The project's aim is not merely to develop the mechanisms for collecting and processing data by scalable methods on analytics and simulation levels. With the strong focus on highly accurate elaborated models, HiDALGO builds up coupled simulations and integrates real-world data in static simulations. The latter yield scientific insights into comprehensive processes and their corresponding phenomena, which occur in the contemporary world and directly affect the quality of human life. This research is driven by three representative case studies: human migration from conflict zones, air pollution levels in the cities, and the spread of malicious information in social media (such as Twitter and telecommunications networks).

In practice, when it comes to coupling of models and data sources for GC simulations in HPC environments, one faces a number of technical challenges, not common for traditional engineering and science applications for HPC. In particular, HPC environments typically operate on static data, which is already available on efficient parallel distributed file systems such as Lustre, BeeGFS, or OrangeFS [5]. Moreover, HPC data centers usually follow a set of security guidelines to isolate users and data which restrict access to incoming data from external sources. However, with the intent to reflect current and upcoming situations for policy-making, GC simulations extensively use recent data from external sources including streaming data coming from physical sensors, social networks, and mobile operators. Consequently, HPC system's operation must be revised in order to allow highly complex simulation, but at the same time, provide the necessary flexibility to incorporate influx data.

Another common challenge stems from the necessity to couple simulations and exchange data across data centers in GC simulation scenarios. This demand is introduced by various reasons including endeavor to reuse time-consuming simulation results in multiple use cases, inability to share massive amounts of data, issues related to licensing the data or simulation software available at different data centers.

1.1   Related Work

Multiscale modeling has a long history of research. Weinan et al. [27] build a preliminary theoretical foundation for the multiscale modeling approaches and highlights key problems in this area. Hoekstra et al. [17] present an epistemological review on the introduction and significance of multiscale modeling and simulation in the interdisciplinary research. They further identify gaps in the current state-of-the-art of multiscale computing and recommend ways to fill these gaps. Zhang et al. [28] presents a study on the complex dynamic bio-system of brain cancer, using a multiscale agent-based modeling (ABM) approach.

Multiscale model coupling is well covered in the literature. Groen et al. [15] present a multiscale survey with a classification of applications in different research domains including: astrophysics, environment, engineering, materials science, energy and biology. They also review a set of multiscale models coupling tools with respect to the domains, approaches, and platforms. Crooks and Castle [9] focused on developing geospatial simulations using the ABM paradigm. They reviewed methods of multiscale coupling of geographical information systems (GIS) with ABM and compared a selection of toolkits which allow such integration. Tao et al. [25] present a multiscale modeling framework (MMF) that deals with the complex issues of weather simulation and discuss several coupling strategies for the sub-systems. Lastly, Gomes et al. [12] review methods and techniques of 'Co-Simulation', a technique that allows different subsystems to be coupled and simulated in a distributed manner.

A full review of individual frequently used *coupling tools* is beyond the scope of this paper, and comprehensively done by Groen et al. [14]. However, by means of example we do highlight a few key tools here. Borgdorff et al. [4] present the Multiscale Coupling Library and Environment (MUSCLE2), which facilitates the coupling and execution of submodels in cyclic multiscale applications. The FabSim [13] toolkit aims to simplify a range of multiscale computational tasks for a diverse range of application domains.

Modern surveys of the *tools for scientific workflows* definition and management [1, 20, 22] cover a broad spectrum of state-of-the-art solution – from mature software, evolved for several decades (like ASKALON, Galaxy, Kepler, Pegasus, or Taverna), to relatively new active developments (like Apache Airavate). Most of the tools stem from the former Grid solutions, where only DIRAC [6] and Airavate [21] support execution on Grid, HPC, and Cloud resources simultaneously. However, DIRAC suffers from poor flexibility, while Airavate does not follow any industrial standard in defining workflows [6]. At the same time, Cloudify [8] is a mature extensible tool for Clouds following industrial strength OASIS TOSCA standard [3]. Croupier orchestrator developed recently extends out-of-the-box Cloudify functionality to the HPC and HPDA environments [6].

1.2   Contributions of the Paper

Despite a broad coverage of different topics, existing literature on the multiscale model coupling barely touches the aspects related to the aforementioned technical challenges of coupling in HPC environments – involving external data sources into the static simulations and coupling across data centers. By introducing a generalized GC simulation workflow, this paper demonstrates a commonality of these technical challenges for various GC and reflects on the approaches to overcome them implemented in the HiDALGO project.

In Sect. 2, we briefly introduce HiDALGO case studies. These representative global challenges comprise a baseline for development of the generalized workflow for GC simulation scenarios, which is defined in Sect. 3. Section 4 is concerned with the implementation of the workflow. This is a core section of the paper, where we present the high level HiDALGO architecture, which implements the generalized workflow, as well as outline HiDALGO approaches to address coupling in HPC environments. Finally, in Sect. 5, we offer a conclusion, including a research and development outlook.

2   Representative Global Challenges

2.1   Human Migration

There are more than 70.8 million people forcibly displaced worldwide. Among them 25.9 million are refugees, half of which fled from Syria, Afghanistan and South Sudan [26]. These fleeing individuals are the unfortunate victims of internal armed conflicts and civil wars, who make decisions to migrate at the times of distress. Their decisions are often based on economic and political push-pull factors in sending and receiving countries. Researchers have mostly investigated why human migration occurs and what effects it has on economies using migration theories and econometric models. However, there is not an appropriate method or model to predict forced displacement counts or movements. In addition, existing models are largely based on regressing existing forced migration data, limiting their predictive power with incomplete or short datasets. Thus, we have created a simulation development approach (SDA) that allows us to forecast movements of forcibly displaced people in conflicts.

We run our simulations using the parallel Python-based Flee code[1]. It relies on an ABM approach with a location network graph, where each refugee in the simulation is represented by a single agent and location network graph reflects geographic information system (GIS) component of the model. In countries with very few roads and in mountainous areas, such as South Sudan, we explore key walking routes, increase the level of detail, and incorporate a broader range of relevant phenomena, such as weather conditions, communications, and food security, through model coupling (cf. Fig. 1).

Our new approach is important due to three main reasons. First, it helps to forecast forced displacement movements when a conflict erupts, guiding decisions on where to provide food and infrastructure. Second, our approach provides forced displacement population estimates in regions where existing data is incomplete, to help prioritize resources to the most important areas. Finally, we can investigate how border closures and other policy decisions are likely

---

[1] http://www.github.com/djgroen/flee-release.

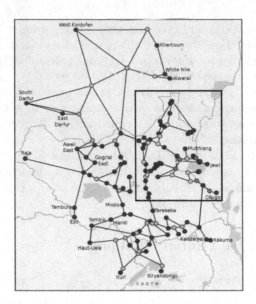

Fig. 1. South Sudan location network graph semi-automatically extracted from Bing and OpenStreetMaps with the HiDALGO GIS toolkit [24]: red vertices stand for conflict zones (refugee depart here), dark green vertices correspond to camps (refugee tend to finish their journeys here), yellow vertices stand for regular locations (refugee may stay or move elsewhere), and light green vertices stand for forced redirection points (representing government-orchestrated movements). This location network includes walking routes allowing to increase resolution of the baseline agent-based model and to couple it with the models for weather conditions, communications, and food security. (Color figure online)

to affect the movements and destinations of forced migration, to provide policy decision-makers with evidence that could support more effective policy and reduce unintended consequences. We have already successfully simulated forced displacement movements in Burundi, Central African Republic, Mali and South Sudan using this prototype, and compared our forecasts with the United Nations High Commissioner for Refugees (UNHCR) refugee camp registrations [24].

2.2   Urban Air Pollution (UAP)

NO_2 is one of the most severe pollutant in urban areas, of which main producer is the vehicular traffic. The level of air contamination can be significantly reduced by proper control on the urban traffic and elaborate architectural development of the cities. UAP case study aims to understand, model, and simulate the spread of vehicular air pollution in the cities.

In order to improve accuracy of the results, we couple detailed computational fluid dynamics (CFD) simulation of the multicomponent air flow in the cities with agent-based traffic simulations, real time traffic and air quality sensor data, and meteorological prediction data. In our experiments with CFD simulation, we use a wide range of software packages including OpenFOAM, NEK5000, Fenics-HPC, and ANSYS Fluent, while traffic simulation part is done with SUMO package [2]. The geometry for meshing of the city is extracted automatically from OpenStreetMaps. For better accuracy, in our CFD simulations, meshes reach the resolution up to 1 m. In this way, we obtained accurate simulation results for the cities of Győr (Hungary, cf. Fig. 2), Graz (Austria), Stuttgart (Germany), Poznan (Poland), and Milwaukee (WI, USA) [18]. These results are further used for model reduction, which allows to lessen computational demands.

Fig. 2. Simulation for the spread of NOx in the city of Győr (Hungary) with the HiDALGO toolchain. NOx concentrations are illustrated via grey-scale point clouds, wind velocity by black arrows, and air path lines by blue lines [18]. (Color figure online)

2.3 Social Network Analysis (SNA)

SNA case study aims to understand, model, and simulate the spread of messages in various social networks. To achieve this goal, several scientific problems have to be addressed, such as the structural and algorithmic properties of the underlying graphs, the stochastic characteristics of information dissemination, and the interplay between these aspects. Furthermore, several programming related points have to be taken care of as well, e.g., the deployment and efficient simulation of the processes described above.

To understand the structure of social networks, we analyzed different types of graphs, which are accessible through SNAP datasets [23] or can be obtained by crawling real world social networks such as Twitter. Apart from well known characteristics, such as degree distribution, clustering coefficient, or distances, one major goal is to understand the size and the structure of the communities in these networks. For this, we apply different clustering algorithms to obtain local partitions of the direct neighborhoods of the nodes. Then, the resulting clusters are merged to obtain an overlapping partitioning of the graph. These results are used to derive synthetic graph models, which are able to properly describe social networks.

To model and simulate the spread of messages, we will analyze different data sources, including tweets from Twitter, and couple them with other real world data such as duration and geographic properties of telephone calls. In our simulation framework, we will integrate the different models from these data sources, and combine them with the synthetic models of the social networks.

3   Generalized Workflow

The GC discussed above were used to construct a generalized workflow for simulation scenarios (cf. Fig. 3), which establishes a common ground and reflects key functionalities for all presented case studies. Here, we briefly outline this workflow. For further details on derivation of the generalized workflow from the individual GC workflows, we refer the interested readers to [11].

The *Data Source* phase denotes the starting point into the workflow including local files as well as external APIs. From these sources data are extracted in the phase of *Data Extraction*. The extracted data can optionally be complemented in a *Synthetic Data Generation* phase which focuses on enriching existing data as well as generating new data. To give an example, in this phase a 3D city model might be constructed based on geospatial information.

The *Data Processing & Feature Extraction* phase prepares the data for the *Model Generation* phase. It transforms the data into standard formats and then seeks to extract relevant information by applying methods from, for instance, signal enhancement, representation learning, or dimensionality reduction.

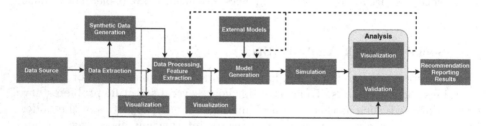

Fig. 3. The generalized workflow in the HiDALGO project.

Visualizations support the validation of data and processing steps at several points of the workflow. The following *Model Generation* phase generates the actual model for the simulation. The generation process is based on the pre-processed data and might optionally incorporate external models, for example, dispersion models for weather forecasts.

The *Simulation* phase employs the model to generate results which are then examined in an *Analysis* phase. *Validation* is performed by comparing the simulation results to a ground truth, if available. In addition, *Visualizations* of the results support the validation process. The *Analysis* phase may directly as well as indirectly influence both the *Data Processing & Feature Extraction* as well as the *Model Generation* phase, for example, a model is updated due to inappropriate predictions. In rare cases the analysis might also impact the data sources, i.e. data sources might be altered (e.g. add a new one), if the results of the analysis are not satisfying.

4   Implementation

4.1   High-Level Architecture

Figure 4 illustrates the high-level HiDALGO architecture. In order to establish computational infrastructure for the implementation of the generalized workflow, HiDALGO involves resources of the three leading European HPC centers – High Performance Supercomputing Center Stuttgart (HLRS), Poznan Supercomputing and Networking Center (PSNC), and European Centre for Medium-Range Weather Forecasts (ECMWF). HLRS and PSNC provide HPC and HPDA clusters and clouds for general purpose applications. Their resources serve to run simulation and data analytics components for various needs of the GC case studies. ECMWF grants access to its cloud facilities to enable distributed, highly efficient computing. While not giving direct access to its HPC systems, which are closed to external access as it is used to produce time-critical twice-daily global weather forecasts, ECMWF provides the migration and UAP case studies with the forecast model's output, as well as climate, atmospheric, and hydrological data via the cloud, which necessitates coupling across data centers.

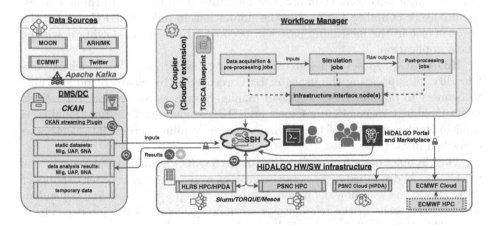

Fig. 4. High-Level HiDALGO architecture

Besides crawled data from Twitter social network for SNA case study, the streaming data for all three case studies comes from MoonStar Communications GmbH (MOON), an international telecommunication company, as well as from ARH Inc. (ARH) and Hungarian Public Road Nonprofit Pte Ltd. Co. (MK), enterprises collecting data from the sensors and traffic lights for the city of Győr.

The overall workflow implementation is orchestrated and monitored by Croupier orchestrator which extends out-of-the-box Cloudify orchestration functionality to the HPC environments [6]. This toolkit provides a unified way to describe complex workflows with YAML files based on OASIS TOSCA standard [3] and to launch them via appropriate scheduling of coupled simulations, data analytics, and data management operations on HPC, HPDA, and cloud resources of multiple data centers at the same time.

Weak and strong coupling mechanisms within the same data center are implemented by means of the tools for coupled simulations (like FabSim) mentioned in Sect. 1.1. Comprehensive Knowledge Archive Network (CKAN) maintains data management operations [7]. In particular, CKAN serves as a single entry point for external static and streaming data, connecting these data sources with HPC and HPDA resources. Apache Kafka enables smooth integration of the streaming data with CKAN and further with the infrastructure of HPC centers [19]. Finally, HiDALGO introduces a custom REST API implementation for coupling with weather and climate data and simulations across data centers. The next subsections explain further details about current state and future plans regarding implementation of coupling mechanisms in HiDALGO.

4.2   Orchestrator and Monitor

The coupling requires that not only the software pieces are synchronized, but also the resource allocation is done in line with the coupling activities. When a ABM simulation needs to be coupled with a CFD simulation, the coupling mechanism must be taken into account in order to facilitate an optimal performance. If we require cyclic coupling, but we run the simulations in different HPC centres, the performance in the messaging part will become a bottleneck, slowing down the whole application.

HiDALGO proposes to use an orchestrator called Croupier [6], which extends Cloudify [8] supporting the usage of HPC resources (by means of plugins able to connect to popular HPC workload managers such as Slurm and Torque). It offers flexibility in combining Cloud and HPC resources. This functionality is used in implementation of the mechanisms for coupling across data centres as described in Sect. 4.5.

The Croupier orchestrator uses the OASIS TOSCA standard [3] to specify the workflows. As described in [6], some extensions were proposed in order to support HPC jobs in the workflow definition. Moreover, it is possible to define dependencies between tasks in the relationships section, by means of the job_depends_on type. This feature allows us to indicate that several tasks represent a coupled simulation and, therefore, the resource allocation should be automated to optimize overall performance. The next step is to adapt that section in the workflow, expressing the type of dependency for specifying the cyclic coupling mechanism: job_mpi_coupled_with, job_data_coupled_with, etc.

It is important to highlight that the coupling has also an impact on the monitoring aspect, since data about the coupling activity itself should be collected, in order to identify potential bottlenecks.

4.3 Coupling Mechanisms for Locally Simulated Models

We distinguish two types of coupling mechanisms:

(i) **Acyclic coupling** refers to the mechanism where the results from the execution of sub-codes are in turn used as input for the execution of the subsequent sub-code. In acyclic coupling sub-codes are not mutually dependent during execution.

(ii) **Cyclic coupling** deals with the simulations where the sub-codes are mutually dependent on each other. Cyclic coupling is further divided into: (a) *sequential cyclic coupling*: where (two or more) sub-codes execute in alternating fashion, such that output of one is used as input by the other and vice versa; and (b) *concurrent cyclic coupling*: where two (or more) sub-codes are executed and exchange their outputs to each other concurrently.

In migration case study, we rely on acyclic coupling for the integration between the conflict model and the migration model, as well as between the migration model and any validation activities that are performed (e.g., by comparing to telecommunications data). We also have several cases of cyclic concurrent coupling, particularly in the integration between a more coarse-grained national model, and the more refined local model (which is in turn coupled to weather data from the Copernicus Project).

UAP case study also draws on acyclic coupling for the integration between CFD model of the multicomponent air flow, traffic model, and meteorological forcasts, as well as for incorporation of streaming data from traffic and air quality sensors into the static simulations.

Likewise, SNA case study exploits acyclic coupling mechanism to couple simulations with streaming data from Twitter and static telecommunications data.

4.4 Coupling with External Data Sources

The Data Management System (DMS) is an indispensable part of the generalized workflow implementation. It is assumed as heavily utilized module of the system by many applications on various stages of scenario execution for reading input data and saving outputs from computing. Furthermore, since parallel computing comes into play where simultaneous operations are executed on the same storage resource, the framework must support distributed and parallel operations as well. It drives us to two main problems which should be solved in the proposed DMS framework: efficient data management and reliable computation.

CKAN is a software of choice for maintaining various data management operations in the HiDALGO system [7]. The CKAN is capable to extend its current functionality by applying a number of plugins. This feature is widely explored by HiDALGO developers in order to offer extra functionality which address specific user demands.

In order to ensure the consistency in the data harvesting and processing method and an adequate level of security in HPC environments, data must be delivered first to the DMS. DMS is connected to the application orchestrator available in the portal and defines appropriate links to the data under workflow definition. In HiDALGO, it is assumed that data can be delivered to the DMS in the three different ways:

- **files** – data prepared by user as standalone files and uploaded to the CKAN from local workstation;
- **links to external data source** – user provides a link to the data located in public area, in the next step data are transferred to the CKAN along with standard procedure;
- **profiled harvester** – in this case the functionality of the CKAN is extended by the plugin which enables access to specific external source using dedicated API and/or providing complementary functionality (e.g., data conversion).

There is only one exception when data do not need to be delivered to the DMS before processing in the HiDALGO project, data are sourced from trusted entity. In this case, only link to external data is created in the CKAN system. One of the example of the trusted entity is ECMWF's data store.

Whenever we need to deal with constantly incoming data (e.g., tweets, traffic data), streaming services must be included as a part of the processing workflow. These types of systems ensure low-latency, high-throughput platform with unified interface, which enables connecting to various external sources. For this purpose, in the HiDALGO project, Apache Kafka [19] system was selected. This is a distributed open-source solution which facilitates building real-time data pipelines for streaming-oriented use cases.

In the HiDALGO project, we are considering streaming data of two different types: (a) *camera* based traffic data, and (b) *air quality monitor* based pollution data. The streaming data will be gathered from the multiple sites around the City of Győr to ARH Data Center Server's GDS middleware software, which will be located at the University of Győr (SZE). SZE will be responsible for coupling these two data streams. For accessing this streaming data from GDS, SZE will be interfacing through GDS's standardized interface: MultiInterface API.

4.5   Coupling with Weather and Climate Data Across HPC Centres

The purpose of coupling with ECMWF's weather forecasts, climate reanalysis, global hydrological, and Copernicus Atmosphere Monitoring Service (CAMS) data within HiDALGO is to enable the migration and UAP case studies to explore improving their simulation models through the inclusion of such data. The overall strategy for the coupling of weather and climate data consists of two stages: (a) static coupling of climate reanalysis data, (b) dynamic coupling via a RESTful API.

To enable coupling with climate reanalysis data each relevant case study has access to the Copernicus Climate Data Store (CDS). Both case studies are using ERA5 [16], ECMWF's latest global atmospheric reanalysis which extends back to 1979. ERA5 is produced daily with a 5 day lag time. Final post-processed data has been manually inserted into CKAN by ECMWF and raw CDS data has been inserted using a custom tool developed by PSNC.

ECMWF will develop a custom Weather and Climate Data API (WCDA) to enable the delivery of weather forecast and climate data to the relevant GC applications running on another HPC centers. The WCDA will be implemented as a RESTful API, using ECMWF's Polytope software, and follow industry norms in terms of GET/POST/DELETE operations. Due to the sizes of data involved, requests to the WCDA may take a considerable time to fulfill (ranging from minutes to days). As a result, download requests to the WCDA will be asynchronous, returning a pollable URL where the client may check the progress and finally retrieve their data.

Due to its RESTful API design the WCDA will enable GC applications to directly submit requests and retrieve the required weather and climate data so long as they can access the WCDA end-points; for the lifetime of the project it is planned to open dedicated ports at HLRS and PSNC to enable this. If this approach is not possible, a fall-back solution is to have a trusted service running at PSNC on a node with internet access which will retrieve data from the WCDA and insert it into CKAN temporarily; the orchestrator will then manage moving this data to the relevant compute node where the GC application is running.

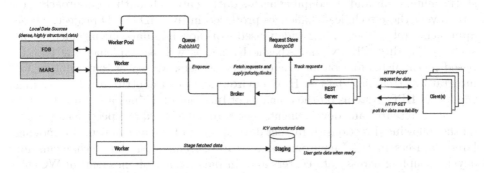

Fig. 5. Preliminary technical design for Polytope, hosting the Weather and Climate REST API

Figure 5 shows a preliminary technical design for Polytope, which will host the Weather and Climate REST API. ECMWF has a flexible object storage solution which is well optimized for weather & climate data, called FDB (Fields Database). The FDB is an internally provided service, used as part of ECMWF's weather forecasting software stack. It operates as a domain-specific object store, designed to store, index and serve meteorological fields produced by ECMWF's forecast model. The FDB serves as a 'hot-object' cache inside ECMWF's high-performance computing facility (HPCF) for the Meteorological Archival and Retrieval System (MARS). MARS makes many decades of meteorological observations and forecasts available to end users. Around 80% of MARS requests are served from the FDB directly, typically for very recently produced data. A subset of this data is later re-aggregated and archived into the permanent archive for long-term availability. Usage of the FDB will allow the WCDA to meet the requirements of data sizes (MiBs to TiBs) and will perform well even with heterogeneous data sizes.

Different collections of data (e.g. forecast data, climate data) will be represented as separate end-points to the WCDA. HTTP requests to each end-point will include authorization metadata in the header and data-specific metadata in the message content. Data will be indexed using multiple scientifically meaningful keys. Endpoints will also be provided to query collection contents and their data schemas.

5   Conclusions and Future Work

Underpinned by the three representative global challenges, the paper derives the generalized workflow targeting these case studies, but also applicable to a wider range of GC. Implementation of the workflow is associated with a number of technical challenges related to coupling. In particular, the paper underlines two aspects of coupling in HPC environments which has not received much attention in the scholarly literature: *i)* involving external data sources into the static simulations and *ii)* coupling across data centers. It outlines approaches to circumvent these technical challenges proposed in the HiDALGO project. These approaches rely on extending widely used workflow and data management tools such as Cloudify, CKAN, and Apache Kafka, as well as implementing REST-based mechanisms for coupling across data centers. Moreover, our solution for integrating streaming data in HPC environments can contribute to expanding the use of HPC systems on the domains of Internet of Things and Industry 4.0.

Future research and developments seen from an applied perspective may focus on the following directions: *i)* further development of the mechanisms for moving large datasets in HPC environments, *ii)* improvement of the mechanisms for acyclic coupling across data centers and, in particular, enhancement of WCDA, *iii)* implementation of the strong coupling mechanisms (via messages passing) in the representative case studies, as well as *iv)* performance evaluation for the proposed solutions.

References

1. Atkinson, M., Gesing, S., Montagnat, J., Taylor, I.: Scientific workflows: past, present and future. Fut. Gener. Comput. Syst. **75**, 216–227 (2017)
2. Behrisch, M., Bieker, L., Erdmann, J., Krajzewicz, D.: SUMO - simulation of urban MObility: an overview. In: 3rd International Conference on Advances in System Simulation, SIMUL 2011, pp. 63–68 (2011)
3. Binz, T., Breiter, G., Leyman, F., Spatzier, T.: Portable cloud services using TOSCA. IEEE Internet Comput. **16**(3), 80–85 (2012)
4. Borgdorff, J., et al.: Distributed multiscale computing with MUSCLE 2, the multiscale coupling library and environment. J. Comput. Sci. **5**(5), 719–731 (2014). https://doi.org/10.1016/j.jocs.2014.04.004

5. Cao, J., et al.: PFault: a general framework for analyzing the reliability of high-performance parallel file systems. In: Proceedings of International Conference on Supercomputing (2018)

6. Carnero, J., Nieto, F.J.: Running simulations in HPC and cloud resources by implementing enhanced TOSCA workflows. In: 2018 International Conference on High Performance Computing & Simulation (HPCS), pp. 431–438 (2018)

7. CKAN: comprehensive knowledge archive network (2020). https://ckan.org/

8. Cloudify: cutting edge orchestration (2020). https://cloudify.co/

9. Crooks, A.T., Castle, C.J.E.: The integration of agent-based modelling and geographical information for geospatial simulation. In: Heppenstall, A., Crooks, A., See, L., Batty, M. (eds.) Agent-Based Models of Geographical Systems, pp. 219–251. Springer, Dordrecht (2012). https://doi.org/10.1007/978-90-481-8927-4_12

10. Dum, R., Johnson, J.: Global systems science and policy. In: Johnson, J., Nowak, A., Ormerod, P., Rosewell, B., Zhang, Y.-C. (eds.) Non-Equilibrium Social Science and Policy. UCS, pp. 209–225. Springer, Cham (2017). https://doi.org/10.1007/978-3-319-42424-8_14

11. Maritsch, M., et al.: Workflow and services definition. Technical Report D6.2, HiDALGO project (2018)

12. Gomes, C., Thule, C., Broman, D., Larsen, P.G., Vangheluwe, H.: Co-simulation: a survey. ACM Comput. Surv. (CSUR) 51(3), 1–33 (2018)

13. Groen, D., Bhati, A.P., Suter, J., Hetherington, J., Zasada, S.J., Coveney, P.V.: FabSim: facilitating computational research through automation on large-scale and distributed e-infrastructures. Comput. Phys. Commun. 207, 375–385 (2016). https://doi.org/10.1016/j.cpc.2016.05.020

14. Groen, D., Knap, J., Neumann, P., Suleimenova, D., Veen, L., Leiter, K.: Mastering the scales: a survey on the benefits of multiscale computing software. Phil. Trans. Roy. Soc. A 377(2142), 20180147 (2019)

15. Groen, D., Zasada, S.J., Coveney, P.V.: Survey of multiscale and multiphysics applications and communities. Comput. Sci. Eng. 16(2), 34–43 (2013)

16. Hersbach, H., et al.: Operational global reanalysis: progress, future directions and synergies with NWP. Technical Report 27, ECMWF (2018). https://doi.org/10.21957/tkic6g3wm, ECMWF ERA report

17. Hoekstra, A., Chopard, B., Coveney, P.: Multiscale modelling and simulation: a position paper. Phil. Trans. Roy. Soc. A Math. Phys. Eng. Sci. 372(2021), 20130377 (2014)

18. Horváth, Z., Liszkai, B., Kovács, Á., Budai, T., Tóth, C.: HiDALGO urban air pollution pilot based on CAMS data (2020). https://atmosphere.copernicus.eu/sites/default/files/2019-09/1_HIDALGO_ZHorvath.pdf

19. Kafka: Kafka web site (2020). https://kafka.apache.org/

20. Kowalik, M., Chiang, H.F., Daues, G., Kooper, R.: DMTN-025: a survey of workflow management systems (2018). https://dmtn-025.lsst.io/

21. Pierce, M.E., et al.: Apache airavata: design and directions of a science gateway framework. Concurrency Comput. Pract. Exp. 27(16), 4282–4291 (2015). https://doi.org/10.1002/cpe.3534

22. da Silva, R.F., Filgueira, R., Pietri, I., Jiang, M., Sakellariou, R., Deelman, E.: A characterization of workflow management systems for extreme-scale applications. Fut. Gener. Comput. Syst. 75, 228–238 (2017)

23. SNAP: stanford large network dataset collection (2014). https://snap.stanford.edu/data/

24. Suleimenova, D., Groen, D.: How policy decisions affect refugee journeys in South Sudan: a study using automated ensemble simulations. J. Artif. Soc. Soc. Simul. **23**(1) (2020). https://doi.org/10.18564/jasss.4193
25. Tao, W.K., et al.: A multiscale modeling system: developments, applications, and critical issues. Bull. Am. Meteorol. Soc. **90**(4), 515–534 (2009)
26. UNHCR: Figures at a Glance (2019). https://www.unhcr.org/figures-at-a-glance.html
27. Weinan, E., Lu, J.: Multiscale modeling. Scholarpedia **6**(10), 11527 (2011)
28. Zhang, L., Wang, Z., Sagotsky, J.A., Deisboeck, T.S.: Multiscale agent-based cancer modeling. J. Math. Biol. **58**(4), 545–559 (2009)

Easing Multiscale Model Design and Coupling with MUSCLE 3

Lourens E. Veen[1]([⊠]) [iD] and Alfons G. Hoekstra[2] [iD]

[1] Netherlands eScience Center, Amsterdam, The Netherlands
l.veen@esciencecenter.nl
[2] University of Amsterdam, Amsterdam, The Netherlands
a.g.hoekstra@uva.nl

Abstract. Multiscale modelling and simulation typically entails coupling multiple simulation codes into a single program. Doing this in an ad-hoc fashion tends to result in a tightly coupled, difficult-to-change computer program. This makes it difficult to experiment with different submodels, or to implement advanced techniques such as surrogate modelling. Furthermore, building the coupling itself is time-consuming. The MUltiScale Coupling Library and Environment version 3 (MUSCLE 3) aims to alleviate these problems. It allows the coupling to be specified in a simple configuration file, which specifies the components of the simulation and how they should be connected together. At runtime a simulation manager takes care of coordination of submodels, while data is exchanged over the network in a peer-to-peer fashion via the MUSCLE library. Submodels need to be linked to this library, but this is minimally invasive and restructuring simulation codes is usually not needed. Once operational, the model may be rewired or augmented by changing the configuration, without further changes to the submodels. MUSCLE 3 is developed openly on GitHub, and is available as Open Source software under the Apache 2.0 license.

Keywords: Multiscale modelling · Model coupling · Software

1 Introduction

Natural systems consist of many interacting processes, each taking place at different scales in time and space. Such multiscale systems are studied for instance in materials science, astrophysics, biomedicine, and nuclear physics [11,16,24,25]. Multiscale systems may extend across different kinds of physics, and beyond into social systems. For example, electricity production and distribution covers processes at time scales ranging from less than a second to several decades, covering physical properties of the infrastructure, weather, and economic aspects [21]. The behaviour of such systems, especially where emergent phenomena are present,

This work was supported by the Netherlands eScience Center and NWO under the e-MUSC project.

may be understood better through simulation. Simulation models of multiscale systems (multiscale models for short), are typically coupled simulations: they consist of several submodels between which information is exchanged.

Constructing multiscale models is a non-trivial task. In addition to the challenge of constructing and verifying a sufficiently accurate model of each of the individual processes in the system, scale bridging techniques must be used to preserve key invariants while exchanging information between different spatiotemporal scales. If submodels that use different domain representations need to communicate, then conversion methods are required to bridge these gaps as well. Multiscale models that exhibit temporal scale separation may require irregular communication patterns, and spatial scale separation results in running multiple instances, possibly varying their number during simulation.

Once verified, the model must be validated and its uncertainty quantified (UQ) [22]. This entails uncertainty propagation (forward UQ) and/or statistical inference of missing parameter values and their uncertainty (inverse UQ). Sensitivity analysis (SA) may also be employed to study the importance of individual model inputs for obtaining a realistic result. Such analysis is often done using ensembles, which is computationally expensive especially if the model on its own already requires significant resources. Recently, semi-intrusive methods have been proposed to improve the efficiency of UQ of multiscale models [18]. These methods leave individual submodels unchanged, but require replacing some of them or augmenting the model with additional components, thus changing the connections between the submodels.

When creating a multiscale model, time and development effort can often be saved by reusing existing submodel implementations. The coupling between the models however is specific to the multiscale model as a whole, and needs to be developed from scratch. Doing this in an ad-hoc fashion tends to result in a tightly coupled, difficult-to-change computer program. Experimenting with different model formulations or performing efficient validation and uncertainty quantification then requires changing the submodel implementations, which in turn makes it difficult to ensure continued interoperability between model components. As a result, significant amounts of time are spent solving technical problems rather than investigating the properties of the system under study.

These issues can be alleviated through the use of a model coupling framework, a software framework which takes care of some of the aspects of coupling submodels together into a coupled simulation. Many coupling frameworks exist, originating from a diversity of fields [2,12,13]. Most of these focus on tightly-coupled scale-overlapping multiphysics simulations, often in a particular domain, and emphasise efficient execution on high-performance computers.

The MUSCLE framework has taken a somewhat different approach, focusing on scale-separated coupled simulation. These types of coupled simulations have specific communication patterns which occupy a space in between tightly-coupled, high communication intensity multiphysics simulations, and pleasingly parallel computations in which there is no communication between components at all. The aforementioned methods for semi-intrusive UQ entail a similar communication style, but require the ability to handle ensembles of (parts of) the coupled simulation. In this paper, we introduce version 3 of the MUltiScale Coupling Library and Environment (MUSCLE 3 [23]), and explain how it helps multiscale model developers in connecting (existing) submodels together, exchanging information between them, and changing the structure of the multiscale model as required for e.g. uncertainty quantification. We compare and contrast MUSCLE 3 to two representative examples: preCICE [5], an overlapping-scale multiphysics framework, and AMUSE [20], another multiscale-oriented coupling framework.

2   Designing Coupled Simulations with the MMSF

MUSCLE 3 is based on the theory of the Multiscale Modeling and Simulation Framework (MMSF, [4,8]). The MMSF provides a systematic method for deriving the required message exchange pattern from the relative scales of the modelled processes. As an example, we show this process for a 2D simulation of In-Stent Restenosis (ISR2D, [6,7,17,19]). This model models stent deployment in a coronary artery, followed by a healing process involving (slow) cell growth and (fast) blood flow through the artery. The biophysical aspects of the model have been described extensively in the literature; here we will focus on the model architecture and communication pattern. Note that we have slightly simplified both the model (ignoring data conversion) and the method (unifying state and boundary condition updates) for convenience.

Simple coupled simulations consist of two or more sequential model runs, where the output of one model is used as an input of the next model. This suffices if one real-world process takes place before the next, or if there is otherwise effectively a one-way information flow between the modeled processes. The pattern of data exchange in such a model may be described as a Directed Acyclic Graph (DAG)-based workflow.

A more complex case is that of cyclic models, in which two or more submodels influence each other's behaviour as the simulation progresses. Using a DAG to describe such a simulation is possible, but requires modelling each executing submodel as a long sequence of state update steps, making the DAG unwieldy and difficult to analyse. Moreover, the number of steps may not be known in advance if a submodel has a variable step size or runs until it detects convergence.

A more compact but still modular representation is obtained by considering the coupled simulation to be a collection of simultaneously executing programs (components) which exchange information during execution by sending and receiving messages. Designing a coupled simulation then becomes a matter of deciding which component should send which information to which other component at which time. Designing this pattern of information exchange between the components is non-trivial. Each submodel must receive the information it needs to perform its next computation as soon as possible and in the correct form. Moreover, in order to avoid deadlock, message sending and receiving should match up exactly between the submodels.

Figure 1 depicts the derivation of the communication pattern of ISR2D according to the MMSF. Figure 1a) shows the spatial and temporal domains

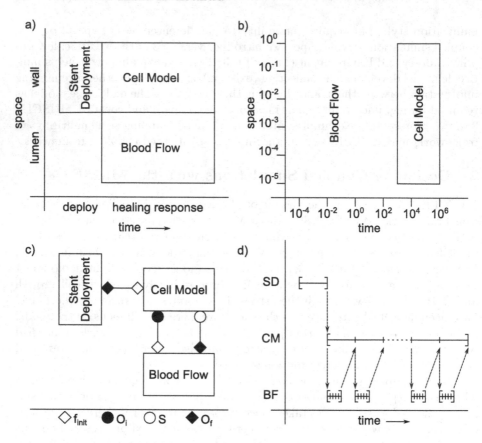

Fig. 1. a) Spatiotemporal domains (ordinal scales), b) Scale Separation Map, c) MML diagram, and d) simulation timeline of the ISR2D model.

in which the three processes comprising the model take place. Temporally, the model can be divided into a deployment phase followed by a healing phase. Spatially, deployment and cell growth act on the arterial wall, while blood flow acts on the lumen (the open space inside the artery). Figure 1b) shows a Scale Separation Map [15] for the healing phase of the model. On the temporal axis, it shows that blood flow occurs on a scale of milliseconds to a second, while cell growth is a process of hours to weeks. Thus, the temporal scales are separated [9]. Spatially, the scales overlap, with the smallest agents in the cell growth model as well as the blood flow model's grid spacing on the order of $10\,\mu\mathrm{m}$, while the domains are both on the order of millimeters.

According to the MMSF, the required communication pattern for the coupled simulation can be derived from the above information. The MMSF assumes that each submodel executes a Submodel Execution Loop (SEL). The SEL starts with an initialisation step (f_{init}), then proceeds to repeatedly observe the state (O_i) and then update the state (S). After a number of repetitions of these two steps, iteration stops and the final state is observed (O_f). During observation steps, (some of the) state of the model may be sent to another simulation component, while during initialisation and state update steps messages may be received.

For the ISR2D model, causality dictates that the deployment phase is simulated before the healing phase, and therefore that the final state of the deployment (O_f) is fed into the initial conditions (f_{init}) of the healing simulation. In the MMSF, this is known as a *dispatch coupling template*. Within the healing phase, there are two submodels which are timescale separated. This calls for the use of the *call* $(O_i$ to $f_{init})$ and *release* $(O_f$ to $S)$ coupling templates.

Figure 1c) shows the resulting connections between the submodels using the Multiscale Modeling Language [10]. In this diagram, the submodels are drawn as boxes, with lines indicating conduits between them through which messages may be transmitted. Decorations at the end of the lines indicate the SEL steps (or *operators*) between which the messages are sent. Note that conduits are unidirectional.

Figure 1d) shows the corresponding timeline of execution. First, deployment is simulated, then the cell growth and blood flow models start. At every timestep of the cell growth submodel (slow dynamics), part of its state is observed (O_i) and used to initialise (f_{init}) the blood flow submodel (fast dynamics). The blood flow model repeatedly updates its state until it converges, then sends (part of) its final state (at its O_f) back to the cell growth model's next state update (S).

3   MUSCLE 3

While the above demonstrates how to design a multiscale model from individual submodels, it does not explain how to implement one. In this section, we introduce MUSCLE 3 and yMMSL, and show how they ease building a complex coupled simulation. MUSCLE 3 is the third incarnation of the MUltiScale Coupling Library and Environment, and is thus the successor of MUSCLE [14] and MUSCLE 2 [3]. MUSCLE 3 consists of two main components: `libmuscle` and the MUSCLE Manager.

Figure 2 shows how `libmuscle` and the MUSCLE Manager work together with each other and with the submodels to enact the simulation. At start-up, the MUSCLE Manager reads in a description of the model and then waits for the submodels to register. The submodels are linked with `libmuscle`, which offers an API through which they can interact with the outside world using *ports*, which are gateways through which messages may be sent and received. To start the simulation, the Manager is started first, passing the configuration, and then the submodels are all started and passed the location of the Manager.

At submodel start-up, `libmuscle` connects to the MUSCLE Manager via TCP, describes how it can be contacted by other components, and then receives a description for each of its ports of which other component it should communicate with and where it may be found. The MUSCLE Manager derives this information from the model topology description and from the registration information sent by the other submodels. The submodels then set up direct peer-to-peer network connections to exchange messages. These connections currently use TCP,

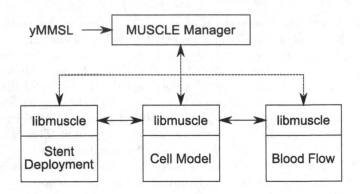

Fig. 2. MUSCLE 3 run-time architecture for ISR2D.

but a negotiation mechanism allows for future addition of faster transports without changes to user code. Submodels may use MPI for internal communication independently of their use of MUSCLE 3 for external communication. In this case, MUSCLE 3 uses a spinloop-free receive barrier to allow resource sharing between submodels that do not run concurrently. In MUSCLE 2, each non-Java model instance is accompanied by a Java minder process which handles communication, an additional complexity that has been removed in MUSCLE 3 in favour of a native `libmuscle` implementation with language bindings.

3.1 Model Description

The Manager is in charge of setting up the connections between the submodels. The model is described to the Manager using yMMSL, a YAML-based serialisation of the Multiscale Modelling and Simulation Language (MMSL). MMSL is a somewhat simplified successor to the MML [10], still based on the same concepts from the MMSF. Listing 1 shows an example yMMSL file for ISR2D. The model is described with its name, the compute elements making up the model, and the conduits between them. The name of each compute element is given, as well as a second identifier which identifies the implementation to use to instantiate this compute element. Conduits are listed in the form `component1.port1: component2.port2`, which means that any messages sent by `component1` on its `port1` are to be sent to `component2` on its `port2`. The components referred to in the `conduits` section must be listed in the `compute_elements` section. MUSCLE 3 reads this file directly, unlike MUSCLE 2 which was configured using a Ruby script that could be derived from the MML XML file.

The yMMSL file also contains settings for the simulation. These can be global settings, like `length` of the simulated artery section, or addressed to a specific submodel, e.g. `bf.velocity`. Submodel-specific settings override global settings if both are given. Settings may be of types float, integer, boolean, string, and 1D or 2D array of float.

Listing 1. yMMSL file for ISR2D

```
ymmsl_version: v0.1

model:
  name: ISR2D
  compute_elements:
    deploy: deploy_stent
    smc: cell_model
    bf: blood_flow
  conduits:
    deploy.final_state_out: smc.initial_state_in
    smc.geom_out: bf.domain_in
    bf.wss_out: smc.wss_in

settings:
  length: 1.5
  lumen_width: 1.0
  deploy.max_depth: 0.11
  bf.velocity: 0.48
  smc.re_recovered_time: 1987200      # 23 days in seconds
  # Many other settings here
```

3.2 Libmuscle

The submodels need to coordinate with the Manager, and communicate with each other. They do this using `libmuscle`, which is a library currently available in Python 3 (via pip), C++ and Fortran. Unlike in MUSCLE 2, whose Java API differed significantly from the native one, the same features are available in all supported languages. Listing 2 shows an example in Python. First, an `Instance` object is created and given a description of the ports that this submodel will use. At this point, `libmuscle` will connect to the Manager to register itself, using an instance name and contact information passed on the command line. Next, the *reuse loop* is entered. If a submodel is used as a micromodel, then it will need to run many times over the course of the simulation. The required number of runs equals the macromodel's number of timesteps, which the micromodel should not have any knowledge of if modularity is to be preserved. A shared setting could solve that, but will not work if the macromodel has varying timesteps or runs until it detects convergence. Determining whether to do another run is therefore taken care of by MUSCLE 3, and the submodel simply calls its `reuse_instance()` function to determine if another run is needed. In most cases, MUSCLE 2 relied on a global end time to shut down the simulation, which is less flexible and potentially error-prone.

Within the reuse loop is the implementation of the Submodel Execution Loop. First, the model is initialised (lines 10–17). Settings are requested from `libmuscle`, passing an (optional) type description so that `libmuscle` can generate an appropriate error message if the submodel is configured incorrectly.

Listing 2. Using libmuscle from Python

```
1   def cell_model() -> None:
2       instance = Instance({
3               Operator.F_INIT: ['initial_state_in'],
4               Operator.O_I: ['geom_out'],
5               Operator.S: ['wss_in'],
6               Operator.O_F: ['final_state_out']})
7
8       while instance.reuse_instance():
9           # F_INIT
10          length = instance.get_setting('length', 'float')
11          width = instance.get_setting('lumen_width', 'float')
12          rec_time = instance.get_setting('re_recovered_time', 'int')
13          t_max = instance.get_setting('t_max', 'float')
14
15          init_msg = instance.receive('initial_state_in')
16          t_cur = msg.timestamp
17          init_model(length, width, init_msg.data)
18
19          while t_cur + dt < t_max:
20              # O_I
21              t_next = t_cur + dt if t_cur + dt < t_max else None
22              msg_out = Message(t_cur, t_next, calc_geometry())
23              instance.send('geom_out', msg_out)
24
25              # S
26              wss_msg = instance.receive('wss_in')
27              update_state(wss_msg.data)
28              t_cur += dt
29
30          # O_F
31          instance.send('final_state', Message(t_cur, None, get_state()))
```

Note that **re_recovered_time** is specified without the prefix; libmuscle will automatically resolve the setting name to either a submodel-specific or a global setting. A message containing the initial state is received on the relevant port (line 15), and the submodel's simulation time is initialised using the corresponding timestamp. The obtained data is then used to initialise the simulation state in a model-specific way, as represented here by an abstract **init_model()** function (line 17).

Next is the iteration part of the SEL, in which the state is repeatedly observed and updated (lines 19–28). In addition to the simulation time corresponding to the current state, the timestamp for the next state is calculated here (line 21). This is unused here, but is required in case two submodels with overlapping timescales are to be coupled [4] and so improves reusability of the model. In ISR2D's O_i operator, the current geometry of the artery is calculated and sent on the **geom_out** port (lines 22–23). Next, the wall shear stress is received and used in the model's state update, after which the simulation time is incremented and the next observation may occur (lines 26–28). Once the final state is reached, it is sent on the corresponding port (line 31). In this example, this port is not connected, which causes MUSCLE 3 to simply ignore the send operation. In practice, a component would be attached which saves this final state to disk, or postprocesses it in some way, possibly via an in-situ/in-transit analysis framework.

Message data may consist of floating point numbers, integers, Booleans, strings, raw byte arrays, or lists or dictionaries containing these, as well as grids of floating point or integer numbers or Booleans, where MUSCLE 2 only supported 1D arrays of numbers. Internally, MUSCLE 3 uses MessagePack for encoding the data before it is sent.

4   Uncertainty Quantification

Uncertainty Quantification of simulation models is an important part of their evaluation. Intrusive methods provide an efficient solution in some cases, but UQ is most often done using Monte Carlo (MC) ensembles. An important innovation in MUSCLE 3 compared to MUSCLE 2 is its flexible support for Monte Carlo-based algorithms. This takes the form of two orthogonal features: instance sets and settings injection.

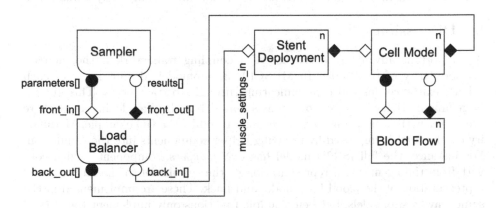

Fig. 3. MMSL diagram for Uncertainty Quantification of ISR2D.

Figure 3 shows an MMSL diagram of a Monte Carlo forward UQ of ISR2D. The simulation has been augmented with a sampler and a load balancer, and there are now multiple instances of each of the three submodels. The sampler samples the uncertain parameters from their respective distributions, and generates a settings object for each ensemble member. These objects are sent to the load balancer, which distributes them evenly among the available model instances. The settings are then sent into a special port on the submodel instances named muscle_settings_in, from where the receiving libmuscle automatically overlays them on top of the centrally provided settings. The settings are then transparently passed on to the corresponding other submodel instances. Final results are passed back via the load balancer to the sampler, which can then compute the required statistics.

To enable communication with sets of instances, MUSCLE 3 offers *vector ports*, recognisable by the square brackets in the name. A vector port allows sending or receiving on any of a number of *slots*, which correspond to instances if the port is connected to a set of them. Vector ports may also be connected to each other, in which case each sending slot corresponds to a receiving slot. In the example, the Sampler component resizes its `parameters[]` port to the number of samples it intends to generate, then generates the settings objects and sends one on each slot. The load balancer receives each object on the corresponding slot of its `front_in` port, and passes it on to a slot on its `back_out` port. It is then received by the corresponding ensemble member, which runs and produces a result for the load balancer to receive on `back_in`. The load balancer then reverses the earlier slot mapping, and passes the result back to the sampler on the same slot on its `front_out` that the sampler sent the corresponding settings object on.

With the exception of the mapping inside the load balancer, all addressing in this use case is done transparently by MUSCLE 3, and components are not aware of the rest of the simulation. In particular, the submodels are not aware of the fact that they are part of an ensemble, and can be completely unmodified.

5   Discussion

An important advantage of the use of a coupling framework is the increase in modularity of the model. In MUSCLE 3, submodels do not know of each other's existence, instead communicating through abstract ports. This gives a large amount of flexibility in how many submodels and submodel instances there are and how they are connected, as demonstrated by the UQ example. Modularity can be further improved by inserting helper components into the simulation. For instance, the full ISR2D model has two *mappers*, components which convert from the agent-based representation of the cell model to the lattice-based representation of the blood flow model and back. These are implemented in the same way as submodels, but being simple functions only implement the F_INIT and O_F parts of the SEL. The use of mappers allows submodels to interact with the outside world on their own terms from a semantic perspective as well as with respect to connectivity. Separate scale bridging components may be used in the same way, except converting between scales rather than between domain representations.

Other coupling libraries and frameworks exist. While a full review is beyond the scope of this paper (see e.g. [12]), we provide a brief comparison here with two other such frameworks in order to show how MUSCLE 3 relates to other solutions.

preCICE is a framework for coupled multiphysics simulations [5]. It comes with adapters for a variety of CFD and finite element solvers, as well as scale bridging algorithms and coupling schemes. Data is exchanged between submodels in the form of variables defined on meshes, which can be written to by one component and read by another. Connections are described in an XML-based configuration file. Like MUSCLE 3, preCICE links submodels to the framework by adding calls to a library to them. For more generic packages, a more extensive adapter is created to enable more configurability. Submodels are started separately, and discover each other via files in a known directory on a shared file system, after which peer-to-peer connections are set up.

preCICE differs from MUSCLE 3 in that it is intended primarily for scale-overlapping, tightly-coupled physics simulations. MUSCLE 3 can do this as well, but is mainly designed for loosely-coupled multiscale models of any kind. For instance, it is not clear how an agent-based cell simulation as used in ISR2D would fit in the preCICE data model. MUSCLE 3's central management of model settings and its support for sets of instances allows it to run ensembles, thus providing support for Uncertainty Quantification. preCICE does not seem to have any features in this direction.

The Astrophysical Multipurpose Software Environment (AMUSE) is a framework for coupled multiscale astrophysics simulations [20, 25]. It comprises a library of well-known astrophysics models wrapped in Python modules, facilities for unit handling and data conversion, and infrastructure for spawning these models and communicating with them at runtime.

Data exchange between AMUSE submodels is in the form of either grids or particle collections, both of which store objects with arbitrary attributes. With respect to linking submodels, AMUSE takes the opposite approach to MUSCLE 3 and preCICE. Instead of linking the model code to a library, the model code is made into a library, and where MUSCLE 3 and preCICE have a built-in configurable coupling paradigm, in AMUSE coupling is done by an arbitrary user-written Python script which calls the model code. This script also starts the submodels, and performs communication by reading and writing to variables in the models.

Linking a submodel to AMUSE is more complex than doing this in MUSCLE 3, because an API needs to be implemented that can access many parts of the model. This API enables access to the model's parameters as well as to its state. AMUSE comes with many existing astrophysics codes however, which will likely suffice for most users. Coupling via a Python script gives the user more flexibility, but also places the responsibility for implementing the coupling completely on the user. Uncertainty quantification could be implemented, although scalability to large ensembles may be affected by the lack of peer-to-peer communication.

6   Conclusions and Future Work

MUSCLE 3, as the latest version of MUSCLE, builds on almost fourteen years of work on the Multiscale Modelling and Simulation Framework and the MUSCLE paradigm. It is mainly designed for building loosely coupled multiscale simulations, rather than scale overlapping multi-physics simulations. Models are described by a yMMSL configuration file, which can be quickly modified to change the model structure. Linking existing codes to the framework can be done quickly and easily due to its library-based design. Other frameworks have more existing integrations however. Which framework is best will thus depend on which kind of problem the user is trying to solve.

MUSCLE 3 is Open Source software available under the Apache 2.0 license, and it is being developed openly on GitHub [23]. Compared to MUSCLE 2, the code base is entirely new and while enough functionality exists for it to be useful, more work remains to be done. We are currently working on getting the first models ported to MUSCLE 3, and we plan to further extend support for Uncertainty Quantification, implementing model components to support the recently-proposed semi-intrusive UQ algorithms [18]. We will also finish implementing semi-intrusive benchmarking of models, which will enable performance measurement and support performance improvements as well as enabling future static scheduling of complex simulations. Other future features could include dynamic instantiation and more efficient load balancing of submodels in order to support the Heterogeneous Multiscale Computing paradigm [1].

References

1. Alowayyed, S., et al.: Patterns for high performance multiscale computing. Fut. Gener. Comput. Syst. **91**, 335–346 (2019). https://doi.org/10.1016/j.future.2018.08.045
2. Babur, O., Verhoeff, T., Van Den Brand, M.G.J.: Multiphysics and multiscale software frameworks: an annotated bibliography. Computer Science Reports, Technische Universiteit Eindhoven (2015)
3. Borgdorff, J., et al.: Distributed multiscale computing with MUSCLE 2, the multiscale coupling library and environment. J. Comput. Sci. **5**(5), 719–731 (2014). https://doi.org/10.1016/j.jocs.2014.04.004
4. Borgdorff, J., Falcone, J.L., Lorenz, E., Bona-Casas, C., Chopard, B., Hoekstra, A.G.: Foundations of distributed multiscale computing: formalization, specification, and analysis. J. Parallel Distrib. Comput. **73**(4), 465–483 (2013). https://doi.org/10.1016/j.jpdc.2012.12.011
5. Bungartz, H.J., et al.: preCICE a fully parallel library for multi-physics surface coupling. Comput. Fluids **141**, 250–258 (2016). http://www.sciencedirect.com/science/article/pii/S0045793016300974
6. Caiazzo, A., et al.: A complex automata approach for in-stent restenosis: two-dimensional multiscale modelling and simulations. J. Comput. Sci. **2**(1), 9–17 (2011). https://doi.org/10.1016/j.jocs.2010.09.002
7. Caiazzo, A., et al.: Towards a complex automata multiscale model of in-stent restenosis. In: Allen, G., Nabrzyski, J., Seidel, E., van Albada, G.D., Dongarra, J., Sloot, P.M.A. (eds.) ICCS 2009. LNCS, vol. 5544, pp. 705–714. Springer, Heidelberg (2009). https://doi.org/10.1007/978-3-642-01970-8_70
8. Chopard, B., Borgdorff, J., Hoekstra, A.G.: A framework for multi-scale modelling. Phil. Trans. Roy. Soc. A Math. Phys. Eng. Sci. **372**(2021), 20130378 (2014). https://doi.org/10.1098/rsta.2013.0378
9. Weinan, E.: Principles of Multiscale Modeling. Cambridge University Press, Cambridge (2011)

10. Falcone, J.L., Chopard, B., Hoekstra, A.: MML: towards a multiscale modeling language. Procedia Comput. Sci. **1**(1), 819–826 (2010). https://doi.org/10.1016/j. procs.2010.04.089. iCCS 2010
11. Gaston, D.R., et al.: Physics-based multiscale coupling for full core nuclear reactor simulation. Ann. Nucl. Energy **84**, 45–54 (2015). https://doi.org/10.1016/ j.anucene.2014.09.060. Multi-Physics Modelling of LWR Static and Transient Behaviour
12. Groen, D., Knap, J., Neumann, P., Suleimenova, D., Veen, L., Leiter, K.: Mastering the scales: a survey on the benefits of multiscale computing software. Phil. Trans. Roy. Soc. A Math. Phys. Eng. Sci. **377**(2142), 20180147 (2019). https://doi.org/ 10.1098/rsta.2018.0147
13. Groen, D., Zasada, S., Coveney, P.: Survey of multiscale and multiphysics applications and communities. Comput. Sci. Eng. **16**, (2012). https://doi.org/10.1109/ MCSE.2013.47
14. Hegewald, J., Krafczyk, M., Tölke, J., Hoekstra, A., Chopard, B.: An agent-based coupling platform for complex automata. In: Bubak, M., van Albada, G.D., Dongarra, J., Sloot, P.M.A. (eds.) ICCS 2008. LNCS, vol. 5102, pp. 227–233. Springer, Heidelberg (2008). https://doi.org/10.1007/978-3-540-69387-1_25
15. Hoekstra, A.G., Lorenz, E., Falcone, J.-L., Chopard, B.: Towards a complex automata framework for multi-scale modeling: formalism and the scale separation map. In: Shi, Y., van Albada, G.D., Dongarra, J., Sloot, P.M.A. (eds.) ICCS 2007. LNCS, vol. 4487, pp. 922–930. Springer, Heidelberg (2007). https://doi.org/ 10.1007/978-3-540-72584-8_121
16. Karakasidis, T.E., Charitidis, C.A.: Multiscale modeling in nanomaterials science. Mater. Sci. Eng. C **27**(5), 1082–1089 (2007). https://doi.org/10.1016/j.msec.2006. 06.029. eMRS 2006 Symposium A: Current Trends in Nanoscience - from Materials to Applications
17. Nikishova, A., Veen, L., Zun, P., Hoekstra, A.G.: Semi-intrusive multiscale metamodelling uncertainty quantification with application to a model of in-stent restenosis. Phil. Trans. Roy. Soc. A Math. Phys. Eng. Sci. **377**(2142), 20180154 (2019). https://doi.org/10.1098/rsta.2018.0154
18. Nikishova, A., Hoekstra, A.G.: Semi-intrusive uncertainty propagation for multiscale models. J. Comput. Sci. **35**, 80–90 (2019). https://doi.org/10.1016/j.jocs. 2019.06.007
19. Nikishova, A., Veen, L., Zun, P., Hoekstra, A.G.: Uncertainty quantification of a multiscale model for in-stent restenosis. Cardiovasc. Eng. Technol. **9**(4), 761–774 (2018). https://doi.org/10.1007/s13239-018-00372-4
20. Pelupessy, F.I., van Elteren, A., de Vries, N., McMillan, S.L.W., Drost, N., Zwart, S.F.P.: The astrophysical multipurpose software environment. Astron. Astrophys. **557**, 84 (2013). https://doi.org/10.1051/0004-6361/201321252
21. Ringkjøb, H.K., Haugan, P.M., Solbrekke, I.M.: A review of modelling tools for energy and electricity systems with large shares of variable renewables. Renew. Sustain. Energy Rev. **96**, 440–459 (2018). https://doi.org/10.1016/j.rser.2018.08. 002
22. Roy, C.J., Oberkampf, W.L.: A comprehensive framework for verification, validation, and uncertainty quantification in scientific computing. Comput. Methods Appl. Mech. Eng. **200**(25), 2131–2144 (2011). https://doi.org/10.1016/j.cma.2011. 03.016
23. Veen, L.: MUSCLE 3 (2019). https://doi.org/10.5281/zenodo.3260941, https:// github.com/multiscale/muscle3

24. Walpole, J., Papin, J.A., Peirce, S.M.: Multiscale computational models of complex biological systems. Ann. Rev. Biomed. Eng. **15**(1), 137–154 (2013). https://doi.org/10.1146/annurev-bioeng-071811-150104. pMID: 23642247
25. Zwart, S.F.P., McMillan, S.L.W., van Elteren, A., Pelupessy, F.I., de Vries, N.: Multi-physics simulations using a hierarchical interchangeable software interface. Comput. Phys. Commun. **184**(3), 456–468 (2013). https://doi.org/10.1016/j.cpc.2012.09.024

Quantum Computing Workshop

Simulations of Quantum Finite Automata

Gustaw Lippa, Krzysztof Makieła, and Marcin Kuta(✉)

Department of Computer Science, Faculty of Computer Science,
Electronics and Telecommunications, AGH University of Science and Technology,
Al. Mickiewicza 30, 30-059 Krakow, Poland
`mkuta@agh.edu.pl`

Abstract. This paper presents a Python library to simulate different kinds of quantum finite automata on a classical computer. The library also provides tools for language generation and visual representation of simulation results. We have conducted experiments to measure the time complexity of the simulation in a function of the automaton size, alphabet size and word length. Examples of library usage are also provided.

Keywords: Quantum finite automata · Acceptance conditions · Cut-point · Automata simulation

1 Introduction

Finite automata are real models of computers, which have only limited amount of memory. Finite automata are also interesting models themselves, due to their simplicity but at the same time rich structure and interesting properties [7]. Since the introduction of finite automata in 1959 by Rabin and Scott [13], the theme has been quite well recognized and approached from various aspects. Connections with different domains, including algebra and logics, have also been established.

The picture for theory of quantum automata and languages generated by them is less clear, and various important problems remain open [2,11]. The task of quantum finite automata is to recognise quantum languages. Studying these languages is useful in establishing the computational and expressive power of quantum machines in general. However, such devices are not yet available and simulators have to be used instead. Because of that, we developed a library written in Python, running on a classical computer and providing implementation of several types of quantum finite automata.

This paper presents the library for simulating quantum finite automata. The library can help in exploring hypotheses on unknown relations between classes of quantum finite automata and quantum languages by providing evidence about accepting probabilities of particular words and sets of words. The library could also be useful for teaching students courses on finite automata in a quantum context.

© Springer Nature Switzerland AG 2020
V. V. Krzhizhanovskaya et al. (Eds.): ICCS 2020, LNCS 12142, pp. 441–450, 2020.
https://doi.org/10.1007/978-3-030-50433-5_34

2   Related Work

Simulation of classical finite automata is a mature area. A comprehensive survey of simulators of classical finite automata is given in [5] and JFLAP emerges as the most mature and popular tool [14].

On the other hand, there have been many libraries focused on bringing quantum computation onto the classical architectures, perhaps the best known being Q#[1]. The low-level libraries include Quirk[2], with graphical interface available through a web browser, or Quantum++[3], a high performance library written in C++11. The high-level libraries provide similar functionalities but focus more on code expressiveness. They often enable using real life simulators or quantum computers as their back-ends. Examples of such libraries include ProjectQ[4], Qiskit[5] and the aforementioned Q#.

However, we have not found a solution focused solely on quantum finite automata. The available pieces of software were either too low-level, focusing on quantum gates and quantum phenomena in micro-scale, or too abstract, providing interfaces for developing quantum algorithms in general, but without tools dedicated specifically for quantum automata.

3   Quantum Finite Automata

In this work we consider only one-way finite automata, i.e., in each step of a simulation the head reads one symbol from the tape and moves forward. Backward or empty moves of the head are forbidden. The paper defines all automata in a uniform framework, which means that for classical finite automata notation differs slightly from the one adopted widely in literature.

Preliminaries. An input alphabet Σ is a finite set of symbols. The working alphabet Γ equals $\Sigma \cup \{\$\}$, where $\$$ denotes a special end-marker symbol outside the input alphabet. Set Q is a finite set of states and $q_I \in Q$ is a distinguished state, called the initial state. Each classical state $q \in Q$ has a quantum counterpart $|q\rangle$. A pure quantum state $|\psi\rangle$ of quantum automaton is defined as

$$|\psi\rangle = \begin{pmatrix} \alpha_1 \\ \vdots \\ \alpha_n \end{pmatrix} = \sum_{i=1}^{n} \alpha_i |q_i\rangle \ , \tag{1}$$

where $\alpha_1, \ldots, \alpha_n \in \mathbb{C}$ and $\sum_{i=1}^{n} |\alpha_i|^2 = 1$.

Vector $\mathbf{1}$ denotes column vector consisting of $|Q|$ ones. Matrix $I_{|Q|}$ denotes square $|Q| \times |Q|$ matrix containing ones on a diagonal and zeros elsewhere.

---

[1] https://docs.microsoft.com/en-us/quantum/language.
[2] https://algassert.com/quirk.
[3] https://github.com/vsoftco/qpp.
[4] https://github.com/ProjectQ-Framework/ProjectQ.
[5] https://qiskit.org.

Definition 1. Nondeterministic Finite Automaton (NFA) [13] is a 5-tuple $\mathcal{A} = (\Sigma, Q, q_I, Q_{\text{acc}}, \{M_\sigma\}_{\sigma \in \Sigma})$, where $Q_{\text{acc}} \subseteq Q$ is a set of accepting states, and for all $\sigma \in \Sigma$ transition matrix M_σ satisfies $M_\sigma \in \{0,1\}^{|Q| \times |Q|}$. If transition matrices M_σ additionally satisfy for all $\sigma \in \Sigma$ condition

$$M_\sigma 1 = 1 , \tag{2}$$

then an automaton is Deterministic Finite Automaton (DFA). Classes of NFAs and DFAs are equivalent as they recognize the same class of languages, i.e., class of regular languages.

Definition 2. Probabilistic Finite Automaton (PFA) [12] is a 5-tuple $\mathcal{A} = (Q, \Sigma, I, F, \{M_\sigma\}_{\sigma \in \Sigma})$, where vector I is a stochastic column vector describing initial distribution of states, i.e., $I \in [0,1]^{|Q| \times 1}$ and $\sum_{i=1}^{|Q|} I_i = 1$. Vector F is a column vector of size $|Q|$ with i-th entry equal 1 if q_i is a accepting state and 0 otherwise. For all $\sigma \in \Sigma$ transition matrix M_σ is Markovian, i.e., its rows define probability distribution. Thus, for all $\sigma \in \Sigma$ we have $M_\sigma \in [0,1]^{|Q| \times |Q|}$ and M_σ satisfies (2).

Definition 3. Measure-Once Quantum Finite Automaton (MO-QFA) [9] is a 5-tuple $\mathcal{A} = (Q, \Sigma, q_I, Q_{\text{acc}}, \{U_\sigma\}_{\sigma \in \Sigma})$, where $Q_{\text{acc}} \subseteq Q$ is a set of accepting states. Transition matrices $\{U_\sigma\}_{\sigma \in \Sigma}$ satisfy $U_\sigma \in \mathbb{C}^{|Q| \times |Q|}$ for all $\sigma \in \Sigma$ and are unitary, i.e., for all $\sigma \in \Sigma$ we have

$$U_\sigma^\dagger U_\sigma = U_\sigma U_\sigma^\dagger = I_{|Q|} . \tag{3}$$

The set of accepting states corresponds to a projective operator:

$$P_{\text{acc}} = \sum_{q \in Q_{\text{acc}}} |q\rangle\langle q| . \tag{4}$$

Definition 4. Measure-Many Quantum Finite Automaton (MM-QFA) [8] is a 6-tuple $\mathcal{A} = (Q, \Sigma, q_I, Q_{\text{acc}}, Q_{\text{rej}}, \{U_\sigma\}_{\sigma \in \Gamma})$, where $Q_{\text{rej}} \subseteq Q$ is a set of rejecting states, and $U_\sigma \in \mathbb{C}^{|Q| \times |Q|}$ are transition matrices satisfying (3).

The automaton partitions set Q into $Q = Q_{\text{acc}} \cup Q_{\text{rej}} \cup Q_{\text{non}}$, where Q_{non} is a set of nonhalting (neutral) states. Sets $Q_{\text{acc}}, Q_{\text{rej}}$ and Q_{non} should be pairwise disjoint.

In a manner analogous to (4), projective operators P_{rej} and P_{non} are defined as follows:

$$P_{\text{rej}} = \sum_{q \in Q_{\text{rej}}} |q\rangle\langle q| , \tag{5}$$

$$P_{\text{non}} = \sum_{q \in Q_{\text{non}}} |q\rangle\langle q| . \tag{6}$$

Definition 5. General Quantum Finite Automaton (GQFA) [10] is a 6-tuple $\mathcal{A} = (Q, \Sigma, q_I, Q_{\text{acc}}, Q_{\text{rej}}, \{U_\sigma\}_{\sigma \in \Gamma})$. The model is similar to MM-QFA, but the transition matrices U_σ are more general – they are a composition of a finite sequence of applications of unitary transformations followed by orthogonal measurements. Note, that there is a more general definition of GQFA, provided by Hirvensalo [6], but it is not considered in our work.

For a broader description of these and other types of quantum automata we refer a reader to [3,7,11].

3.1   Language Acceptance Modes

Let $P_A(x)$ denote probability of accepting word $x \in \Sigma^*$ by automaton A and let λ be a real number such that $\lambda \in [0,1)$. Given $x = \sigma_1 \ldots \sigma_n$, probability $P_A(x)$ is computed as $\|P_{acc}U_{\sigma_n} \ldots U_{\sigma_1}|q_0\rangle\|^2$ for MO-QFA, but for MM-QFA is more complicated.

There exist several modes of language acceptance:

- *with a cut-point* $\lambda \in [0,1)$, if for all $x \in L$, we have $P_A(x) > \lambda$ and for all $x \notin L$, we have $P_A(x) \leq \lambda$. This mode of acceptance is also called *with an unbounded error*.
- *with an isolated cut-point* $\lambda \in [0,1)$, if there exists $\varepsilon \geq 0$, such, that for all $x \in L$, we have $P_A(x) \geq \lambda + \varepsilon$ and for all $x \notin L$, we have $P_A(x) \leq \lambda - \varepsilon$.
- *with a bounded error* $\varepsilon \in [0, \frac{1}{2})$, if for all $x \in L$, we have $P_A(x) \geq 1 - \varepsilon$ and for all $x \notin L$, we have $P_A(x) \leq \varepsilon$. This mode of acceptance is equivalent to acceptance with an isolated cut-point, where cut-point $\lambda = \frac{1}{2}$ is isolated with value $\frac{1}{2} - \varepsilon$.
- *with a positive one-sided unbounded error* if for all $x \in L$, we have $P_A(x) > 0$.
- *with a negative one-sided unbounded error* if for all $x \in L$, we have $P_A(x) = 1$.
- *Monte Carlo acceptance* [4], if there exists $\varepsilon \in (0, \frac{1}{2}]$ such, that for all $x \in L$, we have $P_A(x) = 1$ and for all $x \notin L$, we have $P_A(x) \leq \varepsilon$. Such A is called Monte Carlo *QFA* for L.

Acceptance with a cut-point and acceptance with an isolated cut-point are the most import modes of language acceptance, and are implemented in our library.

4   Library

The simulation library[6] provides its interface through a number of classes, each representing one of the automata models. It also uses modules **LanguageGenerator**, **LanguageChecker** and **Plotter**. Figure 1 presents main components of the simulation library, which offers the following functionality.

Automaton Definition. A new automaton is constructed with an object belonging to a class representing implemented automaton (one of **PFA**, **MO_1QFA**, **MM_1QFA**, **GQFA**) and data defining chosen automaton, such as the alphabet, transition matrices and matrices of projective measurements are passed. A user has to assure unitarity of transition matrices.

---

[6] https://github.com/gustawlippa/QFA.

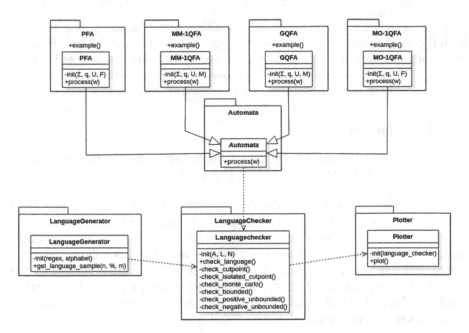

Fig. 1. Model of the quantum finite automata library

Generation of Language Samples. The `LanguageGenerator` module is responsible for this functionality. For finite languages, a user can provide a list of words which entirely define the language. To account for the infinite regular languages, a language can be defined with a regular expression. Generation of samples of stochastic languages is not implemented in the current version of the library.

Automaton Simulation. The `LanguageChecker` module of the library enables simulation of an automaton run on a single word or on a random or user-defined sample of defined language. A simulation is determined by a transition matrix and one simulation of the automaton on a given word is sufficient to obtain all information about word acceptance or rejection. Thus, there is no need to repeat a simulation since transition matrices are stationary. Simulation results are returned with respect to two modes of language acceptance: with a cut-point and with an isolated cut-point.

Results Visualisation. The library has a dedicated `Plotter` module to plot histograms of counts of words accepted and words rejected with a given probability. A cut-point and an isolation interval can also be shown.

5   Simulations and Results

There exist two ways of simulating systems of a probabilistic nature. One way is the *strong* simulation, which requires calculating the exact probability of an outcome. The other way is the *weak* simulation, which is based on a sampling from the output distribution in order to approximate the probability. The latter is the only tractable way of simulating quantum computers because of the complexity of the task. However, our library performs a strong simulation, which is adequate because of the simpler nature of quantum automata.

5.1   Experiments

We have performed a handful of experiments to determine the time complexity of the simulations depending on various parameters.

Automaton size. During our research we concluded that the most important factor in the time complexity of a simulation is the number of automaton states. In a true QFA, the relation between the number of states and computation time would be linear. In our simulator the time complexity is polynomial which agrees with the result proven in [3]. Figure 2 shows simulation time of GQFA with the growing number of its states. The results are quite close to the values predicted theoretically. Simulation time is reported as the arithmetic mean from 5 simulations for each automaton size.

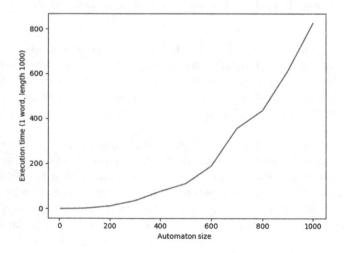

Fig. 2. Simulation time of GQFA in a function of the number of states

Alphabet Size. Figure 3 presents simulation time of GQFA as a function of the size of alphabet, over which automaton is defined. For each alphabet size, simulation time was measured as the arithmetic mean over 500 random words. It should be noted that the scale in ordinate axis (y-axis) does not start from 0. Figure 3 shows that there is no dependence between the size of the alphabet and the time of simulation. This is because the change of the alphabet size influences only the amount of input data (transition matrices and projective measurement matrices) required to define an automaton but does not impact computation time at all.

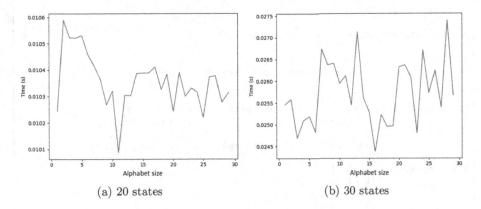

(a) 20 states   (b) 30 states

Fig. 3. Simulation time of GQFA in a function of the alphabet size

Word Length. The relation between the time of computation and the length of an input word is linear. This is not surprising, because with each letter of the input word, an automaton performs a fixed number of matrix multiplications. The size of these matrices is determined by the number of states of automaton. For each letter a transition function must be applied and a measurement may be performed, depending on the type of an automaton. Figure 4 shows simulation time of GQFA as a function of the length of simulated words. For each word length, 100 random words were simulated and simulation time was taken as the arithmetic mean over these 100 words.

5.2   Usage Example

Listing 1.1 presents exemplary code for defining a MM-QFA automaton. This automaton is well-known in literature and was proposed by Ambainis and Freivalds [1] as a part of the proof that there exists a one-way QFA, which recognizes language a^*b^* with probability $p = 0.68..$, where p is the real root of the equation $p^3 + p = 1$. Figure 5 visualizes obtained acceptance probabilities and their corresponding word counts.

```
import numpy as np
from math import sqrt
from QFA.MM_1QFA import MM_1QFA
from QFA.LanguageGenerator import LanguageGenerator
from QFA.LanguageChecker import LanguageChecker
from QFA.Plotter import Plotter

alphabet = 'ab'

p = 0.682327803828019  # Auxillary variable

# Initial state of automaton
initial_state = np.array([[sqrt(1-p)], [sqrt(p)], [0], [0]])

# Transition matrices
U_a = np.array([[1-p,            sqrt(p*(1-p)), 0, -sqrt(p) ],
                [sqrt(p*(1-p)),  p,             0, sqrt(1-p)],
                [0,              0,             1, 0        ],
                [sqrt(p),        -sqrt(1-p),    0, 0        ]])

U_b = np.array([[0, 0, 0, 1],
                [0, 1, 0, 0],
                [0, 0, 1, 0],
                [1, 0, 0, 0]])

U_end = np.array([[0, 0, 0, 1],
                  [0, 0, 1, 0],
                  [0, 1, 0, 0],
                  [1, 0, 0, 0]])

# Accepting and rejecting states are defined with matrices
# representing projective measurements
P_acc = np.array([[0, 0, 0, 0],
                  [0, 0, 0, 0],
                  [0, 0, 1, 0],
                  [0, 0, 0, 0]])

P_rej = np.array([[0, 0, 0, 0],
                  [0, 0, 0, 0],
                  [0, 0, 0, 0],
                  [0, 0, 0, 1]])

qfa = MM_1QFA(alphabet, initial_state,
              [U_a, U_b, U_end], P_acc, P_rej)

language_generator = LanguageGenerator('a*b*', alphabet)
language, not_in_language = language_generator.get_language_sample()

language_checker = LanguageChecker(qfa, language, not_in_language)

plotter = Plotter(language_checker)
plotter.plot()
```

Listing 1.1. Code for defining MM-QFA automaton

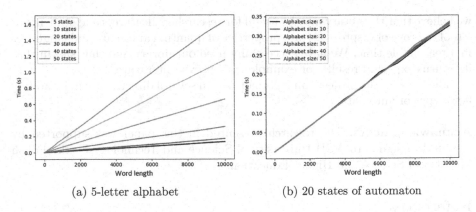

(a) 5-letter alphabet   (b) 20 states of automaton

Fig. 4. Simulation time of GQFA in a function of the length of simulated words

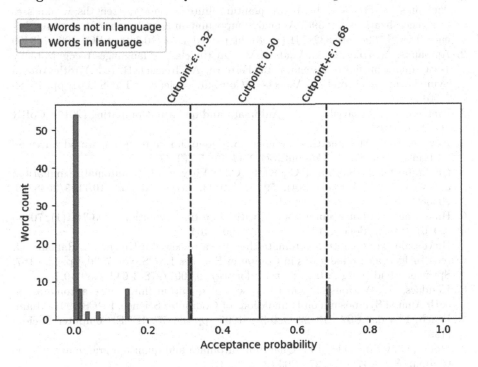

Fig. 5. Visualization of acceptance probabilities and acceptance modes for MM-QFA

6 Conclusions

The library provides a simple API for functionality required to simulate quantum finite automata. We hope the library will encourage research at the intersection of quantum computations and theory of formal languages and automata. We have experimentally shown that the time complexity of simulating a quantum finite automaton is polynomial in relation to the size of the automaton. Nevertheless, we believe that the library may be useful for researchers, lecturers and students as a tool to prove or disprove certain properties of quantum automata and languages in a reasonable time. We have successfully used our library and thus shown that it returns expected results for examples taken from literature.

The scope of the project can be broadened in several directions, e.g. by adding new types of automata.

Acknowledgments. The research presented in this paper was supported by the funds assigned to AGH University of Science and Technology by the Polish Ministry of Science and Higher Education.

References

1. Ambainis, A., Freivalds, R.: 1-way quantum finite automata: strengths, weaknesses and generalizations. In: 39th Annual Symposium on Foundations of Computer Science, FOCS 1998, pp. 332–341 (1998). https://doi.org/10.1109/SFCS.1998.743469
2. Ambainis, A., Kikusts, A., Valdats, M.: On the class of languages recognizableby 1-way quantum finite automata. In: Ferreira, A., Reichel, H. (eds.) 18th Annual Symposium on Theoretical Aspects of Computer Science, STACS 2001, pp. 75–86 (2001)
3. Ambainis, A., Yakaryılmaz, A.: Automata and quantum computing (2015). CoRR abs/1507.01988
4. Bianchi, M.P., Mereghetti, C., Palano, B.: Quantum finite automata: advances on bertoni's ideas. Theor. Comput. Sci. **664**, 39–53 (2017)
5. Chakraborty, P., Saxena, P.C., Katti, C.P.: Fifty years of automata simulation: a review. ACM Inroads **2**(4), 59–70 (2011). https://doi.org/10.1145/2038876.2038893
6. Hirvensalo, M.: Quantum automata with open time evolution. IJNCR **1**(1), 70–85 (2010). https://doi.org/10.4018/jncr.2010010104
7. Hirvensalo, M.: Quantum automata theory – a review. In: Kuich, W., Rahonis, G. (eds.) Algebraic Foundations in Computer Science. LNCS, vol. 7020, pp. 146–167. Springer, Heidelberg (2011). https://doi.org/10.1007/978-3-642-24897-9_7
8. Kondacs, A., Watrous, J.: On the power of quantum finite state automata. In: 38th Annual Symposium on Foundations of Computer Science, FOCS 1997, Miami Beach, Florida, USA, 19–22 October 1997, pp. 66–75. IEEE Computer Society (1997)
9. Moore, C., Crutchfield, J.P.: Quantum automata and quantum grammars. Theor. Comput. Sci. **237**(1–2), 275–306 (2000)
10. Nayak, A.: Optimal lower bounds for quantum automata and random access codes. In: 40th Annual Symposium on Foundations of Computer Science, FOCS 1999, pp. 369–377 (1999)
11. Qiu, D., Li, L., Mateus, P., Gruska, J.: Quantum finite automata. In: Wang, J. (ed.) Handbook of Finite State Based Models and Applications, pp. 113–144 (2012)
12. Rabin, M.O.: Probabilistic automata. Inf. Control **6**(3), 230–245 (1963)
13. Rabin, M.O., Scott, D.S.: Finite automata and their decision problems. IBM J. Res. Dev. **3**(2), 114–125 (1959). https://doi.org/10.1147/rd.32.0114
14. Rodger, S.H., Finley, T.W.: JFLAP: An Interactive Formal Languages and Automata Package. Jones and Bartlett Publishers, Sudbury (2006)

|Lib⟩: A Cross-Platform Programming Framework for Quantum-Accelerated Scientific Computing

Matthias Möller$^{(\boxtimes)}$ ⓘ and Merel Schalkers

Department of Applied Mathematics (DIAM) and 4TU.CEE, Centre for Engineering
Education, Delft University of Technology, Delft, The Netherlands
m.moller@tudelft.nl, merelschalkers@gmail.com
http://ta.twi.tudelft.nl/nw/users/matthias/

Abstract. This paper introduces a new cross-platform programming
framework for developing quantum-accelerated scientific computing
applications and executing them on most of today's cloud-based quantum computers and simulators. It makes use of C++ template meta-
programming techniques to implement quantum algorithms as generic,
platform-independent expressions, which get automatically synthesized
into device-specific compute kernels upon execution. Our software framework supports concurrent and asynchronous execution of multiple quantum kernels via a CUDA-inspired stream concept.

Keywords: Quantum-accelerated scientific computing · Template
meta-programming · Hybrid software development framework

1 Introduction

The development of practically usable quantum computing technologies is in full
swing involving global players like Alibaba, Atos, Google, IBM, and Microsoft
and specialists in this field such as Rigetti Computing and D-Wave. These parties
compete for technology lead and, finally, simply the raw number of qubits they
can provide through their quantum processing units (QPUs), which can be either
hardware quantum computers or quantum computer simulators running on classical high-performance computing hardware. This situation resembles the very
early days of GPU-accelerated computing when the first generation of general-
purpose programmable graphics cards became available but their productive use
in scientific applications was largely hindered by the non-availability of software
development kits (SDKs) and easy-to-use domain-specific software libraries and,
even more severe, the lack of standardized non-proprietary development environments that would lower the dependence on a particular GPU vendor.

Today's quantum software landscape can be grouped into three main categories: *quantum SDKs* [1,6,15,19,22], stand-alone *quantum simulators* [5,11,13],
and *quantum assembly (QASM)* [2,3,12] or *instruction languages (QUIL)* [21].

© Springer Nature Switzerland AG 2020
V. V. Krzhizhanovskaya et al. (Eds.): ICCS 2020, LNCS 12142, pp. 451–464, 2020.
https://doi.org/10.1007/978-3-030-50433-5_35

A recent overview and comparison of gate-based quantum software platforms by LaRosa [14] shows that the field is highly fragmented making it impossible to perform a fair quantitative performance comparison. Moreover, the tools focus on quantum computing experts who are mainly interested in the development of stand-alone quantum algorithms rather than their use as computational building blocks within a possibly hybrid classical-quantum solution procedure.

In our opinion, *practical* quantum computing has the highest chances to become a game-changer for the computational sciences if it is positioned as *special-purpose* accelerator technology that will become available in future heterogeneous compute platforms equipped with GPUs, QPUs and other emerging accelerators like field-programmable gate arrays (FPGAs). Researchers and scientific application developers will then have the free choice between, say, running the HHL-algorithm [9] on a QPU accelerator and adopting one of the many classical numerical methods for solving linear systems of equations on CPUs, GPUs or FPAGs depending on problem sizes and matrix characteristics. In [17] we have outlined a conceptual framework for QPU-accelerated automated design optimization that builds on the HHL-solver as main computational driver.

We believe that end-users from the community of computational science and engineering would be interested in giving QPU-accelerated computing a try with the right software tools at hand. With this vision in mind, we created the |Lib⟩-project [16] (pronounced Lib-Ket), which is a cross-platform programming framework that aims at making QPU-accelerated computing as easily accessible for the masses as GPU computing is today through frameworks like CUDA [18].

The remainder of this paper is structured as follows: Sect. 2 discusses the design principles underlying the |Lib⟩ framework, which is introduced in Sect. 3. Implementation details are discussed in Sect. 4 followed by a brief demonstration of |Lib⟩'s capabilities in Sect. 5. Section 6 completes the paper with a conclusion and an outlook on functionality planned for future releases.

2   Design Principles

To achieve our set-out vision, |Lib⟩ is designed based on the following principles:

- **QPU-accelerated computing**: Quantum computers are used as *special-purpose accelerator devices* within a heterogeneous computer system that can host multiple accelerator technologies (GPUs, FPGAs, ...) side by side.
- **Concurrent task offloading**: Quantum algorithms are implemented as *compute kernels* describing concurrent tasks launched on QPU devices.
- **Single-source quantum-classical programming**: Classical and quantum code is implemented in a single source file, which is compiled into one *hybrid binary executable* executed on the host computer, who offloads certain parts of the computation to the accelerator devices.
- **Write once run anywhere**: Quantum algorithms are implemented once and for all as *generic expressions*, which can be executed on current and future QPU-device types. Support for a particular type is realized by a small set of

conversion functions between |Lib⟩'s unified interface layer and the device-specific low-level application programming interface (API).
- **Standing on the shoulders of giants**: |Lib⟩ is developed on top of existing vendor-specific tools and libraries to exploit their full optimization potential.
- **Seamless integration into status quo**: |Lib⟩ does not create new standards that need to be implemented by others but utilizes the available tools.

The first three principles suggest a conceptual design in the spirit of CUDA [18] or OpenCL [23], which are de-facto standards for GPU computing. To underline the postulated similarity between QPU- and GPU-accelerated computing and to make quantum computing more accessible to experts in classical accelerator technologies, we will utilize a GPU-inspired terminology such as *host* (the CPU and its memory) and *device* (the QPU and its memory), *kernels* and *streams*, as well as *asynchronous execution* and *synchronization* throughout this paper.

The write-once-run-anywhere principle has led us to adopt template meta-programming techniques to implement quantum algorithms as generic expressions, whose evaluation for a particular QPU type is delayed until the program flow has reached the point, where its actual value is really needed. This approach is also known as *lazy evaluation* or *call-by-need* principle in programming language theory and is used successfully in linear algebra libraries [4,7,8,10,20,24].

The last two principles are mainly based on pragmatic considerations. Firstly, introducing yet another approach to quantum programming incompatible to the existing ones would escalate the fragmentation of the quantum software landscape instead of improving the situation for the potential end-users. Moreover, the chosen approach allows for exploiting the expertise and manpower of scientists worldwide working on different aspects of quantum computing and their expert knowledge of non-disclosed technical details of QPU devices to create an open software ecosystem that immediately benefits from any improvement in one of the underling core components. Finally, most human beings are more open to emerging technologies if they come as evolutionary increments of the status quo instead of radical paradigm shifts that call for dumping all previous work.

3   The |Lib⟩ Programming Framework

The open-source, cross-platform |Lib⟩ programming framework is designed as header-only C++14 *quantum expression template library*[1] with minimal external dependencies, namely, an embedded Python interpreter and, possibly, header and/or library files from the respective quantum backends. It can be downloaded free-of-charge from the GitLab repository https://gitlab.com/mmoelle1/LibKet, which provides documentation in form of a wiki and an API documentation and several tutorial examples to get started. In addition to the primary C++ API,

---

[1] In the Dutch language, the word *quantum* is spelled *kwantum*. Hence, the name |Lib⟩ (pronounced Lib-Ket) is an allusion to the *bra-ket notation* introduced in 1939 by Paul Dirac that is widely used for expressing quantum algorithms.

C and Python APIs are being implemented, which adopt just-in-time compilation techniques to exploit the full potential of C++ template meta-programming internally and expose |Lib⟩'s functionality in C and Python-style to the outside.

A comprehensive overview of the |Lib⟩ programming framework is given in Fig. 1. It consists of three layers that provide components for application programmers (high-level (HL) API), quantum algorithm developers (mid-level (ML) API), and QPU providers (low-level (LL) API), respectively.

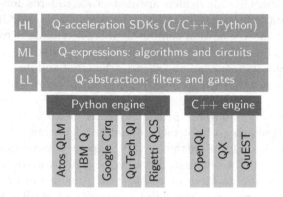

Fig. 1. Overview of the cross-platform |Lib⟩ programming framework.

Before we describe the different software layers in more detail we give a short example on |Lib⟩'s general usage. Consider the C++ code snippet given in Listing 1 which puts the first and third qubit of a quantum register into the maximally entangled first Bell state, where A is qubit 1 and B is qubit 3:

$$|\Phi^+\rangle = \frac{1}{\sqrt{2}}\left(|0\rangle_A \otimes |0\rangle_B + |1\rangle_A \otimes |1\rangle_B\right) = \frac{|00\rangle_{AB} + |11\rangle_{AB}}{\sqrt{2}}. \tag{1}$$

The easiest way to achieve this is to start from the computational basis $|0\rangle$ and apply a Hadamard gate to one qubit followed by a controlled-NOT (CNOT) gate

$$
\begin{array}{c}
|0\rangle_A \;-\boxed{H}\;-\bullet- \\
|0\rangle_B \;-\;-\;-\oplus-
\end{array}
$$

This is realized by the quantum expression that is constructed in lines 8–9 of the code snippet, thereby demonstrating two of |Lib⟩'s most essential components, namely, *Quantum Filters* and *Quantum Gates*, which are implemented in the namespaces `LibKet::filters` and `LibKet::gates`, respectively.

As the name suggests, filters select a subset of the quantum register; see Sect. 4.1 for more details. Here, `sel< 1 >()` selects the first qubit for applying the Hadamard gate. This sub-expression serves as first argument, the control, to the binary CNOT gate, whose action is applied to the third qubit (`sel<3>(...)`). The `init()` gate puts all qubits of the quantum register into the computational basis $|0\rangle$. More information on gates is given in Sect. 4.2. It should be noted

```
1    #include <LibKet.hpp>
2    using namespace LibKet;
3    using namespace LibKet::filters;
4    using namespace LibKet::gates;
5
6    int main()
7    { // Create quantum expression for Bell state between first and third qubit
8      auto expr = cnot( h( sel<1>() ),
9                          sel<3>( init() ) );
10
11     // OPTIONAL: Print quantum expression, cf. Listing 2 (left)
12     show<10>(expr);
13
14     // Create Quantum-Inspire (QI) device with 6 qubits in total
15     QDevice<QDeviceType::qi_26_simulator, 6> device;
16
17     // Populate quantum kernel with quantum expression
18     device(expr);
19
20     // OPTIONAL: Print quantum kernel code, cf. Listing 2 (top-right)
21     std::cout << device << std::endl;
22
23     // Evaluate quantum kernel on Quantum-Inspire platform in the cloud
24     try { utils::json result = device.eval();
25          std::cout << device.get<QResultType::histogram>(result) << std::endl;
26          } catch (const std::exception &e) { std::cerr << e.what() << std::endl; }
27     return 0;
28   }
```

Listing 1: Creation of the first Bell state $|\Phi^+\rangle$ using |Lib⟩'s C++ API.

that the resulting quantum expression is generic, that is, object **expr** holds an abstract syntax tree (AST) representation of the Bell state creation algorithm that can be synthesized to any of |Lib⟩'s quantum backends. For the cloud-based Quantum-Inspire (QI) platform[2], this is accomplished by lines 15 and 18. In short, line 15 creates a **device**object that holds 6 qubits and specializes the generic quantum expression **expr**into common QASM code v1.0 [12], the programming language for the QI backend. The internally stored quantum kernel code as well as the quantum expression **expr**can be printed as illustrated in lines 21 and 12, respectively; see Listing 2. The probability amplitudes resulting from 1024 runs of the quantum algorithm are presented in the same diagram.

The actual execution of the quantum kernel is triggered in line 24, which starts an embedded Python interpreter as sub-process to communicate with the cloud-based quantum simulator platform via the vendor-specific QI-SDK[3]. This call performs blocking execution and returns a JSON object upon successful completion, from which the result can be retrieved. More details on how to customize the execution process, run multiple quantum kernels concurrently and perform non-blocking asynchronous kernel execution are given in Sect. 4.5.

---

[2] https://www.quantum-inspire.com designed and built by the Dutch research center for Quantum Computing and Quantum Internet QuTech (https://qutech.nl). The basic user account only allows utilization of the 26-qubit version of the QI simulator.

[3] https://github.com/QuTech-Delft/quantuminspire.

```
1   BinaryQGate
2   |    gate = QCNOT
3   |  filter = QFilterSelect [ 1 3 ]
4   |   expr0 = UnaryQGate
5   |            |   gate = QHadamard
6   |            | filter = QFilterSelect [ 1 ]
7   |            |   expr = QFilterSelect [ 1 ]
8   |   expr1 = UnaryQGate
9   |            |   gate = QInit
10  |            | filter = QFilterSelect [ 3 ]
11  |            |   expr = QFilter
```

```
1   version 1.0
2   qubits 6
3   h q[1]
4   cnot q[1], q[3]
```

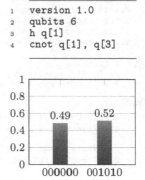

Listing 2: AST of quantum expression (left), resulting QASM code (right-top), and probability amplitudes computed by QuTech's QI simulator (right-bottom).

4 Implementation Details

In what follows, we address the individual |Lib⟩ components and shed some light on their internal realization and ways to extend them to support new backends.

4.1 Quantum Filter Chains

As stated before, |Lib⟩'s quantum filters are meant to select subsets of qubits from the global quantum register to which the following quantum operation is being applied, which is comparable to matrix views in the Eigen library [8].

Since today's and near-future quantum processors have a very limited number of qubits, typically, between 5–50, we consider the assumption of a single global quantum register and the absence of dynamic memory (de)allocation capabilities most practical. Moreover, quantum computing follows the in-memory computing paradigm, that is, data is stored *and* manipulated at fixed locations in memory. This is in contrast to the classical von-Neuman computer architecture, where data is transported between the randomly accessible main memory (RAM) and the central processing unit (CPU), the latter performing the computations.

Table 1 lists all quantum filters supported by |Lib⟩. All filtering operations are applied relative to the given input, which makes it possible combine multiple filters to so-called filter chains. Consider, for instance, the filter chain qubit<2>(shift<2>(range<2, 5>())), which selects the 6-th qubit from the global register, more precisely, the pre-selected set of qubits passed as input.

Thanks to the use of C++ template meta-programming techniques, quantum filters are evaluated at compile time and, hence, even complex filter chains cause no overhead costs at run time. With the aid of gototag<Tag>() it is possible to restore a previously stored filter configuration that has been tagged by the tag<Tag>() function. It is generally recommended to safeguard quantum expressions that should be used as building blocks in larger algorithms by

Table 1. |Lib⟩'s quantum filters.

Class	Function	Example usage	Explanation
QFilterSelectAll	all	all(...)	selects all qubits
QFilterSelect	sel	sel<0,3>(...)	selects q_0 and q_3
QFilterShift	shift	shift<2>(...)	shifts qubit selection by 2
QFilterSelectRange	range	range<2,5>(...)	selects q_2, q_3, q_4, q_5
QRegister	qureg	qureg<2,3>(...)	selects q_2, q_3, q_4
QBit	qubit	qubit<2>	selects q_2
QFilterTag	tag	tag<42>(...)	assigns tagID #42 to current selection
QFilterGotoTag	gototag	gototag<42>(...)	restores selection with tagID #42

tag-gototagpairs to prevent side effects from internal manipulation of the qubit selection.

All components listed in Table 1 come in two flavours, a class whose instantiated objects span the abstract syntax tree (AST) of the expression and a creator function that returns an object of the respective type. Classes are required to implement the operator() for all expressions that should be supported; see Listing 3 for an example. Here and below the universal-reference variant, i.e. operator()(QFilterSelect<_ids...>&&) is omitted due to space limitations but it is implemented for all types to support C++11 move semantics.

Though not foreseen in the current implementation, the just described quantum filter mechanism can be easily extended to support rudimentary stack memory based on a reserved region of the global quantum register. Together with |Lib⟩'s just-in-time (JIT) capabilities (see below) even dynamic memory (de)allocation would be possible with the adopted concept once a sufficiently large number of qubits and circuit depths are reliably supported in quantum hardware to make this feature relevant for practical applications.

4.2 Quantum Gates

|Lib⟩'s implementation of quantum gates follows the same programming paradigm (class with overloaded operator() and gate-creator function) as described above. Additionally, the class provides an overloaded apply(QData<...>& data)method, which is specialized for each supported backend type. Listing 4 illustrates how the application of the Hadamard gate appends QASM code to the data's internal quantum kernel for the cQASMv1backend; see lines 4–13. The static range()method is one of several filter utility functions that returns the actual list of selected qubits based on data's concrete register size at compile time.

Invoking the Hadamard function (lines 16–19) returns a UnaryQGateobject (see below) that stores the current sub-expression, the gate to be applied next, and the filter selection internally. The specialized overload in lines 21–25 ensures that the immediate double-application of the Hadamard gate gets eliminated. |Lib⟩ makes extensive use of this type of rule-based optimization to eliminate

```
1    template<long int _offset>
2    class QFilterShift : public QFilter
3    { public:
4        ...
5        template<std::size_t... _ids>
6        inline constexpr auto operator()(const QFilterSelect<_ids...>&) const noexcept
7        { return QFilterSelect<_offset + _ids...>{}; }
8
9        template<std::size_t... _ids>
10       inline constexpr auto operator()(QFilterSelect<_ids...>&&) const noexcept
11       { return QFilterSelect<_offset + _ids...>{}; }
12       ...
13   };
14
15   template<long int _offset>
16   inline static constexpr auto shift()
17   { return QFilterShift<_offset>{}; }
18
19   template<long int _offset, typename Expr>
20   inline static constexpr auto shift(Expr expr)
21   -> typename std::enable_if<std::is_base_of<QBase, typename std::decay<Expr>::type>::value,
22                              decltype(QFilterShift<_offset>{}(expr))>::type
23   { return QFilterShift<_offset>{}(expr); }
```

Listing 3: Example of an overloaded `operator()` for the `QFilterShift` class.

gate-level expressions of the form `t(tdag(...))` as well as entire quantum circuits followed immediately by their inverse, e.g., `qft(qftdag(...))`.

To orchestrate the interplay of expressions, filters and gates, |Lib⟩ implements unary, binary, and ternary gate containers that hold the aforementioned information as types except for the actual sub-expression which is stored by-value. Instantiations of these nearly stateless classes span the quantum expression's AST (see Listing 2 (left)), whereby an overloaded `operator()` method dispatches between the different variants to apply quantum gates to expressions.

Next to the set of quantum gates that are typically supported by most QPU backends, |Lib⟩ comes with a special hook-gate that can be used to implement common quantum building blocks, e.g., the first Bell state from Listing 1

```
1    QFunctor_alias( Bell, cnot(h(sel<1>()),sel<3>(init())) );
2    auto expr = hook<Bell>();
```

4.3   Quantum Circuit

The main advantage of |Lib⟩'s generic quantum-expression approach becomes visible for circuits, which represent compile-time parametrizable algorithms like the well-known *Quantum Fourier transform*, invoked via the `qft()` function. The implementation follows the same programming paradigms (class with overloaded `operator()` and corresponding creator function with rule-based optimization) but, typically, with a generic `apply()` method, whose synthetization to device-specific instructions is handled by the gates. Our approach makes it, however, possible to also specialize full circuits for selected QPU backends, e.g., to use Qiskit's [1] internal realization of the HHL-solver [9] for the IBM Q platform.

```
1   class QHadamard : public QGate
2   { public:
3     ...
4     // Apply() - overload for common QASM v1.0
5     template<std::size_t _qubits, QBackendType _qbackend, typename _filter>
6     inline static typename std::enable_if<_qbackend == QBackendType::cQASMv1,
7                                 QData<_qubits, QBackendType::cQASMv1>>::type&
8     apply(QData<_qubits, QBackendType::cQASMv1>& data) noexcept
9     { std::string expr = "h q[";
10      for (auto i : _filter::range(data))
11        expr += utils::to_string(i) + (i != *(_filter::range(data).end() - 1) ? "," : "]\n");
12      data.append_kernel(expr);
13      return data; }
14    ... };
15
16    // h() - constant reference
17    template<typename _expr>
18    inline constexpr auto h(const _expr& expr) noexcept
19    { return UnaryQGate<_expr, QHadamard, typename filters::getFilter<_expr>::type>(expr); }
20
21    // h() - constant reference with rule-based optimization
22    template<typename _expr, typename _filter>
23    inline constexpr auto h(const UnaryQGate<_expr, QHadamard,
24                              typename filters::getFilter<_expr>::type>& expr) noexcept
25    { return expr.expr; }
```

Listing 4: Example of an overloaded `apply()` method for the `QHadamard` class.

To ease the development of generic quantum circuits, |Lib⟩ implements a static for-loop that accepts the body as functor being passed as template argument together with loop bounds and step size as illustrated in Listing 5.

Moreover, |Lib⟩ comes with just-in-time (JIT) compilation capabilities making it possible to generate quantum expressions dynamically from user input. Quantum expressions that are given in string format are JIT compiled into dynamically loaded libraries that are cached across multiple program runs.

```
1   template<long int start, long int end, long int step, long int index>
2   struct body {
3     template <typename _expr>
4     static constexpr auto func(_expr&& expr) noexcept
5     {
6       // Apply controlled phase shift on the odd qubits q[1], q[3], ... by the angle
7       // theta = pi/2^k, k = 1, 3, ... controlled by the values of q[0], q[2], ...
8       return crk<index>(sel<index-1>(all()),
9                         sel<index >(all(expr)));
10    }
11  };
12  auto expr = static_for<1,5,2,body>();
```

Listing 5: Example of a static for-loop.

4.4   Quantum Devices

The synthetization of generic quantum expressions into device-dependent quantum instructions that can be executed on a specific QPU is realized by the many specializations of the `QDevice` class, which brings together a particular backend type with device specific details, such as credentials and parameters for connecting to cloud-based services, the maximum number of qubits, the native gate set, and the lattice structure, which might require internal optimization passes.

Lines 15 and 18 of Listing 1 create a device instance for running the quantum algorithm remotely on the Quantum-Inspire simulator platform and populate its internal quantum kernel with the expression given by Eq. (1) for creating the first Bell state, respectively. Next to providing methods for executing the kernel as described in the next section, some device types support extra functionality such as the transpilation of the generic quantum circuit into device-optimized quantum instructions and the export of the resulting circuit to LaTeX. The quantum circuits depicted in Fig. 2 were produced by the following code snippet

```
1   QDevice<QDeviceType::ibmq_london_simulator, 2>  ibmq;
2   QDevice<QDeviceType::cirq_foxtail_simulator, 2> cirq;
3   ibmq(expr); std::cout << ibmq.to_latex() << std::endl;
4   cirq(expr); std::cout << cirq.to_latex() << std::endl;
```

We consider this functionality helpful for getting a better understanding of the actual circuit – possibly with extra swap gates added to enable two-qubit operations on non-neighboring qubits – that is executed on the device rather than its idealized textbook version. The transpilation step can be bypassed by choosing generic simulators such as ibmq_qasm_simulator and cirq_simulator.

(a) Quantum circuit transpiled for IBM's 5-qubit London chip

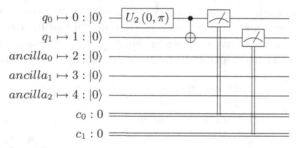

(b) Quantum circuit transpiled for Google's 22-qubit Foxtail chip

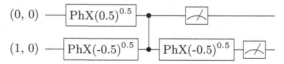

Fig. 2. Quantum circuits for producing the first Bell state, cf. Eq. (1), optimized for (a) IBM's 5-qubit London chip and (b) Google's 22-qubit Foxtail chip.

4.5   Quantum Kernel Execution

Once the generic expression has been synthesized into device-dependent instructions it can be executed on the respective QPU device. As explained before, our aim is to ease the transition from GPU programming to QPU-accelerated computing. |Lib⟩ therefore adopts a CUDA-inspired stream-based execution model, which enables concurrent quantum kernel execution on multiple QPU devices.

The device's `eval()`method called in line 24 of Listing 1 accepts a so-called `QStream<QJobType::Python>` object as optional parameter and so do the methods `execute()`and `execute_async()`as shown in the following code snippet

```
1   QStream<QJobType::Python> stream;
2   auto job = ibmq.execute_async(1024, "", "", "", stream);
3   while(!job->query()) { /* Do other stuff here */ }
4   utils::json result = job->get();
```

While the `eval()`method waits until the execution has finished and returns the result as JSON object or throws an exception upon failure, the `execute()`method returns a pointer to a job object `QJob<QJobType::Python>` that supports `query()`,`wait()` and `get()` operations. Its non-blocking counterpart `execute_async()` can be used to hide the latency stemming from the execution of the quantum kernel on remote QPUs and the overhead costs due to invoking the embedded Python interpreter with other computations on the CPU or other accelerator devices. It is even possible to execute multiple quantum algorithms concurrently on multiple QPUs by launching their kernels in different streams.

Use of an embedded Python interpreter as interface between classical host code and quantum kernels has the advantage that the full potential of vendor-specific SDKs can be exploited to perform circuit optimization and other pre- and post-processing tasks including possible validity checks on the host side before communicating the quantum kernel to the remote QPU device for execution.

The three unused parameters in line 2 of the above code snippet can be used to inject user-defined code preceding the import of Python modules and right before and after the execution of the quantum circuit, respectively. A possible application of this feature is the internal post-processing of measurement results with the functionality provided by a particular SDK[4], e.g., to visualize the measurement outcome as histogram and write it to a graphics file

```
1   auto job = ibmq.execute( /* number of shots   */ 1024,
2   /* script to be run before initialization  */
3   "\tfrom qiskit.visualization import plot_histogram\n",
4   /* script to be run before kernel execution */
5   "",
6   /* script to be run after kernel execution  */
7   "\tplot_histogram(result.get_counts()).savefig('histogram.pdf')\n");
```

While retrieving the outcome of a quantum experiment as JSON object is most flexible it requires backend-specific post-processing steps to extract the

---

[4] Generation of the history plot by the `ibmq` device requires the packages `qiskit` and `matplotlib` to be installed and accessible by the embedded Python interpreter.

desired information. For widely used data such as job identifier and duration, histogram of results, and the state with highest likelihood, each `QDevice` class specialization provides functionality to extract information from the JSON object and convert it into |Lib⟩-specific or intrinsic C++ types, e.g.

```
1   auto duration  = ibmq.get<QResultType::duration >(result);
2   auto histogram = ibmq.get<QResultType::histogram>(result);
```

Let us finally remark that |Lib⟩ also supports the native execution of quantum kernels written in C++, e.g., for quantum simulators like QX [13] and QuEST [11], using the multithreading capabilities that come with C++ 11.

5   Demonstration

|Lib⟩ is a rather young project that is under continuous development. The correct functioning of the core framework described in this paper has been verified by extensive unit tests. A comprehensive presentation of computational examples is beyond the scope of this paper and not possible within the given page limit. We therefore restrict ourselves to a single test case, namely, the quantum expression `qft(init())` and apply it to a quantum register consisting of 1–12 qubits as a first benchmark to measure the performance of different QPU backends.

Figure 3 depicts the run times measured for the following QPU backends: Cirq [6] (v0.7.0, generic simulator), pyQuil [21] (v2.19.0, 9q-square-simulator), QI [13] (v1.1.0), Qiskit [1] (v.0.17.0, qasm-simulator), and QuEST [11] (v3.1.1, CPU-OpenMP simulator). All runs were performed with 1024 shots on a dual-socket Intel Xeon E5-2687W Sandy Bridge EP system with 2×8 cores running at 3.1 GHz with 128 GB of DDR3-1600 memory except for the QI runs, which were executed on a remote system with unknown hardware specification.

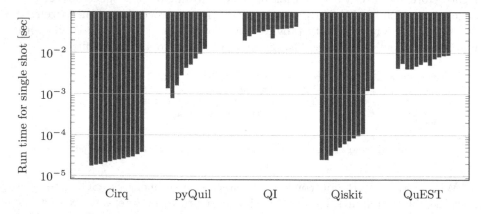

Fig. 3. Run times for the Quantum Fourier transformation executed with 1–12 qubits (per group from left to right) on five different QPU simulator backends.

For some backends, such as pyQuil and Qiskit, increasing the number of qubits and the circuit depth results in significantly longer run times, while others are less sensitive to these parameters. It should be noted that the run times measured for the pyQuil backend include the transformation of the quantum circuit into executable code by the Quil Compiler, which might explain the higher values. The QuEST backend does not allow repeated evaluation of the circuit so that the measured run time might be dominated by overhead costs.

We would like to stress that the presented results are preliminary and should not be considered a comprehensive performance analysis of the QPU backends under consideration. Systematic benchmarking of many more simulator and hardware backends for quantum circuits of different depth and level of entanglement is underway and will be presented in a forthcoming publication.

6   Conclusion

In this paper we have introduced our novel cross-platform programming framework |Lib⟩, which aims at facilitating the use of quantum computers (and their simulators) for accelerating the solution of scientific problems. Primarily addressing today's GPU programmers as early adopters, our framework is largely inspired by Nvidia's CUDA toolkit and offers a similar programming model based on quantum kernels that can be executed concurrently using multiple streams. As a unique feature, |Lib⟩ does not focus on one particular QPU backend but adopts C++ template meta-programming techniques to enable the development of quantum algorithms as generic expressions that can be synthesized to various QPU-backend types, following the write-once-run-anywhere principle.

Ongoing developments focus on the extension of the algorithm library (mid-level API; cf. Fig. 1), especially, variants of the HHL-solver [9] and its computational ingredients such as eigenvalue estimation. Another line of research work addresses the implementation of basic arithmetic routines, which are also used inside the HHL-algorithm to invert eigenvalues. Finally, the extension of the low-level API to support additional QPU backends and to reduce the computational overhead incurred by the use of the embedded Python interpreter and the conversion from JSON objects to C++ types is a permanent quest.

Despite the early development stage of the |Lib⟩ framework, we would like to encourage the scientific computing community to report their experience with it and express feature requests for forthcoming releases to the authors.

Acknowledgments. The authors would like to thank Kelvin Loh and Richard Versluis from TNO for fruitful discussions and financial support of the second author. Moreover, financial support by the 4TU. Centre for Engineering Education is acknowledged. We finally thank the anonymous reviewers for their constructive feedback.

References

1. Abraham, H., et al.: Qiskit: an open-source framework for quantum computing (2019). https://doi.org/10.5281/zenodo.2562110
2. Atos: Atos QLM software stack (2019)
3. Cross, A.W., et al.: Open quantum assembly language (2017)
4. Demidov, D., et al.: Programming CUDA and OpenCL: a case study using modern C++ libraries. SIAM J. Sci. Comput. **35**(5), C453–C472 (2013)
5. Gidney, C.: Quirk: a drag-and-drop quantum circuit simulator that runs in your browser (2019). https://github.com/Strilanc/Quirk
6. Gidney, C., et al.: Cirq: a Python framework for creating, editing, and invoking noisy intermediate scale quantum (NISQ) circuits (2019). https://github.com/quantumlib/Cirq
7. Gottschling, P., et al.: Generic compressed sparse matrix insertion: algorithms and implementations in MTL4 and FEniCS. In: Proceedings of the 8th Workshop on Parallel/High-Performance Object-Oriented Scientific Computing, POOSC 09, pp. 2:1–2:8. ACM, New York (2009)
8. Guennebaud, G., et al.: Eigen v3 (2010). http://eigen.tuxfamily.org
9. Harrow, A.W., et al.: Quantum algorithm for linear systems of equations. Phys. Rev. Lett. **103**, 150502 (2009)
10. Iglberger, K.: Blaze C++ linear algebra library (2012). https://bitbucket.org/blaze-lib
11. Jones, T., et al.: Quest and high performance simulation of quantum computers. Sci. rep. **9**(1), 10736 (2019). https://doi.org/10.1038/s41598-019-47174-9
12. Khammassi, N., et al.: cQASM v1.0: towards a common quantum assembly language (2018)
13. Khammassi, N., et al.: QX: a high-performance quantum computer simulation platform. In: Proceedings of the 2017 Design, Automation & Test in Europe Conference & Exhibition (DATE), pp. 464–469. IEEE, United States (2017)
14. LaRose, R.: Overview and comparison of gate level quantum software platforms. Quantum **3**, 130 (2019)
15. Microsoft: Quantum development kit (2019). https://www.microsoft.com/en-us/quantum/development-kit
16. Möller, M., et al.: LibKet: the quantum expression template library (2019). https://gitlab.com/mmoelle1/LibKet
17. Möller, M., et al.: A conceptual framework for quantum accelerated automated design optimization. Microprocess. Microsyst. **66**, 67–71 (2019)
18. Nickolls, J., et al.: Scalable parallel programming with CUDA. Queue **6**(2), 40–53 (2008)
19. Rigetti Computing: PyQuil: a Python library for quantum programming using Quil (2019). https://github.com/rigetti/pyquil
20. Rupp, K., et al.: ViennaCL – linear algebra library for multi- and many-core architectures. SIAM J. Sci. Comput. **38**(5), S412–S439 (2016)
21. Smith, R.S., et al.: A practical quantum instruction set architecture (2016)
22. Steiger, D.S., et al.: ProjectQ: an open source software framework for quantum computing. Quantum **2**, 49 (2018)
23. Stone, J.E., et al.: OpenCL: a parallel programming standard for heterogeneous computing systems. Comput. Sci. Eng. **12**(3), 66–73 (2010)
24. Yalamanchili, P., et al.: ArrayFire - a high performance software library for parallel computing with an easy-to-use API (2015). https://github.com/arrayfire/arrayfire

Generalized Quantum Deutsch-Jozsa Algorithm

Tomasz Arodz[✉]

Department of Computer Science, Virginia Commonwealth University,
Richmond, VA 23284, USA
tarodz@vcu.edu

Abstract. Quantum computing aims to provide algorithms and hardware that allows for solving computational problems asymptotically faster than on classical computers. Yet, design of new, fast quantum algorithms is not straightforward, and the field faces high barriers of entry for traditional computer scientists. One of the main didactic examples used to introduce speedup resulting from quantum computing is the Deutsch-Jozsa algorithm for discriminating between constant and balanced functions. Here, we show a generalization of the Deutsch-Jozsa algorithm beyond balanced functions that can be used to further illustrate the design choices underpinning quantum algorithms.

Keywords: Quantum speedup · Promise problems · Didactics of quantum computing

1 Introduction

Quantum computing studies algorithms and hardware for performing computation using systems that exploit quantum physics. In the gate model of quantum computing [6], the information storage and processing are done using tools from linear algebra. The information is stored in a quantum register composed of quantum bits. Each quantum bit is represented as a unit-norm vector in a two-dimensional complex Hilbert space. A multi-qubit register is modeled as a tensor product space arising from the individual quantum bits. The information in the quantum register can evolve in time according to invertible, inner product-preserving transformations, modeled mathematically in the gate model of quantum computing as unitary operators. The information in the quantum register can be accessed, to some extent, through quantum measurement, typically modeled as a randomized projection on vectors from the computational basis of the Hilbert space.

The key promise of quantum computing is to achieve speedup compared to classical computers [10], leading to faster algorithms in many domains, including linear algebra [5], database search [4], or machine learning [1,3,9]. The study of quantum algorithms can also lead to more efficient classical methods [7,11].

© Springer Nature Switzerland AG 2020
V. V. Krzhizhanovskaya et al. (Eds.): ICCS 2020, LNCS 12142, pp. 465–472, 2020.
https://doi.org/10.1007/978-3-030-50433-5_36

One of the first, didactic example of a problem for which a quantum computer abstracted using the gate model shows speedup is a promise problem known as the Deutsch-Jozsa problem [2]. Assume that you are given a Boolean function on n-bits: $f : \{0,1\}^n \rightarrow \{0,1\}$ with a black-box access. That is, on a classical computer, you can call it on any bit string and see the result, but you cannot decompile it. On a quantum computer, you have access to an oracle, a unitary transformation U_f that performs the function using some form of input and output encoding.

In the Deutsch-Jozsa problem, we are promised that the function is of one of two types

- *constant*, that is, always returns 0, or always returns 1,
- *balanced*, that is, for half of the 2^n possible inputs it returns 0, for the remaining 2^{n-1} inputs, it returns 1.

The task is to use the ability to execute $f(x)$ for any x to figure out if f is constant, or balanced.

Let $N = 2^n$ be the number of distinct inputs x. On a classical computer, we need at least two calls to f to be able to decide if the function is constant or balanced – in the most optimistic scenario when first call returns 0 and the second returns 1, we know the answer after these two calls to f. But if we see zeros all the time, we need $N/2 + 1$ calls to have the answer – if value number $N/2 + 1$ is also 0, it is a constant function, if the value is 1, it is a balanced function. Thus, pessimistically, we need $O(N) = O(2^n)$ calls to f to solve the Deutsch-Jozsa problem on a classical computer. It is well-known that on a quantum computer, we can solve the Deutsch-Jozsa problem much quicker, using just one call to the unitary oracle U_f.

The Deutsch-Jozsa algorithm has been recently generalized to include discrimination between balanced functions and almost-constant functions [8], with the query complexity increasing with the distance from a constant function. Here, we show that it can be generalized in a different way, to a family of promise problems involving discrimination between a specific constant function f_l, for example an all-zero function, and a family of functions f_k that have fixed level of imbalance, that is, have exactly k outputs equal to one, as long as $k \geq N/2$. We show that this problem can be solved with only one query to the function oracle.

2   Quantum Deutsch-Jozsa Algorithm

To define the Deutsch-Jozsa algorithm, we need to have ability to evaluate Boolean functions using unitary operators, and the ability to use Boolean function evaluation to provide the answer whether the function is constant or balanced.

2.1   Quantum Boolean Function Evaluation

A Boolean function on n bits returning an m bit string is a function $f : \{0,1\}^n \rightarrow \{0,1\}^m$. Consider $n = m$, and a function $f(x) = NOTx$ that negates all bits of x. It is a permutation - a one-to-one mapping - on the set $\{0,1\}^n$. Now consider a function $f_y(x) = y$ XOR $x = y \oplus x$, where $\oplus$ is a modulo-two addition. If a bit in y is 0, the corresponding bit in x is not changed – an identity mapping, a particular form of permutation, on those bits. If a bit in y is 1, the corresponding bit in x is negated - a permutation on those bits. That is, for arbitrary n-bit y, $x \rightarrow y \oplus x$ is a permutation on the set $\{0,1\}^n$. $|y \oplus x\rangle$ vectors for all possible x are mutually orthogonal, same as $|x\rangle$ are. Then, a linear mapping $U_y = \sum_{x \in \{0,1\}^n} |y \oplus x\rangle \langle x|$ will produce $|y \oplus x\rangle$ when we perform $U_y |x\rangle$. Since $y \oplus x$ is a permutation on $\{0,1\}^n$, it is a one-to-one mapping. It also preserves the inner product. The inner product among basis vectors $|x\rangle$, $x \in \{0,1\}^n$ is null, and the inner products among the outputs $U_y |x\rangle$ are also null. This mapping is inner-product-preserving, and thus compatible with the rules of quantum mechanics.

Not all functions $f : \{0,1\}^n \rightarrow \{0,1\}^n$ are permutations; a constant function that always returns all bits set to 0 is not a permutation. Consider an arbitrary function $f : \{0,1\}^n \rightarrow \{0,1\}^m$. Let us have an $n + m$-qubit system $\mathcal{H} = (\mathbb{C}^2)^n \otimes (\mathbb{C}^2)^m$. Consider a state $|x\rangle |z\rangle$, where $|x\rangle$ is one of the 2^n basis states of $(\mathbb{C}^2)^n$, and $|z\rangle$ is one of the 2^m basis states of $(\mathbb{C}^2)^m$. We wish to have a linear transformation U_f, a unitary operator, that takes the vector, and produces, on output, a state $|x\rangle |z \oplus f(x)\rangle$, in particular, for $z = 0^m$, it produces $|x\rangle |f(x)\rangle$. For each x, $z \rightarrow z \oplus f(x)$ is a permutation on the basis states of $(\mathbb{C}^2)^m$, and thus $|x\rangle |z \oplus f(x)\rangle$ is a permutation on the basis states of $\mathcal{H}$. Hence,

$$U_f = \sum_{x,z} (|x\rangle |z \oplus f(x)\rangle)(\langle x| \langle z|),$$

which performs $|x\rangle |0\rangle \rightarrow |x\rangle |f(x)\rangle$, is a one-to-one mapping that preserves the inner product – and thus can describe unitary evolution of a quantum system.

Consider $m = 1$, arbitrary n, and an $n + 1$-qubit system in an arbitrary state of the form $|\psi\rangle |0\rangle$. Let $E = HX$ be a Pauli X gate followed by Hadamard gate; it transforms $|0\rangle$ to $|-\rangle$ and $|1\rangle$ to $|+\rangle$, and is Hermitian. Let $E_n = I^{\otimes n} \otimes E$ denote application of E to the last, $n + 1$ qubit. We have

$$E_n |\psi\rangle |0\rangle = |\psi\rangle |-\rangle, \qquad E_n |\psi\rangle |-\rangle = |\psi\rangle |0\rangle,$$

$$U_f |x\rangle |-\rangle = |x\rangle \frac{|f(x)\rangle - |1 \otimes f(x)\rangle}{\sqrt{2}} = |x\rangle \frac{(-1)^{f(x)}(|0\rangle - |1\rangle)}{\sqrt{2}} = (-1)^{f(x)} |x\rangle |-\rangle$$

We obtained the result by seeing that $|f(x)\rangle - |1 \otimes f(x)\rangle$ differs by a sign depending on the value of $f(x)$; if $f(x) = 0$, it is $|0\rangle - |1\rangle$, if $f(x) = 1$ it is $|1\rangle - |0\rangle$. Then we used phase kickback to the top qubits. Value of $f(x) = 1$ is reflected in change of phase of $|x\rangle$ by π, whereas $f(x) = 0$ results in no change.

We now see two ways of representing the result of applying $f(x)$ as encoded by $U_f = \sum_{x,z} (|x\rangle |z \oplus f(x)\rangle)(\langle x| \langle z|)$:

$$U_f |x\rangle |0\rangle = |x\rangle \otimes |f(x)\rangle, \qquad E_n U_f E_n |x\rangle |0\rangle = \left((-1)^{f(x)} |x\rangle\right) \otimes |0\rangle.$$

The first option is to encode the result in the state of the last qubit, the second is to encode it in the phase of the input qubits.

2.2   The Deutsch-Jozsa Problem

Consider four query states $\frac{|0u\rangle+|1v\rangle}{\sqrt{2}}$ for arbitrary binary u, v. Two are simple, $|FF\rangle = |+\rangle |0\rangle = \frac{|00\rangle+|10\rangle}{\sqrt{2}}$ and $|TT\rangle = |+\rangle |1\rangle = \frac{|01\rangle+|11\rangle}{\sqrt{2}}$, and correspond to two possible one-bit constant functions. Two other are entangled, $|FT\rangle = \frac{|00\rangle+|11\rangle}{\sqrt{2}}$ and $|TF\rangle = \frac{|01\rangle+|10\rangle}{\sqrt{2}}$, and correspond to the two one-bit balanced functions.

If someone prepares a two-qubit system in one of the four query states, can we distinguish whether it is any of these two $|FF\rangle$ and $|TT\rangle$, or any of these two $|TF\rangle$ and $|FT\rangle$? We have $\langle TT | FF\rangle = \langle TF | FT\rangle = 0$, that is, it is easy to distinguish $|TT\rangle$ from $|FF\rangle$, and $|TF\rangle$ from $|FT\rangle$. All other inner products of these four states are equal to $1/2$. Any unitary transformation has to preserve the dimensionality of the space, and the inner products. Thus, there is no mapping that would map $|TT\rangle$ to be orthogonal to $|TF\rangle$ or $|FT\rangle$; same for $|FF\rangle$. If we want to use orthogonality of any member from one group of states to any member from the other group of states to reliably distinguish states from one group from the other, then a group composed of $|TT\rangle$ and $|FF\rangle$ cannot be reliably distinguished from group consisting of $|TF\rangle$ and $|FT\rangle$.

We have seen above two ways of representing the result of applying a binary function $f(x)$ given a state $|x\rangle$: encoding the result as an additional qubit, or as a phase change of the input qubit. Consider a n-bit binary function f. We can construct a state $|+\rangle^{\otimes n} = \sum_{x=0}^{2^n-1} |x\rangle$ using Hadamard transform. Then, we can construct a state $|+^{\otimes n}, 0\rangle = |+\rangle^{\otimes n} |0\rangle$, and apply our two options of evaluating f quantum mechanically to this $n+1$-qubit state. We will get

$$U_f |+^{\otimes n}, 0\rangle = \frac{1}{2^{\frac{n}{2}}} \sum_{x=0}^{2^n-1} (|x\rangle \otimes |f(x)\rangle),$$

$$E_n U_f E_n |+^{\otimes n}, 0\rangle = \left(\frac{1}{2^{\frac{n}{2}}} \sum_{x=0}^{2^n-1} (-1)^{f(x)} |x\rangle\right) \otimes |0\rangle.$$

These two options are not equivalent. The four query states we have seen above can be seen as the first option for $n = 1$ if we equate $u = f(0)$ and $v = f(1)$; indeed $\sum_{x=0}^{1} |x\rangle |f(x)\rangle = \frac{|0u\rangle+|1v\rangle}{\sqrt{2}}$. We have seen that we cannot use this representation to decide, based on orthogonality, if we got a function that is constant, or that returns equal number of 0's and 1's.

On the other hand, consider the second representation, or actually just its first qubit $\frac{1}{\sqrt{2}} \sum_{x=0}^{1} (-1)^{f(x)} |x\rangle = \frac{(-1)^u |0\rangle + (-1)^v |1\rangle}{\sqrt{2}}$. A constant function will have $\pm |+\rangle$, while a balanced function will have $\pm |-\rangle$. We can ignore the global phase, that is, the sign, and we end up with two orthogonal states, $|+\rangle$ for constant and $|-\rangle$ for balanced function. As we can see, while unitary actions after applying the black-box unitary oracle cannot change the distinguishability of states because they must preserve the inner products, differences prior to applying the function oracle can affect our ability to distinguish states.

3   Generalized Deutsch-Jozsa Problem

We show here that the ability to distinguish all-zeros from all-ones that we get from $U_f = U_f(I^{\otimes n} \otimes I)$ and the ability to distinguish all-zeros and all-ones from a balanced function that we get from $U_f E_n = U_f(I^{\otimes n} \otimes E)$ are not the only possibilities for quickly solving promise problems involving two groups of binary functions. We can form a class of single-qubit unitary operators A defined by the condition $|a|^2 + |b|^2 = 1$,

$$
I = \begin{bmatrix} 1 & 0 \\ 0 & 1 \end{bmatrix}, \quad A = \begin{bmatrix} a & -b^* \\ b & a^* \end{bmatrix}, \quad E = \begin{bmatrix} \frac{1}{\sqrt{2}} & \frac{1}{\sqrt{2}} \\ -\frac{1}{\sqrt{2}} & \frac{1}{\sqrt{2}} \end{bmatrix},
$$

and observe that unitaries I and E are at the extreme ends of the family, with a spectrum of other operators in between. These operators can each solve a different promise problem of discriminating between two classes of functions.

Consider two functions f_l and f_k with the corresponding black-box unitaries U_l and U_k, and let $y_j = f_k(j)$ and $x_j = f_l(j)$. Also, let $\xi_j = 1$ if $y_j = x_j$ and $\xi_j = 0$ otherwise, and $\delta_j = 1 - \xi_j$; we will use Ξ to denote the number of outputs on which the functions agree, and Δ to denote the Hamming distance between the outputs, that is, the number of outputs that differ; we have $\Xi + \Delta = 2^n$.

Let $A_n = I^{\otimes n} \otimes A$ denote application of A to the last, $n + 1$-st qubit of an $n + 1$-qubit system. Then, we have

$$
U_k A_n |+^{\otimes n}, 0\rangle = U_k \left(\frac{1}{2^{\frac{n}{2}}} \sum_{j=0}^{2^n - 1} |j\rangle \otimes A |0\rangle \right) = \frac{1}{2^{\frac{n}{2}}} \sum_{j=0}^{2^n - 1} \left(a |j\rangle |y_j\rangle + b |j\rangle |\bar{y}_j\rangle \right)
$$

$$
= \frac{1}{2^{\frac{n}{2}}} \sum_{j=0}^{2^n - 1} \left(a\bar{y}_j |j\rangle |0\rangle + a y_j |j\rangle |1\rangle + b y_j |j\rangle |0\rangle + b\bar{y}_j |j\rangle |1\rangle \right)
$$

$$
= \frac{1}{2^{\frac{n}{2}}} \sum_{j=0}^{2^n - 1} \left((a\bar{y}_j + b y_j) |j\rangle |0\rangle + (a y_j + b\bar{y}_j) |j\rangle |1\rangle \right)
$$

and a similar expression for U_l, with y_j replaced by x_j.

The inner product of the two states is

$$
h_\Delta = \left\langle +^{\otimes n}, 0 \left| A_n{}^\dagger U_l{}^\dagger U_k A_n \right| +^{\otimes n}, 0 \right\rangle = \frac{\Xi - \Delta a}{2^n},
$$

where $\alpha = -2\Re(ab^*)$. Indeed, we have

$$h_\Delta = \frac{1}{2^n} \sum_{j=0}^{2^n-1} \{(a^*\bar{x}_j + b^*x_j)\,\langle j|\,\langle 0| + (a^*x_j + b^*\bar{x}_j)\,\langle j|\,\langle 1|\}$$

$$\{(a\bar{y}_j + by_j)\,|j\rangle\,|0\rangle + (ay_j + b\bar{y}_j)\,|j\rangle\,|1\rangle\}$$

$$= \frac{1}{2^n} \sum_{j=0}^{2^n-1} \Big((a\bar{y}_j + by_j)(a^*\bar{x}_j + b^*x_j)\,\langle j\,|\,j\rangle\,\langle 0\,|\,0\rangle$$

$$+ (a^*x_j + b^*\bar{x}_j)(ay_j + b\bar{y}_j)\,\langle j\,|\,j\rangle\,\langle 1\,|\,1\rangle \Big)$$

$$= \frac{1}{2^n} \sum_{j=0}^{2^n-1} \Big(a\bar{y}_j a^*\bar{x}_j + by_j a^*\bar{x}_j + a\bar{y}_j b^*x_j + by_j b^*x_j$$

$$+ a^*x_j ay_j + b^*\bar{x}_j ay_j + a^*x_j b\bar{y}_j + b^*\bar{x}_j b\bar{y}_j \Big)$$

$$= \frac{1}{2^n} \sum_{j=0}^{2^n-1} \Big((|a|^2 + |b|^2)(\bar{y}_j\bar{x}_j + x_j y_j) + (ab^* + ba^*)(x_j\bar{y}_j + y_j\bar{x}_j) \Big)$$

$$= \frac{1}{2^n} \sum_{j=0}^{2^n-1} \Big(\xi_j + 2\Re(ab^*)\delta_j \Big) = \frac{\Xi + 2\Re(ab^*)\Delta}{2^n} = \frac{\Xi - \Delta\alpha}{2^n}.$$

The last equality comes from the fact that $(\bar{y}_j\bar{x}_j + x_j y_j) = 1$ if and only if $x_j = y_j$, and $(x_j\bar{y}_j + y_j\bar{x}_j) = 1$ if and only if $x_j = \bar{y}_j$.

To achieve perfect distinguishability of the resulting states through quantum measurement, which corresponds to having inner product $h_\Delta = 0$, we need to set a, b such that $\alpha = -2\Re(ab^*)$ becomes $\alpha = \Xi/\Delta = 2^n/\Delta - 1$. Since $\alpha \in [-1, 1]$ for any two unit-norm quantum states, we can do that as long as $\Delta \geq \frac{1}{2}2^n$; the promised functions f_k and f_l must differ on at least half of their outputs to be perfectly distinguishable through orthogonality of the results of $U_k A_n\,|+^{\otimes n}, 0\rangle$ and $U_l A_n\,|+^{\otimes n}, 0\rangle$.

For any $k \in [2^{n-1}, 2^n]$, we can perfectly distinguish functions f_k with exactly k outputs of 1 from the all-zero constant function f_l using just one oracle access by using unitary defined by $\Re(ab^*) = -\frac{2^n/k-1}{2}$; note that here $\Delta = k$. On a classical computer, the same task can be achieved quickly if $k \sim 2^n$, but can take $O(N) = O(2^n)$ function calls if k is close to 2^{n-1}.

As an example, consider a problem involving functions defined over $n = 4$ bits. Let f_l be an all-zero function, that is, for any of the 16 possible input bitstrings, it returns null. Let f_k be an imbalanced function with exactly $k = 12$ out of 16 inputs returning one, and with four null outputs. To discriminate f_l from any possible f_k, we need to find a, b such that $\Re(ab^*) = -(16/12 - 1)/2 = -1/6$ and $|a|^2 + |b|^2 = 1$: we can have $a = 1/\sqrt{6} - 1/\sqrt{3}$ and $b = 1/\sqrt{3} + 1/\sqrt{6}$.

In contrast to applying A prior to U_f to help solve the problem of discriminating an all-zero function f_l and a function f_k with k ones and $N - k$ zeros on outputs, we can consider the result of applying I_n or E_n instead.

For I_n, we arrive at the state $N^{-1/2} \sum_{|\{j\}|=N} |j\rangle |0\rangle$ for the all-zero f_l, while for f_k we arrive at $N^{-1/2} \sum_{|\{j\}|=N-k} |j\rangle |0\rangle + N^{-1/2} \sum_{|\{j\}|=k} |j\rangle |1\rangle$, which has inner product $\frac{N-k}{N}$. Only the all-one function f_k is distinguishable from the all-zero function f_l. Using E_n instead leads to $N^{-1/2} \sum_{|\{j\}|=N} |j\rangle$ for f_l and to $N^{-1/2} \sum_{|\{j\}|=N-k} |j\rangle - N^{-1/2} \sum_{|\{j\}|=k} |j\rangle$ for f_k. These have inner product of $\frac{N-2k}{N}$. Here, only setting $k = N/2$, that is, using a balanced function f_k as in the original Deutsch-Jozsa problem, leads to the ability to distinguish f_l from f_k without error using a single measurement.

With minor changes, we can also, for any $k \in [0, 2^{n-1}]$, distinguish a function with k zero outputs from an all-one constant function. The original Deutsch-Jozsa choice of $k = 2^{n-1}$ is the only case when we can distinguish f_k from both all-zero and all-one functions f_l.

4   Conclusion

The Deutsch-Jozsa problem is often used to illustrate basic concepts in the gate model of quantum computing. Here, we show that the balance-vs-constant function promise problem lies at the edge of a family of promise problems involving discriminating between two subclasses of Boolean functions. The analysis of this broader family of problems can provide improved understanding of the conditions that lead to quantum speedup in Deutsch-Jozsa algorithm.

Acknowledgements. TA is supported by NSF grant CCF-1617710.

References

1. Arodz, T., Saeedi, S.: Quantum sparse support vector machines. arXiv preprint arXiv:1902.01879 (2019)
2. Deutsch, D., Jozsa, R.: Rapid solution of problems by quantum computation. Proc. R. Soc. Lond. Ser. A Math. Phys. Sci. **439**(1907), 553–558 (1992)
3. Dunjko, V., Briegel, H.J.: Machine learning & artificial intelligence in the quantum domain: a review of recent progress. Rep. Prog. Phys. **81**(7), 074001 (2018)
4. Grover, L.K.: A fast quantum mechanical algorithm for database search. arXiv preprint arXiv:quant-ph/9605043 (1996)
5. Harrow, A.W., Hassidim, A., Lloyd, S.: Quantum algorithm for linear systems of equations. Phys. Rev. Lett. **103**(15), 150502 (2009)
6. Nielsen, M.A., Chuang, I.L.: Quantum Computation and Quantum Information. Cambridge University Press, Cambridge (2011)
7. Panahi, A., Saeedi, S., Arodz, T.: word2ket: space-efficient word embeddings inspired by quantum entanglement. In: International Conference on Learning Representations 2020. Preprint. arXiv:1911.04975 (2019)
8. Qiu, D., Zheng, S.: Generalized Deutsch-Jozsa problem and the optimal quantum algorithm. Phys. Rev. A **97**(6), 062331 (2018)
9. Rebentrost, P., Mohseni, M., Lloyd, S.: Quantum support vector machine for big data classification. Phys. Rev. Lett. **113**(13), 130503 (2014)

10. Rønnow, T.F., et al.: Defining and detecting quantum speedup. Science **345**(6195), 420–424 (2014)
11. Tang, E.: A quantum-inspired classical algorithm for recommendation systems. In: Proceedings of the 51st Annual ACM SIGACT Symposium on Theory of Computing, pp. 217–228. ACM (2019)

Revisiting Old Combinatorial Beasts in the Quantum Age: Quantum Annealing Versus Maximal Matching

Daniel Vert, Renaud Sirdey[✉], and Stéphane Louise

CEA, LIST, Palaiseau, France
{daniel.vert2,renaud.sirdey,stephane.louise}@cea.fr

Abstract. This paper experimentally investigates the behavior of analog quantum computers such as commercialized by D-Wave when confronted to instances of the maximum cardinality matching problem specifically designed to be hard to solve by means of simulated annealing. We benchmark a D-Wave "Washington" (2X) with 1098 operational qubits on various sizes of such instances and observe that for all but the most trivially small of these it fails to obtain an optimal solution. Thus, our results suggest that quantum annealing, at least as implemented in a D-Wave device, falls in the same pitfalls as simulated annealing and therefore provides additional evidences suggesting that there exist polynomial-time problems that such a machine cannot solve efficiently to optimality.

Keywords: Quantum computing · Quantum annealing · Bipartite matching

1 Introduction

From a practical view, the emergence of quantum computers able to compete with the performance of the most powerful conventional computers remains highly speculative in the foreseeable future. Indeed, although quantum computing devices are scaling up to the point of achieving the so-called milestone of quantum supremacy [22], these intermediate scale devices, referred to as NISQ [21], will not be able to run mainstream quantum algorithms such as Grover, Shor and their many variants at practically significant scales. Yet there are other breeds of machines in the quantum computing landscape, in particular the so-called analog quantum computers of which the machines presently sold by the Canadian company D-Wave are the first concrete realizations. These machines implement a noisy version of the Quantum Adiabatic Algorithm introduced by Farhi et al. in 2001 [12]. From an abstract point of view, such a machine may be seen as an oracle specialized in the resolution of an NP-hard optimization

© Springer Nature Switzerland AG 2020
V. V. Krzhizhanovskaya et al. (Eds.): ICCS 2020, LNCS 12142, pp. 473–487, 2020.
https://doi.org/10.1007/978-3-030-50433-5_37

problem[1] (of the spin-glass type) with an algorithm functionally analogous to the well-known simulated annealing but with a quantum speedup.

On top of the formal analogies between simulated and quantum annealing, there also appears to be an analogy between the latter present state of art and that of simulated annealing when it was first introduced. So it might be useful to recall a few facts on SA. Indeed, simulated annealing was introduced in the mid-80's [8,18] and its countless practical successes quickly established it as a mainstream method for approximately solving computationally-hard combinatorial optimization problems. Thus, the theoretical computer science community investigated in great depth its convergence properties in an attempt to understand the worst-case behavior of the method. With that respect, these pieces of work, which were performed in the late 80's and early 90's, lead to the following insights. First, when it comes to solving combinatorial optimization problems to optimality, it is necessary (and sufficient) to use a logarithmic cooling schedule [14,15,20] leading to an exponential-time convergence in the worst-case (an unsurprising fact since it is known that $P \neq NP$ in the oracle setting [3]). Second, particular instances of combinatorial problems have been designed to specifically require an exponential number of iterations to reach an optimal solution for example on the (NP-hard) 3-coloring problem [20] and, more importantly for this paper, on the (polynomial) maximum cardinality matching problem [25]. Lastly, another line of works, still active today, investigated the asymptotic behavior of hard combinatorial problems [7,17,26] showing that the cost ratio between best and worst-cost solutions to random instances tends (quite quickly) to 1 as the instance size tend to ∞. These latter results provided clues as to why simple heuristics such as simulated annealing appear to work quite well on large instances as well as to why branch-and-bound type exact resolution methods tend to suffer from a trailing effect (i.e. find optimal or near-optimal solutions relatively quickly but fail to prove their optimality in reasonable time). Despite these results now being quite well established, they can also contribute to the ongoing effort to better understand and benchmark quantum adiabatic algorithms [12] and especially the machines that now implements it in order to determine whether or not they provide a quantum advantage over some classes of classical computations. Still, as it is considered unlikely that any presently known quantum computing paradigm will lead to efficient algorithms for solving NP-hard problems, determining whether or not quantum adiabatic computing yields an advantage over classical computing is most likely an ill-posed question given present knowledge. Yet, as a quantum analogue of simulated annealing, attempting to demonstrate a quantum advantage of adiabatic algorithms over simulated annealing appears to be a better-posed question. At the time of writing, this problem is the focus of a lot of works which, despite claims of exponential speedups in specific cases [11] (which also lead to the development of the promising Simulated Quantum Annealing classical metaheuristic [9]), hint towards a

---

[1] Strictly speaking, to the best of the authors' knowledge, although the general problem is NP-hard, the complexity status of the more specialized instances constrained by the qubit interconnection topology of these machines remains open.

logarithmic decay requirement of the temperature-analog of QA but with smaller constants involved [24] leading to only an $O(1)$ advantage of QA over SA in the general case. Such an advantage has furthermore recently been experimentally demonstrated by Albash and Lidar [1]. The present paper contributes to the study of the QA vs SA issue by experimentally confronting a D-Wave quantum annealer to the pathological instances of the maximum cardinality matching problem proposed by Sasaki and Hajek [25] in order to show that simulated annealing was indeed unable to solve certain polynomial problems in polynomial time. Demonstrating an ability to solve these instances to optimality on a quantum annealer would certainly hint towards a worst-case quantum annealing advantage over simulated annealing whereas failure to do so would tend to demonstrate that quantum annealing remains subject to the same pitfalls as simulated annealing and is therefore unable to solve certain polynomial problems efficiently. As a first step towards this, the present paper experimentally benchmarks a D-Wave "Washington" (2X) with 1098 operational qubits on various sizes of such pathologic instances of the maximum cardinality matching problem and observes that for all but the most trivially small of these it fails to obtain an optimal solution. This thus provides negative evidences towards the existence of a *worst-case* advantage of quantum annealing over classical annealing. As a by-product, our study also provides feedback on using a D-Wave annealer in particular with respect to the size of problems that can be mapped on such a device due to the various constraints of the system. This paper is organized as follows. Section 2 provides some background on quantum annealing, the D-Wave devices and their limitations. Section 3 surveys the maximum cardinality matching problem, introduces the G_n graph family underlying our pathologic instances and subsequently details how we build the QUBO instances to be mapped on the D-Wave from those instances. Then, Sect. 4 extensively details our experimental setup and experimentations and Sect. 5 concludes the paper with a discussion of the results and a number of perspectives to follow up on this work.

2   Quantum Annealing and Its D-Wave Implementation

2.1   The Generalized Ising Problem and QUBO

D-Wave systems are based on a quantum annealing process[2] which goal is to minimize the Ising Hamiltonian:

$$\mathcal{H}(\mathbf{h}, \mathbf{J}, \boldsymbol{\sigma}) = \sum_i h_i \sigma_i + \sum_{i<j} J_{ij} \sigma_i \sigma_j, \tag{1}$$

where the external field $\mathbf{h}$ and spin coupling interactions matrix $\mathbf{J}$ are given, and the vector of spin (or qubit) values $\boldsymbol{\sigma}/\forall i, \sigma_i \in \{-1, 1\}$ is the variable for which the energy of the system is minimized as the process of adiabatic annealing

---

[2] A combinatorial optimization technique functionally similar to conventional (simulated) annealing but which, instead of applying thermal fluctuations, uses quantum phenomena to search the solution space more efficiently [13].

transition the system from a constant coupling with a superposition of spins[3] to the final Hamiltonian as given by Eq. 1. Historically speaking, the Ising Hamiltonian corresponds to the case where only the closest neighbouring spins are allowed to interact (*i.e.* $J_{ij} \neq 0 \iff$ nodes i and j are conterminous). The generalized Ising problem, for which any pair of spins in the system are allowed to interact, is easily transformed into a well known 0–1 optimization problem called QUBO (for Quadratic Unconstrained Binary Optimization) which objective function is given by:

$$O(\mathbf{Q}, \mathbf{x}) = \sum_i Q_{ii} x_i + \sum_{i<j} Q_{ij} x_i x_j, \tag{2}$$

in which the matrix $\mathbf{Q}$ is constant and the goal of the optimization is to find the vector of binary variables $\forall i, x_i \in \{0, 1\}$ that either minimizes or maximizes the objective function $O(\mathbf{Q}, \mathbf{x})$ from Eq. 2. For the minimization problem (but only up to a change of sign for the maximization problem), it is trivial that the generalized Ising problem and the QUBO problem are equivalent given $\forall i, Q_{ii} = h_i, \forall i, j/i \neq j, Q_{ij} = J_{ij}$ and $\forall i, \sigma_i = 2x_i - 1$.

Hence, if quantum annealing can reach a configuration of minimum energy, then the associated state vector solves the equivalent QUBO problem at the same time. As the behavior of each qubit in a quantum annealer allows them to be in a superposition state (a combination of the states "−1" and "+1") until they relax to either one of these eigen-states, it is thought that quantum mechanical phenomena – e.g., quantum tunneling – can help reaching the minimum energy configuration, or at least a close approximation of it, in more cases than with Simulated Annealing (SA). Indeed, when SA only relies on (simulated) temperatures to pass over barriers of potential, in Quantum Annealing, quantum phenomena can help because tunneling is more efficient to pass energy barriers even in the case where the temperature is low. Therefore, this technique is a promising heuristic approach to "quickly" find acceptable solutions for certain classes of complex NP-Hard problems that are easily mapped to these machines, such as optimization, machine learning, or operational research problems.

The physics of the low energies of D-Wave computers [16] is given by a Hamiltonian depending on the time of the form

$$\mathcal{H}(t) = A(t)\mathcal{H}_0 + B(t)\mathcal{H}_P \tag{3}$$

The functions $A(t)$ and $B(t)$ must satisfy $B(t = 0) = 0$ and $A(t = \tau) = 0$ so that, when the state evolution $t = 0$ changes to $t = \tau$, the Hamiltonian $H(t)$ is "annealed" in a purely classical form. Thus, the fundamental state $\mathcal{H}(0) = \mathcal{H}_0$

---

[3] The initial Hamiltonian is proportional to $\sum_{i,j} \sigma_i^x \sigma_j^x$, hence based on Eigen-vectors of operator $\widehat{\sigma^x}$ (on the x-axis) whilst the momentum of spin on Eq. 1 is an Eigenstate of $\widehat{\sigma^z}$ (on the z-axis) for which Eigen-states of $\widehat{\sigma^x}$ are superposition states. The adiabatic theorem allows transitioning from the initial ferromagnetic state on axis x to an eigen-state of the Hamiltonian of Eq. 1 on axis z and hopefully to the lowest energy of it.

evolves to a state $\mathcal{H}(\tau) = \mathcal{H}_P$, the measurements made at time τ give us low energy states of the Ising Hamiltonian (Eq. 1). The adiabatic theorem states that if the time evolution is slow enough (*i.e.* τ is large enough), then the optimal (global) solution $\epsilon(\boldsymbol{\sigma})$ of the system can be obtained with a high probability. $\mathcal{H}_0 = \sum_i \sigma_i^x$ gives the quantum effects, and $\mathcal{H}_P = \sum_i h_i \sigma_i^z + \sum_{(ij)} J_{i,j} \sigma_i^z \sigma_j^z$ is given to encode the problem of the Ising instance.

$$\min \epsilon(\boldsymbol{\sigma}) = \min \left\{ \sum_i h_i \sigma_i + \sum_{i,j} J_{i,j} \sigma_i \sigma_j \right\} \tag{4}$$

2.2 D-Wave Limitations

Nonetheless, it is worth noting that in the case of the current architectures of the D-Wave annealing devices, the freedom to choose the J_{ij} coupling constants is severely restrained by the hardware qubit interconnection topology. In particular, this so-called *Chimera* topology is sparse, with a number of inter-spin couplings limited to a maximum of 6 per qubit (or spin variable). Figure 1 illustrates an instance of the Chimera graph for 128 qubits, $T = (N_T, E_T)$, where nodes N_T are qubits and represent problem variables with programmable weights (h_i), and edges E_T are associated to the couplings J_{ij} between qubits $(J_{ij} \neq 0 \implies (i,j) \in E_T)$. As such, if the graph induced by the nonzero couplings is not isomorphic to the Chimera graph, which is the case most usually, then one must resort to several palliatives among which the duplication of logical qubits onto several physical qubits is the least disruptive one if the corresponding expanded problem can still fit on the target device.

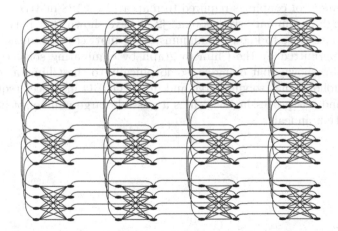

Fig. 1. Representation of a Chimera graph with 4×4 unit cells, each a small 2×4 bipartite graph, for 128 physicals qubits. The links represents all the inter-spin coupling J_{ij} that can be different from 0.

Then, a D-Wave annealer minimizes the energy from the Hamiltonian of Eq. (1) by associating weights (h_i) with qubit spins (σ_i) and couplings (J_{ij}) with couplers between the spins of the two connected qubits (σ_i and σ_j). As an example, the D-Wave 2X system we used has 1098 operational qubits and 3049 operational couplers. As said previously, a number of constraints have an impact on the practical efficiency of this type of machines. In [5], the authors highlight four factors: the precision/control error which is limited by the parameters **h** and **J** which value ranges are also limited[4], the low connectivity[5] in T, and the in fine small number of useful qubits once the topological constraints are accounted for. In [4], the authors show that using large energy gaps in the Ising representation of the model one wants to optimize can greatly mitigate some of the intrinsic limitations of the hardware like precision of the coupling values and noises in the spin measurements. They also suggest using ferromagnetic Ising coupling between qubits (i.e., making qubit duplication) to mitigate the issues with the limited connectivity of the Chimera graph. All these suggestions can be considered good practices (which we did our best to follow) when trying to use the D-Wave machine to solve real Ising or QUBO problems with higher probabilities of outputting the best solution despite hardware and architecture limitations.

Thus, preprocessing algorithms are required to adapt the graph of a problem to the hardware. Pure quantum approaches are limited by the number of variables (duplication included) that can be mapped on the hardware. Larger graphs require the development of hybrid approaches (both classical and quantum) or the reformulation of the problem to adapt to the architecture. For example, for a 128 × 128 matrix size, the number of possible coefficients J_{ij} is 8128 in the worst-case, while the Chimera graph which associates 128 qubits (4 × 4 unit cells) has only 318 couplers. The topology therefore accounts only for ∼ 4% of the total number of couplings required to map a 128 × 128 matrix in the worst case. Although preliminary studies (e.g., [27]) have shown that it is possible to obtain solutions close to known minimums for **Q** matrices with densities higher than those permitted by the Chimera graph by eliminating some coefficients, they have also shown that doing so isomorphically to the Chimera topology is difficult. It follows that solving large and dense QUBO instances requires non-trivial pre and postprocessing as well as a possibly large number of invocations of the quantum annealer.

---

[4] The range of $h_i \in [-2, +2]$ and $J_{i,j} \in [-1, +1]$ is a limitation for all values of the variables to be included in the graph. If the values of h_i and $J_{i,j}$ are outside their respective ranges, then they are unavailable and not mapped.

[5] If the problems to be solved do not match the structure of the T graph architecture, then they cannot be mapped and resolved directly.

3 Solving Maximum Cardinalty Matching on a Quantum Annealer

3.1 Maximum Cardinality Matching and the G_n Graph Family

Given an (undirectered) graph $G = (V, E)$, the maximum matching problem asks for $M \subseteq E$ such that $\forall e, e' \in M^2$, $e \neq e'$ we have that $e \cap e' = \emptyset$ and such that $|M|$ is maximum. The maximum matching problem is a well-known polynomial problem dealt with in almost every textbook on combinatorial optimization (e.g., [19]), yet the algorithm for solving it in general graphs, Edmond's algorithm, is a nontrivial masterpiece of algorithmics. Additionally, when G is bipartite i.e. when there exists two collectively exhaustive and mutually exclusive subsets of E, A and B, such that no edge has both its vertices in A or in B, the problem becomes a special case of the maximum flow problem and can be dealt with several simpler algorithms [19].

It is therefore very interesting that such a seemingly powerful method as simulated annealing can be deceived by special instances of this latter easier problem. Indeed, in a landmark 1988 paper [25], Sasaki and Hajek, have considered the following family of special instances of the bipartite matching problem. Let G_n denote the (undirected) graph with vertices $\bigcup_{i=0}^{n} A^{(i)} \cup \bigcup_{i=0}^{n} B^{(i)}$ where each of the $A^{(i)}$'s and $B^{(j)}$'s have cardinality $n + 1$ (vertex numbering goes from 0 to n), where vertex $A_j^{(i)}$ is connected to vertex $B_j^{(i)}$ and where vertex $B_j^{(i)}$ is connected to all vertices in $A^{(i+1)}$ (for $i \in \{0, \ldots, n\}$ and $j \in \{0, \ldots, n\}$). These graphs are clearly bipartite has neither two vertices in $\bigcup_{i=0}^{n} A^{(i)}$ nor two vertices in $\bigcup_{i=0}^{n} B^{(i)}$ are connected. These graphs therefore exhibit a sequential structure which alternates between sparsely and densely connected subsets of vertices, as illustrated on Fig. 2 for G_3.

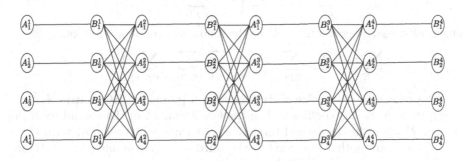

Fig. 2. G_3.

As a special case of the bipartite matching problem, the maximum cardinality matching over G_n can be solved by any algorithm solving the former. Yet, it is even easier as one can easily convince oneself that a maximum matching in G_n is obtained by simply selecting all the edges connecting vertices in $A^{(i)}$ to vertices in $B^{(i)}$ (for $i \in \{0, \ldots, n\}$), i.e. all the edges in the sparsely connected subsets of vertices, and that is the only way to do so. This therefore leads to a maximum matching of cardinality $(n+1)^2$.

We hence have a straightforward special case of a polynomial problem, yet the seminal result of Sasaki and Hajek states that the mathematical expectation of the number of iterations required by a large class of (classical) annealing-type algorithms to reach a maximum matching on G_n is in $O(\exp(n))$. The G_n family therefore provides an interesting playground to study how quantum annealing behaves on problems that are hard for simulated annealing. This is what we do, experimentally, in the sequel.

3.2   QUBO Instances

In order for our results to be fully reproducible we hereafter describe how we converted instances of the maximum matching problem into instances of the Quadratric Unconstrained Boolean Optimization (QUBO) problem which D-Wave machines require as input. Let $G = (V, E)$ denote the (undirected) graph for which a maximum matching is desired. We denote $x_e \in \{0, 1\}$, for $e \in E$, the variable which indicates whether e is in the matching. Hence we have to maximize $\sum_{e \in E} x_e$ subject to the contraints that each vertex v is covered at most once, i.e. $\forall v \in V$,

$$\sum_{e \in \Gamma(v)} x_e \leq 1, \tag{5}$$

where $\Gamma(v)$, in standard graph theory notations, denotes the set of edges which have v as an endpoint. In order to turn this into a QUBO problem we have to move the above constraints into the economic function, for example in maximizing,

$$\sum_{e \in E} x_e - \lambda \sum_{v \in V} \left(1 - \sum_{e \in \Gamma(v)} x_e \right)^2,$$

which, after rearrangements, leads to the following economic function,

$$\sum_{e \in E} x_e + \sum_{v \in V} \sum_{e \in \Gamma(v)} 2\lambda x_e - \sum_{v \in V} \sum_{e \in \Gamma(v)} \sum_{e' \in \Gamma(v)} \lambda x_e x_{e'}$$

Yet we have to reorganize a little to build a proper QUBO matrix. Let $e = (v, w)$, variable x_e has coefficient 1 in the first term, 2λ in the second term (for v) then 2λ again in the second term (for w) then $-\lambda$ in the third term (for v and $e' = e$) and another $-\lambda$ again in the third term (for w and $e' = e$). Hence, the diagonal terms of the QUBO matrix are,

$$Q_{ee} = 1 + 4\lambda - 2\lambda = 1 + 2\lambda.$$

Then, if two distinct edges e and e' share a common vertex, the product of variables $x_e x_{e'}$ has coefficient $-\lambda$, in the third term, when v corresponds to the vertex shared by the two edges, and this is so twice. So, for $e \neq e'$,

$$Q_{ee'} = \begin{cases} -2\lambda \text{ if } e \cap e' \neq \emptyset, \\ 0 \text{ otherwise.} \end{cases}$$

Taking $\lambda = |E|^6$, for example for G_1, we thus obtain an 8 variables QUBO (the corresponding matrix is given in [28]) for which a maximum matching has cost 68, the second best solutions has cost 53 and the worst one (which consist in selecting all edges) has cost -56.

4   Experimental Results

4.1   Concrete Implementation on a D-Wave

In this section, we detail the steps that we have followed to concretely map and solve the QUBO instances associated to G_n, $n \in \{1, 2, 3, 4\}$, on a DW2X operated by the University of South California. Unfortunately (yet unsurprisingly), the QUBO matrices defined in the previous section are not directly mappable on the Chimera interconnection topology and, thus, we need to resort to qubit duplication i.e., use *several* physical qubits to represent *one* problem variable (or "logical qubit"). Fortunately, the D-Wave software pipeline automates this duplication process. Yet, this need for duplication (or equivalently the sparsity of the Chimera interconnection topology) severely limits the size of the instances we were able to map on the device and we had to stop at G_4 which 125 variables required using 951 of the 1098 available qubits. Table 1 provides the number of qubits required for each of our four instances. For G_1, G_2 the maximum duplication is 6 qubits and for G_3, G_4 it is 18 qubits.

Table 1. Number of qubits required to handle the QUBO instances associated to G_1, G_2, G_3 and G_4. See text.

	#var.	#qubits	Average dup.	max. dup.
G_1	8	16	2.0	6
G_2	27	100	3.7	6
G_3	64	431	6.7	18
G_4	125	951	7.6	18

Eventually, qubit duplication leads to an expanded QUBO with more variables and an economic function which includes an additional set of penalty constraints to favor solutions in which qubits representing the same variable indeed end up with the same value. More precisely, each pair of distinct qubits q and q' (associated to the same QUBO variable) adds a penalty term of the form $\varphi q(1-q')$. Where the penalty constant φ is (user) chosen as minus the cost of the worst possible solution to the initial QUBO which is obtained for a vector filled with ones (i.e., a solution that selects all edges of the graph and which therefore maximizes the highly-penalized violations of the cardinality contraints). This

---

[6] As $|E|$ is clearly an upper bound for the cost of any matching, any solution which violates at least one of the constraints 5 cannot be optimal.

therefore guarantees that a solution which violates at least one of these consistency constraints cannot be optimal (please note that we have switched from a maximization problem in Sect. 3.2 to a minimization problem as required by the machine). Lastly, as qubit duplication leads to an expanded QUBO which support graph is trivially isomorphic to the Chimera topology, it can be mapped on the device after a renormalization of its coefficients to ensure that the diagonal terms of Q are in $[-2, 2]$ and the others in $[-1, 1]$.

4.2 Results Summary

This section reports on the experiments we have been able to perform on instances of the previous QUBO problems. As already emphasized, due to the sparsity of the qubit interconnection topology, our QUBO instances were not directly mappable on the D-Wave machine and we had to resort to qubit duplications (whereby one problem variable is represented by several qubits on the D-Wave, bound together to end up with the same value at the end of the annealing process). This need for qubit duplication limited us to G_4 which, with 125 binary variables, already leads to a combinatorial problem of non trivial size. Yet, to solve it, we had to mobilize about 87% of the 1098 qubits of the machine. The results below have been obtained by running 10000 times the quantum annealer with a 20 μs annealing time (although we also experimented with 200 and 2000 μs, which did not appear to affect the results significantly).

Additionally, in order to improve the quality of the results obtained in our experiments, we used different gauges (spin-reversal transformations). The principle of a gauge is to apply a Boolean inversion transformation to operators σ_i in our Hamiltonian (in QUBO terms, after qubit duplication, this just means replacing some variable x_i by $1 - y_i$, with $y_i = 1 - x_i$ and updating the final QUBO matrix accordingly). This transformation has the particularity of not changing the optimal solution of the problem and of limiting the effect of local biases of the qubits, as well as machine accuracy errors [6]. Following common practices (e.g., [2]), we randomly selected 10% of the physical qubits used as gauges for each G_n instance that we mapped to the D-Wave. The results are given in Table 2.

Table 2. Experimental results summary without (left) and with (right) majority voting to fix qubit duplication issues on G_1, G_2, G_3, G_4. See text.

	Opt.	Best	Worst	Mean	Median	Stdev	Best	Worst	Mean	Median	Stdev
G_1	−68	−68	−9	−66.8	−68	4.6	−68	−37	−66.8	−68	4.2
G_2	−495	−495	−29	−398.2	−388	48.1	−495	−277	−400.4	−388	44.6
G_3	−2064	−1810	−505	−1454.8	−1548	157.7	−1810	−911	−1496.5	−1550	111.8
G_4	−6275	−5527	−2507	−4609.9	−4675	346.5	−5527	−3030	−4579.2	−4527	314.1

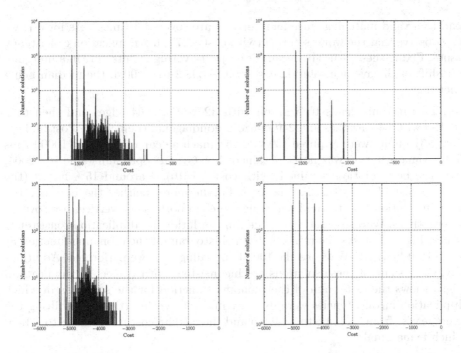

Fig. 3. Histograms on the left represent the economic function over 10000 annealing runs on G_3 and G_4. Histograms on the right represent the economic function over 10000 annealing runs on G_3 and G_4 (with duplication inconsistencies fixed by majority voting).

4.3 Instances Solutions

G_1. This instance leads to a graph with 8 vertices, 8 edges and then (before duplication) to a QUBO with 8 variables and 12 nonzero nondiagonal coefficients[7]; 16 qubits are then finally required. Over 10000 runs, the optimal solution (with a cost of -68) was obtained 9265 times (with correction 9284 times). Interestingly, the worst solution obtained (with a cost of -9) violates duplication consistency as all the 6 qubits representing variable 6 do not have the same value (4 of them are 0, so in that particular case, rounding the solution by means of majority voting gives the optimal solution).

G_2. This instance leads to a graph with 18 vertices, 27 edges and then to a QUBO with 27 variables and 72 nonzero nondiagonal coefficients. Overall, 100 qubits are required. Over 10000 runs the optimal solution (with cost -495) was obtained only 510 times (i.e., a 6% hitting probability). Although the best solution obtained is optimal, the median solution (with cost -388) does not

---

[7] In the Chimera topology the diagonal coefficient are not constraining as there is no limitation on the qubits autocouplings.

lead to a valid matching since four vertices are covered 3 times[8]. As for G_1, we also observe that the worst solution (with cost -277) has duplication consistency issues. Fixing these issues by means of majority voting results only in a marginal left shift of the average solution cost from -398.2 to -400.4, the median being unchanged.

G_3. This instance leads to a graph with 32 vertices, 64 edges and then to a QUBO with 64 variables and 240 nonzero nondiagonal coefficients. Postduplication, 431 qubits were required (39% of the machine capacity). Over 10000 runs the optimal solution was never obtained. For G_3, the optimum value is -2064, thus the best solution obtained (with cost -1810) is around 15% far-off (the median cost of -1548 is 25% far-off). Furthermore, neither the best nor the median solution lead to valid matchings since in both, some vertices are covered several times. We also observe that the worst solution has duplication consistency issues. Figure 3 shows the (renormalized) histogram of the economic function as outputted by the D-Wave for the 10000 annealing runs we performed. Additionally, since some of these solutions are inconsistent with respect to duplication, Fig. 3 shows the histogram of the economic function for the solutions in which duplication inconsistencies were fixed by majority voting (thus left shifting the average cost from -1454.8 to -1496.5 and the median cost from -1548 to -1550 which is marginal).

G_4. This instance leads to a graph with 50 vertices, 125 edges and then to a QUBO with 125 variables and 600 nonzero nondiagonal coefficients. Postduplication, 951 qubits were required (i.e., 87% of the machine capacity). Over 10000 runs the optimal solution was never obtained. Still, Fig. 4 provides a graphic representation of the best solutions obtained, with cost -5527 (median and worst solutions obtained respectively had costs -4675 and -2507). For G_4, the optimum value is -6075, thus the best solution obtained is around 10% far-off (a better ratio than for G_3) and median cost 25%. Furthermore, neither the best nor the median solution lead to valid matchings since in both, some vertices are covered several times. We also observe that the worst solution (as well as many others) has duplication consistency issues. Figure 3 shows the (renormalied) histogram of the economic function as outputted by the D-Wave for the 10000 annealing runs we performed. Additionally, since some of these solutions are inconsistent with respect to duplication, Fig. 3 shows the histogram of the economic function for the solutions in which duplication inconsistencies were fixed by majority voting (thus left shifting the average solution cost from -4609.9 to -4579.2 and the median cost from -4675 to -4527 which is also marginal).

5   Discussion and Perspectives

In this paper, our primary goal was to provide a first study on the behavior of an existing quantum annealer when confronted to old combinatorial beasts

---

[8] Fixing this would require a postprocessing step to produce valid matchings. Of course this is of no relevance for a polynomial problem, but such a postprocessing would thus be required when operationally using a D-Wave for solving non artificial problems.

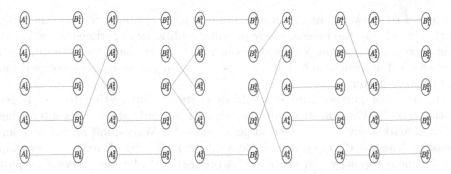

Fig. 4. Graphic representation of the best solution obtained for G_4. See text.

known to defeat classical annealing. At the very least, our study demonstrates that these special instances of the maximum (bipartite) matching problem are not at all straightforward to solve on a quantum annealer and, as such, are worth being included in a standard benchmark of problems for these emerging systems. Furthermore, as this latter problem is polynomial (and the specific instances considered in this paper even have straightforward optimal solutions), it allows to precisely quantify the quality of the solutions obtained by the quantum annealer in terms of distance to optimality.

There also are a number of lessons learnt. First, the need for qubit duplication severely limits the size of the problem which can be mapped on the device leading to a ratio between 5 and 10 qubits for 1 problem variable. Yet, a ≈ 1000 qubits D-Wave can tackle combinatorial problems with a few hundred variables, a size which is clearly nontrivial. Also, the need to embed problem constraints (e.g., in our case, matching constraints requiring that each vertex is covered at most once) in the economic function, even with carefully chosen penalty constants, often lead to invalid solutions. This is true both in terms of qubits duplication consistency issues (i.e., qubits representing the same problem variable having different values) as well as for problem specific constraints. This means that operationally using a quantum annealer requires one or more postprocessing steps (e.g., solving qubit duplication inconsistencies by majority voting), including problem-specific ones (e.g., turning invalid matchings to valid ones).

Of course, the fact that, in our experiments, the D-Wave failed to find optimal solutions for nontrivial instance sizes, does not rule out the existence of an advantage of quantum annealing *as implemented in D-Wave systems* over classical annealing (the existence of which, as previously emphasized, as already been established on specially designed problems [1]). However, our results tends to rule out (or confirm) the absence of an exponential advantage in the general case of quantum over classical annealing.

Also, since the present study takes a worst case (instances) point of view, it does not at all imply that D-Wave machines cannot be practically useful, and, indeed, its capacity to anneal in a few tens of μs makes it inherently very fast compared to software implementations of classical annealing. Stated otherwise, in the line of [23], the present study provides additional experimental evidences that there are (even non NP-hard) problems which are hard for both quantum *and* classical annealing and that on these quantum annealing does not perform significantly better.

In terms of perspectives, it would of course be interesting to test larger instances on D-Wave machines with more qubits. It would also be very interesting to benchmark a device with the next generation of D-Wave qubit interconnection topology (the so-called Pegasus topology [10]) which is significantly denser than the Chimera topology. On the more theoretical side of things, trying to port Sasaki and Hajek proof [25] to the framework of quantum annealing, although easier said than done, is also an insightful perspective. Lastly, bipartite matching over the G_n graphs family also gives an interesting playground to study or benchmark emerging classical quantum-inspired algorithms (e.g. Simulated Quantum Annealing [9]) or annealers.

Acknowledgements. The authors wish to thanks Daniel Estève and Denis Vion, from the Quantronics Group at CEA Paris-Saclay, for their support and fruitul discussions. The authors would also like to warmly thank Pr Daniel Lidar for granting them access to the D-Wave 2X operated at the University of Southern California Center for Quantum Information Science & Technology on which our experiments were run as well as for providing precious feedback and suggestions on an early version of this paper.

References

1. Albash, T., Lidar, D.: Demonstration of a scaling advantage for a quantum annealer over simulated annealing. Phys. Rev. X, 8 (2018)
2. Albash, T., Lidar, D.A.: Demonstration of a scaling advantage for a quantum annealer over simulated annealing. Phys. Rev. X **8**(3), 031016 (2018)
3. Baker, T., Gill, J., Solovay, R.: Relativizations of the $P = ?NP$ question. SIAM J. Comput. **4**, 431–442 (1975)
4. Bian, Z., Chudak, F., Israel, R., Lackey, B., Macready, W.G., Roy, A.: Discrete optimization using quantum annealing on sparse ising models. Frontiers Phys. **2**, 56 (2014)
5. Bian, Z., Chudak, F., Israel, R.B., Lackey, B., Macready, W.G., Roy, A.: Mapping constrained optimization problems to quantum annealing with application to fault diagnosis. Frontiers ICT **3**, 14 (2016)
6. Boixo, S., Albash, T., Spedalieri, F.M., Chancellor, N., Lidar, D.A.: Experimental signature of programmable quantum annealing. Nat. Commun. **4**, 2067 (2013)
7. Burkard, R.E., Fincke, U.: Probabilistic asymptotic properties of some combinatorial optimization problems. Discrete Math. **12**, 21–29 (1985)
8. Cerny, V.: Thermodynamical approach to the traveling salesman problem: an efficient simulation algorithm. J. Optim. Theor. Appl. **5**, 41–51 (1985)
9. Crosson, E., Harrow, A.W.: Simulated quantum annealing can be exponentially faster than classical simulated annealing. In: IEEE FOCS, pp. 714–723 (2016)
10. Dattani, N., Szalay, S., Chancellor, N.: Pegasus: the second connectivity graph for large-scale quantum annealing hardware. arXiv preprint arXiv:1901.07636 (2019)

11. Farhi, E., Goldstone, J., Gutmann, S.: Quantum adiabatic evolution algorithms versus simulated annealing. Technical report 0201031. arXiv:quant-ph (2002)
12. Farhi, E., Goldstone, J., Gutmann, S., Lapan, J., Lundgren, A., Preda, D.: A quantum adiabatic evolution algorithm applied to random instances of an np-complete problem. Science **292**, 472–476 (2001)
13. Farhi, E., Goldstone, J., Gutmann, S., Sipser, M.: Quantum computation by adiabatic evolution. arXiv preprint arXiv:quant-ph/0001106 (2000)
14. Geman, S., Geman, D.: Stochastic relaxation, Gibbs distribution, and the Bayesian restoration of images. IEEE Trans. Pattern Anal. Mach. Intell., 721–741 (1984)
15. Hajek, B.: Cooling schedules for optimal annealing. Math. Oper. Res., 311–329 (1988)
16. Harris, R., et al.: Experimental demonstration of a robust and scalable flux qubit. Phys. Rev. B **81**(13), 134510 (2010)
17. van Houweninge, M., Frenk, J.B.G., Rinnooy Kan, A.H.G.: Asymptotic properties of the quadratic assignment problem. Mathe. Oper. Res. **10**, 100–116 (1985)
18. Kirkpatrick, S., Gelatt Jr., C.D., Vecchi, M.P.: Optimization by simulated annealing. Science (1983)
19. Korte, B., Vygen, J.: Combinatorial Optimization, Theory and Algorithms. Springer, Heidelberg (2012). https://doi.org/10.1007/3-540-29297-7
20. Nolte, A., Schrader, R.: Simulated annealing and its problems to color graphs. In: Diaz, J., Serna, M. (eds.) ESA 1996. LNCS, vol. 1136, pp. 138–151. Springer, Heidelberg (1996). https://doi.org/10.1007/3-540-61680-2_52
21. Preskill, J.: Quantum computing in the NISQ era and beyond. Technical report 1801.00862, arXiv (2018)
22. Google AI Quantum and collaborators. Quantum supremacy using a programmable superconducting processor, September 2019
23. Reichardt, B.E.: The quantum adiabatic optimization algorithm and local minima. In: ACM STOC, pp. 502–510 (2004)
24. Santoro, G.E., Martonak, R., Tosatti, E., Car, R.: Theory of quantum annealing of spin glass. Science **295**, 2427–2430 (2016)
25. Sasaki, G.H., Hajek, B.: The time complexity of maximum matching by simulated annealing. J. ACM **35**, 387–403 (1988)
26. Schauer, J.: Asymptotic behavior of the quadratic knapsack problems. Eur. J. Oper. Res. **255**, 357–363 (2016)
27. Vert, D., Sirdey, R., Louise, S.: On the limitations of the chimera graph topology in using analog quantum computers. In: Proceedings of the 16th ACM International Conference on Computing Frontiers, pp. 226–229. ACM (2019)
28. Vert, D., Sirdey, R., Louise, S.: Revisiting old combinatorial beasts in the quantum age: quantum annealing versus maximal matching. Technical report arXiv:1910.05129 (quant-ph) (2019)

A Quantum Annealing Algorithm
for Finding Pure Nash Equilibria
in Graphical Games

Christoph Roch[✉], Thomy Phan, Sebastian Feld, Robert Müller,
Thomas Gabor, Carsten Hahn, and Claudia Linnhoff-Popien

LMU Munich, Munich, Germany
christoph.roch@ifi.lmu.de

Abstract. We introduce Q-Nash, a quantum annealing algorithm for
the NP-complete problem of finding pure Nash equilibria in graphical
games. The algorithm consists of two phases. The first phase determines
all combinations of best response strategies for each player using classical computation. The second phase finds pure Nash equilibria using
a quantum annealing device by mapping the computed combinations
to a quadratic unconstrained binary optimization formulation based on
the Set Cover problem. We empirically evaluate Q-Nash on D-Wave's
Quantum Annealer 2000Q using different graphical game topologies. The
results with respect to solution quality and computing time are compared
to a Brute Force algorithm and the Iterated Best Response heuristic.

Keywords: Quantum annealing · Game theory · Optimization

1 Introduction

Applications of conventional game theory have played an important role in
many modern strategic decision making processes including diplomacy, economics, national security, and business [6,24,26]. Game theory is a mathematical
paradigm in which such domain-specific decision situations are modeled [9]. Multiple players interact with each other to collectively complete a task or to enforce
their interests. In the classical model of game theory [30] all players choose an
action simultaneously and obtain a certain payoff (*utility*), which depends on
the actions of the other players. The most common solution concept for such a
decision problem is called *Nash equilibrium (NE)* [20], in which no player is able
to unilaterally improve his payoff by changing his chosen action.

There exist many different representations for such simultaneous games. The
most popular is the strategic or standard normal-form game representation,
which is often used for 2-player games, like the prisoner dilemma or battle of
the sexes. However, due to its exponential growth of the representational size
w.r.t the number of players [14], a more compact version, called graphical game,

© Springer Nature Switzerland AG 2020
V. V. Krzhizhanovskaya et al. (Eds.): ICCS 2020, LNCS 12142, pp. 488–501, 2020.
https://doi.org/10.1007/978-3-030-50433-5_38

is increasingly used to model multi-player scenarios [15–17,23]. Here a player's action only depends on a certain number of other players' actions (a so called player's neighborhood). These neighborhoods are visualized by an underlying graph with players as vertices and the dependencies as edges.

While *pure strategies*, where each player unambiguously decides on a particular action, are conceptually simpler than *mixed strategies*, the associated computational problems appear to be harder [10]. This also applies to the compact representation of graphical games, for which the complexity of finding pure strategy Nash equilibria (PNE-GG) was proven to be NP-complete, even in the restricted case of neighborhoods of maximum size 3 with at most 3 actions per player [10].

Although there are many algorithms that find *mixed* or *approximated* NE in graphical games [3,14,21,27], there are only a couple of algorithms that deal with *pure Nash equilibria (PNE)*. In this paper, we focus on the computationally hard problem of finding NE with *pure strategies*, where each player chooses to play an action in a deterministic, non-random manner.

With D-Wave Systems releasing the first commercially available quantum annealer in 2011[1], there is now the possibility to find solutions for such complex problems in a completely different way compared to classical computation. To use D-Wave's quantum annealer, the problem has to be formulated as a *quadratic unconstrained binary optimization (QUBO)* problem [5], which is one possible input type for the annealer. In doing so, the metaheuristic *quantum annealing* seeks to find the minimum of an objective function, i.e., the best solution of the defined configuration space [19].

In this paper, we propose the first quantum annealing algorithm for finding PNE-GG, called *Q-Nash*. The algorithm consists of two phases. The first phase determines all combinations of *best response* strategies for each player using classical computation. The second phase finds pure Nash Equilibria using a quantum annealing device by mapping the computed combinations to a QUBO formulation based on the Set Cover problem. We empirically evaluate Q-Nash on D-Wave's Quantum Annealer 2000Q using different graphical game topologies. The results with respect to solution quality and computing time are compared to a Brute Force algorithm and the Iterated Best Response heuristic.

2   Background

2.1   Graphical Games and Pure Nash Equilibria

In an *n-player game*, each player p $(1 \leq p \leq n)$, has a finite set of strategies or actions, S_p, with $|S_p| \geq 2$. Such a game can be visualized by a set of n matrices M_p. The entry $M_p(s_1, ..., s_n) = M_p(s)$ specifies the payoff to player p when the *joint action* (also, *strategy profile*) of the n players is $s \in S$, with $S = \prod_{i=1}^{n} S_i$ being the set of combined strategy profiles. In order to specify a game with

---

[1] https://www.dwavesys.com/news/d-wave-systems-sells-its-first-quantum-computing-system-lockheed-martin-corporation.

n players and s strategies each, the representational size is ns^n, an amount of information exponential with respect to the number of players. However, players often interact only with a limited number of other players, which allows for a much more succinct representation. In [14] such a compact representation, called *graphical game*, is defined as follows:

Definition 1 (Graphical Game). An *n-player graphical game* is a pair (G, M), where G is an undirected graph with n vertices and M is a set of n matrices M_p with $1 \leq p \leq n$, called the *local game* matrices. Player p is represented by a vertex labeled p in G. We use $N_G(p) \subseteq \{1, ..., n\}$ to denote the set of *neighbors of player p* in G – i.e., those vertices q such that the undirected edge (p, q) appears in G. By convention, $N_G(p)$ always includes p himself. The interpretation is that each player is in a game with only his neighbors in G. Thus, the size of the graphical game representation is only exponential in the maximal node degree d of the graph, ns^d. If $|N_G(p)| = k$ and $s \in \prod_{i=1}^{k} S_i$, $M_p(s)$ denotes the payoff to p when his k neighbors (including himself) play s.

Consider a game with n players and strategy sets $S_1, ..., S_n$. For every strategy profile $s \in S$, the strategy of player p is denoted by s_p and s_{-p} corresponds to the $(n-1)$-tuple of strategies of all players but p. For every $s_p' \in S_p$ and $s_{-p} \in S_{-p}$ we denote by $(s_{-p}; s_p')$ the strategy profile in which player p plays s_p' and all the other players play according to s_{-p}. One has to mention that a strategy profile s is called *global*, if all n players contribute to it, i.e., a global combined strategy s consists of every player playing one of his actions [10].

Definition 2 (Pure Nash Equilibrium). A global strategy profile s is a PNE, if for every player p and strategy $s_p' \in S_p$ we have $M_p(s) \geq M_p(s_{-p}; s_p')$. That is, no player can improve his expected payoff by deviating unilaterally from a Nash equilibrium.

[7] define a *best response* strategy as follows:

Definition 3 (Best Response Strategy). A best response strategy of player p is defined by:

$$BR_{M_p}(s_{-p}) \triangleq \{s_p \mid s_p \in S_p \text{ and } \forall s_p' \in S_p : M_p(s_{-p}; s_p) \geq M_p(s_{-p}; s_p')\}$$

Intuitively, $BR_{M_p}(s_{-p})$ is the set of strategies in S_p that maximize p's payoff if the other players play according to s_{-p}. Thus, a strategy profile s is a pure Nash equilibrium if for every player p, $s_p \in BR_{M_p}(s_{-p})$.

A visual example of a graphical game called *FRIENDS* and its PNE can be found in [10].

2.2   Quantum Annealing

Quantum annealing is a metaheuristic for solving complex optimization and decision problems [12]. D-Wave's quantum annealing heuristic is implemented in hardware, designed to find the lowest energy state of a spin glass system, described by an Ising Hamiltonian,

$$\mathcal{H}(s) = \sum_i h_i x_i + \sum_{i<j} J_{ij} x_i x_j \tag{1}$$

where h_i is the on-site energy of qubit i, J_{ij} are the interaction energies of two qubits i and j, and x_i represents the spin $(-1, +1)$ of the ith qubit. The basic process of quantum annealing is to physically interpolate between an initial Hamiltonian H_I with an easy to prepare minimal energy configuration (or ground state), and a problem Hamiltonian H_P, whose minimal energy configuration is sought that corresponds to the best solution of the defined problem (see Eq. 2). This transition is described by an adiabatic evolution path which is mathematically represented as function $s(t)$ and decreases from 1 to 0 [19].

$$H(t) = s(t)H_I + (1 - s(t))H_P \tag{2}$$

If this transition is executed sufficiently slow, the probability to find the ground state of the problem Hamiltonian is close to 1 [1]. Thus, by mapping the Nash equilibrium decision problem onto a spin glass system, quantum annealing is able to find the solution of it.

For completeness, we map our NE decision problem to an alternative formulation of the Ising spin glass system. The so called QUBO problem [5] is mathematically equivalent and uses 0 and 1 for the spin variables [28]. The quantum annealer is as well designed to minimize the functional form of the QUBO:

$$\min x^t Q x \qquad \text{with } x \in \{0, 1\}^n \tag{3}$$

with x being a vector of binary variables of size n, and Q being an $n \times n$ real-valued matrix describing the relationship between the variables. Given the matrix $Q : n \times n$, the annealing process tries to find binary variable assignments $x \in \{0, 1\}^n$ to minimize the objective function in Eq. 3.

2.3   Set Cover Problem

Since the QUBO formulation of Q-Nash resembles the well known Set Cover (SC) problem, it is introduced here. Within the SC problem, one has to find the smallest possible number of subsets from a given collection of subsets $V_k \subseteq U$ with $1 \le k \le N$, such that the union of them is equal to a global superset U of size n. This problem was proven to be NP-hard [13]. In [18] the QUBO formulation for the Set Cover problem is given by:

$$H_A = A \sum_{\alpha=1}^n \left(1 - \sum_{m=1}^N x_{\alpha,m}\right)^2 + A \sum_{\alpha=1}^n \left(\sum_{m=1}^N m x_{\alpha,m} - \sum_{k:\alpha \in V_k} x_k\right)^2 \tag{4}$$

and

$$H_B = B \sum_{k=1}^{N} x_k \tag{5}$$

with x_k being a binary variable which is 1, if set k is included within the chosen sets, and 0 otherwise. $x_{\alpha,m}$ denotes a binary variable which is 1 if the number of chosen subsets V_k which include element α is $m \geq 1$, and 0 otherwise. The first energy term imposes the constraints that for any given α exactly one $x_{\alpha,m}$ must be 1, since each element of U must be included a fixed number of times. The second term states, that the number of times that we claimed α was included is in fact equal to the number of subsets V_k we have included with α as an element. A is a penalty value, which is added on top of the solution energy, described by $H = H_A + H_B$, if a constraint was not satisfied, i.e. one of the two terms (quadratic differences) are unequal to 0. Therefore adding a penalty value states a solution as invalid. Additionally, the SC problem minimizes over the number of chosen subsets V_k, as stated in Eq. 5. We skip discussing the term due to the fact that it has no impact on our Q-Nash QUBO problem later on.

3   Related Work

Most known algorithms focus on finding *mixed* or *approximated* NE [3,14,21, 27]. However, some investigations in determining PNE in graphical and similar variations of games were made.

Daskalakis and Papadimitriou present a reduction from GG to Markov random fields such that PNE can be found by statistical inference. They use known statistical inference algorithms like belief propagation, junction tree algorithm and Markov Chain Monte Carlo to determine PNE in graphical games [7].

In [11], the authors analyze the problem of computing pure Nash equilibria in action graph games (AGGs), another compact game theoretic representation, which is similar to graphical games. They propose a dynamic-programming approach that constructs equilibria of the game from equilibria of restricted games played on subgraphs of the action graph. In particular, under the premise that the game is symmetric and the action graph has bounded treewidth, their algorithm determines the existence of a PNE in polynomial time.

Palmieri and Lallouet deal with constraint games, for which constraint programming is used to express players preferences. They rethink their solving technique in terms of constraint propagation by considering players preferences as global constraints. Their approach is able to find all pure Nash equilibria for some problems with 200 players and also shows that performance can be improved for graphical games [23].

With quantum computing gaining more and more attention[2] and none of the related work making use of quantum annealing in order to find PNE, we propose a solution approach using a quantum annealer.

---

[2] https://www.gartner.com/smarterwithgartner/the-cios-guide-to-quantum-computing/.

Since some NE algorithms are restricted to a certain graphical game structure, see for example [17], we want to emphasize, that our approach is able to work on every graphical game structure, even if the dependency graph is not connected.

4 Q-Nash

In the following sections we present the concept of Q-Nash. Q-Nash consists of two phases, which are described below.

4.1 Determining Best Response Strategies

In the first phase, we identify each player's *best response* to what the other players might do. That is, for every strategy profile s, we search player p's strategy (or strategies) with the maximum payoff $M_p(s)$. This involves iterating through each player in turn and determining their optimal strategies. An example is given in 1. This can be feasibly done in polynomial time, since one can easily explore each player's matrix M_p representing the utility function, i.e., payoffs [10]. Therefore the first part, which we see as preliminary step for our Q-Nash algorithm, is executed on a classical computer. After doing this, one gets a set of combined strategies with each being a *best response* to the other players' played strategies. This set is denoted by $\mathcal{B} = \{BR_{M_p} \mid (1 \leq p \leq n)\}$ and has the cardinality $C_{\mathcal{B}} = \sum_{p=1}^{n} |BR_{M_p}|$.

Example 1. In Table 1 the local payoff matrix M of player A is visualized. For instance, assume player B plays action 2 and player C chooses action 1. In this case, a *best response* strategy for player A is action 0, due to the fact, that he gets the most payoff in this situation, i.e. 4. This leads to a *best response* strategy combination for player A, denoted by a pointed set $\{A0, B2, C1\}$ with player A being the base point of it.

Table 1. Local dependency payoff matrix M of player A. The payoffs marked in bold correspond to a *best response* strategy of player A.

A	B0 C0	B0 C1	B1 C0	B1 C1	B2 C0	B2 C1
0	**4**	1	2	**2**	1	**4**
1	1	**3**	**2**	1	**2**	2

4.2 Finding PNE Using Quantum Annealing

In the second phase, a PNE is identified, when all players are playing one of their *best response* strategies simultaneously. With the computed set $\mathcal{B}$ of the classical phase, the following question arises:

"Is there a union of the combined best response strategies of $\mathcal{B}$ which results in a global strategy profile, under the premise that every player plays one of his best response actions?"—As stated in Definition 3 this would lead to a PNE.

This question resembles the Set Cover (SC) problem, stated in Sect. 2.3. It asks for the smallest possible number of subsets to cover the elements of a given global set (in our case, this global set would be a feasible global strategy profile and the subsets correspond to our best response strategy profiles of $\mathcal{B}$). However, for our purposes we have to modify the given formulation in Eq. 4 as follows:

$$H = A \sum_{p=1}^{n} \left(1 - \sum_{j=1}^{|S_p|} \sum_{m=1}^{C_\mathcal{B}} x_{p,j,m} \right)^2 + A \sum_{p=1}^{n} \sum_{j=1}^{|S_p|} \left(\sum_{m=1}^{C_\mathcal{B}} m x_{p,j,m} - \sum_{k:p,j \in s_k} x_k \right)^2$$
$$+ A \left(n - \sum_{k=1}^{C_\mathcal{B}} x_k \right)^2$$

(6)

Equation 6 is quite similar to Eq. 4. However, an element α of the superset of Eq. 4 corresponds to a player p and his chosen action j of our global strategy profile. Nevertheless, the intention of those first two energy terms complies with the intention of Eq. 4, stated in Sect. 2.3. Further, another energy term must be added to our QUBO problem as constraint. This last energy term, for which an instance is given in Example 2, states that exactly n sets of $\mathcal{B}$ should be included to form the global strategy profile. This constraint implicitly ensures that every player is playing one of his best response strategies.

A is called penalty value, which is added on top of the solution energy, if a constraint was not satisfied, i.e. one of the three terms (quadratic differences) are unequal to 0. Thus, adding a penalty value states a solution as invalid. Only if the total energy described by $H = 0$ the corresponding solution is a valid *global best response* strategy profile and thus a PNE.

All these energy terms are specified within the QUBO problem matrix Q, which the quantum annealer takes as an input, goes through the annealing process and responds with a binary vector x as a solution, see Sect. 2.2. This vector indicates which best response strategy of each player should be chosen to form a PNE.

Example 2. To demonstrate the function of the last energy term (constraint) of Eq. 6, an excerpt of *best response* strategy combinations of $\mathcal{B}$ (in form of pointed sets, with the bold player-action-combination being the base point) for an arbitrary 4-player game are visualized in Fig. 1. The green union of four *best response* sets leads to a PNE, in which every player is playing one of his *best response* strategies. Although the red union of three sets also leads to a global combined strategy set, in which every player is playing one of his actions, it is not a PNE, due to player B not playing a *best response* strategy.

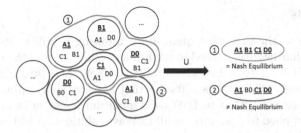

Fig. 1. On the left hand an excerpt of superset $\mathcal{B}$ is given. On the right hand the sets are united, with each (1) & (2) being a global strategy combination, but only one (1) being a PNE.

5 Experiments

5.1 Evaluation Graph-Topologies

For evaluating Q-Nash, we implemented a game generator. It creates graphical game instances of three different popular graphical structures, which were often used in literature [11,15,23,29]. These graphical structures are shown in Fig. 2. As an input for our game generator, one can choose the number of players, the graph-topology and thus the dependencies between the players and the number of actions for each player individually. The corresponding payoffs are sampled randomly from [0, 15]. For our experiments we considered games with three actions per player. The classic theorem of Nash [20] states that for any game, there exists a Nash equilibrium in the space of joint *mixed strategies*. However, in this work we only consider *pure strategies* and therefore there might be (graphical) games without any PNE (see, for instance, [22]). Additionally we want to emphasize, that Q-Nash is able to work on every graphical game structure, even if the dependency graph is not connected, for instance a set of trees (called *forest*).

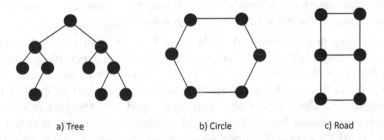

a) Tree   b) Circle   c) Road

Fig. 2. Different graphical game structures (topologies) which indicate the dependencies between players of a game. (a) Tree-Topology with $n = 10$, (b) Circle-Topology with $n = 6$ and (c) Road-Topology with $n = 6$.

5.2   Methods

QBSolv: Due to the fact, that quantum computing is still in its infancy, and corresponding hardware is limited in the number of qubits and their connectivity, we need to fall back to a hybrid method (QBSolv[3]), in order to solve large problem instances. QBSolv is a software that automatically splits instances up into subproblems submitted to D-Wave's quantum annealer, and an extensive tabu search is applied to post-process all D-Wave solutions. Additionally, QBSolv embeds the QUBO problem to the quantum annealing hardware chip. QBSolv further allows to specify certain parameters such as the number of individual solution attempts (*num_repeats*), the subproblem size used to split up instances which do not fit completely onto the D-Wave hardware and many more. For detailed information, see [4].

Brute Force: For evaluating the effectiveness of Q-Nash we implemented a Brute Force (BF) algorithm to compare with. It determines the best response strategy sets of all players in the same way as Q-Nash does and afterwards tries out every possible combination of those sets to form a valid global strategy set which corresponds to a PNE. The number of combinations is exponential with respect to the number of players of the game, $\prod_{p=1}^{n} BR_{M_p}$.

Iterated Best Response: Additionally, an Iterated Best Response (IBR) algorithm was implemented [25]. In each iteration, one player changes his action to the best action that is the best response to the other players their action. This is repeated until a PNE is reached. In that case, the algorithm starts again from an randomly generated global strategy profile, to find other PNE. Furthermore it takes a timespan as an input parameter and terminates after a timeout occurred.

5.3   Computational Times of Q-Nash

With respect to the computational results stated in Table 2 we first introduce the time components of Q-Nash's total computation time. For a better understanding a general overview of Q-Nash is given in Fig. 3.

Determining best responses describes the time Q-Nash classical computing phase takes to identify the *best response* strategies of every player and additionally build up the QUBO matrix. The *Find embedding time* states the time D-Wave's classical embedding heuristic takes to find a valid subproblem hardware embedding. The *QBSolv time* can be divided into the *classical time* and the *quantum annealing time*. The *classical QBSolv time* comprises of not only a tabu search, which iteratively processes all subproblem solutions, it also contains the latency and job queuing time to D-Wave's quantum hardware. The *quantum annealing time* comprises of the number of subproblems times D-Wave's *qpu-access-time*.

---

[3] https://github.com/dwavesystems/qbsolv.

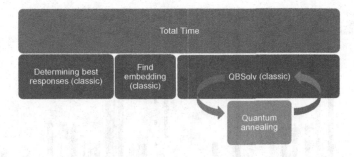

Fig. 3. Overview of the computational time components of Q-Nash

6   Results and Discussion

We investigated the solution quality and computational time of Q-Nash. Figure 4, 5 and 6 show the ability of finding PNE of the three proposed methods (Q-Nash with QBSolv, BF and IBR) in differently structured graphical games. For every graph topology (Tree, Circle and Road) we used games with players ranging from 6 to 30, due to the fact that the road topology needs an even number of players we skipped 15, 21 and 27 player instances. We ran Q-Nash and IBR 20 times on every instance. Additionally, IBR was run as long as Q-Nash (total time) took, to solve the instances.

The results show that Q-Nash with *num_repeats* set to 200, always found the same amount of PNE as the exact BF algorithm for the smaller game instances (6 to 12 Players, except of the 12 Player Tree-Topology instance), while IBR was only able to find all PNE in the Circle-Topology for those instances.

Regarding the larger game instances, one can see that Q-Nash, due being a heuristic, was not always able to find the same amount of PNE per run (20 times) and also was not able to find every PNE in a game, when compared to the exact BF algorithm. Compared to IBR, one can notice, that Q-Nash in general performed better than IBR on Circle-Topology, while IBR outperformed Q-Nash on tree structured games. Regarding the Road-Topology, Q-Nash did better on 12 and 18 Player instances, while IBR surpassed Q-Nash on 24 and 30 Player instances.

As already mentioned in Sect. 5.2 QBSolv splits the QUBO into smaller components (subQUBOS) of a predefined subproblem size, which are then solved independently of each other. This process is executed iteratively as long as there is an improvement and it can be defined using the QBSolv parameter *num_repeats*. This parameter determines the number of times to repeat the splitting of the QUBO problem matrix after finding a better sample. In Fig. 7 the influence of this parameter is shown. We exemplary used a 24 player road game and ran Q-Nash 20 times per parameter setting to show its impact on the effectiveness. As expected one can see, that with increasing number of repeats (*num_repeats*) the inter-quartile range and its median in regard to the number of PNE found, increase. Although an annealing process takes only 20 µs in default,

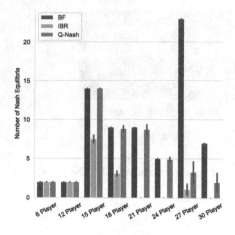

Fig. 4. Circle-Topology games

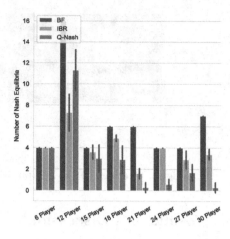

Fig. 5. Tree-Topology games

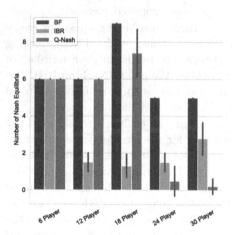

Fig. 6. Road-Topology games

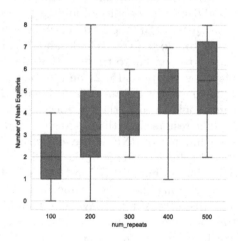

Fig. 7. The impact of *num_repeats* parameter on a 24 player road game.

it adds up with the *num_repeats* parameter and therefore leads to a trade-off between computational time and accuracy.

The computational time results are shown for circle structured graphical games and can be viewed in Table 2. The used game instances differ in the number of players (6–20 Player) and the computing times are given in seconds of our Q-Nash algorithm. As already mentioned in Sect. 5.2 we need to use QBSolv to solve large problem instances and therefore we have to consider the computational results of Q-Nash with caution, as stated in Sect. 5.3.

Due to Q-Nash's immense overhead time (embedding, tabu search, latency, queuing time, etc.) the *total time (1)* is even for the smaller game instances (6 and 12 Players) quite large (85.393 and 125.828 s). Nevertheless the pure Q-Nash

Table 2. Computational times of the Q-Nash algorithm on various circle structured graphical games

		Graphical games (Circle)							
		6 player	8 player	10 player	12 player	14 player	16 player	18 player	20 player
Q-Nash	(1) Total time [s]	**85.393**	**125.828**	**457.906**	**424.701**	**273.283**	**228.609**	**276.593**	**248.547**
	(2) Determining best responses (Classic) [s]	0.391	0.891	1.828	3.063	4.906	7.328	10.704	14.360
	(3) Find embedding (Classic) [s]	4.172	4.062	4.093	5.155	4.813	5.703	4.202	4.391
	(4) QBSolv (Classic) [s]	79.862	120.012	448.834	412.869	260.285	212.849	258.395	226.775
	(5) QA time (Quantum) [s]	0.968	0.863	3.151	3.614	3.279	2.729	3.292	3.021

computational time (*determining best response strategies (2) & QA time (5)*) is comparatively very small (1.359 and 1.754 s).

In general, one can see that the *quantum annealing time (5)* for these game instances is quite constant and the runtime of the *classical Q-Nash phase (4)* is only polynomial, as stated in Sect. 4.1.

With the *embedding time (3)* being constant, the only unpredictable runtime component of Q-Nash is *QBSolv classic (4)*. This is due to the tabu search (including the *num_repeats* parameter), which only terminates after a specified number of iterations in which no improvement of the previously found solutions could be made.

However, with increasing number of qubits and their connectivity it might be possible to map larger game instances directly to the hardware chip such that the hybrid solver QBSolv can be avoided. Thus, the total Q-Nash time would only consist of the *classical algorithm phase time (2)*, the *embedding time (3)* and a fraction of the *quantum annealing time (5)*, since we do not have to split our problem instance into subproblems anymore. At this time, Q-Nash might be a competitive method regarding the computational time.

7   Conclusion

We proposed Q-Nash, to our knowledge the first algorithm that finds PNE-GG using quantum annealing hardware. Regarding the effectiveness of Q-Nash, we showed that for small game instances (ranging from 6–12 players) the algorithm was always able to find all PNE in differently structured graphical games. Anyway we have to mention, that with increasing number of players the variance w.r.t the number of found PNE increased, too.

Due to the fact that quantum computing is still in its infancy and recent hardware is limited in the number of qubits and their connectivity, we had to fall back on a quantum-classical hybrid solver, called QBSolv, which involves additional overhead time. That makes it difficult to draw a fair comparison of Q-Nash and classical state-of-the-art solution methods regarding the computational time. We therefore decomposed the total time into its main components to show their impact. According to the experimental results, the only unpredictable time component is QBSolv's classical tabu search along with latency and job queuing times at D-Wave's cloud computing frontend. However with D-Wave announcing an immense rise of the number of qubits and their connectivity on D-Wave's quantum processors in the next years[4], it might be possible to embed larger game instances directly onto the chip and therefore omit hybrid solvers like QBSolv. Another possibility is using Fujitsu's Digital Annealing Unit (DAU) which also takes a QUBO matrix as input. With DAU being able to solve larger fully connected QUBO problems [2], a shorter total computation time could be achieved.

Regarding future work, it would also be interesting to see, how Q-Nash performs on a gate model quantum computer. For example, one could use the quantum approximate optimization algorithm, proposed by Farhi et al. [8]. This algorithm takes a QUBO Hamiltonian as an input. However since gate model quantum computers are at the moment even more limited in their resources, one has to come up with a clever way to split up large problem instances to fit them on the quantum chip.

References

1. Albash, T., Lidar, D.A.: Adiabatic quantum computation. Rev. Mod. Phys. **90**(1), 015002 (2018)
2. Aramon, M., Rosenberg, G., Valiante, E., Miyazawa, T., Tamura, H., Katzgraber, H.: Physics-inspired optimization for quadratic unconstrained problems using a digital annealer. Bull. Am. Phys. Soc. **7**, 48 (2019)
3. Bhat, N.A., Leyton-Brown, K.: Computing Nash equilibria of action-graph games. In: Proceedings of the 20th Conference on Uncertainty in Artificial Intelligence, pp. 35–42. AUAI Press (2004)
4. Booth, M., Reinhardt, S.P., Roy, A.: Partitioning optimization problems for hybrid classical. quantum execution. Technical report, pp. 01–09 (2017)
5. Boros, E., Hammer, P.L., Tavares, G.: Local search heuristics for quadratic unconstrained binary optimization (QUBO). J. Heuristics **13**(2), 99–132 (2007)
6. Carfi, D., Musolino, F., et al.: Fair redistribution in financial markets: a game theory complete analysis. J. Adv. Stud. Financ. **2**(2), 4 (2011)
7. Daskalakis, C., Papadimitriou, C.H.: Computing pure Nash equilibria in graphical games via Markov random fields. In: Proceedings of the 7th ACM Conference on Electronic Commerce, pp. 91–99. ACM (2006)
8. Farhi, E., Goldstone, J., Gutmann, S.: A quantum approximate optimization algorithm. arXiv preprint arXiv:1411.4028 (2014)
9. Fudenberg, D., Tirole, J.: Game Theory. MIT Press, Cambridge, vol. 392, no. 12, p. 80 (1991)
10. Gottlob, G., Greco, G., Scarcello, F.: Pure nash equilibria: hard and easy games. J. Artif. Intell. Res. **24**, 357–406 (2005)

---

[4] https://www.dwavesys.com/sites/default/files/mwj_dwave_qubits2018.pdf.

11. Jiang, A.X., Leyton-Brown, K.: Computing pure Nash equilibria in symmetric action graph games. In: AAAI, vol. 1, pp. 79–85 (2007)
12. Kadowaki, T., Nishimori, H.: Quantum annealing in the transverse Ising model. Phys. Rev. E **58**(5), 5355 (1998)
13. Karp, R.M.: Reducibility among combinatorial problems. In: Miller, R.E., Thatcher, J.W., Bohlinger, J.D. (eds.) Complexity of computer computations, pp. 85–103. Springer, Boston (1972). https://doi.org/10.1007/978-1-4684-2001-2_9
14. Kearns, M., Littman, M.L., Singh, S.: Graphical models for game theory. arXiv preprint arXiv:1301.2281 (2013)
15. Koller, D., Milch, B.: Multi-agent influence diagrams for representing and solving games. Games Econ. Behav. **45**(1), 181–221 (2003)
16. La Mura, P.: Game networks. In: Proceedings of the 16th Conference on Uncertainty in Artificial Intelligence, pp. 335–342 (2000)
17. Littman, M.L., Kearns, M.J., Singh, S.P.: An efficient, exact algorithm for solving tree-structured graphical games. In: Advances in Neural Information Processing Systems, pp. 817–823 (2002)
18. Lucas, A.: Ising formulations of many NP problems. Front. Phys. **2**, 5 (2014)
19. McGeoch, C.C.: Adiabatic quantum computation and quantum annealing: theory and practice. Synth. Lect. Quantum Comput. **5**(2), 1–93 (2014)
20. Nash, J.: Non-cooperative games. Ann. Math. **54**, 286–295 (1951)
21. Ortiz, L.E., Kearns, M.: Nash propagation for loopy graphical games. In: Advances in Neural Information Processing Systems, pp. 817–824 (2003)
22. Osborne, M.J., Rubinstein, A.: A Course in Game Theory. MIT Press, Cambridge (1994)
23. Palmieri, A., Lallouet, A.: Constraint games revisited. In: International Joint Conference on Artificial Intelligence, IJCAI, vol. 2017, pp. 729–735 (2017)
24. Rabin, M.: Incorporating fairness into game theory and economics. Am. Econ. Rev. **83**, 1281–1302 (1993)
25. Roughgarden, T.: Twenty Lectures on Algorithmic Game Theory. Cambridge University Press, Cambridge (2016)
26. Roy, S., Ellis, C., Shiva, S., Dasgupta, D., Shandilya, V., Wu, Q.: A survey of game theory as applied to network security. In: 2010 43rd Hawaii International Conference on System Sciences (HICSS), pp. 1–10. IEEE (2010)
27. Soni, V., Singh, S., Wellman, M.P.: Constraint satisfaction algorithms for graphical games. In: Proceedings of the 6th International Joint Conference on Autonomous Agents and Multiagent Systems, p. 67. ACM (2007)
28. Su, J., Tu, T., He, L.: A quantum annealing approach for Boolean satisfiability problem. In: Proceedings of the 53rd Annual Design Automation Conference, p. 148. ACM (2016)
29. Vickrey, D., Koller, D.: Multi-agent algorithms for solving graphical games. In: AAAI/IAAI, pp. 345–351 (2002)
30. Von Neumann, J., Morgenstern, O.: Theory of Games and Economic Behavior. Princeton University Press, Princeton (1944)

Hybrid Quantum Annealing Heuristic Method for Solving Job Shop Scheduling Problem

Krzysztof Kurowski[1]([✉]), Jan Węglarz[2], Marek Subocz[1], Rafał Różycki[2], and Grzegorz Waligóra[2]

[1] Poznań Supercomputing and Networking Center, IBCH PAS, Poznań, Poland
{krzysztof.kurowski,marsub}@man.poznan.pl
[2] Poznań University of Technology, Poznań, Poland
{jan.weglarz,rafal.rozycki,grzegorz.waligora}@cs.put.poznan.pl

Abstract. Scheduling problems have attracted the attention of researchers and practitioners for several decades. The quality of different methods developed to solve these problems on classical computers have been collected and compared in various benchmark repositories. Recently, quantum annealing has appeared as promising approach to solve some scheduling problems. The goal of this paper is to check experimentally if and how this approach can be applied for solving a well-known benchmark of the classical Job Shop Scheduling Problem. We present the existing capabilities provided by the D-Wave 2000Q quantum annealing system in the light of this benchmark. We have tested the quantum annealing system features experimentally, and proposed a new heuristic method as a proof-of-concept. In our approach we decompose the considered scheduling problem into a set of smaller optimization problems which fit better into a limited quantum hardware capacity. We have tuned experimentally various parameters of limited fully-connected graphs of qubits available in the quantum annealing system for the heuristic. We also indicate how new improvements in the upcoming D-Wave quantum processor might potentially impact the performance of our approach.

Keywords: Quantum annealing · Job Shop Scheduling Problem · Heuristic

1 Introduction

Quantum computing has attracted the attention of many researchers and provides a new challenge for solving certain classes of combinatorial problems more efficiently than on classical computers. Moreover, quantum computers are starting to approach the limit of classical simulation, and consequently entering an era of unprecedented ways to explore quantum algorithms [1]. There are many ongoing efforts to run new experiments and benchmarks with quantum computing to

© Springer Nature Switzerland AG 2020
V. V. Krzhizhanovskaya et al. (Eds.): ICCS 2020, LNCS 12142, pp. 502–515, 2020.
https://doi.org/10.1007/978-3-030-50433-5_39

discover new applications and solve real-world problems. Many researchers have been trying to find efficient ways of approaching NP-hard scheduling problems over the last decades. A comprehensive study to the theory and applications of scheduling in advanced planning and computer systems as well as various practical industrial, real-time engineering, management science, business administration and information systems use cases were presented in [3]. Recently, there has been much interest in the possibility of using quantum annealing, which is a derivative of adiabatic quantum optimization, e.g. in [4,5,7,9]. In general, the main assumption in adiabatic quantum optimization is that there is a quantum Hamiltonian H_P whose ground state encodes the solution to a considered problem, and another Hamiltonian H_0, whose ground state is easy-to-implement. We first prepare a quantum system to be in the ground state of a known H_0, and then adiabatically change the Hamiltonian for a time T by the following formula:

$$H(t) = (1 - \frac{t}{T})H_0 + \frac{t}{T}H_P \tag{1}$$

Next, if T is large enough, and H_0 and H_P do not interchange, the quantum system will remain in the ground state for all times by the adiabatic theorem of quantum mechanics. At time T, measuring the quantum state will return a solution of a considered problem.

The emerging quantum technologies support mostly 2-local interactions, thus the problem Hamiltonian H_P, containing only 2-local terms between the qubits, can be expressed by the formula:

$$H(\sigma) = \sum_{i=0}^{N-1} h_i \sigma_i + \sum_{i=0}^{N-2} \sum_{j=i+1}^{N-1} J_{ij} \sigma_i \sigma_j \tag{2}$$

The aim is to minimize the energy of a 2-local Ising Hamiltonian function, where $h_i \in \mathbb{R}$, $J_{ij} \in \mathbb{R}$ and $\sigma_i \in \{-1, +1\}$. This physics formula version is often called in short Ising. Various strategies together with useful techniques for mapping a wide variety of NP-hard problems to Ising formulations to benefit from adiabatic quantum optimization have already been demonstrated, e.g., in [8]. In fact, an alternative problem formulation when translated to an objective function, the 2-local condition on the problem Hamiltonian means that the objective function can also be expressed in the form used in Operations Research community:

$$Obj(x) = \sum_{i=0}^{N-1} a_i x_i + \sum_{i=0}^{N-2} \sum_{j=i+1}^{N-1} b_{ij} x_i x_j \tag{3}$$

Thus, the problem is to minimize quadratic pseudo-Boolean objective function which is known as the Quadratic Unconstrained Binary Optimisation (QUBO), where $a_i \in \mathbb{R}$, $b_{ij} \in \mathbb{R}$, and $x_i \in \{1, 0\}$. In a nutshell, to program a quantum annealer we have to provide an appropriate list of a_i and b_{ij} values. One should also note that the conversion between these two formula versions requires only a linear transformation, as $x_i = \frac{(\sigma_i + 1)}{2}$.

Several questions are still open in the field of quantum computing in practice. Technical challenges include the preparation of a quantum physical system and its operation at temperatures close to absolute zero isolated from the surrounding environment in order to behave quantum mechanically. There are also various limits on the quantum system controllability, noise and imperfections. However, in the context of this paper, we address the question what scheduling problems can be solved today using the existing quantum hardware.

We believe that one of the most important applications of quantum annealing is in the category of scheduling. In this paper, however, we have limited our experiments to solve only a certain class of scheduling problems. We wanted to validate experimentally a new quantum annealing heuristic applied for solving a well-known benchmark of the classical Job Shop Scheduling Problem (JSSP). In terms of computational complexity, the JSSP is NP-hard in the strong sense [2]. Consequently, in practice optimal solutions can not be found within a reasonable time. Thus, several polynomial-time heuristics have been developed for finding suboptimal solutions and tested experimentally [16]. Various heuristics have been proposed based on traditional models of computing to find the best-known solutions for JSSP benchmark instances. According to the recent comprehensive literature study in [10], half of them have been based on tabu search algorithms, followed by local search, shifting bottleneck, and branch and bound, and also simulated annealing techniques. The simulated annealing metaheuristic is a probabilistic technique for approximating the global optimum of a given function, successfully applied for solving various scheduling problems, including the JSSP [12] and the MRCPSP (Multi-mode Resource-Constrained Project Scheduling Problem) [13]. In principle, the classical simulated annealing metaheuristic has much in common with a physical process of heating a material and then slowly lowering the temperature to decrease defects, thus minimizing the system energy. There is also a mathematical analogy to the adiabatic theorem of quantum mechanics adopted in quantum annealing processes implemented in the existing quantum hardware.

The rest of this paper is organized as follows. In Sect. 2 we formulate the considered scheduling problem. The procedure of mapping the JSSP to the QUBO formula together with variable pruning techniques are presented in Sect. 3. Our heuristic approach is briefly explained in Sect. 4. The obtained results are presented in Sect. 5. We conclude our paper and present future work in Sect. 6.

2   Problem Formulation

The JSSP can be described by a set of jobs $J = \{\mathbf{j}_1, \dots, \mathbf{j}_N\}$ that must be scheduled on a set of machines $M = \{\mathbf{m}_1, \dots, \mathbf{m}_R\}$. Each job $\mathbf{j}_n$ consists of a sequence of L_n operations that have to be performed in a predefined order:

$$\mathbf{j}_n = \{O_{n1} \to O_{n2} \to \cdots \to O_{nL_n}\} \tag{4}$$

To simplify the notation, we enumerate the operations of all jobs in a lexicographical order, in such a way that:

$$\mathbf{j}_1 = \{O_1 \to O_2 \to \cdots \to O_{k_1}\}$$
$$\mathbf{j}_2 = \{O_{k_1+1} \to O_2 \to \cdots \to O_{k_2}\}$$
$$\cdots$$
$$\mathbf{j}_N = \{O_{k_{N-1}+1} \to O_2 \to \cdots \to O_{k_N}\} \tag{5}$$

The processing time of an operation $O_i, i = 1, 2, \ldots, k_N$, is p_i which is a positive integer. Moreover, each operation requires for its processing a particular machine. Let q_i be the index of the machine $\mathbf{m}_{q_i}$ on which operation O_i is to be executed. By I_m we will denote the set of indices of all of the operations that are to be executed on machine $\mathbf{m}_m$, i.e., $I_m = \{i : q_i = m\}$. There can only be one operation running on any given machine at any given point in time, and each operation of a job needs to be completed before the following one can start. The objective is to schedule all operations in a valid sequence in order to minimize the schedule length (or the makespan), which is the completion time of the last running job.

The JSSP is usually formulated as an optimization problem. However, it can be easily transformed into a decision problem, which is limited to decide whether there exists a feasible solution (schedule) with a makespan smaller than or equal to a given time. The JSSP can be easily mapped to a constraint satisfaction problem (CSP), and the existence of a solution to a CSP can be viewed as a decision problem. The decision version of the JSSP, together with its formulation suitable for a quantum annealing solver was presented in detail in [14]. The obtained results have encouraged us to perform further empirical investigations. We have followed the proposed time-indexed decision version of the JSSP to implement basic steps in our quantum annealing heuristic. However, we have added new conditions and extensions to be able to potentially apply it for practical use. Among many existing JSSP test instances we have decided to focus on the first benchmark set proposed in [15] to analyse various capabilities and properties supported by the D-Wave 2000Q quantum chip. In our preliminary studies, as a proof-of-concept, we have selected a small JSSP test instance denoted as *ft06* (6 jobs and 6 machines), but large enough to investigate several schedule variables and properties relevant in a quantum annealing process aiming at the next generation of D-Wave QPU [17].

Any JSSP instance can be represented as the disjunctive graph $G = (V, C \cup D)$, where V is the set of nodes, representing the operations of the jobs. Each node i has a weight which is equal to the processing time p_i of the corresponding operation O_i, and there are two special nodes, a source 0 and a sink $*$, whereby p_0 and p_* are equal to 0. C is the set of conjunctive arcs which reflect the job-order of all the operations, and the set of these arcs is denoted by D, see Fig. 1 for the selected *ft06* JSSP test instance. Bidirectional connections representing operations executed on the same machine are depicted as nodes with the same colour for better reading. The scheduling decision is to define ordering among all those operations which have to be processed on the same machine. In the disjunctive graph representation it is done by turning undirected arcs into directed ones. A selection S defines a feasible schedule if and only if every undirected arc has been fixed, and the resulting graph $G(S) = (V, C \cup S)$ is acyclic, where S is called a complete selection. For a complete selection S the makespan is equal to the length of the longest weighted path (i.e., critical path) from the source 0 to the sink $*$ in the acyclic graph $G(S) = (V, C \cup S)$.

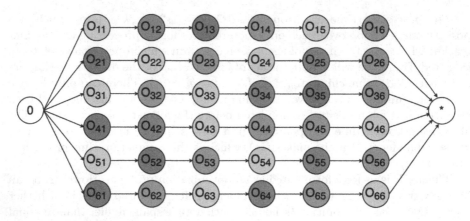

Fig. 1. The example *ft06* (6 jobs, and 6 machines) JSSP test instance represented as the disjunctive graph.

3 JSSP Mapping to QUBO

In this section we present how the considered JSSP can be mapped to QUBO, and then how all the QUBO variables are embedded into a specific quantum hardware, in our case - the D-Wave 2000Q quantum processing unit (QPU). From the QUBO problem formulation perspective, it is essential to note that the QPU is a hardware implementation of an undirected graph with qubits as vertices and couplers as edges among them. For instance, in the D-Wave 2000Q QPU there are 2048 qubits logically mapped into a 16 × 16 matrix of unit cells of eight qubits. Each qubit is connected to at most six other qubits. The D-Wave 2000Q has been designed to solve QUBO problems, where each qubit represents a variable, and couplers between qubits represent the cost associated with qubit pairs. The limited qubits connectivity in the D-Wave 2000Q QPU is a fundamental quantum hardware property which affected the quality of obtained results and in various experiments reduced scheduling problem instances solved by quantum annealing based heuristic methods. However, according to recent publicly available technical specifications, the limited qubits connectivity will be significantly improved in the upcoming D-Wave Pegasus QPU. Therefore, while designing our new heuristic quantum annealing heuristic we have taken this upcoming feature into account.

Following the example JSSP time-indexed QUBO formulation, we define $n_O *$ T binary variables for the JSSP, where n_O is a total number of operations and T is the upper bound for a given JSSP instance. During the initial QUBO mapping phase, we have to assign a set of binary variables for each operation and its various possible discrete starting times:

$$x_{it} = \begin{cases} 1 : \text{operation } O_i \text{ starts at time } t \\ 0 : \text{otherwise} \end{cases} \tag{6}$$

The upper bound T for the JSSP can be simply calculated as the sum of the execution times of all operations, where t is smaller than or equal to T.

3.1 Quadratic Constraints and the Objective Function

Let us introduce a set of penalty functions and corresponding constraints expressed as quadratic constraints, as proposed in [14]. To force the order of operations within a job, the following formula was proposed to count the number of precedence violations among consecutive operations:

$$h_1(x) = \sum_n (\sum_{\substack{k_{n-1} < i < k_n \\ t+p_i > t'}} x_{it} x_{i+1,t'}) \tag{7}$$

Then, there can be only one job running on each machine at any time:

$$h_2(x) = \sum_m (\sum_{(i,t,k,t') \in R_m} x_{it} x_{kt'}) \tag{8}$$

where $R_m = A_m \cup B_m$, and $A_m = \{(i,t,k,t') : (i,k) \in I_m \times I_m, i \neq k, 0 \leq t, t' \leq T, 0 < t' - t < p_i\}$, $B_m = \{(i,t,k,t') : (i,k) \in I_m \times I_m, i < k, t' = t, p_i > 0, p_j > 0\}$. To prevent operation O_j from starting at t' if there is another operation O_i started at time t and $t' - t < p_i$, the set A_m was defined. Then, the set B_m was defined so that two jobs can not start at the same time unless at least one of their execution time is zero.

The last penalty function expressed as the quadratic constraint was defined to force that an operation must start once and only once:

$$h_3(x) = \sum_i (\sum_t x_{it} - 1)^2 \tag{9}$$

The considered JSSP Hamiltonian as the objective function can be expressed simply as the sum of the quadratic constraints defined above:

$$H_T(x) = \eta h_1(x) + \alpha h_2(x) + \beta h_3(x) \tag{10}$$

The penalty constants η, α, and β must be larger than 0 and in practice setup experimentally to ensure that unfeasible solutions do not have a lower energy than ground states. One should also note that the index of H_T indicates the strong dependence of the Hamiltonian on the timespan T, which affects the number of variables, and is one of the critical challenges for running JSSP time-indexed QUBO formulations for reference JSSP benchmark instances on the D-Wave 2000Q QPU.

3.2   Variable Pruning

One of the challenging tasks in our experimental studies was to eliminate as many as possible variables during the variable pruning process. We disabled all of the variables $0 \leq x_{ij} < S$, where S is the sum of execution times of all the operations prior to the considered one in the same job. Then, we also disabled all of the variables $T - S \leq x_{ij} < T$, where S is the sum of execution times of all the operations after the considered one in the same job. Finally, in our public available source code, we introduced a set of parameters for selecting any point in schedule time and corresponding variable or removing a certain time-frame in a schedule.

4   Heuristic

In order to eliminate variables in the JSSP time-indexed QUBO formulation, and consequently be able to run bigger problem instances on the limited number of qubits available, we propose a new hybrid heuristic method which extends the basic variable pruning techniques. The main idea behind the heuristic is to define a processing window and move it in time till the end of a schedule, so only a limited number of operations is considered. In other words, we iterate the processing window in time, and check all the operations if they fit into one of three categories, where:

- s_i is the start time of operation i;
- w_{begin} is the start time of the processing window;
- w_{end} is the end time of the processing window;
- p_i is the execution time of operation i.

A schematic view of the processing window used in our heuristic is presented in Fig. 2. The example operations filled with the red colour are reaching out of the processing window, and therefore they are treated differently. Given that operations A and B belong to the same job, the operation A will not be scheduled when operation B occurs, even though operation B will be removed from the schedule.

A: Inside the processing window

$$\begin{cases} s_i \geq w_{begin} \\ s_i + p_i < w_{end} \end{cases} \tag{11}$$

We include those operations in a new dictionary of jobs, creating a smaller instance.

B: Reaching out of the processing window from the left side

$$\begin{cases} s_i < w_{begin} \\ s_i + p_i < w_{end} \end{cases} \tag{12}$$

Thanks to our modular heuristic implementation we are able to prune additionally variables corresponding to the machines when those operations occupy them and prevent the next operation in the same job from starting before the one ends.

C: Reaching out of the processing window from the right side

$$s_i + p_i \geq w_{end} \tag{13}$$

Respectively to **B**, we are able to disable a machine when those operations take place and prevent the previous operation in the same job from interfering with this one.

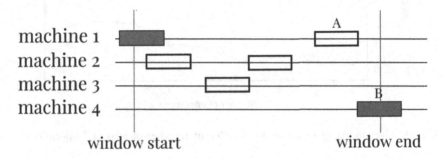

Fig. 2. Dividing a schedule into a set of processing windows for many machines and job operations. (Color figure online)

5   Results

5.1   Jobs vs. Operations

First, we wanted to check which parameters were especially crucial for the quality of generated solutions. In order to ensure that no other parameters interfered with the obtained results, we designed the second simple experiment. During the experiment, we compared schedules of one job consisting of many operations with many jobs consisting of only one operation. Each operation execution time was equal to 1, and it was performed on the same machine, see Fig. 3.

Based on the performed parameters tuning experiments, we discovered that the maximum processing window was around 14 time-units to be able to run the JSSP *ft06* instance on the D-Wave 2000Q QPU. Additionally, during the pre-processing step we defined three classes of job operation length, namely *short, medium, long*, to reduce a number of time-indexed variables and the processing window size down to 5 time-units. The example initial steps in our heuristic improving the scheduling solution by turning undirected arcs into directed ones within the given processing window are presented in Fig. 1.

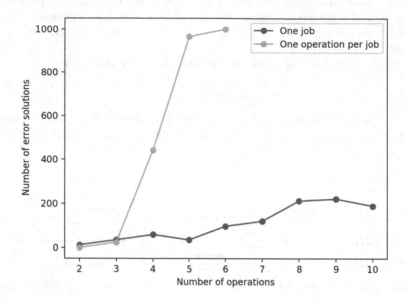

Fig. 3. The quality of solutions for a different number of jobs and operations.

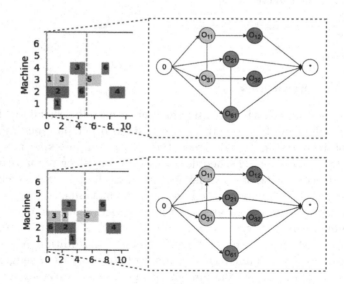

Fig. 4. The initial disjunctive graph representation of the feasible scheduling solution within a processing window with the undirected arcs O_{11}, O_{31} and O_{21}, O_{61} turned by the heuristic into directed arcs during the optimization process.

5.2   Embedding and Qubits Chain Strength

In Sect. 3 we represented the JSSP in the QUBO form as a theoretical graph of qubits and corresponding variables. However, the next challenge in our experimental studies was the embedding procedure of QUBO structures onto the quantum hardware with limited fully-connected graphs of qubits. The graph configuration implemented in the existing D-Wave hardware is called a Chimera structure. In fact, the tested D-Wave 2000Q QPU is a lattice of interconnected qubits. While some qubits connect to others via couplers, the D-Wave QPU is not fully connected, and qubits are arranged in sparsely connected groups of at most six other qubits. Currently, D-Wave provides a set of automated programming tools and APIs to find and perform an embedding automatically. Nevertheless, many quits must be chained together, so the chain is used as a single qubit. The chain strength value c must be set up carefully, as there is no methodology for choosing an optimal value [18]. To evaluate the embedding procedure we designed the following experiment: we randomly generated a set of schedules consisting of 4 jobs, each job consisted of 4 operations, and we assigned relatively short execution times to all operations, so the makespan was $T \leq 7$, and we solved the JSSP problem 100 times. Note that long schedules impose a large number of variables for the JSSP time-indexed QUBO formulation, and then the embedding procedure. Thus, we used different values of the chain strength c to discover what is optimal for our problem, see Fig. 5. Based on the obtained results, we were able to assign the strength value $c = 3$, but we observed a significant impact of its different values on the number of error solutions generated by the D-Wave quantum annealer. To quantify the quality of a generated solution, we simply counted the number of feasible schedules after the quantum optimization procedure.

Additionally, we have performed various experiments with our heuristic to make sure that all the relevant controlling parameters and configurations were used efficiently. In particular, we have selected the $minimum\_classical\_gap = 2.0$ value experimentally to provide enough energy to differentiate the ground state from other states in the QPU. Naturally, we can reduce this value and try to squeeze more problem information onto the QPU, but consequently it will also reduce the accuracy of results. In our case, the biggest constraint is a number of possible variables within a processing window. Therefore, by decreasing the $minimum\_classical\_gap = 1.0$ we had an opportunity to increase the processing window to 6 time-units, and further decreasing of the $minimum\_classical\_gap = 0.5$ value increased the processing window to 7 time-units. However, $minimum\_classical\_gap$ values below 2.0 gave more and more unacceptable solutions and had a negative impact onto the overall quality of obtained solutions.

The existing Chimera D-Wave QPU architecture and available APIs give developers a lot of flexibility to implement and improve strategies by increasing the gap between ground and excited states during the quantum annealing process. A set of interesting error suppression techniques using quantum annealing correction with auxiliary qubits and the energy gap were discussed for instance in [11]. Technically speaking, the D-Wave QPU minimizes the energy of an Ising spin configuration whose pairwise interactions lie on the edges of the Chimera graph. To automatically minor-embed our problem into a structured sampler we used the *EmbeddingComposite* method. However, there are other *dimod composites* with various parameters during pre- and post-processing phases worth considering in our future work. One should note, that the range of coupling strengths available in the D-Wave QPU is finite, so chaining is typically accomplished by setting the coupling strength to the largest allowed negative value and scaling down the input couplings relative to that. Thus, we have also checked various values of the another controlling *extended_j_range* parameter to increase the strength of minor embedding coupling. Typically, using the available larger negative values of J increases the dynamic J range. According to technical specifications of the existing D-Wave 2000Q QPU, strong negative couplings can bias a chain and therefore flux-bias offsets must be applied to recalibrate it to compensate for this effect. However, we have not noticed a significant impact on the quality of obtained solutions by changing the *extended_j_range* parameter. Nevertheless, we argue that additional tests and techniques will be required, in particular the spin-reversal transform can improve results by reducing the impact of possible analog and systematic errors. We plan to compare all the presented controlling parameters on the new Pegasus QPU architecture and explore more precisely the JSSP problem structure and solution space. We will apply new techniques for encoding discrete variables into Ising model qubits, e.g. [6], and try to take advantage of new and more connected Pegasus QPU graph for more efficient embedding of the considered problem.

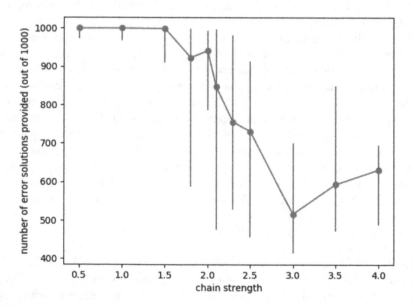

Fig. 5. The chain strength c bars based on the standard deviation and their impact on the JSSP solutions quality.

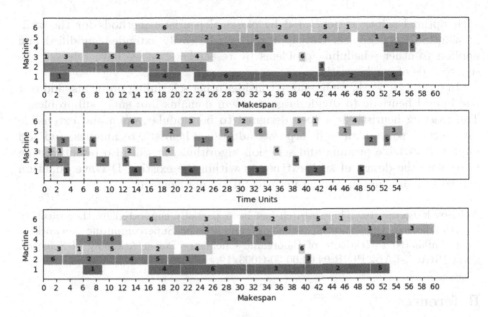

Fig. 6. The incremental optimization process of improving schedule quality within the given processing window generated by the hybrid quantum annealing heuristic for the selected JSSP *ft06* instance.

Finally, our hybrid heuristic method improved the quality of solutions by reducing the makespan of randomly generated schedules down to 60, see Fig. 6. Nevertheless, it is clear that even simple classical heuristics proposed for the JSSP outperform our quantum annealing-based approach reaching a set of many schedules with optimal makespan 55. We expect to use a much bigger processing windows in our heuristic tailored for the upcoming D-Wave Pegasus QPU architecture. The Pegasus graph will allow each qubit to couple to 15 other qubits instead of 6 qubits, so we expect to run the same heuristic successfully for bigger JSSP instances. Our proof-of-concept implementation, including the heuristic source code, has been published in the GitHub repository for reusing and external testing [19].

6   Conclusions

We presented a new quantum annealing heuristic for solving the Job Shop Scheduling Problem (JSSP) on publicly available D-Wave 2000Q QPU. Due to a limited number of available qubits and couplers among qubits implemented in a specific topology, we decomposed the JSSP into a set of smaller optimization problems as window processing slices. We estimated the number of feasible solutions during experimental studies on the well-known scheduling JSSP test instance *ft06*. We also tuned experimentally QUBO parameters proposed for the time-indexed decision version of the JSSP, and compared the obtained results with optimal solutions generated by classical heuristic methods for the well-known JSSP benchmark. Our approach can be easily extended, modified and applied to other scheduling problems by researchers, as we have released the source code of our heuristic.

In our future work, we intend to consider other scheduling problems, and test our hybrid heuristic to divide large problem domains into small subproblems. The existing heuristic has been designed to be modular, open and extensible source code, so we plan to incorporate additional heuristic techniques, and add improved variable pruning and selection algorithms. It will also be interesting to explore the design of such extensions within the existing D-Wave quantum annealers limits and upcoming Pegasus QPU improvements.

Acknowledgements. This research has been partially supported by the statutory funds of Poznan University of Technology and Poznan Supercomputing Networking Center affiliated to Institute of Bioorganic Chemistry Polish Academy of Sciences, grant PRACE-LAB2 POIR.04.02.00-00-C003/19.

References

1. Arute, F., et al.: Quantum supremacy using a programmable superconducting processor. Nature **574**, 505–510 (2019)
2. Błażewicz, J., Lenstra, J.K., Rinnooy, A.H.G.: Scheduling projects subject to resource constraints: classification and complexity. Discrete Appl. Math. **5**, 11–24 (1983)
3. Błażewicz, J., Ecker, K.H., Pesch, E., Schmidt, G., Sterna, M., Węglarz, J.: Handbook on Scheduling: From Theory to Practice. Springer, Cham (2019). https://doi.org/10.1007/978-3-319-99849-7
4. Kazuki, I., Yuma, N., Travis, H.S.: Application of quantum annealing to nurse scheduling problem. Sci. Rep. **9**, 12837 (2019). https://doi.org/10.1038/s41598-019-49172-3
5. Albash, T., Lidar, D.A.: Adiabatic quantum computation. Rev. Mod. Phys. **90**, 015002 (2018)
6. Chancellor, N.: Domain wall encoding of discrete variables for quantum annealing and QAOA. Quantum Sci. Technol. **4**, 4 (2019). https://doi.org/10.1088/2058-9565/ab33c2
7. Humble, T.S., et al.: An integrated programming and development environment for adiabatic quantum optimization. Comput. Sci. Discov. **7**, 015006 (2014)
8. Lucas, A.: Ising formulations of many NP problems. Front. Phys. **2**(5), 1–15 (2014)
9. Boixo, S., Albash, T., Spedalieri, F., et al.: Experimental signature of programmable quantum annealing. Nat. Commun. **4**, 2067 (2013). https://doi.org/10.1038/ncomms3067
10. van Hoorn, J.J.: The current state of bounds on benchmark instances of the job-shop scheduling problem. J. Sched. **21**(1), 127–128 (2017). https://doi.org/10.1007/s10951-017-0547-8
11. Pudenz, K.L., Albash, T., Lidar, D.A.: Quantum annealing correction for random Ising problems. Phys. Rev. A **91**(4), 042302 (2015)
12. van Laarhoven, P.J.M., Aarts, E.H.L., Lenstra, J.K.: Job shop scheduling by simulated annealing. Oper. Res. **40**, 113–125 (1992)

13. Mika, M., Waligóra, G., Węglarz, J.: Overview and state of the art. In: Schwindt, C., Zimmermann, J. (eds.) Handbook on Project Management and Scheduling. IHIS, vol. 1, pp. 445–490. Springer, Cham (2015). https://doi.org/10.1007/978-3-319-05443-8_21

14. Venturelli, D., Marchand, D.J.J., Rojo, G.: Job shop scheduling solver based on quantum annealing, Quantum Artificial Intelligence Laboratory, NASA Ames U.S.R.A. Research Institute for Advanced Computer Science, 1QB Information Technologies (2016). https://arxiv.org/pdf/1506.08479v2.pdf

15. Fisher, H., Thompson, G.L.: Probabilistic learning combinations of local job-shop scheduling rules. In: Muth, J.F., Thompson, G.L. (eds.) Industrial Scheduling, chapter 15, pp. 225–251. Prentice Hall, Englewood Cliffs (1963)

16. Brucker, P., Jurisch, B., Sievers, B.: A branch and bound algorithm for the job-shop scheduling problem. Discrete Appl. Math. **49**, 107–127 (1994)

17. JSSP benchmark instance ft06. http://jobshop.jjvh.nl/instance.php?instance_id=6

18. Coffrin, C.J., Challenges with Chains: Testing the Limits of a D-Wave Quantum Annealer for Discrete Optimization, United States: N.P. (2019). https://doi.org/10.2172/1498001

19. Quantum annealing hybrid heuristic for JSSP. https://github.com/mareksubocz/QuantumJSP

Foundations for Workflow Application Scheduling on D-Wave System

Dawid Tomasiewicz, Maciej Pawlik, Maciej Malawski,
and Katarzyna Rycerz[✉]

Institute of Computer Science, AGH, al. Mickiewicza 30, 30-059 Kraków, Poland
tomasiewicz.dawid@gmail.com, {mapawlik,malawski,kzajac}@agh.edu.pl

Abstract. Many scientific processes and applications can be represented in the standardized form of workflows. One of the key challenges
related to managing and executing workflows is scheduling. As an NP-
hard problem with exponential complexity it imposes limitations on the
size of practically solvable problems. In this paper, we present a solution to the challenge of scheduling workflow applications with the help
of the D-Wave quantum annealer. To the best of our knowledge, there
is no other work directly addressing workflow scheduling using quantum computing. Our solution includes transformation into a Quadratic
Unconstrained Binary Optimization (QUBO) problem and discussion of
experimental results, as well as possible applications of the solution. For
our experiments we choose four problem instances small enough to fit
into the annealer's architecture. For two of our instances the quantum
annealer finds the global optimum for scheduling. We thus show that it
is possible to solve such problems with the help of the D-Wave machine
and discuss the limitations of this approach.

Keywords: D-Wave · QUBO · Workflow scheduling · Serverless

1 Introduction

The paradigm of workflows is commonly used for describing and preserving complex scientific processes and applications [13]. Workflows are usually represented
as Directed Acyclic Graphs (DAG) [23]. Each vertex represents a task, while
edges designate dependencies or data transfers between tasks. By using such
a general representation it is possible to improve application portability and
reusability. The abstract graph representation enables decoupling the application from a specific infrastructure and easily extract parts of the process with
clear understanding of what the extracted part does and what its dependencies
and outputs are. Some examples of scientific applications implemented as workflows include: Montage [4] – image mosaic software used to construct human-
perceptible images of sky features from multiple images captured by telescopes;
software used by the LIGO collaboration, designed to process data related to
detecting gravitational waves [1]; software designed to predict the occurrence
and effects of earthquakes based on geological data [23].

© Springer Nature Switzerland AG 2020
V. V. Krzhizhanovskaya et al. (Eds.): ICCS 2020, LNCS 12142, pp. 516–530, 2020.
https://doi.org/10.1007/978-3-030-50433-5_40

One of the key challenges related to managing and executing scientific workflows is scheduling. The general goal of scheduling is to create a plan of execution with respect to given parameters such as deadline, budget and computing resources. For the scope of this paper we assumed a simple form of workflow scheduling, where we operate on the serverless infrastructure discussed in [3]. In this case the serverless infrastructure is implemented as a set of cloud functions, provided by major cloud service providers, such as Google, Amazon, Azure and IBM. Cloud functions have the potential to be used for compute-intensive tasks, as proposed in [27], with particular emphasis on challenges which require high levels of infrastructure elasticity. Scheduling workflows for serverless infrastructures can be summarized as defining the runtime parameters of a function based on user-supplied deadline and budget. In this paper we assume that the serverless infrastructure consists of an infinite number of machine instances, while the number of machine types is finite, with each machine type associated with specific cost and performance.

In recent years, quantum computing has become popular due to its potential ability to solve problems that are beyond the capabilities of classical computing infrastructures. The research is still in its early stage, however there are many theoretical quantum algorithms already available, such as famous polynomial time Shor factorization [26] or $O(\sqrt{N})$ complexity Grover search in unsorted databases [16]. A good overview of the current status of theoretical quantum algorithms can be found in [18]. There are also various attempts to implement algorithms on the available quantum hardware: IBM-Q[1], Rigetti computing[2] or D-Wave[3]. Scheduling problems are assumed to be one of the challenges which might be efficiently solved by this new approach.

In this paper, we present a solution for the workflow scheduling problem with the use of the D-Wave quantum annealer. In particular, we propose a method of reformulating the problem as Quadratic Unconstrained Binary Optimization (QUBO) [21] required by D-Wave. To achieve this, we developed a Binary Integer Linear Programming (BILP) [19] formulation of the problem, which is then translated to QUBO in a similar way as shown in [14]. Finally, we discuss results obtained on the annealer. We attempt to find the optimal solution for selected instances of workflow scheduling problems, which are constrained by the deadline and have the lowest cost possible.

This paper is organized as follows. In Sect. 3 we provide the overview of quantum computation with the use of the quantum annealer. In Sect. 4 we present the precise formulation of the problem. In Sect. 5 we provide a complete description of transforming the workflow scheduling problem into a QUBO problem. First, the transformation to BILP is presented, which includes all the constraints necessary to be translated. Next comes BILP to QUBO problem translation, using methods from [14]. Finally, Sects. 6 and 7 present full results with a detailed commentary, as well as discussion and suggestions for future work in this matter.

---

[1] httpɔɪ//quantum computing.ibm.com/.

[2] https://www.rigetti.com/.

[3] https://www.dwavesys.com/.

2   Related Work

There is a significant body of knowledge available concerning the topic of scheduling workflows on cloud infrastructures. Due to its widespread adoption, most scheduling solutions operate on Infrastructure as a Service (IaaS) services. To the best of our knowledge, there is no other work directly addressing the problem of workflow scheduling on serverless infrastructures with help of quantum computing. In the presented case, the serverless infrastructure is represented by a Function as a Service (FaaS) type of service. For example, in [2] Arabnejad et al. propose a heuristic list scheduling algorithm with low computational complexity, designed for running workflows on IaaS. Work presented in this paper aims to provide similar features, albeit for FaaS and with help from quantum computing. In particular, we propose a method to create the final execution plan upfront in a short time and at low cost. One of the factors included in the presented algorithm is the sole cost of executing workflow tasks in the cloud. This problem was discussed in more detail in [30]. Zhou et al. addressed the problem of performance offered by cloud services with specific focus on the use case of running workflow applications. The matter of performance offered by serverless infrastructures was studied in [24], with focus on potential parallelism and application of FaaS as an infrastructure for large-scale scientific workflows. The specific topic of scheduling workflows on serverless infrastructures was addressed by Kijak et al. in [20], where they proposed a heuristic algorithm for scheduling workflows on cloud functions.

Although solving scheduling problems with quantum computers is a novelty, some examples of such work are available. In particular, a promising approach is to use the D-Wave quantum annealer based on a chimera graph [5]. A possible method of solving a shop job scheduling problem (JSP) [15] using D-Wave is shown in [29], however the proposed three-dimensional structure of the problem description is a strong limitation for scalability. In general, many NP-hard problems can be formulated as Ising problems [22] which can then be solved using a quantum annealer. For example [8] discusses the problem of finding maximum cliques in a graph. An important factor for solving problems with the use of quantum annealers is small machine size. In [25] the authors discuss heuristics for solving large-scale problems described in [8].

3   D-Wave Computing Overview

D-Wave 2000Q [5] is an adiabatic quantum computer that, unlike its universal counterparts (e.g. IBM-Q or Rigetti), cannot run algorithms implemented with the use of general quantum circuits. Therefore, in spite of its architecture offering a relatively large number of qubits, its usage is limited to certain types of optimisation problems. It is a fully analog machine, the result of which depends on the value of the magnetic field applied to each qubit and the value of the connection between qubits (coupler). Programming the D-Wave 2000Q computer involves determining values of fields and associations. The aim of quantum annealing is to find the minimal energy state for the problem defined by a programmer. Annealing begins with an initial well-known quantum system for which the minimal energy state is known. Then it slowly switches to the quantum system related to the search problem. Ideally, during the whole process of switching, the system remains in its minimal energy state, so it should end up in the minimal energy state of the intended problem.

From a programmer's perspective, the problems to be solved must be formulated as an objective function using the Ising model [22] or, alternatively, a QUBO problem description [21]. Both approaches are isomorphic and can be transformed into each other in polynomial time. In this study, we choose to use QUBO formulation of the problem. In general, a QUBO problem consists of a matrix Q of size NxN, where N is the number of used binary variables and the size of the vector of binary variables that constitute the searched state. It can be converted without loss of generality to the upper triangular matrix as described in [14]. The actual problem solved by D-Wave is to find the final state of n binary variables $X = (x_1, x_2...x_n)$ that minimizes the objective function defined as

$$f(x) = \sum_i Q_{i,i}x_i + \sum_{i,j,i<j} Q_{i,j}x_ix_j. \tag{1}$$

Elements with the same indexes ($Q_{i,i}$) are responsible for "bias", i.e. the initial value of the magnetic field applied to the qubit, and elements with unequal indexes are responsible for the links between qubits (couplers). Minimizing such function is an NP-hard problem which makes it well suited for solving with help from the D-Wave 2000Q quantum computer.

The problem description does not limit the number of connections between qubits. However, the actual hardware is, in fact, a chimera graph (16×16 board of $K(4,4)$ graphs, degree 6); thus, the problem graph must be mapped onto that architecture, which is achieved by minor-embedding. It is a process of mapping each logical binary variable present in a problem description to a group (called a chain) of physical qubits on the actual machine so all required connections are realized. For dense matrices there is probably no better method than minor-embedding based on complete graph embedding described in [11]. For sparse matrices it is reasonable to attempt heuristics to lower the qubit chain length (e.g. as described in [6]).

4   Workflow Problem Formulation

In its general form, the workflow application can be represented as a Directed Acyclic Graph, where each task is represented by a vertex. An example graph is depicted in Fig. 1. We assume that the serverless infrastructure consists of a finite number of machine types, albeit with an infinite number of instances for each type. Each machine type has an associated cost. Each task has an associated running time for each machine type. Our model can be applied to any DAG.

The goal is to assign a machine type to each task with minimal total cost, while respecting the deadline. For simplicity, we will use the term *machine* instead of *machine type* from now on.

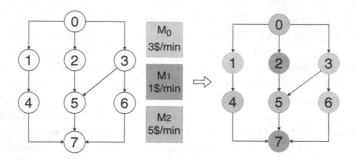

Fig. 1. Example workflow problem as a graph with task count $t = 8$. We search for an assignment of each task to one of the machine types. For example, with machine count $m = 3$ vector $K = [k_0, k_1, k_2]$ describes the cost of time unit execution on each machine while matrix $T = [\tau_{i,j}]_{8\times3}$ describes the execution time of task i running on machine j

The formal definition of the problem addressed in this paper consists of:

- a time matrix $T = [\tau_{i,j}]_{t\times m}$ where t is the number of tasks, m is the number of machines and $\tau_{i,j}$ expresses the execution time of task with number i on the machine with number j;
- a machine cost per time unit vector $K = [k_i]_m$ measured in currency units per time unit. This matrix can be used to calculate the cost of running the workflow;
- a deadline d (an integer, limiting this value is crucial for the size of the resulting QUBO problem);
- a list Θ of paths from the vertex representing the first task to the vertex representing the final task (both inclusive) in a DAG representing the problem.

For the purpose of describing results, we apply the following terminology: a **correct** result meets all the constraints, the **(global) optimum** is a correct result which has the lowest cost possible for the problem instance, while a **wrong** result fails to meet at least one constraint.

5   Transformation to QUBO Problem

The problem addressed in this paper is NP-hard [9]. We provide its translation to a QUBO problem which consists of two steps. First, we propose a transformation of the problem to BILP, which is described in this section. The second step shows how to translate BILP to a QUBO problem using [14].

The general goal of solving a BILP problem (also called "0–1 Integer Programming" [19]) is to find, for a given vector $C = [c_i]_n$, n binary numbers $X = [x_1, ..., x_n]$, for which the function

$$f(x_1, ..., x_n) = C^T X = \sum_{j=1}^{n} c_j x_j \tag{2}$$

is minimal, subject to constraints indicated by the following linear equation:

$$Ax = b, \tag{3}$$

where $A = [a_{i,j}]_{n \times w}$ stores w constraints for n binary numbers.

Translation from Initial Formulation to BILP

In our problem, the BILP binary variables are defined as $x = [x_i]_n$ where $n = t \cdot m$ (see Sect. 4). Variable x_i is set to 1 if the task with number $(i \bmod t)$ runs on a machine with number $(i \operatorname{div} t)$ where div is an integer division. Table 1 contains an example mapping of x_i parameters to task and machine combinations.

Table 1. The binary variables $X = [x_0, x_1, x_2 ... x_{31}]$ for BILP formulation of the sample workflow problem. Task count is $t = 8$ and machine count is $m = 4$. Binary variable x_i is set to 1 when task with number $(i \bmod t)$ runs on a machine with number $(i \operatorname{div} t)$ and to 0 otherwise.

		Task							
		1	2	3	4	5	6	7	8
Machine	0	x_0	x_1	x_2	x_3	x_4	x_5	x_6	x_7
	1	x_8	x_9	x_{10}	x_{11}	x_{12}	x_{13}	x_{14}	x_{15}
	2	x_{16}	x_{17}	x_{18}	x_{19}	x_{20}	x_{21}	x_{22}	x_{23}
	3	x_{24}	x_{25}	x_{26}	x_{27}	x_{28}	x_{29}	x_{30}	x_{31}

The BILP minimization function can subsequently be defined as the total cost of running tasks on selected machines

$$f(X) = \sum_{i=1}^{n} c_i x_i , \tag{4}$$

where $C = [c_i]_n$ is the cost vector and c_i indicates the cost of running task with number $(i \bmod t)$ on the machine with number $(i \operatorname{div} t)$.

The cost vector needed for BILP formulation of the problem is obtained from the problem definition (Sect. 4) as follows: first, the cost matrix $[\gamma_{i,j}]_{t \times m}$ is created by multiplying $\gamma_{i,j} = \tau_{i,j} \cdot k_j$. Next, the cost vector C is obtained by row-by-row vectorization $[\tau_{i,j}]$ using $c_{i+t \cdot j} = \tau_{i,j}$. Finally, the time vector $T = [t_i]_n$ is obtained in an analogous manner by calculating $t_{i+t \cdot j} = \gamma_{i,j}$.

In addition to the minimization function, there are also two types of constraints, the deadline and machine usage, where it is necessary to ensure that only a single machine will be selected for a given task.

Deadline Constraints. We define a matrix $R = [r_{i,j}]_{n \times r}$, where r is the number of all possible paths in the workflow DAG. R stores the time for each machine-based task assignment $i \in [0..n-1]$, but only if the task belongs to the path $\theta \in [0, r-1]$ as follows:

$$r_{i,\theta} = \begin{cases} t_i & \text{if path } \theta \text{ contains task number } i \bmod t \\ 0 & \text{otherwise.} \end{cases} \tag{5}$$

Next, we formulate a set of constraints corresponding to each path θ

$$\sum_{i=0}^{n-1} r_{i,\theta} x_i \le d \,, \tag{6}$$

which we can rewrite as (for $D = [d]_n$)

$$Rx \le D. \tag{7}$$

To describe constraints as part of BILP according to (3) we need to transform an inequality (7) into an equality. To achieve this, we rely on the fact that there always exists a slack variable s such that for natural numbers

$$a \in \mathbb{N}, Z \in \mathbb{N}, a \le Z \implies \exists s \in N : a + s = Z.$$

Therefore it is necessary to extend R with additional columns that represent all necessary slack variables [14]. Generally a slack value s is a natural number, so we represent it using standard binary expansion. If we indicate

$$\sum_{i;task(i) \in \theta} \min_j \tau_{i,j} \tag{8}$$

as the minimum time of running tasks in path θ, then the number of binary variables b for storing slack variables for each path θ equals:

$$b(p) = log_2(|d - \sum_{i;task(i) \in \theta} \min_j \tau_{i,j}|). \tag{9}$$

Machine Usage Constraints. The next category of constraints comprises "one machine per task only" constraints. We define $U = [u_{i,j}]_{n \times t}$ such that for each task $s \in [0, t-1]$ and its machine-based position $i \in [0, n-1]$

$$u_{i,s} = \begin{cases} 1 & \text{if } i \bmod n = s \\ 0 & \text{otherwise.} \end{cases} \tag{10}$$

Therefore, each row of U indicates all machine-based positions of a task corresponding to that row. Then, to assure that only one machine is assigned to each task, the following constraint must be fulfilled:

$$\sum_{i=0}^{n-1} u_{i,t} x_i = 1. \tag{11}$$

If we indicate s as the total number of slack variables, the final BILP constraints matrix (3) is defined as $A = [a_{i,j}]_{(n+s) \times (r+t)}$ and combines all constraints together as follows:

$$a_{i,j} = \begin{cases} r_{i,j} & 0 \le j < r \\ u_{i,j-r} & r \le j < r + t. \end{cases} \tag{12}$$

To perform this operation, we need matrices of compatible sizes. Therefore, the matrix U must also be extended with slack variables, which are set to 0.

Vector b from (3) is defined in a similar manner:

$$b_i = \begin{cases} d & 0 \leq j < r \\ 1 & r \leq j < r + t. \end{cases} \tag{13}$$

Translation from BILP to QUBO Problem

Given matrices A, C and vector b a QUBO matrix can be calculated using the following formula (from [14]):

$$y = x^T C x + P \cdot (Ax - b)^T (Ax - b) = x^T C x + x^T D x + c = x^T Q x + c. \tag{14}$$

By dropping the additive constant c, the exact QUBO problem form, which is minimizing $x^T Q x$, can be formulated. However it is necessary to introduce two additional scalar parameters:

- P - relative strength of all constraints in relation to the objective function, see (14)
- S - weight required for balancing R and U values so that the constraints they represent are efficiently included in the QUBO problem. Namely, it replaces constraints defined by (11) with

$$\sum_{i=0}^{n-1} S u_{i,t} x_i = S. \tag{15}$$

Both mentioned parameters, along with the minor embedding chain strength, need to be balanced, to make sure that (1) the solution meets constraints and (2) the objective function is minimized. Finding proper values for these parameters is not a trivial task. It is important to mention that the resulting QUBO problem might have high resolution, which can result in errors due to the limited resolution of the annealer's Digital to Analog Converter (DAC), so the hardware conversion from QUBO to analog values may not be accurate enough.

6 Results

In this section we describe the experimental results obtained using D-Wave 2000Q, compared with selected classical reference methods. In order to match the limitations of the existing quantum annealer, we considered four sample instances of the problem. Their DAG representation is shown in Fig. 2 and parameters in Table 2.

Table 2. The tested instances of the workflow problem (for definitions of parameters, see Sect. 4)

No.	Binary variable count	T	K	Paths
1	8	$\begin{bmatrix} 6 & 3 & 12 & 9 \\ 2 & 1 & 4 & 3 \end{bmatrix}$	[1,4]	[[0,1,3],[0,2,3]]
2	10	$\begin{bmatrix} 6 & 3 & 12 & 9 & 6 \\ 2 & 1 & 4 & 3 & 2 \end{bmatrix}$	[1,4]	[[0,1,4],[0,2,4],[0,3,4]]
3	15	$\begin{bmatrix} 6 & 3 & 12 & 9 & 6 \\ 2 & 1 & 4 & 3 & 2 \\ 4 & 2 & 8 & 6 & 4 \end{bmatrix}$	[1,5,2]	[[0,1,4],[0,2,4],[0,3,4]]
4	18	$\begin{bmatrix} 12 & 6 & 42 & 18 & 30 & 24 \\ 4 & 2 & 14 & 6 & 10 & 8 \\ 8 & 4 & 28 & 12 & 20 & 16 \end{bmatrix}$	K = [8,18,6]	[[0,1,3,5],[0,1,4,5],[0,2,4,5]]

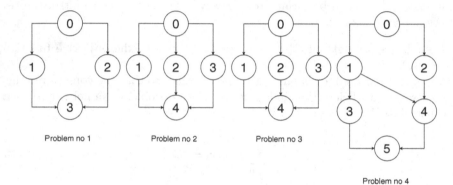

Fig. 2. Graph representations of the tested workflow problems.

Finding Parameters: P, S and Chain Strength. Parameters needed for QUBO problem formulation are dependent on the specific problem instance. For the purpose of this research, parameters were obtained using a metaheuristic. First, (1), we find an initial value of P basing on [14]. The potential initial value of parameter S should be similar to values in vector T in order to achieve proper balance between constraints. Then, (2), we search the discrete space of possible pairs (P, S) for values relatively close to the initial values, using a classical solver – Gurobi[4]. Next, (3), we set the chain strength between physical qubits for the minor-embedded problem basing on the D-Wave guidebook's [11] suggestion that it should be large enough to keep chains and small enough not to cover the actual problem. We have found that it should approximate the largest values of the Q matrix. Finally, (4), the selected chain strength is used as an input for the dwave.embedding library, which performs actual minor embedding. The final parameter values are presented in Table 3. Deadline values for all four

---

[4] http://www.gurobi.com.

problems were selected in such a way that they yield similar percentages of correct solutions.

Table 3. Values of parameters P, S and chain strength found for each size of the workflow problem for the given deadline (which is predefined, but has an impact on the parameters' values). The rightmost column represents the percentage of correct solutions in relation to all possible solutions.

No.	Binary variable number	Deadline	P	S	Chain strength	Percentage of correct solutions
1	8	19	8	10	1200	62,5%
2	10	19	14	25	6650	56,25%
3	15	17	11	10	2800	54,3%
4	18	70	6	40	18000	46,91%

Reference Methods. The problems discussed in this paper are small enough to be solved using classical methods. The following four classical methods have been used to verify D-Wave machine results:

- A brute-force method using initial problem formulation. It was used to calculate exact results, along with minimal energy, through direct use of the objective function (1).
- GNU Linear Programming Kit[5] library for solving BILP problem prior to its translation to QUBO. This method always finds the global optima.
- Gurobi sampler for the QUBO problem without minor-embedding. This method always finds the global optima, provided that parameters P and S are set up properly.
- Gurobi sampler for QUBO problem with minor-embedding, requiring three parameters (P, S, chain strength). The summary of the results is presented in Table 4.

Experiments with D-Wave. We perform experiments using the D-Wave 2000Q 5.0 machine sampled 2000 times with annealing time set to $8\,\mu s$. Figure 3 shows the energy distribution of the actual results. For Problem 1 and Problem 2, the minimal energy corresponds to the global optimum, while for Problem 3 it indicates a correct solution, but not the best possible one. All energies found for the most complex Problem 4 correspond to wrong solutions. It can be noticed that the count of wrong solutions grows along with the size of the solution space. In general, the obtained characteristics of energies remain similar to [17].

---

[5] https://www.gnu.org/software/glpk/.

Table 4. Gurobi solutions table. The rightmost column shows whether the global optimum was found, with the absolute energy error between the lowest Gurobi solution and the brute force global optimum.

No.	Binary variables number	Global optimum found
1	8	YES
2	10	YES
3	15	YES
4	18	NO, wrong (22831)

Details of the results are described in Table 5, where the number of correct solutions found by D-Wave and their relation to the total number of correct solutions is presented. The results are compared to the global optima obtained by the brute force method. It can be noted that D-Wave results are comparable to Gurobi results. In the first two problems, workflow optimization was solved exactly and the global optimum was found. In problem 3, the Gurobi solution was still optimal, while the D-Wave solution was correct, but not the best (12% worse in terms of cost; 33th out of 132 correct solutions). For problem 4 neither sampler found the global optimum.

Table 5. Summary of D-Wave 2000Q results. The fourth column presents the number of D-Wave unique correct solutions in relation to the total number of brute force correct solutions. The fifth column indicates whether the global optimum was found, listing the absolute energy error between the lowest D-Wave solution and the brute-force global optimum. The two rightmost columns refer to the execution time and cost of the lowest D-Wave solution respectively (*min* means global optimum cost).

No.	Binary variables number	Correct solutions samples (from 2000 samples)	Unique correct solutions	Global optimum found	Time	Cost
1	8	98	10 (100%)	YES	19 (d = 19)	34 (min = 34)
2	10	27	14 (77.8%)	YES	16 (d = 19)	40 (min = 40)
3	15	4	4 (3.03%)	NO (6)	16 (d = 17)	45 (min = 40)
4	18	0	0 (0%)	NO (65862)	N/A	N/A

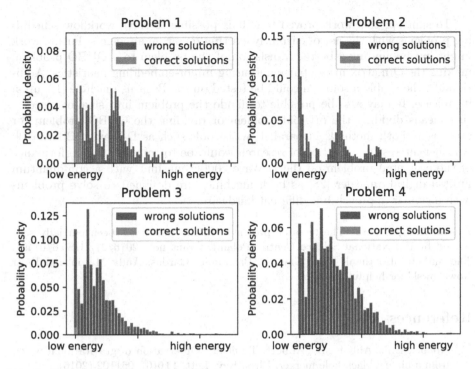

Fig. 3. Histograms of results for the four types of workflow problems tested. The x axis represent values of energies equal to the value of minimized objective function (1). The y axis represents the probability density. For Problems 1–3 the correct solutions found are shown on the left-hand side of the spectrum.

7   Conclusions and Future Work

In this paper we showed that it is possible to translate the workflow scheduling problem into a QUBO problem, execute it on a quantum annealer and achieve not only correct, but also globally optimal results for some of the analyzed problem instances. However, the presented method of adapting the scheduling challenge for D-Wave significantly increases the size of the problem. For example, the 18-binary variable problem required each of 39 QUBO problem variables to be represented by 11 physical qubits; thus the initial problem with a solution space size of $= 3^6 = 729$ ($m^t, m = 3, t = 6$, see Sect. 4) was converted into a problem with size 2^{429}–10^{143} (QUBO problem with 39 variables, 11 qubits each, $39 \cdot 11 = 429$). For such a large QUBO it is difficult for the quantum annealer to find the lowest energy solution. Therefore, in the future we will focus on solutions such as domain wall encoding [7] to reduce the number of binary variables.

However, it is worth noting that for larger problems the brute force method would no longer be usable because of its exponential complexity. This leaves a space for experimenting with a quantum annealer for larger instances, as the annealing process is very fast.

To sum up, this work proved that it is possible to solve workflow scheduling problems with the use of currently existing quantum annealers. Future work might involve finding a better translation of the problem to a QUBO problem, making the Q matrix more sparse and using minor-embedding heuristics. Additionally, the problem solution could be tested on the Pegasus machine [12] upon its release. It may also be possible to divide the problem into smaller pieces – this means dividing the original problem or dividing the QUBO problem (or even using both methods in parallel) with tools such as D-Wave QBSolv [10]. Another interesting direction of research would be to compare the performance of the presented problem on the D-Wave 2000Q machine with a non-quantum Fujitsu digital annealer [28] as both machines are designed to solve problems with a similar approach but different hardware.

Acknowledgements. The research presented in this paper has been partially supported by the National Science Centre, Poland, Grant no. 2016/21/B/ST6/01497. The authors also thank Piotr Gawron, Bartłomiej Gardas, Andy Mason and Piotr Nowakowski for helpful remarks.

References

1. Abbott, B.P., Abbott, R., Abbott, T., et al.: Observation of gravitational waves from a binary black hole merger. Phys. Rev. Lett. **116**(6), 061102 (2016)
2. Arabnejad, H., Barbosa, J.G., Prodan, R.: Low-time complexity budget-deadline constrained workflow scheduling on heterogeneous resources. Future Gener. Comput. Syst. **55**, 29–40 (2016)
3. Baldini, I., et al.: Serverless computing: current trends and open problems. In: Chaudhary, S., Somani, G., Buyya, R. (eds.) Research Advances in Cloud Computing, pp. 1–20. Springer, Singapore (2017). https://doi.org/10.1007/978-981-10-5026-8_1
4. Berriman, G., Good, J., Laity, A., et al.: Montage: a grid enabled image mosaic service for the national virtual observatory. In: Astronomical Data Analysis Software and Systems (ADASS) XIII. vol. 314, p. 593 (2004)
5. Bian, Z., Chudak, F., Macready, W.G., Rose, G.: The Ising model: teaching an old problem new tricks. D-Wave Syst. **2** (2010). https://www.dwavesys.com/resources/publications?type=internal#publication-230
6. Cai, J., Macready, W.G., Roy, A.: A practical heuristic for finding graph minors. arXiv preprint arXiv:1406.2741 (2014)
7. Chancellor, N.: Domain wall encoding of discrete variables for quantum annealing and QAOA. Quantum Sci. Technol. **4**(4), 045004 (2019). https://iopscience.iop.org/article/10.1088/2058-9565/ab33c2
8. Chapuis, G., Djidjev, H., Hahn, G., Rizk, G.: Finding maximum cliques on the D-Wave quantum annealer. J. Signal Process. Syst. **91**(3), 363–377 (2019)
9. Coffman, E.G., Bruno, J.L.: Computer and Job-Shop Scheduling Theory. Wiley, New York (1976)
10. D-Wave Systems: D-Wave Initiates Open Quantum Software Environment (2017). https://www.dwavesys.com/press-releases/d-wave-initiates-open-quantum-software-environment

11. D-Wave Systems Inc.: D'wave problem solving handbook. https://docs.dwavesys. com/docs/latest/_downloads/09-1171A-A_Developer_Guide_Problem_Solving_ Handbook.pdf

12. Dattani, N., Szalay, S., Chancellor, N.: Pegasus: the second connectivity graph for large-scale quantum annealing hardware. arXiv preprint arXiv:1901.07636 (2019)

13. Deelman, E., Gannon, D., Shields, M., Taylor, I.: Workflows and e-science: an overview of workflow system features and capabilities. Future Gener. Comput. Syst. **25**(5), 528–540 (2009)

14. Glover, F., Kochenberger, G., Du, Y.: A Tutorial on Formulating and Using QUBO Models. arXiv preprint arXiv:1811.11538 (2018)

15. Graham, R.L.: Bounds for certain multiprocessing anomalies. ell Syst. Tech. J. **45**(9), 1563–1581 (1966). https://doi.org/10.1002/j.1538-7305.1966.tb01709.x

16. Grover, L.K.: A fast quantum mechanical algorithm for database search. In: Proceedings of the Twenty-Eighth Annual ACM Symposium on Theory of Computing, STOC 1996, pp. 212–219 (1996)

17. Jałowiecki, K., Więckowski, A., Gawron, P., Gardas, B.: Parallel in time dynamics with quantum annealers. arXiv preprint arXiv:1909.0429

18. Jordan, S.: Quantum algorithms zoo web page. https://quantumalgorithmzoo.org/

19. Karp, R.M.: Reducibility Among Combinatorial Problems, pp. 85–103. Springer, Boston (1972). https://doi.org/10.1007/978-1-4684-2001-2_9

20. Kijak, J., Martyna, P., Pawlik, M., Balis, B., Malawski, M.: Challenges for scheduling scientific workflows on cloud functions. In: 11th IEEE International Conference on Cloud Computing, CLOUD 2018, San Francisco, CA, USA, 2–7 July 2018, pp. 460–467. IEEE Computer Society (2018). https://doi.org/10.1109/CLOUD.2018. 00065

21. Lewis, M., Glover, F.: Quadratic unconstrained binary optimization problem preprocessing: theory and empirical analysis. Networks **70**(2), 79–97 (2017)

22. Lucas, A.: Ising formulations of many NP problems. Front. Phys. **2**, 5 (2014). https://doi.org/10.3389/fphy.2014.00005. https://www.frontiersin.org/article/10 3389/fphy.2014.00005

23. Maechling, P., Chalupsky, H., Dougherty, M., et al.: Simplifying construction of complex workflows for non-expert users of the southern california earthquake center community modeling environment. ACM SIGMOD Rec. **34**(3), 24–30 (2005)

24. Pawlik, M., Figiela, K., Malawski, M.: Performance considerations on execution of large scale workflow applications on cloud functions. arXiv preprint arXiv:1909.03555 (2019)

25. Pelofske, E., Hahn, G., Djidjev, H.: Solving large maximum clique problems on a quantum annealer. In: Feld, S., Linnhoff-Popien, C. (eds.) QTOP 2019. LNCS, vol. 11413, pp. 123–135. Springer, Cham (2019). https://doi.org/10.1007/978-3-030-14082-3_11

26. Shor, P.W.: Algorithms for quantum computation: discrete logarithms and factoring. In: Proceedings 35th Annual Symposium on Foundations of Computer Science, pp. 124–134. IEEE (1994)

27. Spillner, J., Mateos, C., Monge, D.A.: FaaSter, better, cheaper: the prospect of serverless scientific computing and HPC. In: Mocskos, E., Nesmachnow, S. (eds.) CARLA 2017. CCIS, vol. 796, pp. 154–168. Springer, Cham (2018). https://doi. org/10.1007/978-3-319-73353-1_11

28. Tsukamoto, S., Takatsu, M., Matsubara, S., Tamura, H.: An accelerator architecture for combinatorial optimization problems. Fujitsu Sci. Tech. J. **53**(5), 8–13 (2017)

29. Venturelli, D., Marchand, D.J.J., Rojo, G.: Quantum annealing implementation of job-shop scheduling. arXiv:1506.08479 (2015)
30. Zhou, A.C., He, B., Liu, C.: Monetary cost optimizations for hosting workflow-as-a-service in IaaS clouds. IEEE Trans. Cloud Comput. 4(1), 34–48 (2015)

A Hybrid Solution Method for the Multi-Service Location Set Covering Problem

Irina Chiscop[1(✉)], Jelle Nauta[1], Bert Veerman[1,2], and Frank Phillipson[1]

[1] The Netherlands Organisation for Applied Scientific Research,
The Hague, The Netherlands
`irina.chiscop@tno.nl`
[2] Vrije Universiteit Amsterdam, Amsterdam, The Netherlands

Abstract. The Multi-Service Location Set Covering Problem is an extension of the well-known Set Covering Problem. It arises in practical applications where a set of physical locations need to be equipped with services to satisfy demand within a certain area, while minimizing costs. In this paper we formulate the problem as a Quadratic Unconstrained Binary Optimization (QUBO) problem, apply the hybrid framework of the D-Wave quantum annealer to solve it, and investigate the feasibility of this approach. To improve the often suboptimal initial solutions found on the D-Wave system, we develop a hybrid quantum/classical optimization algorithm that starts from the seed solution and iteratively creates small subproblems that are more efficiently solved on the D-Wave but often still converge to feasible and improved solutions of the original problem. Finally we suggest some opportunities for increasing the accuracy and performance of our algorithm.

Keywords: D-Wave · MSLSCP · Optimization · Quantum annealing

1 Introduction

The past decade has seen the rapid development of the two paradigms of quantum computing, quantum annealing and gate-based quantum computing. In 2011 D-wave Systems announced the release of the world's first commercial quantum annealer[1] operating on a 128-qubit architecture, which has since been continually extended up to the 2048-qubit version, available from 2017[2]. These technological advances have led to a renewed interest in finding classical intractable problems suited for quantum computing.

---

[1] https://www.dwavesys.com/news/d-wave-systems-sells-its-first-quantum-computing-system-lockheed-martin-corporation.
[2] https://www.dwavesys.com/press-releases/d-wave%C2%A0announces%C2%A0d-wave-2000q-quantum-computer-and-first-system-order.

© Springer Nature Switzerland AG 2020
V. V. Krzhizhanovskaya et al. (Eds.): ICCS 2020, LNCS 12142, pp. 531–545, 2020.
https://doi.org/10.1007/978-3-030-50433-5_41

D-Wave's quantum processor is specifically designed to solve quadratic unconstrained binary optimization (QUBO) problems, and is therefore particularly suited for addressing NP-hard combinatorial optimization problems. Well-known examples that have been implemented on one of D-Wave's quantum processors include maximum clique [4], capacitated vehicle routing [6], minimum vertex cover [10], set cover with pairs [3], traffic flow optimization [9] and integer factorization [8]. These studies have shown that although the current generation of D-Wave annealers may not yet have sufficient scale, precision and connectivity to allow faster or higher quality solutions, they have the suitable infrastructure for modelling real-world instances of these problems, effectively decomposing these into smaller sub-problems and solving these on a real Quantum Processing Unit (QPU).

The paper at hand addresses another NP-hard combinatorial problem, the Multi-Service Location Set Covering Problem (MSLSCP) [11], arising in smart city planning. In smart cities, different services such as Wi-Fi, alarm and air quality or pollution sensors are integrated into street furniture like lamp posts and bus shelters, to create a dense network that can potentially achieve higher transmission rates and thus improve the quality of life. There are costs associated with both enabling a location to be equipped with services, and the actual equipping itself. The goal is then to distribute the services across the existing location in such a way that the demand for each service is satisfied at minimum total cost. As an extension of the Set Covering Problem, which is NP-hard [7], the MSLSCP is also NP-hard. Since for large instances this problem is computationally intractable, several heuristic solution methods have been proposed [11]. The best method found is based on distributing the services one-by-one over the available locations, and is therefore highly dependent on the order in which these services are considered. Given that large instances cannot be optimally solved classically, it is worthwhile to investigate how quantum annealing may be able to provide a better alternative to the current methods.

We present a novel algorithm to solve the MSLSCP, based on a two-phase hybrid approach. In the first phase, an initial solution is obtained from combining classical search heuristics and quantum annealing, whilst in the second phase, an improvement step is applied to reduce the size of the problem. This process is executed iteratively until the user-specified stopping criteria are met. As far as the authors know, this is the first time that the MSLSCP is modeled as a QUBO and solved by an algorithm employing quantum annealing.

The structure of this paper is as follows: Sect. 2 includes the mathematical integer linear programming and QUBO formulations of the problem. The hybrid functionality of the D-Wave system and the newly proposed approach are presented in Sect. 3. Section 4 explains how we generated the test problems to which we applied our method. The results obtained are discussed in Sect. 5. Finally, Sect. 6 provides some conclusions and ideas for further research.

2 Problem Formulation

In mathematical terms, the MSLSCP concerns the distribution of services $\mathcal{F} = \{1, \ldots, F\}$ over a set of locations $\mathcal{L} = \{1, \ldots, L\}$ such that the demand points $\mathcal{G}^u$ for each service $u \in \mathcal{F}$ are covered at minimal cost. Enabling a location $j \in \mathcal{L}$, incurs cost $f_j > 0$, whilst equipping a location $j \in \mathcal{L}$ with service $u \in \mathcal{F}$ incurs cost $c_j^u > 0$. An additional parameter a_{ij}^u is introduced to mark when a demand point falls within the range of a particular service placed at a particular location:

$$a_{ij}^u = \begin{cases} 1 & \text{if demand point } i \in \mathcal{G}^u \text{ is in range of location } j \in \mathcal{L} \text{ for service } u \in \mathcal{F}, \\ 0 & \text{otherwise.} \end{cases}$$

(1)

The following binary decision variables are defined:

$$y_j = \begin{cases} 1 & \text{if location } j \in \mathcal{L} \text{ is open,} \\ 0 & \text{otherwise.} \end{cases}$$

(2)

$$x_j^u = \begin{cases} 1 & \text{if location } j \in \mathcal{L} \text{ is equipped with service } u \in \mathcal{F}, \\ 0 & \text{otherwise.} \end{cases}$$

(3)

The complete integer linear programming formulation of the MSLSCP is then given by:

$$\min \sum_{j \in \mathcal{L}} f_j y_j + \sum_{j \in \mathcal{L}} \sum_{u \in \mathcal{F}} c_j^u x_j^u,$$

(4)

$$\text{s.t.} \sum_{j \in \mathcal{L}} a_{ij}^u x_j^u \geq 1 \quad \forall i \in \mathcal{G}^u, \forall u \in \mathcal{F},$$

(5)

$$x_j^u \leq y_j \quad \forall j \in \mathcal{L}, \forall u \in \mathcal{F},$$

(6)

$$x_j^u \in \{0, 1\} \quad \forall j \in \mathcal{L}, \forall u \in \mathcal{F},$$

(7)

$$y_j \in \{0, 1\} \quad \forall j \in \mathcal{L}.$$

(8)

The objective given by Eq. (4) represents the total sum of costs associated with enabling locations and equipping them with services. Equation (5) is a constraint enforcing all demand points to be satisfied. Constraint Eq. (6) expresses that a location must be enabled if it is equipped with any services. Finally, Eq. (7) and Eq. (8) specify that the opening and equipping variables should (of course) be binary.

To solve this problem on the D-Wave, we need to express the constraints (formulated above as inequalities) as equalities. Based on the formulation in Eq. (4)–(8) the MSLSCP can also be re-written as the following QUBO:

$$\min \quad A \cdot H_A + B \cdot H_B + C \cdot H_C,$$

(9)

where

$$H_A = \sum_{j \in \mathcal{L}} f_j y_j + \sum_{j \in \mathcal{L}} \sum_{u \in \mathcal{F}} c_j^u x_j^u, \tag{10}$$

$$H_B = \sum_{u \in \mathcal{F}} \sum_{i \in \mathcal{G}^u} \left(\sum_{j \in \mathcal{L}} a_{ij}^u x_j^u - \sum_{k=0}^{k_{i,u}^{\max}} 2^k \xi_{i,k}^u - 1 \right)^2, \tag{11}$$

$$H_C = \sum_{j \in \mathcal{L}} \sum_{u \in \mathcal{F}} (x_j^u - x_j^u y_j). \tag{12}$$

H_A is the cost we aim to minimize described by Eq. (4).

H_B corresponds to the inequality constraint Eq. (5). Following the method described in [1], it uses slack variables $\xi_{i,k}^u$, effectively adding equations to capture each possible way of satisfying the inequality. The simplest way in which Eq. (5) can be satisfied is by having only one non-zero $a_{ij}^u x_j^u$ combination, in which case the sum with the $\xi_{i,k}^u$'s should be zero and the entire Eq. (5) is minimized to zero. If there are more non-zero $a_{ij}^u x_j^u$ combinations (i.e. the demand point is serviced multiple times), we can change the appropriate $\xi_{i,k}^u$'s to again minimize the total to zero. The number of additional variables introduced, for each of the $\sum_{u \in \mathcal{F}} |\mathcal{G}^u|$ constraints of type H_B, is given by:

$$k_{i,u}^{\max} = \lfloor \log_2 \left(\sum_{j \in \mathcal{L}} a_{ij}^u - 1 \right) \rfloor. \tag{13}$$

The reason we only need logarithmically many slack variables is that they act as bits in the binary representation of the number of ways in which we can satisfy the inequality Eq. (5).

H_C corresponds to constraint Eq. (6). This term is minimized if services are placed only at open locations.

The penalty coefficients A, B, C should be set such that the minimum of the objective in (9) satisfies the constraints (i.e., we cannot achieve a lower minimum by breaking constraints in favor of lowering the cost in Eq. (10)). In practice there is a delicate trade-off here. Setting lower B and C causes constraints to be violated for many sub-optimal solutions found, while setting higher B and C ensures that the constraints are satisfied but leads to lower accuracy in optimizing the actual cost of Eq. (10). Through some experimental tests we concluded that a suitable parameter setting is:

$$A = 1, \tag{14}$$

$$B = 2 \cdot (\max\{f_j : j \in \mathcal{F}\} + \max\{c_j^u j \in \mathcal{L}, u \in \mathcal{F}\}), \tag{15}$$

$$C = 2 \cdot (\max\{c_j^u : j \in \mathcal{L}, u \in \mathcal{F}\}). \tag{16}$$

3   Solution Approach

To solve the MSLSCP we propose a hybrid iterative approach, which combines a method to reduce the size of the problem together with D-Wave's built-in hybrid framework. First, the process of quantum annealing and D-Wave's hybrid capabilities are presented. Then, a detailed overview of the newly proposed algorithm is given.

3.1 Quantum Annealing on D-Wave

The devices produced by D-Wave Systems are practical implementations of quantum computation by adiabatic evolution [5]. The evolution of a quantum state on D-Wave's QPU is described by a time-dependent Hamiltonian, composed of initial Hamiltonian H_0, whose ground state is easy to create, and final Hamiltonian H_1, whose ground state encodes the solution of the problem at hand:

$$H(t) = \left(1 - \frac{t}{T}\right)H_0 + \frac{t}{T}H_1. \tag{17}$$

The system in Eq. (17) is initialized in the ground state of the initial Hamiltonian, i.e. $H(0) = H_0$. The adiabatic theorem states that if the system evolves according to the Schrödinger equation, and the minimum spectral gap of $H(t)$ is not zero, then for time T large enough, $H(T)$ will converge to the ground state of H_1, which encodes the solution of the problem. This process is known as quantum annealing. Although here we are not concerned with the technical details, it is worthwhile to mention that it is not possible to estimate an annealing time T to ensure that the system always evolves to the desired state. Since there is no estimation of the annealing time, there is also no optimality guarantee.

The D-Wave quantum annealer can accept a problem formulated as an Ising Hamiltonian, corresponding to the term H_1 in Eq. (17), or rewritten as its binary equivalent, in QUBO formulation. Next, this formulation needs to be embedded on the hardware. In the most developed D-Wave 2000Q version of the system, the 2048 qubits are placed in a Chimera architecture: a 16×16 matrix of unit cells consisting of 8 qubits. This allows every qubit to be connected to at most 5 or 6 other qubits. With this limited hardware structure and connectivity, fully embedding a problem on the QPU can sometimes be difficult or simply not possible. In such cases, the D-Wave system employs built-in routines to decompose the problem into smaller sub-problems that are sent to the QPU, and in the end reconstructs the complete solution vector from all sub-sample solutions. The first decomposition algorithm introduced by D-Wave was *qbsolv* [2], which gave a first possibility to solve larger scale problems on the QPU. Although *qbsolv* is the main decomposition approach on the D-Wave system, it does not enable customizations, and therefore is not particularly suited for all kinds of problems.

To alleviate the short-comings of *qbsolv*, D-Wave introduced a hybrid framework which enables users to quickly design and test workflows that iterate over sets of samples through different samplers to solve arbitrarily-sized QUBOs. Large problems can be decomposed in different ways and two or more solution techniques can run in parallel[3]. The schematic representation is shown in Fig. 1. There are four branches shown in this example, more precisely a classical

---

[3] https://readthedocs.com/projects/d-wave-systems-dwave-hybrid/.

Interruptable Tabu search and three decomposition-based methods. A workflow branch has the following structure: decomposer - sampler - composer. We briefly explain the purpose of each these building blocks.

A decomposer is a component that splits up the original problem into sub-problems by selecting only a part of the variables. Classical decomposition approaches are: select first n variables that have the highest impact on your objective (energy-based selection) or select variables that show up in the same set of constraints together (constraint-based selection). Here, *qbsolv* is an example of an energy-based selection algorithm. The sampler is the chosen method to sample (solve) the subproblems coming from the decomposer. This can be simulated annealing or any other QPU-based sampler. Lastly, a composer makes a selection from current samples and updates the complete, final solution according to user-defined criteria.

This workflow allows custom design of the different building blocks and makes it possible to combine these into different methods that can be run in parallel as can be seen in Fig. 1. In our numerical experiments, we use *Kerberos*, the reference hybrid built-in sampler, which combines Tabu search, simulated annealing, and D-Wave sub-problem sampling on problem variables that have high-energy impact.

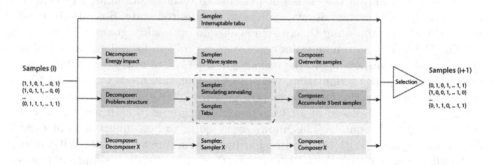

Fig. 1. Schematic representation of D-wave hybrid workflow (see footnote 3).

3.2 Two-Stage Hybrid Algorithm

Using the *Kerberos* sampler from the D-wave hybrid framework, we often cannot find even a feasible - let alone optimal - solution to our problem within a reasonable time. To counter this limitation we propose an hybrid approach in which we make use of the initially found solutions (which may be infeasible) to restrict the solution space and then re-solve the problem. This approach is described in Algorithm 1.

The idea of the algorithm is to construct a first solution by overlapping different samples obtained from Kerberos. In this way we increase the probability of creating a feasible solution. The desired number of solutions to be overlapped

```
1  K_max = number of runs
2  N_max = number of improvement iterations
3  instance = L, F, G^u   ∀u ∈ F
4  solutions = {}
5  for k = 1, ..., K_max do
6  |    solution_k ← KerberosSampler(instance);
7  |    solutions ← solutions ∪ solution_k
8  end
9  solution ← Overlap(solutions)
10 for N = 1, ..., N_max do
11 |   if  solution is feasible  then
12 |   |    L ← {j : y_j = 1}
13 |   |    instance ← L, F, G^u   ∀u ∈ F
14 |   |    solution ← KerberosSampler(instance);
15 |   else
16 |   |    for u ∈ F do
17 |   |    |   G̃_u = {i ∈ G_u : Σ_{j∈L} a^u_{ij} x^u_j = 0}
18 |   |    end
19 |   end
20 |   L ← {j : y_j = 1 or a^u_{ij} = 1   ∀u ∈ F,   ∀i ∈ G̃_u}
21 |   instance ← L, F, G^u   ∀u ∈ F
22 |   solution ← KerberosSampler(instance)
23 end
24 return solution
```

Algorithm 1: Hybrid algorithm pseudocode.

is given as input (step 1). The problem instance to be solved is specified by the sets of locations, services and demand points (step 3). Once the seed solution is created by combining the Kerberos-solutions (steps 5–9), the algorithm moves to the iterated improvement stage. In each iteration, there are two possibilities:

1. If the last solution found was feasible, the problem space is restricted by removing the set of unopened locations from the original input (steps 11–15). This step is motivated by the fact that in a feasible solution, some locations may be opened unnecessarily and the services may be equipped more efficiently on the opened locations.
2. If the last solution found was infeasible, the location set is again reduced (steps 16–18). The infeasible solutions observed were all caused by the violation of demand point coverage constraints. As such, the locations which remain unused in the solution and which cannot cover any of the unsatisfied demand points, can be removed from the location set (step 19).

The newly produced instance can then be solved with Kerberos again and the improvement step process can be repeated (steps 20–21). While these simple strategies of reducing the search space may not always result in optimal solutions, they provide a means to find a reasonably good feasible solution.

4   Problem Generation

To apply our algorithm on a test set, we generate problem instances with the numbers of services, locations and demand points as input parameters. The generation is done in two steps as explained below. For more details on some of the steps, see the code on Github[4].

First we generate the set of coordinates in the unit square at which demand points and locations are situated. This is done in such a way that they are not too close to each other, each demand point is reachable from at least one location, and each location can service at least one demand point.

We then assign each demand point a single requested service, in such a way that each service is requested at least once (see Algorithm 2).

---

1 $\mathcal{F}$ = the set of F services
2 $\mathcal{P}_L$ = Location points
3 $\mathcal{P}_U$ = Demand points
4 triplets $\mathcal{T} = \{\}$
5 f = random service in $\mathcal{F}$
6 **for** $u \in \mathcal{P}_U$ **do**
7 $\qquad L_U = \{l \in \mathcal{P}_L : \text{distance}(u, l) \le d_{max}\}$
8 $\qquad$ **for** $l \in L_U$ **do**
9 $\qquad\qquad \mathcal{T} \leftarrow \mathcal{T} \cup (f, l, u)$
10 $\qquad\qquad$ **if** $\exists (f, l, u) \in \mathcal{T}, l \in \mathcal{P}_L, u \in \mathcal{P}_U \forall f \in \mathcal{F}$ **then**
11 $\qquad\qquad\qquad f \leftarrow$ random service in $\mathcal{F}$
12 $\qquad\qquad$ **else**
13 $\qquad\qquad\qquad f \leftarrow$ some service in $\mathcal{F} : (f, l, u) \notin \mathcal{T} \ \forall l \in \mathcal{P}_L, \forall u \in \mathcal{P}_U$
14 $\qquad\qquad$ **end**
15 $\qquad$ **end**
16 **end**

Algorithm 2: Pseudocode to assign service requirements to demand points by generating triplets $(f, l, u) \in \mathcal{F} \times P_L \times P_U$.

In this way we generate a set of problems of various sizes denoted FLU, where F is the number of services, L is the number of locations and U is the number of demand points. The generated problem and the size of their QUBO formulation, given by the amount of binary variables in the solution vector, is shown in Table 1. Given that on the current 2048 qubit architecture one can embed a fully-connected graph of about 60 nodes, it is clear that most of the problems listed in Table 1 cannot be directly mapped to the QPU and will have to be decomposed.

---

[4] https://github.com/jcnauta/DWave-MSLSCP.

Table 1. Overview of generated problems.

Problem parameters	QUBO size	Problem parameters	QUBO size
F2L50U50	321	F4L50U200	942
F4L50U50	416	F2L100U200	1092
F2L50U100	483	F4L100U200	1274
F2L100U50	498	F2L200U200	1462
F4L50U100	585	F2L50U400	1554
F4L100U50	685	F4L50U400	1626
F2L100U100	691	F2L100U400	1874
F2L50U200	825	F4L100U400	2044
F4L100U100	888	F2L200U400	2340

5 Results

In this section we present the results of our hybrid method described by Algorithm 1. We look first at the quality of the solution obtained in the first sampling phase, and then assess the performance of the iterative improvement step in the second phase.

5.1 First Phase Results

Table 2 shows the results obtained in the first phase of Algorithm 1, in which all problem instances are sampled using the D-Wave's built-in hybrid framework. The Kerberos sampler was run 50 times for each problem, with $time\_out = 60\,s$ and $max\_iter = 10$ as stopping criteria. The rest of the parameters were set to the default values. We assessed the quality of solutions in terms of the deviation from the optimum solution, which was classically obtained from each problem using the CPLEX commercial solver, version 12.8. A striking observation to emerge from these results is that with the exception of the smallest problem, F2L50U50, the average deviation of the 50 runs from optimum is quite large ranging from 20% to over 200%. Moreover, for most problems, the amount of feasible solutions out of the 50 runs was quite small, whilst in a few cases no feasible solution could be found at all. Interestingly, these correspond to problems in which the number of demand points greatly exceeds the number of locations to be equipped with services. A possible explanation is that there are relatively fewer ways to feasibly allocate the services when the number of locations is lower.

Table 2. Phase 1 results for each of the test instances. The entries marked with "-" in the second column indicate that no feasible solution was found for that particular problem.

Problem	Deviation (%) of best feasible solution found of 50 runs	Avg. deviation (%) of 50 runs from optimum	Number of feasible solutions found of 50 runs
F2L50U50	1.87	16.3	41
F4L50U50	54.36	73.8	48
F2L50U100	20.23	30.01	3
F2L100U50	33.36	46.28	50
F4L50U100	123.94	83.98	1
F4L100U50	67.67	97.02	43
F2L100U100	41.13	53.34	8
F2L50U200	-	51.1	0
F4L100U100	87.66	99.84	17
F4L50U200	-	116.21	0
F2L200U100	106.55	66.34	1
F2L100U200	-	30.15	0
F4L200U50	111.61	97.78	1
F4L100U200	-	77.31	0
F4L50U400	-	116.67	0
F2L100U400	-	107.48	0
F4L100U400	-	209.45	0
F2L200U400	256.59	228.83	1

Figure 2 depicts the deviation from the optimal solution of each problem. We see that as the problem size increases, the variance of the results also increases. Once again, there seem to be higher fluctuations in the solutions of problems with fewer locations than demand points. The results obtained so far indicate that the quality of the solutions obtained with Kerberos is rather low. This might be improved by modifying the utilized stopping criteria, but our iterated improvement method (phase 2) addresses this problem automatically and quite reliably.

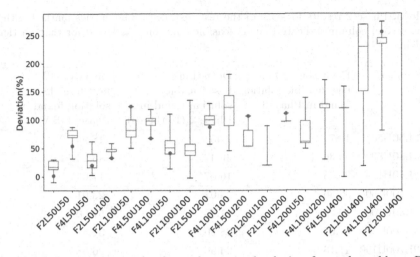

Fig. 2. Deviation of the results from the optimal solution for each problem. Each box corresponds to the area in which the middle 50% of the data reside in with the continuous line being the median. The whiskers are extended to the minimum and maximum values of the data. The red dot marks the best feasible solution found for the respective problem. (Color figure online)

5.2 Improvement Phase Results

In the second phase of Algorithm 1, starting at step 9, the 50 solutions obtained in the first phase were overlapped, resulting in an initial solution to be used as input for the iterated improvement. This second component of the algorithm was run for 10 iteration using both Kerberos and Simulated Annealing (SA) with the same parameter setting for sampling (in step 14 of the algorithm) to ensure a fair comparison between the two. The results of the second phase are summarized in Table 3. The second and third columns of this table show the minimum deviation from the optimum over all iterations resulting in a feasible solution. The results clearly show that with the exception of problem F2L50U50, the improvement step is effective, resulting in either better solutions or finding feasible solutions where Phase 1 did not. Another interesting observation is that Kerberos outperforms SA in Phase 2, almost always achieving solutions with a smaller optimality gap. This is expected, since Kerberos benefits from the additional sampling on the quantum annealer. Exceptional cases are problems F4L50U400, F2L100U400 and F4L100U100 to which Kerberos, unlike SA, fails to find a feasible solution within 10 iterations.

If we now turn to the performance of the second phase in terms of solution improvement, shown in Fig. 3, we observe large variations across different problems. For small-sized problems (Fig. 3a), the initial solution of the second phase is most often feasible, and the optimality gap reduces significantly in the first four or five iterations when using Kerberos. However, it occurs that if the deviation from the optimum becomes small, than the corresponding assignment of services to locations is less flexible, leading to no further improvement in

Table 3. Phase 2 results for each of the test instances. The entries marked with "-" in the second column indicate that no feasible solution was found for that particular problem.

Problem	Deviation (%) of best feasible solution found in Phase 1	Deviation (%) of best feasible solution found in Phase 2 (Kerberos)	Deviation (%) of best feasible solution found in Phase 2 (SA)
F2L50U50	1.87	2.45	18.99
F4L50U50	54.36	40.47	43.24
F2L50U100	20.23	16.07	17.04
F2L100U50	33.36	23.01	23.01
F4L50U100	123.94	21.43	57.60
F4L100U50	67.67	28.80	31.56
F2L100U100	41.13	24.96	26.58
F2L50U200	-	13.95	25.37
F4L100U100	87.66	15.50	42.43
F4L50U200	-	196.34	207.05
F2L200U100	106.55	19.06	36.34
F2L100U200	-	62.04	86.33
F4L200U50	111.61	63.25	57.73
F4L100U200	-	29.33	97.52
F4L50U400	-	-	242.80
F2L100U400	-	-	117.79
F4L100U400	-	-	252.97
F2L200U400	256.60	102.96	156.08

the latter iterations. This behaviour is also observed for some problems when employing SA (Fig. 3b), although in this case, the solution at each iteration may drastically vary, and occasionally become infeasible. For the largest problems, none of the two samplers seem to be very effective since most iterations result in infeasible solutions (Fig. 3c and Fig. 3d). This may be due to the fact that the overlap solution obtained from the 50 solutions in the first phase is infeasible, and does not incorporate enough open locations to allow for an improved search throughout the iterations. To this end, a better initial solution could be obtained in the first phase if we would allow for more than 50 runs.

The results obtained so far suggest that small problems can be solved relatively well with our hybrid approach, when using Kerberos in both phases. The solutions for the larger problems, for which the number of demand points greatly exceeds the number of available locations, are of lower quality. These problems could benefit in particular by a different approach to determine the starting solution. Instead of overlapping all solutions from the first phase, one

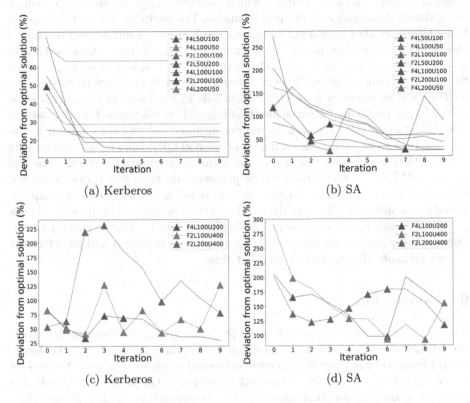

Fig. 3. Phase 2 results per iteration for different problems resulting in feasible solutions (a and b) and infeasible solutions (c and d). The sub-figure captions indicate the sampler used. The triangle markers indicate infeasible solutions.

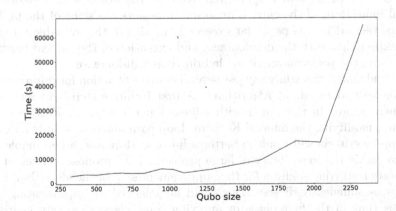

Fig. 4. Wall-clock time taken by Algorithm 1 as a function of the problems' qubo size.

could look into combining solutions which are sufficiently "different", i.e., which have ideally dissimilar sets of open locations. Furthermore, all problems could benefit from further parameter tuning in the Kerberos sampler and increasing the number of iterations in the improvement step. The current stopping criteria for Kerberos and limited number of limitations were chosen so as to provide a first proof of concept of our hybrid approach. Even in the current setting, the wall-clock time recorded for each problem was in the range of 2–5 h with some of the larger problems far exceeding this interval, as shown in Fig. 4. However, the QPU time needed for each of the decomposed subproblems ranged in the interval 10–300 ms, whilst the cumulated recorded QPU time for all experiments was less than 5 min. This is still underperforming in comparison to the classical solver, CPLEX, for which the average problem run time was 0.16 s. Since our experiments were run using a cloud service, it is likely that the high computational times shown in Fig. 4 are due to the latency between submitting problems to the QPU and receiving the solution. In the current D-Wave environment it is difficult to collect timing information for individual problems and therefore, we cannot estimate the queuing or read-out time.

6   Conclusion

This paper introduced a very first attempt to design a hybrid approach to solve the MSLSCP, by modelling the problem as a qubo and employing D-Wave's hybrid framework in combination with a classical improvement step. To the best of our knowledge, this work is the first to address this problem by combining classical means and quantum annealing. It was shown that the two-phase hybrid algorithm leads to improved feasible solutions compared to Kerberos sampling for almost all problems. Although the proposed approach does not yield a direct improvement in terms of solutions quality or running time with respect to classical solvers, it provides a way to model the problem in a suitable way and to reduce the size of the search space iteratively, in the attempt to overcome the physical limitations of the current quantum annealers. As some of the problem instances tackled in this paper far exceeded the size of the available QPU, we would expect that with the development and extension of the current hardware architecture, the performance of the hybrid tools will increase.

The findings of this study suggest several courses of action for improving the performance and results of Algorithm 1. A first intuitive step is to repeat the experiment shown in this paper with adjusted parameters for Kerberos. It is likely that modifying the internal Kerberos loop parameters or simply adjusting the stopping criteria will result in better solutions. Moreover, a reasonable approach to tackle the infeasibility of large problems is to propose a different way to generate a starting solution for the improvement step of the algorithm, which needs to be sufficiently diverse. This could be achieved by implementing more Kerberos runs in the first phase or applying some classical greedy heuristics.

Finally, we expect that intrinsic problem structure given by the spatial distribution of demand points heavily influences the performance of the algorithm. Therefore, it is worthwhile to investigate ways in which the decomposition of the problem on the QPU could be customized to incorporate this aspect.

References

1. D-wave problem-solving handbook; reformulating a problem. https://docs. dwavesys.com/docs/latest/c_handbook_3.html. Accessed 27 Jan 2020
2. Booth, M., Reinhardt, S.P., Roy, A.: Partitioning optimization problems for hybrid classical/quantum execution. Technical report, D-Wave Systems, September 2017
3. Cao, Y., Jiang, S., Perouli, D., Kais, S.: Solving set cover with pairs problem using quantum annealing. Sci. Rep. 6(1), 1–15 (2016)
4. Chapuis, G., Djidjev, H., Hahn, G., Rizk, G.: Finding maximum cliques on the D-Wave quantum annealer. J. Sign. Process. Syst. 91(3), 363–377 (2018). https:// doi.org/10.1007/s11265-018-1357-8
5. Farhi, E., Goldstone, J., Gutmann, S., Sipser, M.: Quantum computation by adiabatic evolution. arXiv:quant-ph/0001106v1 (2000)
6. Feld, S., et al.: A hybrid solution method for the capacitated vehicle routing problem using a quantum annealer. Front. ICT 6, 13 (2019)
7. Garey, M.R., Johnson, D.S.: Computers and Intractability; A Guide to the Theory of NP-Completeness. W. H. Freeman & Co., New York (1990)
8. Jiang, S., Britt, K.A., McCaskey, A.J., Humble, T.S., Kais, S.: Quantum annealing for prime factorization. Sci. Rep. 8(1), 1–9 (2018)
9. Neukart, F., Compostella, G., Seidel, C., Dollen, D.V., Yarkoni, S., Parney, B.: Traffic flow optimization using a quantum annealer. Front. ICT 4, 29 (2017)
10. Pelofske, E., Hahn, G., Djidjev, H.: Solving large minimum vertex cover problems on a quantum annealer. In: Proceedings of the 16th ACM International Conference on Computing Frontiers, CF 2019, pp. 76–84. ACM, New York (2019)
11. Phillipson, F., Vos, T.: Dense multi-service planning in smart cities. In: International Conference on Information Society and Smart Cities (2018)

New Hybrid Quantum Annealing Algorithms for Solving Vehicle Routing Problem

Michał Borowski[ID], Paweł Gora[ID], Katarzyna Karnas[ID], Mateusz Błajda[ID], Krystian Król[ID], Artur Matyjasek[ID], Damian Burczyk[ID], Miron Szewczyk[ID], and Michał Kutwin[(✉)][ID]

Faculty of Mathematics, Informatics and Mechanics, University of Warsaw, Warsaw, Poland
{m.borowski,p.gora,k.karnas,m.blajda,k.krol,a.matyjasek,
d.burczyk,m.szewczyk,m.kutwin}@mimuw.edu.pl

Abstract. We introduce new hybrid algorithms, DBSCAN Solver and Solution Partitioning Solver, which use quantum annealing for solving Vehicle Routing Problem (VRP) and its practical variant: Capacitated Vehicle Routing Problem (CVRP). Both algorithms contain important classical components, but we also present two other algorithms, Full QUBO Solver and Average Partitioning Solver, which can be run only on a quantum processing unit (without CPU) and were prototypes which helped us develop better hybrid approaches. In order to validate our methods, we run comprehensive tests using D-Wave's Leap framework on well-established benchmark test cases as well as on our own test scenarios built based on realistic road networks. We also compared our new quantum and hybrid methods with classical algorithms - well-known metaheuristics for solving VRP and CVRP. The experiments indicate that our hybrid methods give promising results and are able to find solutions of similar or even better quality than the tested classical algorithms.

Keywords: VRP · Vehicle Routing Problem · Quantum annealing

1 Introduction

Vehicle Routing Problem (VRP) is an important combinatorial optimization problem in which the goal is to find the optimal setting of routes for a fleet of vehicles which should deliver some goods from a given origin (depot) to a given set of destinations (customers) [1]. It is a generalization of the Travelling Salesman Problem (TSP) (introduced first as the Truck Dispatching Problem [1]) in which one vehicle has to visit some number of destinations in the optimal way [2]. Both problems are proven to be NP-hard [3]. There exist the exact algorithms able to find optimal solutions in a reasonable time for relatively small instances, but generally, those problems are computationally difficult and the state-of-the-art approaches applied in practice are based on heuristics (constructive, improvement and composite) and metaheuristics [4,5].

© Springer Nature Switzerland AG 2020
V. V. Krzhizhanovskaya et al. (Eds.): ICCS 2020, LNCS 12142, pp. 546–561, 2020.
https://doi.org/10.1007/978-3-030-50433-5_42

Recently, we can observe a noticeable progress in the development of quantum computing algorithms and it turned out that they may be particularly successful in solving combinatorial optimization problems, such as TSP and VRP [6]. The first quantum algorithms for TSP and VRP already exist and in the scientific literature we can find algorithms which can be run on gate-based quantum computers [7–17] as well as quantum annealing algorithms which can be run on adiabatic quantum computers [18–24].

In this paper, we present new methods for solving VRP and its more practical variant, CVRP (Capacitated Vehicle Routing Problem), in which all vehicles have a limited capacity. The algorithms introduced in this paper are based on quantum annealing, because due to the number of available qubits, those algorithms have currently a greater chance to give any practical improvement over classical algorithms.

We developed and present four algorithms: Full QUBO Solver (FQS), Average Partition Solver (AVS), DBSCAN Solver (DBSS) and Solution Partitioning Solver (SPS). The first and second one are designed only for solving VRP, DBSCAN solver can also solve CVRP if capacities of all vehicles are equal, SPS is able to solve CVRP with arbitrary capacities. It is also important to add that the last two methods are hybrid algorithms and they contain important components which should be run on classical processors.

In order to evaluate different algorithms for solving VRP using quantum annealing, we carried out series of experiments using D-Wave's Leap framework [25] which contains implementations of built-in solvers and allows to implement new solvers. We used QBSolv [26] run on quantum processing unit (QPU) and simulating quantum annealing on classical processors (CPU), as well as hybrid solver [27] run on both, QPU and CPU.

Beside quantum algorithms, we also wanted to test and compare several well-known classical algorithms which gave good results in previous studies. Based on a comprehensive literature review [5] and further analysis, we selected 4 metaheuristics: based on simulated annealing [28], bee algorithm [29], evolutionary annealing [30] and recursive DBSCAN with simulated annealing [31], respectively.

In order to reliably compare different algorithms, we conducted experiments on well-established benchmark datasets [32,33], as well as on datasets created by us, with realistic road networks (taken from the OpenStreetMap service) and artificially generated orders.

The rest of the paper is organized as follows: in Sect. 2, we describe in details all the quantum annealing solvers which we used in our experiments. Sections 3 and 4 present the design and results of our experiments, respectively. Section 5 outlines possible future research directions and concludes the paper.

2   CVRP Solvers Based on Quantum Annealing

In this section, we describe QUBO formulations and solvers which we developed for different variants of VRP: general VRP, CVRP with equal capacities and CVRP with arbitrary capacities. Before that, we introduce our notation and assumptions.

2.1 Notation and Assumptions

We assume that in each instance of VRP (or CVRP) we have a road network represented as a directed connected graph with vertices and edges. We also assume that the depots and destinations to which the orders of customers should be delivered are always located in vertices of the road network (in the case of benchmark instances and artificial networks, it may be even assumed that the road network is defined by locations of orders, while in the case of realistic road networks, real locations of orders are usually close enough to vertices determining the road network graph).

Let M be the number of available vehicles and N the number of orders. Let's denote the vehicles as $V = \{v_1, v_2, \ldots, v_M\}$ and the orders as $O = \{o_1, o_2, \ldots, o_N\}$. We assume that there is always a depot located in one of the vertices (we also assume that destinations of orders are not located in the depot - such orders can be just served immediately so are not interesting) and all vehicles are initially located in the depot and should finish all routes back in the depot. Therefore, we have in total $N + 1$ significant vertices and without any loss of generality, we can assume that our graph has exactly $N + 1$ vertices and $N * (N+1)$ directed edges (we can just consider edges built based on the shortest paths between every pair of vertices in the original graph), destination of the order o_i is located in the vertex i and the depot is located in the vertex $N + 1$. We can also denote the cost of the direct travel from the vertex i (destination of the order o_i) to the vertex j (destination of the order o_j) as $C_{i,j}$. We can also define $C_{N+1,i}$ and $C_{i,N+1}$ for $i \in \{1, 2, \ldots, N\}$ as the costs of direct travels from the depot to the destinations of orders and from the destinations of orders to the depot, respectively.

Let's assume that $x_{i,j,k} = 1$ if in a given setting the vehicle number i visits the vertex number j as k-th location on its route (for $j \in \{1, 2, \ldots, N + 1\}$ and $k \in \{0, 1, 2, \ldots, N + 1\}$), otherwise $x_{i,j,k} = 0$. We always have $x_{i,N+1,0} = 1$ and $x_{i,j,0} = 0$ for $j < N + 1$ (the depot is always the first location), and if $x_{i,N+1,K} = 1$ for some K then for $k > K$ $x_{i,j,k} = 1$ (each vehicle stays in the depot after reaching it).

2.2 Full QUBO Solver

First, we defined a basic QUBO formulation used for solving VRP instances. The formulation is based on a similar formulation for TSP in [20].

Let's define the binary function

$$A(y_1, y_2, \ldots, y_n) = \sum_{i=1}^{n} \sum_{j=i+1}^{n} 2y_i y_j - \sum_{i=1}^{n} y_i,$$

where $y_i \in \{0, 1\}$ for $i \in \{1, \ldots, n\}$. It is easy to prove that the minimum value of $A(y_1, y_2, \ldots, y_n)$ is equal to -1 and this value can be achieved only if exactly one of $y_1, y_2, \ldots, y_n$ is equal to 1.

By definition of VRP, the problem of minimizing the total cost can be defined as minimizing the function:

$$C = \sum_{m=1}^{M}\sum_{n=1}^{N} x_{m,n,1}C_{N+1,n} + \sum_{m=1}^{M}\sum_{n=1}^{N} x_{m,n,N}C_{n,N+1} \tag{1}$$

$$+ \sum_{m=1}^{M}\sum_{n=1}^{N-1}\sum_{i=1}^{N+1}\sum_{j=1}^{N+1} x_{m,i,n}x_{m,j,n+1}C_{i,j} \tag{2}$$

The first component of the sum C is a sum of all costs of travels from the depot - the first section of each cars' route. The second is a cost of the last section of a route (to depot) in a special case when a single car serves all N orders (only in such a case the component can be greater than 0). The last part is the cost of all other sections of routes.

To assure that each delivery is served by exactly one vehicle and exactly once, and that each vehicle is in exactly one place at a given time, the following term (in which all A components are equal to -1 only for such desired cases) should be included in our QUBO formulation:

$$Q = \sum_{k=1}^{N} A(x_{1,k,1}, x_{2,k,1}, \ldots, x_{1,k,2}, \ldots, x_{M,k,N}) \tag{3}$$

$$+ \sum_{m=1}^{M}\sum_{n=1}^{N} A(x_{m,1,n}, x_{m,2,n}, \ldots, x_{m,N+1,n}) \tag{4}$$

By definition of VRP, QUBO representation of this optimization problem is

$$QUBO_{VRP} = A_1 \cdot C + A_2 \cdot Q, \tag{5}$$

for some constants A_1 and A_2, which should be set to ensure that the solution found by quantum annealer minimizes Q (which should be $-N - NM$) to ensure satisfiability of the aforementioned constraints (after running initial tests we set $A_1 = 1$, $A_2 = 10^7$).

2.3   Average Partition Solver (APS)

APS is a variation of *Full QUBO Solver* for which we decrease the number of variables for each vehicle by assuming that every vehicle serves approximately the same number of orders. This means, every vehicle can serve up to $A+L$ deliveries, where A is the total number of orders divided by the number of vehicles and L is a parameter (called *"limit radius"*), which controls the number of orders. The QUBO formulation is the same as in case of Full QUBO Solver but the number of variables $x_{i,j,k}$ is lower $(M(A+L)N)$, which simplifies computations.

2.4   DBSCAN Solver (DBSS)

DBSS allows us to use quantum approach combined with a classical algorithm. This particular algorithm is inspired by *recursive DBSCAN* [31]. DBSS uses recursive DBSCAN as a clustering algorithm with limited size of clusters. Then, TSP is solved for each cluster separately by FQS (just by assuming in the QUBO formulation that the number of vehicles equals 1). If the number of clusters is equal to the number of vehicles, the answer is known immediately. Otherwise, the solver runs recursively considering clusters as deliveries, so that each cluster contains orders which in the final result are served one after another without leaving the cluster. What is more, we concluded that by limit the total sum of weights of deliveries in clusters, this algorithm can solve CVRP if all capacities of vehicles are equal.

2.5   Solution Partitioning Solver (SPS)

While adding capacity constraints is not simple, we were looking for the solution that can use results generated by DBSS. Therefore, we developed SPS. It is a simple algorithm which divides TSP solution found by another algorithm (e.g., DBSS) into consecutive intervals, which are the solution for CVRP. The idea is as follows:

Let $d_1, d_2, ..., d_N$ be the TSP solution for N orders, let P_v be a capacity of the vehicle v, let $w_{i,j}$ be the sum of weights of orders $d_i, d_{i+1}, d_{i+2}, ..., d_j$ (in the order corresponding to TSP solution) and let $cost_{i,j}$ be the total cost of serving only orders $d_i, d_{i+1}, ..., d_j$. Also, let $dp_{i,S}$ be the cost of the best solution for orders $d_1, d_2, d_3, ..., d_i$ and for the set of vehicles S. Now, the dynamic programming formula for solving CVRP is given by:

$$dp_{i,S} = \min_{v \in S, 0 \leq j \leq i, w_{j+1,i} \leq P_v} \{dp_{j,S\setminus\{v\}} + cost_{j+1,i}\}, \tag{6}$$

where $cost_{i,j} = 0$ and $w_{i,j} = 0$ for $i > j$. Formula (6) returns a plenty of possible routes, but it also finds the optimal solution. We can speed it up by noticing that if two vehicles have the same capacity, it doesn't matter which one of them we choose, but pessimistically, capacities can be pairwise distinct. We propose the following heuristic to optimize this solution:

1. Instead of set S of vehicles, consider a sequence $v_1, v_2, \ldots, v_M$ of vehicles and assume that we attach them to deliveries in such an order.
2. Now, our dynamic programming formula is given by:

$$dp_{i,\{v_1,v_2,...,v_k\}} = \min_{0 \leq j \leq i, w_{j+1,i} \leq P_{v_k}} \{dp_{j,\{v_1,...,v_{k-1}\}} + cost_{j+1,i}\} \tag{7}$$

3. To count this dynamic effectively, we can observe that:

$$\forall_{j<i} cost_{j,i} = C_{N+1,j} + C_{i,N+1} + \sum_{k=j}^{i-1} C_{k,k+1} \tag{8}$$

$$\forall_{j<i} cost_{j,i} = cost_{j,i-1} + C_{i-1,i} + C_{i,N+1} - C_{i-1,N+1} \tag{9}$$

$$\forall_{j<i} cost_{j,i} - cost_{j,i-1} = C_{i-1,i} + C_{i,N+1} - C_{i-1,N+1} \tag{10}$$

$$\forall_{j<i,1\leq k\leq M}(dp_{j-1,\{v_1,v_2,...,v_k\}} + cost_{j,i}) - (dp_{j-1,\{v_1,v_2,...,v_k\}}$$
$$+ cost_{j,i-1}) = C_{i-1,i} + C_{i,N+1} - C_{i-1,N+1} \tag{11}$$

So if we have counted dp for fixed k, then for counting dp for $k+1$ we can store all dp values for k and increase them, one by one (starting from $i = j+1$), by a right side of Eq. 10. Using monotonic queue, we can get minimum in $O(1)$ time.

We can now select some random permutations of vehicles and perform dynamic programming for each of them. The number of permutations can be regulated by additional parameter. With optimization of dynamic programming, the complexity of this algorithm is $O(NMR)$, where R is the number of permutations.

The greatest limitation of SPS is that it considers only one TSP solution. Nonetheless, we observed that DBSS for more than one vehicle works in a similar way.

3   Design of Experiments

The goal of our experiments was to test and compare different formulations of QUBO (solving different variants of VRP) on different datasets and with different solvers and settings (number of qubits and quota of time on quantum processor). We ran them using D-Wave's Leap platform [25] and its 2 solvers: qbsolv [26] and hybrid solver [27]. To run comprehensive and comparable experiments, we prepared several datasets:

- Christofides1979 - a standard benchmark dataset for CVRP, well-known and frequently investigated by the scientific community [32,33],
- A dataset built by us based on a realistic road network of Belgium, acquired from the OpenStreetMap service.

Christofides1979 consists of 14 tests, where each test instance is described by three files. The first one provides the number of vehicles and their capacity (the same for all vehicles). The second file describes the orders, i.e. their coordinates in 2−dimensional plane and the demand. The last file reports the time matrix (times of travel between various vertices in a graph). For a purpose of running our experiments and compare the results, we selected only 9 out of 14 tests because in case of other tests some hybrid or classical algorithms were not able to find any good solutions. All the important parameters describing Christofides1979 instances are given in Table 1.

Table 1. Parameters of instances of Christofides1979 used in our experiments.

Test name	Nr of vehicles	Capacity	Nr of orders
CMT11	7	200	120
CMT12	10	200	100
CMT13	11	200	120
CMT14	11	200	100
CMT3	8	200	100
CMT6	6	160	50
CMT7	11	140	75
CMT8	9	200	100
CMT9	14	200	150

In the case of the second dataset, we generated in total 51 tests. Each test was characterized by the number of orders. Table 2 presents a description of this dataset. Basically, it consists of 4 groups of test cases: small test (small number of orders), medium tests (medium number of orders), big tests (large number of order), mixed tests (various number of orders with some additional conditions).

Table 2. Parameters and descriptions of tests

Test	Number of orders	Description
small-0	2	No further conditions
small-1	2	
small-2	2	
small-3	1	
small-4	2	
small-5	5	
small-6	6	
small-7	5	
small-8	4	
small-9	6	
medium-0	20	No further conditions
medium-1	26	
medium-2	27	
medium-3	24	
medium-4	25	
medium-5	25	
medium-6	20	
medium-7	14	
medium-8	17	
medium-9	15	
big-0	52	No further conditions
big-1	42	
big-2	48	
big-3	48	
big-4	50	

(*continued*)

Table 2. (*continued*)

Test	Number of orders	Description
group1-1	42	No further conditions
group1-2	54	
range-6-1	47	Magazines are at most 6 km from city center
range-6-2	50	
range-8-12-1	50	Magazines are at least 8 km and at most 12 km from city center
range-8-12-2	50	
range-8-12-3	46	
range-8-12-4	51	
range-8-12-5	50	
range-8-12-6	50	
range-5-1	50	Orders are at most 5 km from city center. Vehicles have capacity greater than total demand
range-5-1	50	
range-3-1	37	Orders are within 3 km from city center
range-3-2	29	
range-4-1	9	Orders are within 4 km from city center
range-4-2	7	
range-4-75-1	75	Orders are within 4 km from city center. We have 75 orders
range-4-75-2	75	
range-4-100-1	100	Orders are within 4 km from city center. We have 100 orders
range-4-100-2	100	
range-4-150-1	150	Orders are within 4 km from city center. We have 150 orders
range-4-150-2	150	
range-4-200-1	200	Magazines and orders are within 4 km from city center. We have 200 orders
range-4-200-2	200	
clustered1-1	57	In each one of four 1-km circles spread across the map, there is between 6 and 20 orders
clustered1-2	55	

In every experiment, our programs computed the minimal cost of serving all orders. D-Wave's quantum annealing machine is naturally nondeterministic, so are the returned results, so for every algorithm and on every test case we ran 5 experiments. The code of programs used in our experiments is publicly available at [34].

4 Results of Experiments

In this section, we present results of experiments conducted using QBSolv and hybrid solver built-in D-Wave's Leap framework and using algorithms described in Sect. 3.

4.1 Full QUBO Solver (FQS)

First, we investigated Full QUBO Solver (FQS) on test cases small-0 - small-9. On every test except small-0, we ran experiments for 3 different numbers of vehicles $(1, 2, 3)$ on quantum processor (FQS QPU [26]), its classical simulator

(FQS CPU) and using a hybrid solver (FQS Hybrid [27]). On small-0 there were only 2 orders so we tested only 1, 2 vehicles.

As we can see in Table 3, QBSolv (FQS CPU and FQS QPU) exacerbates final results in test cases with more vehicles. For more vehicles, it can potentially generate the same solution as for less vehicles, because some vehicles can be just ignored. Solutions generated with hybrid solver (FQS Hybrid) confirm that. However, the size of QUBO makes the solutions with more vehicles unavailable for QBSolv. In hybrid solver, we have such a problem in only one case (small-9). However, in only 1 test case (small-3) QBSolv was able to improve the solution returned for smaller number of vehicles. In addition, in most cases QBSolv was not able to find a solution on QPU, the size of the instance and the number of the required variables and qubits was just too large. Also the required time of computations on QPU was worse than in case of CPU or hybrid approach. Therefore, we concluded that it doesn't make sense to run more experiments on QPU for larger test cases (with more cars and more orders) and we conducted next tests only using QBSolv on CPU and using a hybrid solver.

For larger VRP instances (medium-0 - medium-9), we observed that the transition from one vehicle to two vehicles is difficult. QBSolv usually returns much worse results (there is only 1 exception, test case medium-8). For the hybrid solver, in only one case the result for two vehicles is better (medium-6) but the results are usually still better than in case of QBSolv. We also noticed that the order of deliveries in tests with one vehicle was not optimal for majority of test cases. Only the least instances - with up to 15 orders - seem to be solved optimally. An interesting thing is that differences between results for two vehicles and one vehicle are very discrepant and it is not caused by the number of orders. By analyzing full results, we concluded that for 2 vehicles the solvers divided deliveries evenhandedly and for some tests it is a good way to build the optimal solution. We came up with an idea that since solvers found only these solutions, we can ask them to optimize only that kind of solutions, so we implemented Average Partition Solver, which demands less qubits.

4.2 Average Partition Solver (APS)

We extended Full QUBO Solver with an option of changing the maximum difference between the number of deliveries attached to the vehicles, i.e., a deflection from the average number of deliveries per one vehicle. We found out experimentally that it should be $\frac{1}{10}$ of the number of deliveries, which gives maximum difference in our test cases equal to 5. Having 1 vehicle, APS works exactly the same as Full QUBO Solver, so we ran experiments only for more vehicles (but we also included the results for 1 vehicle in Table 3, just for comparison).

In most test cases, the results found using APS were better than results found by FQS. We can also notice that differences between results for 3 vehicles and results for 2 vehicles generated by APS are lower than the differences between results for 2 vehicles and 1 vehicle generated by FQS. However, in case of 3 vehicles, QBSolv on CPU still can't find better solutions with only 2 vehicles. The hybrid solver can find better solutions in cases with 3 vehicles than in cases with only 2 vehicles in 4 (out of 10) test cases.

Table 3. Results on small and medium datasets

Test	Vehicles	FQS CPU	FQS QPU	FQS Hybrid	APS CPU	APS Hybrid	DBSS CPU
small-0	1, 2	11286	11286	11286	11286	11286	–
small-1	1	10643	10643	10643	10643	10643	–
	2	10643	10643	10643	12379	12379	–
	3	10643	–	10643	–	–	–
small-2	1	21311	21311	21311	21311	21311	–
	2	21311	–	21311	24508	24508	–
	3	22192	–	21311	–	–	–
small-3	1	18044	18044	18044	18044	18044	–
	2	20819	–	18033	22193	22193	–
	3	22843	–	18033	–	–	–
small-4	1	15424	15424	15424	15424	15424	–
	2	17364	–	15424	19472	19472	–
	3	17364	–	15424	–	–	–
small-5	1	10906	10906	10906	10906	10906	–
	2	11676	–	10906	13480	13480	–
	3	11754	–	10906	–	–	–
small-6	1	20859	20859	20859	20859	20859	–
	2	26735	–	20859	26735	26735	–
	3	27110	–	20859	–	–	–
small-7	1	18117	18117	18117	18117	18117	–
	2	18710	–	18117	23114	23114	–
	3	21666	–	18117	–	–	–
small-8	1	12198	12198	12198	12198	12198	–
	2	12494	–	12198	13282	13282	–
	3	13282	–	12198	–	–	–
small-9	1	19184	19184	19184	19184	19184	–
	2	19848	–	19184	21438	21438	–
	3	21438	–	19848	–	–	–
medium-0	1	20774	–	21775	20774	21775	24583
	2	36966	–	29879	25737	25217	27994
	3				28226	27237	34185
medium-1	1	29868	–	29423	29868	29423	27606
	2	50639	–	39485	30820	31129	31346
	3	–	–	–	33376	32018	32588
medium-2	1	37045	–	35208	37045	35208	29442
	2	55579	–	36511	33235	33163	32947
	3	–	–	–	36600	32569	34480
medium-3	1	30206	–	29422	30206	29422	31092
	2	51787	–	35774	31428	30273	33790
	3	–	–	–	35994	33627	33712
medium-4	1	21257	–	20762	21257	20762	21435
	2	34379	–	25470	22410	22722	22885
	3	–	–	–	23599	22176	25446
medium-5	1	23013	–	21642	23013	21462	21737
	2	36149	–	22041	22775	23076	23403
	3	–	–	–	24899	22386	24336

(*continued*)

Table 3. (*continued*)

Test	Vehicles	FQS CPU	FQS QPU	FQS Hybrid	APS CPU	APS Hybrid	DBSS CPU
medium-6	1	23804	–	24664	23804	23804	23926
	2	35826	–	24490	24265	25178	25510
	3	–	–	–	27032	23364	25122
medium-7	1	22847	–	22847	22847	22847	28308
	2	33441	–	26550	24331	24460	30482
	3	–	–	–	27156	27156	34064
medium-8	1	23843	–	14566	23843	14566	15575
	2	20804	–	15931	14256	14808	15829
	3	–	–	–	15815	15466	16930
medium-9	1	12228	–	12395	12228	12395	12842
	2	16606	–	13950	12321	12830	14926
	3	–	–	–	13221	13178	14619

4.3   DBSCAN Solver (DBSS)

We can see in Table 3 that DBSS usually gives worse results than the APS, but we expected that it may change in case of tests with more orders thanks to utilizing the power of recursive DBSCAN.

Indeed, on big test cases with a larger number of orders, DBSS gives much better results than APS (Table 4). Additionally, DBSS can be run on larger instances and don't need assumption that every vehicle serves approximately the same number of deliveries (as it is in case of APS).

Table 4. Comparison of results for Average Partition Solver and DBSCAN Solver on big test cases.

	Vehicles	APS CPU	DBSS CPU
big-0	1	80084	71594
	2	97286	71051
big-1	1	157660	146828
	2	206782	149200
big-2	1	168646	154105
big-3	1	85873	62236
big-4	1	156411	129279

4.4   Solution Partitioning Solver (SPS)

At the beginning, we tested SPS on test cases where all capacities are equal, in order to compare results with DBSS which can solve this problem. The results are presented in Table 5. In some cases, our solvers were not able to find the proper solutions (we mark such cases as "Not valid") but in general, SPS outperformed DBSS.

Table 5. Comparison of DBSCAN Solver and Solution Partitioning Solver (SPS) run on CPU on big test cases with various capacities.

	Vehicles	Capacity	SPS (CPU)	DBSS (CPU)
big-0	2	100	70928	73508
	2	85	72295	73189
	2	80	75150	Not valid
	3	100	71320	76717
	3	70	71251	78012
	3	55	Not valid	76807
	5	100	71740	Not valid
	5	50	78726	91066
	5	40	85976	Not valid
big-1	2	100	150608	158631
	2	80	150608	152946
	2	65	150804	156188
	3	100	151525	153673
	3	60	153190	152854
	3	45	164055	Not valid
	5	100	151930	168789
	5	40	156242	165271
	5	30	174519	176935

Based on those experiments, we decided to test further only SPS and compare it with 4 classical algorithms - simulated annealing (SA), bee algorithm (BEE), evolutionary annealing (EA) and recursive DBSCAN with simulated annealing (DBSA). We ran next experiments with even more orders on mixed test cases generated by us (Table 2) and on benchmark datasets Christofides1979 (Table 1). The results are presented in Table 6 and Table 7.

Table 6. Comparison of results achieved by Solution Partitioning Solver (SPS) and classical algorithms (SA - simulated annealing, BEE - Bee algorithm, EA - evolutionary annealing, DBSA - DBSCAN with simulated annealing) on a benchmark dataset Christofides79.

Test name	SPS	SA	BEE	EA	DBSA
CMT11	25.54	23.62	36.18	16.52	19.94
CMT12	26.84	53.06	20.24	20.68	21.37
CMT13	25.97	86.72	34.66	35.05	19.44
CMT14	26.83	52.52	20.23	20.23	22.8
CMT3	25.13	48.3	28.38	28.82	–
CMT6	17.58	48.3	15.42	28.82	15.82
CMT7	29.42	41.4	27.89	31.68	23.18
CMT8	26.5	51.16	26.67	28.09	19.4
CMT9	34.14	76.34	44.25	42.81	–

Table 7. Results of Solution Partitioning Solver compared with results for classical algorithms run on artificially generated test cases.

	Type	Deliveries	SPS	Simul. Ann.	Bee	Evolution
clustered1-1	Average	57	69850	66379	60876	48923
	Best	57	69080	52119	56358	48152
clustered1-2	Average	55	77173	74341	81438	54719
	Best	55	75530	59947	68772	53490
group1-1	Average	42	158919	156217	153495	137989
	Best	42	155388	146526	142774	135593
group1-2	Average	54	171732	145380	145325	137626
	Best	54	165043	141065	140947	136307
range-6-1	Average	47	71670	68003	67234	59937
	Best	47	68459	62312	64404	59827
range-6-2	Average	50	80490	84380	83915	73651
	Best	50	79640	79574	85917	73051
range-8-12-1	Average	50	142008	146553	142835	129069
	Best	50	140170	136369	127372	126555
range-8-12-2	Average	50	146798	137628	145332	129048
	Best	50	143598	135493	136776	128803
range-8-12-3	Average	46	105544	105051	98366	92792
	Best	46	101577	99004	94423	91921
range-8-12-4	Average	51	147993	143309	148900	128316
	Best	51	145559	140088	128575	124405
range-8-12-5	Average	50	146719	143516	145685	134162
	Best	50	143993	139784	139796	133245
range-8-12-6	Average	50	146984	148194	150121	136326
	Best	50	141467	138781	139400	134692
range-5-1	Average	50	81728	68900	69052	67896
	Best	50	72527	67984	68022	67691
range-5-2	Average	50	81759	69342	68564	67981
	Best	50	76868	67958	67780	67716
range-3-1	Average	37	39790	37268	36260	29326
	Best	50	36851	32877	35650	29180
range-3-2	Average	29	34361	39336	34068	30497
	Best	50	33548	35340	32908	30466
range-4-1	Average	50	21559	21604	21604	21604
	Best	50	21317	21604	21604	21604
range-4-2	Average	50	18044	18498	18640	18498
	Best	50	18044	18498	18497	18498
range-4-100-1	Average	100	84916	106625	118550	85346
	Best	50	81303	98522	112389	84514
range-4-100-2	Average	100	91527	105538	127744	86538
	Best	50	88566	97312	111513	84750
range-4-150-1	Average	150	90394	98711	119547	101126
	Best	50	88040	91972	108442	100195
range-4-150-2	Average	150	112539	118351	171620	125444
	Best	50	110104	110401	170164	121462
range-4-200-1	Average	200	112618	124269	179239	139991
	Best	50	111259	120510	171530	137684
range-4-200-2	Average	200	135243	158634	223262	202373
	Best	50	131349	135931	203352	194707
range-4-75-1	Average	75	62439	60423	65381	52701
	Best	50	60283	56337	62051	51846
range-4-75-2	Average	75	72077	76964	85849	60753
	Best	50	70403	71164	84140	60168

5   Conclusion and Future Research Directions

We introduced new hybrid algorithms for solving VRP and CVRP and ran tests using D-Wave's Leap framework on well-established benchmark test cases and on our own test scenarios built based on realistic road networks. We also compared our new quantum and hybrid methods with classical algorithms - well-known metaheuristics for solving VRP and CVRP. The results indicate that our hybrid methods give promising results and are able to find solutions of a similar quality to the tested classical algorithms.

Our primary future research direction is extending QUBO formulations to solve even more realistic variant of VRP - the Vehicle Routing Problem with Time Windows (VRPTW). Also, we are planning to compare our hybrid algorithms with even more classical algorithms for solving VRP and its variants.

Acknowledgment. The presented research was carried out within the frame of the project "Green LAst-mile Delivery" (GLAD) realized at the University of Warsaw with the project partners: Colruyt Group, University of Cambridge and Technion. The project is supported by EIT Food, which is a Knowledge and Innovation Community (KIC) established by the European Institute for Innovation & Technology (EIT), an independent EU body set up in 2008 to drive innovation and entrepreneurship across Europe.

References

1. Dantzig, G.B., Ramser, J.H.: The truck dispatching problem. Manage. Sci. **6**(1), 80–91 (1959)
2. Kirkman, T.: XVIII. On the representation of polyedra. Philos. Trans. R. Soc. Lond. **146**, 413–418 (1856)
3. Karp, R.M.: Reducibility among combinatorial problems. In: Miller, R.E., Thatcher, J.W., Bohlinger, J.D. (eds.) Complexity of Computer Computations. IRSS, pp. 85–103. Springer, Boston (1972). https://doi.org/10.1007/978-1-4684-2001-2_9
4. Laporte, G., Toth, P., Vigo, D.: Vehicle routing: historical perspective and recent contributions. EURO J. Transp. Logist. **2**, 1–4 (2013). https://doi.org/10.1007/s13676-013-0020-6
5. Gora, P., et al.: On a road to optimal fleet routing algorithms: a gentle introduction to the state-of-the-art. In: Smart Delivery Systems, Solving Complex Vehicle Routing Problems, Intelligent Data-Centric Systems, pp. 37–92 (2020)
6. Zahedinejad, E., Zaribafiyan, A.: Combinatorial optimization on gate model quantum computers: a survey (2017). https://arxiv.org/abs/1708.05294
7. Feng, X., Wang, Y., Ge, H., Zhou, C., Liang, Y.: Quantum-inspired evolutionary algorithm for travelling salesman problem. In: Liu, G., Tan, V., Han, X. (eds.) Computational Methods, pp. 1363–1367. Springer, Dordrecht (2006). https://doi.org/10.1007/978-1-4020-3953-9_55
8. Beheshti, A.K., Hejazi, S.R.: A novel hybrid column generation-metaheuristic approach for the vehicle routing problem with general soft time window. Inf. Sci. **316**, 598–615 (2015)

9. Beheshti, A.K., Hejazi, S.R.: A quantum evolutionary algorithm for the vehicle routing problem with delivery time cost. Int. J. Ind. Eng. Prod. Res. **25**(4), 287–295 (2014)

10. Greenwood, G.W.: Finding solutions to NP problems: philosophical differences between quantum and evolutionary search algorithms. In: Proceedings of the 2001 Congress on Evolutionary Computation (2001)

11. Zeng, K., Peng, G., Cai, Z., Huang, Z., Yang, X.: A hybrid natural computing approach for the VRP problem based on PSO, GA and quantum computation. In: Yeo, S.S., Pan, Y., Lee, Y., Chang, H. (eds.) Computer Science and its Applications. LNEE, vol. 203, pp. 23–28. Springer, Dordrecht (2012). https://doi.org/10.1007/978-94-007-5699-1_3

12. Srinivasan, K., Satyajit, S., Behera, B.K., Panigrahi, P.K.: Efficient quantum algorithm for solving travelling salesman problem: an IBM quantum experience (2018). https://arxiv.org/abs/1805.10928

13. Cui, L., Wang, L., Deng, J., Zhang, J.: A new improved quantum evolution algorithm with local search procedure for capacitated vehicle routing problem. Math. Probl. Eng. (2013). https://doi.org/10.1155/2013/159495. Article ID 159495

14. Zhang, J., Wang, W., Zhao, Y., Cattani, C.: Multiobjective quantum evolutionary algorithm for the vehicle routing problem with customer satisfaction. Math. Probl. Eng. (2012). https://doi.org/10.1155/2012/879614. https://www.hindawi.com/journals/mpe/2012/879614, Article ID 879614

15. Dai, H., Yang, Y., Li, H., Li, C.: Bi-direction quantum crossover-based clonal selection algorithm and its applications. Expert Syst. Appl. **41**(16), 7248–7258 (2014)

16. You, X., Miao, X., Liu S.: Quantum computing-based Ant Colony Optimization algorithm for TSP. In: 2nd International Conference on Power Electronics and Intelligent Transportation System (PEITS), Shenzhen, 2009, pp. 359–362 (2009). https://doi.org/10.1109/PEITS.2009.5406879

17. Wang, Y., et al.: A novel quantum swarm evolutionary algorithm and its applications. Neurocomputing **70**, 633–640 (2007)

18. Martoňák, R., Santoro, G.E., Tosatti, E.: Quantum annealing of the traveling-salesman problem. Phys. Rev. E **70**, 057701 (2004)

19. Santoro, G.E., Tosatti, E.: Optimization using quantum mechanics: quantum annealing through adiabatic evolution. J. Phys. A Math. Gen. **39**(36), R393 (2006)

20. Lucas, A.: Ising formulations of many NP problems. Front. Phys. **2**, 5 (2014)

21. Smelyanskiy, V.N., et al.: A Near-Term Quantum Computing Approach for Hard Computational Problems in Space Exploration (2012)

22. Kieu, T.D.: Quantum adiabatic computation and the travelling salesman problem (2006). https://arxiv.org/abs/quant-ph/0601151

23. Feld, S., et al.: A hybrid solution method for the capacitated vehicle routing problem using a quantum annealer. In: Frontiers in ICT, vol. 6 (2019). https://www.frontiersin.org/articles/10.3389/fict.2019.00013/full

24. Feld, S., et al.: A Hybrid Solution Method for the Capacitated Vehicle Routing Problem Using a Quantum Annealer (2019)

25. D-Wave's Leap project. https://www.dwavesys.com/take-leap. Accessed 7 Feb 2020

26. https://docs.ocean.dwavesys.com/projects/qbsolv/en/latest. Accessed 7 Feb 2020

27. https://docs.ocean.dwavesys.com/projects/hybrid/en/latest. Accessed 7 Feb 2020

28. Tavakkoli-Moghaddam, R., Safae, N., Kah, M.M.O., Rabbani, M.: A new capacitated vehicle routing problem with split service for minimizing fleet cost by simulated annealing. J. Franklin Inst. **344**(5), 406–425 (2007)

29. Szeto, W.Y., Yongzhong, W.Y., Ho, S.C.: An artificial bee colony algorithm for the capacitated vehicle routing problem. Eur. J. Oper. Res. **215**(1), 126–135 (2011)
30. Bañosa, R., Ortega, J., Gil, C., Márquez, A.L., Toroc, F.: A hybrid meta-heuristic for multi-objective vehicle routing problems with time windows. Comput. Ind. Eng. **65**(2), 286–296 (2013)
31. Bujel, K., Lai, F., Szczecinski, M., So, W., Fernandez, M.: Solving high volume capacitated vehicle routing problem with time windows using recursive-DBSCAN clustering algorithm. arXiv:1812.02300v2
32. http://www.vrp-rep.org/datasets/item/2014-0002.html. Accessed 7 Feb 2020
33. Christofides, N., Mingozzi, A., Toth, P.: The vehicle routing problem. In: Christofides, N., Mingozzi, A., Toth, P., Sandi, C. (eds.) Combinatorial Optimization, pp. 315–338. Wiley, Chichester (1979)
34. Code used in our experiments. https://github.com/xBorox1/D-Wave-Leap---CVRP/tree/master/vrp. Accessed 7 Feb 2020

Multi-agent Reinforcement Learning Using Simulated Quantum Annealing

Niels M. P. Neumann[✉], Paolo B. U. L. de Heer, Irina Chiscop,
and Frank Phillipson

The Netherlands Organisation for Applied Scientific Research,
Anna van Buerenplein 1, 2595 DA The Hague, The Netherlands
{niels.neumann,paolo.deheer,irina.chiscop,frank.phillipson}@tno.nl

Abstract. With quantum computers still under heavy development, already numerous quantum machine learning algorithms have been proposed for both gate-based quantum computers and quantum annealers. Recently, a quantum annealing version of a reinforcement learning algorithm for grid-traversal using one agent was published. We extend this work based on quantum Boltzmann machines, by allowing for any number of agents. We show that the use of quantum annealing can improve the learning compared to classical methods. We do this both by means of actual quantum hardware and by simulated quantum annealing.

Keywords: Multi-agent · Reinforcement learning · Quantum computing · D-Wave · Quantum annealing

1 Introduction

Currently, there are two different quantum computing paradigms. The first is gate-based quantum computing, which is closely related to classical digital computers. Making gate-based quantum computers is difficult, and state-of-the-art devices therefore typically have only a few qubits. The second paradigm is quantum annealing, based on the work of Kadowaki and Nishimore [17]. Problems have already been solved using quantum annealing, in some cases much faster than with classical equivalents [7,23]. Applications of quantum annealing are diverse and include traffic optimization [23], auto-encoders [18], cyber security problems [24], chemistry applications [12,28] and machine learning [7,8,21].

Especially the latter is of interest as the amount of data the world processes yearly is ever increasing [14], while the growth of the classical computing power is expected to stop at some point [27]. Quantum annealing might provide the necessary improvements to tackle these upcoming challenges.

One specific type of machine learning is reinforcement learning, where an optimal action policy is learnt through trial and error. Reinforcement learning can be used for a large variety of applications, ranging from autonomous robots [29] to determining optimal social or economical interactions [3]. Recently, reinforcement learning has seen many improvements, most notably the use of

V. V. Krzhizhanovskaya et al. (Eds.): ICCS 2020, LNCS 12142, pp. 562–575, 2020.
https://doi.org/10.1007/978-3-030-50433-5_43

neural networks to encode the quality of state-action combinations. Since then, it has been successfully applied to complex games such as Go [25] and solving a Rubik's cube [2].

In this work we consider a specific reinforcement learning architecture called a Boltzmann machine [1]. Boltzmann machines are stochastic recurrent neural networks and provide a highly versatile basis to solve optimisation problems. However, the main reason against widespread use of Boltzmann machines is that the training times are exponential in the input size. In order to effectively use Boltzmann machines, efficient solutions for complex (sub)routines must be found. One of the complex subroutines is finding the optimal parameters of a Boltzmann machine. This task is especially well suited for simulated annealing, and hence for quantum annealing.

So far, little research has been done on quantum reinforcement learning. Early work demonstrated that applying quantum theory to reinforcement learning problems can improve the algorithms, with potential improvements to be quadratic in learning efficiency and exponential in performance [10,11]. Only recently, quantum reinforcement learning algorithms are implemented on quantum hardware, with [8] one of the first to do so. They demonstrated quantum-enabled reinforcement learning through quantum annealer experiments.

In this article, we consider the work of [8] and implement their proposed quantum annealing algorithm to find the best action policy in a gridworld environment. A gridworld environment, shown in Fig. 2, is a simulation model where an agent can move from cell to cell, and where potential rewards, penalties and barriers are defined for certain cells. Next, we extend the work to an arbitrary number of agents, each searching for the optimal path to certain goals. This work is, to our knowledge, the first simulated quantum annealing-based approach for multi-agent gridworlds. The algorithm can also be run on quantum annealing hardware if available.

In the following section, we will give more details on reinforcement learning and Boltzmann machines. In Sect. 3 we will describe the used method and the extensions towards a multi-agent environment. Results will be presented and discussed in Sect. 4, while Sect. 5 gives a conclusion.

2 Background

A reinforcement learning problem is described as a *Markov Decision Process* (MDP) [6,15], which is a discrete time stochastic system. At every timestep t the agent is in a state s_t and chooses an action a_t from its available actions in that state. The system then moves to the next state s_{t+1} and the agent receives a reward or penalty $R_{a_t}(s_t, s_{t+1})$ for taking that specific action in that state. A policy π maps states to a probability distribution over actions and, when used as $\pi(s)$ it returns the highest-valued action a for state s. The policy will be optimized over the cumulative rewards attained by the agent for all state-action combinations. To find the optimal policy π^*, the Q-function $Q(s, a)$ is used which defines for each state-action pair the Q-value, denoting the expected cumulative reward, or the *quality*.

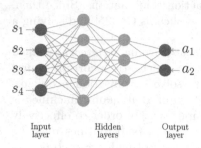

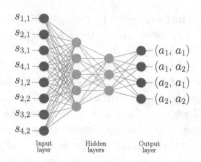

(a) A single-agent Boltzmann machine with four states and two actions.

(b) A multi-agent Boltzmann machine with two agents and per agent four states and two actions .

Fig. 1. Examples of restricted Boltzmann machines for reinforcement learning environments with one or more agents.

The Q-function is trained by trial and error, by repeatedly taking actions in the environment and updating the Q-values using the Bellman equation [5]:

$$Q^\pi(s_t, a_t) = \mathcal{E}_{s_{t+1}} \left[R_{a_t}(s_t, s_{t+1}) + \gamma Q^\pi(s_{t+1}, \pi(s_{t+1})) \right]. \qquad (1)$$

Different structures can be used to represent the Q-function, ranging from a simple but very limited tabular Q-function to a (deep) neural network which encodes the values with the state vector as input nodes, and all possible actions as output nodes. In such deep neural networks, the link between nodes i and j is assigned a weight w_{ij}. These weights can then be updated using, for example, gradient descent, which minimizes a loss function. If a multi-layered neural network is used, it is called deep reinforcement learning (DRL). A special type of DRL is given by Boltzmann machines and their restricted variants.

A Boltzmann machine is a type of neural network that can be used to encode the Q-function. In a general Boltzmann machine, all nodes are connected to each other. In a restricted Boltzmann machine (RBM), nodes are divided into subsets of visible nodes v and hidden nodes h, where nodes in the same subset have no connections. The hidden nodes can be further separated in multiple hidden node subsets, resulting in a multi-layered (deep) RBM, an example of which can be seen in Fig. 1a with two hidden layers of 5 and 3 nodes respectively. There are also two visible layers. Connections between distinct nodes i and j are assigned a weight w_{ij}. Additionally, each node i is assigned a bias w_{ii}, indicating a preference to one of the two possible values ± 1 for that node. All links are bidirectional in RBMs, meaning $w_{ij} = w_{ji}$. Hence, they differ from feed-forward neural networks, where the weight of one direction is typically set to 0.

Using v_i for visible nodes and h_j for hidden ones, we can associate a global energy configuration to an RBM using

$$E(v, h) = -\sum_i w_{ii} v_i - \sum_j w_{jj} h_j - \sum_i \sum_j v_i w_{ij} h_j. \qquad (2)$$

The probability for nodes being plus or minus one depends on this global energy and is given by

$$p_{\text{node } i=1} = \frac{1}{1 + \exp(-\frac{\Delta E_i}{T})}.$$

Here $\Delta E_i = E_{\text{node } i=1} - E_{\text{node } i=-1}$ is the difference in global energy if node i is 1 or -1 and T is an internal model parameter, referred to as the temperature.

Simulated annealing can be used to update the weights w_{ij} quickly. In this approach, a subset of visible nodes are fixed (clamped) to their current values after which the network is sampled. During this process the anneal temperature is decreased slowly. This anneal parameter affects (decreases) the probability that the annealing process moves to a worse solution than the current one to avoid potential local minima. This sampling results in the convergence of the overall probability distribution of the RBM where the global energy of the network fluctuates around the global minimum.

3   Method

First, we will explain how the restricted quantum Boltzmann machine can be used to learn an optimal traversal-policy in a single-agent gridworld setting. Next, in Sect. 3.2 we will explain how to extend this model to work for a multi-agent environment.

3.1   Single-Agent Quantum Learning

In [8], an approach to a restricted quantum Boltzmann machine was introduced for a gridworld problem. In their approach, each state is assigned an input node and each action an output node. Additional nodes in the hidden layers are used to be able to learn the best state-action combinations. The topology for the hidden layers is a hyperparameter that is set before the execution of the algorithm. The task presented to the restricted Boltzmann machine is to find the optimal traversal-policy of the grid, given a position and a corresponding action.

Using a Hamiltonian associated to a restricted Boltzmann machine, we can find its energy. In its most general form, the Hamiltonian $\mathcal{H}_{\mathbf{v}}$ is given by

$$\mathcal{H}_{\mathbf{v}} = -\sum_{\substack{v \in V \\ h \in H}} w_{vh} v \sigma_h^z - \sum_{\{v,v'\} \subseteq V} w_{vv'} vv' - \sum_{\{h,h'\} \subseteq H} w_{hh'} \sigma_h^z \sigma_{h'}^z - \Gamma \sum_{h \in H} \sigma_h^x \quad (3)$$

with $\mathbf{v}$ denoting the prescribed fixed assignments of the visible nodes, i.e. the input and output nodes. Here V is the set of all visible nodes, while H is the set of all hidden nodes. Note that setting $w_{hh'} = 0$ has the same effect as removing the link between nodes h and h'. Also, Γ is an annealing parameter, while σ_i^z and σ_i^x are the spin-values of node i in the z- and x-direction, respectively. Note that in Eq. (3) no σ_v^z variables occur, as the visible nodes are fixed for a given sample, indicated by the v-terms. Note the correspondence between this Hamiltonian and the global energy configuration given in Eq. (2). The optimal traversal-policy is found by training the restricted Boltzmann machine, which means that the weights $w_{vv'}$, w_{vh} and $w_{hh'}$ are optimized based on presented training samples.

For each training step i, a random state s_i^1 is chosen which, together with a chosen action a_i^1, forms the tuple (s_i^1, a_i^1). Based on this state-action combination, a second state is determined: $s_i^2 \leftarrow a_i^1(s_i^1)$. The corresponding optimal second action a_i^2 can be found by minimizing the free energy of the restricted Boltzmann machine given by the Hamiltonian of Eq. (3). As no closed expression exists for this free energy, an approximate approach based on sampling $\mathcal{H}_\mathbf{v}$ is used.

For all possible actions a from state s_i^2, the Q-function corresponding to the RBM is evaluated. The action a that minimizes Q, is taken as a_i^2. Ideally, one would use the Hamiltonian $\mathcal{H}_\mathbf{v}$ from Eq. (3) for the Q-function. However, $\mathcal{H}_\mathbf{v}$ has both σ_h^x and σ_h^z terms that correspond to the spin of variable h in the x- and z-direction. As these two directions are perpendicular, measuring the state of one direction destroys the state of the other. Therefore, instead of $\mathcal{H}_\mathbf{v}$, we use an effective Hamiltonian $\mathcal{H}_\mathbf{v}^{eff}$ for the Q-function. In this effective Hamiltonian all σ_h^x terms are replaced by σ^z terms by using so-called *replica stacking* [20], based on the Suzuki-Trotter expansion of Eq. (3) [13,26].

With replica stacking, the Boltzmann machine is replicated r times in total. Connections between corresponding nodes in adjacent replicas are added. Thus, node i in replica k is connected to node i in replica $k \pm 1$ modulo r. Using the replicas, we obtain a new effective Hamiltonian $\mathcal{H}_{\mathbf{v}=(s,a)}^{eff}$ with all σ^x variables replaced by σ^z variables. We refer to the spin variables in the z-direction as $\sigma_{i,k}$ for node i in replica k and we identify $\sigma_{h,0} \equiv \sigma_{h,r}$. All σ^z variables can be measured simultaneously. Additionally, the weights in the effective Hamiltonian are scaled by the number of replicas. In its clamped version, i.e. with $v = (s,a)$ fixed, the effective resulting Hamiltonian $\mathcal{H}_{\mathbf{v}=(s,a)}^{eff}$ is given by

$$
\mathcal{H}_{\mathbf{v}=(s,a)}^{eff} = - \sum_{\substack{h \in H \\ h-s \text{ adjacent}}} \sum_{k=1}^{r} \frac{w_{sh}}{r} \sigma_{h,k} - \sum_{\substack{h \in H \\ h-a \text{ adjacent}}} \sum_{k=1}^{r} \frac{w_{ah}}{r} \sigma_{h,k}
$$

$$
- \sum_{\{h,h'\} \subseteq H} \sum_{k=1}^{r} \frac{w_{hh'}}{r} \sigma_{h,k} \sigma_{h',k} - J^+ \sum_{h \in H} \sum_{k=0}^{r} \sigma_{h,k} \sigma_{h,k+1}. \qquad (4)
$$

Note that J^+ is an annealing parameter that can be set and relates to the original annealing parameter Γ. Throughout this paper, the values selected for Γ and J^+ are identical to those in [8].

For a single evaluation of the Hamiltonian and all corresponding spin variables, we get a specific spin configuration $\hat{h}$. We evaluate the circuit n_{runs} times for a fixed combination of s and a, which gives a multi-set $\hat{h}_{s,a} = \{\hat{h}_1, \ldots, \hat{h}_{n_{runs}}\}$ of evaluations. From $\hat{h}_{s,a}$, we construct a set of configurations $C_{\hat{h}_{s,a}}$ of unique spin combinations by removing duplicate solutions and retaining only one occurrence of each spin combination. Each spin configuration in $C_{\hat{h}_{s,a}}$ thus corresponds to one or more configurations in $\hat{h}_{s,a}$, and each configuration in $\hat{h}_{s,a}$ corresponds to precisely one configuration in $C_{\hat{h}_{s,a}}$.

The quality of $\hat{h}_{s,a}$, and implicitly of the weights of the RBM, is evaluated using the Q-function

$$Q(s,a) = -\left\langle \mathcal{H}^{eff}_{\mathbf{v}=(s,a)} \right\rangle - \frac{1}{\beta} \sum_{c \in C_{\hat{h}_{s,a}}} \mathbb{P}(c|s,a) \log \mathbb{P}(c|s,a), \tag{5}$$

where the Hamiltonian is averaged over all spin-configurations in $\hat{h}_{s,a}$. Furthermore, β is an annealing parameter and the frequency of occurrence of c in $\hat{h}_{s,a}$ is given by the probability $\mathbb{P}(c|s,a)$. The effective Hamiltonian $\mathcal{H}^{eff}_{\mathbf{v}=(s,a)}$ from Eq. (4) is used.

Using the Q-function from Eq. (5), the best action a_i^2 for state s_i^2 is given by

$$a_i^2 = \operatorname{argmin}_a Q(s,\,a) \tag{6}$$

$$= \operatorname{argmin}_a \left(-\left\langle \mathcal{H}^{\text{eff}}_{\mathbf{v}=(s,a)} \right\rangle - \frac{1}{\beta} \sum_{c \in C_{\hat{h}_{s,a}}} \mathbb{P}(c|s,a) \log \mathbb{P}(c|s,a) \right). \tag{7}$$

Once the optimal action a_i^2 for state s_i^2 is found, the weights of the restricted Boltzmann machine are updated following

$$\Delta w_{hh'} = \epsilon \left(R_{a_i^1}\left(s_i^1, s_i^2\right) + \gamma Q\left(s_i^2, a_i^2\right) - Q\left(s_i^1, a_i^1\right) \right) \langle hh' \rangle, \tag{8}$$

$$\Delta w_{vh} = \epsilon \left(R_{a_i^1}\left(s_i^1, s_i^2\right) + \gamma Q\left(s_i^2, a_i^2\right) - Q\left(s_i^1, a_i^1\right) \right) v \langle h \rangle, \tag{9}$$

where, v is one of the clamped variables s_i^1 or a_i^1. The averages $\langle h \rangle$ and $\langle hh' \rangle$ are obtained by averaging the spin configurations in $\hat{h}_{s,a}$ for each h and all products hh' for adjacent h and h'. Based on the gridworld, a reward or penalty is given using the reward function $R_{a_i^1}(s_i^1, s_i^2)$. The learning rate is given by ϵ, and γ is a discount factor related to expected future rewards, representing a feature of the problem.

If the training phase is sufficiently long, the weights are updated such that the restricted Boltzmann machine gives the optimal policy for all state-action combinations. The required number of training samples depends on the topology of the RBM and the specific problem at hand. In the next section we will consider the extensions on this model to accommodate multi-agent learning.

3.2   Multi-agent Quantum Learning

In the previous section we considered a model with only a single agent having to learn an optimal policy in a grid, however, many applications involve multiple agents having conjoined tasks. For instance, one may think of a search-and-rescue setting where first an asset must be secured before a safe-point can be reached.

This model can be solved in different ways. First and foremost, different models can be trained for each task/agent involved. In essence, this is a form of multiple independent single-agent models. We will however focus on a model including *all* agents and *all* rewards simultaneously. This can be interpreted as one party giving orders to all agents on what to do next, given the states.

We consider the situation where the number of target locations is equal to the number of agents and each agent has to reach a target. The targets are not preassigned to the agents, however, each target can only be occupied by one agent simultaneously.

For this multi-agent setting, we extend the restricted Boltzmann machine as presented before. Firstly, each action of each agent is considered as an input state. For M agents, each with N actions, this gives MN input states. The output states are all possible combinations of the different actions of all agents. This means that if each agent has k possible actions, there are k^M different output states.

When using a one-hot encoding of states to nodes, the number of nodes in the network increases significantly compared to using a binary qubit encoding which allows for a more efficient encoding. Training the model using a binary encoding, however, is more complex than with one-hot encoding since for the former only a few nodes carry information on which states and actions are of interest while for binary encoding all nodes are used to encode the information. Therefore, we chose one-hot encoding similar to [8].

The Boltzmann machine for a multi-agent setting is closely related to that of a single-agent setting. An example is given in Fig. 1b for two agents. Here, input $s_{i,j}$ represents state i of agent j and output (a_m, a_n) means action m for the first agent and action n for the second.

Apart from a different RBM topology, also the effective Hamiltonian of Eq. (4) changes to accommodate the extra agents and the increase in possible action combinations for all agents. Again, all weights are initialized and state-action combinations denoted by tuples $(s_{i_1,1}, \ldots, s_{i_M,M}, (a_{i_1}, \ldots, a_{i_M}))$, are given as input to the Boltzmann machine. Let $a = (a_{i_1}, \ldots, a_{i_M})$ and $\mathcal{S} = \{s_{i_1,1}, \ldots, s_{i_M,M}\}$ and let r be the number of replicas. Nodes corresponding to these states and actions are clamped to 1, and other visible nodes are clamped to 0. The effective Hamiltonian is then given by

$$\mathcal{H}_{\mathbf{v}}^{eff}(v = (\mathcal{S}, a)) = - \sum_{\substack{s \in \mathcal{S}, h \in H \\ h-s \text{ adjacent}}} \sum_{k=1}^{r} \frac{w_{sh}}{r} \sigma_{h,k} - \sum_{\substack{h \in H \\ h-a \text{ adjacent}}} \sum_{k=1}^{r} \frac{w_{ah}}{r} \sigma_{h,k}$$

$$- \sum_{\{h,h'\} \subseteq H} \sum_{k=1}^{r} \frac{w_{hh'}}{r} \sigma_{h,k} \sigma_{h',k} - J^{+} \sum_{h \in H} \sum_{k=0}^{r} \sigma_{h,k} \sigma_{h,k+1}.$$

$$(10)$$

In each training iteration a random state for each agent is chosen, together with the corresponding action. For each agent, a new state is determined based on these actions. The effective Hamiltonian is sampled n_{runs} times and the next best actions for the agents are found by minimizing the Q-function, with Eq. (10) used as effective Hamiltonian in Eq. (5). Next, the policy and weights of the Boltzmann machine are updated.

To update the weights of connections, Eq. (8) and Eq. (9) are used. Note that this requires the reward function to be evaluated for the entirety of the choices, consisting of the new states of all agents and the corresponding next actions.

4 Numerical Experiments

In this section the setup and results of our experiments are presented and discussed. First we explain how the models are sampled in Sect. 4.1, then the used gridworlds are introduced in Sect. 4.2. The corresponding results for the single-agent and multi-agent learning are presented and discussed in Sect. 4.3 and Sect. 4.4, respectively.

4.1 Simulated Quantum Annealing

There are various annealing dynamics [4] that can be used to sample spin values from the Boltzmann distribution resulting from the effective Hamiltonian of Eq. (3). The case $\Gamma = 0$ corresponds to purely classical simulated annealing [19]. Simulated annealing (SA) is also known as thermal annealing and finds its origin in metallurgy where the cooling of a material is controlled to improve its quality and correct defects.

For $\Gamma \neq 0$, we have quantum annealing (QA) if the annealing process starts from the ground state of the transverse field and ends with a classical energy corresponding to the ground state energy of the Hamiltonian. The ground energy corresponds to the minimum value of the cost function that is optimized. No replicas are used for QA. The devices made by D-Wave Systems physically implement this process of quantum annealing.

However, we can also simulate quantum annealing using the effective Hamiltonian with replicas (Eq. (4)) instead of the Hamiltonian with the transverse field (Eq. (3)). This representation of the original Hamiltonian as an effective one, corresponds to simulated quantum annealing (SQA). Theoretically, SQA is a method to classically emulate the dynamics of quantum annealing by a quantum Monte Carlo method whose parameters are changed slowly during the simulation [22]. In other words, by employing the Suzuki-Trotter formula with replica stacking, one can simulate the quantum system described by the original Hamiltonian in Eq. (3).

Although SQA does not reproduce quantum annealing, it provides a way to understand phenomena such as tunneling in quantum annealers [16]. SQA can have an advantage over SA thanks to the capability to change the amplitudes of states in parallel, as proven in [9]. Therefore, we opted for SQA in our numerical experiments. We implemented the effective Hamiltonian on two different back-ends. The first using classical sampling given by simulated annealing (SQA SA). The second by implementing the effective Hamiltonian on the D-Wave 2000Q (SQA D-Wave 2000Q), a 2048 qubits quantum processor. Furthermore, we implemented a classical DRL algorithm for comparison.

(a) Gridworld problems considered in single-agent experiments. (b) Gridworld problems considered in multi-agent experiments.

Fig. 2. All grids used to test the performance. On the left the three grids used for the single-agent scenario. On the right the grids used for multi-agent learning. The 3×3-grid is used with and without wall.

4.2 Gridworld Environments

We illustrate the free energy-based reinforcement learning algorithm proposed in Sect. 3.2 by applying it on the environments shown in Fig. 2a and Fig. 2b for one and two agents, respectively. In each gridworld, agents are allowed to take the actions up, down, left, right and stand still. We consider only examples with deterministic rewards. Furthermore, we consider environments both with and without forbidden states (walls) or penalty states. The goal of the agent is to reach the reward while avoiding penalty states. In case of multiple agents and rewards, each of the agents must reach a different reward. The considered multi-agent gridworlds focus on two agents. These environments are however easily extendable to an arbitrary number of agents.

The discount factor γ, explained in Sect. 3.1, was set to 0.8, similar to [8]. An agent reaching a target location is rewarded a value of 200, while ending up in a penalty state is penalized by -200. An extra penalty of -10 is given for each step an agent takes. As the rewards propagate through the network, the penalty assigned to taking steps is overcome. In the multi-agent case, a reward of 100 is given to each agent if each is at a different reward state simultaneously.

To assess the results of the multi-agent QBM-based reinforcement learning algorithm, we compare the learned policy for each environment with the optimal one using a fidelity measure. The optimal policy for this measure was determined logically thanks to the simple nature of these environments. As fidelity measure for the single-agent experiments, the formula from [20] is used. The fidelity at the i-th training sample for the multi-agent case with n agents is defined as

$$fidelity(i) = (T_r \times |S|^n)^{-1} \sum_{k=1}^{T_r} \sum_{s \in S^n} \mathbf{1}_{A(s,i,k) \in \pi^*(s)}. \tag{11}$$

Here, T_r denotes the number of independent runs for the method, $|S|$ denotes the total amount of states in the environment, π^* denotes the optimal policy and $A(s,i,k)$ denotes the action assigned at the k-th run and i-th training sample to the state pair s. Each state pair s is an n-tuple consisting of the state of each agent. This definition of fidelity for the multi-agent case essentially records the amount of state pairs in which all agents took the optimal actions over all runs.

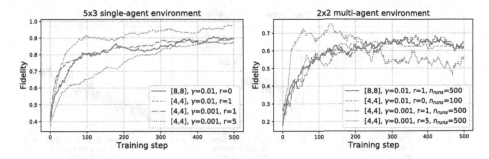

Fig. 3. Fidelity scores corresponding to hyperparameter choices with the highest average fidelity for the 5×3 single-agent and 2×2 multi-agent gridworld. The hyperparameters considered are the hidden layer size (described as an array of the number of nodes per layer), the learning rate γ and the number of replicas r. For the multi-agent grid search we also considered the number of samples per training step. The legends indicate the used values. (Color figure online)

4.3   Single-Agent Results

Before running the experiment, a grid search is performed to find the best setting for some hyperparameters. The parameters considered are: structure of the hidden layers, learning rate γ and number of replicas r used. These replicas are needed for the Suzuki-Trotter expansion of Eq. (3). The SQA SA reinforcement learning algorithm was run $T_r = 20$ times on the 5×3 grid shown in Fig. 2a for $T_s = 500$ training samples each run. In total, 18 different hyperparameter combinations are considered. For each, an average fidelity over all training steps is computed. The four best combinations are shown in the left plot of Fig. 3. Based on these results, the parameters corresponding to the orange curve (i.e. hidden layer size $= [4, 4], \gamma = 0.01, r = 1$) have been used in the experiments. These settings are used for all single-agent environments. The three different sampling approaches explained in Sect. 4.1 are used for each of the three environments. The results are all shown in Fig. 4.

We achieved similar results compared to the original single-agent reinforcement learning work in [8]. Our common means of comparison is the 5×3 gridworld problem, which in [8] also exhibits the best performance with SQA. Despite the fact that we did not make a distinction on the underlying graph of the SQA method, in our case the algorithm seems to achieve a higher fidelity within the first few training steps (~ 0.9 at the 100-th step in comparison to ~ 0.6 in [8]) and to exhibit less variation in the fidelity later on in training. This may be due to the different method chosen for sampling the effective Hamiltonian.

Comparing sampling using SQA simulated annealing with SQA D-Wave 2000Q, we see the latter shows more variance in the results. This can be explained by the stochastic nature of the D-Wave system, the limited availability of QPU time in this research and the fact that only 100 D-Wave 2000Q samples are used at every training step. We expect that increasing the number of D-Wave 2000Q samples per training iteration increases the overall fidelity and results in a

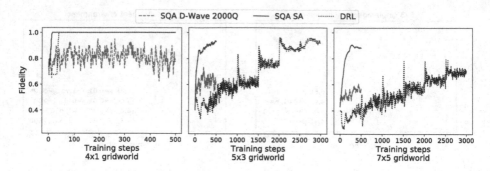

Fig. 4. The performance of the different RL implementations for the three single-agent gridworlds. All algorithms have been run $T_r = 10$ times.

smoother curve. However, it could also stem from the translation of the problem to the D-Wave 2000Q architecture. A problem that is too large to be directly embedded on the QPU is decomposed in smaller parts and the result might be suboptimal. This issue can be resolved by a richer QPU architecture or a more efficient decomposition.

Furthermore, the results could also be improved by a more environment specific hyperparameter selection. We now used the hyperparameters optimized for the 5×3 gridworld for each of the other environments. A gridsearch for each environment separately will probably improve the results. Increasing the number of training steps and averaging over more training runs will likely give a better performance and reduce variance for both SQA methods. Finally, adjusting the annealing schedule by optimizing the annealing parameter Γ could also lead to significantly better results.

Comparing the DRL to the SQA SA algorithm, we observe that the SQA SA algorithm achieves a higher fidelity using fewer training samples than the DRL for all three environments. Even SQA D-Wave 2000Q, with the limitations listed above, outperforms the classical reinforcement learning approach with exception of the 4×1 gridworld, the simplest environment. It is important to note that the DRL algorithm will ultimately reach a fidelity similar to both SQA approaches, but it does not reach this performance for the 5×3 and 7×5 gridworlds until having taken about six to twenty times as many training steps, respectively. Hence, the simulated quantum annealing approach on the D-Wave system learns more efficiently in terms of timesteps.

4.4   Multi-agent Results

As the multi-agent environments are fundamentally different from the single-agent ones, different hyperparameters might be needed. Therefore, we again run a grid search to find the optimal values for the same hyperparameters as in the single-agent case. Additionally, due to the complexity of the multi-agent environments, the number of annealing samples per training step $n_{runs} \in \{100, 500\}$ is also considered in the grid search.

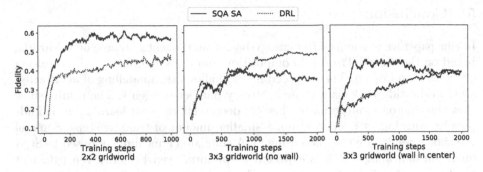

Fig. 5. Different RL methods for the three multi-agent gridworlds. All results are averaged over $T_r = 5$ runs.

For each combination of hyperparameters, the algorithm was run $T_r = 15$ times for $T_s = 500$ training samples each run for the 2×2 gridworld problem shown in Fig. 2b. In total, 36 combinations are considered and the performance of the best four combinations is given in the right plot in Fig. 3.

Based on the results from this gridsearch, suitable choices for the model parameters would either be given by the parameter sets corresponding to the green fidelity curve or the blue one. We opt for the blue fidelity curve, corresponding to a hidden layer topology of $[8, 8]$, a learning rate $\gamma = 0.01$, one replica and $n_{runs} = 500$ samples per training step, since that one is expected to be generalize better due to the larger hidden network and increased sampling.

The same hyperparameters found in the grid search conducted on the 2×2 gridworld problem are used for the two other environments. In Fig. 5, the results for the multi-agent environments are shown. As the available D-Wave 2000Q QPU time was limited in this research, only the results for the multi-agent SQA simulated annealing and the multi-agent DRL method are shown. An aspect that immediately stands out from the performance plots is the fast learning rate achieved by SQA SA within the first 250 training steps. In the case of classical DRL, learning progresses slower and the maximum fidelity reached is still lower than the best values achieved by SQA in the earlier iterations. We also see that the overall achieved fidelity is rather low for each of the environments compared to the single agent environments. This indicates that the learned policies are far from optimal. This can be due to the challenging nature of the small environments where multiple opposing strategies can be optimal, for instance, agent 1 moving to target 1 and agent 2 to target 2, and vice versa.

We expect the results for SQA D-Wave 2000Q to be better than the classical results, as SQA D-Wave 2000Q excels at sampling from a Boltzmann distribution, given sufficiently large hardware and sufficiently long decoherence times. We see that for two of the three environments, SQA SA learns faster and achieves at least a similar fidelity as classical methods. This faster learning and higher achieved fidelity is also expected of SQA D-Wave 2000Q.

5   Conclusion

In this paper we introduced free energy-based multi-agent reinforcement learning based on the Suzuki-Trotter decomposition and SQA sampling of the resulting effective Hamiltonian. The proposed method allows the modelling of arbitrarily-sized gridworld problems with an arbitrary number of agents. The results show that this approach outperforms classical deep reinforcement learning, as it finds policies with higher fidelity within a smaller amount of training steps. Some of the shown results are obtained using SQA simulated annealing, opposed to SQA quantum annealing which is expected to perform even better, given sufficient hardware and sufficiently many runs. Hence, a natural progression of this work would be to obtain corresponding results for SQA D-Wave 2000Q. The current architecture of the quantum annealing hardware is rather limited in size and a larger QPU is needed to allow fast and accurate reinforcement learning algorithm implementations of large problems.

Furthermore, implementing the original Hamiltonian without replicas on quantum hardware, thus employing proper quantum annealing, might prove beneficial. This takes away the need for the Suzuki-Trotter expansion and thereby a potential source of uncertainty. Moreover, from a practical point of view, it is worthwhile to investigate more complex multi-agent environments, where agents for instance have to compete or cooperate, or environments with stochasticity.

References

1. Ackley, D.H., Hinton, G.E., Sejnowski, T.J.: A learning algorithm for Boltzmann machines. Cogn. Sci. **9**(1), 147–169 (1985)
2. Agostinelli, F., McAleer, S., Shmakov, A., Baldi, P.: Solving the Rubik's cube with deep reinforcement learning and search. Nat. Mach. Intell. **1**, 356–363 (2019)
3. Arel, I., Liu, C., Urbanik, T., Kohls, A.: Reinforcement learning-based multi-agent system for network traffic signal control. IET Intell. Transp. Syst. **4**(2), 128–135 (2010)
4. Bapst, V., Semerjian, G.: Thermal, quantum and simulated quantum annealing: analytical comparisons for simple models. J. Phys. Conf. Ser. **473**, 012011 (2013)
5. Bellman, R.: On the theory of dynamic programming. Proc. Natl. Acad. Sci. **38**(8), 716–719 (1952)
6. Bellman, R.: A Markovian decision process. Indiana Univ. Math. J. **6**, 679–684 (1957)
7. Benedetti, M., Realpe-Gómez, J., Perdomo-Ortiz, A.: Quantum-assisted Helmholtz machines: a quantum-classical deep learning framework for industrial datasets in near-term devices. Quantum Sci. Technol. **3**(3), 034007 (2018)
8. Crawford, D., Levit, A., Ghadermarzy, N., Oberoi, J.S., Ronagh, P.: Reinforcement learning using quantum Boltzmann machines. CoRR 1612.05695
9. Crosson, E., Harrow, A.W.: Simulated quantum annealing can be exponentially faster than classical simulated annealing. In: 2016 IEEE 57th Annual Symposium on Foundations of Computer Science (FOCS). IEEE, October 2016
10. Dong, D., Chen, C., Li, H., Tarn, T.J.: Quantum reinforcement learning. IEEE Trans. Syst. Man Cybern. Part B (Cybern.) **38**(5), 1207–1220 (2008)

11. Dunjko, V., Taylor, J.M., Briegel, H.J.: Quantum-enhanced machine learning. Phys. Rev. Lett. **117**(13), 130501 (2016)
12. Finnila, A., Gomez, M., Sebenik, C., Stenson, C., Doll, J.: Quantum annealing: a new method for minimizing multidimensional functions. Chem. Phys. Lett. **219**(5), 343–348 (1994)
13. Hatano, N., Suzuki, M.: Finding exponential product formulas of higher orders. In: Das, A., Chakrabarti, B.K. (eds.) Quantum Annealing and Other Optimization Methods. LNP, vol. 679, pp. 37–68. Springer, Heidelberg (2005). https://doi.org/10.1007/11526216_2
14. Hilbert, M., López, P.: The world's technological capacity to store, communicate, and compute information. Science **332**(6025), 60–65 (2011)
15. Howard, R.A.: Dynamic Programming and Markov Processes. Wiley for The Massachusetts Institute of Technology, Cambridge (1964)
16. Isakov, S.V., et al.: Understanding quantum tunneling through quantum Monte Carlo simulations. Phys. Rev. Lett. **117**(18), 180402 (2016)
17. Kadowaki, T., Nishimori, H.: Quantum annealing in the transverse Ising model. Phys. Rev. E **58**, 5355–5363 (1998)
18. Khoshaman, A., Vinci, W., Denis, B., Andriyash, E., Amin, M.H.: Quantum variational autoencoder. CoRR (2018)
19. Kirkpatrick, S., Gelatt, C.D., Vecchi, M.P.: Optimization by simulated annealing. Science **220**(4598), 671–680 (1983)
20. Levit, A., Crawford, D., Ghadermarzy, N., Oberoi, J.S., Zahedinejad, E., Ronagh, P.: Free energy-based reinforcement learning using a quantum processor. CoRR (2017)
21. Li, R.Y., Felice, R.D., Rohs, R., Lidar, D.A.: Quantum annealing versus classical machine learning applied to a simplified computational biology problem. NPJ Quantum Inf. **4**(1), 1–10 (2018)
22. Mbeng, G.B., Privitera, L., Arceci, L., Santoro, G.E.: Dynamics of simulated quantum annealing in random Ising chains. Phys. Rev. B **99**(6), 064201 (2019)
23. Neukart, F., Compostella, G., Seidel, C., Dollen, D.V., Yarkoni, S., Parney, B.: Traffic flow optimization using a quantum annealer. Front. ICT **4**, 29 (2017)
24. Neukart, F., Dollen, D.V., Seidel, C.: Quantum-assisted cluster analysis on a quantum annealing device. Front. Phys. **6**, 55 (2018)
25. Silver, D., et al.: Mastering the game of go with deep neural networks and tree search. Nature **529**(7587), 484–489 (2016)
26. Suzuki, M.: Generalized Trotter's formula and systematic approximants of exponential operators and inner derivations with applications to many-body problems. Commun. Math. Phys. **51**(2), 183–190 (1976). https://doi.org/10.1007/BF01609348
27. Waldrop, M.M.: The chips are down for Moore's law. Nature **530**(7589), 144–147 (2016)
28. Xia, R., Bian, T., Kais, S.: Electronic structure calculations and the Ising Hamiltonian. J. Phys. Chem. B **122**(13), 3384–3395 (2018)
29. Zhu, Y., et al.: Target-driven visual navigation in indoor scenes using deep reinforcement learning. In: 2017 IEEE International Conference on Robotics and Automation (ICRA), pp. 3357–3364. IEEE (2017)

Quantum Hopfield Neural Networks: A New Approach and Its Storage Capacity

Nicholas Meinhardt[1,2], Niels M. P. Neumann[1(✉)], and Frank Phillipson[1]

[1] The Netherlands Organisation for Applied Scientific Research,
Anna van Buerenplein 1, 2595 DA The Hague, The Netherlands
{niels.neumann,frank.phillipson}@tno.nl
[2] Department of Physics, ETH Zurich, 8093 Zurich, Switzerland
nmeinhar@student.ethz.ch

Abstract. At the interface between quantum computing and machine learning, the field of quantum machine learning aims to improve classical machine learning algorithms with the help of quantum computers. Examples are Hopfield neural networks, which can store patterns and thereby are used as associative memory. However, the storage capacity of such classical networks is limited. In this work, we present a new approach to quantum Hopfield neural networks with classical inputs and outputs. The approach is easily extendable to quantum inputs or outputs. Performance is evaluated by three measures of error rates, introduced in this paper. We simulate our approach and find increased storage capacity compared to classical networks for small systems. We furthermore present classical results that indicate an increased storage capacity for quantum Hopfield neural networks in large systems as well.

Keywords: Hopfield neural networks · Gate-based quantum computing · Storage capacity · Quantum machine learning

1 Introduction

While conventional computers are restricted to classical operations, quantum computers implement the rules of quantum mechanics to process information [5], using quantum principles such as superpositions. The basic units to store information on quantum computers are two-level quantum bits, or qubits. Due to superpositions of both levels, qubits allow for a more flexible representation of information than classical bits. One widely accepted premise is that quantum computers have computational advantages over classical processing [6], giving rise to the notion of 'quantum supremacy' [13], which only recently has been claimed for the first time in experiments [1].

A candidate to show a quantum advantage is believed to be quantum machine learning (QML) [4,12], a field of research at the interface between quantum information processing and machine learning. Even though machine learning is an important tool that is widely used to process data and extract information

© Springer Nature Switzerland AG 2020
V. V. Krzhizhanovskaya et al. (Eds.): ICCS 2020, LNCS 12142, pp. 576–590, 2020.
https://doi.org/10.1007/978-3-030-50433-5_44

from it [4], it also faces its limits. The amount of data processed worldwide each year is steadily increasing, while the limits of computing power are rapidly approaching [7]. Therefore, more efficient algorithms, such as found in the quantum domain, are crucial.

We consider neural networks (NN), a subclass of machine learning algorithms consisting of nodes that can be connected in various configurations and interact with each other via weighted edges. As special case, Hopfield neural networks (HNN) consist of a single layer of nodes, all connected with each other via symmetric edges and without self-connections [8]. In an HNN, nodes are updated using the updating rule $x_i \leftarrow \text{sign}\left(\sum_{j=1}^{n} W_{ij} x_j\right)$, where $\text{sign}\,(\cdot)$ refers to the sign-function, W_{ij} is the weight between node i and j and $W_{ii} = 0$. A graphical representation of an HNN is given in Fig. 1, where k an indicator for the number of updating iterations in the direction of the dashed arrows.

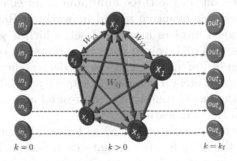

Fig. 1. Schematic overview of a fully-connected Hopfield neural network with 5 neurons. First, the neurons are initialized (orange nodes), then the network evolves in time or number of iterations k according to the weight matrix with entries W_{ij} (blue plane). The final configuration is read out (green nodes). The dashed arrows indicate the direction of updating or time. (Color figure online)

Due to this connectivity, HNNs can be used as associative memories, meaning that they can store a set of patterns and associate noisy inputs with the closest stored pattern. Memory patterns can be imprinted onto the network by the use of training schemes, for instance Hebbian learning [15]. Here, the weights are calculated directly from all memory patterns, and thereby only a low computational effort is required. It is possible to store an exponential number of stable attractors in an HNN if the set of attractors is predetermined and fixed [11]. In general, however, fewer patterns can be stored if they are randomly selected, resulting in a very limited storage capacity of HNNs. For Hebbian learning $n/(4 \log n)$ patterns can be stored asymptotically in an HNN with n nodes [9].

Translating HNNs to counterparts in the quantum domain is assumed to offer storage capacities beyond the reach of classical networks [14,18]. For example, in [18] a quantum HNN is proposed that could offer an exponential capacity when qutrits are used. When using qubits however, no increased capacity has been demonstrated yet for quantum HNNs.

In this work, we provide a new approach for hybrid quantum-classical HNNs, which 1) allows for classical and quantum inputs and outputs; 2) is able to store classical bit strings as attractors; and 3) fulfills three minimum requirements to allow an analogy to classical HNNs as posed in Ref. [16]. The first requirement is that the quantum HNN must comply with quantum theory while respecting the structure of NNs. The second requirement is that the quantum model should solve the discrepancy between unitary quantum evolution and dissipative dynamics of NNs. Thirdly, it should offer the feature of an associative memory, meaning that inputs are mapped to the closest stable outputs that encode the learned bit patterns. We furthermore provide numerical evidence that the capacity indeed increases for gate-based quantum HNNs, when storing randomly chosen bit strings.

Previously proposed implementations of HNNs either deal with non-random memory patterns [3], or do not account for the discrepancy between dissipative and unitary dynamics, one of the three minimum requirements [14]. We follow the recent proposal of deep quantum neural networks in Ref. [2] for our HNN-development. Our model involves a training set, which is generated based on the chosen memories, and all involved gate operations are optimized using the training scheme given in Ref. [2]. We test the model's ability to store randomly chosen bit strings and thereby estimate its capacity. While limited to small system sizes due to the model complexity, the results are compared to those of a classical HNN with Hebbian learning.

The remainder of this work is organized as follows: We present our quantum model in Sect. 2 and the setup for the simulations in Sect. 3. The results of these simulations are given in Sect. 4. Finally, we provide a summary of the results in Sect. 5 and a conclusion in Sect. 6.

2   Quantum Hopfield Neural Networks

We first present a feed-forward interpretation of quantum HNNs in Sect. 2.1 and then explain how to train these feed-forward quantum HNNs in Sect. 2.2.

2.1   A Feed-Forward Interpretation of Quantum HNNs

HNNs can be implemented as feed-forward NNs by regarding each update step as a new layer of neurons. In the feed-forward interpretation, the weight matrix is layer depended and can be written as $W^{(l)}$. Depending on whether the HNN is updated synchronously or asynchronously, the weights might differ between layers. In the former case, the weights $W_{ij}^{(l)}$ are exactly as W_{ij} of the usual HNN. Hence, the weights are symmetric in both the subscripts and the layers and the superscript l can be omitted. Note that HNNs have no self-connections, such that $W_{ii} = 0$ for all i. Therefore, the interpretation of an HNN with synchronous updating as a feed-forward NN is valid. The number of layers l can be seen as a time parameter. Figure 2a shows an HNN with three neurons and a feed-forward interpretation of this network is given in Fig. 2b.

Note that we are not restricted to synchronous updating. In principle any updating rule may be applied and the weights of the feed-forward interpretation may differ drastically from the ones of the single-layer scheme in general. The weights do not necessarily need to agree with the ones of Hebbian learning. Note that the fundamental properties of HNNs of storing and retrieving patterns are retained.

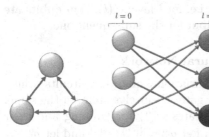

(a) A Hopfield neural network with three neurons and without self-connections.

(b) A feed-forward interpretation of Fig. 2a. A single updating step l, without connections between the i-th neuron of adjacent layers. The colors indicate three different outputs.

(c) Quantum circuit of the network of Fig. 2b. Each unitary U_i is associated with the output neuron of the same color. The U_i act on a fresh ancilla qubit illustrated in gold and all input qubits except for the i-th qubit shown in gray. The overall quantum operation U is a product of all U_i.

Fig. 2. Two interpretations of the updating process in classical HNNs and the corresponding quantum model. (Color figure online)

One important advantage of the feed-forward interpretation is that we can use existing proposals to translate classical NNs to a gate-based quantum analog. To implement quantum analogs of feed-forward NNs, neurons can be implemented directly as qubits and weights between two neurons as operators [14,17]. We use another promising approach to feed-forward NNs, where unitary operations U acting on a quantum register are associated with the classical perceptrons [2]. In the following, we will only consider a single synchronous update. More updating steps can be added by repeating the presented approach.

Using the qubit encoding scheme, a classical bit string $(x_1, \ldots, x_n)$ is encoded in the corresponding computational basis state $|x_1 \ldots x_n\rangle$. In HNNs, neurons can only take the values ± 1 and we identify $+1 \leftrightarrow |0\rangle$ and $-1 \leftrightarrow |1\rangle$. Consequently, the classical input layer is replaced by state initialization of the quantum input register. The neurons of each subsequent layer of the classical feed-forward NN model are replaced by unitaries U_i, which act on the input register and each on an additional, freshly prepared ancilla qubit. Figure 2c gives an example of this quantum analogue for three neurons and a single update. The colors correspond with those of the classical neurons of the classical network in Fig. 2b and the golden lines represent the ancilla qubits.

Note that input qubit i is not affected due to the absence of self-connections. The only output qubit affected by unitary U_i is ancilla qubit i and the output state corresponds to the output of classical neuron i. To retrieve a classical output from the quantum system, ancilla qubits are measured at the end of the circuit. Using a majority vote over multiple measurement rounds, the most likely outcome is chosen as updated state. The original input qubits are discarded after an update round of applying all U_i, meaning the input register is traced out. For a single update, $2n$ qubits are needed. For l updates, i.e. for l layers, $(l+1)n$ qubits are needed and the output of a layer is used as input for the subsequent one.

2.2 Training a Quantum Hopfield Neural Network

The goal is to train the unitaries, opposed to variational quantum circuits, where classical gate parameters are trained and the gates themselves remain the same. Assume we have a training set of N input states $|\phi_k^{\text{in}}\rangle$ for training and their desired output states $|\phi_k^{\text{out}}\rangle$, for $k = 1, \ldots, N$. Let $\rho_k^{\text{in}} = |\phi_k^{\text{in}}\rangle\langle\phi_k^{\text{in}}|$ and let $\rho_k^{\text{out}} = U |\phi_k^{\text{in}}\rangle\langle\phi_k^{\text{in}}| U^\dagger$ be the actual output of the quantum circuit U with input $|\phi_k^{\text{in}}\rangle$. Furthermore, let the fidelity of the circuit be given by

$$\mathcal{F}\left(U |\phi_k^{\text{in}}\rangle\langle\phi_k^{\text{in}}| U^\dagger, |\phi_k^{\text{out}}\rangle\langle\phi_k^{\text{out}}|\right) = \langle\phi_k^{\text{out}}| U |\phi_k^{\text{in}}\rangle\langle\phi_k^{\text{in}}| U^\dagger |\phi_k^{\text{out}}\rangle. \tag{1}$$

This fidelity corresponds with how well the output state after applying the unitary gates matches the desired output state. The cost function C is defined as

$$C = \frac{1}{N} \sum_{k=1}^{N} \langle\phi_k^{\text{out}}| \rho_k^{\text{out}} |\phi_k^{\text{out}}\rangle. \tag{2}$$

To optimize C, we train the unitary operations $U_j(s)$, which are parametrized by s as a measure of the training iterations or the training duration. After a time step ε, the unitaries are then updated according to

$$U_j(s + \varepsilon) = e^{i\varepsilon K_j(s)} U_j(s), \tag{3}$$

where the $K_j(s)$ are Hermitian matrices given by

$$K_j(s) = \frac{i 2^{n+1}}{2N\lambda} \sum_k \text{Tr}_{\text{rest}}[M_j^k(s)]. \tag{4}$$

Here $1/\lambda$ is a learning rate. These $K_j(s)$ can be estimated by taking a partial trace of matrices M_j^k that act on the whole space $\mathcal{H}_{2^{2n}}$ of all input and output qubits. This partial trace Tr_{rest} traces out all qubits not related to the unitary U_j. These qubits are all other ancilla qubits and the j-th input qubit if self-connections are removed. The M_j^k can be calculated from all unitaries, input and output training states as

$$M_j^k(s) = \Big[U_j(s)\cdots U_1(s)\rho_k^{\text{in}} \otimes |0...0\rangle\langle0...0| U_1^\dagger(s)\cdots U_j^\dagger(s),$$
$$U_{j+1}^\dagger(s)\ldots U_n^\dagger(s)\mathbb{1} \otimes |\phi_k^{\text{out}}\rangle\langle\phi_k^{\text{out}}| U_n(s)\cdots U_{j+1}(s)\Big], \tag{5}$$

where $[A, B] = AB - BA$ is the commutator of two operators A and B.

This updating scheme can be applied and implemented directly. In each iteration, all M_j^k are estimated and K_j is obtained by tracing out all unrelated qubits. Using Eq. (3), the unitaries are consequently updated in small time steps ε. The derivation of Eq. (4) and (5) involves Tayloring the exponential in Eq. (3) around $\varepsilon = 0$ to the first order and is provided in [2].

3   Simulating HNNs

In this section, we present the setup of our simulations of the HNNs. First we introduce the training set in Sect. 3.1, then we discuss the scaling of the simulations in Sect. 3.2 and afterwards we explain how to evaluate the performance of the HNNs in Sect. 3.3. Finally, Sect. 3.4 explains how we implemented the simulations and how we ran them.

3.1   Creating a Training Set

Let S be a set of m classical memory patterns $S = \{x^{(p)}\}_p$. We generate a training set T of input and output states from S using the qubit encoding. First, we add all memory patterns $x^{(p)}$ as both input and output patterns to T. Additionally, we add noisy memory patterns to the training set to prevent the unitaries from simply swapping input and output registers, without actually acting as an associative memory. All bit strings at a Hamming distance smaller or equal to d around each memory pattern in S are used as input states and are denoted by $|x_{p_k}^{in}\rangle$. These $|x_{p_k}^{in}\rangle$ states are noisy versions of the memory state $x^{(p)}$. Hence, for each memory state $x^{(p)}$, the respective $|x_{p_k}^{in}\rangle$ are associated with $|x_p^{out}\rangle$ as output states.

The number of training samples depends on the number of patterns at distance at most d to a given pattern. For m memories, the total number of generated training samples N_{train} depends on the binomial coefficients $\binom{n}{d}$ and is given by

$$N_{train}(m, d) = m \sum_{c=0}^{d} \binom{n}{c}. \tag{6}$$

Note that the order of training samples does not influence the results, as K_j is estimated as a sum over all training samples in Eq. (4). Also note that for large enough d, one noisy input pattern may be associated with different output states. For example, for $n = 3$, $m = 2$ and $d = 1$, the string $(1, 0, 1)$ is at distance one from both $(1, 0, 0)$ and $(0, 0, 1)$, yielding two contradicting training pairs. Consequently, the cost function in Eq. (2) cannot be exactly one, but takes smaller values. Clearly, the larger m and d are with respect to n, the more contradicting training pairs there are and the smaller the maximum of the cost function is.

3.2   Model Complexity

Let us consider the complexity of our model. To estimate K_j using Eq. (4), in each iteration, $\mathrm{Tr}_{\mathrm{rest}}[M_j^k]$ must be estimated for each training pair k. Hence, the duration of training is linear in the number of training samples $N_{\mathrm{train}}(m, d)$ and the time required to estimate $\mathrm{Tr}_{\mathrm{rest}}[M_j^k]$, denoted by $t_{M_j^k}$. The time to update U_j according to Eq. (3) is denoted as t_{upd}. This is repeated for all n unitaries and N_{iter} iteration steps. The total training duration is thus given by

$$t_{\mathrm{tot}} = N_{\mathrm{iter}}n\left(t_{\mathrm{upd}} + N_{\mathrm{train}}(m, d)t_{M_j^k}\right). \tag{7}$$

To estimate both terms in Eq. (5), we need $2(n+1)$ multiplications of $2^{2n} \times 2^{2n}$-matrices. As most matrices in Eq. (5) do not depend on k, the result of the multiplication can be reused. Therefore, the second term in Eq. (7) can be rewritten as $N_{\mathrm{train}}(m, d)t_{M_j^k} = \mathcal{O}\left((2n - 4)2^{3(2n)} + 4N_{\mathrm{train}}2^{3(2n)}\right)$, where we used that multiplying two complex $a \times a$ matrices requires $\mathcal{O}(a^3)$ multiplications of complex numbers in general. Neglecting the computational costs for the partial trace and matrix exponential and assuming a constant time for each multiplication, the total time complexity can be summarized as

$$t_{\mathrm{tot}} = \mathcal{O}\left(N_{\mathrm{iter}}mn^2 2^{6n}\right), \tag{8}$$

where only the samples at distance $d = 1$ are included in the training set. This complexity is independent of whether or not self-connections are removed. It does however restrict us to classical simulations of small systems with $n \leq 5$ only.

3.3   Evaluating the Performance

Different HNNs with different training and updating schemes can be compared by the capacity of the HNN, an important measure to estimate the performance as an associative memory. The capacity relates to the maximum number of storable patterns, which requires some measure of the number of retrieval errors. We give three types of errors, each decreasingly strict in assigning errors. The proposed error rates are the *strict*, *message* and *bit* error rates and are given by:

$$\mathrm{SER}_{n,m} := 1 - \mathbb{1}\left[\forall p \in \{1, \ldots, m\}, \forall j \in \{1, \ldots, n\} : x_j^{(p)} = f(x^{(p)})_j\right], \tag{9}$$

$$\mathrm{MER}_{n,m,\eta} := \frac{1}{mN_{\mathrm{vic}}} \sum_{p=1}^{m} \sum_{p_k=1}^{N_{\mathrm{vic}}} \left(1 - \mathbb{1}[\forall j \in \{1, \ldots, n\} : x_j^{(p)} = f(x_\eta^{(p_k)})_j]\right), \tag{10}$$

$$\mathrm{BER}_{n,m,\eta} := \frac{1}{mN_{\mathrm{vic}}} \sum_{p=1}^{m} \sum_{p_k=1}^{N_{\mathrm{vic}}} \frac{1}{n} H\left(x^{(p)}, f(x_\eta^{(p_k)})\right). \tag{11}$$

Here, n is the input size and m the number of distinct stored patterns, with $m \leq 2^n$. The memory patterns are chosen randomly. Furthermore, $\mathbb{1}[\cdot]$ is the indicator function, which is one if its argument is true, and zero otherwise.

The SER (Eq. (9)) only considers the patterns the HNN should memorize and equals one if at least one bit of any memory pattern cannot be retrieved. This definition corresponds to the one given in [9]. The MER (Eq. (10)) is less strict and uses N_{vic} noisy probe vectors $x_\eta^{(p_k)}$ for each memory $x^{(p)}$. These probe vectors are random noisy versions of the memory patterns, generated with noise parameter η. The MER equals the fraction of the probe vectors from which $x^{(p)}$ cannot be recovered exactly. Finally, the BER (Eq. (11)) also uses the probe vectors $x_\eta^{(p_k)}$. The BER considers all bits separately that cannot be retrieved correctly. For $\eta = 0$, these three error rates are decreasingly strict: $\text{SER} \geq \text{MER} \geq \text{BER}$.

For error rates $\text{ER} \in \{\text{SER}, \text{MER}, \text{BER}\}$ and threshold t, we estimate the number of storable patterns in an HNN using

$$m_{\max}^t(n, ER) := \max\{m \in \{1, \ldots, 2^n\} \mid ER(n, m) \leq t\}. \qquad (12)$$

The capacity C_t of an HNN is now given by normalizing $m_{\max}^t$ by n. For large system sizes, $m_{\max}^t(n, \text{SER})$ is independent of t. For classical HNNs with Hebbian learning the capacity is given by $n/(4 \log n)$.

The capacity of an HNN cannot be determined accurately with only a single set of random memory patterns. Therefore, the error rate for r different random sets of memory patterns are averaged to better approximate the error rates. Note that we require the learned patterns to be stable states of the network and that other states in the vicinity should be attracted. Furthermore, SER is not an appropriate measure for the attractiveness of memory patterns. This follows as the noisy probe samples are randomly generated and might therefore not be part of the memories basin of attraction.

Memories may contain some patterns multiple times, for instance due to small system sizes. For such memories, effectively fewer patterns are stored. Therefore, we generate memories at random, but require them to be distinct.

3.4   Simulation Methods

We simulate both the classical and the quantum HNN using matrix multiplications. The classical simulation is implemented straightforwardly by applying the updating rule for neurons and the Hebbian learning scheme to estimate the weights. For the quantum HNN, the unitaries U_j are initialized uniformly at random and updated according to Eq. (3), where the matrices K_j are estimated from Eq. (4) and (5). Quantum measurements are implemented by choosing the most likely outcome based on the probabilities calculated from the output states, in case of several equally likely outcomes an error is assigned in general. The code used for our simulations is available at [10].

The learning rate $1/\lambda$ introduced in Eq. (4) can be chosen freely and controls the step width of updates. We chose $\lambda = 1$ based on an estimation of the MER with varying $\lambda \in [0.01, 50]$ for system size $n = 4$, $N_{\text{train}} = 50$ training iterations and $r = 100$ repetitions.

We train the unitaries in 50 training iterations on r randomly generated sets of memory patterns. For each set, we estimate the three error rates of retrieval when presenting the memories as input to the trained quantum model. The training sets include all samples at a distance $d \leq 1$ around the respective memory patterns. We repeat this estimation with systems of size $n \leq 5$ and $m \leq 2^n$ memories for all r runs. We sample the error rates $r = 500$ times for $n \in \{3, 4\}$, and up to 1200 times for $n = 5$, to reduce the confidence intervals to a reasonable level.

4   Results

We present the results for the error rates for noisy input retrieval and the capacity of both the quantum HNN (Sect. 4.1 and Sect. 4.2) and the classical HNN (Sect. 4.3 and Sect. 4.4). We end by comparing the results for both in Sect. 4.5. All results are presented with 99% confidence intervals (CI).

4.1   Error Rates of Retrieving Memory Patterns

The error rates when presenting stored memories as input states are displayed in Fig. 3 for system sizes $n \in \{3, 4, 5\}$. The error rates are averaged over the corresponding r rounds. In all simulations, the error rates are zero for $m = 1$ and increase monotonically with m. The SER increases quickly for small m and reaches one at $m = 5$ $(n = 3)$, $m = 7$ $(n = 4)$ and $m = 9$ $(n = 5)$. The MER increases moderately and does not reach one, but settles at around 0.7 for $n = 3$ and $n = 4$. The BER increases the least of all rates and remains below 0.2 for all considered systems.

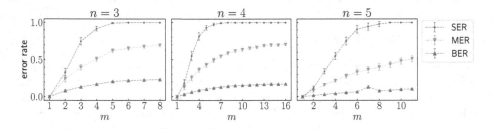

Fig. 3. Estimated SER, MER and BER versus the number of stored patterns m for different system sizes $n = 3, 4, 5$ for a quantum HNN.

The noisy input samples are generated with noise rates $\eta \in \{0.1, 0.2, 0.3\}$ and performance is evaluated for BER and MER. The results are shown in Fig. 4, together with the noiseless results for $\eta = 0$. We find that both the MER and BER monotonically increase with m. Even for $m = 2^n$ and noise rate $\eta = 0.3$, the BER remains below 0.3 and the MER below 0.85 in all considered cases.

For all m, the differences between the error rates for different noise rates remain approximately constant. We notice that the MER for $\eta = 0$ and $\eta = 0.1$ are within the range of each other's confidence intervals for almost all m. For $n = 5$, the CIs are increasingly large due to the varying number of repetitions.

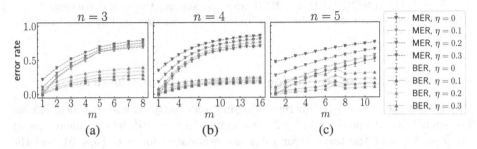

(a)   (b)   (c)

Fig. 4. Estimated MER and BER of retrieving the correct memory patterns from noisy inputs versus the number of stored patterns m. The considered system sizes are (a) $n = 3$, (b) $n = 4$ and (c) $n = 5$.

4.2 Capacity of the Quantum Model

Based on the error rate estimations, we estimate the ability of our quantum model to store and retrieve patterns. The estimated maximum numbers of storable patterns m_{max}^t are given in Table 1 for error rates SER and MER and thresholds $t = 0$ and $t = 0.1$. For this, the point estimates of both error rates and their CIs are compared to the thresholds. Only for $n = 5$, there are several m-values with confidence intervals that contain error rates below the threshold values. Hence, not all m_{max}^t can be estimated with certainty, and therefore all possible values are indicated by curly brackets.

Table 1. Maximum number of storable patterns $m_{\mathrm{max}}^t(\cdot)$ when presenting memories as inputs, for SER and MER and thresholds $t \in \{0, 0.1\}$. To obtain all m_{max}^t, the 99% confidence intervals around error rates are considered.

n	$m_{\mathrm{max}}^0(\mathrm{SER})$	$m_{\mathrm{max}}^{0.1}(\mathrm{SER})$	$m_{\mathrm{max}}^0(\mathrm{MER})$	$m_{\mathrm{max}}^{0.1}(\mathrm{MER})$
3	1	1	1	1
4	1	1	1	1
5	1	$\{1, 2\}$	1	$\{1, 2\}$

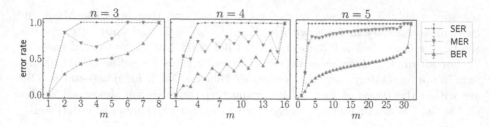

Fig. 5. Estimated SER, MER and BER when presenting the stored patterns to the classical HNN versus their number m for different system sizes n. The stored patterns are required to be distinct.

4.3 Error Rates of Retrieving Memory Patterns Classically

We estimate the error rates for retrieving memory patterns with classical HNNs. For each fixed n and $1 \leq m \leq 2^n$, we generate $r = 10^4$ sets of memories at random. Each of the three error rates are estimated for $n \in \{3, 4, 5\}$ and the memory patterns as inputs, the results are shown in Fig. 5. We find that with increasing number of patterns m, the error rates increase as well. All error rates are exactly zero for $m = 1$ and one for $m = 2^n$. For even n, both the MER and SER fluctuate for different m and are higher if m is even. In contrast, for odd n we see a smooth evolution. The SER increases to 1 rapidly for all n. The results for MER are similar to those for the SER. The BER stays well below the other error rates and increases only moderately, before reaching unity for $m = 2^n$.

When presenting noisy input states to the HNN, we see different behavior. As in the quantum case, only the MER and BER are estimated. For each memory pattern, we generate $N_{vic} = 100$ noisy samples with the same noise rates η as before. The results for different system sizes n are shown in Fig. 6. The different noise rates are indicated by different colors. Again we see less fluctuations for increasing n. Errors increase earlier in the noisy case than in the noiseless case, as expected.

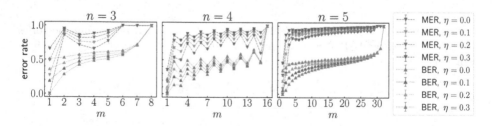

Fig. 6. Estimated MER and BER versus the number of stored patterns m for system sizes $n \in \{3, 4, 5\}$ when presenting noisy test samples as inputs to the classical HNN. The test samples are generated with noise rates $\eta \in \{0.1, 0.2, 0.3\}$. Additionally, the error rates for presenting the noiseless memories are shown.

4.4 Capacity of Classical HNNs with Hebbian Learning

We evaluate $m_{\max}^t$ for 100 iterations and in each iteration we estimate the error
rates using $r = 10^4$ randomly chosen sets of distinct memories for different m
and n. We consider the strict error rates in this analysis. In Fig. 7a and 7b the
results are shown for thresholds $t = 0$ and $t = 0.1$. The results for the MER are
similar to those of the SER. The theoretical capacity $n/(4 \log n)$ is shown as an
orange line. We see a step-wise behavior for all shown results and we see that
the results for $t = 0$ correspond relatively well with the theoretical limit.

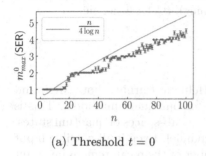
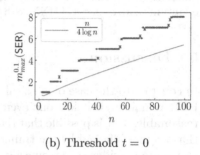

(a) Threshold $t = 0$ (b) Threshold $t = 0$

Fig. 7. Estimated $m_{\max}^t(n, \mathrm{SER})$ for thresholds (a) $t = 0$ and (b) $t = 0.1$. The obtained
values are based on $r = 10^4$ runs with random sets of memories to estimate the SER for
a classical HNN with Hebbian learning. The asymptotic limit of retrievable patterns is
displayed by the orange curve $n/(4 \log n)$. (Color figure online)

4.5 Comparison of Classical and Trained Quantum HNN

We compare the error rates, which are estimated when presenting the memory
patterns to the respective model, for the classical HNN with Hebbian learning
and our trained quantum model in Fig. 8. The considered system sizes are $n \in
\{3, 4, 5\}$. For $n = 5$, we have only few data for the quantum model due to the
computational cost, such that a comparison can be only made for $m \leq 11$.

For all n, the MER and BER of the quantum model are smaller than for the
classical HNN. The only exception is for $n = 4$ and $m = 15$, where the MER of
the classical model is smaller. We also find that the SER of the quantum model
is smaller than the classical SER for small m and reaches one only for higher
values of m. While the MER and BER fluctuate like a saw-tooth for even n for
the classical HNN, we do not find this behavior for the quantum model.

In Table 1 the maximum number of stored patterns without errors is given
for the trained quantum model. Based on the results in Fig. 8, we see that the
classical HNN with Hebbian learning can only store one pattern reliably.

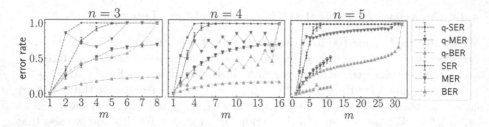

Fig. 8. Comparison of estimated error rates versus the number of stored patterns m for different system sizes $n \in \{3, 4, 5\}$. The blue (red) data are achieved using our trained quantum model (classical HNN with Hebbian learning). (Color figure online)

5 Discussion

In contrast to the case of classical HNNs with Hebbian learning, both MER and BER remain well below one even for $m = 2^n$ for our quantum model. This is reasonable, as it is possible that there are invariant sub-spaces of quantum states, that are not affected by the trained quantum channel. Even if all possible input states are considered as memories, a small number of them can remain invariant under the channel action and thus yield a retrieval error rate less than one.

The estimated error rates for noisy inputs for a quantum HNN stay well below the results for the classical HNN with Hebbian learning and they increase slower. However, when comparing the relative increase in error rate for noiseless and noisy patterns, the classical and the quantum HNN score roughly the same.

Within the level of confidence obtained with the results, we can conclude that our quantum model can store more memories than the classical HNN using Hebbian learning. Already for $n > 4$ it is likely that the quantum model can store more than one memory given that SER or MER are below $t = 0.1$, whereas the classical model can only store a single memory reliably.

The capacity estimates for the classical HNN with Hebbian learning follow the theoretical optimal curve. Due to the high computational costs of the simulations, these results are unavailable for the quantum HNN. Based on the shown results, we do expect capacity improvements for the quantum model over the classical theoretical optimum.

The high computational costs of the simulations of the quantum model originate from the exponential complexity given in Sect. 3.2. This in turn results in very limited system sizes we can simulate. Nonetheless, simulating larger systems in sufficiently many repetitions is valuable, because it allows us to compare the number of stored patterns to other implementations of HNNs.

The presented model can be implemented on general quantum devices and an implementation would require $3n + 1$ qubits and $n4^n$ multi-qubit gates.

6   Conclusion and Outlook

In this work we consider classical HNNs with Hebbian learning and quantum HNNs, where the unitaries are explicitly trained. Based on the presented results, we conclude that the quantum HNN can indeed be used to store classical bit strings as stable attractors with a higher capacity than classical HNNs.

Using a numerical analysis, we consider the number of randomly chosen bit strings that can be stored by an associative model. For $n = 5$ we found that the number of storable patterns is one or two, given an error rate threshold of 0.1, whereas only a single pattern can be stored using a classical HNN with Hebbian learning. For threshold zero, the storage capacity for small system sizes is equal for both classical and quantum HNNs.

It is possible to implement the trained quantum model on actual quantum devices, requiring $3n + 1$ qubits. This might even allow for faster evaluation of the training scheme due to fast execution times on quantum devices. This would allow testing of the trained quantum model on larger systems than in our simulations. However, the number of required gate parameters of the algorithm has a similar scaling as the time complexity when implemented straightforwardly. Therefore, we expect that the scaling prevents experimental realizations of much larger systems.

We conclude that the trained quantum model of our work should be understood as a toy example on the path towards a quantum algorithm for associative memories with possibly larger capacity. The achievement of a quantum advantage by increasing the storage capacity of quantum neural networks beyond classical limits is far from obvious, and more research is required.

Although only classical inputs have been considered, the presented quantum models can also be used for quantum data as inputs and outputs. The ability of our model to store and retrieve quantum states should be studied in future research. We suggest comparing our trained quantum model to classical algorithms that involve non-static training schemes for HNNs, i.e., where the weights are optimized on a training set with respect to a cost-function. In this way, it can be clarified experimentally, whether the better performance of the quantum model originates purely from the fact that it is trained, or from an actual quantum advantage over classical schemes. Moreover, we propose to analyze the storage capacity of our model theoretically, both for quantum and classical memory states. In this way, we hope to find an answer to the ultimate question of whether a quantum advantage can be achieved in the storage capacity of neural networks.

References

1. Arute, F., et al.: Quantum supremacy using a programmable superconducting processor. Nature **574**, 505–510 (2019)
2. Beer, K., Bondarenko, D., Farrelly, T., Osborne, T.J., Salzmann, R., Wolf, R.: Efficient learning for deep quantum neural networks. arXiv:1902.10445, February 2019

3. Cabrera, E., Sossa, H.: Generating exponentially stable states for a Hopfield neural network. Neurocomputing **275**, 358–365 (2018)
4. Dunjko, V., Briegel, H.J.: Machine learning and artificial intelligence in the quantum domain: a review of recent progress. Rep. Prog. Phys. **81**(7), 074001 (2018)
5. Feynman, R.P.: Quantum mechanical computers. Opt. News **11**(2), 11 (1985)
6. Harrow, A.W., Montanaro, A.: Quantum computational supremacy. Nature **549**(7671), 203–209 (2017)
7. Hilbert, M., Lopez, P.: The world's technological capacity to store, communicate, and compute information. Science **332**(6025), 60–65 (2011)
8. Hopfield, J.J.: Neural networks and physical systems with emergent collective computational abilities. Proc. Natl. Acad. Sci. **79**(8), 2554–2558 (1982)
9. McEliece, R., Posner, E., Rodemich, E., Venkatesh, S.: The capacity of the Hopfield associative memory. IEEE Trans. Inf. Theory **33**(4), 461–482 (1987)
10. Meinhardt, N.: NMeinhardt/QuantumHNN 1.0 (Version 1.0). Zenodo (2019), 11 April 2020. https://doi.org/10.5281/zenodo.3748421
11. Mitarai, K., Negoro, M., Kitagawa, M., Fujii, K.: Quantum circuit learning. Phys. Rev. A **98**(3), 032309 (2018)
12. Neumann, N., Phillipson, F., Versluis, R.: Machine learning in the quantum era. Digitale Welt **3**(2), 24–29 (2019). https://doi.org/10.1007/s42354-019-0164-0
13. Preskill, J.: Quantum computing and the entanglement frontier. In: 25th Solvay Conference on Physics, March 2012
14. Rebentrost, P., Bromley, T.R., Weedbrook, C., Lloyd, S.: Quantum Hopfield neural network. Phys. Rev. A **98**(4), 042308 (2018)
15. Rojas, R.: Neural Networks. Springer, Heidelberg (1996). https://doi.org/10.1007/978-3-642-61068-4
16. Schuld, M., Sinayskiy, I., Petruccione, F.: The quest for a quantum neural network. Quantum Inf. Process. **13**(11), 2567–2586 (2014). https://doi.org/10.1007/s11128-014-0809-8
17. Schuld, M., Sinayskiy, I., Petruccione, F.: Simulating a perceptron on a quantum computer. Phys. Lett. A **379**(7), 660–663 (2015)
18. Ventura, D., Martinez, T.: Quantum associative memory with exponential capacity. In: 1998 IEEE International Joint Conference on Neural Networks Proceedings. IEEE World Congress on Computational Intelligence (Cat. No. 98CH36227), vol. 1, pp. 509–513. IEEE (2002)

A Variational Algorithm for Quantum Neural Networks

Antonio Macaluso[1]([✉]), Luca Clissa[2,3], Stefano Lodi[1], and Claudio Sartori[1]

[1] Department of Computer Science and Engineering, University of Bologna,
Bologna, Italy
{antonio.macaluso2,stefano.lodi,claudio.sartori}@unibo.it
[2] Department of Physics and Astronomy, University of Bologna, Bologna, Italy
luca.clissa2@unibo.it
[3] Istituto Nazionale di Fisica Nucleare (INFN), Bologna, Italy

Abstract. Quantum Computing leverages the laws of quantum mechanics to build computers endowed with tremendous computing power. The field is attracting ever-increasing attention from both academic and private sectors, as testified by the recent demonstration of quantum supremacy in practice. However, the intrinsic restriction to linear operations significantly limits the range of relevant use cases for the application of Quantum Computing. In this work, we introduce a novel variational algorithm for quantum Single Layer Perceptron. Thanks to the universal approximation theorem, and given that the number of hidden neurons scales exponentially with the number of qubits, our framework opens to the possibility of approximating any function on quantum computers. Thus, the proposed approach produces a model with substantial descriptive power, and widens the horizon of potential applications already in the NISQ era, especially the ones related to Quantum Artificial Intelligence. In particular, we design a quantum circuit to perform linear combinations in superposition and discuss adaptations to classification and regression tasks. After this theoretical investigation, we also provide practical implementations using various simulation environments. Finally, we test the proposed algorithm on synthetic data exploiting both simulators and real quantum devices.

Keywords: Quantum AI · Quantum Machine Learning · Quantum computing · Quantum variational algorithms · Machine Learning · Neural Networks

1 Background and Motivation

The field of quantum computing (QC) has recently achieved a historic milestone with quantum supremacy [1], thus attracting increasing interest and fostering future research. One of the topics in which QC may have a higher impact is Quantum Machine Learning (QML), i.e. a sub-discipline of quantum information processing whose intent is developing quantum algorithms that learn from

© Springer Nature Switzerland AG 2020
V. V. Krzhizhanovskaya et al. (Eds.): ICCS 2020, LNCS 12142, pp. 591–604, 2020.
https://doi.org/10.1007/978-3-030-50433-5_45

data. However, the ability to deliver a significant boost in performance through quantum algorithms on near-term devices is still to be demonstrated. Given these premises, Neural Networks (NN) are among the most desired targets when coming to transposing classical models into their quantum counterpart. In fact, NN have demonstrated remarkable performances in many real-world applications and multiple learning tasks, including clustering, classification, regression and pattern recognition.

In this work, we introduce a general model framework that reproduces a quantum state equivalent to the output of a classical Single Layer Perceptron (SLP). This is achieved by implementing an efficient variational algorithm that performs linear combinations in superposition. The results are then passed altogether through an activation function with just one application. Importantly, the framework supports pluggable activation function routines, thus allowing an easy way to adapt the approach to different use cases. In Sect. 2, we design a quantum circuit that generates a quantum SLP (qSLP) with two hidden neurons. Section 3 is devoted to practical experiments to test our model as a linear classifier. Finally, Sect. 4 describes how our approach can be extended to the case of more hidden neurons.

1.1   Quantum Variational Algorithms

The construction of full-scale, error-corrected quantum devices still poses many technical challenges. At the same time, significant progress has been made in the development of small-scale quantum computers, thus giving rise to the so-called Noisy Intermediate-Scale Quantum (NISQ) era. For this reason, many researchers are currently focusing on algorithms for NISQ machines that may have an immediate impact on real-world applications, e.g. chemistry [2] and optimisation [3, 4].

Such machines, however, are still not sufficiently powerful to be a credible alternative to the classical ones. For this reason, *hybrid computation* was proposed to exploit near-term devices to benefit from the performance boost expected from quantum technologies. Quantum variational algorithms [5, 6] represent the most promising attempt in this direction, and they are designed to tackle optimisation problems using both classical and quantum resources. The latter component is referred to as variational circuit, and it presents three ingredients: *i*) a parametrised quantum circuit $U(x; \theta)$, *ii*) a quantum output $f(x; \theta)$ and *iii*) an updating rule for the parameters θ.

The general hybrid approach is illustrated in Fig. 1. The data, x, are initially pre-processed on a classical device to determine the input quantum state. The quantum hardware then prepares a quantum state $|x\rangle$ and computes $U(x; \theta)$ with randomly initialised parameters θ. After multiple executions of $U(x; \theta)$, the classical component post-processes the measurements and generates a prediction $f(x; \theta)$. Finally, the parameters are updated, and the whole cycle is run multiple times in a closed loop between the classical and quantum hardware.

Interestingly, the first practical demonstration of quantum advantage over classical supercomputers is related precisely to variational algorithms [7]. Other

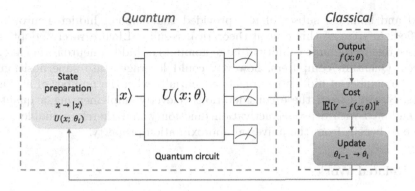

Fig. 1. Scheme of a hybrid quantum-classical algorithm for supervised learning. The quantum variational circuit is depicted in green, while the classical component is represented in blue. (Color figure online)

applications related to Machine Learning (ML) problems were also explored [8,9]. More recently, Schuld et al. [10] proposed a low-depth variational algorithm for classification. The strengths of this approach are two-fold. On one side, the possibility of learning gate parameters enables the adaptation of the architecture for different use cases. On the other hand, the choice of amplitude encoding allows obtaining model predictions with a single-qubit measurement. Importantly, simulations on standard benchmark datasets showed good performances, with the advantage of requiring fewer parameters than classical alternatives.

1.2   Neural Network as Universal Approximator

A Single Hidden Layer Neural Network (or Single Layer Perceptron - SLP) [11] is a two-stage model suitable for both classification and regression. Given a training point (x_i, y_i), the output of a feedforward NN with a single hidden layer containing H neurons can be expressed as:

$$f(x_i) = \sigma_{\text{out}} \left(\sum_{j=1}^{H} \beta_j \sigma_{\text{hid}} \left(L(x_i; \theta_j) \right) \right). \tag{1}$$

Each hidden neuron j computes a linear combination, $L(\cdot)$, of the input features $x_i \in \mathbb{R}^p$ with coefficients given by the p-dimensional vector θ_j. This operation is performed for all neurons, and the results are individually fed into the inner activation function σ_{hid}. The outputs of the previous operation are then linearly combined with coefficients β_j. Finally, a task-dependent outer activation function, σ_{out}, is applied.

Despite being more straightforward than the deep architectures proposed in recent years, the SLP model can be very expressive. According to the *universal approximation theorem* [12], in fact, a SLP with a non-constant, bounded and continuous activation function can approximate any continuous function on a closed and bounded subset of $\mathbb{R}^n$, provided that enough hidden neurons are specified. In spite of this crucial theoretical result, SLP are rarely adopted in practice due to the unfeasibility of large amounts of hidden neurons on classical devices. Quantum computers, however, could leverage state superposition to scale the number of hidden neurons exponentially with the number of available qubits. Starting from these considerations, cleverly implementing a quantum SLP endowed with a proper activation function would therefore enable a real chance to benefit from the universal approximation property.

1.3 Related Works

Several attempts for building a quantum perceptron unit were discussed in the literature [13–15]. A concrete implementation in near-term processors is illustrated in [16], where the authors introduced a model for binary classification using a modified version of the perceptron updating rule. A key characteristic of their architecture is the theoretical exponential advantage in storage resources over classical alternatives. This constitutes the first step towards the efficient implementation of quantum NN on near-term quantum processing hardware.

To the best of our knowledge, however, there are no trainable algorithms that efficiently reproduce a quantum state encoding the output of a classical SLP. Also, the available approaches rely on the introduction of severe constraints on the input data in order to reproduce non-linear activation functions, which makes the algorithms hardly useful in practice.

1.4 Contribution

In this work, we propose a new variational algorithm reproducing a quantum Single Layer Perceptron, whose output is equivalent to the classical counterpart. In particular, building on top of the approach described in [10], we design a general framework that allows efficient computation using just mild constraints on the input. Also, the flexible architecture enables to plug in custom implementations of the activation function routine, thus adapting to different use cases. Thanks to the possibility of learning the parameters for a given task, the proposed framework allows training models that can potentially approximate any function.

However, we do not address the problem of implementing a non-linear activation function. Our goal is to provide a framework that generates multiple linear combinations in superposition entangled with a control register. In this way, instead of executing a given activation function for each hidden neuron, a single application is needed to propagate it to all of the quantum states. This allows scaling the number of hidden neurons exponentially with the number of qubits, thus enabling the qSLP to be a concrete alternative for approximating complex and diverse functions.

2   Variational Algorithm for Single Hidden Layer Neural Network

2.1   Encode Data in Amplitude Encoding

The first issue to address when using a quantum computer for data analysis is *state preparation*, i.e. the design of a process that loads the data from a classical memory to a quantum system. The most general encoding adopted in QML is amplitude encoding [17]. This strategy associates quantum amplitudes with real vectors of observations at the cost of introducing just normalisation constraints. Formally, a normalised vector $x \in \mathbb{R}^{2^n}$ can be described by the amplitudes of a quantum state $|x\rangle$ as:

$$|x\rangle = \sum_{k=1}^{2^n} x_k |k\rangle \longleftrightarrow x = \begin{pmatrix} x_1 \\ \vdots \\ x_{2^n} \end{pmatrix}. \tag{2}$$

In this way, it is possible to use the index register to indicate the k-th feature. The main advantage of this encoding is that we only need n qubits for a vector of $p = 2^n$ elements. This means that, if a quantum algorithm is polynomial in n, then it will have a polylogarithmic runtime dependency on the data size. A possible strategy for amplitude encoding has been proposed by Möttönen et al. [18], which is the one used for experiments in this work. The goal of this approach is to map an arbitrary state $|x\rangle$ to the ground $|0\ldots0\rangle$. Once the circuit is obtained, then all of the operations are inverted and applied in the reversed order.

2.2   Activation Function

The implementation of a proper activation function – in the sense of the Universal Approximation Theorem – is one of the major issues for building a complete quantum Neural Network. This is due to the restrictions to linear and unitary operations imposed by the laws of quantum mechanics [19]. The most promising attempt to solve this problem is described in [20], where the authors use the repeat-until-success technique to achieve non-linearity. The most significant limitation is the requirement of inputs in the range $\left[0, \frac{\pi}{2}\right]$, which is a severe constraint for real-world problems.

In this work, we do not discuss how to implement a non-linear activation function. However, we provide an framework that permits to train a quantum SLP for a given activation function Σ. Our architecture is naturally capable of incorporating any implementation of an activation function whose parameters are learned, like the one described in [21]. Indeed, we can think of extending the circuit that trains the qSLP to also learn the activation parameters. For this reason, new implementations of non-linear activation functions are naturally pluggable in the proposed framework as long as they fit in a learning paradigm.

2.3   Gates as Linear Operators

A variational circuit $U(\theta)$ is composed of a series of gates, each one possibly parametrised by a set of parameters $\{\theta_l\}_{l=1,\ldots,L}$. Formally, $U(\theta)$ is the product of matrices:

$$U(\theta) = U_L \cdots U_l \cdots U_1, \tag{3}$$

where each U_l is composed of a single-qubit or a two-qubit quantum gate. In order to make the single-qubit gate trainable it is necessary to formulate U_l in terms of parameters that can be learned. This is possible by adopting a single-qubit gate G which is defined as the following 2×2 unitary matrix [22]:

$$G(\alpha, \beta, \gamma) = \begin{pmatrix} e^{i\beta}cos(\alpha/2) & e^{i\gamma}sin(\alpha/2) \\ -e^{-i\gamma}sin(\alpha/2) & e^{-i\beta}cos(\alpha/2) \end{pmatrix}. \tag{4}$$

Thus, we can now express each U_l in terms of single-qubit gates, G_i, acting on the i-th qubit:

$$U_l = \mathbb{1}_1 \otimes \cdots \otimes G_i \otimes \cdots \otimes \mathbb{1}_n, \tag{5}$$

where n is the total number of qubits of the quantum system. This representation of $U(\theta)$ is convenient since it allows computing the gradient analytically, as shown in [10].

Alternatively, we can express Eq. (4) using complex numbers $z, u \in \mathbb{C}$ instead of trigonometric functions:

$$G(z, v) = \begin{pmatrix} z & v \\ -v^* & z^* \end{pmatrix}, \tag{6}$$

where $|z|^2 + |v|^2 = 1$. This parametrisation avoids non-linear dependencies between the circuit parameters and the model output. Notice that the definition of linear operator given in Eq. (6) involves complex coefficients. Therefore, it describes a more general operation with respect to the classical counterpart adopted in an SLP, that only allows for linear combinations with real-valued coefficient. Nonetheless, one can still parametrise the circuit using Pauli-Y rotation in case one wants to restrict the computation to the real domain.

2.4   Quantum Single Hidden Layer Network with Two Neurons

In this section we introduce the basic idea of a quantum Single Layer Perceptron with two neurons in the hidden layer. The generalisation of the algorithm is then discussed in Sect. 4.

Intuitively, a qSLP can be implemented into a quantum computer in two steps. Firstly, we generate different linear operations in superposition, each one having different parameters θ_j, entangled with a control register. Secondly, we propagate the activation function to all the linear combinations in superposition. Notice that, thanks this approach, instead of executing a given activation function for each hidden neuron, we need only one application to obtain the output of all the neurons in the hidden layer. To this end, three quantum registers are necessary: *control, data* (denoted by $|\psi\rangle$) and *temporary* register ($|\phi\rangle$). The latter is responsible for generating the linear combinations of the input data in superposition. Also, it can be in any arbitrary state, possibly even unknown.

The algorithm is composed of five main steps: *state preparation, entangled linear operators in superposition, application of the activation function, read-out step, post-processing.*

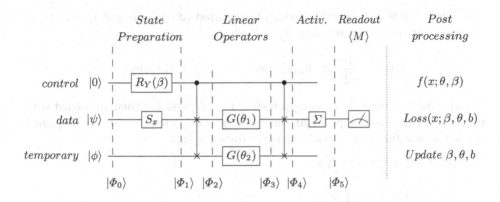

Fig. 2. Quantum circuit for training a qSLP.

(Step 1) The state preparation includes encoding the data, x, in the amplitude of $|\psi\rangle$ and applying a parametrised Y-rotation $R_y(\beta)$ to the control qubit:

$$|\Phi_1\rangle = \big(R_y(\beta) \otimes S_x \otimes \mathbb{1}\big)|\Phi_0\rangle = \big(R_y(\beta) \otimes S_x \otimes \mathbb{1}\big)|0\rangle|0\rangle|\phi\rangle$$
$$= (\beta_1|0\rangle + \beta_2|1\rangle) \otimes |x\rangle \otimes |\phi\rangle = \beta_1|0\rangle|x\rangle|\phi\rangle + \beta_2|1\rangle|x\rangle|\phi\rangle), \quad (7)$$

where S_x indicates the routine that encodes the data, $|\beta_1|^2 + |\beta_2|^2 = 1$ and $\beta_1, \beta_2 \in \mathbb{R}$.

(Step 2) We exploit the idea of quantum forking [23] to generate two different linear operations in superposition, each entangled with the control qubit.

2.1 The first controlled-swap is applied to swap $|x\rangle$ with $|\phi\rangle$ if the control qubit is equal to $|1\rangle$:

$$|\Phi_2\rangle = \frac{1}{\sqrt{E}}\Big(\beta_1|0\rangle|x\rangle|\phi\rangle + \beta_2|1\rangle|\phi\rangle|x\rangle\Big) \quad (8)$$

where E is a normalisation constant.

2.2 Two linear operations parametrised by two different sets (θ_1 and θ_2) act on $|\psi\rangle$ and $|\phi\rangle$ respectively:

$$
\begin{aligned}
|\Phi_3\rangle &= \Big(\mathbb{1} \otimes G(\theta_1) \otimes G(\theta_2)\Big)|\Phi_2\rangle \\
&= \frac{1}{\sqrt{E}}\Big(\beta_1 |0\rangle\, G(\theta_1)\,|x\rangle\,|\phi\rangle + \beta_2 |1\rangle\,|\phi\rangle\, G(\theta_2)\,|x\rangle \Big) \\
&= \frac{1}{\sqrt{E}}\Big(\beta_1 |0\rangle\,|L(x;\theta_1)\rangle\,|\phi\rangle + \beta_2 |1\rangle\,|\phi\rangle\,|L(x;\theta_2)\rangle \Big).
\end{aligned}
\tag{9}
$$

2.3 Then, the second controlled-swap is executed to swap $|L(x;\theta_2)\rangle$ with $|\phi\rangle$ if the control qubit is equal to $|1\rangle$:

$$
|\Phi_4\rangle = \frac{1}{\sqrt{E}}\Big(\beta_1 |0\rangle\,|L(x;\theta_1)\rangle\,|\phi\rangle + \beta_2 |1\rangle\,|L(x;\theta_2)\rangle\,|\phi\rangle \Big).
\tag{10}
$$

Finally, the two linear operations are stored in $|\psi\rangle$ and are then entangled with one state of the control qubit. At this point, a routine is necessary to propagate the activation function in both the trajectories of $|\psi\rangle$.

(Step 3) Activation function:

$$
\begin{aligned}
|\Phi_5\rangle &= \Big(\mathbb{1} \otimes \Sigma \otimes \mathbb{1}\Big)|\Phi_4\rangle \\
&= \frac{1}{\sqrt{E}}\Big(\beta_1 |0\rangle\, \Sigma\,|L(x;\theta_1)\rangle\,|\phi\rangle + \beta_2 |1\rangle\, \Sigma\,|L(x;\theta_2)\rangle\,|\phi\rangle \Big) \\
&= \frac{1}{\sqrt{E}}\Big(\beta_1 |0\rangle\,\big|\sigma_{hid}\big[L(x;\theta_1)\big]\big\rangle\,|\phi\rangle + \beta_2 |1\rangle\,\big|\sigma_{hid}\big[L(x;\theta_2)\big]\big\rangle\,|\phi\rangle \Big).
\end{aligned}
\tag{11}
$$

At the end of Step 3 the two linear operations, $L(\cdot)$, are put through the same activation function, σ_{hid}, represented by the gate Σ. The results are then encoded in the quantum register $|\psi\rangle$. Each output is finally weighed by the parameters of the control qubit (β), i.e. the coefficients attached to the hidden neurons in the linear combination that produces the output of the NN. This is exactly the quantum version of the two-neurons classical SLP presented in Eq. (1).

(Step 4) The measurement of $|\psi\rangle$ can be expressed as the expected value of the Pauli-Z operator acting on the quantum state $|x\rangle$:

$$
\langle M \rangle = \langle \Phi_0|\, U^\dagger(\beta,\theta)(\mathbb{1} \otimes \sigma_z \otimes \mathbb{1})U(\beta,\theta)\,|\Phi_0\rangle = \pi(x;\beta,\theta),
\tag{12}
$$

where $U(\beta,\theta)$ represents the qSLP circuit. In order to get an estimate of $\pi(\cdot)$, we have to run the entire circuit multiple times.

(Step 5) The post-processing is performed classically and is task-dependent. For classification models we need four steps: (i) adding a learnable bias term b to produce a continuous output, (ii) applying a thresholding operation, (iii) computing the loss function and (iv) updating the parameters. Notice that all these steps are customisable and can be adapted to the particular needs of the application. In the case of the experiments presented in Sect. 3 we adopt the following thresholding operation:

$$f(x_i; \beta, \theta, b) = \begin{cases} 1 & \text{if } \pi(x_i; \beta, \theta) + b > 0.5 \\ 0 & \text{else} \end{cases}, \tag{13}$$

where b is the bias term and $f(x_i; \beta, \theta, b)$ gives us the predicted class for observation x. As loss function we choose the *Sum of Squared Errors* (SSE) between the predictions and the true values y:

$$SSE = Loss(\Theta; D) = \sum_{i=1}^{N} [y_i - f(x_i; \Theta)]^2, \tag{14}$$

where N is the total number of observations in the sample and $\Theta = \{\beta, \theta, b\}$. Finally, we exploit the Nesterov accelerated gradient method for updating the parameters, although many alternative optimisation strategies can be adopted to update the parameters [24] .

To summarise, the variational algorithm described above allows reproducing a classical Neural Network with one hidden layer on a quantum computer. In particular, it includes a variational circuit adopted for encoding the data, performing the linear combinations of input neurons and applying the same activation function to their results with just one execution. A single iteration during the learning process is then completed using classical resources to measure the output of the network, compute the loss function and update the parameters. The whole process is then repeated iteratively until convergence, as for classical Neural Networks.

As a final remark, notice that having a post-processing step that is extremely flexible enables the adoption of this model both for regression and classification problems, thus enhancing the impact of such algorithm.

3   Experiments

To test the performances of the qSLP, we implemented the circuit illustrated in Fig. 2 using PennyLane [25], a software framework for optimisation and Machine Learning. This library can be used for both quantum and hybrid computations, and allows using quantum objects (e.g. qubits, gates) in conjunction with classical elements (e.g. variables, functions). It can handle many learning tasks such as training a hybrid ML model in a supervised fashion. In addition, we also implemented a version of the qSLP on the Qiskit framework. In this way, we were able to execute the pre-trained algorithm obtained with PennyLane both on QASM simulators and on a real device.

In our case, the goal is to find the parameters of the quantum circuit (β, θ) plus the additional bias term b. In absence of a gate Σ which implements a non-linear activation function, the final quantum state of $|\psi\rangle$ is:

$$|\Phi_5\rangle = \frac{1}{\sqrt{E}} \Big(\beta_1 |0\rangle |L(x; \theta_1)\rangle |\phi\rangle + \beta_2 |1\rangle |L(x; \theta_2)\rangle |\phi\rangle \Big), \tag{15}$$

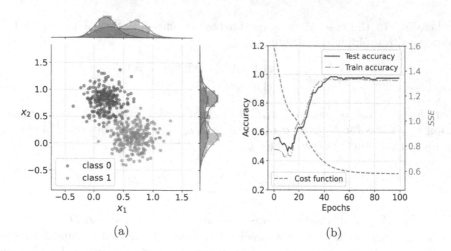

(a)   (b)

Fig. 3. The plot on the left illustrates the distributions of generated data in the two classes $(0, 1)$. The plot on the right shows the trends over training epochs of the cost function and the accuracy.

which is a linear transformation of the input data and defines a linear classifier. Notice that $Pr[y_i = 1|x_i]$ for a given observation x_i corresponds to the square of the linear transformation of hidden neurons with coefficients β_j plus a bias term, b.

In practice, we generated linearly separable data to test our classifier. In particular, we drew a random sample of 500 observations (250 per class) from two independent bivariate Gaussian distributions, with different mean vectors and the same covariance matrix (Fig. 3a). Then, we used the 75% of the data for training and the remaining 25% for testing. The training metrics for the model trained on the PennyLane simulator are illustrated in Fig. 3b. The results demonstrate that the quantum SLP is able to classify correctly the observations, as testified by the high classification accuracy in both training and test sets, 0.97 and 0.95 respectively. After the model was trained, the variational algorithm was also implemented using Qiskit, and its performance was tested on 50 newly-generated observations. In this way, it was possible to test the pre-trained model on both the QASM simulator – which emulates the execution of a quantum circuit on a real device, also including highly configurable noise models – and a real device. Results are reported in Table 1. The PennyLane implementation was in line with the training results, and was the most accurate (94% accuracy), as expected since the framework assumes a perfect device. The effects of introducing the intrinsic noise due to quantum computations, instead, can be appreciated in the Qiskit implementations. Both alternatives showed lower performances, although the decrease in accuracy was certainly smaller for QASM. The real device, instead, presented a significant deterioration. This may be due to the depth of the implemented circuit, especially regarding the encoding part, that seems to be prohibitive considering the actual quantum devices.

Table 1. Test accuracy of multiple implementations. The performance deteriorates as we introduce intrinsic quantum noise (QASM) and current technology limits (IBM – Vigo).

PennyLane	QASM	IBM (Vigo)
94%	90%	64%

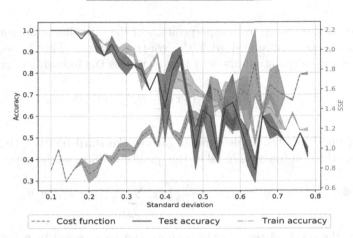

Fig. 4. Assessment metrics trend as a function of distributions overlapping. Larger standard deviations cause the two distributions to overlap, so that observations belonging to the two classes are mixed together and, hence, harder to separate. As a consequence, model performances decrease and non-linearity is required.

In addition, we investigated how the performance of the qSLP implemented in PennyLane changes as the generated distributions get closer and less separated. To this end, we drew multiple samples from the two distributions, each time increasing the common standard deviation so to force reciprocal contamination. As expected, the accuracy showed a decreasing trend as the overlap of the distributions increased (Fig. 4). In conclusion, the experiments show that the proposed architecture works well for linearly separable data. However, performance decreases as we add to the problem a level of complexity that cannot be solved by linear classifiers.

4   Generalisation to H Hidden Neurons

In this section we discuss the generalisation of the quantum SLP to the case of $H > 2$ hidden neurons.

In order to extend the quantum state in Eq. (11), we can consider a *data* register whose size depends on the number of input features, a *control* register made by d qubits, and another register (*output*) that stores the output of the Neural Network. Intuitively, the algorithm can be summarised into three steps. First, the control register is turned into a non-uniform superposition parameterised by the 2^d-dimensional vector β by means of an oracle B:

$$|\Phi_1\rangle = (\mathbb{1} \otimes B \otimes \mathbb{1})\,|x\rangle_{\text{data}}\,|0\rangle_{\text{control}}\,|0\rangle_{\text{output}}$$
$$\rightarrow \frac{1}{\sqrt{E}}\Big(|x\rangle \otimes \sum_j \beta_j\,|j\rangle \otimes |0\rangle \Big). \tag{16}$$

The second step generates a superposition of the same linear operation with different parameters entangled with the control register. This is possible by assuming to have a quantum oracle Λ that performs the following operation:

$$|\Phi_2\rangle = \Lambda\,|\Phi_1\rangle \rightarrow \frac{1}{\sqrt{E}}\Big(|x\rangle \sum_j \beta_j\,|j\rangle\,|L(x;\theta_j)\rangle \Big). \tag{17}$$

Finally, the third step applies the Σ gate to the third register, thus propagating the activation function in all of the quantum states of the superposition:

$$|\Phi_3\rangle = (\mathbb{1} \otimes \mathbb{1} \otimes \Sigma)\,|\Phi_2\rangle \rightarrow \frac{1}{\sqrt{E}}\Big(|x\rangle \sum_j \beta_j\,|j\rangle\,|\sigma[L(x;\theta_j)]\rangle \Big). \tag{18}$$

In this way, the result of the algorithm above can be accessed by a single-qubit measurement. Regarding the parameters, β and $\{\theta_j\}_{j=1,\dots,H}$ can be randomly initialised and the same hybrid optimisation process presented in Sect. 2.4 can be exploited.

As a final remark, it is important to notice that our algorithm entangles linear combinations to the states of the control register. As a consequence, the number of linear combinations that can be performed is equal to the number of possible states of the quantum system. This, in turn, implies that the number of hidden neurons H scales exponentially with the number of states of the control register, 2^d. This is a consequence of each hidden neuron being represented by a single linear combination. Thus, the exponential scaling property enables the construction of quantum Neural Networks with an arbitrary large number of hidden neurons as the amount of available qubits increases. In other terms, we can build qSLP with an incredible descriptive power that may be really capable of being an universal approximator.

5   Conclusions and Outlook

In this work, we proposed an implementation of a quantum version of the Single Layer Perceptron. The key idea is to use a single state preparation routine and apply different linear combinations in superposition, each entangled with a control register. This allows propagating the routine of a generic activation function to all of the states with only one operation. As a result, a model trained through our algorithm is potentially able to approximate any desired function as long as enough hidden neurons and a non-linear activation function are available.

Furthermore, we provided a practical implementation of our variational algorithm that reproduces a quantum SLP for classification with two hidden neurons and an identity function as activation.

In addition, we tested our algorithm on synthetic data and demonstrated that the model works well in case of linearly separable observations, with a test accuracy of 95%. However, the performance deteriorates when facing the intrinsic noise due to quantum computations and current technology limits. On the other hand, experiments showed how the performance of the model deteriorates as the distributions of the two classes overlap so to contaminate each other, thus testifying the necessity of introducing non-linearity into the model. For this reason, the main challenge to tackle in the near future is the design of a routine that reproduces a non-linear activation function.

Another natural follow-up of this work is the implementation of a generalisation of the quantum SLP to the case of $H > 2$ hidden neurons. This would be beneficial for more hands-on experimentation, including, for instance, the discussion of a regression task.

In conclusion, we are still far from proving that Machine Learning can benefit from Quantum Computing in practice. However, thanks to the flexibility of variational algorithms, we believe that the hybrid quantum-classical approach may be the ideal setting to make universal approximation possible in quantum computers.

References

1. Arute, F., et al.: Quantum supremacy using a programmable superconducting processor. Nature **574**(7779), 505–510 (2019)
2. Lanyon, B.P., et al.: Towards quantum chemistry on a quantum computer. Nature Chem. **2**(2), 106 (2010)
3. Farhi, E., Goldstone, J., Gutmann, S.: A quantum approximate optimization algorithm. arXiv preprint arXiv:1411.4028 (2014)
4. Kandala, A., et al.: Hardware-efficient variational quantum eigensolver for small molecules and quantum magnets. Nature **549**(7671), 242 (2017)
5. Wecker, D., Hastings, M.B., Troyer, M.: Progress towards practical quantum variational algorithms. Phys. Rev. A **92**, 042303 (2015)
6. Moll, N., et al.: Quantum optimization using variational algorithms on near-term quantum devices. Quantum Sci. Technol. **3**(3), 030503 (2018)
7. Peruzzo, A.: A variational eigenvalue solver on a photonic quantum processor. Nature Commun. **5**, 4213 (2014)
8. Ristè, D.: Demonstration of quantum advantage in machine learning. NPJ Quantum Inf. **3**(1), 16 (2017)
9. Biamonte, J., Wittek, P., Pancotti, N., Rebentrost, P., Wiebe, N., Lloyd, S.: Quantum machine learning. Nature **549**(7671), 195–202 (2017)
10. Schuld, M., Bocharov, A., Svore, K., Wiebe, N.: Circuit-centric quantum classifiers. arXiv preprint arXiv:1804.00633 (2018)
11. Hastie, T., Tibshirani, R., Friedman, J.: The Elements of Statistical Learning. SSS. Springer, New York (2000). https://doi.org/10.1007/978-0-387-84858-7
12. Hornik, K., Stinchcombe, M., White, H.: Multilayer feedforward networks are universal approximators. Neural Netw. **2**(5), 359–366 (1989)

13. Schuld, M., Sinayskiy, I., Petruccione, F.: The quest for a quantum neural network. Quantum Inf. Process. **13**(11), 2567–2586 (2014)
14. Schuld, M., Sinayskiy, I., Petruccione, F.: Simulating a perceptron on a quantum computer. Phys. Lett. A **379**(7), 660–663 (2015)
15. Faber, J., Giraldi, G.A.: Quantum models of artificial neural networks, **5**(7.2), 5–7 (2002). http://arquivosweb.lncc.br/pdfs/QNN-Review.pdf
16. Tacchino, F., Macchiavello, C., Gerace, D., Bajoni, D.: An artificial neuron implemented on an actual quantum processor, zak1998quantum. NPJ Quantum Inf. **5**(1), 26 (2019)
17. Schuld, M., Petruccione, F.: Supervised Learning with Quantum Computers. QST. Springer, Cham (2018). https://doi.org/10.1007/978-3-319-96424-9
18. Mottonen, M., Vartiainen, J.J., Bergholm, V., Salomaa, M.M.: Transformation of quantum states using uniformly controlled rotations. arXiv preprint arXiv:quant-ph/0407010 (2004)
19. Nielsen, M.A., Chuang, I.: Quantum computation and quantum information. Am. J. Phys. **70**, 558 (2002)
20. Cao, Y., Guerreschi, G.G., Aspuru-Guzik, A.: Quantum neuron: an elementary building block for machine learning on quantum computers. arXiv preprint arXiv:1711.11240 (2017)
21. Wei, H.: Towards a real quantum neuron. Nat. Sci. **10**(3), 99–109 (2018)
22. Barenco, A., et al.: Elementary gates for quantum computation. Phys. Rev. A **52**, 3457–3467 (1995)
23. Park, D.K., Sinayskiy, I., Fingerhuth, M., Petruccione, F., Kevin Rhee, J.-K.: Quantum forking for fast weighted power summation. arXiv preprint arXiv:1902.07959 (2019)
24. Ruder, S.: An overview of gradient descent optimization algorithms. arXiv preprint arXiv:1609.04747 (2016)
25. Bergholm, V., et al.: Pennylane: automatic differentiation of hybrid quantum-classical computations. arXiv preprint arXiv:1811.04968 (2018)

Imperfect Distributed Quantum Phase Estimation

Niels M. P. Neumann[✉], Roy van Houte, and Thomas Attema

The Netherlands Organisation for Applied Scientific Research,
Anna van Buerenplein 1, 2595 DA The Hague, The Netherlands
{niels.neumann,roy.vanhoute,thomas.attema}@tno.nl

Abstract. In the near-term, the number of qubits in quantum computers will be limited to a few hundreds. Therefore, problems are often too
large and complex to be run on quantum devices. By distributing quantum algorithms over different devices, larger problem instances can be
run. This distributing however, often requires operations between two
qubits of different devices. Using shared entangled states and classical
communication, these operations between different devices can still be
performed. In the ideal case of perfect fidelity, distributed quantum computing is a solution to achieving scalable quantum computers with a
larger number of qubits. In this work we consider the effects on the output fidelity of a quantum algorithm when using noisy shared entangled
states. We consider the quantum phase estimation algorithm and present
two distribution schemes for the algorithm. We give the resource requirements for both and show that using less noisy shared entangled states
results in a higher overall fidelity.

Keywords: Noisy quantum computing · Phase estimation ·
Distributed quantum computing · Entanglement fidelity · Quantum
Fourier transform

1 Introduction

The field of quantum computation already contains an extensive amount of theoretical knowledge and has found more applications in the last decades [10]. The
combination of quantum computing and quantum networks opens a whole new
world of information and communication technology as new applications emerge.
Applications of quantum computing and quantum networks are being developed
that are not feasible using classical computers and classical communication, such
as applications for security [1,13], telescopy [5] and clock-synchronization [8].

Current quantum computers are far from solving large practical problems and
implementing such quantum computers still comes with many challenges [11].
One of the hurdles of such a universal quantum computer is the number of
qubits. The required number of qubits depends on both the application and
implementation of the corresponding algorithm. This means that a single quantum computer with only a few qubits will in general not be able to solve larger

© Springer Nature Switzerland AG 2020
V. V. Krzhizhanovskaya et al. (Eds.): ICCS 2020, LNCS 12142, pp. 605–615, 2020.
https://doi.org/10.1007/978-3-030-50433-5_46

problems. However, using a quantum network to link together multiple quantum computers, each with a handful of qubits, larger problem instances can be solved. This concept is called *distributed quantum computing* (DQC) [3,14].

Using a network of computers to solve large problems is not unique to quantum computers and quantum algorithms. In general, distributed (quantum) computing combines different (quantum) computers, where each machine performs part of the computation. This gives either a speed-up (parallelism) or allows to solve large problem instances (larger computers). These (quantum) computers may be physically separated. In this work, we focus on quantum computers and we consider the effects on the output quality, when distributing a quantum algorithm over multiple devices. The total number of usable qubits depends on the sum of the qubits of each quantum computer separately. Additionally, each device has a communication qubit, used for the shared entanglement with other quantum computers.

Different quantum computers can be linked by shared entangled qubit pairs. This entanglement allows for physically separated machines. Operations involving qubits from a single quantum computer are called *local*, whereas operations using qubits from different devices are called *non-local*. Non-local operations require shared entangled qubit pairs and ideally these pairs are in one of the Bell states [15], often referred to as EPR-pairs. These are used, together with classical communication bits, to perform non-local operations.

Due to noise in the quantum gates and qubit decoherence, the output of a quantum algorithm might differ from the theoretical output. A measure on how well the output quantum state matches with the theoretically expected output, is the fidelity and is given in Eq. (3.2). The lower the fidelity, the more the two states differ. When distributing a quantum algorithm, imperfections in the shared EPR-pair form another potential source of uncertainty. These imperfections can for instance occur due to an imperfect generation process or imperfections in the used quantum channels.

Non-local gates were first described by Eisert et al. [3] in 2000. Later, Yimsiriwattana and Lomonaco showed a modular approach to distributed quantum computing and suggest the use of quantum teleportation to decrease the number of EPR-pairs required [15]. In 2004 they extended their approach to a distributed version of Shor's algorithm [14]. A distributed version of Grover's algorithm was presented by Exman and Levy in 2012 [4]. Only recently, in 2018, Moghadam et al. designed an algorithm to optimize the teleportation cost of distributed quantum circuits [16].

In each of these works, however, only the perfect setting is considered with no gate- and qubit errors. We relax this assumption by allowing imperfect EPR-pairs. Local quantum operations are still assumed to be noiseless and qubits are assumed to not decohere. Hence, errors can only be introduced by the imperfect shared entanglement used for non-local operations. We present two different distribution schemes, one standard implementation and one implementation where operations are combined, thereby requiring less imperfect EPR-pairs.

In Sect. 2, we will explain the quantum phase estimation algorithm (Sect. 2.1), non-local controlled operations (Sect. 2.2) and the distributed quantum Fourier transform (Sect. 2.3). Afterwards we introduce the setup of our simulations in Sect. 3.1 and we present the corresponding results in Sect. 3.2. Conclusions are given in Sect. 4.

2 Distributed Quantum Computing and Phase Estimation

We will first briefly explain the quantum phase estimation algorithm in Sect. 2.1. Then we explain how to perform non-local controlled U-gates in Sect. 2.2. We end this section by giving two implementation schemes for the distributed quantum Fourier transform in Sect. 2.3.

2.1 Quantum Phase Estimation

The phase estimation algorithm, first presented by Kitaev [9], returns an approximation of an eigenvalue of a given unitary U and a corresponding eigenvector. It has a wide range of applications, the most famous of which is Shor's algorithm [12].

More formally, if U is a unitary operation on m qubits, and $|\psi\rangle$ is an eigenstate of U, then $U |\psi\rangle = \exp(2\pi i \varphi) |\psi\rangle$ for some phase $\varphi \in [0, 1)$. Let

$$\varphi = \sum_{i=1}^{\infty} \varphi_i 2^{-i} = 0.\varphi_1 \varphi_2 \ldots \tag{2.1}$$

be the binary representation of φ. If we truncate the sum of Eq. (2.1) to n, we have an n-bit approximation of φ given by $0.\varphi_1 \varphi_2 \ldots \varphi_n$. This n-bit approximation of φ is found using the quantum phase estimation algorithm.

A quantum circuit implementation of the quantum phase estimation is given in Fig. 1, with two registers of n and m qubits, respectively. If φ can be represented exactly in at most n-bits, this will be the output of the algorithm with

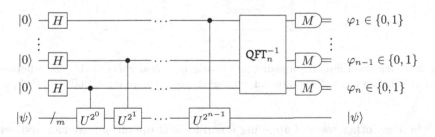

Fig. 1. The quantum phase estimation circuit for a unitary U acting on m qubits. The result is an n-bit approximation of the eigenvalue φ of eigenvector $|\psi\rangle$. The block QFT_n^{-1} is the inverse quantum Fourier transform on n qubits.

certainty. Otherwise, the approximation will round the phase and the correct result is given with probability at least $4/\pi^2$ [2].

First a Hadamard gate is applied on the first n qubits. Afterwards, controlled-$U^{2^{n-i}}$ gates are applied, with control qubit i in the first register and the qubits in the second register as target. Then an inverse quantum Fourier transform on the first register is applied and the qubits are measured. This gives the n-bit phase approximation of φ.

2.2 Distributed Controlled U-gate

A universal gate-set for local operations is given by a CNOT-gate and single qubit rotations [9]. A universal gate-set for non-local operations is thus obtained by a combination of the local universal gate set and non-local CNOT gates. By combining non-local CNOT-gates and local operations, arbitrary non-local operations are constructed. This is similar to how one would construct arbitrary local operations.

Suppose we want to apply a controlled U-gate between two qubits $|\psi\rangle$ and $|\phi\rangle$ on two different devices. Furthermore, let there be two extra qubits, one on each device, that share an entangled state in the Bell state $\frac{1}{\sqrt{2}}(|00\rangle + |11\rangle)$ and assume the two devices can communicate classically. The quantum circuit given in Fig. 2 performs this non-local controlled U-gate. Here, the two dotted boxes indicate the two quantum devices. The operation E_2 entangles $|0\rangle_1$ and $|0\rangle_2$ in a Bell-state and M indicates a measurement of the corresponding qubit. The double lines indicate classical control by the measured value. This quantum circuit is treated in more detail in [3, 15].

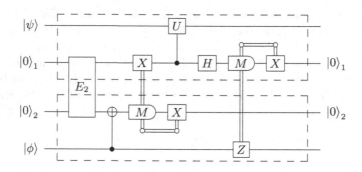

Fig. 2. A quantum circuit implementation of a non-local controlled U-gate between $|\phi\rangle$ and $|\psi\rangle$. The block E_2 creates an entangled qubit pair in state $(|00\rangle + |11\rangle)/\sqrt{2}$.

There are other ways of applying a non-local U operation. We can first teleport $|\psi\rangle$ to the other device, do all operations locally, and then teleport the resulting state back. This, however, requires one extra qubit per device, more operations and two shared EPR-pairs instead of one.

2.3   Distributed Quantum Fourier Transform

The quantum Fourier transform maps an n-qubit state $|k\rangle$ to $\sum_{j=0}^{2^n-1} e^{2\pi ijk/2^n} |j\rangle$, with i the complex unit. A recursive implementation of the quantum Fourier transform is given in Fig. 3. We see that the implementation can be decomposed in Hadamard gates and controlled R_k-gates. These rotation gates R_k are given by

$$R_k = \begin{pmatrix} 1 & 0 \\ 0 & e^{2\pi i/2^k} \end{pmatrix}. \tag{2.2}$$

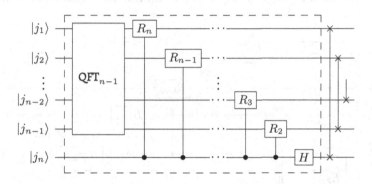

Fig. 3. A recursive quantum circuit for an n-qubit quantum Fourier transform. The dotted rectangle represents the QFT$_n$. By definition QFT$_1 = H$.

Table 1. The resources requirements for the local quantum Fourier transform on $2n$ qubits and the non-local quantum Fourier transform distributed over two devices of n qubits each.

Method	Number of qubits	Gates	Communication (EPR-pairs)	Communication (classical bits)
Local gates	$2n$	$2n^2 + n$	0	0
Standard implementation	$2n + 2$	$10n^2 + n$	n^2	$2n^2$
Combined implementation	$2n + 2$	$2n^2 + 9n$	n	$2n$

Note that the last operations in Fig. 3 can be omitted, as they only swap the order of the qubits. Performing non-local SWAP-gates gives a high computational overhead, whereas reversing the order of the measurement results is easily accounted for classically. Even if the output of the quantum Fourier transform is used as input for further computations, these operations can be accounted for without using SWAP gates.

A non-local implementation of the quantum Fourier transform is obtained by combining the quantum circuits shown in Fig. 2 and Fig. 3, with $U = R_k$. We refer to the approach of simply replacing each controlled gate by a non-local one if necessary, as the *standard* approach. Instead of replacing each controlled operation by a non-local one, we can also use a single shared entangled state to perform multiple non-local gates, by grouping all operations on one computer that are controlled by a single qubit from another quantum computer. This quantum circuit is given in Fig. 4, where only a single qubit of the second device is shown. This *combined* approach uses less shared entangled states and has less communication overhead.

In Table 1 we show the resource requirements to run a quantum Fourier transform on $2n$ qubits for both the local implementation and the two presented non-local implementations. In the non-local situation, two additional qubits are used for the shared entanglement, as well as additional classical communication bits.

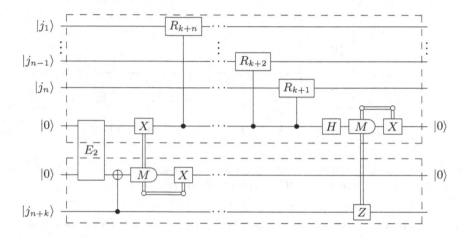

Fig. 4. Part of a distributed quantum Fourier transform, where n non-local operations, on n qubits are performed using a single shared entangled state. The control is $|j_{n+k}\rangle$, the targets are $|j_1\rangle, \ldots, |j_n\rangle$. Only a single qubit of the second quantum computer is shown, others are omitted. The dashed boxes indicate the quantum computers and the double lines indicate classical communication.

3   Non-local Quantum Circuits with Imperfect Entanglement

In this section we will first explain the setup of our simulations (Sect. 3.1) and then present the results of these simulations (Sect. 3.2).

3.1   The Simulation Setup

We implemented the two distributed quantum circuits presented in the previous section, as well as an implementation of a local circuit. For these implementations we used Python 3.6 and the QuTiP Python package [6,7]. Simulations are run in the density state formalism. Depolarizing noise is applied to the shared entangled states using noise parameter α. The density representation of the noisy EPR-pair is given by

$$\eta(\alpha) = \frac{(1-\alpha)}{2}(|00\rangle + |11\rangle)(\langle 00| + \langle 11|) + \frac{\alpha}{4}I_2 \otimes I_2. \tag{3.1}$$

If $\alpha = 0$, the state corresponds to one of the Bell-states, whereas for increasing α, the state becomes more ideally mixed. In our simulations we consider k quantum computers, each with n_i qubits. Each device has one additional qubit used for the shared entanglement.

For different topologies we compare the output of the quantum circuit $\eta_{out}(\alpha)$ with the output in a noiseless situation $\eta_{out}(0)$. The quality is expressed in terms of the fidelity between a pair of density matrices ρ and σ and is given by

$$F(\rho, \sigma) = \left[\text{Tr}\sqrt{\rho^{1/2}\sigma\rho^{1/2}}\right]^2. \tag{3.2}$$

3.2   Results

We run our simulations for the unitary operation

$$R_\varphi = \begin{pmatrix} 1 & 0 \\ 0 & \exp(2\pi i \varphi) \end{pmatrix}$$

with eigenvector $|\psi\rangle = |1\rangle$. The corresponding eigenvalue is $\exp(2\pi i \varphi)$ which has phase φ. We consider different noise rates $\alpha \in [0, 1]$.

First, we consider a random fixed angle $\varphi = 72/128$, and two quantum computers of 4 qubits each, 7 qubits of which we use for the approximation. We run the simulations for different noise rates and the results for $\alpha = 0.1$ and 0.5 are shown in Fig. 5. Note that φ translates to a fraction of 2π and hence to an angle of $\varphi * 2\pi = 202.5°$ in these plots. Results are presented using log-radar plots for both the standard and the combined implementation. The shown results are the log-values of the output probabilities. The results for $\alpha = 0$ and $\alpha = 1$ are not shown. For $\alpha = 0$, no errors occur and φ is retrieved with certainty. For $\alpha = 1$, the result is uniform for all states.

The results for both implementations are similar and show a repetitive pattern, with spikes every $45°$. The largest spike is found at $202.5°$, corresponding to the phase φ to be found. These effects were also found for quantum computers of different sizes and when distributing over more than two devices.

For the 7-bit approximation, 128 different measurement outcomes are possible. For both the standard and the combined implementation, we found that the results are independent from the initial angle up to rotations with steps of $1/128$. Therefore, the probability distribution of $\varphi = 72/128$, is the rotated probability distribution of $\varphi = 0$. More generally, we also found that the probability distributions for m-bit approximations are equivalent up to rotation for $m > n$. For example, the 9-bit phases $\varphi = 1/512$ and $\varphi = 3/512$, give the same probability distribution in a 7-bit approximation.

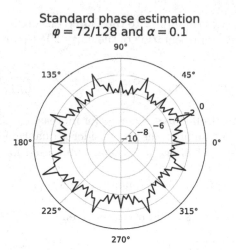

(a) Standard implementation with noise parameter $\alpha = 0.1$.

(b) Standard implementation with noise parameter $\alpha = 0.5$.

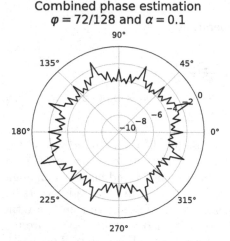
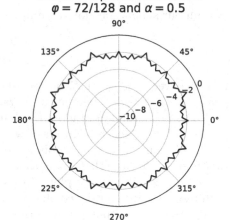

(c) Combined implementation with noise parameter $\alpha = 0.1$.

(d) Combined implementation with noise parameter $\alpha = 0.5$.

Fig. 5. The probability distributions for the standard and combined distributed phase estimation circuit for $\alpha = 0.1$ and $\alpha = 0.5$

Even though the probability distributions for the standard and combined approach seem similar, they are not the same. We found that the probability of correct retrieval of angle φ was highest for the combined implementation.

In Fig. 6 we show the fidelity for both implementations for varying α-values for the same network as before: two quantum computers with 4 qubits each. As expected, the results are the same for $\alpha = 0$. For $\alpha = 1$, the probability distributions are uniform and hence equal for both the standard and the combined implementation. Note that with increasing noise rate, the fidelity drops off quickly. However, also note that the fidelity will not become zero, due to the uniform distribution obtained for $\alpha = 1$.

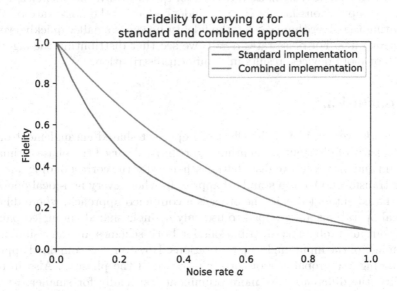

Fig. 6. Comparison between the fidelity for the standard and combined approach for varying noise rates α for two quantum computers of 4 qubits each.

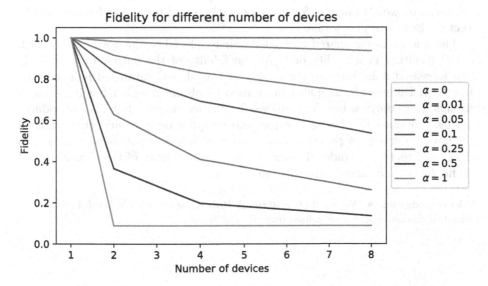

Fig. 7. The fidelity for a varying number of devices the phase estimation algorithm is distributed over. Different noise rates α are shown.

Finally, we consider the effects of distributing the algorithm over more devices. We consider 8 qubits in total and distributed the quantum phase estimation algorithm over k quantum computers for $k \in \{1, 2, 4, 8\}$. The results are shown in Fig. 7 for varying noise rates $\alpha \in [0, 1]$. Naturally, the fidelity is 1 when doing all computations locally ($k = 1$), independent of the noise rate α. When distributing the algorithm ($k > 1$), the fidelity becomes smaller quickly even for small error rates. For noise rates $\alpha = 1$, we see that distributing the algorithm over two or more devices, results in a uniform distribution.

4   Conclusions

In this work, we considered the effect of imperfect shared entanglement on the output fidelity of distributed quantum algorithms. We used the phase estimation algorithm and proposed two distribution schemes for the corresponding quantum Fourier transform. One is a standard approach, where every non-local operation uses a shared entangled pair; the other is a combined approach, where different non-local operations are grouped to use only a single shared entangled pair.

The output probability distributions for both schemes are very similar and independent of the input angle, up to rotations. However, the combined approach gives the highest probability of correct retrieval of the phase φ. Also in terms of fidelity, the differences are more prominent, especially for smaller α-values. Again the combined approach shows the highest fidelity. For high noise rates, the fidelity of both is very similar and near uniform.

We thus found that using less shared entangled states is beneficial for the output in terms of fidelity. Note however, that the results presented in this paper are based on simulations and hence a formal proof of the result is still needed. Furthermore, we assumed perfect local operations and not qubit decoherence. In practice, both will play a role.

The fidelity of the shared entangled pair is related to the noise rate α, with $\alpha = 0$ resulting in a fidelity of 1. As the fidelity of the output drops quickly with increased noise rates α, the fidelity of this shared entangled pair must be close to 1. Different techniques can be used to obtain a higher fidelity, such as entanglement purification. This allows for higher fidelity, but also introduces overhead. In our case of no gate errors and no qubit decoherence, this overhead will have no effect. In practical cases, we may however not neglect these two effects and there is a trade-off between the output fidelity of the algorithm and the fidelity of the shared entangled qubit pair.

Acknowledgments. We would like to thank Rob Knegjens for his helpful advice and insightful discussions. These helped improve the paper.

References

1. Bennett, C.H., Brassard, G.: Quantum cryptography: public key distribution and coin tossing. Theor. Comput. Sci. **560**, 7–11 (2014)
2. Cleve, R., Ekert, A., Macchiavello, C., Mosca, M.: Quantum algorithms revisited. Proc. R. Soc. London. Ser. A Math. Phys. Eng. Sci. **454**(1969), 339–354 (1998)
3. Eisert, J., Jacobs, K., Papadopoulos, P., Plenio, M.B.: Optimal local implementation of nonlocal quantum gates. Phys. Rev. A **62**, 052317 (2000)
4. Exman, I., Levy, E.: Quantum probes reduce measurements: application to distributed Grover algorithm (2012)
5. Gottesman, D., Jennewein, T., Croke, S.: Longer-baseline telescopes using quantum repeaters. Phys. Rev. Lett. **109**, 070503 (2012)
6. Johansson, J., Nation, P., Nori, F.: QuTiP: an open-source Python framework for the dynamics of open quantum systems. Comput. Phys. Commun. **183**(8), 1760–1772 (2012)
7. Johansson, J., Nation, P., Nori, F.: QuTiP 2: a Python framework for the dynamics of open quantum systems. Comput. Phys. Commun. **184**(4), 1234–1240 (2013)
8. Jozsa, R., Abrams, D.S., Dowling, J.P., Williams, C.P.: Quantum clock synchronization based on shared prior entanglement. Phys. Rev. Lett. **85**, 2010–2013 (2000)
9. Kitaev, A.Y.: Quantum computations: algorithms and error correction. Russ. Math. Surv. **52**(6), 1191–1249 (1997)
10. Nielsen, M.A., Chuang, I.L.: Quantum Computation and Quantum Information: 10th Anniversary Edition, 10th edn. Cambridge University Press, New York (2010)
11. Preskill, J.: Quantum computing in the NISQ era and beyond. Quantum **2**, 79 (2018)
12. Shor, P.W.: Polynomial time algorithms for discrete logarithms and factoring on a quantum computer. In: Adleman, L.M., Huang, M.-D. (eds.) ANTS 1994. LNCS, vol. 877, pp. 289–289. Springer, Heidelberg (1994). https://doi.org/10.1007/3-540-58691-1_68
13. Stucki, D., et al.: Long-term performance of the SwissQuantum quantum key distribution network in a field environment. New J. Phys. **13**(12), 123001 (2011)
14. Yimsiriwattana, A., Lomonaco Jr., S.J.: Distributed quantum computing: a distributed Shor algorithm. In: Donkor, E., Pirich, A.R., Brandt, H.E. (eds.) Quantum Information and Computation II, vol. 5436, pp. 360–372. International Society for Optics and Photonics, SPIE (2004)
15. Yimsiriwattana, A., Lomonaco, Jr., S.J.: Generalized GHZ states and distributed quantum computing (2004)
16. Zomorodi-Moghadam, M., Houshmand, M., Houshmand, M.: Optimizing teleportation cost in distributed quantum circuits. Int. J. Theor. Phys. **57**(3), 848–861 (2018)

Optimal Representation of Quantum Channels

Paulina Lewandowska[✉], Ryszard Kukulski, and Łukasz Pawela

Institute of Theoretical and Applied Informatics, Polish Academy of Sciences,
ul. Bałtycka 5, 44-100 Gliwice, Poland
plewandowska@iitis.pl
https://iitis.pl/en/person/plewandowska

Abstract. This work shows an approach to reduce the dimensionality of matrix representations of quantum channels. It is achieved by finding a base of the cone of positive semidefinite matrices which represent quantum channels. Next, this is implemented in the `Julia` programming language as a part of the `QuantumInformation.jl` package.

Keywords: Julia programming language · Computational quantum information · Quantum channels · Convex cones · Base of hermiticity preserving maps

1 Introduction

Nowadays the fields of quantum information processing and machine learning are coming together leading to the emergence of quantum machine learning [1,2]. This area can be broadly divided into three, depending whether the data, algorithms or both are of quantum or classical nature. In this work we are interested in the case of quantum data being processed by a classical algorithm. The natural question arises: how this data should be represented and loaded into our algorithm? To be more precise, we are interested how to represent quantum channels in a succinct manner so that it can be an input into a classical neural network.

The goal of such a network would be to approximate, up to a reasonable error the distance between two channels Φ and Ψ. As Φ and Ψ are linear mappings transforming matrices into matrices it may not seem obvious how to define the distance between them. Turns out, there exists one notion of distance between channels which has an operational interpretation. The distance between Φ and Ψ can be expressed in the terms of so called diamond norm

$$\|\Phi - \Psi\|_\diamond = \max_{\|X\|_1=1} \| ((\Phi - \Psi) \otimes \mathbb{1}) (X)\|_1. \tag{1}$$

This quantity plays a central role in the problem of quantum operation discrimination which has gained a lot of traction recently. This is due to the fact that

© Springer Nature Switzerland AG 2020
V. V. Krzhizhanovskaya et al. (Eds.): ICCS 2020, LNCS 12142, pp. 616–626, 2020.
https://doi.org/10.1007/978-3-030-50433-5_47

this distance provides an upper bound on the probability of discrimination of Φ and Ψ.

Consider a following setup. We are given a black box which is said to contain, with equal probability, either Φ or Ψ. What is the probability of guessing which of these is in the box if we are allowed to use the box only once? Turns out that this probability p is connected with the distance between Φ and Ψ [3]

$$p = \frac{1}{2} + \frac{1}{4}\|\Phi - \Psi\|_\diamond. \tag{2}$$

However the explicit form of the diamond norm contains an optimization over all input matrices X. In principle this can be solved via semidefinite programming, but regrettably this quickly becomes intractable with the growing dimension of the input matrix. That is why it would desirable to have the possibility to train a classical algorithm, like a neural network, on a relatively small set of quantum channels and have the ability to quickly approximate the distance between arbitrary channels utilizing this network.

That is why this paper aims at finding an optimal representation of quantum channels for the purposes of machine learning. By *optimal* we understand the lowest possible number of real parameters needed to define a quantum channel [4]. Further, we would like this representation to be technically usable so that we could train, for instance, neural networks to approximate functions of this objects. This approach could provide a large speed boost in the problem of quantum channel discrimination [3,5].

Our work is naturally divided into three parts. In the first part we show the mathematical structures needed to find the optimal representation. This involves dealing with cones of positive semidefinite matrices. The second part we present the example of whereas the last part presents the implementation of this example in the `Julia` language. This implementation is now a part of the `QuantumInformation.jl` [6,7] numerical library available on-line at https://github.com/iitis/QuantumInformation.jl. Surprisingly, despite the complex mathematical structure and quite technical proofs, the implementation is relatively simple and therefore useful.

2 Mathematical Framework

2.1 Quantum Channels

Let $\mathcal{X}, \mathcal{Y}$ be complex finite-dimensional vector spaces, let $L(\mathcal{X}, \mathcal{Y})$ be the set of all linear operators transforming vectors from $\mathcal{X}$ to $\mathcal{Y}$ and denote $L(\mathcal{X}) := L(\mathcal{X}, \mathcal{X})$. Further, consider mappings of the form

$$\Phi : L(\mathcal{X}) \to L(\mathcal{Y}). \tag{3}$$

The set of all such mappings will be denoted $T(\mathcal{X}, \mathcal{Y})$ and $T(\mathcal{X}) := T(\mathcal{X}, \mathcal{X})$. Quantum channels are such $\Phi \in T(\mathcal{X}, \mathcal{Y})$ which are trace preserving and completely positive. The former means that

$$\forall A \in L(\mathcal{X}) \quad \mathrm{Tr}(\Phi(A)) = \mathrm{Tr}(A). \tag{4}$$

The latter is a bit more complicated. Formally this condition can be written as

$$\forall \mathcal{Z} \ \forall A \in \mathrm{L}(\mathcal{X} \otimes \mathcal{Z}) \ \ A \geq 0 \implies \left(\Phi \otimes \mathbb{1}_{\mathrm{L}(\mathcal{Z})}\right)(A) \geq 0. \tag{5}$$

The intuitive explanation is as follows. First, consider a $\rho \in \mathrm{L}(\mathcal{X})$ such that $\mathrm{Tr}(\rho) = 1$ and $\rho \geq 0$. Such an operator is called a quantum state. We would like our channels not only to transform states into states, but also we would like the ability to perform a channel on only a part of the system. In other words we would like the output of $\left(\Phi \otimes \mathbb{1}_{\mathrm{L}(\mathcal{Z})}\right)(\rho)$ to also be a proper quantum state for an arbitrary space $\mathcal{Z}$ and all $\rho \in \mathrm{L}(\mathcal{X} \otimes \mathcal{Z})$. This can only be fulfilled when we introduce the need for completely positivity. We will denote the set of all quantum channels as $\mathrm{C}(\mathcal{X}, \mathcal{Y})$ and $\mathrm{C}(\mathcal{X}) = \mathrm{C}(\mathcal{X}, \mathcal{X})$.

The mappings $\mathrm{T}(\mathcal{X}, \mathcal{Y})$ may be represented in a number of ways. For our purposes only the Choi-Jamiołkowski isomorphism [8,9] will be relevant. This representation states that there exists a bijection J between the sets $\mathrm{T}(\mathcal{X}, \mathcal{Y})$ and $\mathrm{L}(\mathcal{Y} \otimes \mathcal{X})$. This bijection can be explicitly written as

$$J(\Phi) = \sum_{i,j}^{\dim(\mathcal{X})} \Phi(|i\rangle\langle j|) \otimes |i\rangle\langle j|. \tag{6}$$

Φ is completely positive if and only if $J(\Phi) \geq 0$; Φ is trace preserving if and only if $\mathrm{Tr}_{\mathcal{Y}} J(\Phi) = \mathbb{1}_{\mathcal{X}}$. Finally, Φ is Hermiticity preserving if and only if $J(\Phi) \in \mathrm{Herm}(\mathcal{Y} \otimes \mathcal{X})$, where $\mathrm{Herm}(\mathcal{X})$ denotes the set of all Hermitian matrices in $\mathrm{L}(\mathcal{X})$.

2.2   Convex Cone Structures

Consider $\mathcal{X}$ is a real finite-dimensional vector space and $\mathcal{C} \subset \mathcal{X}$ is a closed convex cone. We assume that $\mathcal{C}$ is pointed, i.e. $\mathcal{C} \cap -\mathcal{C} = \{0\}$ and generating, i.e. for each $x \in \mathcal{X}$ there exists $u, w \in \mathcal{C}$ such that $x = u - w$. Such a cone $\mathcal{C}$ is called a proper cone in the space $\mathcal{X}$. The proper cone $\mathcal{C}$ becomes a partially ordered vector space $x \geq y \iff x - y \in \mathcal{C}$ for each $x, y \in \mathcal{X}$. Let $\mathcal{X}^*$ be the space dual to $\mathcal{X}$ defined by the inner product $\langle \cdot | \cdot \rangle$. Then, we may introduce a partial order in $\mathcal{X}^*$ as well with the dual cone

$$\mathcal{C}^* = \{f \in \mathcal{X}^* : \langle f | z \rangle \geq 0, \forall z \in \mathcal{C}\}. \tag{7}$$

The cone $\mathcal{C}^*$ is also closed and convex cone. If $\mathcal{C}$ is generating in space $\mathcal{X}$, then $\mathcal{C}^*$ is pointed and we may introduce partial order in $\mathcal{X}^*$ given by

$$f \geq g \iff f - g \in \mathcal{C}^* \tag{8}$$

for all $f, g \in \mathcal{X}^*$.

An interior point $e \in \mathrm{int}(\mathcal{C})$ of a cone $\mathcal{C}$ is called an order unit [10] if for each $x \in \mathcal{X}$, there exists $\lambda > 0$ such that $\lambda e - x \in \mathcal{C}$ whereas a base of $\mathcal{C}$ is defined as compact and convex subset $B \subset \mathcal{C}$ such that for every $z \in \mathcal{C} \setminus \{0\}$, there exists unique $t > 0$ and an element $b \in B$ such that $z = tb$. The following theorem shows there exists relation between the order unit e and a base of cone $\mathcal{C}$.

Theorem 1. *The set $B_e = \{z \in \mathcal{C} : \langle e|z \rangle = 1\}$ is the base of $\mathcal{C}$ (determined by element e) if and only if an element e is an order unit and $e \in \text{int}(\mathcal{C}^*)$.*

The proof of this theorem is presented in Appendix A.

2.3 Base of Hermiticity Preserving Maps

Let us now define the finite-dimensional linear space

$$\{\Phi \in \mathrm{T}(\mathcal{X}, \mathcal{Y}) : \Phi - \text{Hermiticity preserving}\}. \tag{9}$$

Due to the Choi–Jamiolkowski isomorphism, the set of all Hermiticity preserving linear maps of a finite-dimensional space is mathematically closely related to the set

$$\mathcal{V} = \{J(\Phi) : J(\Phi) \in \text{Herm}(\mathcal{Y} \otimes \mathcal{X})\}, \tag{10}$$

of all Choi matrices of Hermiticity preserving maps.

In every linear space of Hermitian matrices $\text{Herm}(\mathcal{Z})$ we can introduce an orthonormal basis $\mathcal{B}(\mathcal{Z})$. The basis $\mathcal{B}(\mathcal{Z})$ is a collection of $\dim(\mathcal{Z})^2$ matrices. The standard orthonormal basis is denoted by the set

$$\mathcal{B}(\mathcal{Z}) =$$

$$\left\{ \frac{\mathbb{1}_{\mathcal{Z}}}{\sqrt{\dim(\mathcal{Z})}}, \right.$$

$$\frac{\sum_{a=1}^{k} |a\rangle\langle a| - k\,|k+1\rangle\langle k+1|}{\sqrt{k + k^2}}, \text{ for } k = 1, \ldots, \dim(\mathcal{Z}) - 1,$$

$$\left. \frac{|a\rangle\langle b| + |b\rangle\langle a|}{\sqrt{2}}, \frac{i\,|a\rangle\langle b| - i\,|b\rangle\langle a|}{\sqrt{2}}, \text{ for } a, b = 1, \ldots, \dim(\mathcal{Z}) \text{ and } a \neq b \right\}. \tag{11}$$

If we consider the space $\mathcal{V}$ of all Choi matrices of Hermiticity preserving maps we receive the $\dim(\mathcal{X})^2 \dim(\mathcal{Y})^2$ dimensional space. To reduce the number of dimensions of $\mathcal{V}$ we introduce the concept of a cone in this space and the base of cone.

Now we introduce a proper cone in the space $\mathcal{V}$ as

$$\mathcal{C} = \{J(\Phi) \in \mathcal{V} : J(\Phi) \geq 0\}, \tag{12}$$

and a subspace $\mathcal{S} \subset \mathcal{V}$ such that

$$\mathcal{S} = \{J(\Phi) \in \mathcal{V} : \text{Tr}_{\mathcal{Y}}\, J(\Phi) = c\mathbb{1}_{\mathcal{X}}, c \in \mathbb{R}\}. \tag{13}$$

Fact 1. *The set of Choi matrices of quantum channels $\mathrm{C}(\mathcal{X}, \mathcal{Y})$ is the intersection of sets*

$$\mathcal{C} \cap \{J(\Phi) \in \mathcal{V} : \text{Tr}_{\mathcal{Y}}\, J(\Phi) = \mathbb{1}_{\mathcal{X}}\}. \tag{14}$$

We can also introduce the orthogonal complement $\mathcal{S}^\perp$ of $\mathcal{S}$ which is given by

$$\mathcal{S}^\perp := \{X \in \mathcal{V} : \mathrm{Tr}\,(XY) = 0, Y \in \mathcal{S}\}. \tag{15}$$

Fact 2. *The set $\mathcal{S}^\perp$ is given by*

$$\mathcal{S}^\perp = \{\mathbb{1}_\mathcal{Y} \otimes H : H \in \mathrm{Herm}(\mathcal{X}), \mathrm{Tr}(H) = 0\}. \tag{16}$$

The proof of this fact is presented in Appendix B.

We can also consider a proper cone $\mathcal{C}_\mathcal{S}$ in space $\mathcal{S}$ given by $\mathcal{C}_\mathcal{S} = \mathcal{S} \cap \mathcal{C}$ and a base $B_\mathcal{S} \subset \mathcal{C}_\mathcal{S}$ of the cone $\mathcal{C}_\mathcal{S}$. We can prove, using Theorem 1, that the set $B_\mathcal{S}$ is the base of cone $\mathcal{C}_\mathcal{S}$ if and only if $B_\mathcal{S} = \mathcal{S} \cap B_E$ for some order unit $E \in \mathrm{int}(\mathcal{C}^*)$. The base $B_\mathcal{S}$ determined by an order unit E will be denoted as $B_\mathcal{S}^E$ and is given by

$$B_\mathcal{S}^E = \{X \in \mathcal{C}_\mathcal{S} : \langle X|E\rangle = 1\}. \tag{17}$$

One can easily see that identity matrix $\mathbb{1}_\mathcal{Y} \otimes \mathbb{1}_\mathcal{X}$ is an order unit in cone $\mathcal{C}$. Thus we have the following observation.

Fact 3. *For $E := \frac{\mathbb{1}_\mathcal{Y} \otimes \mathbb{1}_\mathcal{X}}{\dim(\mathcal{X})}$ the base $B_\mathcal{S}^E$ is determined by the set of Choi matrices of quantum channels $\Phi \in \mathrm{C}(\mathcal{X}, \mathcal{Y})$ i.e.*

$$B_\mathcal{S}^E = \{J(\Phi) : \Phi \in \mathrm{C}(\mathcal{X}, \mathcal{Y})\}. \tag{18}$$

We are ready to establish the main result of our work.

Theorem 2. *The linear space $\mathcal{S}$ is the smallest linear subspace containing the set of quantum channels $\mathrm{C}(\mathcal{X}, \mathcal{Y})$ with orthonormal basis $\mathcal{B}(\mathcal{S})$ given by*

$$\left\{\frac{\mathbb{1}_\mathcal{Y} \otimes \mathbb{1}_\mathcal{X}}{\sqrt{\dim(\mathcal{X})\dim(\mathcal{Y})}}\right\} \cup \left\{G \otimes H : G \in \mathcal{B}(\mathcal{Y}) \backslash \left\{\frac{\mathbb{1}_\mathcal{Y}}{\sqrt{\dim(\mathcal{Y})}}\right\}, H \in \mathcal{B}(\mathcal{X})\right\}. \tag{19}$$

Moreover,

$$\dim(\mathcal{S}) = \dim(\mathcal{X})^2 \dim(\mathcal{Y})^2 - \dim(\mathcal{X})^2 + 1. \tag{20}$$

The proof of this theorem is presented in Appendix C.

Combining Theorem 2 with Fact 1 we obtain the following corollary.

Corollary 1. *Every quantum channel $\Phi \in \mathrm{C}(\mathcal{X}, \mathcal{Y})$ can be uniquely determined by $\dim(\mathcal{X})^2 \dim(\mathcal{Y})^2 - \dim(\mathcal{X})^2$ real numbers.*

Moreover, there exists extra, single non-zero coefficient which is fixed for all quantum channels $\mathrm{C}(\mathcal{X}, \mathcal{Y})$. Existence of this coefficient is a consequence of trace preserving condition $\mathrm{Tr}_\mathcal{Y}\, J(\Phi) = \mathbb{1}_\mathcal{X}$ and it can be calculated via

$$\left\langle \frac{\mathbb{1}_\mathcal{Y} \otimes \mathbb{1}_\mathcal{X}}{\sqrt{\dim(\mathcal{X})\dim(\mathcal{Y})}} \middle| J(\Phi) \right\rangle = \sqrt{\frac{\dim(\mathcal{X})}{\dim(\mathcal{Y})}}. \tag{21}$$

As a conclusion, we reduced the dimension of computational space by $\dim(\mathcal{X})^2$,

3   Example

In this section we present how one can use the `Julia` language and `QuantumInformation.jl` library in order express quantum channels as vectors in the space $\mathcal{S}$.

Let us consider $\mathcal{X} = \mathbb{C}^2$ and $\mathcal{Y} = \mathbb{C}^3$ along with quantum channels $\Phi \in C(\mathcal{X})$ given by

$$\Phi(X) = \frac{1}{2}\begin{bmatrix} 1 & 1 \\ 1 & -1 \end{bmatrix} X \begin{bmatrix} 1 & 1 \\ 1 & -1 \end{bmatrix}, \quad X \in L(\mathcal{X}), \tag{22}$$

and $\Psi \in C(\mathcal{Y})$ defined as

$$\Psi(Y) = \begin{bmatrix} 1 & 0.92 - 0.14i & 0.84 - 0.19i \\ 0.92 + 0.14i & 1 & 0.81 + 0.06i \\ 0.84 + 0.19i & 0.81 - 0.06i & 1 \end{bmatrix} \odot Y, \quad Y \in L(\mathcal{Y}), \tag{23}$$

where $\odot$ denotes the Hadamard product.

First we calculate the Choi matrices of Φ given by

$$J(\Phi) = \begin{bmatrix} 0.5 & 0.5 & 0.5 & -0.5 \\ 0.5 & 0.5 & 0.5 & -0.5 \\ 0.5 & 0.5 & 0.5 & -0.5 \\ -0.5 & -0.5 & -0.5 & 0.5 \end{bmatrix}. \tag{24}$$

Analogously for Ψ we have

$$J(\Psi) = \begin{bmatrix} 1 & 0\,0\,0 & 0.92 - 0.14i & 0\,0\,0 & 0.84 - 0.19i \\ 0 & 0\,0\,0 & 0 & 0\,0\,0 & 0 \\ 0 & 0\,0\,0 & 0 & 0\,0\,0 & 0 \\ 0 & 0\,0\,0 & 0 & 0\,0\,0 & 0 \\ 0.92 + 0.14i & 0\,0\,0 & 1 & 0\,0\,0 & 0.81 + 0.06i \\ 0 & 0\,0\,0 & 0 & 0\,0\,0 & 0 \\ 0 & 0\,0\,0 & 0 & 0\,0\,0 & 0 \\ 0 & 0\,0\,0 & 0 & 0\,0\,0 & 0 \\ 0.84 + 0.19i & 0\,0\,0 & 0.81 - 0.06i & 0\,0\,0 & 1 \end{bmatrix}. \tag{25}$$

Now we use the function `channelbasis`. The inputs of this function are the dimensions of spaces $\mathcal{X}$ and $\mathcal{Y}$ of channels Φ, Ψ. The function returns an orthonormal basis of $\mathcal{S}$. Then, we are able to use the function `represent` which factor out Choi matrices $J(\Phi), J(\Psi)$ on basis elements and returns a vector representations $v_{J(\Phi)}, v_{J(\Psi)}$ of basis coefficients. In our examples we have

$$v_{J(\Phi)} = \begin{bmatrix} 1.0 \\ 0.70711 \\ 0.70711 \\ -0.70711 \\ -0.70711 \\ 1.0 \end{bmatrix} \oplus \mathbf{0}_7, \quad v_{J(\Psi)} = \begin{bmatrix} 0.70711 \\ -0.70711 \\ -0.20356 \\ -0.26978 \\ 1.29553 \\ 0.40825 \\ 0.40825 \\ -0.8165 \\ 0.08119 \\ 1.18231 \\ 1.14206 \\ 1.0 \end{bmatrix} \oplus \mathbf{0}_{61} \tag{26}$$

where $\mathbf{0}_i$ denotes vector of zeros of length i. If we want to reverse vector representation process, we can use function `combine`. The output matrix elements shall be accurate with original Choi matrix elements to 10^{-16} or better.

The explicit code of implementation in Julia language is presented in Appendix D.

4   Conclusion

In this work we find a matrix basis for quantum channels and provide strict mathematical proofs supporting our result. This basis allows us to reduce the dimensionality of the matrix which represents a quantum channel. This, in turn, allows us to speed up computation of a class of functions of these channels, which is applicable in, for instance, the study of quantum channel discrimination. Our analytical results are accompanied by functions written in the `Julia` language which decompose a given quantum channel in our basis. This implementation is now a part of the `QuantumInformation.jl` package [6,7].

Acknowledgements. This work was supported by the Foundation for Polish Science (FNP) under grant number POIR.04.04.00-00-17C1/18-00.

A   Proof of Theorem 1

Proof. "$\Longrightarrow$" Consider that B_e is a base of $\mathcal{C}$. An element $e \in \text{int}(\mathcal{C}^*)$ if and only if there exists $r > 0$ such that the ball $K(e, r) \subset \mathcal{C}^*$, which is equivalent above condition

$$\exists_{r>0} \forall_{f \in \mathcal{X}^*} \left(||f - e|| < r \implies \forall_{z \in \mathcal{C}} \langle f|z \rangle \geq 0 \right). \tag{27}$$

By using the fact that in finite-dimensional spaces all norms are equivalent, we use the definition of induced norm given by

$$||f - e|| = \inf \left\{ M \in [0, \infty) : \forall_{x \in \mathcal{X}} | \langle f|x \rangle - \langle e|x \rangle | \leq M ||x|| \right\}. \tag{28}$$

Then, we have

$$\|f - e\| < r \iff \exists_{0 < M < r} \forall_{x \in \mathcal{X}} |\langle f|x\rangle - \langle e|x\rangle| \leq M\|x\|. \tag{29}$$

Assume that $f \in K(e, r)$, $M := \max\{\|b\| : b \in B_e\}$ and $0 < r < \frac{1}{M}$. Then, we have

$$|\langle f|b\rangle - \langle e|b\rangle| \leq \|f - e\| \cdot \|b\| \leq r\|b\| \leq rM < 1. \tag{30}$$

If $b \in B_e$, then $\langle e|b\rangle = 1$. Hence $|\langle f|b\rangle - 1| < 1$. That entails that $\langle f|b\rangle > 0$. By using the assumption we have $\langle f|z\rangle = t\langle f|b\rangle$, which implies that $\langle f|z\rangle > 0$.

"$\impliedby$" Now consider that $e \in \text{int}(\mathcal{C}^*)$ is an order unit. It easy to see that B_e is a convex set. First we prove that $\langle e|z\rangle \neq 0$. Let $z \in \mathcal{C} \setminus \{0\}$ and $e \in \text{int}(\mathcal{C}^*)$. If $e \in \mathcal{C}^*$, then $\langle e|z\rangle \geq 0$. It suffices to show that $\langle e|z\rangle \neq 0$. We will show this fact by contradiction. Assume that $\langle e|z\rangle = 0$ and let $\epsilon > 0$. By the Hahn–Banach theorem [11], there exists $z^* \in \mathcal{X}^*$ such that $\langle z^*|z\rangle = \|z\|$. Then,

$$\langle e - \epsilon z^*|z\rangle = -\epsilon\|z\| < 0 \tag{31}$$

It implies that $K(e, \epsilon) \notin \mathcal{C}^*$, which is contradiction with the assumption $e \in \text{int}(\mathcal{C}^*)$. Therefore, $\langle e|z\rangle > 0$.

Let us see that if $b := \frac{z}{\langle e|z\rangle}$ and $t := \langle e|z\rangle$, then each element $z \in \mathcal{C} \setminus \{0\}$ can be written as $z = tb$. To prove that B_e is compact we note that $\mathcal{X}$ is a finite-dimensional space. Then the set B_e is compact if and only if B_e is closed and bounded. To prove that B_e is closed, we take any sequence $(z_n)_{n \in \mathbb{N}} \in B_e$ such that $z_n \xrightarrow{n \to \infty} z$. By the inner product continuity, we get

$$1 = \lim_{n \to \infty} \langle e|z_n\rangle = \left\langle e\middle| \lim_{n \to \infty} z_n \right\rangle = \langle e|z\rangle. \tag{32}$$

It implies that $z \in B_e$. Therefore B_e is closed. To prove that B_e is bounded we show there exists $M \in [0, \infty)$ such that $\|z\| \leq M$ for every $z \in B_e$. Let us take a compact sphere $S(0, 1)$ and closed cone $\mathcal{C}$. Then $S = S(0, 1) \cap \mathcal{C}$ is also compact. Notice the function $f : S \to \mathbb{R}_+$ given by $f(x) = \langle e|x\rangle$, where e is an order unit. By the Weierstrass theorem, a function f attains infimum and supremum. Therefore there exists $x_0 \in S$ such that $0 \leq f(x_0) = \inf_{x \in S} f(x)$. Consider by contradiction that $f(x_0) = \langle e|x_0\rangle = 0$. We have $0 = \langle e|x_0\rangle = \langle e|tb_0\rangle = t$, where $b_0 \in B_e$, which is a contradiction with the assumption $t > 0$. Thus there exists $\lambda := \langle e|x_0\rangle > 0$ such that $\langle e|z\rangle \geq \lambda\|z\|$ for every $z \in B_e$, hence $\|z\| \leq \frac{1}{\lambda}$. Taking $M := \frac{1}{\lambda}$, we get thesis.

B   Proof of Fact 2

Proof. It is clear that $\dim(\mathcal{V}) = (\dim(\mathcal{X})\dim(\mathcal{Y}))^2$. Consider a linear space $\mathcal{V} \oplus \mathbb{R}$ which is $(\dim(\mathcal{X})\dim(\mathcal{Y}))^2 + 1$ dimensional. Let $J(\Phi) \in \mathcal{S}$. The condition $\text{Tr}_{\mathcal{Y}} J(\Phi) = c\mathbb{1}_{\mathcal{X}}, c \in \mathbb{R}$ in the space $\mathcal{V} \oplus \mathbb{R}$ is equivalent to

$$\sum_{k=1}^{\dim(\mathcal{Y})} \Re\left(J(\varPhi)_{j+(k-1)\dim(\mathcal{X}),i+(k-1)\dim(\mathcal{X})}\right) = 0 \quad i > j,$$

$$\sum_{k=1}^{\dim(\mathcal{Y})} \Im\left(J(\varPhi)_{j+(k-1)\dim(\mathcal{X}),i+(k-1)\dim(\mathcal{X})}\right) = 0 \quad i < j, \qquad (33)$$

$$\sum_{k=1}^{\dim(\mathcal{Y})} J(\varPhi)_{j+(k-1)\dim(\mathcal{X}),i+(k-1)\dim(\mathcal{X})} - c = 0 \quad i = j,$$

for all $i, j \in \{1, \ldots, \dim(\mathcal{X})\}$. This homogeneous system of $\dim(\mathcal{X})^2$ linear equations is linearly independent. By rank–nullity theorem [12], we have

$$\dim(\mathcal{S}) = (\dim(\mathcal{X})\dim(\mathcal{Y}))^2 + 1 - \dim(\mathcal{X})^2. \qquad (34)$$

Therefore, $\dim(\mathcal{S}^{\perp}) = \dim(\mathcal{X})^2 - 1$. To complete the proof, note that

$$\dim\left(\{\mathbb{1}_{\mathcal{Y}} \otimes H : H \in \mathrm{Herm}(\mathcal{X}), \mathrm{Tr}(H) = 0\}\right) = \dim(\mathcal{X})^2 - 1. \qquad (35)$$

C   Proof of Theorem 2

Proof. According to Fact 3 the set $\mathrm{C}(\mathcal{X}, \mathcal{Y})$ is the base of a proper cone $\mathcal{C}_S$. That means

$$\mathrm{span}\left(\{J(\varPhi) \in \mathcal{V} : J(\varPhi) \geq 0, \mathrm{Tr}_{\mathcal{Y}} J(\varPhi) = \mathbb{1}_{\mathcal{X}}\}\right) = \mathcal{S}. \qquad (36)$$

Now we fix an orthonormal basis of the space $\mathcal{V}$. Let it be given as the collection

$$\mathcal{B}(\mathcal{V}) = \{G \otimes H : G \in \mathcal{B}(\mathcal{Y}), H \in \mathcal{B}(\mathcal{X})\}. \qquad (37)$$

By using Fact 2, if we take $X \in S^{\perp}$, than there exists $H \in \mathrm{Herm}(\mathcal{X})$, $\mathrm{Tr}(H) = 0$ such that $X = \mathbb{1}_{\mathcal{Y}} \otimes H$. Let us set up the basis of $S^{\perp}$

$$\mathcal{B}(S^{\perp}) = \left\{\frac{\mathbb{1}_{\mathcal{Y}}}{\sqrt{\dim(\mathcal{Y})}} \otimes H : H \in \mathcal{B}(\mathcal{X}) \backslash \left\{\frac{\mathbb{1}_{\mathcal{X}}}{\sqrt{\dim(\mathcal{X})}}\right\}\right\}. \qquad (38)$$

Bearing in mind the relation $\mathcal{V} = \mathcal{S} \oplus \mathcal{S}^{\perp}$, we conclude that basis of $\mathcal{S}$ can be chosen as $\mathcal{B}(\mathcal{S}) = \mathcal{B}(\mathcal{V}) \backslash \mathcal{B}(\mathcal{S}^{\perp})$, namely

$$\left\{\frac{\mathbb{1}_{\mathcal{Y}} \otimes \mathbb{1}_{\mathcal{X}}}{\sqrt{\dim(\mathcal{X})\dim(\mathcal{Y})}}\right\} \cup \left\{G \otimes H : G \in \mathcal{B}(\mathcal{Y}) \backslash \left\{\frac{\mathbb{1}_{\mathcal{Y}}}{\sqrt{\dim(\mathcal{Y})}}\right\}, H \in \mathcal{B}(\mathcal{X})\right\},$$
$$(39)$$

which completes the proof.

D   Julia implementation

Here, we present the code structure for the basis representation of Choi matrix of a qubit unitary channel Φ given by Eq. (22).

```
julia> using QuantumInformation

julia> H=hadamard(2)
2×2 Array{Float64,2}:
0.707107    0.707107
0.707107   -0.707107

julia> # defining Choi Matrix
J_Φ=res(H)*res(H)'
4×4 Array{Float64,2}:
 0.5    0.5    0.5   -0.5
 0.5    0.5    0.5   -0.5
 0.5    0.5    0.5   -0.5
-0.5   -0.5   -0.5    0.5

julia> # representing Choi matrix in the basis of the subspace S
v_J_Φ=represent(channelbasis(Matrix{ComplexF64}, 2, 2),J_Φ)
13-element Array{Float64,1}:
0.0
0.9999999999999996
0.0
0.0
0.0
0.0
0.0
0.0
0.7071067811865474
0.7071067811865474
-0.7071067811865474
-0.7071067811865474
0.9999999999999998

julia> # recovering the original Choi matrix from its basis representation
J_Φ_recovered=combine(channelbasis(Matrix{ComplexF64}, 2,2),v_J_Φ).matrix
4×4 Array{Complex{Float64},2}:
 0.5+0.0im    0.5+0.0im    0.5+0.0im   -0.5+0.0im
 0.5+0.0im    0.5+0.0im    0.5+0.0im   -0.5+0.0im
 0.5+0.0im    0.5+0.0im    0.5+0.0im   -0.5+0.0im
-0.5+0.0im   -0.5+0.0im   -0.5+0.0im    0.5+0.0im

julia> # checking accuracy of recovery process using trace norm
print(norm_trace(J_Φ-J_Φ_recovered))
8.881784197001252e-16
```

References

1. Biamonte, J., Wittek, P., Pancotti, N., Rebentrost, P., Wiebe, N., Lloyd, S.: Quantum machine learning. Nature **549**, 195–202 (2017)
2. Wittek, P.: Quantum Machine Learning: What Quantum Computing Means to Data Mining. Academic Press, Cambridge (2014)
3. Helstrom, C.W.: Quantum Detection and Estimation Theory. Academic Press, Cambridge (1976)
4. Holbrook, J.A., Kribs, D.W., Laflamme, R.: Noiseless subsystems and the structure of the commutant in quantum error correction. Quantum Inf. Process. **2**(5), 381–419 (2003)
5. Jenčová, A.: Base norms and discrimination of generalized quantum channels. J. Math. Phys. **55**(2), 022201 (2014)
6. QuantumInformation.jl. https://github.com/iitis/QuantumInformation.jl

7. Gawron, P., Kurzyk, D., Pawela, Ł.: QuantumInformation.jl–a Julia package for numerical computation in quantum information theory. PLoS ONE **13**(12), e0209358 (2018)

8. Choi, M.-D.: Completely positive linear maps on complex matrices. Linear Algebra Appl. **10**(3), 285–290 (1975)

9. Jamiołkowski, A.: Linear transformations which preserve trace and positive semidefiniteness of operators. Rep. Math. Phys. **3**(4), 275–278 (1972)

10. Fuchssteiner, B., Lusky, W.: Convex Cones. Elsevier, Amsterdam (2011)

11. Rudin, W.: Principles of Mathematical Analysis, vol. 3. McGraw-Hill, New York (1964)

12. Meyer, C.D.: Matrix Analysis and Applied Linear Algebra, vol. 71. SIAM, Philadelphia (2000)

Perturbation of the Numerical Range
of Unitary Matrices

Ryszard Kukulski[✉], Paulina Lewandowska, and Łukasz Pawela

Institute of Theoretical and Applied Informatics, Polish Academy of Sciences, ul.
Bałtycka 5, 44-100 Gliwice, Poland
rkukulski@iitis.pl, https://www.iitis.pl/en/person/rkukulski

Abstract. In this work we show how to approach the problem of manipulating the numerical range of a unitary matrix. This task has far-reaching impact on the study of discrimination of quantum measurements. We achieve the aforementioned manipulation by introducing a method which allows us to find a unitary matrix whose numerical range contains the origin where at the same time the distance between unitary matrix and its perturbation is relative small in given metric.

Keywords: Numerical range · Perturbation of the numerical range · Unitary matrices · Quantum unitary channels

1   Introduction

One of the most important tasks in quantum information theory is a problem of distinguishability of quantum channels [1,2]. Imagine we have an unknown device, a black-box. The only information we have is that it performs one of two channels, say Φ and Ψ. We want to tell whether it is possible to discriminate Φ and Ψ perfectly, i.e. with probability equal to one. Helstrom's result [3] gives the analytical formula for upper bound of probability of discrimination quantum channels using the special operators norm called diamond norm or sometimes referred to as completely bounded trace norm [4]. The Holevo-Helstrom theorem says that the quantum channels Ψ and Φ are perfectly distinguishable if and only if the distance between them is equal two by using diamond norm. In general, numerical computing of diamond norm is a complex task. Therefore, researchers were limited to smaller classes of quantum channels. One of the first results was the study of discrimination of unitary channels $\Phi_U : \rho \rightarrow U\rho U^\dagger$ where ρ is a quantum state. The sufficient condition for perfect discrimination of unitary channels Φ_U and Φ_1 is that zero belongs to the numerical range of unitary matrix U [5].

The situation in which zero belongs to numerical range of unitary matrix U paves the way toward simple calculating of probability of discrimination unitary channels without the necessity of computing the diamond norm. Now consider the following scenario. We have two quantum channels Φ_U and Φ_1 such that

© Springer Nature Switzerland AG 2020
V. V. Krzhizhanovskaya et al. (Eds.): ICCS 2020, LNCS 12142, pp. 627–637, 2020.
https://doi.org/10.1007/978-3-030-50433-5_48

zero does not belong to the numerical range of U. Hence, we know that we cannot distinguish between Φ_U and Φ_1 perfectly. Therefore, we can assume some kind of noise and consider the unitary channel Φ_V instead of Φ_U such that the distance between unitary matrix V and U is relative small where at the same time zero belongs in numerical range of V. Such a unitary matrix V will be called perturbation of U.

In this work we are interested in determining the perturbation form of V. Our motivation is two-fold. On the one hand considering the unitary channels Φ_V and Φ_1 we know that they will be perfectly distinguishable. On the other hand our method of computing V does not change the measurement result in standard basis.

Our work is naturally divided into three parts. In the first part we show the mathematical preliminaries needed to present our main result. The second part presents the theorem which gives us the method of manipulation of numerical range of unitary matrices. In third part we show the example illustrative our theorem. Concluding remarks and some future work are presented in the end of our work.

2   Mathematical Preliminaries

Let us introduce the following notation. Let $\mathbb{C}^d$ be complex d-dimensional vector space. We denote the set of all matrix operators by $\mathrm{L}(\mathbb{C}^{d_1}, \mathbb{C}^{d_2})$ while the set of isometries by $\mathrm{U}(\mathbb{C}^{d_1}, \mathbb{C}^{d_2})$. It easy to see that every square isometry is a unitary matrix. The set of all unitary matrices we will be denoted by $U \in \mathrm{U}(\mathbb{C}^d)$. We will be also interested in diagonal matrices and diagonal unitary matrices denoted by $\mathrm{Diag}(\mathbb{C}^d)$ and $\mathrm{DU}(\mathbb{C}^d)$ respectively. Next classes of matrices that will be used in this work are Hermitian matrices denoted by $\mathrm{Herm}(\mathbb{C}^d)$. All of the above-mentioned matrices are normal matrices i.e. $AA^\dagger = A^\dagger A$. Every normal matrix A can be expressed as a linear combination of projections onto pairwise orthogonal subspaces

$$A = \sum_{i=1}^{k} \lambda_i \, |x_i\rangle\langle x_i| , \tag{1}$$

where scalar $\lambda_i \in \mathbb{C}$ is an eigenvalue of A and $|x_i\rangle \subset \mathbb{C}^d$ is an eigenvector corresponding to the eigenvalue λ_i. This expression of a normal matrix A is called a spectral decomposition of A [6]. Many interesting and useful norms, not only for normal matrices, can be defined on spaces of matrix operators. In this work we will mostly be concerned with a family of norms called Schatten [7] p-norms defined as

$$||A||_p = \left(\mathrm{tr} \left(\left(A^\dagger A \right)^{\frac{p}{2}} \right) \right)^{\frac{1}{p}} \tag{2}$$

for any $A \in \mathrm{L}(\mathbb{C}^{d_1}, \mathbb{C}^{d_2})$. The Schatten ∞-norm is defined as

$$||A||_\infty = \max \left\{ ||A\,|u\rangle\,|| : |u\rangle \in \mathbb{C}^{d_1}, ||\,|u\rangle\,|| \leq 1 \right\}. \tag{3}$$

For a given square matrix A the set of all eigenvalues of A will be denoted by $\lambda(A)$ and $r(\lambda_i)$ will denote the multiplicity of each eigenvalue $\lambda_i \in \lambda(A)$. For any square matrix A, one defines its numerical range [8,9] as a subset of the complex plane

$$W(A) = \{z \in \mathbb{C} : z = \langle \psi | A | \psi \rangle, |\psi \rangle \in \mathbb{C}^d, \langle \psi | \psi \rangle = 1\}. \tag{4}$$

It is easy to see that $\lambda(A) \subseteq W(A)$. One of the most important properties of $W(A)$ is its convexity which was shown by Hausdorff and Toeplitz [10,11]. For any normal matrix A the set $W(A)$ is a convex hull of spectrum of A which will be denoted by $\mathrm{conv}(\lambda(A))$. Another well-known property of $W(U)$ for any unitary matrix $U \in U(\mathbb{C}^d)$ is the fact that its numerical range forms a polygon whose vertices are eigenvalues of U lying in unit circle on complex plane. In our work we introduce the counterclockwise order of eigenvalues of unitary matrix U [12] such that we choose any eigenvalue named $\lambda_1 \in \lambda(U)$ on the unit circle and next eigenvalues are labeled counterclockwise.

In our setup we consider the space $L(\mathbb{C}^d)$. Imagine that the matrices are points in space $L(\mathbb{C}^d)$ and the distance between them is bounded by small constant $0 < c \ll 1$. We will take two unitary matrices - matrix $U \in U(\mathbb{C}^d)$ and its perturbation $V \in U(\mathbb{C}^d)$ i.e. $||U - V||_\infty \leq c$ by using ∞-Schatten norm. We want to determine the path connecting these points given by smooth curve. To do so, we fix continuous parametric (by parameter t) curve $U(t) \in U(\mathbb{C}^d)$ for any $t \in [0,1]$ with boundary conditions $U(0) := U$ and $U(1) := V$. The most natural and also the shortest curve connecting U and V is geodesic [13] given by

$$t \to U \exp\left(t \operatorname{Log}\left(U^\dagger V\right)\right), \tag{5}$$

where Log is the matrix function such that it changes eigenvalues $\lambda \in \lambda(U)$ into $\log(\lambda(U))$, where $-i \log(\lambda(U)) \subset (-\pi, \pi]$.

We will study how the numerical range $W(U(t))$ will be changed depending on parameter t. Let $H := -i \operatorname{Log}\left(U^\dagger V\right)$. Let us see that $H \in \operatorname{Herm}(\mathbb{C}^d)$ and $W(H) \subset (-\pi, \pi]$ for any $U, V \in U(\mathbb{C}^d)$. Taking the spectral decomposition $H = Y D Y^\dagger, Y \in U(\mathbb{C}^d), D \in \operatorname{Diag}(\mathbb{C}^d)$ we get

$$\begin{aligned}
W\left(U \exp\left(itH\right)\right) &= W\left(U \exp\left(itY D Y^\dagger\right)\right) = W\left(U Y \exp\left(itD\right) Y^\dagger\right) \\
&= W\left(Y^\dagger U Y \exp\left(itD\right)\right) = W\left(\widetilde{U} \exp\left(itD\right)\right),
\end{aligned} \tag{6}$$

where $\widetilde{U} := Y^\dagger U Y \in U(\mathbb{C}^d)$. Hence, without loss of generality in our analysis we will assume that H is a diagonal matrix. Moreover, we can assume that $D \geq 0$ which follows from simple calculations

$$\begin{aligned}
W\left(U \exp\left(itD\right)\right) &= W\left(U \exp\left(itD_+\right)\left(\exp\left(it\alpha \mathbb{1}\right)\right)\right) = W\left(e^{it\alpha} U \exp\left(itD_+\right)\right) \\
&\equiv W\left(U \exp\left(itD_+\right)\right).
\end{aligned} \tag{7}$$

Let us see that the numerical range of $U(t)$ for any $t \in [0,1]$ is invariant to above calculations although the trajectory of $U(t)$ is changed. Therefore, we will consider the curve

$$t \to U \exp\left(itD_+\right), \tag{8}$$

where $t \in [0,1]$ and $U \in U(\mathbb{C}^d), D_+ \in \operatorname{Diag}(\mathbb{C}^d)$ such that $D_+ \geq 0$.

3   Main Result

In this section we will focus on the behavior of the spectrum of the unitary matrices $U(t)$, which will reveal the behavior of $W(U(t))$ for relatively small parameter t. Without loss of generality we can assume that $\mathrm{tr}\,(D_+) = 1$. Together with the fact that $D_+ \in \mathrm{Diag}(\mathbb{C}^d)$ and $D_+ \geq 0$ we can note that $D_+ = \sum_{i=1}^d p_i \, |i\rangle\langle i|$, where $p \in \mathbb{C}^d$ is a probability vector. Let us also define the set

$$S_\lambda^M = \left\{ |x\rangle \in \mathbb{C}^d : (\lambda \mathbb{1}_d - M) \, |x\rangle = 0, \| \, |x\rangle \, \|_2 = 1 \right\} \tag{9}$$

for some matrix $M \in \mathrm{L}(\mathbb{C}^d)$ which consists of unit eigenvectors corresponding to the eigenvalue λ of the matrix M. We denote by $k = r(\lambda)$ the multiplicity of eigenvalue λ whereas by $I_{M,\lambda} \in \mathrm{U}(\mathbb{C}^k, \mathbb{C}^d)$ we denote the isometry which columns are formed by eigenvectors corresponding to eigenvalue λ of a such matrix M. Let $\lambda(t), \beta(t) \in \mathbb{C}$ for $t \geq 0$. We will write $\lambda(t) \approx \beta(t)$ for relatively small $t \geq 0$, whenever $\lambda(0) = \beta(0)$ and $\frac{\partial}{\partial t}\lambda(0) = \frac{\partial}{\partial t}\beta(0)$.

Theorem 1. *Let $U \in \mathrm{U}(\mathbb{C}^d)$ be a unitary matrix with spectral decomposition*

$$U = \sum_{j=1}^d \lambda_j \, |x_j\rangle\langle x_j|. \tag{10}$$

Assume that the eigenvalue $\lambda \in \lambda(U)$ is such that $r(\lambda) = k$. Let us define a matrix $E(t)$ given by

$$E(t) = \exp(itD_+) = \sum_{i=1}^d e^{ip_i t} \, |i\rangle\langle i| \in \mathrm{DU}(\mathbb{C}^d), \quad t \geq 0. \tag{11}$$

Let $\lambda(t) := \lambda(UE(t))$ and let every $\lambda_j(t) \in \lambda(t)$ corresponds to eigenvector $|x_j(t)\rangle$. Assume that $\lambda_1(t), \ldots, \lambda_k(t)$ are such eigenvalues that $\lambda_j(t) \to \lambda$, as $t \to 0$. Then:

(a) *If $\min_{|x\rangle \in S_\lambda^U} \sum_{i=1}^d p_i |\langle i|x\rangle|^2 = 0$, then λ is an eigenvalue of $UE(t)$.*

(b) *If $|\{p_i : p_i > 0\}| = l < k$, then λ is an eigenvalue of $UE(t)$ and $r(\lambda) \geq k - l$.*

(c) *Each eigenvalue of product $UE(t)$ moves counterclockwise or stays in the initial position as parameter t increases.*

(d) *If $k = 1$, then*

$$\lambda_1(t) \approx \lambda \exp\left(it \sum_{i=1}^d p_i |\langle i|x_1\rangle|^2 \right)$$

for small $t \geq 0$.

(e) *Let $Q := I_{U,\lambda}^\dagger D_+ I_{U,\lambda}$ and $\lambda_1(Q) \leq \lambda_2(Q) \leq \ldots \leq \lambda_k(Q)$. Then we have*

$$\lambda_j(t) \approx \lambda \exp\left(i\lambda_j(Q)t \right)$$

for small $t \geq 0$ and eigenvector $|x_j\rangle$ corresponding to $\lambda_j \in \lambda(U)$ is given by

$$|x_j\rangle = I_{U,\lambda} \, |v_j\rangle,$$

where $|v_j\rangle \in S_{\lambda_j(Q)}^Q$.

(f) For each $j = 1, \ldots, d$ we have

$$\frac{\partial}{\partial t} \lambda_j(t) = i\lambda_j(t) \sum_{i=1}^{d} p_i |\langle i | x_j(t) \rangle|^2.$$

Moreover,

$$\sum_{j=1}^{d} \left| \frac{\partial}{\partial t} \lambda_j(t) \right| = 1.$$

This theorem gives us equations which one can use to predict behavior of $W(UE(t))$. Observe the postulate (f) fully determines the movement of the spectrum. However, this is a theoretical statement and in practice determining the function $t \mapsto |x_j(t)\rangle$ is a numerically complex task. The postulates $(a) - (e)$ play a key role in numerical calculations of $W(UE(t))$. The most important fact comes from (c) which says that all eigenvalues move in the same direction or stay in the initial position. The instantaneous velocity of a given eigenvalue in general case is given in (e), while in the case of eigenvalue with multiplicity equal one, the instantaneous velocity is determined by (d). We can see that whenever the spectrum of the matrix U is not degenerated, calculating these velocities is easy. What is more, when some eigenvalue is degenerated, the postulate (e) not only gives us method to calculate the trajectory of this eigenvalue, but also determines the form of corresponding eigenvector. It is worth noting that the postulates $(d), (e)$ give us only an approximation of the velocities, so despite being useful in numerical calculations, these expressions are valid only in the neighborhood of $t = 0$. Moreover, sometimes we are able to precisely specify this velocities. This happens in the cases presented in $(a), (b)$. Whenever the calculated velocity is zero we know for sure that this eigenvalue will stay in the initial position. According to the postulate (b) the same happens when the multiplicity of the eigenvalue is greater than the number of positive elements of vector p.

4 Example

We start with sampling some random unitary matrix $U \in U(\mathbb{C}^3)$ such that $0 \notin W(U)$

$$U = \begin{bmatrix} 0.267868 + 0.026891i & 0.752935 - 0.510663i & -0.314404 - 0.0313982i \\ -0.83413 - 0.0693252i & 0.245915 - 0.275811i & 0.34174 - 0.214685i \\ 0.472125 + 0.0635826i & 0.0211772 - 0.18793i & 0.795835 - 0.322391i \end{bmatrix}. \quad (12)$$

We would like to construct matrix $E \in \mathcal{DU}(\mathbb{C}^d)$ to obtain $0 \in W(UE)$. The first feature we need to analyze is the numerical range of U given as $\text{conv}(\{\lambda_1, \lambda_2, \lambda_3\})$ for $\lambda_1, \lambda_2, \lambda_3 \in \lambda(U)$. The Fig. 1 presents numerical range $W(U)$, which boundary is determined by dashed triangle. The most distant pair of eigenvalues, which determine the distance between $W(U)$ and the origin is the pair (λ_1, λ_3). Therefore, our analysis will be focused on eigenvectors corresponding to λ_1 and λ_3. Let us calculate eigenvectors of matrix U and then according to the postulate (d)

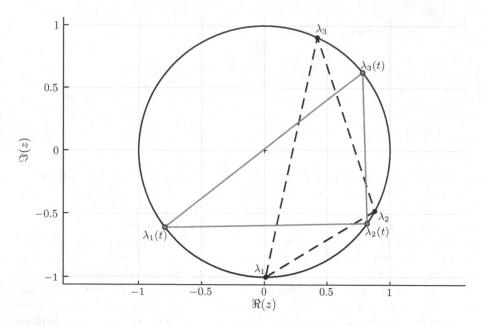

Fig. 1. The figure consists two triangles representing boundaries of the numerical ranges: $W(U)$ (dashed black) and $W(UE(t))$ for $t = 1.5$ (solid red). (Color figure online)

choose appropriate probability vector p. The squared modules of eigenvectors entries form the matrix

$$Q = \begin{bmatrix} 0.426542 & 0.543517 & 0.0299407 \\ 0.0480551 & 0.105588 & 0.846357 \\ 0.525403 & 0.350895 & 0.123702 \end{bmatrix} \tag{13}$$

Rows of the matrix Q correspond to the considered eigenvectors, i.e. first row corresponds to λ_1, second to λ_2 and third to λ_3. Analyzing the first and the third row, we can see the greatest difference in values is in the second column, namely between values $Q_{1,2} = 0.543517$ and $Q_{3,2} = 0.350895$. If we consider probability vector $p = (0, 1, 0)$ with associated matrix $E(t) = \sum_{i=1}^{3} e^{ip_i t} |i\rangle\langle i|$, then according to Theorem 1, the values $Q_{1,2}$ and $Q_{3,2}$ will indicate the instantaneous velocities of eigenvalues λ_1 and λ_3, respectively. Therefore, the first eigenvalue will move faster than the third, and hence, in order to minimize the distance between the numerical range and the origin, we have to rotate spectrum clockwise. This can be done by substituting $E(t) \leftarrow E(t)^\dagger$, namely $E(t) = \sum_{i=1}^{3} e^{-ip_i t} |i\rangle\langle i|$. The postulate (d) of Theorem 1 provides our calculations accurate only in the neighborhood of $t = 0$, but this may be not sufficient to satisfy $0 \in W(UE(t))$. Therefore, we utilize the postulate (f) and calculate instantaneous velocities $|\langle 2|x_1(t)\rangle|^2$ and $|\langle 2|x_3(t)\rangle|^2$ of eigenvalues $\lambda_1(t)$ and $\lambda_3(t)$ accurately for each $t > 0$. As long as $|\langle 2|x_1(t)\rangle|^2 > |\langle 2|x_3(t)\rangle|^2$ the eigenvalue $\lambda_1(t)$ will be faster than $\lambda_3(t)$, which will progressively minimize the value $\min|W(UE(t))|$. The Fig. 2 shows velocities $|\langle 2|x_1(t)\rangle|^2$ and $|\langle 2|x_3(t)\rangle|^2$ calculated for $t \in [0, 1.5]$.

As we can see, for $t \in [0, 1.5]$ the relation $|\langle 2|x_1(t)\rangle|^2 > |\langle 2|x_3(t)\rangle|^2$ is satisfied, so it remains to check the motion of $W(U(t))$, where we define $U(t) := UE(t)$. First of all, in the Fig. 1 we showed the numerical range $W(UE(1.5))$ and we can see that $0 \in W(UE(1.5))$. The behavior of function $t \mapsto W(U(t))$ is presented in Fig. 3. At last, numerical calculations reveal that the time t when zero enters numerical range $W(U(t))$ is approximately $t \approx 1.45$ (see Fig. 4 for more details).

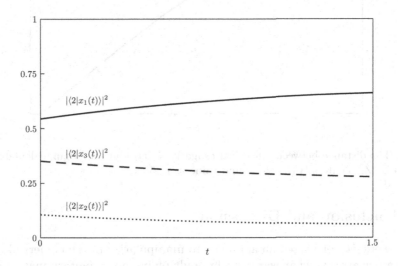

Fig. 2. The instantaneous velocities of eigenvalues $\lambda_1(t)$ (solid line), $\lambda_2(t)$ (dotted line) and $\lambda_3(t)$ (dashed line) of matrices $U(t)$. The eigenvectors $|x_1(t)\rangle, |x_2(t)\rangle, |x_3(t)\rangle$ correspond to eigenvalues $\lambda_1, \lambda_2, \lambda_3$, respectively.

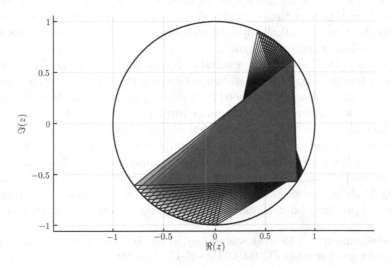

Fig. 3. Time dependence of $W(U(t))$ for $t \in [0, 1.5]$.

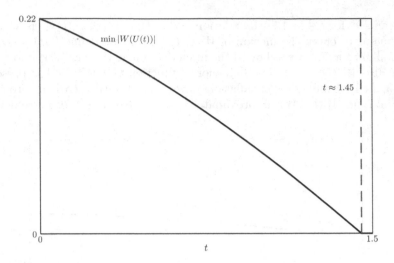

Fig. 4. The distance between numerical range $W(U(t))$ and the origin, calculated for $t \in [0, 1.5]$. Zero enters $W(U(t))$ around value ≈ 1.45.

5   Conclusion and Discussion

In this work we considered an approach to manipulation of the numerical range of unitary matrices. That was done by multiplying given unitary matrix U by some unitary matrix E which is diagonal in the fixed computational basis (we took the standard basis) and is relatively close to identity matrix. We established differential equations describing behavior of eigenvalues and presented their approximated solutions, which we find useful in numerical calculations. Our motivation was to find for given unitary matrix U the closest unitary matrix of the form $V = UE$ such that channels Φ_V and Φ_1 are perfectly distinguishable. It is important to stress that applying channel Φ_V to the quantum states leaves their classical distribution unchanged.

The results in Theorem 1 are suitable to solve various tasks. For example, one would like to maximize the distance between the numerical range and the point zero. Such task was introduced in [14] and plays crucial role in calculating diamond norm of difference of two von Neumann measurements.

In the end, we would like to note that solving numerical tasks with the use of postulates (d) and (e) of Theorem 1 requires one to find the bounds on maximal value of the parameter t still allowing to apply these results. Such bounds should depend on variability of functions $t \mapsto |\langle i|x_j(t)\rangle|^2$. Numerical analysis (e.g. Fig. 2) shows smoothness of functions of this type, giving an argument for existence of practical bounds and hence providing direction of future research.

Acknowledgements. This work was supported by the Foundation for Polish Science (FNP) under grant number POIR.04.04.00-00-17C1/18-00.

A   Proof of Theorem 1

Proof. (a) This fact is implicated by equation $\sum\limits_{i=1}^{d} p_i |\langle i|x\rangle|^2 = 0$ for some eigenvector $|x\rangle$ of eigenvalue λ. Eventually, we obtain $UE(t)|x\rangle = U|x\rangle = \lambda|x\rangle$.

(b) We will show that there are at least $k - l$ orthogonal eigenvectors $|x\rangle$ of eigenvalue λ, for which $\sum\limits_{i=1}^{d} p_i |\langle i|x\rangle|^2 = 0$, so (a) will imply (b). W.l.o.g. assume $p_i > 0$ for $i = 1, \ldots, l$. For each matrix $W \in U(\mathbb{C}^k)$ the columns of isometry $I_{U,\lambda}W$ consist of eigenvectors of eigenvalue λ. We can choose such a matrix W for which first $k - l$ columns are orthogonal to each of vectors $I_{U,\lambda}^\top |1\rangle, \ldots, I_{U,\lambda}^\top |l\rangle$. One can note that for $|x\rangle \in \{I_{U,\lambda}W|1\rangle, \ldots, I_{U,\lambda}W|k-l\rangle\}$ we obtain $\sum\limits_{i=1}^{d} p_i |\langle i|x\rangle|^2 = 0$.

(c) Fix some eigenvalue λ with $r(\lambda) = k$. We introduce the notation of $\Pi = I_{U,\lambda} I_{U,\lambda}^\dagger$. We consider the subspace $\{|x\rangle : D_+ |x\rangle = 0, \Pi|x\rangle = |x\rangle\}$ along with projection Π_0 on this subspace. Denote $\Pi_+ = \Pi - \Pi_0$. Let

$$c := \min_{|x\rangle : \Pi_+ |x\rangle = |x\rangle, \||x\rangle\|_2 = 1} \langle x| D_+ |x\rangle . \tag{14}$$

One can note that $c > 0$. Take unit vector $|x\rangle$ and define $|v\rangle = (\mathbb{1}_d - \Pi)|x\rangle$. First of all, we will show that $\operatorname{tr}(\mathbb{1}_d - \Pi)|x\rangle\langle x| \in O(\epsilon)$ if $|\lambda - \langle x| U|x\rangle| = \epsilon$. Direct calculations reveal that

$$\epsilon = |\lambda - \langle x| U|x\rangle| = |\lambda \operatorname{tr}|x\rangle\langle x| - \operatorname{tr} U\Pi|x\rangle\langle x| - \operatorname{tr} U(\mathbb{1}_d - \Pi)|x\rangle\langle x||$$
$$= |\lambda \operatorname{tr}(\mathbb{1}_d - \Pi)|x\rangle\langle x| - \operatorname{tr}(\mathbb{1}_d - \Pi)U(\mathbb{1}_d - \Pi)|x\rangle\langle x||. \tag{15}$$

If $\|v\| = 0$ the statement is true, so assume $\|v\| \neq 0$. We obtain

$$\epsilon = |\lambda \langle v| - \rangle \langle v| U|v\rangle| \geq \langle v|v\rangle \operatorname{dist}(\lambda, \operatorname{conv}(\lambda_{k+1}, \ldots, \lambda_d)) > 0, \tag{16}$$

which finishes this part of proof.

In the second part we will check the behavior of points $\langle x| U|x\rangle$ in the neighborhood of point λ. We assume that $|\lambda - \langle x| U|x\rangle| = \epsilon$ for relatively small $\epsilon \geq 0$, so $\operatorname{tr}(\mathbb{1}_d - \Pi)|x\rangle\langle x| \in O(\epsilon)$. The derivative of trajectory of such a point is

$$\frac{\partial}{\partial t}(\langle x| UE(t)|x\rangle)(0) = i\langle x| UD_+ |x\rangle . \tag{17}$$

We can rewrite the above as

$$i\langle x| UD_+ |x\rangle = i\langle x| (\Pi_+ + \Pi_0 + \mathbb{1}_d - \Pi)UD_+(\Pi_+ + \Pi_0 + \mathbb{1}_d - \Pi)|x\rangle$$
$$= i\lambda \langle x| \Pi_+ D_+ \Pi_+ |x\rangle + i\langle x| \Pi_+ UD_+(\mathbb{1}_d - \Pi)|x\rangle$$
$$+ i\langle x| (\mathbb{1}_d - \Pi)UD_+ \Pi_+ |x\rangle + i\langle x| (\mathbb{1}_d - \Pi)UD_+(\mathbb{1}_d - \Pi)|x\rangle). \tag{18}$$

The above equation means that the instantaneous velocity of point $\langle x| U|x\rangle$ is the sum of the velocity

$$i\lambda \langle x| \Pi_+ D_+ \Pi_+ |x\rangle \tag{19}$$

which is responsible for counterclockwise movement and which speed is $\langle x|\, \Pi_+ D_+ \Pi_+ \,|x\rangle$ and the "noise" velocity which for the most pessimistic scenario can be rotated in any direction and which speed is at most

$$|\langle x|\, \Pi_+ U D_+ (\mathbb{1}_d - \Pi) \,|x\rangle\,|+|\,\langle x|\, (\mathbb{1}_d - \Pi) U D_+ \Pi_+ \,|x\rangle\,|+|\,\langle x|\, (\mathbb{1}_d - \Pi) U D_+ (\mathbb{1}_d - \Pi) \,|x\rangle\,|. \quad (20)$$

The speed of velocity with direction $i\lambda$ can be lower bounded by

$$c\,\langle x|\, \Pi_+ \,|x\rangle, \quad (21)$$

while the speed of the second velocity can be upper bounded by

$$2\sqrt{\langle x|\, \Pi_+ \,|x\rangle}\sqrt{\langle x|\, (\mathbb{1}_d - \Pi)\,|x\rangle} + \langle x|\, (\mathbb{1}_d - \Pi)\,|x\rangle \le 2\sqrt{\langle x|\, \Pi_+ \,|x\rangle}\sqrt{\epsilon} + \epsilon. \quad (22)$$

There exists constant $d > 0$ depending on the geometry of the numerical range of the matrix U, such that if

$$c\,\langle x|\, \Pi_+ \,|x\rangle \ge d(2\sqrt{\langle x|\, \Pi_+ \,|x\rangle}\sqrt{\epsilon} + \epsilon), \quad (23)$$

then the point $\langle x|\, U \,|x\rangle$ moves counterclockwise. This is true if

$$\langle x|\, \Pi_+ \,|x\rangle \ge \frac{cd + 2d^2 + 2d\sqrt{d^2 + cd}}{c^2}\epsilon. \quad (24)$$

In the case when the above inequality does not hold, the speed of the second velocity is upper bounded by a some linear function of variable ϵ. That means there exists t_0 such that for $t \le t_0$ there can not exists eigenvalue $\widetilde{\lambda} \in UE(t)$ which $\widetilde{\lambda} \to \lambda$ as $t \to 0$ and $\widetilde{\lambda}$ is before λ in the counterclockwise order. To finish the proof we can see the above holds for any $t \ge 0$ due to fact that $E(t_0 + t) = E(t_0)E(t)$.

(d) To see this we first need to describe local dynamics of point $\beta(t) = \langle x_1|\, UE(t)\,|x_1\rangle = \lambda\,\langle x_1|\, E(t)\,|x_1\rangle$. One can note that $\beta(0) = \lambda$ and $\frac{\partial}{\partial t}\beta(0) = i\lambda \sum_{i=1}^{d} p_i |\langle i|x_1\rangle|^2$. That means $\beta(t) \approx \lambda \exp(it \sum_{i=1}^{d} p_i |\langle i|x_1\rangle|^2)$. To see that $\beta(t) \approx \lambda_1(t)$ we need to utilize the following facts:

- Eigenvalues of $UE(t)$ are continuous functions.
- $\beta(t) \in W(UE(t))$.
- Trajectory of $\beta(t)$ is curved in such a way that holds $\frac{1-|\beta(t)|}{|\beta(t)-\lambda|} \to 0$.

The above means that if $R(t) \subset \{|z| = 1\}$ is an arch in which we can potentially find eigenvalue $\lambda_1(t)$ according to the fact that $\beta(t) \in W(UE(t))$, then it is true that $\frac{|R(t)|}{|\beta(t)-\lambda|} \to 0$ and consequently $\lambda_1(t) \approx \beta(t)$ for small $t \ge 0$.

(e) The see this point we need to utilize the postulate (c) along with the proof of the postulate (d) for eigenvectors $|x\rangle \in \{I_{U,\lambda}\,|v_1\rangle, \ldots, I_{U,\lambda}\,|v_k\rangle\}$, where $Q := I_{U,\lambda}^\dagger D_+ I_{U,\lambda}$ and $|v_j\rangle \in S_{\lambda_j(Q)}^Q$ for $j = 1, \ldots, k$.

(f) This relation follows from (e).

References

1. Helstrom, C.W.: Quantum Detection and Estimation Theory. Academic press, New York (1976)
2. Jenčová, A.: Base norms and discrimination of generalized quantum channels. J. Math. Phys. **55**(2), 022201 (2014). https://doi.org/10.1063/1.4863715
3. Helstrom, C.W.: Quantum detection and estimation theory. J. Stat. Phys. **1**(2), 231–252 (1969). https://doi.org/10.1007/BF01007479
4. Aharonov, D., Kitaev, A., Nisan, N.: Quantum circuits with mixed states. In: Proceedings of the Thirtieth Annual ACM Symposium on Theory of Computing, pp. 20–30 (1998)
5. Watrous, J.: The Theory of Quantum Information. Cambridge University Press, Cambridge (2018)
6. Hall, B.C.: Quantum Theory for Mathematicians, vol. 267. Springer, New York (2013). https://doi.org/10.1007/978-1-4614-7116-5
7. Franklin, J.N.: Matrix theory. Courier Corporation (2012)
8. Murnaghan, F.D.: On the field of values of a square matrix. Proc. Natl. Acad. Sci. U.S.A. **18**(3), 246 (1932). https://doi.org/10.1073/pnas.18.3.246
9. Gustafson, K., Rao, D.: The field of values of linear operators and matrices, universitext (1997)
10. Toeplitz, O.: Das algebraische analogon zu einem satze von fejér. Math. Z. **2**(1–2), 187–197 (1918). https://doi.org/10.1007/BF01212904
11. Hausdorff, F.: Der wertvorrat einer bilinearform. Math. Z. **3**(1), 314–316 (1919). https://doi.org/10.1007/BF01292610
12. Bhatia, R.: Matrix Analysis, vol. 169. Springer Science & Business Media, Alemania (2013)
13. Antezana, J., Larotonda, G., Varela, A.: Optimal paths for symmetric actions in the unitary group. Commun. Math. Phys. **328**(2), 481–497 (2014). https://doi.org/10.1007/s00220-014-2041-x
14. Puchała, Z., Pawela, Ł., Krawiec, A., Kukulski, R.: Strategies for optimal single-shot discrimination of quantum measurements. Phys. Rev. A **98**(4), 042103 (2018). https://doi.org/10.1103/PhysRevA.98.042103

Design of Short Codes for Quantum Channels with Asymmetric Pauli Errors

Marco Chiani[(⊠)] and Lorenzo Valentini

DEI/CNIT, University of Bologna, Via Università 50, Cesena, Italy
{marco.chiani,lorenzo.valentini13}@unibo.it

Abstract. One of the main problems in quantum information systems is the presence of errors due to noise. Many quantum error correcting codes have been designed to deal with generic errors. In this paper we construct new stabilizer codes able to correct a given number e_g of generic Pauli X, Y and Z errors, plus a number e_z of Pauli errors of a specified type (e.g., Z errors). These codes can be of interest when the quantum channel is asymmetric, i.e., when some types of error occur more frequently than others. For example, we design a $[[9,1]]$ quantum error correcting code able to correct up to one generic qubit error plus one Z error in arbitrary positions. According to a generalized version of the quantum Hamming bound, it is the shortest code with this error correction capability.

1 Introduction

The possibility to exploit the unique features of quantum mechanics is paving the way to new approaches for acquiring, processing and transmitting information, with applications in quantum communications, computing, cryptography, and sensing [1–10]. In this regard, one of the main problem is the noise due to unwanted interaction of the quantum information with the environment. Error correction techniques are therefore essential for quantum computation, quantum memories and quantum communication systems [11–14]. Compared to the classical case, quantum error correction is made more difficult by the laws of quantum mechanics which imply that qubits cannot be copied or measured without perturbing state superposition [15]. Moreover, there is continuum of errors that could occur on a qubit. However, it has been shown that in order to correct an arbitrary qubit error it suffices to consider error correction on the discrete set of Pauli operators, i.e., the bit flip X, phase flip Z, and combined bit-phase flip Y [11, 16–18]. Hence, we can consider in general a channel introducing qubit errors X, Y, and Z with probabilities p_x, p_y, and p_z, respectively, and leaving the qubit intact with probability $1 - \rho$, where $\rho = p_x + p_y + p_z$. A special case of this model is the so-called depolarizing channel for which $p_x = p_y = p_z = \rho/3$. Quantum error correcting codes for this channel are naturally designed to protect against equiprobable Pauli errors [19–21].

This work was supported in part by the Italian Ministry for Education, University and Research (MIUR) under the program "Dipartimenti di Eccellenza (2018–2022)".

ⓒ Springer Nature Switzerland AG 2020
V. V. Krzhizhanovskaya et al. (Eds.): ICCS 2020, LNCS 12142, pp. 638–649, 2020.
https://doi.org/10.1007/978-3-030-50433-5_49

However, not all channels exhibit this symmetric behaviour of Pauli errors as, in some situations, some types of error are more likely than others [22]. In fact, depending on the technology adopted for the system implementation, the different types of Pauli error can have quite different probabilities of occurrence, leading to asymmetric quantum channels [23,24]. An example is the Pauli-twirled channel associated to the combination of amplitude damping and dephasing channels [23]. This model has $p_x = p_y$ and $p_z = A\rho/(A+2)$, where ρ is the error probability, and the asymmetry is accounted for by the parameter $A = p_z/p_x$. This parameter is a function of the relaxation time, T_1, and the dephasing time, T_2, which are in general different, leading to $A > 1$ [24,25].

Owing to this considerations, it can be useful to investigate the design of quantum codes with error correction capabilities tailored to specific channel models. For example, codes for the amplitude damping channel have been proposed in [26–31], while quantum error correcting codes for more general asymmetric channels are investigated in [22–25]. In particular, asymmetric Calderbank Shor Steane (CSS) codes, where the two classical parity check matrices are chosen with different error correction capability (e.g., Bose Chaudhuri Hocquenghem (BCH) codes for X errors and low density parity check (LDPC) codes for Z errors), are investigated in [22,23]. Inherent to the CSS construction there are two distinct error correction capabilities for the X and the Z errors; the resulting asymmetric codes, denoted as $[[n, k, d_x/d_z]]$, can correct up to $t_x = \lfloor (d_x - 1)/2 \rfloor$ Pauli X errors and $t_z = \lfloor (d_z - 1)/2 \rfloor$ Pauli Z errors per codeword. In fact, due to the possibility of employing tools from classical error correction, many works have been focused on asymmetric codes based on the CSS construction, which, however, may not lead to the shortest codes (e.g., for the symmetric channel compare the $[[7,1]]$ CSS Steane code with the shortest $[[5,1]]$ code [20,21]).

In this paper, we relax the CSS constraint in order to obtain the shortest asymmetric stabilizer codes able to correct a given number e_g of generic Pauli errors, plus a number e_Z of Pauli errors of a specified type (e.g., Z errors). We denote these as the asymmetric $[[n, k]]$ codes with (e_g, e_Z). To this aim we first derive a generalized version of the quantum Hamming bound, which was developed to correct generic errors, for an asymmetric error correction capability (e_g, e_Z). Then, we construct a $[[9,1]]$ code with $(e_g = 1, e_Z = 1)$ which, according to the new quantum Hamming bound, is the shortest possible code. Finally, we extend the construction method to the class of $[[n,1]]$ codes with $e_g = 1$ and arbitrary e_Z.

2 Notation

A qubit is an element of the two-dimensional Hilbert space $\mathcal{H}^2$, with basis $|0\rangle$ and $|1\rangle$ [17]. An n-tuple of qubits (n qubits) is an element of the 2^n-dimensional Hilbert space, $\mathcal{H}^{2^n}$, with basis composed by all possible tensor products $|i_1\rangle |i_2\rangle \cdots |i_n\rangle$, with $i_j \in \{0,1\}, 1 \le j \le n$. The Pauli operators, denoted as I, X, Z, and Y, are defined by $I |a\rangle = |a\rangle$, $X |a\rangle = |a \oplus 1\rangle$, $Z |a\rangle = (-1)^a |a\rangle$, and $Y |a\rangle = i(-1)^a |a \oplus 1\rangle$ for $a \in \{0,1\}$. These operators either commute or anticommute.

With $[[n,k]]$ we indicate a quantum error correcting code (QECC) that encodes k data qubits $|\varphi\rangle$ into a codeword of n qubits $|\psi\rangle$. We use the stabilizer formalism, where a stabilizer code $\mathcal{C}$ is generated by $n-k$ independent and commuting operators $\boldsymbol{G}_i \in \mathcal{G}_n$, called generators [17,32,33]. The code $\mathcal{C}$ is the set of quantum states $|\psi\rangle$ satisfying

$$\boldsymbol{G}_i |\psi\rangle = |\psi\rangle , \quad i = 1, 2, \ldots, n-k . \tag{1}$$

Assume a codeword $|\psi\rangle \in \mathcal{C}$ affected by a channel error described by the operator $\boldsymbol{E} \in \mathcal{G}_n$. For error correction, the received state $\boldsymbol{E}|\psi\rangle$ is measured according to the generators $\boldsymbol{G}_1, \boldsymbol{G}_2, \ldots, \boldsymbol{G}_{n-k}$, resulting in a quantum error syndrome $\boldsymbol{s}(\boldsymbol{E}) = (s_1, s_2, \ldots, s_{n-k})$, with each $s_i = 0$ or 1 depending on the fact that $\boldsymbol{E}$ commutes or anticommutes with $\boldsymbol{G}_i$, respectively. Note that, due to (1), the syndrome depends on $\boldsymbol{E}$ and not on the particular q-codeword $|\psi\rangle$. Moreover, measuring the syndrome does not change the quantum state, which remains $\boldsymbol{E}|\psi\rangle$. Let $\mathcal{S} = \{\boldsymbol{s}^{(1)}, \boldsymbol{s}^{(2)}, \ldots, \boldsymbol{s}^{(m)}\}$ be the set of $m = 2^{n-k}$ possible syndromes, with $\boldsymbol{s}^{(1)} = (0, 0, \ldots, 0)$ denoting the syndrome of the operators $\boldsymbol{E}$ (including the identity $\boldsymbol{I}$, i.e., the no-errors operator) such that $\boldsymbol{E}|\psi\rangle$ is still a valid q-codeword. A generic Pauli error $\boldsymbol{E} \in \mathcal{G}_n$ can be described by specifying the single Pauli errors on each qubit. We use $\boldsymbol{X}_i, \boldsymbol{Y}_i, \boldsymbol{Z}_i$ to denote the Pauli $\boldsymbol{X}, \boldsymbol{Y}, \boldsymbol{Z}$ error on the i-th qubit.

3   Hamming Bounds for Quantum Asymmetric Codes

The standard quantum Hamming bound (QHB) gives a necessary condition for the existence of non-degenerate error correcting codes: a quantum code which encodes k qubits in n qubits can correct up to t generic errors per codeword only if [17,34]

$$2^{n-k} \geq \sum_{j=0}^{t} \binom{n}{j} 3^j . \tag{2}$$

The bound is easily proved by noticing that the number of syndromes, 2^{n-k}, must be at least equal to that of the distinct errors we want to correct. Since for each position there could be three Pauli errors ($\boldsymbol{X}$, $\boldsymbol{Y}$ or $\boldsymbol{Z}$), the number of distinct patterns having j qubits in error is $\binom{n}{j} 3^j$, and this gives the bound (2).

In this paper we investigate non-degenerate QECCs which can correct some generic errors ($\boldsymbol{X}$, $\boldsymbol{Y}$ or $\boldsymbol{Z}$), plus some fixed errors (e.g., $\boldsymbol{Z}$ errors). We derive therefore the following generalized quantum Hamming bound (GQHB).

Theorem 1 (Generalized Quantum Hamming Bound). *A quantum code which encodes k qubits in n qubits can correct up to e_{g} generic errors plus up to e_Z fixed errors per codeword only if*

$$2^{n-k} \geq \sum_{j=0}^{e_{\mathrm{g}}+e_Z} \binom{n}{j} \sum_{i=0}^{e_{\mathrm{g}}} \binom{j}{i} 2^i . \tag{3}$$

Proof. For the proof we need to enumerate the different patterns of error. The number of patterns of up to e_g generic errors is given by (2) with $t = e_g$. Then, we have to add the number of configurations with $e_g < j \leq e_g + e_z$ errors, composed by e_g generic errors and the remaining $j - e_g$ Pauli Z errors. We can write

$$2^{n-k} \geq \sum_{j=0}^{e_g} \binom{n}{j} 3^j + \sum_{j=e_g+1}^{e_g+e_z} \binom{n}{j} [3^j - f(j; e_g)] \tag{4}$$

where $f(j; e_g)$ is a function that returns the number of non-correctable patterns of j errors. This is the solution of the following combinatorial problem: given j positions of the errors, count the number of all combinations with more than e_g symbols from the set $\mathcal{P}_{XY} = \{X, Y\}$ and the remaining from the set $\mathcal{P}_Z = \{Z\}$. We have therefore

$$f(j; e_g) = \sum_{i=0}^{j-e_g-1} \binom{j}{i} 2^{j-i} \tag{5}$$

which allows to write

$$\begin{aligned}
g(j; e_g) &= 3^j - f(j; e_g) \\
&= \sum_{i=0}^{j} \binom{j}{i} 2^{j-i} - \sum_{i=0}^{j-e_g-1} \binom{j}{i} 2^{j-i} \\
&= \sum_{i=j-e_g}^{j} \binom{j}{i} 2^{j-i} = \sum_{i=0}^{e_g} \binom{j}{i} 2^i.
\end{aligned} \tag{6}$$

It is easy to see that $g(j; e_g)$ is equal to 3^j if $j \leq e_g$, so substituting and incorporating the summation in (4) we finally obtain

$$2^{n-k} \geq \sum_{j=0}^{e_g+e_z} \binom{n}{j} g(j; e_g) = \sum_{j=0}^{e_g+e_z} \binom{n}{j} \sum_{i=0}^{e_g} \binom{j}{i} 2^i. \tag{7}$$

The GQHB can be used to compare codes which can correct t generic errors with codes correcting a total of t errors with e_g of them generic and the others fixed. In Table 1 we report the minimum code lengths n_{min} resulting from the Hamming bounds, for different values of the total number of errors t, and assuming $e_g = 1$ for the GQHB. From the table we can observe the possible gain in qubits for the asymmetric case.

Table 1. Comparison between the minimum code lengths n_{min}^{QHB}, n_{min}^{GQHB} according to the Hamming bounds (2), (3). For the GQHB $t = e_g + e_Z$ with $e_g = 1$.

	$t = 1$	$t = 2$	$t = 3$	$t = 4$
$k = 1$	5,5	10,9	15,12	20,15
$k = 2$	7,7	12,10	16,14	21,17
$k = 3$	8,8	13,12	18,15	23,19

4   Construction of Short Asymmetric Codes by Syndrome Assignment

In this section we present a construction of short stabilizer asymmetric codes with $k = 1$ and $e_g = 1$, i.e., for $[[n, 1]]$ QECCs with error correction capability $(1, e_Z)$. The design is based on the error syndromes: specifically, we proceed by assigning different syndromes to the different correctable error patters.

Let us first observe that the vector syndrome of a composed error $E = E_1 E_2$, with $E_1, E_2 \in \mathcal{G}_n$, can be expressed as $s(E) = s(E_1 E_2) = s(E_2 E_1) = s(E_1) \oplus s(E_2)$ where $\oplus$ is the elementwise modulo 2 addition. Moreover, $XZ = iY$, and for the syndromes we have $s(X_i Z_i) = s(Y_i)$, $s(X_i Y_i) = s(Z_i)$, and $s(Y_i Z_i) = s(X_i)$. Hence, once we have assigned the syndromes for the single error patterns X_i and Z_i, with $i = 1, \ldots, n$, the syndromes for all possible errors are automatically determined.

The key point in the design is therefore to find an assignment giving distinct syndromes for all correctable error patterns.

In the following, if not specified otherwise, the indexes i, j will run from 1 to n, and the index ℓ will run from 1 to $n - 1$. Also, the weight of a syndrome is the number of non-zero elements in the associated vector.

4.1   Construction of $[[n, 1]]$ QECCs with $e_g = 1, e_Z = 1$

For this case we need to solve the following problem: assign $2n$ syndromes $s(X_i)$ and $s(Z_i)$ such that the syndromes of the errors I, X_i, Y_i, Z_i, $X_i Z_j$, $Y_i Z_j$, $Z_i Z_j$, $\forall i \neq j$, are all different.

Now, we aim to construct the shortest possible code according to the GQHB, i.e., a code with $n = 9$ (see Table 1). We start by assigning the syndromes of Z_i as reported in the following table.

With this choice we have assigned all possible syndromes of weight 1 and 8. Also, the combinations of $Z_i Z_j$ with $i \neq j$, cover all possible syndromes of weight 2 and 7.

To assign the syndromes of X_i we then use a Monte Carlo approach. To reduce the search space, i.e., the set of possible syndromes, we observe the following:

– The weight of $s(X_i)$ cannot be 3 or 6. This is because otherwise $s(Z_j X_i)$ would have weight 2 or 7 for some i and j, which are already assigned for

Table 2. Assigned syndromes for single Pauli Z errors.

	s_8	s_7	s_6	s_5	s_4	s_3	s_2	s_1
Z_1	0	0	0	0	0	0	0	1
Z_2	0	0	0	0	0	0	1	0
Z_3	0	0	0	0	0	1	0	0
Z_4	0	0	0	0	1	0	0	0
Z_5	0	0	0	1	0	0	0	0
Z_6	0	0	1	0	0	0	0	0
Z_7	0	1	0	0	0	0	0	0
Z_8	1	0	0	0	0	0	0	0
Z_9	1	1	1	1	1	1	1	1

errors of the type $Z_i Z_j$. Therefore the possible weights for $s(X_i)$ are only 4 and 5. The same observation applies to $s(Y_i)$. We then fix the weight for $s(X_i)$ equal to 4.

- We can obtain $s(Y_\ell)$ with weight 5 for $\ell = 1, \ldots, 8$, by imposing to "0" the ℓ-th element of the syndrome of X_ℓ. Note that Y_9 has weight 4 since X_9 has weight 4.

By following the previous rules, a possible assignment obtained by Monte Carlo is reported in Table 3.

Table 3. Possible syndromes for single Pauli X errors.

	s_8	s_7	s_6	s_5	s_4	s_3	s_2	s_1
X_1	1	0	1	1	1	0	0	0
X_2	1	0	0	1	0	1	0	1
X_3	0	0	1	0	1	0	1	1
X_4	1	1	1	0	0	1	0	0
X_5	0	1	0	0	1	1	0	1
X_6	1	1	0	0	0	0	1	1
X_7	0	0	1	1	0	1	1	0
X_8	0	1	0	1	1	0	1	0
X_9	1	0	0	0	1	1	1	0

From Tables 2 and 3 we can then build the stabilizer matrix with the following procedure, where $s_j(X_i)$ indicates the j-th elements of the X_i's syndrome:

- if $s_j(X_i) = 0$ and $s_j(Z_i) = 0$ put the element I in position (j, i) of the stabilizer matrix because it is the only Pauli operator which commutes with both.

- if $s_j(X_i) = 1$ and $s_j(Z_i) = 0$ put the element Z in position (j,i) of the stabilizer matrix because it is the only Pauli operator which commutes with Z and anti-commute with X.
- if $s_j(X_i) = 0$ and $s_j(Z_i) = 1$ put the element X in position (j,i) of the stabilizer matrix because it is the only Pauli operator which commutes with X and anti-commute with Z.
- if $s_j(X_i) = 1$ and $s_j(Z_i) = 1$ put the element Y in position (j,i) of the stabilizer matrix because it is the only Pauli operator which anti-commutes with both.

The resulting stabilizer matrix, after checking the commutation conditions, is represented in Table 4.

Table 4. Stabilizer for a $[[9,1]]$ QECC with $e_{\mathrm{g}} = 1$ and $e_{\mathrm{Z}} = 1$.

	1	2	3	4	5	6	7	8	9
G_1	X	Z	Z	I	Z	Z	I	I	X
G_2	I	X	Z	I	I	Z	Z	Z	Y
G_3	I	Z	X	Z	Z	I	Z	I	Y
G_4	Z	I	Z	X	Z	I	I	Z	Y
G_5	Z	Z	I	I	X	I	Z	Z	X
G_6	Z	I	Z	Z	I	X	Z	I	X
G_7	I	I	I	Z	Z	Z	X	Z	X
G_8	Z	Z	I	Z	I	Z	I	X	Y

4.2   Construction of $[[n,1]]$ QECCs with $e_{\mathrm{g}} = 1$ and $e_{\mathrm{Z}} \geqslant 1$

The construction presented in the previous section can be generalized to the case of more fixed errors, $e_{\mathrm{Z}} \geqslant 1$. In this section we indicate $\tilde{t} = e_{\mathrm{g}} + e_{\mathrm{Z}}$. By adopting the same assignment proposed in Table 2, it is easy to see that we use all possible syndromes with weight in the range $[0, \tilde{t}]$ and $[n - \tilde{t}, n - 1]$, covering all possible error operators with up to $\tilde{t}$ errors of type Z. For the assignment of the syndromes $s(X_i)$ we can generalize the previously exposed arguments, as follows:

- The weight of $s(X_i)$ cannot be less than $2\tilde{t}$ or greater than $n - 2\tilde{t}$. This is because otherwise $s(Z_{j_1} \ldots Z_{j_L} X_i)$ would have weight in the range $[0, \tilde{t}]$ or $[n - \tilde{t}, n - 1]$ for some $L \leq e_{\mathrm{Z}}$ and some choices of $j_1, \ldots, j_L$. These syndromes are already assigned for errors of the type $Z_{j_1} \ldots Z_{j_M}$ for some $M \leq e_{\mathrm{Z}}$ and some choices of $j_1, \ldots, j_M$. Therefore the possible weights for $s(X_i)$ are in the range $[2\tilde{t}, n - 2\tilde{t}]$. The same observation applies to $s(Y_i)$.

– Setting the ℓ-th element of the syndrome of $\boldsymbol{X}_\ell$ to "0" we obtain that $\boldsymbol{s}(\boldsymbol{Y}_\ell)$ has the weight of $\boldsymbol{s}(\boldsymbol{X}_\ell)$ increased by 1, with $\ell = 1,\ldots,n-1$. Hence, in order to have both $\boldsymbol{s}(\boldsymbol{X}_\ell)$ and $\boldsymbol{s}(\boldsymbol{Y}_\ell)$ in the permitted range, we must have $n - 4\tilde{t} \geq 1$. Note that this constraint can be stricter than the GQHB. For example, we cannot construct the $[[12,1]]$ code with $e_g = 1, e_Z = 2$.

– With the previous choice, the sum of the weights of the syndromes $\boldsymbol{s}(\boldsymbol{Y}_n)$ and $\boldsymbol{s}(\boldsymbol{X}_n)$ is $n-1$. Then, a good choice is to assign to $\boldsymbol{s}(\boldsymbol{X}_n)$ a weight $\lceil (n-1)/2 \rceil$ or $\lfloor (n-1)/2 \rfloor$. In this case, if n is odd $\boldsymbol{s}(\boldsymbol{Y}_n)$ would have the same weight, which is in the correct range because $n - 4\tilde{t} \geq 0$ is guaranteed by the previous point; if n is even the weights are still in the correct range because $n - 4\tilde{t} \geq 1$.

The resulting algorithm is reported below.

Result: Stabilizer matrix from Assignment Construction
Choose n and $\tilde{t}$ to satisfy the constraint $n - 4\tilde{t} \geq 1$;
Assign $\boldsymbol{s}(\boldsymbol{Z}_i)$ as in Table 2;
Pick a random syndrome for $\boldsymbol{s}(\boldsymbol{X}_n)$ with weight $\lfloor (n-1)/2 \rfloor$;
Assign $\boldsymbol{s}(\boldsymbol{Y}_n)$, $\boldsymbol{s}(\boldsymbol{X}_n \boldsymbol{Z}_{j_1} \ldots \boldsymbol{Z}_{j_L})$ and $\boldsymbol{s}(\boldsymbol{Y}_n \boldsymbol{Z}_{j_1} \ldots \boldsymbol{Z}_{j_L})$ for each
    $L = 1,\ldots,e_Z$ and for each possible combination of $j_1,\ldots,j_L \neq n$;
for $\ell = 1$ **to** $n - 1$ **do**
    goodPick $= 0$;
    while *goodPick $== 0$* **do**
        Pick a random syndrome for $\boldsymbol{s}(\boldsymbol{X}_\ell)$ with weight in $\left[2\tilde{t}, n - 2\tilde{t} - 1\right]$
           and $s_\ell(\boldsymbol{X}_\ell) = 0$;
        if $\boldsymbol{s}(\boldsymbol{Y}_\ell)$, $\boldsymbol{s}(\boldsymbol{X}_\ell \boldsymbol{Z}_{j_1} \ldots \boldsymbol{Z}_{j_L})$ *and* $\boldsymbol{s}(\boldsymbol{Y}_\ell \boldsymbol{Z}_{j_1} \ldots \boldsymbol{Z}_{j_L})$ *are not already*
        assigned for all possible combinations **then**
           goodPick $= 1$;
           Assign $\boldsymbol{s}(\boldsymbol{Y}_\ell)$ and all $\boldsymbol{s}(\boldsymbol{X}_\ell \boldsymbol{Z}_{j_1} \ldots \boldsymbol{Z}_{j_L})$, $\boldsymbol{s}(\boldsymbol{Y}_\ell \boldsymbol{Z}_{j_1} \ldots \boldsymbol{Z}_{j_L})$;
        end
        if *No more possible syndromes* **then**
           Restart the algorithm;
        end
    end
end
Construct the Stabilizer Matrix from $\boldsymbol{s}(\boldsymbol{X}_i)$ and $\boldsymbol{s}(\boldsymbol{Z}_i)$;
Check if all of the generators commute with each other.

Algorithm 1: Construction by syndrome assignment, $k = 1, e_g = 1$.

---

5 Performance Analysis

It is well known that the *Codeword Error Rate* (CWER) for a standard $[[n,k]]$ QECC which corrects up to t generic errors per codeword is

$$P_e = 1 - \sum_{j=0}^{t} \binom{n}{j} (1-\rho)^{n-j} \rho^j \tag{8}$$

where $\rho = p_x + p_y + p_z$ is the error probability.

We now generalize this expression to an $[[n, k]]$ QECC which corrects up to e_g generic errors and up to e_Z Pauli Z errors per codeword. By weighting each pattern of correctable errors with the corresponding probability of occurrence, it is not difficult to show that the performance in terms of CWER is

$$P_e = 1 - \sum_{j=0}^{e_g + e_Z} \binom{n}{j} (1 - \rho)^{n-j} \xi(j; e_g) \tag{9}$$

where

$$\xi(j; e_g) = \begin{cases} \rho^j & \text{if } j \leq e_g \\ \sum_{i=j-e_g}^{j} \binom{j}{i} p_z^i \sum_{\ell=0}^{j-i} \binom{j-i}{\ell} p_x^\ell p_y^{j-i-\ell} & \text{otherwise} . \end{cases} \tag{10}$$

In the case of asymmetric channels with $p_x = p_y = \rho/(A+2)$, $p_z = A\rho/(A+2)$, and $A = p_z/p_x$ [35], the expression in (9) can be simplified to

$$P_e = 1 - \sum_{j=0}^{e_g + e_Z} \binom{n}{j} (1 - \rho)^{n-j} \rho^j \left(1 - 2^{j+1} \frac{(A/2)^{j-e_g} - 1}{(A-2)(A+2)^j} u_{j-e_g-1} \right) \tag{11}$$

where $u_i = 1$ if $i \geq 0$, otherwise $u_i = 0$.

Using the previous expressions, we report in Fig. 1 the performance in terms of CWER for different codes, assuming an asymmetric channel. The parameter A accounts for the asymmetry of the channel, and for $A = 1$ we have the standard depolarizing channel. In the figure we plot the CWER for the new asymmetric $[[9, 1]]$ code specified in Table 4 with $e_g = 1$ and $e_Z = 1$, over channels with asymmetry parameter $A = 1, 3$ and 10. For comparison, in the same figure we report the CWER for the known 5-qubits code, the Shor's 9-qubits code, both correcting $t = 1$ generic errors, and a $[[11, 1]]$ code with $t = 2$ [36].

First, we note that for the symmetric codes the performance does not depend on the asymmetry parameter A, but just on the overall error probability ρ. For these codes, for a given t the best CWER is obtained with the shortest code. As expected, the performance of the new asymmetric $[[9, 1]]$ code improves as A increases. In particular, for the symmetric channel, $A = 1$, the 5-qubits code performs better than the new one, due to its shorter codeword size. However, already with a small channel asymmetry, $A = 3$, the new code performs better than the 5-qubits code. For $A = 10$ the new code performs similarly to the $[[11, 1]]$ symmetric code with $t = 2$. Asymptotically for large A, the channel errors tend to be of type Z only, and consequently the new code behaves like a code with $t = 2$.

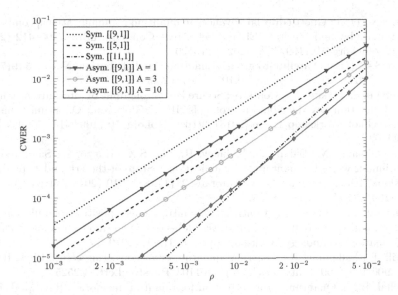

Fig. 1. Performance of short codes over an asymmetric channel, $k = 1$. Symmetric codes: 9-qubits code and 5-qubits code with $t = 1$, 11-qubits code with $t = 2$. Asymmetric 9-qubits code with $e_g = 1, e_Z = 1$.

6 Conclusions

We have investigated a new class of stabilizer short codes for quantum asymmetric Pauli channels, capable to correct up to e_g generic errors plus e_Z errors of type Z. We generalized the quantum Hamming bound and derived the analytical expressions for the performance for the new codes. Then, we designed a [[9,1]] QECC capable to correct up to 1 generic error plus 1 Z error, which is the shortest according to the new bound. The comparisons with known symmetric QECCs confirm the advantage of the proposed code in the presence of channel asymmetry.

References

1. Zoller, P., et al.: Quantum information processing and communication. Eur. Phys. J. D Atomic Mol. Opt. Plasma Phys. **36**(2), 203–228 (2005). https://doi.org/10. 1140/epjd/e2005-00251-1
2. Kimble, H.J.: The quantum internet. Nature **453**(7198), 1023 (2008). https://doi. org/10.1038/nature07127
3. Wehner, S., Elkouss, D., Hanson, R.: Quantum internet: a vision for the road ahead. Science **362**(6412), 1–9 (2018). https://doi.org/10.1126/science.aam9288
4. Grumbling, E., Horowitz, M. (eds.): Quantum Computing: Progress and Prospects. The National Academies Press, Washington, DC (2019)
5. Quantum Networks for Open Science Workshop: Office of Science US Department of Energy. Rockville, MD (2018)

6. Ng, S.X., et al.: Guest editorial advances in quantum communications, computing, cryptography, and sensing. IEEE J. Sel. Areas Commun. **38**(3), 405–412 (2020). https://doi.org/10.1109/JSAC.2020.2973529

7. Liao, S.-K., et al.: Satellite-to-ground quantum key distribution. Nature **549**(7670), 43–47 (2017). https://doi.org/10.1038/nature23655

8. Guerrini, S., Chiani, M., Conti, A.: Secure key throughput of intermittent trusted-relay quantum key distribution protocols. IEEE Globecom: Quantum Commun. Inf. Technol. Workshop **1**, 1–6 (2018). https://doi.org/10.1109/GLOCOMW.2018. 8644402

9. Hosseinidehaj, N., Babar, Z., Malaney, R., Ng, S.X., Hanzo, L.: Satellite-based continuous-variable quantum communications: State-of-the-art and a predictive outlook. IEEE Commun. Surv. Tutorials **21**(1), 881–919 (2018). https://doi.org/ 10.1109/COMST.2018.2864557

10. Guerrini, S., Chiani, M., Conti, A.: Quantum pulse position modulation with photon-added coherent states. In: IEEE Globecom: Quantum Communications and Information Technology Workshop, pp. 1–5. IEEE (2019)

11. Knill, E., Laflamme, R.: Theory of quantum error-correcting codes. Phys. Rev. A **55**, 900–911 (1997). https://doi.org/10.1103/PhysRevLett.84.2525

12. Terhal, B.M.: Quantum error correction for quantum memories. Rev. Mod. Phys. **87**, 307–346 (2015). https://doi.org/10.1103/RevModPhys.87.307

13. Munro, W.J., Stephens, A.M., Devitt, S.J., Harrison, K.A., Nemoto, K.: Quantum communication without the necessity of quantum memories. Nat. Photonics **6**(11), 777 (2012). https://doi.org/10.1038/nphoton.2012.243

14. Muralidharan, S., Li, L., Kim, J., Lütkenhaus, N., Lukin, M.D., Jiang, L.: Optimal architectures for long distance quantum communication. Sci. Rep. **6**, 20463 (2016). https://doi.org/10.1038/srep20463

15. Wootters, W.K., Zurek, W.H.: A single quantum cannot be cloned. Nature **299**(5886), 802 (1982). https://doi.org/10.1038/299802a0

16. Bennett, C.H., DiVincenzo, D.P., Smolin, J.A., Wootters, K.W.: Mixed state entanglement and quantum error correction. Phys. Rev. A **54**(5), 3824–3851 (1996). https://doi.org/10.1103/PhysRevA.54.3824

17. Nielsen, M.A., Chuang, I.L.: Quantum Computation and Quantum Information. Cambridge University Press, Cambridge (2010)

18. Gottesman, D.: An introduction to quantum error correction and fault-tolerant quantum computation. arXiv preprint quant-ph/0904.2557 (2009)

19. Shor, P.W.: Scheme for reducing decoherence in quantum computer memory. Phys. Rev. A **52**, R2493–R2496 (1995). https://doi.org/10.1103/physreva.52.r2493

20. Steane, A.M.: Error correcting codes in quantum theory. Phys. Rev. Lett. **77**(5), 793 (1996). https://doi.org/10.1103/PhysRevLett.77.793

21. Laflamme, R., Miquel, C., Paz, J.P., Zurek, W.H.: Perfect quantum error correcting code. Phys. Rev. Lett. **77**(1), 198 (1996). https://doi.org/10.1103/PhysRevLett. 77.198

22. Ioffe, L., Mézard, M.: Asymmetric quantum error-correcting codes. Phys. Rev. A **75**(3), 032345 (2007). https://doi.org/10.1103/PhysRevA.75.032345

23. Sarvepalli, P.K., Klappenecker, A., Rötteler, M.: Asymmetric quantum codes: constructions, bounds and performance. Proc. Roy. Soc. Math. Phys. Eng. Sci. **465**(2105), 1645–1672 (2009). https://doi.org/10.1098/rspa.2008.0439

24. Gyongyosi, L., Imre, S., Nguyen, H.V.: A survey on quantum channel capacities. IEEE Commun. Surv. Tutorials **20**(2), 1149–1205 (2018). https://doi.org/10.1109/ COMST.2017.2786748

25. Evans, Z.W.E., Stephens, A.M., Cole, J.H., Hollenberg, L.C.L.: Error correction optimisation in the presence of x/z asymmetry. arXiv preprint arXiv:0709.3875 (2007)
26. Fletcher, A.S., Shor, P.W., Win, M.Z.: Channel-adapted quantum error correction for the amplitude damping channel. IEEE Trans. Inf. Theory **54**(12), 5705–5718 (2008). https://doi.org/10.1109/TIT.2008.2006458
27. Fletcher, A.S., Shor, P.W., Win, M.Z.: Structured near-optimal channel-adapted quantum error correction. Phys. Rev. A **77**, 012320 (2008). https://doi.org/10.1103/PhysRevA.77.012320
28. Lang, R., Shor, P.W.: Nonadditive quantum error correcting codes adapted to the ampltitude damping channel. arXiv preprint arXiv:0712.2586 (2007)
29. Leung, D.W., Nielsen, M.A., Chuang, I.L., Yamamoto, Y.: Approximate quantum error correction can lead to better codes. Phys. Rev. A **56**(4), 2567 (1997). https://doi.org/10.1103/PhysRevA.56.2567
30. Shor, P.W., Smith, G., Smolin, J.A., Zeng, B.: High performance single-error-correcting quantum codes for amplitude damping. IEEE Trans. Inf. Theory **57**(10), 7180–7188 (2011). https://doi.org/10.1109/TIT.2011.2165149
31. Jackson, T., Grassl, M., Zeng, B.: Codeword stabilized quantum codes for asymmetric channels. In: 2016 IEEE International Symposium on Information Theory (ISIT), pp. 2264–2268. IEEE (2016)
32. Gottesman, D.: Class of quantum error-correcting codes saturating the quantum Hamming bound. Phys. Rev. A **54**, 1862 (1996). https://doi.org/10.1103/PhysRevA.54.1862
33. Gottesman, D.: An introduction to quantum error correction and fault-tolerant quantum computation. Proc. Symp. Appl. Math. **68**, 13–58 (2009)
34. Ekert, A., Macchiavello, C.: Quantum error correction for communication. Phys. Rev. Lett. **77**(12), 2585 (1996)
35. Sarvepalli, P.K., Klappenecker, A., Rotteler, M.: Asymmetric quantum LDPC codes. In: 2008 IEEE International Symposium on Information Theory, pp. 305–309, July 2008
36. Grassl, M.: Bounds on the minimum distance of linear codes and quantum codes (2007). http://www.codetables.de. Accessed on 20 Dec 2019

Simulation Methodology for Electron Transfer in CMOS Quantum Dots

Andrii Sokolov[1]([✉]), Dmytro Mishagli[1], Panagiotis Giounanlis[1], Imran Bashir[2], Dirk Leipold[2], Eugene Koskin[1], Robert Bogdan Staszewski[1,2], and Elena Blokhina[1]

[1] University College Dublin, Belfield Dublin 4, Ireland
andrii.sokolov@ucdconnect.ie
[2] Equal1 Labs, Fremont, CA 94536, USA

Abstract. The construction of quantum computer simulators requires advanced software which can capture the most significant characteristics of the quantum behavior and quantum states of qubits in such systems. Additionally, one needs to provide valid models for the description of the interface between classical circuitry and quantum core hardware. In this study, we model electron transport in semiconductor qubits based on an advanced CMOS technology. Starting from 3D simulations, we demonstrate an order reduction and the steps necessary to obtain ordinary differential equations on probability amplitudes in a multi-particle system. We compare numerical and semi-analytical techniques concluding this paper by examining two case studies: the electron transfer through multiple quantum dots and the construction of a Hadamard gate simulated using a numerical method to solve the time-dependent Schrödinger equation and the tight-binding formalism for a time-dependent Hamiltonian.

Keywords: CMOS quantum dots · Charge qubits · Position-based charge qubits · Tight binding formalism · Split-operator method · Electron transfer · X-rotation gate

1 Introduction

This work is motivated by the development of Complementary Metal-Oxide-Semiconductor (CMOS) charge qubits in one of the most advanced technologies, 22FDX by GlobalFoundries employing a 22 nm Fully-Depleted Silicon-On-Insulator (FDSOI) process. Building charge, spin or hybrid qubits by exploiting the fine-feature lithography of CMOS devices is currently a dominant trend towards a large-scale quantum computer [3,8,13,19,20,23,25]. Although of all the mentioned CMOS silicon qubits, charge qubits have quite short decoherence time [17,24], they are revisited now in light of the high material interface purity, fine feature size and fast speed of operation of the latest mainstream nanometer-scale CMOS process technology (see the review section in Ref. [5]). The speed

© Springer Nature Switzerland AG 2020
V. V. Krzhizhanovskaya et al. (Eds.): ICCS 2020, LNCS 12142, pp. 650–663, 2020.
https://doi.org/10.1007/978-3-030-50433-5_50

of operation of quantum gates can now be ultra short due to the transistor cut-off frequency reaching half-terahertz in advanced CMOS. Thus, the number of quantum gate operations can be on the same order as with spin-based or hybrid qubits. For this reason, the paper is focused on the modelling of charge qubits.

The development of quantum computer emulators requires high-level software representing the physics of quantum structures under study with high accuracy. At the same time, it should incorporate the control of quantum states of a large number of qubits, the realization of quantum gates, the act of measurement and transport of quantum information including the interface between classical circuitry and quantum registers, mentioning some of the major components needed for such an attempt. To facilitate an accurate modeling of interacting qubits, one is required to solve a many-body quantum system, which is a very resource intense task for classical simulators. There are some methods from computational quantum physics that could be used as a first approach for such problems. The density function theory (DFT) is one of the most known and, in principle, exact *ab initio* method [18, 26]. However, it is not applicable in practice for strongly correlated electron problems, since the exact exchange correlation functional is not known. In principle, the electrons in the position-based charge qubits described in this study are strongly interacting by Coulomb interaction [22]. From this point of view, any single-electron method such as DFT cannot be reliably used. Configuration interaction methods from quantum chemistry may also be considered exact *ab initio* methods, but of course they are limited to extremely small system sizes due to the well-known exponential scaling of the Hilbert space with system size, making such methods essentially useless for our present purposes. By contrast, virtually all methods in quantum physics employ effective (or reduced) models that capture the essential physics of interest while throwing away irrelevant details [7]. Such approaches can even be quantitative semi-empirical, by fitting model parameters to experimental data.

In this study, we will focus on the modelling of a specific type of CMOS silicon qubits known as position-based charge qubits (or simply as charge qubits) [5]. Starting from 3D simulations, we demonstrate an order reduction and the steps necessary to obtain ordinary differential equations on probability amplitudes in a multi-particle system. We compare numerical and semi-analytical techniques concluding this paper by examining two case studies: electron transfer through multiple quantum dots and construction of a Hadamard gate simulated using the time-dependent Schrödinger equation and the tight-binding formalism.

2   Coupled Quantum Dot Chains and Structures in 3D

In this section, we provide the general description of the structures containing coupled quantum dots (QDs) under study and key results of their Fine-Element Method (FEM) modelling employing COMSOL Multiphysics. As we aim to penetrate the behaviour of a quantum processor fabricated in a commercial technology [3], FEM simulations have been carried out on the structures whose dimensions and composition are inspired by the 22FDX technology of GlobalFoundries.

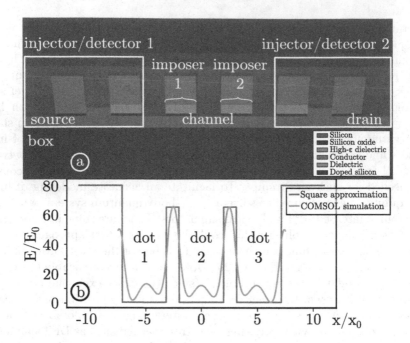

Fig. 1. (a) Schematic diagram of a CMOS structure implementing three QDs including two imposers between the QDs and two injectors/detectors at the edges. (b) Normalised potential energy as a function of the position obtained from FEM electromagnetic simulations (orange line) and a simplified piece-wise potential energy function (blue line). (Color figure online)

Table 1. Parameters used for simulations in this study

Elementary charge, e	1.602×10^{-19}	C
Effective mass, m_e^*	$1.08 \times 9.109 \times 10^{-31}$	kg
Length unit, x_0	20	nm
Energy unit, E_0	$\hbar^2/2m_e^* x_0^2 = 1.41 \times 10^{-23} = 87.6$	J μeV
Time unit, t_0	$2\pi\hbar/E_0 = 47.3$	ps

The schematic 3D structure under study is shown in Fig. 1(a). The parameters and normalisation units used are presented in Table 1. Each structure can be seen as transistor-like devices arranged in a chain. Each 'transistor' contains a control gate (which we call an imposer) made of a very thin SiO$_2$ layer, high-k dielectric layer and a thick heavily doped polysilicon layer. Beneath a gate, there is a thin depleted silicon channel and a thin buried oxide (BoX) layer deposited on a thick lightly doped silicon wafer. An insulating coat is usually deposited on the top of the entire structure for electric isolation. The edges of the structure are connected to classic circuit devices known as single-electron injectors (detectors).

Semiconductor equilibrium simulations using the Semiconductor module of COMSOL Multiphysics have been carried out for temperatures ranging from 10 to 100 K to confirm that the silicon channel is completely depleted and there are no thermally generated carriers or carriers diffused from the interconnections with the heavily doped regions belonging to the injectors and detectors [3]. Electromagnetic simulations have been carried out to investigate the electric potential and electric field propagation through the structure when a given voltage is applied at the imposers. The electric potential developed in the silicon channel allows us to calculate the modulation of the edge of the conduction band and the formation of potential energy wells in the channel. An example of the potential energy on the surface of the silicon channel as a function of the coordinate along the channel is shown in Fig. 1(b) where one can observe three potential wells forming in the region between the imposers.

In general, there exist specific imposer voltages so that potential energy "barriers" are formed under the imposers and "wells" are formed in between the imposers. We refer to the region between the imposers where a potential energy well is formed as a quantum dot. In the figure, the minimum of the potential energy is conventionally placed at 0 meV. In a typical scenario, barriers of 1 to 4 meV can be formed when a sub-threshold voltage is applied at the imposers. The resulting potential energy can be effectively approximated by an equivalent piece-wise linear function. The time-independent Schrödinger equation (TISE) with a piece-wise linear potential energy can be solved analytically or numerically in one, two or three dimensions to find a set of eigenfunctions and eigenenergies. The time-dependent Schrödinger equation (TDSE) can also be solved (as we will show later in the paper), but even in the 1D case, it requires substantial computational resources. The next section outlines the problem overview and the assumptions taken in this study.

3   Outline of Quantum Mechanical Modelling

As the main model, we consider an isolated quantum system without external fluctuations. However, we will also present some results including effects of decoherence, which arise from the coupling between the quantum system under consideration with the environment. In this paper, the coupling appears as time-dependent noise terms in the effective Hamiltonian, but it should be noted that there are a number of approaches to model decoherence effects. Coupling to a fermionic bath leads to Kondo-type physics and the low-energy quenching of spin qubits, while coupling to a bosonic environment, leads to the localization through the Caldeira-Legget mechanism [15]. In general, we note that both kinds of processes occur at some characteristic timescale (not instantaneously). In addition, we neglect other degrees of freedom (spin, valley and orbital) since we are interested in electron transfer in charge qubits (also called position-based qubits, as stated earlier).

We start from the fundamental equation in Quantum Mechanics, the Schrö-dinger equation. We have made some assumptions, which reduce the complexity of the problem:

1. We are interested in the transfer of an electron through the silicon channel of a quantum register containing multiple QDs. Since the channel is very thin, the electron wave function is "shallow" and we can neglect one dimension. In addition, FEM simulations show the symmetry of the wave function, hence we use the reduction of the wave function along the symmetry line. In this case, it is reasonable to use the 1D modelling [10].
2. The complexity of the multi-electron model is significantly higher than that of a single-electron one. Our first step is to consider the transfer of one electron through the channel by changing the imposer voltages.
3. In general, any changes of imposer voltages may cause non-trivial changes to the potential energy which can cause second-order effects on the wave function evolution. At the moment we do not take into account second- and third-order effects associated with abrupt imposer voltage changes of electronic noise.

The starting point is the 1D time-dependent Schrödinger equation:

$$i\hbar\frac{\partial \Phi(x,t)}{\partial t} = \left[-\frac{\hbar^2}{2m_e^*}\frac{\partial^2}{\partial x^2} + V(x,t) \right] \Phi(x,t), \tag{1}$$

where $\hbar$ is the reduced Plank's constant, m_e^* is the effective mass of the elec-tron, $\Phi(x,t)$ is the wave function of the particle, and $V(x,t)$ is the 1D potential energy function. The the potential energy of the electron is obtained from the simulations presented in Sect. 2, see Fig. 1. The potential energy calculated from the FEM simulations and its changes due to the variation of imposer voltages are exported as a table function from the COMSOL electrostatic simulator and is used as a 'pre-generated' input parameter of the quantum simulator.

In order to have a useful simulator of the studied quantum structure, one needs to define a possible "localised" state of an electron injected in the structure, simulate its evolution with time at a given potential energy along the structure and calculate the probability of the electron to be measured at the edges of the structure by a detector device. The eigenenergies and eigenfunctions of the time independent case are particularly useful since they define the frequency of transitions, and, as the next step, we address the calculation eigenfunctions for an arbitrary shaped potential energy.

4   Eigenenergy and Eigenfunction Calculation Using the Matrix Diagonalisation Method

In this section, we discuss a method for obtaining an approximate solution to a one-dimensional TISE [14]. A distinguishable feature of this method is that it allows one to obtain all the bound states and their corresponding wave functions at once. The method is closely related but not equivalent to the perturbation theory [12]. Assume that an electron is confined by a finite potential $V(x)$ and this finite potential is located *inside* an infinite potential well $V_\infty(x)$, so that the Hamiltonian $\hat{H}$ of such a system reads

$$\hat{H} = \hat{H}_0 + \hat{V}. \tag{2}$$

Here $\hat{H}_0$ is the Hamiltonian operator that corresponds to the problem of a particle (an electron) in an infinite potential well, $\hat{H}_0 = -\hbar^2/2m \cdot d^2/dx^2 + V_\infty(x)$. The eigenvalues $E_n^{(0)}$ and eigenfunctions $\psi_n^{(0)}$ of such an operator are well known. The eigenfunctions ψ_n and eigenvalues E_n of the operator (2) are the solutions to the equation

$$(\hat{H}_0 + \hat{V})\psi_n = E_n\psi_n. \tag{3}$$

Since the functions $\psi_n^{(0)}$ form a complete basis set, for the required functions ψ_n we can write the expansion

$$\psi_n = \sum_m c_{nm}\psi_m^{(0)}, \tag{4}$$

where c_{nm} are unknown coefficients. Thus, the initial Eq. (3) yields

$$\sum_m c_{nm}(\hat{H}_0 + \hat{V})\psi_m^{(0)} = E_n \sum_m c_{nm}\psi_m^{(0)}. \tag{5}$$

Multiplying both sides of the latter equation by $\psi_k^{(0)*}$ and integrating over the variable x, we obtain

$$\sum_m c_{nm}H_{km} = E_n c_{nk}, \qquad \text{where} \qquad H_{km} = E_k^{(0)}\delta_{km} + V_{km}. \tag{6}$$

The matrix element V_{km} is determined by the potential $V(x)$:

$$V_{km} = \int \psi_k^{(0)*}V(x)\psi_m^{(0)}dx.$$

$$= \frac{2}{L}\int_0^L \sin\left(\frac{k\pi x}{L}\right)V(x)\sin\left(\frac{m\pi x}{L}\right)dx \tag{7}$$

Thus, the problem of solving the differential equation (3) is reduced to the square matrix equation problem (6) that involves the integrals (7). The coefficients c_{nm} form the eigenvectors of the Hamiltonian matrix H_{km} where n stays for the index of a given bound energy level. Such a problem is usually solved by a matrix diagonalisation procedure, hence the name of the method. Below we discuss different potentials $V(x)$ and their impact on the solution.

In case of a square well (potential barrier) structure, the integration in (7) is trivial. Indeed, for a piece-wise potential that takes either zero or some constant value V_i for $x \in [a, b]$, we have

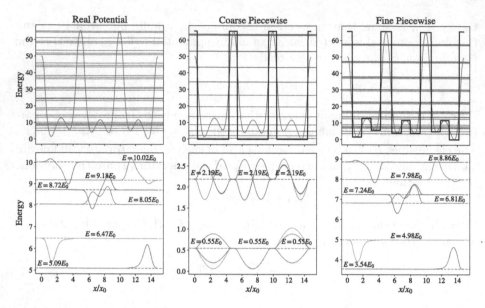

Fig. 2. Energy spectra of the potential generated from FEM simulations (COM-SOL) and the six lowest energy levels with the corresponding wave functions for the three cases: realistic potential function integrated with splines (left), coarse piece-wise approximation (middle) and fine piece-wise approximation (right). *Left, realistic potential*: the lowest energy levels (as well as the corresponding wave functions) exhibiting bound states 'localised' in each well. *Middle, coarse piece-wise approximation*: the lowest energy levels are grouped in three, exhibiting 'non-localised' bound states. *Right, fine piece-wise approximation*: the lowest energy levels are of the same order as those of the realistic potential energy; the wave functions match qualitatively.

$$
V_{km} = V_i \times
\begin{cases}
\dfrac{1}{2}\left(x - \dfrac{L}{\pi k^2}\sin\dfrac{k^2\pi}{L}x\right)\Big|_a^b, & k = m; \\[2ex]
\dfrac{L}{2\pi}\left(\dfrac{\sin(k-m)\pi x/L}{k-m} + \dfrac{\sin(k+m)\pi x/L}{k+m}\right)\Big|_a^b, & k \neq m.
\end{cases}
\tag{8}
$$

When the potential $V(x)$ is given by some continuous but known function, the integration in (7) should not be problematic. While it can be tedious, in principle this is a solvable problem.

In the present study, we use the smooth potential energy generated by FEM simulations as shown in Fig. 1. For comparison, we create its coarse piece-wise approximation and also its fine piece-wise approximations as shown in Fig. 2. The realistic smooth potential energy function is given in form of data points, and thus the easiest approach to the calculation of (7) is numerical integration (e.g., using splines). Placing the potential inside an infinite well of length L and separating it from the edges with a distance h, as described above, we obtain equivalent formulation of the problem within the Matrix Diagonalisation Method.

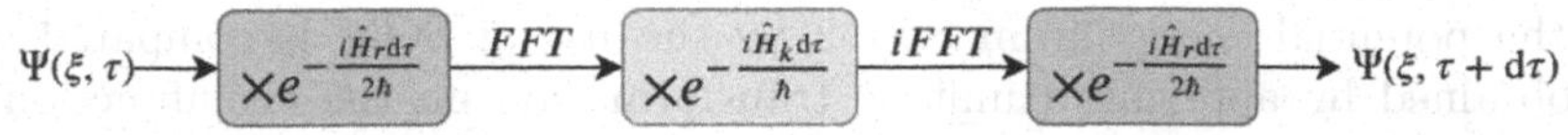

$$\Psi(\xi, \tau) \longrightarrow \boxed{\times e^{-\frac{i\hat{H}_r d\tau}{2\hbar}}} \xrightarrow{FFT} \boxed{\times e^{-\frac{i\hat{H}_k d\tau}{\hbar}}} \xrightarrow{iFFT} \boxed{\times e^{-\frac{i\hat{H}_r d\tau}{2\hbar}}} \longrightarrow \Psi(\xi, \tau + d\tau)$$

Fig. 3. High-level algorithm illustrating the steps of the split-operator method.

The result of the eigenenergy and eigenfunction calculations for the realistic potential energy[1] is shown in Fig. 2 in the left column. We have used 320 basis functions, and the width of the surrounding infinite well was $L = 44.472$ of dimensionless units, which gave 32 bound states. One can see that such a potential has a nontrivial asymmetric allocation of the wave functions. The coarse and fine piece-wise approximations to the realistic function are shown in the same figure in the middle and right columns correspondingly. In the case of the coarse approximation, when the fine features of the potential energy such as "double-wells" are ignored, the lowest energy levels are grouped so that they are hardly distinguishable. There is also a significant difference in the values of the eigenenergies compared to the realistic case. On the other hand, the fine piece-wise approximation allows one to preserve both qualitative and quantitative properties of the real potential energy.

Solving the time independent problem gives a good understanding of the system's properties at early stages of modelling. The time-dependent solution is discussed next.

5 Split-Operator Method

Since we would like to model the evolution of an electron injected into a register of coupled QDs and subject to a variable potential energy, we employ a Split-Operator Method (SOM) to solve the time-dependent problem [2,11]. This method has been chosen due to its effectiveness at a moderate computational cost. Since the method is well documented, we provide only its brief description, discuss its computational cost and show its application to the studied system.

The split-operator method is based on the fact that in conventional cases it is possible to split the Hamiltonian operator into two components, one being position dependent $\hat{H}_r$ and the other being momentum dependent $\hat{H}_k$ [16]. Moreover, it is possible to show that in the position space, the $\hat{H}_r$ operator is reduced to the multiplication of the wave function by a position dependent function. The same reduction can be done for the momentum dependent component of the Hamiltonian operator in the momentum space. This concept is illustrated in Fig. 3.

This method allows one to observe the evolution of the wave function with time, which is particularly useful in the case of a time-dependent Hamiltonian. It also allows one to calculate the ground state of a system (in this case one can use the imaginary time step idt). In our case, when we deal with multiple potential wells (or as we say coupled quantum dots), it allows us to obtain corresponding localised states. The construction of the localised states in a system

---

[1] The source code for this section is available from www.github.com/mishagli/qsol.

with the potential energy from Sect. 2 by the use of SOM is compared with that obtained by applying a unitary transformation on the eigenfunction set (obtained in Sect. 4). Both methods show a good agreement. However, it is more convenient to use the matrix diagonalisation method in order to obtain the full spectrum of possible states.

With regard to the computational cost of this method, it has two contributing aspects:

☐ The calculation of the forward and inverse Fast Fourier Transforms (FFT and iFFT correspondingly) is proportional to the number of discretisation points n as $n \ln(n)$. A higher number n provides a higher resolution for the wave function, and therefore for the probability density function. Note that in order to use the FFT algorithm, the number of coordinate points should be a power of 2 (in our case, 512).

☐ This operation is repeated over a desired number of time steps $d\tau$. The accuracy of the SOM is $O(dt^3)$, and thus the time step operation should be done with a relatively small time step (in our case 10^{-4} of dimensionless units of time).

After we have implemented the split-operator method and ensured that it gives consistent results, our next task is to simulate the behaviour of an electron in a quantum register. We will show two examples: the transport of an electron from the first to the last QD in the register combined with electron probability oscillations and "spliting" electron's wave function between a number of QDs. These simulations are presented in a later section using the coarse piece-wise approximation and by adjusting the height of the potential barriers separating the QDs.

6   Multiple Quantum Dot Model in the Tight Biding Formalism

In this section, we will introduce the tight binding formalism. This will allow one to capture the time dependent dynamics assuming an ideal quantum transport in the quantum structure (register) and can be easily extended to multi-particle and multi-energy level systems [5,9]. We use this approach for additional verification of the SOM modelling.

In the tight binding formalism, we assume that electrons can be represented by wave functions associated with localised states in a discrete lattice. We can visualize the quantum register of Fig. 1(a) as a pseudo-1D lattice. For each QD, we consider one effective quantum state, which can be represented as $|j\rangle$, where $j = 1, 2, 3...$ is an integer denoting the position of each QD. Then, the Hamiltonian of the system can be written as:

$$\hat{\mathbf{H}} = -t_s \left(\sum_{jk} \hat{c}_j^\dagger \hat{c}_k + \hat{c}_k^\dagger \hat{c}_j \right) \tag{9}$$

where $\hat{c}_j^\dagger$ $(\hat{c}_j)$ are the creation (annihilation) operators, which create (annihilate) a particle at site j. The terms $t_{s,jk}$ are describing tunnelling, i.e., the hopping of electrons from site j to site k in the lattice. Here, we will consider the symmetrical case, and therefore we will omit the potential energy of each QD (since it only shifts the total energy of the system by a constant and has no physical contribution to its evolution). We will also assume that sites k and j are immediate neighbors, and we will disregard any probability of a particle hopping to a distant QD. It should be noted that for single-electron tunneling processes through potential barriers, the tunneling rate is exponentially suppressed in the barrier height and width, therefore this is a reasonable assumption. In this representation, the wave function can be expressed as a linear combination in the position basis:

$$|\Psi\rangle = \sum_k c_k(t)\,|k\rangle \tag{10}$$

where $c_k(t)$ are the complex probability amplitudes of the states $|k\rangle$, and the normalisation restriction applies $\sum_k |c_k(t)|^2 = 1$. Furthermore, by considering the time-dependent Schrödinger equation [5], one can write:

$$i\hbar\frac{d\mathbf{C}(t)}{dt} = \hat{\mathbf{H}}(t)\mathbf{C} \tag{11}$$

where $\mathbf{C}(t) = \{c_1(t), c_2(t), \ldots, c_k(t)\}$ is the vector of the probability amplitudes.

In this work, we are interested in describing the dynamics of the system for a time-dependent case where the hopping coefficients $t_{s,jk}$ do not remain constant with time (and can be increased or decreased by means of controlling the potential energy function). In order to perform quantum operations, one needs to apply a correct sequence of voltage pulses at the imposers (gates) at specific time instances, changing the tunnelling probabilities. In such a case, the Hamiltonian of the system is changing with time. In the system under study we can consider a sudden change in the Hamiltonian [21]. Also, the applied external fields (square voltage pulses) will be assumed of small magnitudes. As a consequence, the Hamiltonian of the system takes the form:

$$\hat{\mathbf{H}}(t) = \begin{bmatrix} 0 & t_{h,21}(t) & 0 \\ t_{h,12}(t) & 0 & t_{s,32}(t) \\ 0 & t_{s,23}(t) & 0 \end{bmatrix} \tag{12}$$

where each of the $t_{h,jk}$ hopping terms is a piece-wise time dependent-function of the form:

$$t_{h,jk} = \begin{cases} t_{h,\text{low}} & t < t_{jk,0} \\ t_{h,\text{high}} & t_{jk,0} \leq t_{jk} \leq t_{jk,0} + t_{jk,\text{width}} \\ t_{h,\text{low}} & t > t_{jk,0} + t_{jk,\text{width}}. \end{cases} \tag{13}$$

where $t_{jk,0}$ is the initial time instance when the pulse is applied at the imposer separating the quantum dots j and k, and $t_{jk,\text{width}}$ is the time duration for which the pulse is applied. Note that one needs to match the initial conditions between the solutions of (11), before and after the application of the pulse. The values used in this simulation are: $t_{h,\text{high}} = 2.82E_0$, $t_{h,\text{low}} = 0.0013E_0$.

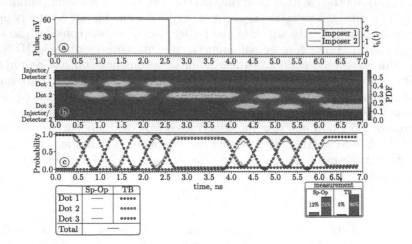

Fig. 4. Transport of a single injected electron from dot 1 to dot 3. (a) Control pulses applied to imposers and corresponding hopping terms. (b) Probability heatmap generated by the SOM showing electron transfer and Rabi oscillations between pairs of dots. (c) Probability of detecting an electron in each quantum dot (lines correspond to the SOM, circles — to the tide-binding formalism).

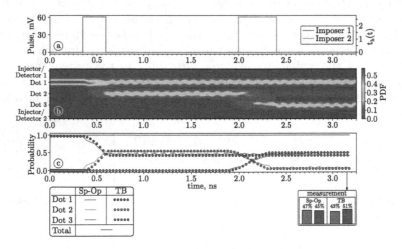

Fig. 5. Rotation gate operation applied onto a single injected electron in dot 1 and detected at both detectors. (a) Control pulses applied to imposers and corresponding hopping terms. (b) Probability heatmap generated using the SOM showing the reaction of the electron to the pulses. (c) Probability of detecting the electron in each quantum dot.

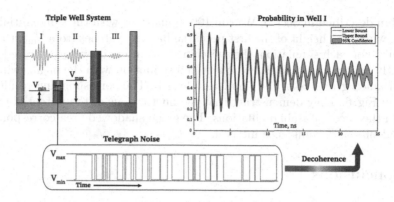

Fig. 6. Simulation of decoherence on example of a triple well system, where the left and right walls are 200 units tall. The barrier that separates wells II and III is fixed and equals 64 units. The barrier between wells I and II switched randomly between $V_{min} = 4$ and $V_{max} = 5$ units in accordance to the telegraph noise. The average switching time is ≈ 0.2 ns. (Color figure online)

7   Discussion of Simulation Results

Figure 4 demonstrates the transfer of an electron injected to dot 1 and detected by two detectors (1 and 2) placed at the opposite edges of the structure. Two pulses of 2.6 ns duration and 60 mV amplitude are applied at imposers 1 and 2 respectively (Fig. 4(a)). The heatmap (Fig. 4(b)) demonstrates the PDF of the injected electron as a function of time (the horizontal axis) and coordinate along the register (the vertical axis), and its reaction to the applied pulses. The most informative is Fig. 4(c) which shows the probability of electron being in each of the dots, and the final measurement in form of a probability histogram. It is obvious that in this case it is possible to transport the electron faster, but this graph was also made to demonstrate the feasibility of Rabi oscillations with period T_0 between the two dots when the barrier separating the dots is lowered. Thus, to make the transport operation, we need to have the pulse duration $T_0/2 + kT_0$ with integer k.

An X-rotation gate is an operation which allows one to control the probability of detecting an electron. In order to "split" the electron's wavefunction into two equal parts, one needs to apply a pulse of duration equal to $T_0/4 + kT_0$ (Fig. 5(a)). Note that the second pulse applied is two times longer since we aimed at transporting a 'part' of the wave function to the third well. As one can see, the shown pulses allow one to realise an X-rotation gate. However, it is evident that each operation leaves some residual wave function in the dots, and thus it affects the accuracy (fidelity) of each operation.

The presented results do not take into account decoherence effects or non-zero temperatures. However, it would have been unfair not to mention the decoherence that results from fluctuations, which always take place in a real system. Using a simple two-levels telegraph noise model [1,4,6], we modified the algorithm described in Sect. 5. We run 100 simulations with the same initial conditions, where the height of the first potential barrier switches from V_{min} to V_{max} randomly. For each simulation, the switching time intervals were generated independently in accordance to the exponential distribution with the mean switching interval much grater than the simulation length. The corresponding results are shown in Fig. 6. They demonstrate that the fluctuations of the potential energy result in the decay of Rabi oscillations. The green shadowed area corresponds to the Student's 95% confidence interval.

8   Conclusions

In this paper, we discussed the the development of a quantum simulator for charge qubits based on FDSOI CMOS technology. We started from the description of the system and proceeded with the discussion and comparison of numerical and semi-analytical techniques to model the behaviour of a single electron in such a structure. We presented a high-level multi-particle model to simulate the evolution of the various quantum states in such a system using the tight binding approach. We demonstrated two case studies: the electron transport through multiple QDs and the construction of an X-rotation gate, and showed the effect of decoherence due to potential energy fluctuations.

Acknowledgement. A.S., D.M. and P.G. contributed equally to this work. This work was supported in part by the Science Foundation of Ireland under Grant 13/RC/2077 and under Grant 14/RP/I2921.

References

1. Abel, B., Marquardt, F.: Decoherence by quantum telegraph noise: a numerical evaluation. Phys. Rev. B **78**(20), 201302 (2008)
2. Bandrauk, A.D., Shen, H.: Improved exponential split operator method for solving the time-dependent schrödinger equation. Chem. Phys. Lett. **176**(5), 428–432 (1991)
3. Bashir, I., et al.: A mixed-signal control core for a fully integrated semiconductor quantum computer system-on-chip. In: Proceedings of IEEE ESSDERC/ESSCIRC, Krakow, Poland, 23–26 September 2019 (2019)
4. Bergli, J., Galperin, Y.M., Altshuler, B.: Decoherence in qubits due to low-frequency noise. New J. Phys. **11**(2), 025002 (2009)
5. Blokhina, E., Giounanlis, P., Mitchell, A., Leipold, D.R., Staszewski, R.B.: CMOS position-based charge qubits: theoretical analysis of control and entanglement. IEEE Access **8**, 4182 (2020)
6. Cai, X.: Quantum dephasing induced by non-markovian random telegraph noise. Sci. Rep. **10**(1), 1–11 (2020)
7. Cubitt, T.S., Montanaro, A., Piddock, S.: Universal quantum hamiltonians. Proc. Natl. Acad. Sci. **115**(38), 9497–9502 (2018)
8. Fujisawa, T., Hayashi, T., Cheong, H., Jeong, Y., Hirayama, Y.: Rotation and phase-shift operations for a charge qubit in a double quantum dot. Phys. E Low-Dimen. Syst. Nanostruct. **21**(2-4), 1046–1052 (2004)

9. Giounanlis, P., Blokhina, E., Leipold, D., Staszewski, R.B.: Photon enhanced interaction and entanglement in semiconductor position-based qubits. Appl. Sci. **9**(21), 4534 (2019)
10. Giounanlis, P., Blokhina, E., Pomorski, K., Leipold, D.R., Staszewski, R.B.: Modeling of semiconductor electrostatic qubits realized through coupled quantum dots. IEEE Access **7**, 49262–49278 (2019)
11. Hermann, M.R., Fleck Jr., J.: Split-operator spectral method for solving the time-dependent schrödinger equation in spherical coordinates. Phys. Rev. A **38**(12), 6000 (1988)
12. Landau, L.D., Lifshitz, E.M.: Quantum Mechanics - Non-Relativistic Theory, Course of Theoretical Physics, vol. 3. Pergamon Press, Oxford (1981)
13. Li, Y.C., Chen, X., Muga, J., Sherman, E.Y.: Qubit gates with simultaneous transport in double quantum dots. New J. Phys. **20**(11), 113029 (2018)
14. Marsiglio, F.: The harmonic oscillator in quantum mechanics: a third way. Am. J. Phys. **77**, 253–258 (2009). https://doi.org/10.1119/1.3042207
15. Mitchell, A.K., Logan, D.E.: Two-channel kondo phases and frustration-induced transitions in triple quantum dots. Phys. Rev. B **81**(7), 075126 (2010)
16. Pahlavani, M.R.: Theoretical Concepts of Quantum Mechanics. BoD-Books on Demand, Norderstedt (2012)
17. Petersson, K., Petta, J.R., Lu, H., Gossard, A.: Quantum coherence in a one-electron semiconductor charge qubit. Phys. Rev. Lett. **105**(24), 246804 (2010)
18. Remediakis, I., Kaxiras, E.: Band-structure calculations for semiconductors within generalized-density-functional theory. Phys. Rev. B **59**(8), 5536 (1999)
19. Sarma, S.D., Wang, X., Yang, S.: Hubbard model description of silicon spin qubits: charge stability diagram and tunnel coupling in si double quantum dots. Phys. Rev. B **83**(23), 235314 (2011)
20. Shinkai, G., Hayashi, T., Ota, T., Fujisawa, T.: Correlated coherent oscillations in coupled semiconductor charge qubits. Phys. Rev. Lett. **103**, 056802 (2009). https://doi.org/10.1103/PhysRevLett.103.056802
21. Song, H., Yang, F., Wang, X.: Condition for sudden approximation and its application in the problem of compression of an infinite well. Eur. J. Phys. **36**(3), 035009 (2015)
22. Szafran, B.: Paired electron motion in interacting chains of quantum dots. Phys. Rev. B **101**, 075306 (2020). https://doi.org/10.1103/PhysRevB.101.075306. https://link.aps.org/doi/10.1103/PhysRevB.101.075306
23. Veldhorst, M., et al.: A two-qubit logic gate in silicon. Nature **526**(7573), 410 (2015)
24. Weichselbaum, A., Ulloa, S.: Charge qubits and limitations of electrostatic quantum gates. Phys. Rev. A **70**(3), 032328 (2004)
25. Yang, Y.C., Coppersmith, S., Friesen, M.: Achieving high-fidelity single-qubit gates in a strongly driven charge qubit with 1/f charge noise. npj Quantum Inf. **5**(1), 12 (2019)
26. Zheng, H., Changlani, H.J., Williams, K.T., Busemeyer, B., Wagner, L.K.: From real materials to model hamiltonians with density matrix downfolding. Front. Phys. **6**, 43 (2018)

Author Index